Methods in Enzymology

Volume IX

CARBOHYDRATE METABOLISM

METHODS IN ENZYMOLOGY

EDITORS-IN-CHIEF

Sidney P. Colowick   Nathan O. Kaplan

Methods in Enzymology

Volume IX

Carbohydrate Metabolism

EDITED BY

Willis A. Wood

MICHIGAN STATE UNIVERSITY
DEPARTMENT OF BIOCHEMISTRY
EAST LANSING, MICHIGAN

1966

ACADEMIC PRESS New York and London

ACADEMIC PRESS, INC.
111 Fifth Avenue, New York, New York 10003

United Kingdom Edition published by
ACADEMIC PRESS, INC. (LONDON) LTD.
Berkeley Square House, London W1X 6BA

LIBRARY OF CONGRESS CATALOG CARD NUMBER: 54-9110

Third Printing, 1970

PRINTED IN THE UNITED STATES OF AMERICA

Contributors to Volume IX

Article numbers are shown in parentheses following the names of contributors.
Affiliations listed are current.

Robert H. Abeles (123), *Brandeis University, Graduate Dept. of Biochemistry, Waltham, Massachusetts*

William S. Allison (43), *Brandeis University, Graduate Dept. of Biochemistry, Waltham, Massachusetts*

D. Amaral (19), *Instituto de Bioquimica, Universidade de Parana, Curitiba-Parana, Brazil*

Donald E. Anderson (115), *Michigan State University, Dept. of Biochemistry, East Lansing, Michigan*

R. L. Anderson (11, 70b, 71, 82, 105), *Michigan State University, Dept. of Biochemistry, East Lansing, Michigan*

A. A. App (112c), *Boyce Thompson Institute for Plant Research, Inc., Yonkers, New York*

William J. Arion (111), *University of North Dakota School of Medicine, Dept. of Biochemistry, Grand Forks, North Dakota*

Carlos Asensio (76b), *Instituto Marañón, Centro de Investigaciones Biológicas, CSIC, Madrid, Spain*

Anita J. Aspen (47), *National Institutes of Health, National Institute of Arthritis and Metabolic Diseases, Laboratory of Biochemical Pharmacology, Bethesda, Maryland*

A. L. Baker (62), *Worthington Biochemical Corporation, Freehold, New Jersey*

R. L. Baldwin (122), *University of California, Dept. of Animal Husbandry, Davis, California*

S. S. Barkulis (76a), *Ciba Pharmaceutical Company, Research Department, Summit, New Jersey*

Harold J. Blumenthal (13a, 13b, 93, 118, 119), *Loyola University, Stritch School of Medicine, Dept. of Microbiology, Hines, Illinois*

Oscar Bodansky (55, 99), *Sloan-Kettering Institute for Cancer Research and Memorial Hospital for Cancer and Allied Diseases, Division of Biochemistry, New York, New York*

Myron Brin (89), *Upstate Medical Center, State University of New York, Syracuse, New York*

L. A. Burgoyne (57), *University of Adelaide, Dept. of Biochemistry, Adelaide, South Australia*

John Calder (87), *University of Texas, Dept. of Chemistry, Austin, Texas*

M. Chakravorty (33a), *Banaras Hindu University, College of Medical Sciences, Dept. of Biochemistry and Biophysics, Varanasi-5, India*

Frixos Charalampous (125), *University of Pennsylvania School of Medicine, Dept. of Biochemistry, Philadelphia, Pennsylvania*

I-Wen Chen (125), *University of Pennsylvania School of Medicine, Dept. of Biochemistry, Philadelphia, Pennsylvania*

C. Chiang (37), *Squibb Institute for Medical Research, New Brunswick, New Jersey*

T. H. Chiu (84), *University of Pittsburgh, Dept. of Biology, Pittsburgh, Pennsylvania*

Terenzio Cremona (56, 59), *Laboratorio Internazionale di Genetica e Biofisica, Naples, Italy*

E. A. Davidson (126), *Duke University, Dept. of Biochemistry, Durham, North Carolina*

J. De Ley (34, 39, 40, 64), *Laboratory of Microbiology, University of Ghent, Ghent, Belgium*

Goetz F. Domagk (101), *Physiologisch-Chemisches, Institut der Universität Göttingen, West Germany*

N. L. EDSON (29), *University of Otago Medical School, Dept. of Biochemistry, Dunedin, New Zealand*

ELLIS ENGLESBERG (3, 80), *University of California, Dept. of Biological Sciences, Santa Barbara, California*

DAVID SIDNEY FEINGOLD (17, 84), *University of Pittsburgh, Dept. of Biology, Pittsburgh, Pennsylvania*

JUNE M. FESSENDEN (90), *Public Health Research Institute of the City of New York, Inc., New York, New York*

GRACE M. FIMOGNARI (54), *University of California School of Medicine, Dept. of Biochemistry, San Francisco, California*

DONALD C. FISH (13a, 13b, 93), *U.S. Army Biological Laboratories, Fort Detrick, Frederick, Maryland*

D. D. FOSSITT (35), *University of Michigan Medical Center, Dept. of Industrial Health, Ann Arbor, Michigan*

OTHMAR GABRIEL (5), *National Institutes of Health, Department of Health, Education and Welfare, Bethesda, Maryland*

J. GAZITH (69b), *Vanderbilt University School of Medicine, Dept. of Microbiology, Nashville, Tennessee*

MOHAMMAD A. GHALAMBOR (14, 83, 94, 95), *Pahlavi University, Department of Biochemistry, Faculty of Medicine, Shiraz, Iran*

L. N. GIBBINS (75), *Cawthorn Institute, P.O. Box 175, Nelson, New Zealand*

MELVIN GOLDBERG (90), *Univ. of California Medical Center, Dept. of Pathology, San Francisco, California*

R. H. GOODING (69b), *Vanderbilt University School of Medicine, Dept. of Microbiology, Nashville, Tennessee*

A. M. GOTTO (49), *The Massachusetts General Hospital, Dept. of Medicine, Boston, Massachusetts*

ENRICO GRAZI (26), *Universitá degli Studi, Istituto di Chimica Biologica, Ferrara, Italy*

EUGENE E. GREBNER (17), *National Institutes of Health, Bethesda, Maryland*

D. P. GROTH (97b), *Emory University, Dept. of Biochemistry, Atlanta, Georgia*

WILLIAM E. GROVES (87), *St. Jude's Hospital, Memphis, Tennessee*

NABA K. GUPTA (124), *University of Wisconsin, Institute for Enzyme Research, Madison, Wisconsin*

LOWELL P. HAGER (51), *University of Illinois, Dept. of Chemistry and Chemical Engineering, Urbana, Illinois*

P. HANDLER (68), *Duke University School of Medicine, Dept. of Biochemistry, Durham, North Carolina*

R. G. HANSEN (127, 128), *Michigan State University, Dept. of Biochemistry, East Lansing, Michigan*

JENS G. HAUGE (20a, 21b), *University of Bergen, Dept. of Biochemistry, Bergen, Norway*

OSAMU HAYAISHI (16), *Kyoto University, Dept. of Medical Chemistry, Kyoto, Japan*

EDWARD C. HEATH (14, 83, 94, 95), *The Johns Hopkins University School of Medicine, Dept. of Physiological Chemistry, Baltimore, Maryland*

M. R. HEINRICH (74), *National Aeronautics and Space Administration, Ames Research Center, Moffett Field, California*

KLAUS HERRMANN (109), *Chemisches Laboratorium der Universität Freiburg, Freiburg i. Breisgau, Germany*

P. HOFFEE (97a), *Yeshiva University, Albert Einstein College of Medicine, Dept. of Molecular Biology, Bronx, New York*

B. L. HORECKER (19, 28, 33, 88, 97a, 112b, 113), *Yeshiva University, Albert Einstein College of Medicine, Dept. of Molecular Biology, Bronx, New York*

SUSAN B. HORWITZ (30, 31), *Emory University, Dept. of Pharmacology, Atlanta, Georgia*

SALLY M. HOWARD (74), *University of Southern California School of Medicine, Dept. of Biochemistry, Los Angeles, California*

JAMES R. HUNSLEY (87), *Michigan State*

University, Dept. of Biochemistry, East Lansing, Michigan

RONALD E. HURLBERT (121), *Washington State University, Dept. of Bacteriology and Public Health, Pullman, Washington*

JORDAN M. INGRAM (36, 91), *Microbiology Research Institute, Central Experimental Farm, Canada Department of Agriculture Research Branch, Ottawa, Canada*

V. JAGANNATHAN (69a), *National Chemical Laboratory, Biochemistry Division, Poona, India*

WILLIAM B. JAKOBY (38, 46, 47, 48, 49, 65, 121), *National Institutes of Health, National Institute of Arthritis and Metabolic Diseases, Bethesda, Maryland*

THERESSA JEPSON (13b, 119), *University of Michigan, Dept. of Botany, Ann Arbor, Michigan*

MIRA D. JOSHI (69a), *National Chemical Laboratory, Biochemistry Division, Poona, India*

ELLIOT JUNI (4), *University of Michigan, Dept. of Microbiology, Ann Arbor, Michigan*

M. Y. KAMEL (70b, 71), *Michigan State University, Dept. of Biochemistry, East Lansing, Michigan*

F. KELLY-FALCOZ (19), *Centre Nationale de la Recherche Scientifique, Gif-sur-Yvette, France*

K. KERSTERS (34, 64), *University of Ghent, Laboratory of Microbiology, Ghent, Belgium*

TSOO E. KING (20b, 32), *Oregon State University, Laboratory for Respiratory Enzymology, Corvallis, Oregon*

LEONARD D. KOHN (46, 48, 49), *National Institutes of Health, National Institute of Arthritis and Metabolic Diseases, Laboratory of Biochemical Pharmacology, Bethesda, Maryland*

H. L. KORNBERG (49), *University of Leicester, Dept. of Biochemistry, Leicester, England*

EMILIE KOVACH (17), *Wayne State University, Dept. of Microbiology, Detroit, Michigan*

S. G. KNIGHT (37), *University of Wisconsin, College of Agriculture, Dept. of Bacteriology, Madison, Wisconsin*

L. O. KRAMPITZ (15), *Western Reserve University School of Medicine, Dept. of Microbiology, Cleveland, Ohio*

STEPHEN A. KUBY (23), *University of Utah School of Medicine, Laboratory of Metabolic and Hereditary Diseases, Salt Lake City, Utah*

WERNER KUNDIG (72), *Johns Hopkins University, Dept. of Biology, Baltimore, Maryland*

G. KURZ (22), *Chemisches Laboratorium der Universität Freiburg, Freiburg i. Breisgau, Germany*

ROBERT G. LANGDON (24), *The Johns Hopkins University School of Medicine, Dept. of Physiological Chemistry, Baltimore, Maryland*

HENRY A. LARDY (77a), *University of Wisconsin, Institute for Enzyme Research, Madison, Wisconsin*

N. L. LEE (80), *University of California, Dept. of Biological Sciences, Santa Barbara, California*

K. H. LING (77a), *National Taiwan University College of Medicine, Dept. of Biochemistry, Taipei, Taiwan, China*

M. MALAMY (113), *Princeton University, Program in Biochemical Sciences, Princeton, New Jersey*

T. MANN (32), *Cambridge University, Molteno Institute and Institute for Reproduction Biochemistry, Cambridge, England*

TAG E. MANSOUR (77b), *Stanford University School of Medicine, Dept. of Pharmacology, Palo Alto, California*

FRANK MARCUS (77a), *University of Chile, Faculty of Medicine, Dept. of Biochemistry, Santiago, Chile*

PAUL A. MARKS (25, 27), *College of Physicians and Surgeons of Columbia University, Dept. of Medicine, New York, New York*

VINCENT MASSEY (52), *The University of Michigan Medical School, Dept. of*

Biological Chemistry, Ann Arbor, Michigan

JARY S. MAYES (127, 128), *State University of New York Children's Hospital, Dept. of Biochemical Genetics, Buffalo, New York*

H. PAUL MELOCHE (12, 91, 116), *The Institute for Cancer Research, Biochemistry Division, Philadelphia, Pennsylvania*

J. M. MERRICK (117), *University of Buffalo, Dept. of Biochemistry, Buffalo, New York*

O. NEAL MILLER (61, 86), *Tulane University School of Medicine, Metabolism Research Laboratory, Dept. of Biochemistry and Nutrition, New Orleans, Louisiana*

R. P. MORTLOCK (8, 79, 102), *University of Massachusetts, Dept. of Microbiology, Amherst, Massachusetts*

HENRY I. NAKADA (100), *University of California, Dept. of Biological Sciences, Santa Barbara, California*

STUART A. NARROD (65), *Women's Medical College, Dept. of Biochemistry, Philadelphia, Pennsylvania*

ERNST A. NOLTMANN (23, 98a), *University of California, Dept. of Biochemistry, Riverside, California*

ROBERT C. NORDLIE (111), *University of North Dakota School of Medicine, Dept. of Biochemistry, Grand Forks, North Dakota*

VERNER PAETKAU (77a), *University of Wisconsin, Institute for Enzyme Research, Madison, Wisconsin*

M. J. PARRY (70a), *University of Birmingham, Dept. of Biochemistry, Birmingham, England*

JOHN H. PAZUR (18), *Pennsylvania State University, Dept. of Biochemistry, College Park, Pennsylvania*

A. S. PERLIN (6), *Prairie Regional Laboratory, National Research Council, Saskatoon, Saskatchewan, Canada*

BURTON M. POGELL (2), *The Albany Medical College of Union University, Dept. of Biochemistry, Albany, New York*

JACK PREISS (41, 42, 107), *University of California, Dept. of Biochemistry and Biophysics, Davis, California*

SANDRO PONTREMOLI (10, 26, 112a), *Universitá degli Studi, Istituto di Chimica Biologica, Ferrara, Italy*

J. R. QUAYLE (63, 67), *The University, Dept. of Biochemistry, Sheffield, England*

EFRAIM RACKER (66, 90), *Cornell University, Dept. of Biochemistry, Ithaca, New York*

K. V. RAJAGOPALAN (68), *Duke University School of Medicine, Dept. of Biochemistry, Durham, North Carolina*

T. V. RAJKUMAR (87), *University of Washington, Dept. of Biochemistry, Seattle, Washington*

N. APPAJI RAO (114), *Indian Institute of Science, Dept. of Biochemistry, Bangalore, India*

LESTER J. REED (5, 10), *University of Texas, Clayton Foundation, Biochemical Institute, Dept. of Chemistry, Austin, Texas*

W. J. REEVES, JR. (54), *University of California School of Medicine, Dept. of Biochemistry, San Francisco, California*

F. J. REITHEL (98b), *University of Oregon, Dept. of Chemistry, Eugene, Oregon*

WILLIAM G. ROBINSON (60), *New York University School of Medicine, NYU Medical Center, Dept. of Biochemistry, New York, New York*

ZELDA B. ROSE (66), *The Institute for Cancer Research, Dept. of Biochemistry, Philadelphia, Pennsylvania*

SAUL ROSEMAN (72, 110, 117), *The Johns Hopkins University, McCollum-Pratt Institute, Dept. of Biology, Baltimore, Maryland*

O. M. ROSEN (97a, 112b), *Yeshiva University, Albert Einstein College of Medicine, Dept. of Medicine, Bronx, New York*

S. M. ROSEN (112b), *Yeshiva University, Albert Einstein College of Medicine, Dept. of Medicine, Bronx, New York*

Manuel Ruiz-Amil (76b), *Instituto de Biología Celular, Centro de Investigaciones Biológicas, CSIC, Madrid, Spain*

William J. Rutter (87), *University of Washington, Dept. of Biochemistry, Seattle, Washington*

H. L. Sadoff (21a), *Michigan State University, Dept. of Microbiology, East Lansing, Michigan*

María L. Salas (77c), *Instituto "G. Marañón," Centro de Investigaciones Biológicas, Madrid, Spain*

H. J. Sallach (44, 45), *University of Wisconsin, Dept. of Physiological Chemistry, Madison, Wisconsin*

B. M. Scher (33b), *New York University School of Medicine, Dept. of Microbiology, New York, New York*

I. T. Schulze (69b), *The Public Health Research Institute of The City of New York, Inc., New York, New York*

Morton K. Schwartz (55, 99), *Sloan-Kettering Institute for Cancer Research and Memorial Hospital for Cancer and Allied Diseases, Division of Biochemistry, New York, New York*

D. R. D. Shaw (29), *University of Otago Medical School, Dept. of Biochemistry, Dunedin, New Zealand*

C. W. Shuster (92), *Western Reserve University School of Medicine, Dept. of Microbiology, Cleveland, Ohio*

A. S. Sieben (6), *Prairie Regional Laboratory, National Research Council, Saskatoon, Saskatchewan, Canada*

F. J. Simpson (6, 9, 75, 81), *Prairie Regional Laboratory, National Research Council, Saskatoon, Saskatchewan, Canada*

Thomas P. Singer (56, 59), *Division of Molecular Biology, Veterans Administration Hospital, San Francisco, California*

Roberts A. Smith (73), *University of California, Dept. of Chemistry, Los Angeles, California*

Alan M. Snoswell (58), *University of Melbourne, School of Veterinary Science, Biochemistry Unit, Victoria, Australia*

Alberto Sols (77c), *Instituto "G. Marañón," Centro de Investigaciones Biológicas, Madrid, Spain*

Chava Telem Spivak (110), *University of Michigan, Dept. of Biochemistry, Ann Arbor, Michigan*

Francis Stolzenbach (53), *Brandeis University, Graduate Dept. of Biochemistry, Waltham, Massachusetts*

R. H. Symons (57), *University of Adelaide, Dept. of Biochemistry, Adelaide, South Australia*

Yasuyuki Takagi (96), *Kyushu University, Faculty of Medicine, Dept. of Biochemistry, Sukuoka, Japan*

Kouichi R. Tanaka (85), *University of California School of Medicine, Dept. of Medicine, Los Angeles, California*

O. Tchola (88), *Yeshiva University, Albert Einstein College of Medicine, Dept. of Molecular Biology, Bronx, New York*

Myna C. Theisen (73), *University of California, Dept. of Chemistry, Los Angeles, California*

N. E. Tolbert (62, 115), *Michigan State University, Dept. of Biochemistry, East Lansing, Michigan*

C. S. Vaidyanathan (114), *Indian Institute of Science, Department of Biochemistry, Bangalore, India*

William N. Valentine (85), *University of California School of Medicine, Dept. of Medicine, Los Angeles, California*

Birgit Vennesland (124), *University of Chicago, Dept. of Biochemistry, Chicago, Illinois*

Wesley A. Volk (7, 78, 103), *University of Virginia School of Medicine, Dept. of Microbiology, Charlottesville, Virginia*

Robert Votaw (15), *Western Reserve University School of Medicine, Dept. of Microbiology, Cleveland, Ohio*

D. G. Walker (70a), *University of Birmingham, Dept. of Biochemistry, Birmingham, England*

KURT WALLENFELS (22, 109), *Chemisches Laboratorium der Universität, Freiburg i. Br., Germany*

E. W. WESTHEAD (120), *Dartmouth Medical School, Dept. of Biochemistry, Hanover, New Hampshire*

F. ROBERT WILLIAMS (51), *Service de Biochimie, Centre d'Etudes Nucleaires de Saclay, Gif-sur-Yvette, France*

W. T. WILLIAMSON (108), *Dairy Techniks Corporation, Kalamazoo, Michigan*

CHARLES R. WILLMS (50), *Southwest Texas State College, Dept. of Chemistry, San Marcos, Texas*

NORRIS P. WOOD (1, 129), *University of Rhode Island, Dept. of Microbiology, Kingston, Rhode Island*

W. A. WOOD (8, 12, 35, 36, 79, 82, 91, 106, 108, 116, 122), *Michigan State University, Dept. of Biochemistry, East Lansing, Michigan*

B. M. WOODFIN (87), *University of Michigan, Dept. of Biological Chemistry, Ann Arbor, Michigan*

K. YAMANAKA (104, 106), *Kagawa University, Faculty of Agriculture, Dept. of Agricultural Chemistry, Miki-tyô, Kagawa-ken, Japan*

RONALD ZECH (101), *Physiologisch.-Chemisches, Institut der Universität, Göttingen, West Germany*

Preface

The growth of enzymology over the past decade may be seen in the fact that two volumes, VIII and IX, are now needed to update the material formerly covered by Volume I of this series. In Volume IX, "Carbohydrate Metabolism," enzymes involved in metabolic reactions between disaccharides and pyruvate, together with appropriate procedures and methods for preparing substrates, have been covered. In contrast to earlier volumes, the organization groups together enzymes of like reaction type. Within these large divisions, the organization follows metabolic sequences. In recognition of the importance of subtle differences in properties of an enzyme, preparations of the same enzyme from several sources have been included. Where applicable, additional contributions were secured which deal with enzymes of importance in medical diagnosis. The format of these chapters departs from the typical in that description of assay procedures used in clinical laboratories, together with general information and interpretation of results, are presented.

The problem of identification and availability of microorganisms used as sources of enzymes has been somewhat bothersome. It is fairly common to find recorded in the literature the preparation of an enzyme from a mutant which is not generally available. Important preparations of this type have been included in this volume, and attempts have been made to identify properly the organism involved by its number in the American Type Culture Collection or other similar collection. It should be noted, however, that securing a microorganism in the proper genetic and physiological form is not always easy, and reliance upon the culture collections to provide organisms in the condition used by the authors in their contributions unfortunately cannot be assured. It is also unfortunate that in most cases it is not possible to define the steps to be taken to prepare or convert cultures from diverse sources to the form which yields the enzymes as described. Nevertheless, in the editor's judgment, most, if not all, of the procedures involving microorganisms are straightforward, and the species and strains usually are typical.

It has been a pleasure to serve as editor of this volume. The cooperation and productivity of the contributors have been magnificent. It is also a pleasure to recognize the services of Mr. Donald Schneider who painstakingly prepared the index for this volume.

East Lansing, Michigan
November, 1966

W. A. Wood

Table of Contents

Section I. General Analytical and Preparative Procedures

Section II. Preparation of Substrates

Section III. Dehydrogenases and Oxidases

Section IV. Kinases and Transphosphorylases

Section V. Aldolases

Section VI. Isomerases and Epimerases

Section VII. Phosphatases

Section VIII. Dehydrases

Section IX. Miscellaneous Enzymes

METHODS IN ENZYMOLOGY

EDITED BY

Sidney P. Colowick and Nathan O. Kaplan

VANDERBILT UNIVERSITY
SCHOOL OF MEDICINE
NASHVILLE, TENNESSEE

DEPARTMENT OF CHEMISTRY
UNIVERSITY OF CALIFORNIA
AT SAN DIEGO
LA JOLLA, CALIFORNIA

I. Preparation and Assay of Enzymes
II. Preparation and Assay of Enzymes
III. Preparation and Assay of Substrates
IV. Special Techniques for the Enzymologist
V. Preparation and Assay of Enzymes
VI. Preparation and Assay of Enzymes (*Continued*)
   Preparation and Assay of Substrates
   Special Techniques
VII. Cumulative Subject Index

METHODS IN ENZYMOLOGY

EDITORS-IN-CHIEF

Sidney P. Colowick   Nathan O. Kaplan

VOLUME VIII. Complex Carbohydrates
Edited by ELIZABETH F. NEUFELD AND VICTOR GINSBURG

VOLUME IX. Carbohydrate Metabolism
Edited by WILLIS A. WOOD

VOLUME X. Oxidation and Phosphorylation
Edited by RONALD W. ESTABROOK AND MAYNARD E. PULLMAN

VOLUME XI. Enzyme Structure
Edited by C. H. W. HIRS

VOLUME XII. Nucleic Acids (in two volumes)
Edited by LAWRENCE GROSSMAN AND KIVIE MOLDAVE

VOLUME XIII. Citric Acid Cycle
Edited by J. M. LOWENSTEIN

VOLUME XIV. Lipids
Edited by J. M. LOWENSTEIN

VOLUME XV. Steroids and Terpenoids
Edited by RAYMOND B. CLAYTON

VOLUME XVI. Fast Reactions
Edited by KENNETH KUSTIN

In Preparation

Metabolism of Amino Acids and Amines
Edited by HERBERT TABOR AND CELIA WHITE TABOR

Vitamins and Coenzymes
Edited by DONALD B. McCORMICK AND LEMUEL D. WRIGHT

Proteolytic Enzymes
Edited by GERTRUDE E. PERLMANN AND LASZLO LORAND

Photosynthesis
Edited by A. SAN PIETRO

Enzyme Purification and Related Techniques
Edited by WILLIAM B. JAKOBY

Section I

General Analytical and Preparative Procedures

[1] Control of Oxidation-Reduction Potential during Purification

By NORRIS P. WOOD

Control of the oxidation-reduction (O-R) potential during purification has become an important procedure for enzymes that are rapidly and irreversibly inactivated by oxygen. The method to be described has certain advantages over other techniques for preventing oxidation of extracts such as the use of chelating agents to prevent catalysis of oxidation by metal ions, or the removal of oxygen from solution by evacuation, boiling, or bubbling.

Reductants

A reductant is needed that will react rapidly with oxygen at low temperatures, in high salt concentrations, and with sufficient capacity that the potential will remain steady throughout purification. The reductant or its oxidized form should not react with or bind to the enzyme.

Ascorbic acid and sulfhydryl compounds, such as glutathione, cysteine, mercaptoethanol, or 2,3-dimercaptopropanol, when used alone react too slowly for practical use and accurate potentiometric measurements are difficult to obtain.[1-3] Also, potentials of thiols measured with a platinum electrode do not follow theory for perfectly reversible systems, but in practice the potential observed is dependent on the concentration of the reduced form RSH, not on the oxidized form RSSR.[2]

Ferrothiol Complex. When ferrous sulfate and a thiol are added together, a "ferrothiol" complex is formed which can be rapidly oxidized by oxygen to a ferricomplex.[2,4] The ferricomplex undergoes spontaneous reduction at the expense of RSH to form a freely autoxidizable ferrous complex, and the regeneration may be repeated until the RSH has been converted to RSSR. Regeneration of ferric complex by RSH provides a certain amount of O-R "buffer" action, but the regeneration is rate limiting. The potential of the system, however, should remain steady provided that the entrance of oxygen does not exceed the rate of Fe^{3+} reduction by RSH. In practice, the amount of O-R "buffering" and the rate of reduction by RSH will depend on the amount of complex and RSH that

[1] M. O. Oster and N. P. Wood, *J. Bacteriol.* **87,** 104 (1964).

[2] R. K. Cannon and G. M. Richardson, *Biochem. J.* **23,** 1242 (1929).

[3] N. Tanaka, L. M. Kolthoff, and W. Stricks, *J. Am. Chem. Soc.* **77,** 2004 (1955).

[4] L. Michaelis and E. S. G. Barron, *J. Biol. Chem.* **83,** 191 (1929).

can be added to the system. For enzyme assay, excess RSH can be dialyzed or diluted to a noninhibitory level.

The ferrithiol-ferrothiol system is reversible and obtains rapid equilibrium with other reversible O-R systems. The potential of the iron present in the dissolved complex is determined by the ratio of the concentration of ferrous-ferric iron, and the concentration of the active ferrothiol is proportional to the concentration of RSH and ferrithiol. The oxidation of the ferrothiol is rapid at pH 8 and above, but below pH 6, very little or no oxidation of the ferrothiol is demonstrable.[2] Hydrogen peroxide, an intermediate of the reaction, is oxidized further by Fe^{++} or RSH.

The more negative the potential of the complex, the more rapid is the oxidation by molecular oxygen.[5] In practice, a complex of ferrous sulfate and 2,3-dimercaptopropanol will produce a voltage response at pH 7.0 more negative than the ferrocomplex of phosphate, Versene, cysteine, glutathione, mercaptoethanol, or ascorbate. The voltage response is rapid until a maximal negative value is approached.[6] Upon admission of oxygen, the potential rises from the initial low value to a second plateau which apparently represents the normal potential (potential at 50% oxidation) of the system. Ferrous sulfate (0.001 *M*) and 2,3-dimercaptopropanol (0.003 *M*) in 0.06 *M* potassium phosphate, pH 7.0, 25° produced this second plateau at —435 mv, and mercaptoethanol (0.002 *M*) and ferrous sulfate (0.001 *M*) produced the plateau at —275 mv (measured against the saturated calomel electrode, 25°, pH 7.0).

The complex of ferrous sulfate (0.001 *M*) and 2,3-dimercaptopropanol (0.003 *M*) will provide a steady potential during purification of extracts provided that no more than traces of oxygen enter the system. This concentration has been used routinely for purifying enzymes of the formate-pyruvate exchange system of *Streptococcus faecalis*[6] (see this volume [129]).

Cobalt–Thiol Chelates. The cobaltous-cysteine chelate is a powerful reductant and absorbant of oxygen. The reducing ability is highest at pH 7.8, and the potential equals the potential of the hydrogen electrode. Once cobalt has formed a chelate with cysteine, the cobaltic state cannot be readily reduced to the cobaltous state and forms the end product of the reaction.[7] Maximum rate of oxygen consumption is obtained when the ratio of cysteine to cobalt is 3:1.[4]

[5] L. Michaelis and C. V. Smythe, *J. Biol. Chem.* **94,** 329 (1931).

[6] N. P. Wood, P. Paolella, and D. Lindmark, *J. Bacteriol.* (in press).

[7] L. Michaelis, *in* "Currents in Biochemical Research" (D. E. Green, ed.), p. 220. Wiley (Interscience), New York, 1946.

The chelate of cobaltous sulfate and 2,3-dimercaptopropanol (ratio 2:1) combines with the extract and then forms a nondialyzable gelatinous coating on dialysis tubing.[6] The cysteine-cobalt chelate is tightly bound and does not combine with cell extract. The cysteine-cobalt chelate can be used at maximal potential with very little loss in voltage, provided that the chelate is in excess.

Ferrous–Phosphate Complex. Phosphate and pyrophosphate buffers have an accelerating influence on Fe^{++} —O_2 reaction.[8] In acid phosphate buffer, the rate of oxidation is first order in ferrous ion and is independent of ferric ion concentration.[9] Solutions of ferrous sulfate containing sodium pyrophosphate were found to oxidize about 1000 times faster than ferrous sulfate, and rates are about 5 times faster than predicted for phosphate catalysis.[8,10]

Solutions of ferrous sulfate (0.001–0.01 *M*) in phosphate buffer can be used to lower the O-R potential during purification, but the potential changes rapidly in the presence of trace amounts of air and no buffering action at a negative potential range is observed.

Measure of the O-R Potential

The many problems involved in controlling the O-R potential make it of major importance to have a means of measuring the potential of the extract at any time and at any stage of purification. A measure of the potential will not only show the effectiveness of reducing agents, but will indicate when air is entering the system. Two methods are used for measuring O-R potential, (1) the dye indicator method and (2) the direct electrometric determination of the potential.

O-R Indicators. A choice of indicators for measuring low O-R potentials is available.[11] Those most commonly used are:

	E'_0, pH 7
Methylene blue	+0.011
Indigo trisulfonate	−0.081
Indigo disulfonate	−0.125
Janus green	−0.225
Phenosafranine	−0.252
Safranine T	−0.289
Neutral red	−0.325
Benzyl viologen	−0.359
Methyl viologen	−0.440

[8] J. King and N. Davidson, *J. Am. Chem. Soc.* **80**, 1542 (1958).
[9] M. Cher and N. Davidson, *J. Am. Chem. Soc.* **77**, 793 (1955).
[10] A. B. Lamb and L. W. Edler, Jr., *J. Am. Chem. Soc.* **53**, 137 (1931).
[11] W. M. Clark, "Oxidation-Reduction Potentials of Organic Systems," p. 131. Williams & Wilkins, Baltimore, Maryland, 1960.

Each has a range of about 0.1 volt and is used in concentrations of 10^{-4} to 10^{-5} *M*. Indicators have been used in reaction mixtures and for preparing reagents free of oxygen, but all indicators have definite limitations. Many biological systems have small effective O-R capacity, and are not well poised. In these systems, the oxidized form of the dye is capable of oxidizing the enzymes without the dye being reduced appreciably so that the potential of the system approximates to that of the oxidized form of the dye.[12]

Direct Electrometric Determination. A more reliable method of measuring the potential of an extract is with a null point potentiometer or more conveniently with an electronic pH-millivolt meter. The potentiometric method has its limitations. Biological systems are slow to equilibrate with the electrode and accurate measurements may be difficult to obtain. Thiols often show irregular drifts in potential when measured with a platinum electrode. However, with the addition of ferrous ion there is an immediate improvement in the electrode response and discrepancies are generally eliminated.[2]

To measure the potential, an unattackable electrode (platinum) is immersed in the extract to form a half-cell. The potential of the half-cell is measured against a standard half-cell (calomel electrode) which makes contact with the extract through a salt bridge.

PLATINUM ELECTRODES. Bright platinum electrodes are commonly used for measuring the O-R potential of biological systems. An electrode can be prepared by sealing a wire spiral (No. 26) into soft glass tubing.[12] Greater surface area is provided when platinum foil (1 cm^2 and 0.125 mm thick) is spot-welded to a 2 cm length of platinum wire (No. 26),[13] commercial platinum electrodes of the thimble type (Beckman No. 39271) are also satisfactory and eliminate the problems of working with mercury.

Platinum electrodes can be cleaned by an alternate immersion in alcoholic KOH solution and in warm cleaning solution prepared by combining three volumes of 12 *M* HCl with one volume of 16 *M* HNO_3 and four volumes of water. After a slight etching, the electrodes are washed in water. They will retain their activity for several hours.[13]

STANDARD HALF-CELL. A commercial calomel electrode (Beckman No. 39170) used for pH determinations can be used for this purpose. The salt bridge connection is made through a fiber in the tip of the electrode. When a number of flasks or test tubes are used, it is more convenient to keep the calomel electrode in a reservoir of KCl kept at the same tem-

[12] F. Hewitt, "Oxidation-Reduction Potentials in Bacteriology and Biochemistry." Livingstone, Edinburgh and London, 1950.

[13] R. G. Bates, "Determination of pH." Wiley, New York, 1964.

perature as the platinum electrode and make salt bridge connections to the flasks or test tubes.

The potential of the saturated KCl electrode is not as reproducible as that of the unsaturated types and is subject to large hysteresis errors. The 0.1 N and 3.5 N calomel electrodes are reproducible, and the potential changes very little with changes of temperature.[13]

SALT BRIDGE. The salt bridge ensures electrical contact without introducing errors such as those encountered due to liquid:liquid junction potential. The bridge consists of an inverted glass U-tube of internal bore about 3 mm and filled with KCl (of the same concentration as in the calomel electrode) in 2% agar.[12] One end of the bridge is turned up at the tip to minimize downward diffusion of KCl (a sintered glass tip can be used for the purpose). The melted agar-KCl solution is drawn into the bridge by suction and is cooled before the vacuum is released. In use the agar-filled limb with the turned-up tip is immersed in the cell extract and the other limb makes contact with a reservoir of KCl.

Purification of Cell Extracts

The critical requirement of all the work is the rigorous exclusion of oxygen from all reagents and apparatus. A trial run should be made with reduced methylene blue-buffer solution to indicate when oxygen is entering the system. One apparatus found to be successful is shown in Fig. 1.

Preparation of Extracts. The French pressure cell is suitable for preparing extracts under anaerobic conditions because the extract can be extruded into a vessel which is being continuously flushed with an inert gas. Air, trapped inside the cell, must be expelled completely before pressure is applied. Active extracts can be prepared without reducing agents when stringent methods are used to remove oxygen from solutions by bubbling (40–60 minutes), evacuation, and boiling. Since oxygen will ultimately enter the system, a greater margin of safety is provided by adding to the extract and boiled solutions a reducing agent that will react rapidly with oxygen. Low activity usually reflects inactivation of the enzyme by oxygen.

Protamine and Ammonium Sulfate Treatment. For adding ammonium sulfate and protamine, a container such as a graduated cylinder cut to a convenient height, is fitted with a rubber stopper with holes for a platinum electrode, salt bridge, gas inlet and outlet, and a glass tube extending to the bottom of the cylinder to facilitate the removal of cell extract by gas pressure. Ammonium sulfate, protamine, and reducing agents are added by syringe through the gas exit port.

Air trapped in powdered ammonium sulfate makes the use of dry powder difficult. To remove oxygen from saturated ammonium sulfate,

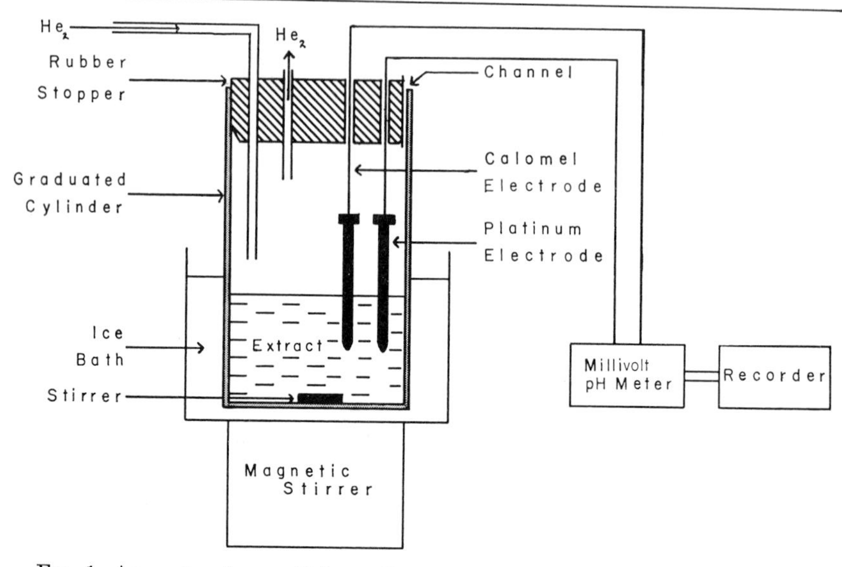

FIG. 1. Apparatus for purifying cell extracts. A container, such as a graduate cylinder cut to a convenient height, is fitted with a 4-holed rubber stopper which is grooved for inserting a syringe needle. The electrodes are suspended below the stopper to facilitate a positioning of the electrodes.

the solution is boiled, adjusted to the desired pH with concentrated NH_4OH, and cooled under gas. The reducing solution is added to a slurry of saturated ammonium sulfate and crystals. The supernatant solution is used to produce 60–70% saturation, and a slurry of crystals is added for complete saturation of the extract.

Saturated ammonium sulfate can also be reduced by hydrogen and palladium asbestos. The reduction takes an unusually long time, but the solution can be reduced very quickly when a small trace of reversible dye of sufficient negative potential range, such as safranine or rosinduline,[14] is added. Dyes of less negative potential ranges, such as methylene blue or the indophenols, are very much less, or not at all, effective in this respect.

Oxygen-Free Gases. Nitrogen, helium, argon, and hydrogen have been used to provide an anaerobic atmosphere. "Prepurified" nitrogen and sometimes helium need further purification, and this can be accomplished by passing the gas through a vanadious sulfate solution, pyrogallol, or over bright copper turnings heated to 600–700° in a combustion tube.[15]

[14] L. Michaelis and H. Eagle, *J. Biol. Chem.* **87**, 713 (1930).

[15] W. W. Umbreit, R. H. Burris, and J. F. Stauffer, "Manometric Techniques." Burgess, Minneapolis, Minnesota, 1964.

Column chromatography. Buffer solutions and Sephadex preparations are boiled and cooled under gas atmosphere. Mercaptopropanol should be added to the Sephadex and buffer before the column is packed and should be replaced with fresh buffer-reducing solution at the end of 24 and 48 hours (or until the O-R potential remains steady). The column is then developed under an oxygen-free atmosphere.

Reaction mixtures. To test for enzyme activity, a modification of the Thunberg method is recommended. Ingredients of the reaction are added except enzyme. A train of test tubes or flasks connected with two-holed rubber stoppers and glass tubing is flushed with an inert gas and then evacuated (twice) by a suitable oil pump or water aspirator to remove the gases in solution. This is accomplished readily by connecting one end of the train to the gas outlet and the other end of the train to a water aspirator. A vacuum is created when the gas is shut off. During a 3-minute interval, the tubes are held in a slanted position to minimize bumping and are tapped occasionally to facilitate the escape of bubbles. After the evacuation process, the cell extract or enzyme solution is added to the tubes while the tubes are being flushed with gas. Glass tubing should be used to prevent diffusion of oxygen. However, the diffusion of oxygen through Tygon tubing did not affect the activity significantly during purification.

[2] Enzyme Purification by Specific Elution Procedures with Substrate

By BURTON M. POGELL

In a discussion of enzyme purification in the first volume of this series, Colowick[1] referred to Heppel's procedure for elution of inorganic pyrophosphatase from alumina gel with dilute pyrophosphate[2] and suggested that utilization of the specificity of enzyme-substrate binding might be of general usefulness in the purification of enzymes. Actually, as early as 1910, Starkenstein[3] reported the tight binding of amylase by "insoluble" starch. A number of reports have since appeared on the purification of enzymes and separation of different enzyme activities by elution procedures with substrates. Although it is still far from being a general tool, application of this type of procedure in many specific

[1] S. P. Colowick, Vol. I [11].
[2] L. A. Heppel, Vol. II [91].
[3] E. Starkenstein, *Biochem. Z.* **24**, 210 (1910).

cases has proved to be of great practical value and has given purifications of a different order of magnitude over those obtained with classical fractionation steps.

Purification of Liver Fructose 1,6-Diphosphatase[4,5]

Assay Method.[6] Two micromoles of sodium fructose diphosphate (FDP) (Sigma Chemical Co.), 1 micromole of $MgSO_4$, 0.1 micromole of $MnCl_2$, 2 micromoles of mercaptoethanol, 10 micromoles of sodium borate (pH 9.5), and enzyme were incubated in a final volume of 0.2 ml at 38°. The reaction was terminated by the addition of 0.2 ml of 10% trichloroacetic acid, and inorganic phosphate was determined in the supernatant by the method of Gomori.[7] *N*-Phenyl-*p*-phenylenediamine hydrochloride was used as reducing agent in place of Elon. This reduced the time necessary for maximal color development to 10 minutes (Eastman Kodak abstract 432).

Specific activities are expressed as micromoles of inorganic phosphate released per hour per milligram of protein.

A detailed description of our development of a method for purification of rabbit liver fructose diphosphatase by substrate elution will serve to illustrate one approach to this problem. The rabbit liver enzyme has an isoelectric point of about 8, and therefore negatively charged substituted celluloses were tested for retention of the enzyme. Both CM-cellulose and P-cellulose were found to bind the enzyme at pH values below its isoelectric point. The CM-cellulose was chosen for more detailed studies, with the assumption that the weaker ionic binding would facilitate elution by substrate. Since CM-cellulose has a pK of about 4,[8] columns were run at pH 6 in order to have optimal charge on both adsorbant and enzyme. Optimal conditions for both adsorption and elution of enzyme were determined with small columns prepared from CM-cellulose (0.5–0.6 g of standard Selectacel from C. Schleicher and Schuell Co., Keene, New Hampshire packed in columns of 1–1.2 cm diameter). The CM-cellulose was first suspended in about 10 ml of 5 m*M* sodium malonate (pH 6.0) and readjusted to pH 6.0 by addition of NaOH; the slurry was poured without pressure into the column. Glass wool was placed at both the bottom and top of the cellulose. All column

[4] B. M. Pogell, *Biochem. Biophys. Res. Commun.* **7**, 225 (1962).

[5] B. M. Pogell, *in* "Fructose-1,6-diphosphatase and Its Role in Gluconeogenesis" (R. W. McGilvery and B. M. Pogell, eds.), p. 20. Am. Inst. Biol. Sci., Washington, D. C., 1964. (See also this volume [112a].)

[6] R. W. McGilvery, Vol. II [84].

[7] G. Gomori, *J. Lab. Clin. Med.* **27**, 955 (1941–1942).

[8] E. A. Peterson and H. A. Sober, *J. Am. Chem. Soc.* **78**, 751 (1956).

experiments were carried out at 0–4°. Flow rates were about 3 ml per minute, but this was not critical.

First, conditions were determined so that 5–10% of the applied enzyme (in 5 m*M* sodium malonate, pH 6.0) was not retained by the cellulose. This procedure ensured saturation of the column with enzyme for subsequent facilitation of elution. With a preparation of specific activity of 12.5, a 0.6 g column was saturated by 10 mg of enzyme. The amount of enzyme required for this purpose must be determined empirically, since it varied with different preparations, their degree of purification, and the particular batch of CM-cellulose being employed. The column was then washed with malonate until there was no more protein in the effluent (i.e., $\Delta A_{215 m\mu} - A_{225 m\mu} = <0.01$ with malonate as a blank).[9] This was followed by elution with malonate containing the specific eluant being tested and at pH 6.0.

The high selectivity of the substrate, FDP, as eluant for fructose diphosphatase is shown in the table. In the experiments summarized in Part A of the table, the minimal concentration of NaCl necessary to remove the bulk of the enzyme was first determined: 15 m*M* NaCl eluted 83% of the fructose diphosphatase and gave material of specific activity of 114, representing a ninefold purification. In contrast, concentrations of either the dimagnesium or sodium salts of FDP as low as 0.05 m*M* eluted the bulk of the enzyme activity with much lower amounts of contaminating protein. Since a divalent metal is necessary for phosphatase activity, elution by the sodium salt indicates either interaction of the substrate and enzyme in the absence of divalent metal or the presence of small amounts of magnesium in the enzyme preparation. The specificity of the FDP elution was shown by the fact that other salts, including tenfold higher concentrations of the hexose monophosphates, only partially removed the enzyme from the column and gave much lower specific activities. These experiments have been extended starting with material of lower specific activity, and purifications of fructose diphosphatase as high as 290-fold have been obtained in a single adsorption and elution from CM-cellulose with sodium FDP (Part B of the table). FDP eluted most of the enzyme in a sharp peak, whereas the activity came off the column in a much more diffuse manner with either 3 m*M* glucose 6-phosphate or glucose 1-phosphate.

It should be noted as seen in experiment 6 of Part A and experiment 5 of Part B, that, although elution with FDP after hexose monophosphate application removed a large percentage of the remaining enzyme, the specific activity was much lower than that found by elution with FDP

[9] J. B. Murphy and M. W. Kies, *Biochim. Biophys. Acta* **45**, 382 (1960).

SPECIFICITY OF ELUTION OF FRUCTOSE 1,6-DIPHOSPHATASE FROM CM-CELLULOSE COLUMNS[a]

Experiment	Eluant	Concentration (mM)	Per cent eluted	Specific activity	Purification factor
A. 1	NaCl	15	83	114	9
2	Mg_2FDP	0.14	97	818	65
3	Na_4FDP	0.05	80	1040	83
4a	$MgCl_2$	20	0	—	—
4b	$MgCl_2$	100	70	31	5.5
5	Na_2F-6-P	0.46	51	123	9.8
6a	Na_2G-6-P	0.50	18	157	12.6
6b	K_2G-1-P	0.50	4	—	—
6c	Na_4FDP	0.05	36	102	8.1
B. 1	Mg_2FDP	0.05	37	1770	290
2	Na_4FDP	0.15	58	935	153
3	Na_4FDP	0.30	90	692	122
4a	Na_2G-6-P	0.30	0	—	—
4b	Na_2G-6-P	3.0	>87	446	79
5a	K_2G-1-P	0.30	0	—	—
5b	K_2G-1-P	3.0	38	190	32
5c	Na_4FDP	0.30	24	437	74

[a] All eluants were dissolved in 5 mM sodium malonate (pH 6.0), and the pH was readjusted to pH 6.0. Where more than one eluant is listed in a single experiment, the respective solutions were applied in the order shown. Part A. The starting material (specific activity of 12.5) was rabbit liver enzyme partially purified by high speed centrifugation to remove particulate material, acid precipitation to remove further insoluble protein, and ammonium sulfate fractionation followed by dialysis and lyophilization. Of the lyophilized sample, 10 mg was dissolved in 1 ml of malonate buffer and applied to a 0.6 g CM-cellulose column. At least five 10-ml fractions were collected before changing to the eluants above. Protein was determined by either the 280 mμ absorbancy or the absorbancy difference at 215 mμ and 225 mμ (see text footnote 9). In experiments 1 and 2, 20 mg of lyophilized sample was applied to a 1.2 g column. Part B. Starting material (specific activity of 5.7–6.1 based on protein content and 3.5–3.7 on a dry weight basis) was prepared by dialyzing a high speed rabbit liver supernatant, removing any precipitate formed on adjustment to pH 6.0, and lyophilizing; 31 mg of dialyzed sample was dissolved in 1 ml of malonate buffer and applied to a 0.5 g CM-cellulose column. The column then was washed with buffer until the absorbancy difference ($A_{215m\mu}$-$A_{225m\mu}$) was zero. This took 150–200 ml. Protein was determined by the micro Lowry procedure (Vol. III [73]).

alone. In this type of column experiment, pretreatment with any eluant apparently alters the effect of the second eluant. This type of purification procedure has been modified and extended to large-size columns.[5] Also, substrate elution was recently utilized in the purification and crystallization of rabbit liver fructose diphosphatase.[10]

[10] S. Pontremoli and E. Grazi, this volume [26].

Purification of Aldolase

Sodium FDP was found to be much more specific than NaCl for the elution of rabbit liver aldolase from CM-cellulose columns.[4] Under similar conditions to those for fructose diphosphatase purification, 2.5 m*M* FDP eluted more than 90% of the adsorbed aldolase. The peak fraction contained enzyme 29-fold purer than the initial material. With 30 m*M* NaCl as eluant, the aldolase was eluted in a much more diffuse band and with much less purification. In these experiments, FDP and NaCl elution gave at least two peaks of aldolase activity, suggesting that the liver enzyme was composed of more than one molecular species.

Blostein and Rutter[11] were able to resolve crystalline muscle aldolase into two components by elution from CM-cellulose columns with FDP. Purification and crystallization of two different forms of rabbit liver aldolase has been attained utilizing elution from CM-cellulose columns with FDP.[12] Russell[13] has purified an aldolase from *Chlamydomonas mundana* grown on CO_2 from an initial specific activity of 10–20 micromoles/hr/mg protein to 3000–4000 micromoles/hr/mg protein by adsorption on P-cellulose and elution with 2 m*M* FDP.

Purification of Enzymes Bound to Polysaccharide Substrates

Numerous reports have appeared on the purification of amylases by taking advantage of their property of being tightly bound to starch granules.[14] Holmbergh[15] in 1933 reported the separation of malt α-amylase from β-amylase by adsorption on starch. Thayer[16] found that the exoamylase (an α-amylase) of *Pseudomonas saccharophila* was adsorbed by "insoluble" starch, whereas the endoamylase (a β-amylase) was not. In his purification procedure, the exoamylase was first removed by adsorption on a column of potato starch and Celite and then eluted by a gradient of "soluble" starch.

Successful purifications have also been attained for several other enzymes involved in either the synthesis or degradation of polysaccharide molecules using similar procedures. Leloir and Goldemberg[17] observed that the glycogen synthetase of liver was bound to "particulate" glycogen and were able to obtain a 300-fold purification of this enzyme

[11] R. Blostein and W. J. Rutter, *J. Biol. Chem.* **238,** 3280 (1963).
[12] W. J. Rutter, private communication.
[13] G. K. Russell, private communication.
[14] E. H. Fischer and E. A. Stein, *in* "The Enzymes" (P. D. Boyer, H. Lardy, and K. Myrbäck, eds.), 2nd ed., p. 313. Academic Press, New York, 1960.
[15] O. Holmbergh, *Biochem. Z.* **258,** 134 (1933).
[16] P. S. Thayer, *J. Bacteriol.* **66,** 656 (1953).
[17] L. F. Leloir and S. H. Goldemberg, *J. Biol. Chem.* **235,** 919 (1960). See also Vol. V [14].

(with respect to protein) by first separating the glycogen fraction from homogenates of livers of well-fed rats by differential centrifugation and then washing the "particulate" glycogen with saline containing soluble starch. This latter step removed most of the phosphorylase from the particles. Leloir and Goldemberg were able also to elute the enzyme from the particulate fraction with "soluble" glycogen.

The starch synthetase of beans has been found to be tightly bound to the starch granules and was purified by isolation and washing of these particles.[18] Evidence has recently been presented[19] that this enzyme is specifically bound to the linear polysaccharide, amylose, rather than to the branched-chain amylopectin. These studies stemmed from observations that starch synthetase from extracts of glutinous rice grain, which contains mostly amylopectin, was soluble, whereas it was particle bound in nonglutinous rice grain, where a considerable amount of amylose is present.

de la Haba[20] devised one of the more elegant procedures for isolation of crystalline phosphorylase *a* utilizing substrate adsorption and elution. The enzyme from crude rabbit muscle extracts was adsorbed on a starch column, washed, and then eluted with "soluble" glycogen. Advantage was taken of the facts that the enzyme was more tightly bound to the starch at 3° and more easily eluted by glycogen after the temperature of the column was raised to 14°. In a large-scale experiment, 100% recovery of activity was attained and the enzyme was purified 33-fold, to a specific activity at least as great as that of the twice-crystallized enzyme. Dissociation of the phosphorylase-glycogen complex was attained by allowing the enzyme to degrade the glycogen to completion.

Purification of Tyrosinase

A somewhat different approach to the utilization of specific enzyme-substrate binding sites was developed by Lerman for the purification of mushroom tyrosinase.[21] Azo dyes, which were known inhibitors of tyrosinase, were coupled through ether linkages to cellulose. Several of these cellulose adsorbants specifically retained the tyrosinase and very little other protein. A 61-fold purification with an overall recovery of 56% was obtained by adsorbing the enzyme and washing the column at pH 6 and then eluting at more alkaline pH values.

[18] L. F. Leloir, M. A. Rongine de Fekete, and C. E. Cardini, *J. Biol. Chem.* **236**, 636 (1961).
[19] T. Akazawa and T. Murata, *Biochem. Biophys. Res. Commun.* **19**, 21 (1965).
[20] G. de la Haba, *Biochim. Biophys. Acta* **59**, 672 (1962).
[21] L. S. Lerman, *Proc. Natl. Acad. Sci. U.S.* **39**, 232 (1953).

Other Enzymes

A combination of malate and DPN in Tris buffer (pH 8.5) was found to be more specific than NaCl for the elution of the malic dehydrogenase of *Ascaris suum* from DEAE-cellulose columns.[22] Our attempts to elute crude yeast hexokinase from DEAE-cellulose with various possible combinations of substrates and products were unsuccessful.[23]

Note Added in Proof: Arsenis and McCormick[24] have recently reported the selective purification of several enzymes by chromatography on flavin-substituted cellulose. This procedure was particularly effective for the purification of rat liver flavokinase on a column of flavin-cellulose, 7-celluloseacetamido-6,9-dimethylisoalloxazine, and of the apoenzyme from spinach glycolate oxidase on FMN-cellulose.

[22] M. B. Rhodes, C. L. Marsh, and G. W. Kelley, Jr., *Exptl. Parasitol.* **15**, 403 (1964).
[23] J. Gazith and B. M. Pogell, unpublished observations.
[24] C. Arsenis and D. B. McCormick, *J. Biol. Chem.* **239**, 3093 (1964); **241**, 330 (1966).

[3] Isolation of Mutants in the L-Arabinose Gene-Enzyme Complex

By Ellis Englesberg

The L-arabinose gene–enzyme complex in *Escherichia coli* B/r[1-12] is illustrated in Fig. 1. The structures of L-arabinose permease, L-arabinose isomerase, L-ribulokinase, and L-ribulose 5-phosphate 4-epimerase are determined by genes E, A, B, and D, respectively, and all four proteins

[1] J. Gross and E. Englesberg, *Virology* **9**, 314 (1959).
[2] E. Englesberg, *J. Bacteriol.* **81**, 996 (1961).
[3] E. Englesberg, R. L. Anderson, R. Weinberg, N. Lee, P. Hoffee, G. Huttenhauer, and H. Boyer, *J. Bacteriol.* **84**, 137 (1962).
[4] Nancy Lee and E. Englesberg, *Proc. Natl. Acad. Sci. U.S.* **48**, 335 (1962).
[5] Nancy Lee and E. Englesberg, *Proc. Natl. Acad. Sci. U.S.* **50**, 696 (1963).
[6] R. B. Helling and R. Weinberg, *Genetics* **48**, 1397 (1963).
[7] R. Cribbs and E. Englesberg, *Genetics* **49**, 95 (1964).
[8] E. Englesberg, J. Irr, and J. Power, *Bacteriol. Proc.* p. 19 (1964).
[9] D. Isaacson and E. Englesberg, *Bacteriol. Proc.* p. 113 (1964).
[10] C. Novotny and E. Englesberg, *Proc. 6th Intern. Congr. Biochem., New York, 1964, III-46.*
[11] E. Englesberg, J. Irr, J. Power, and N. Lee, *J. Bacteriol.* **90**, 946 (1965).
[12] E. Englesberg and E. Lieberman, unpublished data.

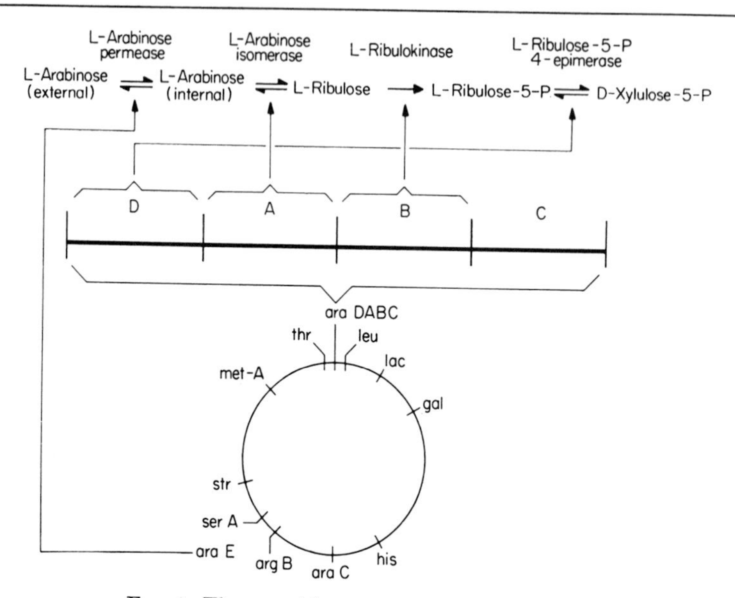

FIG. 1. The L-arabinose gene–enzyme complex.

are inducible by L-arabinose. Gene C, the activator gene, is a regulatory gene of a new type exhibiting positive control over structural gene expression.[8,11] Since expression of the four structural genes is under control of only gene C, induction is not sequential but essentially coincident; mutants deficient in any one enzyme, as a result of mutation in a particular structural gene, can be induced to produce the other enzymes in this system.

In utilizing mutants in the L-arabinose pathway for biochemical studies, it should be noted that mutations in the structural genes may have various and profound pleiotropic effects. Some of these effects can be extremely helpful,[13] and in other cases they can be a nuisance to the investigator, depending upon the particular mutant concerned and the intended use. Two examples will illustrate this point. (1) Mutants in the B gene, besides being deficient in L-ribulokinase, have increased or decreased rates of synthesis (coordinate) of isomerase, epimerase, and (in the case of a CRM former) L-ribulokinase CRM.[2,5] A hyperinducible mutant can be of great value in a study of the isomerase and epimerase, whereas, use of hypoinducible mutants would be a disadvantage in such a case. (2) Growth of L-arabinose negative mutants in the D gene is inhibited by L-arabinose. So although extracts of such mutants may be free of this enzyme, it is difficult to fully induce the other enzymes in the

[13] N. Lee and E. Englesberg, see this volume [80].

L-arabinose pathway because of this inhibition. However, such D mutants are valuable in providing a means of directly selecting for mutants in genes A, B, and D, as described below.[3,14]

General

Media[1]

Synthetic medium (mineral base). KH_2PO_4–K_2HPO_4, pH 7.0, 0.1%; $MgSO_4$ $7H_2O$, 0.01%; $(NH_4)_2SO_4$, 0.1%. Sterile concentrated carbohydrate(s) are added and amino acids when required. Agar (Bacto-Difco), when added, is at a final concentration of 0.2%.

Casein hydrolyzate mineral medium. This medium contains the mineral base plus 0.05% casamino acids (Difco). L-Arabinose, when added, is at a final concentration of 0.2%.

EMB L-arabinose made up according to the Difco formula with 1.0% L-arabinose as sole carbohydrate

Nutrient broth and nutrient agar (Difco)

Bacteria and Virus[1]

Escherichia coli B/r and a leucine requiring (leu^-) derivative

Bacteriophage Plbt

Growth of Phage and Bacteria and Transduction[1,2,14] Lysates are prepared by a modified soft-agar technique[15] using minimal Tris glucose agar medium[16] supplemented with 20% L broth.[14,17] Plates are incubated for 4–5 hours, lids up. To each dish 2–3 ml of eluting fluid (NaCl, 0.9%; peptone, 0.3%) is added, the top agar layer is thoroughly broken up, transferred to a centrifuge cup, and centrifuged at 4000–5000 rpm for 10–15 minutes to remove the agar. The supernatant is poured into a screw-capped test tube, and 0.3 ml of chloroform is added per 10 ml of suspension. The tube is shaken, and the suspension is checked for "sterility."

Transduction is carried out as previously described.[1]

Cultures are incubated at 37°.

Maintenance of Mutants. Mutants are maintained in one-half strength nutrient agar deeps and by lyophilization. Constitutive mutants are lyophilized soon after isolation since there is selection against hyperconstitutive mutants in nutrient agar deeps (or slants).

[14] H. Boyer, E. Englesberg, and R. Weinberg, *Genetics* **47**, 417 (1962).
[15] M. Swanstrom and M. H. Adams, *Proc. Soc. Exptl. Biol. Med.* **78**, 372 (1951).
[16] A. Hershey, *Virology* **1**, 108 (1955).
[17] E. S. Lennox, *Virology* **1**, 190 (1955).

Isolation of Arabinose Negative Mutants

General Procedure.[18] An overnight nutrient broth culture (5 ml) of *Escherichia coli* B/r is irradiated in a petri dish on a slow shaker with a dose of ultraviolet irradiation,[19] giving a survival of about 10^{-6} cells per milliliter. Aliquots (0.2 ml) of the irradiated suspension are distributed into tubes containing 5 ml of nutrient broth. The tubes are incubated overnight with aeration, and 0.1-ml samples of a 0.5×10^{-5} dilution of each culture are smeared on the surface of each of five EMB L-arabinose plates.[20] The plates are incubated in the dark; after 24 hours of incubation, negative clones, appearing as pink colonies, are isolated in pure culture. Those mutants which grow on mineral glucose agar, but fail to grow on mineral L-arabinose medium or grow poorly on the latter, are transferred to deeps and lyophilized. To ensure the isolation of mutants produced as a result of independent mutational events, only one of each type of negative colony is isolated from the five plates from one broth culture. Usually about half the cultures will contain an L-arabinose negative colony.

L-Arabinose negative mutants in the A, B, C, and E genes produce, on EMB L-arabinose, pink colonies about the size of, or somewhat larger than, the purple colony produced by the L-arabinose positive parent. Mutants in the D gene appear in 24–48 hours as minute (0.5 mm or smaller) pink colonies due to the L-arabinose inhibition.[3]

Presumptive identification of permeaseless mutants (E gene mutants) from mutants in the A, B, or C genes can be ascertained by the slow growth of the former on mineral-arabinose agar. Mutants in the A, B, and C genes (except for leaky ones) fail to grow on this medium.

Identification of the Type of Mutants Isolated Using Intact Cells.[2] The growth from an overnight nutrient agar slant culture is harvested, centrifuged, and resuspended in saline to a turbidity that is equivalent to approximately 1.7×10^9 cells/ml. A 0.2-ml sample of this suspension is used to inoculate 4.8 ml of casein hydrolyzate (0.05%), mineral medium, with and without 0.2% L-arabinose, in optically tared test tubes (18×150 mm). The tubes are incubated with aeration, and turbidity is measured directly at half-hour intervals for 4–5 hours in a Fisher electrophotometer modified to take the test tubes. A 425-mμ filter is employed, uninoculated medium being used as the blank. Sub-

[18] For the sake of simplification, the procedure described is for a prototroph. If a leucine auxotroph is employed in the isolation of L-arabinose negative mutants, leucine should be added to the synthetic medium.

[19] Chemical mutagens may be employed in place of ultraviolet irradiation.

[20] The penicillin selection technique has invariably proved unsuccessful in our hands for the isolation of L-arabinose negative mutants.

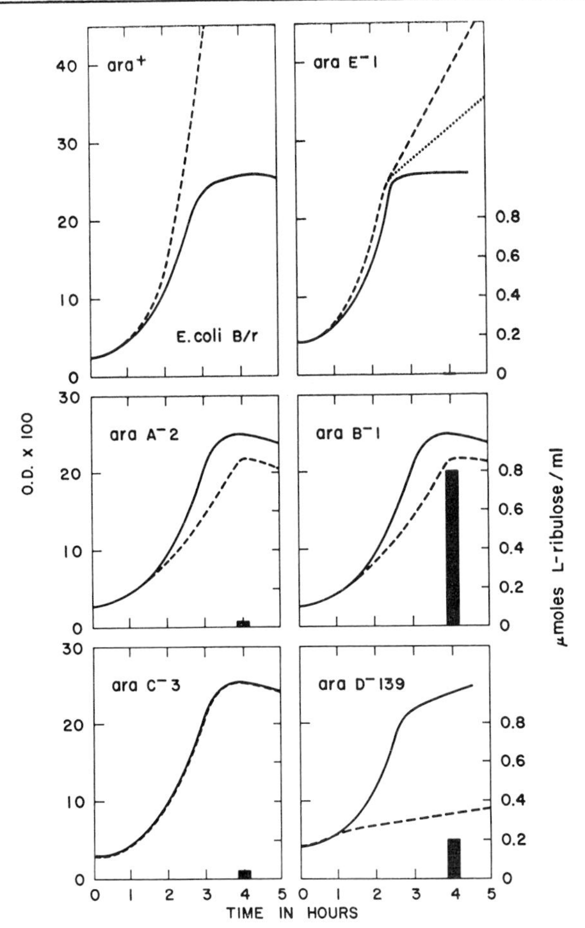

FIG. 2. Comparison of the growth (lines) and L-ribulose production (bars) of *Escherichia coli* B/r and various L-arabinose negative mutants in a casein hydrolysate (0.05%), mineral medium in the presence or absence of 0.2% L-arabinose (-----). The growth rate of the E⁻ mutant (permeaseless) is dependent upon the concentration of L-arabinose employed. Notice reduction in growth rate with 0.1% L-arabinose (.....).

sequent to full growth (4 hours of incubation), 0.1 ml of each culture is assayed for ribulose by the cysteine-carbozole test, the appropriate medium being used as the blank.[2,21] For rapid scanning of a large number of mutants, turbidity readings may be taken only at 4 hours of incubation. The results of a typical assay are shown in Fig. 2.

The permease assay,[11] the preparation of cell free extracts, and the assay procedure for the individual enzymes[2,5,7] have been described.

[21] Z. Dische and E. Borenfreund, *J. Biol. Chem.* **192**, 583 (1951).

Positive Selection of Arabinose Negative Mutants in the A, B, and C Genes by the Use of D^- Mutants.[3,14] An overnight culture of a D^- mutant started from a small inoculum (approximately 1×10^3 cells) is streaked on to the surface of an EMB L-arabinose plate. The growth of the parent culture is inhibited, and resistant negative colonies appear in 24 hours and continue to appear for several days.[22] L-Arabinose positive revertants are also present. The L-arabinose resistant, L-arabinose negative clones are double L-arabinose negative mutants of three types; D^-A^-, D^-B^-, and D^-C^-. The new mutants may be characterized by procedures described above.

Isolation of the new mutant site from D^- is accomplished in the following manner: Phage Plbt grown on the double mutant is used to transduce an ara^+ leu^- strain. Selection is made for leu^+ transductants on mineral glucose agar, and the transductants that are also ara^- are isolated. To test for the presence of the original D^- site, phage grown on the original D^- mutant are used to infect each of the isolated transductants and selection is for ara^+. Those cultures that yield ara^+ transductants are free of the original D^- mutant site. On the average, one out of five leu^+ transductants will have only the new ara^- mutant site.

Positive Selection of Arabinose Negative Mutants in the C and E Genes by Use of A^- Mutants.[9] An overnight culture of an A^- mutant, started from a small inoculum (approximately 1×10^3 cells) is streaked on to the surface of a L-arabinose (0.02–0.2%), glycerol (0.1%), mineral medium with the potassium phosphate content reduced to one-tenth its normal amount. Mutants in the A gene are inhibited by L-arabinose in this medium, and resistant colonies appear in 24–48 hours and continue to appear for several days.[22] These mutants contain the original A^- mutant site and an additional mutant site in either the C or E genes (A^-C^- or A^-E^-). It is a simple matter to separate the E^- mutant site from A^- since the E and A genes are unlinked.[9,12] Phage grown on the wild type are used to transduce the double mutants. Selection is for L-arabinose utilization on mineral L-arabinose plates. If the culture is A^-E^-, small colonies (A^+E^-) will appear in 48 hours. No wild-type recombinants are possible and A^-E^+ recombinants will not grow on this medium. On the other hand, if the double mutant is A^-C^-, wild-type colonies will appear in 24 hours since the A and C genes are cotransducible. In this case, the C^- mutant site can be separated from A^- by similar procedures as outlined for D^- double mutants.

[22] To ensure the isolation of mutants of independent origin, only one resistant colony appearing in 24 hours and those colonies appearing in 48 hours or later should be selected.

Isolation of Constitutive Mutants

General.[8,11] D-Fucose inhibits the growth of *E. coli* B/r in mineral L-arabinose medium, and selection occurs for D-fucose resistant mutants that produce various but coordinate constitutive levels of enzymes (including permease) involved in the initial steps in L-arabinose metabolism.

Isolation Using Liquid Medium. This procedure is designed to conserve D-fucose and to isolate, at a maximum, 10 constitutive mutants derived from independent mutational events. An overnight nutrient broth culture of *E. coli* is diluted into fresh nutrient broth to yield approximately 10^3 viable cells per milliliter. The broth suspension is divided into 10 separate 1-ml portions and incubated overnight. Ten tubes containing 2.5 ml of mineral L-arabinose (0.04%), D-fucose (0.08%), liquid medium (MAF) are each inoculated with 0.05 ml of these overnight cultures, incubated until they become turbid (2–7 days), and then streaked onto nutrient agar. Colonies that appear are tested by picking and streaking on to MAF agar (0.1% L-arabinose and 0.2% D-fucose). Most of the colonies appearing on the nutrient agar plates are resistant to fucose inhibition, so that it is only necessary to test two or three colonies per culture. (One MAF plate is sufficient to test 100 colonies.) One fucose-resistant clone from each of the nutrient broth cultures is isolated in pure culture by restreaking on nutrient agar, then lyophilized, after again verifying that each of the isolates are resistant to D-fucose.

Isolation from MAF Agar Plates. MAF agar plates are smeared with 0.1 ml of a nutrient broth culture of *E. coli* started from small inocula as described above. D-Fucose-resistant colonies appear in 24 hours and continue to arise for several days.[22] Isolation is carried out on MAF agar plates.

[4] Periodate Methods: Determination of Compounds Oxidized to Carbonyl Acids

By ELLIOT JUNI

Analytical Method[1]

Principle. Periodate oxidations have been widely used for structural analysis, particularly for the study of carbohydrates.[2] Methods have been

[1] E. Juni and G. A. Heym, *Anal. Biochem.* **4,** 143 (1962).

[2] J. M. Bobbitt, *Advan. Carbohydrate Chem.* **11,** 1 (1956). See also J. R. Dyer, *in* "Methods of Biochemical Analysis" (D. Glick, ed.), Vol. III, p. 111. Wiley (Inter-

described for estimation of the oxidation products, those most frequently encountered including formic acid,[3] formaldehyde,[4] and carbon dioxide.[2] When carbon dioxide is formed during periodate oxidation of sugars and sugar acids, this product usually arises as a result of oxidation of α-carbonyl acid intermediates.[2] Under conditions of strong acidity, periodic acid will not oxidize α-carbonyl acids.[1] The procedure to be described makes possible the quantitative determination of the following compounds by analysis of the α-carbonyl acids formed upon periodate oxidation:

Sugar acids

$$CH_2OH{-}(CHOH)_4{-}CO_2H \xrightarrow{+\ 4\ HIO_4} HCHO + 3\ HCO_2H + CHO{-}CO_2H \qquad (1)$$

Disaccharides (1,3-or 1,4-linked)

$$\begin{array}{l}\ulcorner CH(OH)\\ HCOH\\ ROCH \quad O\\ HCOH\\ \llcorner HC\\ CH_2OH\end{array} \xrightarrow{+\ HIO_4} \begin{array}{l}\ulcorner CHO\\ CHO\\ ROCH \quad O\\ HCOH\\ \llcorner HC\\ CH_2OH\end{array} \xrightarrow{+\ H_2O} \begin{array}{l}HCO_2H\\ CHO\\ ROCH\\ HCOH\\ HCOH\\ CH_2OH\end{array} \xrightarrow{+\ 2\ HIO_4}$$

$$\begin{array}{l}CHO\\ ROCH\\ CHO\\ HCO_2H\\ HCHO\end{array} \xrightarrow{+\ HIO_4} \begin{array}{l}CHO\\ ROCOH\\ CHO\end{array} \xrightarrow{+\ HIO_4} \begin{array}{l}HCO_2H\\ ROCO\\ CHO\end{array} \xrightarrow{+\ H_2O} ROH + \begin{array}{l}CO_2H\\ CHO\end{array} \qquad (2)$$

α-Acetolactic acid

$$\begin{array}{l}CH_3{-}COH{-}CO_2H\\ \quad\ CO\\ \quad\ CH_3\end{array} \xrightarrow{+\ NaBH_4} \begin{array}{l}CH_3{-}COH{-}CO_2H\\ \quad\ CHOH\\ \quad\ CH_3\end{array} \xrightarrow{+\ HIO_4} \begin{array}{l}CH_3{-}CO{-}CO_2H\\ \quad\ CHO\\ \quad\ CH_3\end{array} \qquad (3)$$

science), New York, 1956. See also R. D. Guthrie, *in* "Methods in Carbohydrate Chemistry" (R. L. Whistler and M. L. Wolfrom, eds.), Vol. I, pp. 432, 435. Academic Press, New York, 1962.

[3] See Vol. III [52].

[4] See Vol. III [43]. See also J. C. Speck, *in* "Methods in Carbohydrate Chemistry" (R. L. Whistler and M. L. Wolfrom, eds.), Vol. I, p. 441, Academic Press, New York, 1962.

Reagents

Sulfuric acid, 10 *N* aqueous solution. To 72.2 ml of distilled water is added 27.8 ml of reagent-grade sulfuric acid.

Periodic acid or sodium periodate, 0.1 *M* aqueous solution. Dissolve 2.28 g of paraperiodic acid (H_5IO_6) or 2.14 g of sodium meta periodate ($NaIO_4$) in distilled water to 100 ml. Store in a brown bottle in the dark.

Thioacetamide, 0.867 *M* aqueous solution. Dissolve 650 mg of thioacetamide (Distillation Products Industries) in distilled water to 10 ml. Prepare fresh each day.

p-Nitrophenylhydrazine reagent, 0.1 *M* solution in 10 *N* sulfuric acid. Dissolve 380 mg of *p*-nitrophenylhydrazine hydrochloride (Distillation Products Industries) in distilled water to 10 ml and filter through Whatman No. 1 filter paper. Add a volume of 20 *N* H_2SO_4 equal to the volume of the filtrate. Prepare fresh each day and store at 25–30° to prevent crystallization of *p*-nitrophenylhydrazine hydrosulfate.

Ethyl acetate, reagent grade

Tris, 0.5 *M* aqueous solution. Dissolve 12.1 g of Tris in distilled water to 200 ml.

Sodium borohydride, 1.0 *M* aqueous solution. Dissolve 380 mg of $NaBH_4$ in distilled water to 10 ml. Filter twice through Whatman No. 1 filter paper. Prepare fresh each day.

Sodium gluconate standard, 0.0001 *M* aqueous solution. Stock 0.1 *M* solution is prepared by dissolving 1.09 g of sodium gluconate in distilled water to 50 ml. The 0.0001 *M* solution of sodium gluconate is prepared daily by dilution from the 0.1 *M* stock solution. The stock solution is stored at 3°.

Sodium pyruvate standard, 0.0001 *M* aqueous solution. Stock 0.1 *M* solution is prepared by dissolving 550 mg of sodium pyruvate (Nutritional Biochemicals Corporation) in distilled water to 50 ml. The 0.0001 *M* solution of sodium pyruvate is prepared daily by dilution from the 0.1 *M* stock solution. The stock solution is stored at 3°.

Procedure

Step 1. Periodate Oxidation. To an 18 × 150 mm Pyrex test tube are added 1.0 ml of 10 *N* H_2SO_4 and exactly 1.0 ml of 0.1 *M* paraperiodic acid or 0.1 *M* sodium meta periodate. The contents are mixed thoroughly. The maximum sample volume employed is 2.0 ml. When smaller samples are used, sufficient water is first placed in the tube to make a total volume of 4.0 ml when the sample is finally added. All samples are run in duplicate, and duplicate reagent blanks, containing 2.0 ml of water in place of sample, are always prepared for each assay. For samples giving rise to glyoxylate, sodium gluconate may be used as a standard. Sodium pyruvate is the standard when pyruvate is the carbonyl acid expected after periodate oxidation. After addition of a neutral or slightly alkaline sample, containing 0.02–0.2 micromole of a carbonyl acid precursor, the contents of each tube are mixed thoroughly and placed in a 30°C water bath.

Step. 2. Termination of Periodate Oxidation. After incubation at 30°C for 45 minutes all tubes are removed from the water bath and exactly 0.5 ml of 0.867 *M* thioacetamide is added to each tube. Immediately after addition of thioacetamide, each tube is mixed gently by hand and then allowed to stand at room temperature for 5–10 minutes. The tubes are than mixed individually (orbital mixer[5]) for 30 seconds.

When periodate reacts with thioacetamide some of the iodine formed as an intermediate of the reduction combines with sulfur to yield a brown precipitate. This sulfur-iodine complex does not interfere with any of the further steps in the assay procedure.

Step 3. Formation of p-Nitrophenylhydrazones. After all tubes in a given assay are mixed, 0.5 ml of *p*-nitrophenylhydrazine reagent is added to each tube; after mixing again, all tubes are returned to the 30° water bath.

Step 4. Extraction of p-Nitrophenylhydrazones from Aqueous Phase. After incubation for 45 minutes at 30° the tubes are removed from the water bath, 5.0 ml of ethyl acetate is added to each tube, followed by mixing (orbital mixer) for 30 seconds.[6] As much as possible of the upper ethyl acetate layer, which separates fairly rapidly after mixing is stopped, is removed using a 5.0-ml pipet with the aid of a pipettor,[7] and the ethyl acetate is transferred to a dry 18 × 150 mm Pyrex tube. A pellet, consisting largely of sulfur, usually appears at the water–ethyl acetate interface. Approximately 4.5 ml of ethyl acetate can be removed without contamination by liquid from the aqueous phase. In order to recover most of the carbonyl acid derivative still remaining in the small amount of ethyl acetate covering the aqueous phase, 2 ml of ethyl acetate is added to the residual ethyl acetate layer and, without further

[5] Vortex, Jr., Scientific Industries Inc., Springfield, Massachusetts.

[6] In order to avoid possible loss of liquid during mixing, each tube is first covered with an inverted polyethylene thimble (Size No. 2, Cat. No. 9314-F, Arthur H. Thomas Co., Philadelphia), which is kept in position by the index finger of the hand used to hold the tube.

[7] Propipette, Instrumentation Associates, New York.

mixing, this ethyl acetate is removed and placed in the tube containing the bulk of the original ethyl acetate used for extraction. This procedure is repeated once more to make a total of two 2-ml rinses.

Step 5. Extraction of Carbonyl Acid p-Nitrophenylhydrazones. To the ethyl acetate extract is added 4.0 ml of 0.5 *M* Tris. The carbonyl acid *p*-nitrophenylhydrazones are extracted into the Tris by mixing (orbital mixer) for 30 seconds. After separation of a clear bottom layer of Tris, as much as possible of the upper ethyl acetate layer, which may be somewhat turbid, is removed with a 10-ml pipette (with pipettor) and discarded. The Tris extract is covered with 5 ml of ethyl acetate and washed by mixing (orbital mixer) for 30 seconds.

Step. 6 Determination of Concentration of Extracted Carbonyl Acid. For each tube, approximately 3 ml of Tris extract is removed from beneath the ethyl acetate wash with a 5-ml pipette (with pipettor) and transferred to a 1-cm rectangular absoption cell. The absorbance of reagent blanks and samples are read at 390 mμ in a Beckman Model DU Spectrophotometer against a water or 0.5 *M* Tris blank. The absorbancies of duplicate tubes, which usually differ by less than 3%, are averaged and substracted from the average reading for the reagent blanks. For all glyoxylate precursors that yield equivalent amounts of glyoxylate upon periodate oxidation the net absorbance for 0.1 micromole does not differ by more than 2% from a value of 0.585. The corresponding absorbance for pyruvate precursors is 0.564. The net absorbance is a linear function of concentration for compounds that give rise to glyoxylate or pyruvate upon periodate oxidation. Linearity is still maintained using as large a sample as 1.0 micromole of gluconate or 1.0 micromole of pyruvate and diluting the derivative-containing Tris extract tenfold with 0.5 *M* Tris before determining the absorbance.

Step 7. Reduction prior to Periodate Oxidation. To 1.9 ml of neutral or slightly alkaline sample in an 18 × 150 mm Pyrex tube is added 0.1 ml of 1.0 *M* sodium borohydride solution. After mixing, all tubes treated with the reducing agent, including reagent blanks, are incubated in a 30° water bath for 30 minutes. The tubes are then removed from the water bath and to each is added 1.0 ml of 10 *N* H_2SO_4. After thorough mixing the tubes are incubated at 30° for 10 minutes followed by mixing (orbital mixer) for 30 seconds. To each tube is then added 1.0 ml of 0.1 *M* paraperiodic acid, or of 0.1 *M* sodium meta periodate. The tubes are mixed thoroughly and incubated at 30° for 45 minutes. The remaining procedure is the same as steps 2–6 outlined above.

Destruction of excess borohydride with acid results in lactonization of any sugar acids present. The following procedure is therefore employed, when sugar acids are present, to reduce and delactonize. To a

5-ml neutral or slightly alkaline sample, containing 0.15–1.5 micromoles of carbonyl acid precursor, is added 0.5 ml of 1.0 *M* sodium borohydride. After mixing, the solution is incubated at 30° for 30 minutes. Excess borohydride is then destroyed by addition of 1.0 ml of 1.0 *N* H_2SO_4. The acidified sample is incubated at 30° for 10 minutes followed by mixing (orbital mixer) for 30 seconds. Delactonization is accomplished by adding 0.5 ml of 2.5 *N* NaOH, mixing, and allowing the solution to stand for several minutes. The standard assay (steps 1–6) is made using 1.0 ml of the delactonized solution. The small amount of alkali in this sample does not cause any interference.

Step 8. Determination of Carbonyl Acids prior to Periodate Oxidation (Modified Friedemann and Haugen Test). To an 18 × 150 mm Pyrex tube are added 3.0 ml of sample or water (blank) and 1.0 ml of 10 *N* H_2SO_4, and the contents are thoroughly mixed. *p*-Nitrophenylhydrazones of carbonyl acids are prepared and assayed as outlined in steps 3–6.

Applicability of Assay

The following sugar acids have been found to yield equimolar amounts of glyoxylic acid upon periodate oxidation using the above assay procedure: gluconic, galactonic, glucoheptonic, gulonic, glucuronic, saccharic (two equivalents), and 6-phosphogluconic acids.[8] Periodate oxidation of sugar acid lactones gives rise to unpredictable small excesses of glyoxylate as a result of an "overoxidation" reaction characteristic of this reagent.[8] For this reason it is necessary that samples containing sugar acid lactones be neutralized or made slightly alkaline before oxidation with periodic acid in order to hydrolyze any lactone present. Reduction of 2-ketogluconate with sodium borohydride prior to periodate oxidation makes possible the quantitative determination of this compound.

Periodate oxidation of maltose, lactose, and cellobiose results in equimolar yields of glyoxylic acid [Eq. (2)]. Since only 1,3- or 1,4-linked disaccharides yield glyoxylic acid upon periodate oxidation,[8] it is possible to use this assay procedure for linkage studies of disaccharides and oligosaccharides. Periodate oxidation of disaccharides is a slower process than for sugar acids, and the assay procedure described above should be altered to provide for 2 hours of incubation with periodic acid rather than 45 minutes. Detailed time curves for all the compounds studied have been published.[8] Lactobionic and cellobionic acids are each oxidized by periodate to 2 moles of glyoxylic acid, oxidation being complete in 3 hours.

Some aldoses, such as glucose, give an approximate 1% yield of

[8] E. Juni and G. A. Heym, *Anal. Biochem.* **4**, 159 (1962).

glyoxylic acid upon periodate oxidation. Prior reduction with sodium borohydride diminishes this small interference to 0.2% or less. Reduction of disaccharides before periodate oxidation serves to lower the glyoxylic acid yield to 2% or less. Possible explanations for these effects have been reported.[8] Another advantage of reduction with sodium borohydride prior to periodate oxidation is obtained when free carbonyl acids are present since these are also reduced and will not interfere in the assay for carbonyl acids formed by periodate oxidation even when the free carbonyl acids are present in 100-fold excess.

Periodate oxidation of ketoses is known to produce a mixture of glyoxylic and glycolic acids, since initially some molecules are cleaved between carbons 1 and 2 while others are cleaved between carbons 2 and 3. By the present assay procedure, glyoxylic acid yields from fructose and sorbose were found to be 38.8% and 48.9%, respectively.

Other compounds shown to give quantitative glyoxylic acid yields in this assay are tartaric (two equivalents) and glyceric acids, a 3-hour oxidation period being required for complete oxidation of tartaric acid. Under the conditions of the assay, 3-phosphoglyceric acid is not oxidized to glyoxylic acid. Using the present procedure it has been possible to distinguish hydroxypyruvate from its isomer tartronic semialdehyde. Hydroxypyruvate is oxidized to glyoxylic acid only after prior reduction to glycerate whereas tartronic semialdehyde is oxidized to glyoxylate both before and after reduction with sodium borohydride.

Application of the standard assay procedure preceded by sodium borohydride reduction makes it possible to determine α-acetolactate in the presence of as much as a 100-fold excess of pyruvate, the enzymatic precursor of α-acetolactate, as well as a 100-fold excess of acetoin, the decarboxylation product of α-acetolactate. Prior reduction with sodium borohydride [Eq. (3)] under slightly alkaline conditions, where α-acetolactate is stable, has the additional advantage of producing a dihydroxy acid; this acid, unlike α-acetolactate, is stable in strong acid and is then oxidized to pyruvic acid and acetaldehyde. Acetaldehyde and other nonacidic carbonyl compounds do not interfere in the assay. Another compound which should give rise to pyruvic acid upon periodate oxidation is saccharinic acid.

Since it is possible to control accurately the time of incubation of sample with periodic acid, the assay procedure may be used to study the kinetics of oxidation of compounds for which carbonyl acids occur as intermediates. It has been shown, for example, that periodate oxidation of various disaccharides proceeds with a lag before the appearance of glyoxylic acid.[8] The lag observed is characteristic for the particular disaccharide studied, being shortest for maltose (15 minutes) and extending for almost an hour for the case of cellobiose. Kinetic analysis of periodate oxidation of aldobionic acids also shows a unique behavior for each of the compounds studied.[8]

The assay procedure has been used for the estimation of hydroxypyruvic aldehyde and its differentiation from glyceraldehyde and dihydroxyacetone, since only hydroxypyruvic aldehyde is cleaved, at least in part, to glyoxylic acid and formaldehyde.[9] It has been possible to distinguish 2-ketogluconate from 5-ketogluconate by this method since only the latter compound will be oxidized to glyoxylic acid.[10]

The method has also been shown to be suitable for use in biological systems. Protein, most of which may be removed by conventional procedures, does not cause any interference. Serine and threonine are known to be oxidized by periodate with the intermediary formation of glyoxylate. These amino acids are not oxidized under the strongly acidic conditions of this assay. The presence in the sample to be oxidized of compounds susceptible to periodate oxidation has been shown not to interfere if as much as 25 micromoles of periodic acid (25% of the periodic acid added to a sample) is used up during oxidation of such compounds.

Use of the Procedure as a General Assay for Carbonyl Acids. The application of the assay procedure for analysis of carbonyl acids in the absence of periodate, as described in step 8, has several advantages over methods involving use of 2,4-dinitrophenylhydrazine as the carbonyl reagent.[11]

Carboxylic acids such as formate and acetate cause interference in assays using 2,4-dinitrophenylhydrazine, but not when *p*-nitrophenylhydrazine is used.[1] *p*-Nitrophenylhydrazine hydrochloride is considerably more soluble than 2,4-dinitrophenylhydrazine hydrochloride. It is therefore possible to use larger concentrations of the former reagent, a fact that is of considerable advantage when assaying for carbonyl acids in the presence of large excesses of nonacidic carbonyl compounds, which can complete with carbonyl acids for reagent. It has been shown that Tris, unlike sodium carbonate solutions usually used, completely extracts the *p*-nitrophenylhydrazones of glyoxylate and pyruvate from ethyl acetate.[1]

It has been customary to determine the concentration of carbonyl acid hydrazones by measuring the absorbance of the colored solution obtained upon addition of strong alkali.[11] Since the color obtained with glyoxylic acid fades whether 2,4-dinitrophenylhydrazine or *p*-nitro-

[9] H. C. Reeves and S. J. Ajl, *J. Biol. Chem.* **240,** 569 (1965).
[10] J. L. Goddard and J. R. Sokatch, *J. Bacteriol.* **87,** 844 (1964).
[11] See Vol. III [66].

phenylhydrazine is used, it has been found more advantageous to determine the absorbancies of solutions of the hydrazones directly without addition of alkali, these derivatives being stable in either carbonate or Tris.

The use of *p*-nitrophenylhydrazine in place of 2,4-dinitrophenylhydrazine also makes it possible to compare directly results of carbonyl acid determinations with and without prior periodate oxidation. The orbital mixer used for extractions between aqueous and organic solvent phases has proved to be extremely efficient and eliminates the need for an aeration mixing apparatus.

[5] Determination of Formate

By OTHMAR GABRIEL

General Method

Principle.[1] Formate is converted to *p*-bromophenacyl formate by reaction with 2,4′-dibromoacetophenone. The *p*-bromophenacyl formate is isolated from the reaction mixture by chromatography on silica gel. The ethanol solution of the pure compound has an absorption maximum at 256 mμ which permits fast and accurate quantitation. An aliquot of the same solution is used for the radioactivity measurement. For alternate methods see Vol. III [52], [60], and [61].

Reagents and Materials. Two grams 2,4′dibromoacetophenone (Eastman Organic Chemicals) is recrystallized from 8–10 ml ethanol before use. Addition of 100 mg Norit A to the boiling alcoholic solution and filtration on a sintered glass funnel results in a colorless solution. Upon cooling in an ice bath, pure white crystals are obtained which are washed with a small volume of pentane.

Silica gel G for thin layer chromatography, according to Stahl, can be obtained from Brinkmann Instruments Inc. and silica gel 200 mesh from Davison Chemical Company.

Measurements of radioactivity are carried out in a liquid scintillation counter using Bray's system.[2]

Preparation and Isolation of *p*-Bromophenacyl Formate

Procedure. To 1 ml of an aqueous solution containing 50–1000 micromoles of sodium formate, adjusted to pH 5.0, is added an equimolar

[1] O. Gabriel, *Anal. Biochem.* **10**, 143 (1965).
[2] G. A. Bray, *Anal. Biochem.* **1**, 279 (1960).

amount of 2,4′-dibromoacetophenone dissolved in 3.0 ml of ethanol. After refluxing for 30 minutes, the ethanol is removed by distillation *in vacuo*. On cooling in an ice bath the aqueous solution solidifies to form a white crystalline mass. The material dissolves readily in 5–10 ml ether. The ether solution is washed 3 times with 5 ml of water and the water washes are discarded. The ether is evaporated *in vacuo* at room temperature. The residue is recrystallized from 1–5 ml of ethanol and washed with cold 20% ethanol and pentane. The dry product is dissolved in 3 ml benzene–ethylacetate (95:5 v/v) and added to a column of silica gel (20 g, 2 × 11 cm) which had been previously washed with the benzene–ethylacetate solvent. The column is eluted with 200 ml of the same solvent, and 10-ml fractions are collected at a flow rate of approximately 1 ml per minute. Each fraction is checked for radioactivity, and a small aliquot is subjected to thin layer chromatography (see below).

The tubes containing the formyl ester are pooled (Table I), the solvent is removed *in vacuo* at room temperature, and the residue is recrystallized from absolute ethanol. The pure formyl ester is dissolved in ethanol, and the concentration is determined by spectrophotometry at 256 mμ (Table I). An aliquot of this solution is pipetted into a glass

TABLE I
PROPERTIES OF *p*-BROMOPHENACYL DERIVATIVES

Compound	Test tube number in silica gel column chromatography	R_f Value in thin layer chromatography	Melting point (Kofler) (°C)	$\epsilon_{256} \times 10^4$ in ethanol
Br—C_6H_4—$COCH_2Br$	3	0.65	113	—
Br—C_6H_4—$COCH_2OH$	12–18	0.10	120 subl.[a] 130 dec.[b] 140–142	1.668
Br—C_6H_4—$COCH_2OOCH$	6–9	0.33	99 dec.[b] 102–103	1.736

[a] Start of sublimation.
[b] Start of decomposition.

vial and the ethanol is evaporated by a stream of air. Bray's solution (10 ml) is added to the vial, and the radioactivity is determined in a liquid scintillation counter. Several recrystallizations from ethanol, establishing constancy of the specific activity, the melting point, and the demonstration of a single spot on thin layer chromatography, are sufficient to prove the identity in most instances.

Thin Layer Chromatography. Silica gel G (25 g) is suspended in 62 ml water and spread evenly over glass plates at a thickness of 0.25 mm. After activation of the gel by heating at 100° for 30 minutes, ascending chromatography was carried out for 20–30 minutes in a vapor-saturated chamber containing benzene–ethylacetate (95:5 v/v). Samples of *p*-bromophenacyl formate dissolved in this solvent must not be allowed to dry when applied to the chromatogram; because of lability, the compound decomposes partially to *p*-bromophenacyl alcohol, and double spot formation may result.

The widely separated *p*-bromophenacyl derivatives are located in ultraviolet light with a sensitivity of about 0.01 micromole. A somewhat more sensitive method of visualization entails spraying the plate evenly with a reagent containing 100 mg 2,4-dinitrophenylhydrazine and 5 ml concentrated HCl made up to 100 ml with ethanol; the spots appear as dark yellow areas which increase in intensity upon standing. Detection of radioactivity on thin layer chromatography can be achieved by direct scanning employing a Vanguard Strip Scanner with attachment for thin layer chromatography. In cases of insufficient radioactivity and/or low counting efficiency (tritium), the chromatogram is divided into zones, and the silica gel is scraped off and counted directly in the liquid scintillation counter by the system of Snyder and Stephens.[3]

Application of the Method

A reaction mixture containing 1 micromole, or less, of formate with a sufficiently high specific activity may be diluted with 500–1000 micromoles of nonlabeled sodium formate, and the *p*-bromophenacyl formate be prepared as described above. However, in cases where the radioactive formate is of low specific activity and is present in amounts of 50 micromoles or more, in a mixture containing other reaction products, separation of the formate is necessary prior to the formation of the *p*-bromophenacyl derivative. The separation techniques used must be quantitative since isotope discrimination may cause appreciable changes in the isotope distribution. A convenient method, applicable in the presence of inorganic salts, is liquid–liquid extraction of the aqueous acidic solution (pH 2.0 or less) with ether. The ether reservoir flask should contain sufficient $NaHCO_3$ dissolved in a small volume of water to convert the extracted formic acid to its sodium salt.

[3] F. Snyder and N. Stephens, *Anal. Biochem.* **4**, 128 (1962).

Corrections. A correction factor to eliminate errors inherent in the method as well as to account for possible effects due to isotope discrimination should be established by running a sample containing radioactive formic acid of known specific activity. The formic acid used for this purpose has to be chromatographically homogeneous (Whatman No. 1; solvent isopropanol–ammonium hydroxide 70:30 v/v).

TABLE II
PROPERTIES OF *p*-BROMOPHENACYL DERIVATIVES

Compound	Migration in 90 minutes[a]	Crystalline form	Melting point (Kofler) (°C)
p-Bromophenacyl acetate	12.4	Hexahedron	86
p-Bromophenacyl lactate	4.5	Long needles	112
p-Bromophenacyl alcohol	8.0	—	140–142
2-4′-Dibromoacetophenone	16.5	—	113

[a] Continuous thin layer chromatography on silica gel G was carried out in a vapor-saturated container. In these experiments the B-N chamber (Brinkmnan Instrument Company) is used. The solvent system consists of benzene–ethylacetate (95:5 v/v).

Specificity; Properties of p-Bromophenacyl Esters. Many carboxylic acids behave similarly to formate and give rise to the corresponding *p*-bromophenacyl esters. However, they are readily distinguished by their physical constants (examples: see Table II) as well as by their behavior on isotope dilution. *p*-Bromophenacyl esters of carboxylic acids (except α-keto acids) are generally more stable than the formyl ester. *p*-Bromophenacyl formate readily decomposes to *p*-bromophenacyl alcohol. Thus, drying of the compound on thin layer plates, prolonged exposure to heat during recrystallization or storage of the crystalline material over a period of weeks, will result in partial saponification.

Section II

Preparation of Substrates

[6] Erythrose 4-Phosphate[1]

By F. J. SIMPSON, A. S. PERLIN, and A. S. SIEBEN

Principles. The procedure consists of the oxidation of glucose 6-phosphate with lead tetraacetate,[2] separation of the product, erythrose 4-phosphate from glucose 6-phosphate by chromatography on Dowex 1 formate, and preparation of the stable hydrazone that is recovered as the barium salt.[3] Less than the theoretical amount of lead tetraacetate is used to avoid overoxidation and production of glyceraldehyde 3-phosphate.[3] The chemical synthesis of erythrose 4-phosphate is described by C. E. Ballou.[4]

Reagents

Barium glucose 6-phosphate heptahydrate (mol. wt. 521.6). The water of crystallization and purity varies somewhat with the source. The salt may be assayed before use with glucose phosphate dehydrogenase.[5]

Lead tetraacetate. The commercial product is stored moistened with acetic acid. The salt is dissolved in glacial acetic acid using about 0.8 g per 40 ml. The exact concentration is determined by adding a portion (e.g., 1.0 ml) to 15 ml of a solution containing 100 g of KI and 500 g of sodium acetate per liter,[6] then titrating the iodine released with 0.02 N sodium thiosulfate (starch end point).

Dowex 1 formate. Dowex 1-X8 chloride is suspended in 4 N HCl and washed with water by decantation until the fines are removed. The resin is placed in a column and washed with water to remove the excess HCl, followed by a solution of 3 M sodium formate in 1 N formic acid until the resin is free from chloride ion, then with one-quarter bed volume of 50% formic acid, and finally with water until the effluent is neutral. Columns for chromatography are prepared by suspending the resin in 2 volumes of 0.2 N formic acid, degassing and then pouring the suspension into a suitable column and allowing the resin to

[1] Issued as N.R.C. No. 9023.

[2] J. N. Baxter, A. S. Perlin, and F. J. Simpson, *Can. J. Biochem. Physiol.* **37**, 199 (1959).

[3] A. S. Sieben, A. S. Perlin, and F. J. Simpson, *Can. J. Biochem.* **44**, 663–69 (1966).

[4] See Vol. VI [70].

[5] See Vol. III [19]; see also this volume [23].

[6] J. P. Cordner and K. H. Pausacker, *J. Chem. Soc.*, p. 102 (1953).

settle by gravity. The resin is then equilibrated by passing 500 ml of 0.2 N formic acid through the column. The column should be approximately 45 cm long, and the ratio of the cross sectional area in square centimeters to the millimoles of sugar phosphate maintained at about 1.2:1.

Procedure. Barium glucose 6-phosphate heptahydrate (1 millimole, 0.53 g) is placed in a 500-ml beaker and moistened with 2 ml of water. Then 5 ml of glacial acetic acid, reagent grade, is added, and the suspension is warmed to 50–60° for a few minutes to dissolve the salt. Sulfuric acid (0.35 ml of 6 N) is added with stirring and is followed by 245 ml of glacial acetic acid.

A solution containing 1.7 millimoles of lead tetraacetate and 0.6 ml of 6 N sulfuric acid in approximately 40 ml of glacial acetic acid is prepared just prior to use and placed in a dropping funnel (a fine precipitate that need not be removed develops if the lead tetraacetate contains divalent lead). This reagent is added dropwise into the rapidly stirred solution of glucose 6-phosphate at about 25° at such a rate that the reaction mixture never contains more than a slight excess of oxidant. This may be checked readily with moistened starch-iodide paper. The addition takes about 30 minutes.

The mixture is filtered through a thin layer of Celite "filter-aid," and the filtrate is concentrated in a rotary evaporator at 35–40° to about 20 ml. The Celite is washed with 100 ml of water, and the washings are combined with the filtrate. Evaporation is continued until a volume of 15–20 ml is again reached, thus removing most of the acetic acid. This concentrate, containing 0.2 millimole of G-6-P and 0.77 millimoles of E-4-P, is transferred with rinsing to an ether extractor. The remainder of the acetic acid is removed by extracting continuously with ether for 15–20 hours (overnight). The dissolved ether is evaporated from the aqueous layer which is then passed through a small column of Amberlite 1R-120 (H^+) (2 g) to remove remaining traces of cations. The volume of the eluate and rinse from the column is reduced to 5–10 ml by evaporation *in vacuo* at 20–30°.

The above solution (0.17 millimole G-6-P, 0.72 millimole E-4-P) adjusted to 0.2 N total acid by dilution or by addition of formic acid is adsorbed on the Dowex-1 formate column (1.2 × 45 cm) and rinsed in with 25 ml of 0.2 N formic acid. Elution of the glucose 6-phosphate is accomplished by applying a linear gradient from 0.05 M ammonium formate in 0.2 N formic acid (500 ml) to 0.1 M ammonium formate in 0.2 N formic acid (500 ml).[7] Fractions of 5 ml are collected. The presence

[7] R. M. Bock and N. Ling, *Anal. Chem.* **26**, 1543 (1954).

of glucose 6-phosphate in the eluate is detected by withdrawing a sample (0.5 ml) from every fifth tube and assaying with phenol–sulfuric acid.[8] The method consists of adding to 1 ml of suitably diluted sample (0–0.3 micromole), 1 ml of 5% phenol and 4 ml of 95% sulfuric acid with a syringe or Conway pipette. Within 10 minutes an orange color is produced and the density may be measured at 490 mμ. Where $l =$ 1 cm, the glucose 6-phosphate, micromoles per sample = (O.D. sample —O.D. blank) (0.365).

When all the glucose 6-phosphate is eluted, a solution of 2.0 *M* ammonium formate in 0.2 *N* formic acid is applied to the column to remove the erythrose 4-phosphate. This sugar produces a pink color in the phenol–sulfuric acid reaction and may be readily detected in the column eluate. Since other sugars are not present, the amount of erythrose 4-phosphate may be determined by measuring the optical density at 525 mμ. The micromoles per sample = (O.D. sample — O.D. blank) (2.19). The erythrose 4-phosphate also may be determined enzymatically.[2,3]

The eluate containing the erythrose 4-phosphate and 2 *M* ammonium formate (0.67 millimole E-4-P) is pooled and passed through a column of Amberlite 1R-120 (H^+) to remove the cations. The eluate and washings are combined and extracted continuously with ether for 36–48 hours to remove the formic acid. The solution is then concentrated to approximately 20 ml at temperatures below 20° in a rotary evaporator. The pH of this concentrated solution is adjusted to 8.6 with 2 *N* hydrazine, which usually is in excess of that required to form the hydrazone.[9] In this mixture 1.3 g of barium bromide is dissolved. Any precipitate that forms is removed by centrifugation. The barium salt of the erythrose 4-phosphate hydrazone is precipitated with 4 volumes of ethanol and held at 0° for a few hours. The precipitate is recovered by centrifugation, washed once with 70% ethanol and once with absolute alcohol, and then dried *in vacuo*. The dry salt is dissolved in 10 ml of water with, if necessary, the aid of a drop of 2 *N* HBr. One milliliter of 2 *M* barium bromide is added, and the pH of the solution is adjusted to 8.6 with hydrazine. Any precipitate that forms is removed by centrifugation. The barium salt is again precipitated with 3 volumes of ethanol, recovered, washed and dried as before. Yield = 0.5 millimole of erythrose 4-phosphate. The product does not contain glucose 6-phosphate or glyceraldehyde 3-phosphate. The barium salt of erythrose 4-phosphate hydrazone is quite stable and may be stored for long periods of time in the refrigerator.

[8] M. Dubois, K. A. Gilles, J. K. Hamilton, P. A. Rebers, and F. Smith, *Anal. Chem.* **28**, 350 (1956).

[9] L. M. Hall, *Biochem. Biophys. Res. Commun.* 3, 239 (1960).

The erythrose 4-phosphate may be recovered from the barium salt of the hydrazone by dissolving the salt, adjusting to pH 3–4, and passing the solution through a small column of Dowex 50W H^+ to remove the hydrazine and barium.[9]

[7] D-Arabinose 5-Phosphate[1]

By WESLEY A. VOLK

Principle

D-Arabinose 5-phosphate is prepared by allowing ninhydrin to react with D-glucosamine 6-phosphate, then purifying the D-arabinose 5-phosphate on a Dowex 1 formate column.

Preparation

Phosphorylation of D-Glucosamine. A modification of the method of Brown[2] (eliminating the use of buffer and thereby simplifying the final purification of the glucosamine 6-phosphate) was used to phosphorylate D-glucosamine. The reaction mixture consisted of 4.0 g of glucosamine HCl, 1.6 g of ATP, and 1 millimole of $MgCl_2$ in a total volume of 200 ml. The pH was adjusted to 8.0, and the solution was placed on a magnetic stirrer. To the reaction mixture was added 100 mg of Sigma type III hexokinase, dissolved in 50 ml of water, and the pH was kept at 8.0 by manually adding 1.0 N KOH from a burette. The reaction was complete after 2 hours at room temperature, and the glucosamine 6-phosphate was purified as outlined by Brown.[2] Final yield of purified glucosamine 6-phosphate from the above experiment was 3.5 millimoles.

Ninhydrin Reaction. The reaction mixture for the synthesis of D-arabinose 5-phosphate contained 400 micromoles of glucosamine 6-phosphate, 1000 micromoles of ninhydrin, and 1500 micromoles of citrate buffer, pH 4.7, in a final volume of 60 ml. The yield of D-arabinose 5-phosphate remained constant between pH 4.2 and 5.2, but decreased markedly on either side of this pH range. The reaction mixture was placed in a boiling water bath and pentose formation was followed by means of the orcinol test. Maximum yield of pentose was obtained after a 40-minute heating period.

Purification of Arabinose 5-Phosphate. At the end of the heating period the dark blue mixture was cooled and adsorbed on a Dowex 1

[1] W. A. Volk, *Biochim. Biophys. Acta* **37**, 365, (1960).

[2] D. H. Brown, *Biochim. Biophys. Acta* **7**, 487, (1951).

formate column (2 × 20 cm). It should be noted that considerably larger amounts could be handled if the arabinose 5-phosphate is precipitated at this point as its alcohol insoluble barium salt and then redissolved and adsorbed on a Dowex 1 formate column. The column was eluted with 0.2 N formic acid containing 0.02 M ammonium formate. Two well separated orcinol-positive fractions were eluted. The first comprised less than 1% of the total orcinol-positive material and is believed to contain both ribose and fructose phosphates. The second fraction contained 334 micromoles of D-arabinose 5-phosphate. This fraction was pooled and neutralized to pH 7.5 with KOH, and barium chloride (1000 micromoles) and ethanol (to a final concentration of 80%) were added. After standing overnight at 2° the barium salt was collected by centrifugation and dissolved in water. The total yield of D-arabinose 5-phosphate was 197 micromoles.

[8] 2-Ketopentoses

By R. P. MORTLOCK and W. A. WOOD

Enzymatic Method

Principle. Pentitol is oxidized by pentitol dehydrogenases to the corresponding 2-ketopentose. Crystalline lactic dehydrogenase and pyruvate are added to the reaction mixture to accomplish the oxidation of reduced NAD. The nonspecific ribitol dehydrogenase of *Aerobacter aerogenes* has been utilized for the preparation of D-ribulose from ribitol and L-xylulose from L-arabitol. D-Arabitol dehydrogenase may be utilized to oxidize D-arabitol to D-xylulose. The purification of both of these dehydrogenases has been described.[1] In general, the ketopentoses prepared by this procedure are of excellent purity (< 0.1% contamination of other 2-ketopentoses).

Method. The reaction mixture consists of phosphate buffer, pH 7.5 (0.5 millimole), potassium pyruvate (4.0 millimoles), NAD (10 micromoles), pentitol (2.0 millimoles), pentitol dehydrogenase (1000 units), and lactic dehydrogenase (10 units)[2] in a total volume of 10 ml. The mixture is incubated at 37° with slow stirring. Ketopentose formation is followed by assay of aliquots by the cysteine-carbazole test.[3]

When the reaction has ceased, 5 ml of 1 M $BaCl_2$ is added. The

[1] This volume [35].
[2] Crystalline suspension, 36 units/ml, Worthington Biochemical Corporation.
[3] Vol. III [12].

mixture is centrifuged and the supernatant solution is added to 9 volumes of cold ethanol. After several hours, the ethanol mixture is filtered. The ethanol is removed from the filtrate under vacuum, and the remaining solution is deionized by passage through Dowex 50 (H^+) and Dowex 3 (CO_3^{2-}) columns. The remaining traces of pentitol may be separated by use of a Dowex bisulfite column.[4]

Alternate Enzymatic Methods. Aldopentose isomerases have also been utilized for the preparation of ketopentoses, but yields are normally low (5%). Cohen, using D-arabinose isomerase from *Escherichia coli,* reported a 70–90% yield of D-ribulose when borate was present in the reaction mixtures.[5] *Acetobacter suboxydans* has been utilized for the preparation of D-xylulose from D-arabitol,[6] and L-ribulose from ribitol.

On the basis of an enzymatic assay, the ketopentoses prepared by the chemical method have lower purity than those prepared enzymatically. Thus, L-xylulose prepared by refluxing L-xylose with pyridine is normally 10–20% pure based upon an assay with xylitol (→ L-xylulose) dehydrogenase, whereas L-xylulose prepared by the pentitol dehydrogenase oxidation of L-arabitol possesses complete biological activity.

Chemical Method

Principle. The aldopentose corresponding in structure is refluxed with dry pyridine.[7]

Method. The aldopentose (30 g) is refluxed for 4 hours with 250 ml of dry pyridine. The pyridine is then removed by distillation under reduced pressure, and the residue is diluted with water and concentrated under vacuum to remove traces of pyridine.[8] The sirup is refrigerated overnight to crystallize excess aldopentose, and the crystal cake is washed with cold absolute ethanol. The ethanol washings are concentrated under vacuum, and the crystallization procedure is repeated. When no further crystallization of aldopentose is observed, a small portion of distilled water is added and the ethanol is removed under vacuum.

A Dowex 1 borate column may be utilized to separate the desired ketopentose from other sugars remaining in the sirup.[9] The sirup is dissolved in sufficient 0.004 *M* potassium tetraborate to complex the pentoses,[10] and the solution is passed through a Dowex 1 borate column.

[4] S. Adachi and H. Sugawara, *Arch. Biochem. Biophys.* **100,** 468 (1963).
[5] S. S. Cohen, *J. Biol. Chem.* **201,** 71 (1953).
[6] J. McCorkindale and N. L. Edson, *Biochem. J.* **57,** 518 (1954).
[7] C. Glatthaar and T. Reichstein, *Helv. Chim. Acta* **18,** 80 (1935).
[8] During this and subsequent steps, the temperature should not exceed 40°.
[9] Vol. III [15].
[10] Determined by the orcinol test: Vol. III [13].

The column is eluted with a linear gradient of from 0.005 *M* to 0.25 *M* potassium tetraborate.[11] The fractions containing the desired ketopentose are pooled and passed through a cationic exchanger in the hydrogen form. Methanol is added, and the borate is removed as methyl borate by vacuum distillation. The methanol treatment is repeated several times to remove the borate completely. The ketopentose is recovered as a concentrated sirup and diluted to the concentration desired as determined by the cysteine-carbazole test.[3] All four of the 2-ketopentose epimers have been prepared in this manner. Preparation of the *o*-nitrophenylhydrazone (ribulose) or *p*-bromophenylhydrazone (xylulose) also have been utilized to recover the ketopentose.

Assay

D- or L-xylulose may be differentiated from D- or L-ribulose by rate of color development during the cysteine-carbazole test of Dische and Borenfreund.[3] Maximum color is obtained after 15 minutes' incubation with ribulose and after 100 minutes with xylulose. Purity of preparations can be determined by paper chromatography[12] and by gradient elution from Dowex borate.[9] Biological activity and purity may be tested by use of the four ketopentokinases of *Aerobacter aerogenes*.[13]

[11] When 80 fractions (800 ml total volume) are collected, the xylulose peak is normally between fractions 14 and 19, while the ribulose peak is between fractions 40 and 45.

[12] Vol. III [11].

[13] This volume [79–82].

[9] Ketopentose 5-Phosphates

By F. J. SIMPSON

All four pentulose 5-phosphates have been prepared by methods employing preparations of specific enzymes such as kinases, isomerases, epimerases, transketolase, and phosphogluconate dehydrogenase. The products can be isolated as the barium, calcium, or lithium salts or as the more stable barium salts of the hydrazones from which they are readily recovered.

D- and L-Xylulose 5-Phosphate (D- and L-*threo*-Pentulose 5-Phosphate)

D-Xylulose 5-phosphate has been prepared by a number of methods, including phosphorylation of D-xylulose with a specific kinase,[1-3] by the

[1] P. K. Stumpf and B. L. Horecker, *J. Biol. Chem.* **218**, 753 (1956).

[2] J. Hickman and G. Ashwell, *J. Biol. Chem.* **232**, 737 (1958).

[3] F. J. Simpson and B. K. Bhuyan, *Can. J. Microbiol.* **8**, 663 (1962).

transfer of a two-carbon fragment from hydroxypyruvate to glyceraldehyde 3-phosphate with transketolase,[4] and by the action of 3-epimerase on D-ribulose 5-phosphate by a procedure that also yields ribulose diphosphate.[5]

The simplest procedure is the phosphorylation of D-xylulose with a specific kinase, free of interfering enzymes, obtained from *Aerobacter aerogenes* grown on D-xylose.[3,6]

Likewise, L-xylulose 5-phosphate is prepared with a specific L-xylulokinase produced by the same organism when grown on L-xylose.[7]

D-Xylulose is easily prepared in good yield by the oxidation of D-arabinitol with *Acetobacter suboxydans*, ATCC 621,[8] with either resting cells[9] or growing cells.[10] The crude sirup obtained after removal of the cells, desalting, and clarification with charcoal, may be stored at —20°, pH 2–3, and used without purification by the formation of a crystalline derivative since the D-xylulokinase is specific for D-xylulose. The small amounts of xylose and arabinitol present do not interfere. D-Arabinitol can be obtained commercially, prepared by reduction of D-arabinose with sodium borohydride,[11] or recovered from fermentations of glucose by certain yeasts.[12] Alternatively, D-xylulose can be obtained by epimerization of D-xylose with hot pyridine[13,14] or with sodium hydroxide in the presence of borate and isolation on a Dowex 1 borate column.[15] The D-xylose isomerases also may be employed to carry out this conversion.[16–19] L-Xylulose is usually prepared by the chemical epimerization of L-xylose[13–15] but the L-xylose isomerase could be used.[20]

[4] P. Srere, J. R. Cooper, M. Tabachnick, and E. Racker, *Arch. Biochem. Biophys.* **74,** 295 (1958).

[5] B. L. Horecker, J. Hurwitz, and P. K. Stumpf, Vol. III, p. 193.

[6] F. J. Simpson, this volume [81].

[7] R. L. Anderson and W. A. Wood, *J. Biol. Chem.* **237,** 1029 (1962).

[8] L. A. Underkofler, A. C. Bantz, and W. H. Peterson, *J. Bacteriol.* **45,** 183 (1943).

[9] V. Moses and R. J. Ferrier, *Biochem. J.* **83,** 8 (1962).

[10] R. Prince and T. Reichstein, *Helv. Chim. Acta* **20,** 101 (1937).

[11] H. L. Frush and H. S. Isbell, *J. Am. Chem. Soc.* **78,** 2844 (1956).

[12] G. J. Hajny, *Appl. Microbiol.* **12,** 87 (1964).

[13] L. Hough and R. S. Theobald, *in* "Methods in Carbohydrate Chemistry" (R. L. Whistler and M. L. Wolfrom, eds.), Vol. I, p. 94. Academic Press, New York, 1962.

[14] O. Touster, *in* "Methods in Carbohydrate Chemistry" (R. L. Whistler and M. L. Wolfrom, ed.), Vol. I, p. 94. Academic Press, New York, 1962.

[15] J. F. Mendicino, *J. Am. Chem. Soc.* **82,** 4975 (1960).

[16] R. L. Anderson and D. P. Allison, *J. Biol. Chem.* **240,** 2367 (1965).

[17] D. P. Burma and B. L. Horecker, *J. Biol. Chem.* **231,** 1053 (1958).

[18] S. Mitsuhashi and J. O. Lampen, *J. Biol. Chem.* **204,** 1011 (1953).

[19] K. Yamanaka, *Agr. Biol. Chem.* **26,** 175 (1962).

[20] R. L. Anderson and W. A. Wood, *J. Biol. Chem.* **237,** 296 (1962).

The preparation of D-xylulose 5-phosphate with the purified D-xylulokinase prepared from *Aerobacter aerogenes*[6] is described below. Other pentulose 5-phosphates can be prepared by phosphorylating the pentulose with the appropriate kinase.

The reaction mixture (100 ml) contains 2 millimoles of D-xylulose, 2 millimoles of ATP, 0.1 millimole of EDTA, and 5 millimoles of $MgCl_2$, adjusted to pH 6.9 with 1 N hydrazine. The reaction is begun by the addition of 200 units (200 micromoles/minute),[6] of D-xylulokinase, and the mixture is maintained at pH 6.6–6.9 by the periodic addition of 1 N hydrazine until the reaction is complete (90–100 minutes at 30°). Hydrazine at low concentrations does not inhibit the D-xylulokinase of *A. aerogenes,* but other enzymes may be more sensitive.

The reaction mixture is cooled to 0°, 2 ml of 70% perchloric acid is added; after it has been mixed for 1 minute, the solution is neutralized by the cautious addition of approximately 6 ml of 4 N KOH. The resultant precipitate is removed by filtering the cold solution through glass fiber filter paper. The precipitate is rinsed with a little cold water and discarded. The combined filtrates are adjusted to pH 9 by the addition of hydrazine. Nucleotides are precipitated from this solution by the slow addition with stirring of 5 ml of 2 M barium bromide. The precipitate is removed by centrifugation and washed three times with 20 ml of water. The combined liquors are diluted with 4 volumes of cold ethanol, and the precipitate is allowed to settle at 0° for 2 hours or more. The clear supernatant fluid is decanted, and the precipitate is collected by centrifugation.

The precipitate is suspended in 50 ml of cold water and adjusted to pH 2.2–2.5 with HBr. Material that will not dissolve is removed by centrifugation and washed with 20 ml of water. The combined liquors are placed in a beaker in ice and are treated with 1 g of charcoal (Norit A, acid washed, prepared by boiling for 10 minutes in 1 N HCl, then washing with distilled water until neutral, and drying in air) for 10 minutes; then the charcoal is filtered off on glass fiber paper in a Büchner funnel and rinsed with a little water. This treatment is repeated two to four times until the absorption at 260 mμ attains a minimum value. One milliliter of 2 M barium bromide is added to the clarified solution and sufficient hydrazine to raise the pH to 8.5.[21] The barium salt of the D-xylulose 5-phosphate hydrazone is precipitated by the addition of 4 volumes of ethanol. After it has been allowed to settle for 2 hours at 0°, the precipitate is collected by centrifugation, washed twice with 80% ethanol, and dried *in vacuo* over calcium chloride and sodium

[21] L. M. Hall, *Biochem. Biophys. Res. Commun.* 3, 239 (1960).

hydroxide. The yield is 0.7–0.8 g containing 62–80% barium D-xylulose 5-phosphate hydrazone as determined by enzymatic analysis.

D-Xylulose 5-phosphate may be recovered from the salt by dissolving the required amount in a little water containing an equivalent amount of HBr, then removing the barium and most of the hydrazine by percolation through a column of cationic ion exchange resin (e.g., four equivalents of Dowex 50W H^+).[21] For use with enzymes, such as transketolase, that are sensitive to hydrazine, the solution should be extracted twice with 0.2 volume of benzaldehyde (freshly distilled) followed by three extractions with ether. The ether may be removed by gently bubbling nitrogen gas through the solution.

If the phosphorylation is made with small amounts of substrate (e.g., 200 μmoles) the pentulose-phosphate may be recovered from the reaction mixture by column chromatography on Dowex 1 formate resin employing gradient elution.[1,22–24] The eluate containing the pentulose phosphate is adjusted to pH 8.5 with hydrazine, barium bromide is added, then the salt is precipitated with ethanol and recovered by centrifugation.

L-Ribulose 5-Phosphate (L-*erythro*-Pentulose 5-Phosphate)

This pentulose phosphate has been prepared by the enzymatic phosphorylation of L-ribulose with the L-ribulokinases prepared from cells of *Aerobacter aerogenes*,[25] *Escherichia coli*,[26] and *Lactobacillus plantarum*.[23] The procedure described for D-xylulose 5-phosphate may be used to obtain L-ribulose 5-phosphate from L-ribulose with the L-ribulokinase preferably prepared from a mutant of *E. coli* lacking L-ribulosephosphate 4-epimerase.[27]

L-Ribulose is obtained in high yield and purity from the oxidation of ribitol (adonitol) with *Acetobacter suboxydans*.[9,10] The pentulose can also be prepared by the epimerization of L-arabinose with pyridine[28] or with sodium hydroxide in the presence of borate.[15] Similarly, in the presence of borate, L-arabinose isomerases[19,26,29] will convert L-arabinose

[22] G. R. Bartlett, *J. Biol. Chem.* **234**, 459 (1959).

[23] D. P. Burma and B. L. Horecker, *J. Biol. Chem.* **231**, 1039 (1958).

[24] B. L. Horecker, P. Z. Smyrniotis, and J. E. Seegmiller, *J. Biol. Chem.* **193**, 383 (1951).

[25] W. A. Wood and F. J. Simpson, Vol. V [32a].

[26] N. Lee and E. Englesberg, *Proc. Natl. Acad. Sci. U.S.* **48**, 335 (1962).

[27] E. Englesberg, R. L. Anderson, R. Weinberg, N. Lee, P. Hoffee, G. Huttenhauer, and H. Boyer, *J. Bacteriol.* **84**, 137 (1962).

[28] P. A. Levene and R. S. Tipson, *J. Biol. Chem.* **115**, 731 (1936).

[29] E. C. Heath, B. L. Horecker, P. Z. Smyrniotis, and Y. Takagi, *J. Biol. Chem.* **231**, 1031 (1958).

to L-ribulose in good yield. The ribulose may be separated from arabinose by chromatography on Dowex 1 borate,[15] or by oxidation of the arabinose with bromine and recovery of the ribulose as the *o*-nitrophenyl hydrazone.[24,30,31]

D-Ribulose 5-Phosphate (D-*erythro*-Pentulose 5-Phosphate)

This ester may likewise be prepared by the phosphorylation of D-ribulose with a suitable kinase.[32,33] D-Ribulose can be obtained by chemical epimerization of D-arabinose[15,31] or by employing isomerases in the presence of borate.[34,35] The D-ribose 5-phosphate isomerases[4,36-38] and the D-arabinose 5-phosphate isomerase[39] may be used, but the mixture of sugar phosphates has to be separated.[24] The best method for preparing D-ribulose 5-phosphate is the enzymatic oxidation of D-gluconate 6-phosphate with crystalline D-gluconate 6-phosphate dehydrogenase.[40] This crystalline enzyme is readily prepared from *Candida utilis* (PRL Y56 or ATCC 9950[41] may be used). The procedure described by Pontremoli *et al.*[42] requires a thirtyfold concentration by evaporation at low temperature before the final precipitation and crystallization with ammonium sulfate. We found that the concentration could be done conveniently by lyophilization. Approximately 250 ml portions of the dilute preparation are placed in 1-liter round-bottom flasks and rapidly frozen as an even "shell," then while frozen taken to near dryness in a high vacuum system. The recovery of enzyme activity is nearly quantitative. The D-ribulose 5-phosphate obtained from D-gluconate 6-phosphate by using this crystalline enzyme in conjunction with crystalline lactic acid dehydrogenase and pyruvate for the regeneration of NADP does not contain other pentulose or pentose phosphates.[40] We have prepared the

[30] R. L. Barker and W. G. Overend, *Chem. Ind. (London)*, p. 1298 (1960).
[31] C. Glatthaar and T. Reichstein, *Helv. Chim. Acta* **18**, 80 (1935).
[32] H. J. Fromm, *J. Biol. Chem.* **234**, 3097 (1959).
[33] T. Kameyama and N. Shimazono, *J. Biochem. (Tokyo)* **57**, 339 (1965).
[34] S. S. Cohen, *J. Biol. Chem.* **201**, 71 (1953).
[35] M. Di Girolamo, L. Frontali, and G. Tecce. *Quaderni Nutr.* **18**, 131 (1959).
[36] B. Axelrod and R. Jang, *J. Biol. Chem.* **209**, 847 (1954).
[37] J. Hurwitz, A. Weissbach, B. L. Horecker, and P. Z. Smyrniotis, *J. Biol. Chem.* **218**, 769 (1956).
[38] M. Urivetsky and K. K. Tsuboi, *Arch. Biochem. Biophys.* **103**, 1 (1963).
[39] W. A. Volk, *J. Biol. Chem.* **235**, 1550 (1960).
[40] S. Pontremoli and G. Mangiarotti, *J. Biol. Chem.* **237**, 643 (1962).
[41] C. O. Reiser, *J. Agr. Food Chem.* **2**, 70 (1954). Dried cells may be purchased from Lake States Yeast and Chemical Division, St. Regis Paper Co., Rhinelander, Wisconsin.
[42] S. Pontremoli, A. De Flora, E. Grazi, G. Mangiarotti, A. Bonsignore, and B. L. Horecker, *J. Biol. Chem.* **236**, 2975 (1961).

pentulose ester on a scale ten times that described by the authors with excellent yields. The hydrazone is readily obtained by adding an excess of hydrazine before precipitating the barium salt with ethanol.

[10] D-Ribulose 5-Phosphate

By SANDRO PONTREMOLI

Preparation

Principle. D-Ribulose 5-phosphate is prepared by the TPN-coupled enzymatic oxidation of D-gluconate 6-phosphate followed by treatment with charcoal and precipitation of the phosphorylated product as barium or lithium salt.[1]

Reagents

Glycylglycine buffer, 0.3 *M* pH 7.6
D-Gluconate 6-phosphate (Na salt), 0.2 *M*
Sodium pyruvate, 0.3 *M*
TPN, 0.1 *M*
Crystalline lactic dehydrogenase (from rabbit muscle, purchased from Boehringer and Söhne, Germany)
Crystalline D-gluconate 6-phosphate dehydrogenase (specific activity 190 units per milligram of protein). For the preparation of the enzyme, see this volume [26].

Both enzymes must be centrifuged before use and dissolved in water to avoid the introduction of excessive amounts of ammonium sulfate.

Procedure. The reaction mixture (8.0 ml) contains: 3.7 ml of water, 1.2 ml of D-gluconate 6-phosphate, 1.0 ml of glycylglycine buffer, 1.6 ml of sodium pyruvate, 0.2 ml of TPN, 0.24 ml (1.2 mg) of lactic dehydrogenase, and 0.06 ml (0.6 mg) of D-gluconate 6-phosphate dehydrogenase. The reaction is started by the addition of the D-gluconate 6-phosphate dehydrogenase. The contents are mixed and incubated at 30°. Aliquots are assayed at intervals for D-gluconate 6-phosphate. After 45 minutes when less than 1% remains, the reaction mixture is rapidly chilled in an ice bath and treated with 0.78 ml of 45% trichloroacetic acid and 100 mg of acid-washed charcoal,[2] shaken, and left 10 minutes in ice. The suspension is filtered through a sintered glass filter, and the residue is washed with 2.5 ml of water.

Preparation of the Barium Salt. To the combined filtrate and wash-

[1] S. Pontremoli and G. Mangiarotti, *J. Biol. Chem.* **237**, 643 (1962).
[2] For the preparation of the charcoal, see this volume [26].

ings (11 ml) is added 1 ml of 1 *M* barium acetate. After 20 minutes in ice the turbid suspension is centrifuged, the precipitate is washed with 1 ml of water, and the supernatant solution and washing are combined (12.2 ml). This solution is treated with 3.6 ml of saturated $Ba(OH)_2$, adjusted to pH 6.6 with saturated KOH, and precipitated with 4 volumes of cold absolute ethanol. After 30 minutes the precipitate is collected by centrifugation, washed with 80% ethanol, and dried *in vacuo*.

The yield of dried barium salt is 104.5 mg (first barium salt precipitate). An aliquot of the barium salt (55 mg) is dissolved in 5.0 ml of 0.1 *M* CH_3COOH, the solution is adjusted to pH 6.6 with 0.01 ml of saturated KOH and 0.5 ml of saturated $Ba(OH)_2$ and is precipitated with 4 volumes of cold absolute ethanol. After 45 minutes the precipitate is collected by centrifugation, washed with 80% ethanol, and dried *in vacuo*. The yield of dried barium salt is 44 mg (second barium salt precipitate).

Preparation of the Lithium Salt. To obtain the lithium salt, 49.5 mg of the first barium salt precipitate is dissolved in 4 ml of 0.1 *M* CH_3COOH, an aliquot (0.5 ml) is removed for analysis, and the remaining solution is passed through a Dowex 50 H^+ column to remove Ba^{++}. The effluent and washings (4.1 ml) are combined and brought to pH 6.3 with 4.5 ml of 0.1 *N* LiOH. The lithium salt is precipitated at 0° with 100 ml of 10% methanol in acetone. After 30 minutes the precipitate is collected by centrifugation, washed with 10 ml of the acetone–methanol mixture, and dried *in vacuo*. The yield of the dried lithium salt is 19 mg.

Properties and Purity of the Product. The barium and lithium salts of the product, prepared as described, have been characterized as D-ribulose 5-phosphate by chemical and enzymatic analyses. The properties of the phosphorylated sugar are summarized in the table. On the basis of

CHEMICAL AND ENZYMATIC PROPERTIES OF D-RIBULOSE 5-PHOSPHATE LITHIUM AND BARIUM SALTS

Test	Ba salt, 1st ppt.	Ba salt, 2nd ppt.	Li salt
Orcinol test,[a] 870 mμ (total micromoles)	212	86.2	64
Cysteine carbazole,[b] 540 mμ (total micromoles)	211	85.5	62
P_i (micromoles)	0	0	0
Bromine oxidation[c]	No	No	No
D-Ribose 5-phosphate[d] formed at equilibrium (%)	Not done	70	71

[a] F. Dickens and D. H. Williamson, *Biochem. J.* **64,** 657 (1956).

[b] G. Ashwell and J. Hickman, *J. Biol. Chem.* **226,** 65 (1957).

[c] At 28°, pH 6.0 in 6 minutes.

[d] Incubated with D-ribose 5-phosphate isomerase prepared according to J. Hurwitz, A. Weissbach, B. L. Horecker, and P. Z. Smyrniotis, *J. Biol. Chem.* **218,** 769 (1956).

these data the product appears to be D-ribulose 5-phosphate in almost pure form. The presence of D-ribose 5-phosphate is excluded since the product is not oxidized by bromine in a period of time sufficient to oxidize 70% of the aldopentose. By this procedure quantities of D-ribose 5-phosphate approaching 10% of the total would be readily detected. The presence of D-xylulose 5-phosphate is excluded by the equilibrium ratio obtained with D-ribose 5-phosphate isomerase. The D-ribulose 5-phosphate preparation reacts with TPNH + H^+ in the presence of CO_2 and D-gluconate 6-phosphate dehydrogenase to produce D-gluconate 6-phosphate and TPN.

The yield based on D-gluconate 6-phosphate was 80%. The product is stable if stored at 2° in a vacuum desiccator.

Comments. Previous methods for the enzymatic preparation of D-ribulose 5-phosphate by enzymatic oxidation of D-gluconate 6-phosphate,[3] isomerization of D-ribose 5-phosphate[4] or phosphorylation of D-ribulose,[5] required separation of the phosphorylated sugar by ion exchange chromatography since other phosphate esters or inorganic phosphate were present in the reaction mixture. The yield was low, and the product was contaminated by other phosphate esters. With the present method, pure ribulose 5-phosphate may be obtained as the lithium or the barium salt with an overall yield of 80%.

[3] B. L. Horecker and P. Z. Smyrniotis, *J. Biol. Chem.* **193,** 383 (1951).
[4] P. Srere, J. R. Cooper, M. Tabachnick, and E. Racker, *Arch. Biochem. Biophys.* **74,** 295 (1958).
[5] F. J. Simpson and W. A. Wood, *J. Biol. Chem.* **230,** 473 (1958).

[11] L-Ribulose 5-Phosphate

By R. L. ANDERSON

The enzymatic procedure described here is based on the use of L-ribulokinase from an L-ribulose 5-phosphate 4-epimeraseless mutant of *Escherichia coli*.[1] The use of this mutant obviates the need to purify L-ribulokinase to obtain a product free from D-xylulose 5-phosphate and D-ribulose 5-phosphate. Crystalline L-ribulokinase has been prepared,[2] however, and could be used if desired. L-Ribulose 5-phosphate has also

[1] E. Englesberg, R. L. Anderson, R. Weinberg, N. Lee, P. Hoffee, G. Huttenhauer, and H. Boyer, *J. Bacteriol.* **84,** 137 (1962).
[2] See this volume [80].

been prepared with L-ribulokinases purified from *Aerobacter aerogenes*[3] and *Lactobacillus plantarum*.[4]

Preparation of L-Ribulose. L-Ribulose may be prepared by the microbiological oxidation of ribitol,[5] or by the chemical isomerization of L-arabinose,[6] as described here. Twenty-five grams of L-arabinose is refluxed with 200 ml of pyridine (dried by distillation over KOH) for 4 hours. The pyridine is removed by evaporation under reduced pressure at 35° in a rotary evaporator, and the precipitated L-arabinose is removed by vacuum filtration. The resultant sirup is dissolved in a small volume of ethanol and stored in a refrigerator overnight. This usually results in the further precipitation of L-arabinose, which is removed by vacuum filtration. This procedure may be repeated until no more precipitated L-arabinose is obtained. The recovery of L-arabinose is about 19 g. The final sirup, which contains a mixture of L-ribulose and L-arabinose, is dissolved in about 30 ml of water. The L-ribulose concentration may be determined with the cysteine-carbazole reagents described by Dische and Borenfreund,[7] using D-ribulose-*o*-nitrophenylhydrazone (available commercially) as a standard. The yield of L-ribulose is about 10%. It can be purified and separated from L-arabinose by chromatography on Dowex 1 borate[8] or by preparation of the *o*-nitrophenylhydrazone,[6] but this is unnecessary because L-arabinose is not phosphorylated by L-ribulokinase.

Production and Selection of L-Ribulose 5-Phosphate 4-Epimeraseless (D-Gene) mutants of E. coli B/r. The procedure for obtaining mutants of *E. coli* B/r deficient in L-ribulose 5-phosphate 4-epimerase (D-gene) has been detailed by Englesberg.[9]

Growth of Organism and Induction of L-Ribulokinase. The mutant used in the procedure described here was ara-139,[1] but any D-gene mutant with a suitable level of L-ribulokinase can be used. D-gene mutants of *E. coli* B/r must be induced by L-arabinose to form L-ribulokinase, but because the cells are inhibited by L-arabinose, they must first be grown in its absence. Cells are grown aerobically in a casein hydrolyzate–mineral medium[10] at 37° in Fernbach flasks. The inoculum is a 1% volume of an overnight nutrient broth culture. When the optical

[3] F. J. Simpson and W. A. Wood, *J. Biol. Chem.* **230**, 473 (1958).
[4] See Vol. V [28b].
[5] T. Reichstein, *Helv. Chim. Acta,* **17**, 996 (1934).
[6] C. Glatthaar and T. Reichstein, *Helv. Chim. Acta* **18**, 80 (1935).
[7] Z. Dische and E. Borenfreund, *J. Biol. Chem.* **192**, 583 (1951). See also Vol. III [12].
[8] F. J. Simpson, M. J. Wolin, and W. A. Wood, *J. Biol. Chem.* **230**, 457 (1958).
[9] See this volume [3].
[10] E. Englesberg, *J. Bacteriol.* **81**, 996 (1961).

density at 425 mμ reaches about 0.6–0.8 (measured in 18-mm test tubes), L-arabinose is added to a concentration of 0.4%, and the incubation is continued for an additional 3–4 hours.[1] The cells are then harvested by centrifugation. The cell pellet may be stored frozen prior to preparing extracts without a loss of L-ribulokinase activity.

Preparation of L-Ribulokinase. Although it should be possible to use a crude cell extract for the phosphorylation of L-ribulose, it is desirable to purify L-ribulokinase somewhat to partially remove competing enzymes such as adenosine triphosphatase. The procedure described here used 9 g wet weight of cells and involved precipitating the nucleic acids with protamine sulfate followed by fractionation with ammonium sulfate at 60–80% of saturation. The procedures were performed at 0–4°.

The cells were suspended in about 30 ml of 1 m*M* sodium glutathione–1 m*M* EDTA (pH 7.0) and sonicated for 10–15 minutes in a Raytheon 10-kc sonic oscillator. The broken-cell suspension was then centrifuged at 31,000 *g* for 10 minutes, yielding 35 ml of supernatant with a protein concentration[11] of 48 mg/ml. The extract was diluted to 167 ml with 1 m*M* sodium glutathione–1 m*M* EDTA (pH 7.0). Ammonium sulfate (2.2 g) was added with stirring, followed by 33 ml of 2% protamine sulfate. The precipitate was removed by centrifugation and discarded. To the supernatant solution (188 ml) was added 39.8 g of ammonium sulfate, and the precipitate that formed was centrifuged and discarded. To the supernatant solution was added 22.5 g of ammonium sulfate, and the resulting precipitate (L-ribulokinase fraction) was collected by centrifugation and dissolved in water. The activity of L-ribulokinase in the fraction may be assayed by any of several described procedures.[2]

Preparation of L-Ribulose 5-Phosphate. A typical reaction mixture contains 4 millimoles of L-ribulose, 5 millimoles of ATP, 5 millimoles of $MgCl_2$, 5 millimoles of NaF, 1 millimole of reduced sodium glutathione, 2.5 micromoles of EDTA, and L-ribulokinase in a volume of 200 ml and at a pH of 7.5. The amount of L-ribulokinase in the reaction should be great enough to allow the reaction to reach completion in 3–6 hours. The reaction is followed and maintained at pH 7.5 by titration with NaOH using a recording pH stat. When the reaction has stopped, acetic acid is added to give a concentration of 0.2 *M*. The precipitate is filtered and discarded, and the supernatant solution is adjusted to pH 6.7 with NaOH. The L-ribulose 5-phosphate may then be purified and separated from nucleotides by chromatography on Dowex-1 formate[3] or by the procedure described here. Barium acetate (20 millimoles) is added, and the precipi-

[11] O. Warburg and W. Christian, *Biochem. Z.*, **310**, 384 (1941).

tate is centrifuged and discarded. The supernatant solution is adjusted to pH 2.0 with HBr, and the 260 mμ absorbance is measured (usually a 1:100 dilution is required). Activated charcoal (Darco G-60, about 10 g) is added and removed by filtration, and the 260 mμ absorbance is again measured (usually no dilution is required). The charcoal treatment is repeated with smaller amounts of charcoal until the 260 mμ absorbance reaches a minimum. The solution is then adjusted to pH 6.7 with NaOH, and 4 volumes of ethanol are added. The mixture is chilled overnight and the precipitate is collected by centrifugation, washed twice with 80–90% ethanol, and dried in a vacuum. The product (about 900 mg) is an amorphous powder consisting of barium L-ribulose 5-phosphate free from D-ribulose 5-phosphate and D-xylulose 5-phosphate.

[12] 2-Keto-3-deoxy-6-phosphogluconate[1]

By H. PAUL MELOCHE and W. A. WOOD

D-Glyceraldehyde 3-phosphate + pyruvate ⇄
2-keto-3-deoxy-6-phosphogluconate + pyruvate

There is no convenient chemical means available for synthesizing KDPG. However, two enzymatic methods have been used. In the first method 6-phosphogluconate is dehydrated with the enzyme 6-phosphogluconic dehydrase. This has the disadvantage that the dehydrase must be free of KDPG aldolase. The method described herein is based upon the fact that the KDPG aldolase catalyzed condensation of pyruvate and G-3-P would be driven far to the right if one of the reactants were in excess. A prime advantage of this method is that it can be carried out with crude enzyme preparations. By employing isotopic pyruvate or G-3-P, one could synthesize KDPG with a label in any desired position. (The one exception would involve hydrogen isotopes at carbon 3 of KDPG.)

Synthesis

Reagents

DL-Glyceraldehyde-3-phosphate diethylacetal, barium salt, commercial

[1] Abbreviations used: KDPG, 2-keto-3-deoxy-6-phosphogluconate(ic); G-3-P, D-glyceraldehyde 3-phosphate; NADH, reduced nicotinamide adenine dinucleotide.

Dowex 50, hydrogen form
Sodium pyruvate, crystalline
KDPG aldolase
Barium acetate
$Ba(OH)_2$ solution, saturated

Procedure. One gram of diethylacetal of DL-glyceraldehyde 3-phosphate Ba salt (1.3 millimoles of G-3-P) is put in solution in 15 ml of water using Dowex 50 (hydrogen form). The resin is added until all the salt is dissolved and the supernatant is acid to Congo red paper. This preparation is filtered, and the filtrate is incubated at 40° for 48 hours to hydrolyze the acetal. To this solution is added 550 mg of crystalline sodium pyruvate (4 millimoles), and the pH is adjusted to 7 with alkali. The volume of this solution should be about 25 ml. Then, 200 units of KDPG aldolase[2] are added, and the mixture is incubated 1 hour at room temperature. After incubation, 2 g of barium acetate (8 millimoles) is added to the reaction mixture and the pH is adjusted to 3.5 with concentrated HCl. Any precipitate present is removed by centrifugation. To the supernatant is added 5 volumes of ethanol, and the precipitate is allowed to settle overnight in the cold.

The precipitate is collected by centrifugation and dissolved in dilute HCl. The pH of the solution is adjusted to 3.5 with saturated $Ba(OH)_2$ solution, and the barium salt is precipitated by the addition of 5 volumes of ethanol and standing in the cold. The precipitate is collected by centrifugation and dried *in vacuo*. The resulting KDPG is obtained in 72% yield (relative to the initial G-3-P) and is about 50% pure (the theoretical molecular weight of mono-barium KDPG, assuming it exists as a ketal, is 395.20). This material is adequate for use as substrate in the KDPG aldolase assay.

The monobarium KDPG isolated can be further purified by ion exchange chromatography. The salt is put in solution with Dowex 50 (hydrogen form). The resin is removed by filtration, and the filtrate containing the KDPG (pH about 2) is adjusted to pH 6 with alkali and applied to a Dowex 1 (chloride form) column. The column is eluted with a linear gradient of 0–0.1 *N* HCl. Those tubes positive for KDPG are pooled and adjusted to pH 3.5 with saturated $Ba(OH)_2$. After the addition of 5 volumes of ethanol, the salt is allowed to settle out in the cold. The salt is collected by centrifugation, washed with ethanol and recentrifuged, and finally dried *in vacuo*. The recovery of KDPG on chromatography is excellent, and the salt assays 70–75% pure (weight basis) without correction for moisture.

[2] See this volume [91].

Determinations

KDPG can be determined either chemically or enzymatically. The chemical method is based upon the spectrophotometric determination of its semicarbazone at 250 mμ.[3] Under the conditions described the extinction of the semicarbazone is 2.0×10^{-3}.

The enzymatic method is based upon cleavage of KDPG by the aldolase and determining the pyruvate or G-3-P formed. A convenient assay results from the addition of 0.05 ml of the stock assay solution[2] and 0.15 ml of water to a microcuvette. The initial optical density at 340 mμ is read against a water blank. After the addition of 1 μl of an appropriate dilution of neutralized KDPG, 1 O.D. change is equivalent to 0.0322 micromole of the keto acid. A similar assay based upon the G-3-P formed by KDPG cleavage can also be used. In this assay, NADH oxidation in the presence of crystalline triosephosphate isomerase and α-glycerophosphate dehydrogenase is determined.

[3] J. MacGee and M. Doudoroff, *J. Biol. Chem.* **210,** 617 (1954).

[13a] D-Glucaric and Some Related Acids[1]

By DONALD C. FISH and HAROLD J. BLUMENTHAL

D-Glucaric (saccharic) acid, when isolated as the potassium acid salt following the oxidation of D-glucose with nitric acid,[2] is contaminated with small amounts of potassium acid oxalate and potassium acid D-tartrate.[3] Some commercial samples of potassium acid D-glucarate vary considerably in their properties, such as by having 26% potassium instead of the 15.8% required by theory. Although recrystallization of the potassium acid salt from water yields a product whose potassium content is close to theory, other criteria of purity are needed, especially since boiling water can lactonize potassium acid glucarate.[4] Although crystalline 1,4- and 6,3-lactones of D-glucaric acid with sharp melting points can be prepared,[4] potassium acid glucarate itself does not have a usable melting point and free glucaric acid is not convenient to prepare.[5]

The cyclohexylamine salts can conveniently be used to purify and to

[1] D. C. Fish, Ph.D. Thesis, Univ. of Michigan, 1964.

[2] C. L. Mehltretter and C. E. Rist, *J. Agr. Food Chem.* **1,** 779 (1953).

[3] S. Kiyooha, *J. Agr. Chem. Soc. Japan* **34,** 170 (1960).

[4] R. J. Bose, T. L. Hullar, B. A. Lewis, and F. Smith, *J. Org. Chem.* **26,** 1300 (1961).

[5] K. Rehorst, *Ber.* **61,** 163 (1928).

assess the purity of D-glucarate, a substrate for D-glucarate dehydrase.[6] The cyclohexylamine salts of D-glucaric and galactaric (mucic) acids and of four organic acids that could arise as contaminants during preparation of potassium acid glucarate or calcium glucarate, namely, oxalic, D-gluconic, D-glucuronic, and D-tartaric acids, are easily prepared. Unlike the diphenylhydrazides of D-glucaric[7] and galactaric acids, both of which have melting points accompanied by decomposition, the dicylohexylamine salts of the two naturally occurring hexaric acids have sharp melting points that are distinguishable from each other and from the melting points of the salts of related organic acids. Furthermore, the free acid or the sodium salt can be prepared easily and quantitatively by shaking an aqueous solution of the cyclohexylamine salt for 30 minutes at room temperature with an excess of Dowex 50 cation exchange resin, H^+ or Na^+ form, respectively.

Reagents

Monopotassium or calcium salts of D-glucarate, galactarate, oxalate, tartrate, D-glucuronate, D-gluconate
Dowex 50 (H^+) cation exchange resin
Cyclohexylamine, freshly distilled
Isopropanol, methanol, ethyl acetate, absolute ethanol

Procedure

DICYCLOHEXYLAMMONIUM D-GLUCARATE. A sample (2.0 g) of commercial potassium acid D-glucarate or calcium D-glucarate tetrahydrate is placed in a 50-ml erlenmeyer flask containing 20 ml of water, and the free acid is generated by shaking with a 20% excess of Dowex 50 cation exchange resin (H^+ form) for 30 minutes. The liquid is filtered into a 250-ml flask, the resin is washed with several volumes of water, and the combined filtrates are adjusted to a pH 9 by the addition of cyclohexylamine. The solution is then concentrated to dryness under reduced pressure in a rotary evaporator, bath temperature 40°. The completely dry salt is recrystallized from isopropanol–water, 12:1 (v/v), and the filtered crystals are allowed to dry in air. The yield of salt from potassium acid glucarate is 1.8 g (60% of theory), with a melting point of 185–188°. After a second recrystallization, the yield is 1.2 g (40% of theory), and the melting point rises to 191–192°. There is no further increase in the melting point upon subsequent recrystallizations or upon drying *in vacuo*.

OTHER CYCLOHEXYLAMMONIUM SALTS. The salts of organic acids must

[6] See this volume [118].
[7] E. Fischer and P. Passmore, *Ber.* **22**, 2728 (1889).

PROPERTIES OF THE CYCLOHEXYLAMINE SALTS OF D-GLUCARIC AND RELATED ACIDS

Acid salt	Solvent for recrystallization	Melting point, corrected (°C)	Yield (%)[b]	Elemental analysis[a]						
				Calculated for	Calculated (%)			Found (%)		
					C	H	N	C	H	N
Dicyclohexylammonium D-glucarate	Isopropanol–H_2O, 12:1 (v/v)	191–192	40[c]	$C_{18}H_{36}N_2O_8$ (408.50)	52.93	8.88	6.86	53.22[c]	8.80	7.24
			51[d]					52.50[d]	8.88	6.90
Dicyclohexylammonium galactarate	Isopropanol–H_2O, 3:1 (v/v)	209–210	63	$C_{18}H_{36}N_2O_8$ (408.50)	52.93	8.88	6.86	52.82	8.78	6.95
Dicyclohexylammonium oxalate	Methanol–ethyl acetate–H_2O, 25:50:3 (v/v/v)	248–251 (sublimes)	55	$C_{14}H_{28}N_2O_4$ (288.39)	—	—	9.71	—	—	9.72
Dicyclohexylammonium (+)-tartrate	Isopropanol–H_2O, 10:1 (v/v)	208–209	61	$C_{16}H_{32}N_2O_6$ (348.45)	—	—	8.04	—	—	8.00
Cyclohexylammonium D-glucuronate	Ethanol–H_2O, 20:1 (v/v)	161–162 (decomp.)	*ca.* 25	$C_{12}H_{23}NO_7$ (293.32)	—	—	4.78	—	—	4.84
Cyclohexylammonium D-gluconate	Isopropanol	114–116 (sublimes)	*ca.* 35	$C_{12}H_{25}NO_7$ (295.34)	—	—	4.74	—	—	4.65

[a] All salts, except that of D-glucuronate, were dried over magnesium perchlorate *in vacuo* for 4 hours at 78° before analysis. The salt of D-glucuronate was dried *in vacuo* for 5 hours at room temperature. Analyses were performed by the Spang Microanalytical Laboratory, Ann Arbor, Michigan.

[b] After two recrystallizations.

[c] From potassium acid glucarate, Nutritional Biochemicals Corp.

[d] From calcium glucarate tetrahydrate, Mann Research Laboratories.

first be converted to the free acid by treatment of the aqueous solution with Dowex 50 (H^+ form). The crude cyclohexylamine salts of these acids are prepared in the same manner as the dicyclohexylammonium glucarate. The solvents used to recrystallize the crude salts twice are listed in the table along with the yields, melting points, and elemental analyses. The cyclohexylammonium D-gluconate crystals have to be filtered within 36 hours and immediately dried in a vacuum desiccator or they become gummy and require recrystallization.

[13b] Preparation and Properties of α-Keto-β-deoxy-D-glucarate

By HAROLD J. BLUMENTHAL, DONALD C. FISH, and THERESSA JEPSON

The initial intermediate in the microbial catabolism of either of the naturally occurring hexaric acids is an α-keto-β-deoxy-D-glucarate. When D-glucarate dehydrase[1] acts upon D-glucaric acid as the substrate, an 85:15 mixture of 5-keto-4-deoxy-D-glucarate and 2-keto-3-deoxy-D-glucarate results.[2] With galactarate dehydrase acting upon its substrate galactaric acid, the product is entirely 5-keto-4-deoxy-D-glucarate.[3,4] This report describes the enzymatic preparation and properties of the α-keto-β-deoxy-D-glucarates, substrates for ketodeoxyglucarate aldolase,[5] and intermediates in the formation of α-ketoglutaric acid.[6]

Procedure I: 5-Keto-4-deoxy-D-glucarate by Action of Galactarate Dehydrase on Galactarate

Reagents

Disodium galactarate, 0.1 *M*. Commercial galactaric (mucic) acid is neutralized with sodium hydroxide.

Tris-HCl, pH 8.0, 0.8 *M*

Disodium EDTA, 0.02 *M*

Crude *Escherichia coli* galactarate dehydrase[1] in 0.02 *M* disodium galactarate

[1] See H. J. Blumenthal, this volume [118].

[2] D. C. Fish and H. J. Blumenthal, *Bacteriol. Proc.* p. 192 (1961).

[3] H. J. Blumenthal and T. Jepson, *Biochem. Biophys. Res. Commun.* **17**, 282 (1964).

[4] See H. J. Blumenthal and T. Jepson, this volume [119].

[5] See D. C. Fish and H. J. Blumenthal, this volume [93].

[6] H. J. Blumenthal and T. Jepson, *Bacteriol. Proc.* p. 82 (1965).

Procedure. The incubation mixture (182 ml) contains 10 millimoles of disodium galactarate, 9.6 millimoles of Tris-HCl, pH 8.0, 0.6 millimole of EDTA, and 40 ml of crude *E. coli* galactarate dehydrase in 0.02 *M* disodium galactarate.[4] The labile galactarate dehydrase must be prepared immediately prior to use from either fresh or frozen cells. The flask is incubated at 30° and occasionally stirred. Since EDTA inhibits the action of the ketodeoxyglucarate aldolase also present in the crude extract,[5] the galactarate can be converted into ketodeoxyglucarate almost quantitatively if an excess of enzyme is employed. The progress of the conversion can be followed by removing small samples and measuring product formation by the periodate-thiobarbiturate procedure.[7] Ordinarily, the reaction is terminated after 6 hours by placing the flask in a boiling water bath for 2 minutes. The denatured protein is removed by high speed centrifugation, and the supernatant fluid is stored at —20° until ready for purification. For certain studies, this supernatant solution may be used without further purification. By adding an excess of Mg^{++}, for example, the inhibitory effect of the EDTA can be reversed and the solution will serve as a substrate for the ketodeoxyglucarate aldolase.[5] Furthermore, a number of buffers may be substituted for the Tris-HCl used in the incubation.[4] Since galactose can easily be converted to galactaric acid by oxidation with nitric acid,[8] specifically ^{14}C-labeled 5-keto-4-deoxy-D-glucaric acid can also be prepared by this enzymatic procedure from the corresponding ^{14}C-galactaric acids.

For further purification, one half of the above incubation is placed on a 3.5 (diameter) × 30 cm column of Dowex 1-formate, 8% cross-linked, 200–400 mesh. After the sample has been washed in, a plug of wet glass wool and water to a height of 10 cm are placed over the resin. A gradient of increasing formic acid concentration is then applied to the column. The reservoir contains 6 *N* formic acid, and the mixing chamber 175 ml of water. Fractions of about 20 ml are collected at 3-minute intervals, and samples are analyzed for the presence of ketodeoxyglucarate by the periodate-thiobarbiturate procedure.[7] A small peak of unknown periodate thiobarbiturate-positive material that appears early is discarded. The tubes containing the major peak of periodate-thiobarbiturate positive material are pooled and lyophilized. This peak appears when the formic acid concentration is about 5.4 *N*, immediately after the elution of pyruvic acid (when present). The material in the lyophile flasks is dissolved in the minimum quantity of distilled water, and finely powdered calcium hydroxide is added slowly and carefully to bring the pH to 8.0.

[7] A. Weissbach and J. Hurwitz, *J. Biol. Chem.* **234**, 705 (1959).
[8] C. L. Mehltretter and C. E. Rist, *J. Agr. Food Chem.* **1**, 779 (1953).

Additional quantities of calcium hydroxide are added over a period of approximately 4 hours to keep the pH at 8.0 and complete the delactonization. After this, three volumes of absolute ethanol are added to the completely clear solution and the flask is placed at 4° overnight. The white flocculent precipitate is removed by filtration and dried *in vacuo* over silica gel and paraffin shavings at room temperature for 3 days. The salt is then dried *in vacuo* at 78° over magnesium perchlorate for 4 hours. Elemental analysis calculated for $CaC_6H_6O_7$: C—31.33; H—3.25; Ca—17.43; values found were C—30.59; 30.64; H—3.07, 3.12; Ca—17.50. Yields of the calcium salt in excess of 50% of theory can be obtained in the first crop of precipitate.

The dry calcium salt is stable at room temperature. It is converted to the disodium salt, using Dowex 50-Na^+, and these solutions are stored at −20°.

Procedure II: 5-Keto-4-deoxy-D-glucarate and 2-Keto-3-deoxy-D-glucarate from Action of D-Glucarate Dehydrase on D-Glucarate

Reagents

Disodium D-glucarate (saccharate), 0.1 *M*. Commercial sources of glucarate are frequently impure. See this volume [13a] for the purification of this substrate as the crystalline dicyclohexylammonium salt. However, even the unpurified commercial potassium acid glucarate can be used.

Tris-HCl, pH 7.5, 0.8 *M*

$MgSO_4$, 0.08 *M*

Sodium arsenite, 0.012 *M*

Crude D-glucarate dehydrase[1]

Procedure. The incubation mixture (182 ml) contains 10 millimoles of disodium D-glucarate, 9.6 millimoles of Tris-HCl, pH 7.5, 1.2 millimoles of $MgSO_4$, 0.18 millimole of sodium arsenite, and 40 ml crude of *E. coli* D-glucarate dehydrase. Since the crude enzyme is relatively stable, it can be prepared and stored frozen for long periods prior to use. The flask is incubated at 30° for 4 hours or until product formation, determined by the periodate-thiobarbiturate procedure,[7] is at a maximum. Crude enzyme extracts contain ketodeoxyglucarate aldolase,[5] which slowly cleaves the ketodeoxyglucarate under the conditions employed. The reaction is terminated and the ketodeoxyglucarate is purified and converted to the calcium salt, as described in Procedure I. Elemental analysis of the calcium salt calculated for $CaC_6H_6O_7 \cdot H_2O$: C—29.03; H—3.25; Ca—16.15; values found were C—29.05, 29.11; H—3.30, 3.19; Ca—16.00, 16.08. The melting point (corrected) of this product is 163–

164°. The yield of the calcium salt in the first crop of precipitate is about 30% of theory.

Extinction Coefficient. When an analytical sample of the calcium salt of mixed deoxyglucarates formed by Procedure II is converted to the disodium salt and analyzed by the periodate-thiobarbiturate procedure,[7] an extinction coefficient of 60,000 is found at 551 mμ. Not all α-keto-β-deoxy sugar acids yield such values. A number of 2-keto-3-deoxy-onic acids analyzed by the same procedure yield extinction coefficients of 40,000–44,000[7,9] whereas values about twice these are also reported.[6] It is known that the rate of release of formyl pyruvate in the presence of periodate is different for epimers.[7,10] With the ketodeoxyglucarate, the extinction coefficient does not vary even when the time for periodate oxidation is lengthened to 40 minutes. However, to ensure that the isolation and purification of the ketodeoxyglucarate does not alter its properties, as far as analysis by the periodate-thiobarbiturate procedure is concerned, a sample of D-glucarate can be quantitatively converted to ketodeoxyglucarate and assayed directly. When a 0.186 micromole sample of D-glucarate (from crystalline dicyclohexylammonium D-glucarate[11]), in 160 micromoles of Tris-HCl buffer, pH 7.5, 7 micromoles $MgSO_4$, and 0.073 unit of purified D-glucarate dehydrase[1] free of ketodeoxyglucarate aldolase, in a final volume of 0.8 ml, is incubated at 30° and samples are removed at intervals and assayed directly by the periodate-thiobarbiturate procedure, the extinction coefficient at the maximum is found to be 59,000. We choose to use the initial value of 60,000, determined on the isolated compound, as the extinction coefficient.

It should be mentioned, however, that when the amount of ketodeoxyglucarate added to ketodeoxyglucarate aldolase[5] is based on the extinction coefficient of 60,000, the recovery of products is only 80–85% of theory when the reaction is allowed to proceed to completion.[12] In these experiments, performed in three different buffers, purified ketodeoxyglucarate aldolase[5] is used and the products formed are determined directly by coupling with lactic dehydrogenase and/or tartronate semialdehyde reductase and measuring the DPNH utilization.[12] When similar experiments are performed in parallel with D-glucarate and D-glucarate dehydrase[1] generating the ketodeoxyglucarate, recoveries of products are 90–100% of theory.[12,13] The reasons for the incomplete recoveries from ketodeoxyglucarate are presently unknown.

[9] D. B. Sprinson, J. Rothschild, and M. Sprecher, *J. Biol. Chem.* **238**, 3170 (1963).

[10] J. Preiss and G. Ashwell, *J. Biol. Chem.* **238**, 1571 (1963).

[11] See D. C. Fish and H. J. Blumenthal, this volume [13a].

[12] D. C. Fish, Ph.D. Thesis, Univ. of Michigan, 1964.

[13] H. J. Blumenthal and D. C. Fish, *Biochem. Biophys. Res. Commun.* **11**, 239 (1963).

[14] 3-Deoxyoctulosonate[1]

By MOHAMMAD A. GHALAMBOR and EDWARD C. HEATH

3-Deoxyoctulosonate may be prepared by any of the following procedures: (A) isolation from the cell wall lipopolysaccharide (or cells) of *Escherichia coli;* (B) enzymatic synthesis with crude preparations of 3-deoxyoctulosonate 8-phosphate synthetase[2] from *E. coli;* or (C) chemical synthesis. Neither of the materials prepared by isolation or enzymatic synthesis has been obtained in crystalline form, and therefore, if it is desirable to prepare a sample that would provide an analytically pure sample of 3-deoxyoctulosonate, the chemical procedure should be used. 3-Deoxyoctulosonate may be conveniently analyzed with the thiobarbituric acid reagents.[3] The preparation of 3-deoxyoctulosonate-1-^{14}C is described in Vol. VIII [14].

Isolation from *Escherichia Coli*[4]

Escherichia coli 0111-B_4 (ATCC 12015) is grown in Trypticase Soy broth (Baltimore Biological Laboratories) at 37° on a rotary shaker (1 liter of medium in a 2-liter Erlenmeyer flask) for 10–18 hours. If larger quantities of cells are required, the organism may be grown in the same medium in a fermentor with forced aeration; in this case, the growth cycle usually requires 4–6 hours. The cells are harvested and washed with cold water; they may be frozen until needed.

The cells (300 g) are resuspended in 600 ml of water with the aid of a Waring blendor. The cell suspension is added slowly to 4 liters of ethanol while the mixture is being stirred vigorously. The suspension is heated in a boiling water bath (while being stirred) until the temperature of the mixture reaches 70°; the suspension is then allowed to stand at room temperature overnight. The insoluble residue is harvested by centrifugation, resuspended in 1 liter of 0.2 N H_2SO_4, and placed in a boiling water bath for 30 minutes. The insoluble material is removed by centrifugation and discarded. The supernatant solution is adjusted to pH 6 by

[1] This presentation is based on work previously published in preliminary form: M. A. Ghalambor and E. C. Heath, *Biochem. Biophys. Res. Commun.* **11,** 288 (1963).

[2] D. H. Levin and E. Racker, *J. Biol. Chem.* **234,** 2532 (1959).

[3] D. Aminoff, *Biochem. J.* **81,** 384 (1961).

[4] 3-Deoxyoctulosonate is a constituent of the cell wall lipopolysaccharide of this and other microorganisms and can be prepared from these isolated polymers. However, a somewhat higher yield is obtained, and it is more convenient to isolate the compound from cells.

the addition of saturated barium hydroxide solution, and the barium sulfate precipitate is removed by centrifugation and discarded.[5] The supernatant solution is applied to a column (2.5×30 cm) of Dowex 1 (HCO_3^-) resin (200–400 mesh, 8% crosslinked) and the column is washed with 1 liter of water. The resin is eluted with a linear gradient consisting of 500 ml of water in the mixing vessel and 500 ml of 0.2 *M* ammonium bicarbonate in the reservoir; 10-ml fractions are collected on an automatic fraction collector. 3-Deoxyoctulosonate is eluted from the column usually between the 30th and the 50th fractions, as determined by assay of aliquots from the various fractions. After the fractions containing 3-deoxyoctulosonate have been located, they are pooled and treated with Dowex 50 (H^+) resin (excess water is removed from the resin by filtration just prior to use). The resin is added in a batchwise manner with vigorous shaking after each addition. Care must be taken during this procedure to avoid loss of sample by excessive foaming. After sufficient resin has been added to adjust the solution to pH 3–5, the resin is removed by filtration, and the filtrate is concentrated to one-third volume by evaporation under reduced pressure. The solution is then adjusted to pH 6.5 by the addition of potassium hydroxide, and concentration is continued until the desired volume is attained. The yield of 3-deoxyoctulosonate is approximately 500 micromoles. Upon analysis of this material by paper chromatography in a variety of solvent systems,[6] a single thiobarbituric acid-positive[7] component is obtained which corresponds to authentic 3-deoxyoctulosonate prepared enzymatically (see procedure outlined below).

Enzymatic Preparation

This procedure utilizes a crude protein fraction from *E. coli* 0111-B_4 that contains most of the 3-deoxyoctulosonate 8-phosphate synthetase activity (see Vol. VIII [14]); the fraction also contains phosphatase activity which rapidly converts the initial product of the reaction to 3-deoxyoctulosonate and inorganic phosphate. The combination of these

[5] It is necessary to remove the dilute acid insoluble residue before precipitation of barium sulfate to prevent loss of 3-deoxyoctulosonate.

[6] The following chromatographic solvent systems were employed: A, water-saturated phenol; B, ethyl acetate, acetic acid, water (3:1:3); C, methyl ethyl ketone, acetic acid, water (8:1:1); D, *n*-butanol, pyridine, 0.1 *N* HCl (5:3:2). All solvent systems clearly separated 3-deoxyoctulosonate from 3-deoxy-D-*erythro*-hexulosonate, 3-deoxy-D-*arabino*-heptulosonate, and acetyl neuraminate. The mobility of 3-deoxyoctulosonate in the various solvent systems relative to the mobility of authentic 3-deoxy-D-*erythro*-hexulosonate is as follows: A, 0.56; B, 0.59; C, 0.58; and D, 0.52, respectively. Chromatography is conducted on Whatman No. 1 paper.

[7] L. Warren, *Nature* **186**, 237 (1960).

activities in this fraction permit essentially stoichiometric conversion of phosphopyruvate and D-arabinose 5-phosphate to free 3-deoxyoctulosonate. The assay procedure is that described in Vol. VIII [14].

The organism is grown and the crude cell-free extract is prepared as described in Vol. VIII [14]. The crude extract (130 ml) prepared from 30 g of cells is treated with 14 ml of 2% protamine sulfate solution, the precipitate is removed by centrifugation and discarded. The supernatant fluid (140 ml) is treated with 140 ml of saturated (0°) ammonium sulfate solution and stirred for 5 minutes; the precipitate is removed by centrifugation and discarded. To the supernatant fluid is added 35 g of solid ammonium sulfate; the mixture is stirred for 5 minutes and centrifuged at 25,000 *g* for 15 minutes. The supernatant fluid is discarded, and the precipitate is dissolved in 15 ml of 0.01 *M* Tris buffer, pH 7.3; the latter fraction contains most of the synthetase and phosphatase activities. An incubation mixture is prepared as follows (millimoles in a final volume of 80 ml): phosphopyruvate, 0.8; D-arabinose 5-phosphate, 0.35; Tris buffer, pH 7.3, 10; and 4 ml of the ammonium sulfate fraction. The mixture is incubated at 37° for 1 hour, chilled, and treated with 10 ml of 50% trichloroacetic acid; the precipitate is removed by centrifugation. The supernatant fluid is extracted with ether repeatedly until the aqueous phase is about pH 5, and 3-deoxyoctulosonate is isolated by ion exchange chromatography as described in the preceding part of this paper. The yield of 3-deoxyoctulosonate obtained in this preparation is 0.32 millimole (approximately 90%) of chromatographically homogeneous material.

Chemical Synthesis

This procedure is based on reaction, at pH 11, of oxalacetic acid with one of a variety of aldoses (with elimination of CO_2) to yield an aldol condensation product of pyruvic acid and the corresponding aldose.[1,8] As the reaction involves strong alkaline conditions, the product(s) of the reaction usually consists of a mixture of isomers, and procedures must be employed that permit the resolution of the mixture. The chemical synthesis of a single isomer of 3-deoxyoctulosonate that corresponds by all criteria tested to authentic, natural 3-deoxyoctulosonate, is obtained by the following series of reactions:

Reaction A. Oxalacetic acid + D-arabinose $\xrightarrow{\text{pH 11}}$ mixture of 3-deoxyoctulosonate isomers

Reaction B. 3-Deoxyoctulosonate isomers $\xrightarrow[\text{3. Fractional crystallization}]{\text{1. Acetylation, 2. Methylation}}$ crystalline 2,4,5,7,8-penta-acetyl-3-deoxyoctulosonate methyl ester

[8] J. W. Cornforth, M. E. Firth, and A. Gottschalk, *Biochem. J.* **68**, 57 (1958).

Reaction C. 2,4,5,7,8-penta-acetyl-3-deoxyoctulosonate-methyl ester $\xrightarrow{\text{dil. NaOH}}$ 3-deoxyoctulosonate (natural isomer)

Reaction A. A 100-millimole sample (13 g) of oxalacetic acid (California Biochemical Corp.) is dissolved in 100 ml of water, chilled to 5°, and rapidly adjusted to pH 11 with 20% potassium hydroxide. Immediately, 250 millimoles (37.5 g) of D-arabinose (Pfanstiehl), dissolved in 150 ml of water, is added and the mixture is allowed to stand at room temperature for 2 hours; occasionally during this period, the pH is checked and readjusted to pH 11 if necessary. The mixture is adjusted to pH 5 with glacial acetic acid, diluted with water to a total volume of 1 liter, and applied to a column (4.5 × 40 cm) containing Dowex 1-HCO_3^- resin (200–400 mesh, 8% cross-linked). The column is washed with 2 liters of water and then eluted with a linear gradient consisting of 2 liters of water in the mixing vessel and 2 liters of 0.2 *M* ammonium bicarbonate in the reservoir; 25-ml fractions are collected on an automatic fraction collector. The fractions (usually between the 50th and the 100th fractions) containing thiobarbituric acid-reactive material are pooled[9] and treated with an excess of Dowex 50-H^+ resin as described above. After removal of the resin by filtration, the filtrate is concentrated to 200 ml under reduced pressure. The sample contains approximately 85 millimoles of thiobarbituric acid-reactive (natural 3-deoxyoctulosonate standard) material. However, qualitative analysis of the product by paper chromatography indicates the presence of at least four thiobarbituric acid-reactive components: one major component (approximately 70%) corresponds to natural 3-deoxyoctulosonate; two minor components with mobilities less than authentic 3-deoxyoctulosonate and one minor component exhibiting a relative mobility faster than authentic material.

Reaction B. An aliquot (15 millimoles) of the crude synthetic mixture of the free acids obtained in the preceding reaction is concentrated to a sirup under reduced pressure. The sirup is dehydrated by first dissolving it in 25 ml of methanol, then adding 100 ml of ethanol; the mixture is concentrated to dryness. This procedure is repeated several times, and the resultant, pale yellow powder is dried under reduced pressure over calcium chloride for 18 hours. The dried sample is dissolved in 70 ml of anhydrous pyridine and chilled to 0°; 50 ml of cold acetic anhydride is added. The flask is stoppered tightly (glass, standard taper) and shaken in an ice slurry for 48 hours. The clear, brown sirup is poured over about 500 ml of finely crushed ice and stirred until all the ice is melted. The clear, tan-colored solution is concentrated to a sirup

[9] This fractionation procedure does not separate significantly the isomeric mixture obtained in this reaction.

under reduced pressure; the sirup is dissolved in ethanol and concentrated to a sirup several times to remove the last traces of pyridine. The sample is dissolved in 100 ml of ethanol, 50 ml of water is added, and the mixture is shaken for a few minutes with 25 ml (bed volume) of Dowex 50-H^+ resin. The resin is removed by filtration, and the filtrate is concentrated to a sirup under reduced pressure. The sirup is suspended in 30 ml of water and extracted five times with 2-volume portions of ether. The pooled ether extracts are concentrated under reduced pressure to a sirup; the sample is dissolved repeatedly (3 times) in ethanol followed by concentration to dryness. The sample is dried under reduced pressure over calcium chloride and solid potassium hydroxide pellets overnight. The dry sirup (approximately 3 g) is dissolved in approximately 50 ml of cold ether and mixed with 150 ml of a cold solution of diazomethane (prepared from 22 g of *N*-methyl-*N*-nitroso-*p*-toluenesulfonamide.)[10] The mixture is tightly stoppered and permitted to stand in an ice bath overnight. The yellowish solution is concentrated to a sirup under reduced pressure, and the sample is dissolved in approximately 50 ml of ethanol. Water is added to the ethanolic solution until a slight turbidity appears, the solution is treated with 25 ml (bed volume) of mixed-bed resin,[11] the resin is removed by filtration, and the filtrate is concentrated to a sirup under reduced pressure. The sirup is dissolved in a minimum volume of boiling ethanol, and the mixture is allowed to cool slowly on the surface of the hot plate; crystals appear after a short time, and crystallization is allowed to continue at room temperature for 18 hours. The suspension is then placed in a refrigerator and allowed to stand for an additional period of 24 hours. The crystals are harvested on a fritted glass funnel, washed with a small amount of ice cold ethanol, and dried. The mother liquor is concentrated, and a second crop of crystals may be obtained. The crystalline material is pooled, recrystallized from ethanol as described, and dried under reduced pressure over phosphorus pentoxide. The total yield of recrystallized material is 368 mg (0.8 millimole).

Properties of 2,4,5,7,8-pentaacetyl-3-deoxyoctulosonate methyl ester. Molecular weight = 462; melting point = 155–156°; $[\alpha]_D^{23} = +109.7 \pm 0.5°$ (c = 1.387 in methanol). The elementary analysis is tabulated.

	C	H	Acetyl	Methoxyl
Theoretical	49.35	5.67	47.60	6.71
Found	49.22	5.73	48.57	6.23

[10] Prepared as described in Vol. VI [76], p. 515.

[11] Equal parts of Dowex 50-H^+ resin and Dowex 1-HCO_3^- resin. Both resins are 200–400 mesh and 8% cross-linked. Just prior to use, water is removed by suction filtration on a fritted glass funnel.

Reaction C. Dissolve 10 mg of the pentaacetyl methyl ester derivative in 1 ml of methanol, add 1 ml of 0.2 *N* NaOH, and incubate the mixture at 37° for 15 minutes. Treat the solution with just enough Dowex 50-H^+ resin (dried with acetone immediately prior to use) to adjust the pH to 5 or lower. Remove the resin by filtration, adjust the filtrate to pH 6.5 with sodium hydroxide, and concentrate it to dryness. Dissolve the residue in 1 ml of water, yielding a solution of 0.022 *M* sodium 3-deoxyoctulosonate. Analysis of this solution by paper chromatography[6] indicates the presence of a single thiobarbituric acid-reactive component which corresponds in all systems to the natural compound. In addition, the compound obtained in this manner is reactive with 3-deoxyoctulosonate aldolase and CMP-3-deoxyoctulosonate synthetase in a manner indistinguishable from authentic 3-deoxyoctulosonate obtained from natural sources.

[15] α-Hydroxyethylthiamine Diphosphate and α,β-Dihydroxyethylthiamine Diphosphate

By L. O. Krampitz and Robert Votaw

Thiamine diphosphate (TDP) with an α-hydroxyethyl group substitution at position 2 of the thiazolium ring has been found to be the active intermediate in many enzymatic reactions involving pyruvate metabolism.[1-3] An α,β-dihydroxyethyl group substitution at position two of the thiazolium ring is the intermediate involved in the transketolase and phosphoketolase reactions.[1,4] In order to establish proof of structure by synthesis, the α-hydroxyethyl derivative of thiamine was synthesized employing conventional methods for the synthesis of thiamine or thiamine-like compounds.[5] Inasmuch as the pyrophosphate ester of thiamine is the active form of the coenzyme, only the preparation of the α-hydroxyethyl and α,β-dihydroxyethyl derivatives of thiamine diphosphate will be described.

[1] L. O. Krampitz, I. Suzuki, and G. Greull, *Federation Proc.* **20,** 971 (1961).
[2] L. O. Krampitz, I. Suzuki, and G. Greull, *Ann. N.Y. Acad. Sci.* **98,** 466 (1962).
[3] H. Holzer, *Angew. Chem.* **73,** 721 (1961).
[4] R. Votaw, W. T. Williamson, L. O. Krampitz, and W. A. Wood, *Biochem. Z.* **338,** 756 (1963).
[5] L. O. Krampitz, G. Greull, C. S. Miller, J. B. Bicking, H. R. Skeggs, and J. M. Sprague. *J. Am. Chem. Soc.* **80,** 5893 (1958).

α-Hydroxyethylthiamine Diphosphate (HETDP)

Principle. Mizuhara *et al.*[6] have found that under slightly alkaline conditions thiamine or thiamine diphosphate catalyzes the nonenzymatic conversion of pyruvate to acetoin. Acetaldehyde is also converted to acetoin, although at a somewhat slower rate. Following the proposal of Breslow[7] that carbon atom 2 of the thiazolium ring of thiamine is involved in the catalysis, the presence of α-hydroxyethylthiamine as an intermediate in the nonenzymatic system employed by Mizuhara was sought. Comparison of the intermediate with the authentic α-hydroxyethylthiamine obtained by synthesis[5] was made. Under proper conditions α-hydroxyethylthiamine accumulates in this system to the extent of 30–40% of the amount of thiamine present. Employing thiamine diphosphate, comparable yields of α-hydroxyethylthiamine diphosphate can be obtained.

Reagents

Thiamine diphosphate
K_2CO_3, solid
K_2CO_3, 2.0 *M*
Acetaldehyde, redistilled
Nitrogen, gaseous
Nitrogen, liquid
Cationic exchange resin, IRC 50 (H^+), Amberlite
Ethanol, absolute
Diethyl ether, absolute
Whatman 3 MM paper
Whatman 31 extra thick paper
n-Propanol
Ammonium formate, 1.0 *M*, pH 5.0

Procedure. The following description employs 1 g of thiamine diphosphate as starting material; however, if a smaller-scale preparation is desired, the procedure may be scaled down proportionately.

One gram (~ 2.1 millimoles) thiamine diphosphate is partially dissolved in 5 ml of H_2O in a small two-necked flask. The solution is cooled to 0° and solid K_2CO_3 (~ 590 mg) is added in small portions until pH 8.8 is obtained. Two milliliters of 2.0 *M* K_2CO_3 and 1 ml H_2O are added; 42 millimoles of acetaldehyde are added. The temperature of the mixture is maintained at 0°, and the flask is flushed with N_2. The flask is stop-

[6] S. Mizuhara, R. Tamura, and H. Arata, *Proc. Japan Acad.* **27**, 302 (1951).
[7] R. Breslow, *J. Am. Chem. Soc.* **80**, 3719 (1958).

pered tightly and securely. The solution is incubated at 50° for 3 hours; after this time it has a light yellow color and the pH is 7.9. The flask and its contents are cooled to 0° and attached to a liquid nitrogen trap. Gaseous nitrogen is passed through the flask, and the temperature is gradually brought to 50°. Much of the excess acetaldehyde is thus removed and trapped. If radioactive acetaldehyde is employed in the synthesis, a 2,4-dinitrophenylhydrazone derivative of the trapped acetaldehyde is made in order to determine its specific activity. The solution is again cooled to 0°; 55 milliequivalents of IRC 50 (H^+) are added, and the solution is allowed to stand at 0° for 6 hours. Under these conditions thiamine diphosphate and its derivatives are only slightly cationic, if at all. The solution containing the resin is placed on a fritted glass funnel and washed with small aliquots of water until the total volume of the eluate is 200 ml. Recovery of thiamine diphosphate compounds is followed by testing aliquots of the washings for absorption at 265 mμ, pH 7.0.

The eluate which has a pH of 5.4–5.8 is lyophilized to dryness. A glassy, slightly yellow residue is obtained which is triturated with 20 ml of absolute ethyl alcohol. This is a tedious step. The residue is mostly soluble; however, a white residue remains which contains inorganic salts. After separation of the solution, the precipitate is washed with 20 ml of absolute ethyl alcohol and the washings are combined with the previous solution of 20 ml of ethyl alcohol. The alcoholic solution is slightly yellow and contains all the ultraviolet-absorbing material. Absolute diethyl ether is added to the alcoholic solution until precipitation no longer occurs (approximately 3 volumes). The precipitate is collected and dried *in vacuo* (1.25 g).

The dried precipitate obtained above contains decomposition products of TDP as well as HETDP and TDP. In order to determine qualitatively whether HETDP accumulated in the mixture, an amount of the dried precipitate equivalent to 5 μg TDP (based upon a molar extinction coefficient of 8.1×10^3 at 265 mμ, pH 7.0 for TDP) is dissolved in a small amount of water and chromatographed descendingly on Whatman 3 MM paper at 4° with the following solvent: *n*-propanol–1.0 *M* ammonium formate (pH 5.0)–water, 65:15:25. In this system thiamine has an R_f of 0.65, thiamine monophosphate 0.45, thiamine diphosphate 0.33, and α-hydroxyethyl thiamine diphosphate 0.415 when viewed with ultraviolet light. If the reaction has proceeded properly, spots containing unreacted TDP and HETDP will be distinctly discernible. Faint spots of other thiamine derivative will also be seen, but distinct from TDP and HETDP.

For the preparative separation, 50 mg of the dried precipitate are dissolved in a minimum amount of water and spotted on Whatman 31 extra thick paper (18.25 × 22.5 inches) and chromatographed descendingly at 4° employing the *n*-propanol–ammonium formate solvent cited above. A reference spot containing approximately 0.1 micromole TDP is also placed on each sheet. Several sheets are employed. In order to obtain excellent separation the solvent front is permitted to leave the paper for 3 hours. The sheets are dried at room temperature and the ultraviolet absorbing material immediately ahead of the TDP spot is cut from the sheets and eluted with water. The combined water eluates from all the sheets are lyophilized to dryness, and the residue is exposed to vacuum for an additional 6 hours to remove the volatile ammonium formate. The dry residue is triturated with absolute ethanol at 0°, and absolute ethyl ether is added to precipitate the HETDP. The precipitate has a very faint yellow color and upon rechromatography exhibited only slight traces of TDP.

α,β-Dihydroxyethylthiamine Diphosphate (diHETDP)

Principle. Employing the same rationale as for the synthesis of α-hydroxyethylthiamine diphosphate described in the previous section, the expectation was that substitution of glycolaldehyde for acetaldehyde in the procedure would result in the preparation of α,β-dihydroxyethylthiamine diphosphate. Although thiamine diphosphate rapidly catalyzed the nonenzymatic condensation of glycolaldehyde to form a variety of sugars, α,β-dihydroxyethylthiamine diphosphate did not accumulate as an intermediate. During the preparation of α-hydroxyethylthiamine by the procedure outlined in the previous section, an adduct of acetoin and thiamine was also isolated. Presumably α-hydroxyethylthiamine will condense with another mole of acetaldehyde to form the acetoin adduct which has a degree of stability prior to forming acetoin and thiamine. By analogy formaldehyde should condense with thiamine or thiamine diphosphate to form the *hydroxymethyl* adduct at position 2 of the thiazolium ring followed by a further condensation with another mole of formaldehyde to form α,β-dihydroxyethyl thiamine diphosphate. The following procedure utilizes this principle for the preparation of α,β-dihydroxyethylthiamine diphosphate. Jaenicke and Koch[8] have prepared *hydroxymethylthiamine* diphosphate by reacting thiamine diphosphate with formaldehyde at pH 5.0–6.0. Apparently the condensation of a second mole of formaldehyde with hydroxymethylthiamine diphosphate does not occur under slightly acid conditions.

Reagents. Same as those required for the preparation of α-hydroxy-

[8] L. Jaenicke and J. Koch, *Biochem. Z.* **336**, 432 (1962).

ethylthiamine diphosphate described in the previous section, but with the following substitutions:

Formaldehyde for acetaldehyde
Ammonium formate 0.5 *M*, pH 5.0 for ammonium formate 1.0 *M*, pH 5.0

Procedure. The following description employs 1 g of thiamine diphosphate as starting material; however, if a smaller-scale preparation is desired, the procedure may be scaled down proportionately.

One gram ($\sim$2.1 millimoles) thiamine diphosphate is partially dissolved in 5 ml H_2O in a small two-necked flask. The solution is cooled to 0° and solid K_2CO_3 ($\sim$590 mg) is added in small portions until pH 8.8 is obtained. Two milliliters of 2.0 *M* K_2CO_3 and 1 ml H_2O are added; 42 millimoles formaldehyde are added. The temperature of the mixture is maintained at 0°, and the flask is flushed with N_2. The flask is stoppered, the stoppers being securely fastened. The solution is incubated at 50° for 2 hours; after this time it has a light yellow color and the pH is 8.0. The flask and its contents are cooled to 0° and attached to a liquid nitrogen trap. Gaseous nitrogen is passed through the flask, and the temperature is gradually brought to 23°. Much of the excess formaldehyde is thus removed and trapped. If radioactive formaldehyde is employed in the synthesis, a dimedon derivative of the trapped formaldehyde is made in order to determine its specific activity.

The solution is again cooled to 0°, and 55 milliequivalents of IRC 50 (H^+) are added. The flask is agitated in order to facilitate the removal of CO_2. The incubation mixture and resin are transferred to a small column containing a layer of IRC 50 (H^+) and washed with 5-ml aliquots of water until the total volume of the eluate is 200 ml. Recovery of thiamine diphosphate compounds is followed by testing aliquots of the eluates for absorption at 265 mμ, pH 7.0.

The eluate which has a pH of 5.4 to 5.8 is lyophilized to approximately 7 ml. *Caution must be taken not to permit drying of the material.* The flask is frequently thawed and shell frozen during the lyophilization in order that drying on the sidewalls of the flask does not occur. A procedure has not been effected by which α,β-dihydroxyethylthiamine diphosphate is stable upon drying. Solutions containing diHETDP which are dried show mostly the presence of TDP when rechromatographed.

Ten volumes (70 ml) of cold absolute ethanol are added to the 7 ml of lyophilized solution. To this, 400 ml of diethyl ether is added slowly, and the solution is kept at −20° for 12 hours. The precipitate is collected by centrifugation, allowed to drain well, but not permitted to dry. The precipitate is dissolved in 4 ml of H_2O preparatory to chromatography on paper. From 30 to 60 micromoles of thiamine diphosphate-containing compounds are applied on each sheet of Whatman No. 31 extra thick paper. The application of the material should be performed quickly in order that it does not dry. Development of the chromatogram is begun before the sample dries. Descending chromatography is performed at 4° employing the following solvent: *n*-propanol–0.5 *M* ammonium formate (pH 5.0)–water, 65:15:25. Development of the chromatogram is permitted to proceed for 14 hours with the solvent flowing off the paper. A reference spot containing approximately 0.1 micromole of TDP is also placed on each sheet. At the conclusion of the chromatography period, the excess solvent is drained from each sheet but drying of the sheet is prevented. The α,β-dihydroxyethylthiamine diphosphate band can be observed with an ultraviolet light source immediately ahead of the thiamine diphosphate. This band is cut out from the several sheets, and the α,β-dihydroxyethylthiamine diphosphate is eluted with water. After elution the eluate is concentrated by lyophilization to the desired concentration; however, caution is taken to prevent drying of the material on the sides of the flask. The concentration of α,β-dihydroxyethylthiamine diphosphate may be determined by assuming that the molar extinction coefficient at 265 mμ at pH 7.0 is identical to thiamine diphosphate, i.e., 8.1×10^3.

It may be necessary in some cases to rechromatograph the material a second time. Employing radioactive formaldehyde for the synthesis of α,β-dihydroxyethylthiamine diphosphate, preparations have been obtained whose specific activity ratio to the specific activity of the formaldehyde employed ranged from 1.70:1.0 to 1.91:1.0. The theoretical ratio is 2.0:1.0.

The final lyophilized solutions of α,β-dihydroxyethylthiamine diphosphate have been found to be stable for several weeks when kept frozen. The α,β-dihydroxyethylthiamine diphosphate present in the alcohol-ether precipitate described above is relatively stable and may be kept in suspension with the alcohol and ether for several months. Aliquots of the precipitate are usually chromatographed as needed.

Section III

Dehydrogenases and Oxidases

[16] Lactose Dehydrogenase

By OSAMU HAYAISHI

$$\text{Lactose} \xrightarrow{-2\text{ H}} \text{lactobionic-}\delta\text{-lactone} \xrightarrow{+\text{H}_2\text{O}} \text{lactobionic acid}$$

Lactose dehydrogenase of *Pseudomonas graveolens* catalyzes the stoichiometric oxidation of α- or β-lactose to lactobionic-δ-lactone in the presence of an appropriate hydrogen acceptor. The lactobionic-δ-lactone thus produced is hydrolyzed by another enzyme, lactonase, to lactobionic acid. This method is based on the procedure of Nishizuka and Hayaishi.[1]

Assay Method

Principle. The reduction of 2,6-dichlorophenolindophenol was followed spectrophotometrically at 590 mμ at 23°.

Reagents

Potassium phosphate buffer, pH 5.6, 0.1 *M*
Lactose, 0.01 *M*
2,6-Dichlorophenolindophenol, 0.0025 *M*

Procedure. A standard reaction mixture contains 100 micromoles of potassium phosphate buffer, pH 5.6, 10 micromoles of lactose, 0.25 micromole of the dye, and the enzyme in a total volume of 3.0 ml. Absorbancy at 590 mμ is determined at half-minute intervals.

Definition of Unit and Specific Activity. One unit of enzyme activity was defined as that amount which caused the reduction of 0.1 micromole of the dye in 5 minutes under the above conditions, and the specific activity was expressed as units per milligram of protein. ϵm*M* of 2,6-dichlorophenolindophenol is 19.1 at 590 μM. The decrease in absorbancy was linear with time at least for 5 minutes and was proportional to the amount of enzyme within the range of 0.2–1.5 units of enzyme.

Purification Procedure

Growth of Cells. Pseudomonas graveolens (ATCC-4683) was subcultured monthly on an agar slant containing 3 g of lactose, 1 g of Difco yeast extract, 0.2 g of $MgCl_2$, 5 g of $(NH_4)_2SO_4$, 1.5 g of K_2HPO_4, 0.5 g of KH_2PO_4 and 20 g of Difco agar per liter of tap water. The cells were grown in a liquid medium of the same composition with the omission of agar. Inoculation was carried out by transferring the cells from a slant to

[1] Y. Nishizuka and O. Hayaishi, *J. Biol. Chem.* **237**, 2721 (1962).

1 liter of the medium. The bacteria were grown at 22° for approximately 17 hours with moderate mechanical shaking and then transferred to 10 liters of the same medium. Further growth was allowed to take place at 22° for 10 hours with vigorous aeration. The cells were then harvested with the aid of a Sharples centrifuge and washed twice with 0.85% NaCl solution. The yield of wet packed cells was about 3 g per liter of the medium.

Step 1. Crude Extract and Particulate Fraction. All manipulations were performed at 0–4°. Of the cell paste, 35 g was suspended in 100 ml of 0.02 *M* phosphate buffer, pH 7.0. The suspension was subjected to the action of a Kubota 9-kc sonic oscillator for 7 minutes, followed by centrifugation in an International centrifuge for 20 minutes at 10,000 *g*. The extract (120 ml) was further centrifuged in a Spinco Model L centrifuge for 60 minutes at 40,000 *g*. The supernatant solution was decanted and saved for the preparation of lactonase. The precipitate was collected, washed once with 100 ml of the same buffer, centrifuged once again above, and suspended in 20 ml of buffer. The suspension, referred to as the particulate fraction, had more than 90% of the activity of the crude extract. The activity of this fraction was stable for at least a month at —20°.

Step 2. Solubilization. To the particulate fraction (24 ml), 8 ml of 2% sodium deoxycholate, pH 7.5, and 2 ml of *n*-butanol were added with continuous stirring. The mixture was then homogenized for 5 minutes in a glass homogenizer and centrifuged for 30 minutes at 25,000 *g*.

Step 3. Chloroform Fraction. To the brownish yellow supernatant solution (32 ml), an equal volume of cold chloroform was added. The mixture was shaken vigorously for 5 minutes, and the resulting milky solution was centrifuged for 30 minutes at 25,000 *g*. The clear yellow supernatant solution (aqueous phase) was removed and dialyzed for 3 hours against a large volume of 0.02 *M* phosphate buffer, pH 7.0.

Step 4. Protamine Treatment. To the dialyzed enzyme solution (29 ml), 12 ml of 0.4% protamine solution, pH 7.0, were added with stirring. After approximately 10 minutes, the resulting precipitate was removed by centrifugation.

Step 5. Ammonium Sulfate Fractionation. To the supernatant solution (40 ml), 8.4 g of ammonium sulfate was added (35% saturation). After approximately 15 minutes, the precipitate was centrifuged, and 6.8 g of ammonium sulfate was added to the supernatant solution (60% saturation). The brownish yellow precipitate was collected by centrifugation and dissolved in 20 ml of cold water.

Step 6. Ethanol Fractionation. To the enzyme solution (20 ml), an equal volume of cold ethanol (—20°) was added slowly with stirring, and the precipitate was removed by centrifugation. To the supernatant solution, 60 ml of cold ethanol was added in the same manner. The orange-yellow precipitate was immediately collected by centrifugation and dissolved in 3 ml of 0.02 *M* phosphate buffer, pH 6.5. Insoluble material was removed by centrifugation.

Summary of the purification procedure is shown in Table I.

TABLE I
PURIFICATION OF LACTOSE DEHYDROGENASE

			Lactose		D-Glucose		
Fraction	Volume (ml)	Total protein (mg)	Specific activity (a)	Yield (%)	Specific activity[a] (b)	Yield (%)	Ratio of specific activities (a:b)
1. Crude extract	120	2925	0.7	100	3.8	100	0.18
2. Deoxycholate-*n*-butanol	32	282	4.0	57	9.1	23	0.44
3. Chloroform	29	189	5.1	47	6.2	10	0.82
4. Protamine	40	124	7.6	46	8.6	9	0.86
5. Ammonium sulfate	20	32	15.6	24	15.8	5	0.96
6. Ethanol	3	7	36.0	12	42.3	2	0.85

[a] The conditions of assay of the activity and the units of specific activity were the same as that of lactose dehydrogenase, except that 10 micromoles of D-glucose was employed as substrate.

Properties

Stability. The purified enzyme was unstable and lost about 20% of its activity during storage for 12 hours at 0°. Freezing and thawing caused complete loss of activity.

Effects of pH and Substrate Concentration. When the activity of the solubilized enzyme was assayed spectrophotometrically with 2,6-dichlorophenolindophenol, an optimal pH of 5.6 was found. Similarly when the enzyme activity in the particulate fraction was followed manometrically, the optimal pH was 5.8. The K_m value for lactose, as calculated from the Lineweaver-Burk plot, was approximately 1.1×10^{-2} *M*.

Hydrogen Acceptor. Whereas the particulate bound enzyme reacted directly with oxygen, the solubilized enzyme preparation required an artificial electron acceptor such as 2,6-dichlorophenolindophenol, methylene blue, or ferricyanide. Neither oxygen, pyridine nucleotides, nor heart muscle cytochrome *c* could replace these acceptors. Under anaerobic conditions, the addition of lactose to the particulate fraction caused the appearance of at least three main absorption maxima at 420, 525, and 555 mμ. These bands disappeared in the presence of either oxygen or ferricyanide. Apparently the enzyme is closely associated with some hemoprotein system in the cell.

The Possible Role of Flavin Adenine Dinucleotide as the Prosthetic Group. The purified enzyme preparation displays the absorption spectrum of flavin nucleotide as well as of some heme components. The difference spectra of the preparation exhibited peaks in the range of 350–360 mμ and 465–475 mμ on reduction with either lactose or hydrosulfite. All attempts to resolve the prosthetic group from the enzyme protein, such as repeated precipitation by ammonium sulfate at pH 3–5 and prolonged dialysis against an acidic solution, resulted in an irreversible loss of activity. When the enzyme preparation was treated in the cold with perchloric acid at a final concentration of 1.5%, a compound which showed greenish fluorescence (maximum of the fluorescence emission spectrum; 540 mμ) was liberated, almost quantitatively, from the acid-denatured enzyme preparation. The liberated flavin was tentatively identified as FAD by paper chromatography, and could serve as a prosthetic group for D-amino acid oxidase, which is specific for FAD. Since flavin-containing proteins other than lactose dehydrogenase may be present in the enzyme preparation, precise identification of the prosthetic group must await further purification of the enzyme.

Activations and Inhibitors. No requirement for a metal ion could be demonstrated. Several metal-binding agents such as ethylenediamine tetraacetate, 8-hydroxyquinoline, α,α′-dipyridyl, and metals including Fe^{3+}, Cu^{++}, and Co^{++}, at a concentration of 10^{-2} *M* did not affect the enzyme activity. The activity of the enzyme was inhibited about 50% by 1 *M* sodium chloride.

Substrate Specificity. Several naturally occurring aldoses, such as D-glucose, D-galactose, D-mannose, D-talose, D-ribose, D-xylose, L-arabinose, maltose, cellobiose, and D-fucose, were also oxidized by the purified enzyme preparation as shown in Table II. D-Glucosamine, D-galactosamine, and 2-deoxyribose were not oxidized to a measurable extent, but 3-deoxy-D-glucose was rapidly oxidized.

Several experiments were performed to ascertain whether or not a single enzyme was responsible for the oxidation of these sugars. The results were compatible with the supposition that lactose dehydrogenase was a separate entity. The ratio of the specific activities for several aldoses to that for D-glucose varied during enzyme purification (Table II). For example, the ratio of specific activity for lactose to that for D-glucose increased about 4- to 5-fold (Table I). As shown in Table III, when cells were grown in a medium in which lactose was replaced

TABLE II
SUBSTRATE SPECIFICITY[a]

Substrate	Ethanol fraction	Chloroform fraction	Particulate fraction
D-Glucose	100	100	100
D-Galactose	97	92	85
D-Mannose	19	22	17
D-Talose	54	55	14
D-Ribose	97	95	68
D-Xylose	95	91	104
L-Xylose	<1	<1	<1
L-Arabinose	106	102	110
L-Lyxose	<1	<1	<1
α-Lactose	85	82	18
β-Lactose	87	85	22
Maltose	95	87	35
Cellobiose	96	84	22
L-Rhamnose	<1	<1	<1
L-Fucose	<1	<1	<1
D-Fucose	97	99	104
D-Glucosamine	4	4	5
D-Galactosamine	3		
2-Deoxy-D-ribose	<1	<1	<1
3-Deoxy-D-glucose	113	102	87
D-Fructose	0	0	0
L-Sorbose	0	0	0

[a] Enzyme activity was measured spectrophotometrically under the standard assay conditions in the presence of 10 micromoles of various substrates. All values represent the percentage of activity as compared to the value obtained with D-glucose, and are averages of triplicate experiments.

by either D-glucose, D-galactose, D-ribose, or L-arabinose, the specific activities of the enzyme preparations for several aldoses as well as their ratios to that for D-glucose also varied. Similar results were obtained when D-mannose, D-xylose, or maltose was employed as the substrate in growth media. The ratios of the activities for lactose, maltose, cellobiose, and D-ribose were constant during enzyme purification and in various growth conditions. The activities with D-glucose, D-galactose, L-arabinose, and D-fucose paralleled each other and appeared to form a separate group. When cells were grown with D-glucose or D-galactose, lactose and D-ribose were not oxidized to a measurable extent. Lactose dehydrogenase was induced by either D-ribose or L-arabinose, however. The evidence indicates that the best preparation of lactose dehydrogenase obtained to date contained at least two different aldose dehydrogenases. Preliminary attempts to separate these enzyme by either starch block

TABLE III
SUBSTRATE SPECIFICITY OF ENZYME PREPARATION OBTAINED FROM CELLS GROWN WITH VARIOUS CARBON SOURCES[a]

Substrate enzyme	Carbon source for growth				
	Lactose	D-Glucose	D-Galactose	D-Ribose	L-Arabinose
D-Glucose	100 (5.8)	100 (13.2)	100 (8.6)	100 (7.0)	100 (7.4)
D-Galactose	92 (5.4)	84 (11.0)	94 (8.0)	92 (6.3)	99 (7.4)
D-Mannose	22 (1.2)	15 (2.0)	32 (2.8)	23 (1.6)	26 (1.9)
D-Ribose	95 (5.6)	10 (1.4)	23 (2.0)	83 (5.7)	93 (6.8)
D-Xylose	91 (5.2)	—	41 (3.6)	92 (6.4)	100 (7.4)
L-Arabinose	106 (6.2)	105 (13.8)	109 (9.4)	100 (7.0)	110 (8.2)
α-Lactose	82 (4.8)	5 (0.6)	15 (1.2)	62 (4.4)	73 (5.4)
β-Lactose	85 (4.9)	6 (0.8)	23 (2.0)	70 (4.9)	77 (5.7)
Maltose	87 (5.0)	8 (1.0)	33 (2.8)	75 (5.3)	87 (6.4)
Cellobiose	84 (4.9)	—	26 (2.2)	65 (4.6)	80 (5.9)
D-Fucose	99 (5.7)	—	94 (8.1)	90 (6.3)	93 (6.9)
3-Deoxy-D-glucose	102 (5.9)	—	70 (6.0)	—	—

[a] The activities were measured spectrophotometrically by the standard assay conditions in the presence of 10 micromoles of various substrates. Chloroform-treated fractions, the fourth step, were employed as the enzyme preparations. All values represent the percentage of activity as compared to the value obtained with D-glucose. Figures in parentheses express the units of activity per milligram of protein. All the values are averages of triplicate experiments.

electrophoresis at pH 8.0, or diethylaminoethyl cellulose column chromatography were unsuccessful. The precise nature of the enzyme and its substrate specificity can be elucidated only after further purification.

[17] D-Aldohexopyranoside Dehydrogenase[1]

By EUGENE E. GREBNER, EMILIE KOVACH, and DAVID SIDNEY FEINGOLD

D-Glucopyranoside + acceptor → D-*ribo*-hexopyranosid-3-ulose + reduced acceptor
D-Galactopyranoside + acceptor → D-*xylo*-hexopyranosid-3-ulose + reduced acceptor

Assay

Principle. The assay is based on the reduction of 2,6-dichlorophenol-indophenol to its leuco form, coupled to the oxidation of methyl-α-D-glucopyranoside.

[1] E. E. Grebner and D. S. Feingold, *Biochem. Biophys. Res. Commun.* **19**, 37 (1965).

Reagents

2,6-Dichlorophenolindophenol (DIP). Sufficient DIP is dissolved in 0.1 *M* Na phosphate buffer, pH 6.0, to give a solution with an absorbance of approximately 0.6 at 600 mμ.

1.0 *M* methyl α-D-glucopyranoside (MG)

Enzyme solution

Procedure. (All reagents are tempered at 28°.) To 2.8 ml of DIP in a cuvette are added 0.1 ml of MG and 0.1 ml of enzyme. The change of absorbance at 600 mμ is followed for at least 2 minutes at 28°. A control cuvette contains H_2O in place of MG.

Definition of Unit and Specific Activity. One unit of enzyme catalyzes a decrease of 1 absorbance unit per minute at 28°. Specific activity is expressed as units per milligram protein.[2]

Purification Procedure

Organism. Agrobacterium tumefaciens strain B-6 (virulent) or II BNV-6 (avirulent) is maintained on potato dextrose agar slants containing 0.5% yeast extract.

Medium. The medium[3] is made up in three parts: (A) 20 g MG, 9 g Na_2HPO_4, 13 g KH_2PO_4, and 4 g $(NH_4)_2SO_4$ in 500 ml H_2O; (B) 0.8 g $MgSO_4$, 1.0 mg H_3BO_3, 2.0 mg $CuSO_4 \cdot 5\ H_2O$, 2.0 mg $MnSO_4 \cdot H_2O$, 2.0 mg $Na_2MoO_4 \cdot 2\ H_2O$, 2.0 mg $CaCl_2$, 10.0 mg $ZnCl_2$ or 21.1 mg $ZnSO_4$ in 480 ml H_2O; and (C) 20.0 mg $FeCl_3 \cdot 6\ H_2O$ in 20 ml H_2O. A and B are autoclaved separately, C is sterilized by filtration. The solutions are mixed at room temperature and if necessary made up to a final volume of 1 liter with sterile distilled H_2O.

Growth of Cells. A loopful of *A. tumefaciens* is transferred from a slant into 100 ml of medium contained in 500-ml Erlenmeyer flasks equipped with two stainless-steel baffles and a gauze closure. The flasks are shaken (300 oscillations per minute) at 25° on a New Brunswick Gyrotory Shaker,[4] for approximately 54 hours, and are harvested at a Klett reading of 650–700 [filter 42 (420 mμ)]. This corresponds to an absorbancy value of 1.2–1.4. In order to assure high yield of enzyme, adequate aeration and good temperature control are essential during growth of cells.

Resting Cell Suspensions. (All following operations are performed at

[2] O. H. Lowry, N. J. Rosebrough, A. L. Farr, and R. J. Randall, *J. Biol. Chem.* **193,** 265 (1951).

[3] A. J. Kraght and M. P. Starr, *Arch. Biochem. Biophys.* **42,** 271 (1953).

[4] New Brunswick Scientific Co., New Brunswick, New Jersey.

0–4° unless otherwise stated.) The cells are harvested by centrifugation, washed twice by suspension in H_2O and centrifugation, and then resuspended in H_2O. Aerated resting cell preparations (in buffer) can be used to prepare a number of glycosid-3-uloses from D-glucosides and D-galactosides.[1]

Cell-Free Extracts. The resting cell suspension is held at 0–4° overnight; the cells are spun down, resuspended in sufficient 0.1 *M* sodium phosphate buffer, pH 6.0, to yield a suspension, a 1:10 dilution of which gives a Klett reading of 650–750 (filter 42). The cells are disrupted in a French Pressure Cell[5] at 20,000 psi. The broken-cell suspension is spun at 40,000 *g* for 10 minutes, the pellet is discarded, and the supernatant fluid (crude extract) is retained.

$MnCl_2$ Precipitation. Sufficient 0.5 *M* $MnCl_2$ is added to the crude extract with stirring to give a final concentration of 0.025 *M*. The suspension is gently stirred for 1 hour, and the precipitate is then removed by centrifugation and discarded. The supernatant fluid ($MnCl_2$ supernatant) is retained.

Dialysis. The $MnCl_2$ supernatant is dialyzed for 20 hours against 100 volumes of 0.1 *M* sodium phosphate buffer, pH 6.0. The precipitate which forms is discarded (dialysis supernatant).

Heat Step. The dialysis supernatant is rapidly brought to 55°, and after 5 minutes is cooled in an ice bath. The precipitate is spun out and discarded (heat-step supernatant).

Ammonium Sulfate Precipitation. The heat-step supernatant is brought to pH 7.0 with 0.5 *M* NH_4OH. Solid $(NH_4)_2SO_4$ is added to 50% saturation, the pH being maintained at 7.0 with 0.5 *M* NH_4OH. The precipitate is discarded, and the supernatant fluid is made 75% saturated with $(NH_4)_2SO_4$, the pH being held at 7.0 as previously. The precipitate is taken up in a minimal volume of 0.01 *M* sodium phosphate buffer pH 7.0 [$(NH_4)_2SO_4$ fraction].

Calcium Phosphate Gel Treatment. One volume of $(NH_4)_2SO_4$ fraction is passed through a column containing 10 volumes of Sephadex G-25 equilibrated with 0.01 *M* sodium phosphate buffer, pH 7.0, the same buffer being used to elute the column. Active fractions are pooled and Ca_3PO_4 gel[6] (approximately 1.5 mg dry weight per milligram protein) is added with stirring. (The exact amount of gel required to adsorb about 10% of activity should be determined in a preliminary small-scale trial.) The suspension is stirred for 10 minutes and the gel is spun down and discarded. To the supernatant fluid more gel is added with stirring

[5] Vol. I [9].
[6] Vol. I [11].

(approximately 16 mg dry weight per milligram protein; the exact amount required to just adsorb all enzyme activity should be determined in a preliminary small-scale trial). After it has been stirred for 10 minutes, the suspension is centrifuged; the supernatant fluid is discarded. The gel is eluted with 10% $(NH_4)_2SO_4$ in 0.1 *M* Na phosphate buffer, pH 7.0. Two elutions, each with one-half the volume of the supernatant fluid in the adsorption step, suffice to elute the bulk of the enzyme. The enzyme is concentrated by addition of 0.47 g $(NH_4)_2SO_4$ per milliliter eluate. The precipitated protein is collected by centrifugation and dissolved in a minimal volume of 0.1 *M* sodium phosphate buffer, pH 7.0 (gel eluate). An overall purification of 32-fold is achieved, with recovery of about half the initial activity.

PURIFICATION PROCEDURE

Fraction	Volume (ml)	Total units	Specific activity (units/mg)	Recovery (%)
Crude extract	54	203	0.11	100
$MnCl_2$ supernatant	50	200	0.15	99
Dialysis supernatant	43	200	0.19	99
Heat supernatant	40	190	0.34	94
$(NH_4)_2SO_4$ fraction	3	187	1.30	93
Gel eluate	13	114	3.50	56

Properties

Stability. Crude extract and purified fractions up to and including the heat-step supernatant are stable for at least 48 hours at 0–4°. More highly purified enzyme is less stable and loses about 50% of its activity over a 24-hour period.

pH Optimum. The enzyme has a sharp pH optimum at 6.0.

Specificity. In general the enzyme is specific for the D-glucopyranosyl and D-galactopyranosyl moieties. Relative rates of reduction of DIP by the enzyme in the presence of different substrates is as follows: cellobiose, 100; methyl α-D-glucopyranoside, 89; lactose, 81; maltose, 81; D-glucose, 76; sucrose, 65; D-galactose, 58; melibiose, 55; α-D-glucopyranosyl phosphate, 43; α-D-galactopyranosyl phosphate, 43; D-glucose 6-phosphate, 23; trehalose, 23; raffinose, 19. Aldopentoses, D-fructose, L-rhamnose, L-fucose, D-fucose, D-galacturonic acid, D-glucitol, *m*-inositol, and L-galactose are not substrates.

Reaction Product. The D-glucopyranoside or D-galactopyranoside moieties are oxidized to the 3-oxo derivatives (3-uloses). Thus, D-glucose is converted to D-*ribo*-hexos-3-ulose, sucrose to β-D-fructofuranosyl-α-D-*ribo*-hexopyranosid-3-ulose, lactose to 4-O-β-D-*xylo*-hexopyranosyl-3-ulose-D-glucose. For preparation of glycosid-3-uloses the enzyme can be used with catalytic quantities of the autoxidizable H-acceptor phenazine methosulfate. Reaction mixtures contain 100 micromoles of substrate (i.e., sucrose), 0.1 micromole of phenazine methosulfate, 0.5 mg of enzyme, and 100 micromoles of 0.1 *M* Na phosphate buffer, pH 6.0, in a total volume of 0.1 ml. After incubation in air for 4 hours at 25°, the reaction components are separated by chromatography on Whatman 3 MM paper, using butanone–acetone–acetic acid–H_2O (20:10:6:9, v/v) as developing solvent. The reaction product (if sucrose is the substrate, β-D-fructofuranosyl-α-D-*ribo*-hexopyranosid-3-ulose) usually has a slightly higher R_f than the parent sugar. It reduces alkaline triphenyltetrazolium[7] instantly in the cold, yielding the red formazan, and gives a characteristic purple-red color with urea phosphate spray.[8]

[7] Made by mixing equal volumes of 1% triphenyltetrazolium chloride and 1 *N* NaOH immediately before spraying. See K. Wallenfels, *Naturwissenschaften* **34**, 491 (1950).

[8] Made by dissolving 3 g urea in 100 ml H_2O-saturated *n*-butanol (1 *M* in phosphoric acid) then adding 95% ethanol until one phase is obtained. After spraying, the paper is heated at 100° for 5–10 minutes. See C. S. Wise, F. J. Dimler, H. A. Davis, and C. E. Rist, *Anal. Chem.* **27**, 33 (1955).

[18] Glucose Oxidase from *Aspergillus niger*

By JOHN H. PAZUR

$$\beta\text{-D-Glucose} + H_2O + O_2 \rightarrow \text{D-gluconic acid} + H_2O_2$$

Assay Method

Principle. Glucose oxidase activity can be measured conveniently by determining the rate of oxygen uptake during the oxidation of D-glucose by the enzyme in a Warburg manometer in the presence of excess oxygen and catalase. Since catalase is often present in crude preparations of glucose oxidase, the catalase is added to ensure comparable reaction conditions in the assay procedure. The method[1] described in the following section represents a modification of the procedure originally reported by Scott.[2] An alternate method of assay based on the measurement of the

[1] J. H. Pazur and K. Kleppe, *Biochemistry* **3**, 578 (1964).

[2] D. Scott, *J. Agr. Food Chem.* **1**, 729 (1953).

D-gluconic acid also has been described, and it may be advantageous to use it under certain conditions.[3]

Reagents

D-Glucose, 0.2 *M*, in 0.1 *M* potassium phosphate buffer of pH 5.9
Sodium hydroxide, 2.5 *M*

Procedure. Conventional Warburg flasks and apparatus were employed in the assay procedure. Two milliliters of 0.2 *M* D-glucose solution was placed in the main chamber of the flask, 0.5 ml of enzyme solution containing from 0.02 to 0.5 unit of glucose oxidase activity was placed in the side arm, and 0.1 ml of 2.5 *M* sodium hydroxide was placed in the center well. For glucose oxidase samples devoid of catalase, a few units of crystalline catalase were added during the dilution of the glucose oxidase sample in the range of activity desirable in the assay. The flasks were attached to the manometer and equilibrated at 30° for 15 minutes. At the end of this time the enzyme solution was introduced into the main compartment and the rate of oxygen uptake was determined as a function of time. Appropriate controls and a thermobar were always run simultaneously with the assays, and all assays were performed in triplicate.

Definition of Enzyme Unit and Specific Activity. The unit of glucose oxidase activity is defined as that quantity of enzyme which will cause the uptake of 11.2 μl of oxygen per minute at standard temperature and pressure in the Warburg manometer at 30° and atmospheric pressure with a substrate concentration of 0.16 *M* D-glucose in 0.1 *M* potassium phosphate buffer of pH 5.9 containing excess oxygen and catalase. The unit as defined above represents the oxidation of 1 micromole of glucose per minute and is in accord with the recommendation of the Commission on Enzymes of the International Union of Biochemistry.

Specific activities are expressed as units of enzyme activity per milligram of protein. The protein in the enzyme preparations is determined by the method of Lowry *et al.*,[4] crystalline ovalbumin being used as the standard.

Purification Procedure

Step 1. Preparation of Crude Enzyme Concentrate. The glucose oxidase from *Aspergillus niger* is an intracellular enzyme present in the mycelium of the organism. For the preparation of crude enzyme

[3] L. A. Underkofler, *Proc. Intern. Symp. Enzyme Chem., Tokyo-Kyoto, 1957*, p. 486 (I.U.B. Vol. 2). Maruzen, Tokyo, 1958.

[4] O. H. Lowry, N. J. Rosebrough, A. L. Farr, and R. J. Randall, *J. Biol. Chem.* **193**, 265 (1951).

concentrates, *Aspergillus niger* was grown in submerged culture in a mineral medium containing a source of organic nitrogen and carbohydrate. The mycelium was collected by filtration and extracted with water.

The glucose oxidase along with other enzymes and proteins in the extract were precipitated from the solution by addition of several volumes of ethyl or propyl alcohol. The precipitate was collected by centrifugation or by filtration and air dried. A small sample of the preparation was dissolved in water and assayed for glucose oxidase activity. In general, considerable variation in activity is noted with such preparations. A commercial product can be obtained at a standard level of activity from several enzyme producers (e.g., Miles Chemical Co., Elkhart, Indiana). Such preparations are convenient starting material for the isolation of pure glucose oxidase.

Step 2. Solubilization and Dialysis of the Enzyme. One hundred grams of glucose oxidase concentrate, prepared as described above or obtained from an enzyme supplier, was shaken in 150 ml of water for 2 hours. The insoluble material was removed by filtration or centrifugation, and the clear solution was dialyzed in the cold against distilled water for approximately 30 hours. During the dialysis procedure a transfer of the tube contents to new dialysis tubes was necessary every 6–10 hours to avoid disintegration of the cellulose tube due to the action of cellulases also present in the crude enzyme preparation.

Step 3. Ammonium Sulfate Fractionation. The enzyme sample from the dialysis tube was transferred to a graduated cylinder and ammonium sulfate (40 g per 100 ml of solution) was added to yield approximately 60% ammonium sulfate saturation. The solution was refrigerated overnight, and a yellow-colored precipitate which formed was collected by centrifugation. Most of the glucose oxidase activity was in the supernatant; the precipitate was accordingly discarded. Additional ammonium sulfate (20 g/100 ml) was added to the supernatant to yield a solution of approximately 85% ammonium sulfate saturation. The precipitate obtained on refrigeration of the solution overnight was collected by centrifugation and dissolved in 100 ml of distilled water. The glucose oxidase activity was found to be in the solution of the precipitate. This solution was dialyzed against 0.01 M sodium acetate buffer of pH 4.5 for 12 hours at 3°.

Step 4. Chromatography on DEAE-Cellulose. Thirty grams of DEAE-cellulose (Brown Co., Berlin, New Hampshire) was stirred in water and transferred into a glass column (450 $\times$ 35 mm). The column was washed with 0.3 liter of 0.1 N sodium hydroxide followed by 0.3 liter of 0.5 N acetic acid, and finally with 3 liter of 0.05 M sodium acetate buffer of pH 4.5. At completion of the washing the pH of the eluate was also 4.5.

The solution of glucose oxidase from step 3 was introduced from a separatory funnel slowly onto the column. After absorption of the sample on the DEAE-cellulose was complete, the column was washed with approximately 0.5 liter of 0.07 *M* sodium acetate buffer of pH 4.5. During the washing some protein material visible by its brown color was removed from the column, and the glucose oxidase, visible by its yellow color, diffused into a broad band. This band of glucose oxidase was then eluted with 0.1 *M* sodium acetate buffer of pH 3.7. The pH of the enzyme solution obtained from the column was 4.1, and this pH was adjusted to 4.5 with acetic acid. The solution of glucose oxidase was subjected to rechromatography on a column containing 10 g of DEAE-cellulose. The washing of this smaller column and elution of the enzyme was performed in a comparable manner to that described above.

The glucose oxidase solution obtained from the second column contained 0.7% protein and assayed approximately 900 units per milliliter of solution. A summary of the purification procedure is presented in Table I.

TABLE I
GLUCOSE OXIDASE ACTIVITIES OF FRACTIONS OBTAINED AT VARIOUS STAGES OF PURIFICATION

Fraction	Volume (ml)	Units per milliliter	Total units	Units per milligram protein
Initial enzyme sample	205	541	112,000	16
After dialysis	563	186	105,000	51
After ammonium sulfate fractionation	100	680	68,000	—
1st DEAE-cellulose chromatography	77	766	59,000	—
2nd DEAE-cellulose chromatography	60	902	54,000	130

Properties

The glucose oxidase from *Aspergillus niger* has not been obtained in crystalline form although many different crystallization procedures have been attempted. Ultracentrifuge data, electrophoresis, and chromatographic behavior indicate that the glucose oxidase prepared by the above procedure is in a high state of purity.[1] A crystalline glucose oxidase has been isolated by similar procedures from *Penicillum amagasakiense* by other workers.[5] Perhaps since the enzyme from the latter

[5] K. Kusai, I. Sekuzu, B. Hagihara, K. Okunuki, S. Yamauchi, and M. Nakai, *Biochim. Biophys. Acta* **40**, 555 (1960).

organism is extracellular, its physical form may facilitate crystallization.

The *Aspergillus niger* glucose oxidase contains two moles of flavin adenine dinucleotide firmly bonded to the protein. The molecular weight of the enzyme is approximately 150,000 as determined from sedimentation and diffusion data. The pH optimum for enzyme activity is 5.5, and the isolectric point is 4.2. Recently it has been reported that the glucose oxidase contains carbohydrate residues as integral structural units of the enzyme molecule.[6] The carbohydrate residues that have been identified in hydrolyzates of the enzyme include D-mannose, D-galactose, and D-glucosamine, the last most probably occurring in the *N*-acetyl form in the native enzyme.

Pure glucose oxidase when maintained at cold temperatures is very stable. Little loss of activity has been noted with enzyme samples stored for several years at 3°.

In view of the widespread use of glucose oxidase as a specific reagent for the detection and determination of D-glucose, it should be pointed out that several derivatives and epimers of D-glucose are also oxidized by the enzyme at an appreciable rate. Data on the relative rate of action of the glucose oxidase on monosaccharides are available from several laboratories.[1,7,8] Some representative values for oxidation rates with various hexoses are recorded in Table II. The rate of oxidation of D-glucose under the conditions of the assay procedure has been assigned an

TABLE II
RELATIVE RATE OF OXIDATION OF HEXOSES BY GLUCOSE OXIDASE

Compound	Apparent R_f value	Relative rate
D-Glucose	0.52	100
L-Glucose	0.52	0
2-Deoxy-D-glucose	0.71	20
4-*O*-Methyl-D-glucose	0.70	15
6-Deoxy-D-glucose	0.71	10
4-Deoxy-D-glucose	0.63	2
D-Mannose	0.59	1
3-Deoxy-D-glucose	0.70	1
6-*O*-Methyl-D-glucose	0.68	1
D-Galactose	0.47	0.5
5-Deoxy-D-glucose	0.68	0.05
D-Allose	0.54	0.02
3-*O*-Methyl-D-glucose	0.71	0.02

[6] J. H. Pazur, K. Kleppe, and E. M. Ball, *Arch. Biochem. Biophys.* **103**, 515 (1963).
[7] D. Keilin and E. F. Hartree, *Biochem. J.* **50**, 331 (1952).
[8] A. Sols and G. De La Fuente, *Biochem. Biophys. Acta* **24**, 206 (1957).

arbitrary value of 100. The rates of oxidation of the other substrates has been compared to that for D-glucose. Also recorded in Table II are the apparent R_f values[1] which were obtained from paper chromatograms of the compounds developed by two ascents of the solvent system of *n*-butyl alcohol–pyridine–water (9:5:7 by volume).

[19] Galactose Oxidase of *Polyporus circinatus*[1-4]

By D. AMARAL, F. KELLY-FALCOZ, and B. L. HORECKER

D-Galactose and α- or β-D-galactopyranosides $+ O_2 \longrightarrow$ D-Galacto-hexodialdose and α- or β-D-galacto-hexodialdosides $+ H_2O_2$

Assay Method

Principle. The hydrogen peroxide formed in the oxidation of the substrate by molecular oxygen is utilized for the coupled oxidation of a suitable chromagen to yield a colored product:

$$H_2O_2 + \text{chromagen} \rightarrow \text{dye} + 2H_2O$$

In the present procedure horseradish peroxidase is added to catalyze the peroxidation of the chromagen *o*-dianisidine.[5,6]

[1] There has been considerable uncertainty as to the identification of the organism. It was originally designated as *Polyporus circinatus*,[2,3] but in a later publication Nobles and Madhosingh[4] suggested that this was a misnomer. Recent studies by Dr. L. B. Lockwood, Miles Chemical Company, Elkhart, Indiana (personal communication) indicate that only cultures of *Polyporus circinatus* produced the enzyme and that cultures of *Dactylium dendroides* from a number of sources are inactive. Until this question is resolved it is considered advisable to use the original designation.

[2] J. A. D. Cooper, W. Smith, M. Bacila, and H. Medina, *J. Biol. Chem.* **234,** 445 (1959).

[3] G. Avigad, D. Amaral, C. Asensio, and B. L. Horecker, *J. Biol. Chem.* **237,** 2736 (1962).

[4] M. K. Nobles and C. Madhosingh, *Biochem. Biophys. Res. Commun.* **12,** 146 (1963).

[5] A. S. G. Huggett and D. A. Nixon, *Biochem. J.* **66,** 12P (1957).

[6] W. Fischer and J. Zapf [*Biochem. Z.* **337,** 186 (1964)] report that better results can be obtained if *o*-cresol is used in place of *o*-dianisidine.

Reagents

Chromagen-peroxidase mixture. Dissolve 5 mg of *o*-dianisidine in 0.5 ml of methanol. Add to about 30 ml of water. Add 5 ml of 0.1 *M* phosphate buffer, pH 7.0, and 5.0 ml of 1% horseradish peroxidase and dilute to 50 ml with water. Store in a brown bottle.

Galactose. 10% (w/v) in water

Enzyme. The enzyme solutions are diluted with water where necessary. Samples containing ammonium sulfate are dialyzed before assay.

Procedure. To 1.0 ml of chromagen-peroxidase mixture add 0.05 ml of galactose and 0.01–0.05 ml of enzyme solution. Incubate for 10 minutes at 30° and read the absorbance at 420 mμ. Add enzyme to give readings of 0.2–0.4 units in the test. Run a control solution lacking galactose each day to correct for color in the reagents. The test is not proportional throughout the entire range, and a calibration curve should be run for each batch of chromagen.

Definition of Unit and Specific Activity. One unit of enzyme is the amount which causes an absorbance change of 1.000 per 10 minutes under the conditions of the assay. Specific activity is expressed as units per milligram of protein. Protein is determined by the Folin-Ciocalteu reagent.[7]

Preparation of Enzyme

Growth of Polyporus circinatus and Preparation of Culture Filtrate. Stock cultures of the organism are maintained on Sabouraud-glucose agar slants.[8] The culture medium for preparation of enzyme has the following composition:

KH_2PO_4 (anhydrous)	0.9%	54 g
$Na_2HPO_4 \cdot 7\ H_2O$	0.8%	90.5 g
$(NH_4)_2SO_4$	0.2%	12 g
$MgSO_4$	0.02%	1.2 g
$MnSO_4$	0.0002%	0.012 g
NH_4NO_3	0.1%	6.0 g
Yeast extract	0.1%	6.0 g
Distilled water	—	to 6 liters

To each of 12 2-liter flasks add 500 ml of medium. Autoclave at 10 pounds per square inch (psi) for 20 minutes. To each flask add 0.1 ml

[7] O. H. Lowry, N. J. Rosebrough, A. L. Farr, and R. J. Randall, *J. Biol. Chem.* 193, 265 (1951).

[8] Difco Manual, Difco Laboratories, Detroit, 9th ed., 1953, p. 238.

of 0.5 *M* $CuSO_4$ (final concentration 10^{-4} *M*) and 12.5 ml of 40% glucose solution (final concentration 1%). These solutions are sterilized separately for 15 minutes at 15 psi before addition. Inoculate each flask with a portion of mycelium taken directly from a fully grown Sabouraud-glucose agar slant. Shake the flasks in a rotary shaker (New Brunswick Model G-25) at 25°, at a speed of 200 rpm and a shaking radius of 2 inches. Excessive shaking or aeration with sparging results in loss of activity from the culture media. Follow the appearance of enzyme in the growing cultures by testing 0.01-ml aliquots of culture filtrates obtained by filtration through a pad of glass wool. Harvest the culture after 48–96 hours when the enzyme activity reaches 20–50 units per milliliter. Filter through 4 layers of cheesecloth (culture filtrate). Carry out the following steps at room temperature, except where otherwise specified.

Purification Procedure[9]

Step 1. Ammonium Sulfate Fractionation. To 4800 ml of culture filtrate, stirred mechanically, add 48 g of powdered cellulose (Whatman reagent grade) and 3200 g of ammonium sulfate. During the addition of the solid ammonium sulfate maintain the pH at 7.0 by the addition of concentrated ammonium hydroxide. Continue mechanical stirring for 1 hour at room temperature. Collect the cellulose powder together with the protein precipitate on a medium sintered glass filter. Suspend the cake in 480 ml of 60% saturated ammonium sulfate solution (adjusted to pH 7 with ammonium hydroxide) and pack it into a column 4 cm in diameter. Wash the column with 500 ml of 0.1 *M* phosphate buffer, pH 7.0. When the filtrate becomes colored (after about 480 ml have been collected) begin collecting fractions of 25 ml each. The first 6–8 colored fractions (150–200 ml) can be expected to contain the bulk of the enzyme activity. To these combined fractions add solid ammonium sulfate with mechanical stirring until the solution is saturated and a small excess remains undissolved. Collect the precipitate by centrifugation at 25° in the GSA head of the Servall RC2 centrifuge and suspend it in 0.1 *M* phosphate buffer, pH 7.0. To the suspension (42 ml) add 0.1 ml (0.5 mg) of crystalline catalase suspension (Mann Research Laboratories) and dialyze it overnight at 0° against 0.02 *M* phosphate buffer, pH 7.0 (ammonium sulfate fraction, 48 ml).

Step 2. DEAE-Cellulose Chromatography. Prepare a column of DEAE-cellulose (Schleicher and Schuell Company), 2.0 × 20 cm, which has been equilibrated with the 0.02 *M* phosphate buffer. Pass the dialyzed solution (48 ml) through the column at room temperature and elute with the same buffer. Collect fractions of about 20 ml each and analyze for

[9] The procedure is a modification of those previously reported.[3]

galactose oxidase activity. The enzyme should appear in the second, third, and fourth fractions (DEAE-cellulose eluate).

Step 3. Ammonium Sulfate Precipitation. To the DEAE-cellulose eluate (63 ml) add an equal volume of saturated ammonium sulfate solution. Centrifuge the solution and discard any precipitate which may have formed. To the supernatant solution add solid ammonium sulfate to 85% saturation (250 grams per liter of solution). Collect the precipitate by centrifugation at top speed for 1 hour at 0° in the SS-3A head of the Servall RC2 centrifuge. Dissolve the precipitate in 2.5 ml of 0.02 *M* phosphate buffer, pH 7. Dialyze the solution overnight at 0° against 0.1 *M* phosphate buffer, pH 7.0, and discard any precipitate which has formed (second ammonium sulfate fraction, 2.5 ml).

A summary of the purification procedure is given in Table I.

TABLE I
PURIFICATION PROCEDURE

Fraction	Volume (ml)	Total units	Specific activity (units/mg protein)	Yield (%)
Culture filtrate	4800	109,000	140	—
Ammonium sulfate fraction	48	76,800	1190	71
DEAE-cellulose eluate	63	69,000	2900	63
Second ammonium sulfate fraction	2.5	56,000	8000	51

Properties

Physical and Chemical Properties. The purified enzyme is homogeneous in gradient electrophoresis[10] and in sucrose density gradient centrifugation.[11] The molecular weight, based on equilibrium centrifugation,[12] is 42,400. It contains one equivalent of copper per 40–48,000 g and is inhibited by cyanide (10^{-4} *M*) and diethyldithiocarbamate (10^{-4} *M*). The cyanide-treated enzyme is fully reactivated by dialysis. Amino acid analysis and determination of half-cysteine by the method of Crestfield *et al.*[13] indicates the presence of 7 half-cysteine residues.[14] The native protein contains no SH group which can be titrated amperometrically,

[10] A. W. Bernheimer, *Arch. Biochem. Biophys.* **96**, 226 (1962).

[11] D. Amaral, L. Bernstein, D. Morse, and B. L. Horecker, *J. Biol. Chem.* **238**, 2281 (1963).

[12] D. A. Yphantis, *Ann. N.Y. Acad. Sci.* **88**, 586 (1960).

[13] A. M. Crestfield, S. Moore, and W. H. Stein, *J. Biol. Chem.* **238**, 622 (1963).

[14] F. Kelly-Falcoz, H. Greenberg, and B. L. Horecker, *J. Biol. Chem.* **240**, 2966 (1965).

but the apoenzyme (see below) in 8 M urea contains one titrable SH group. The enzyme is inactivated by reduction with mercaptoethanol and reactivated by incubation in air.[14]

The enzyme is colorless, even in the most concentrated solutions examined (several milligrams per milliliter). Electron paramagnetic resonance (EPR) studies[15] indicate that it contains cupric copper in a weakly bonded complex. The EPR signal is not changed by addition of substrate under anaerobic or aerobic conditions.

The purified enzyme preparation contains no other detectable enzyme activities.

Specificity.[3] The enzyme shows significant activity with the following monosaccharides: D-galactose, 1,5-anhydrogalactitol, 2-deoxy-D-galactose, D-talose, D-galactosamine, *N*-acetyl-D-galactosamine. It is not active with other common monosaccharides, including D-fructose, D-glucose, and D-mannose. It is active with galactosides, including methyl-2-D-galactopyranoside and methyl-β-D-galactopyranoside, and also with melibiose and, to a lesser extent, lactose. Oligosaccharides and polysaccharides containing galactose, such as raffinose, stachyose, planteose, and guaran are among the best substrates. It shows no activity with galactose 1-phosphate, UDP-galactose, or α- or β-galactofuranosides. The relative activities with several substrates are given in Table II.

The enzyme has been used for the determination of free galactose[3,6,16]

TABLE II
RELATIVE ACTIVITIES AND AFFINITY CONSTANTS[a]

Substrate	Relative velocity	K_m
D-Galactose	100	0.24
D-Galactosamine	75	0.45
N-Acetyl-D-galactosamine	92	—
2-Deoxy-D-galactose	32	—
Methyl α-D-galactopyranoside	125	—
Methyl β-D-galactopyranoside	340	—
Raffinose	180	0.025
Melibiose	80	0.045
Stachyose	610	0.013
Guaran	180	0.0003
β-Thiodigalactoside	154	—
Lactose	2	>0.5

[a] G. Avigad, D. Amaral, C. Asensio, and B. L. Horecker, *J. Biol. Chem.* **238** (1963).

[15] W. E. Blumberg, B. L. Horecker, F. Kelly-Falcoz, and J. Peisach, *Biochim. Biophys. Acta* **96**, 336 (1965).
[16] E. S. Rorem and J. C. Lewis, *Anal. Biochem.* **3**, 230 (1962).

and galactose in cerebrosides[17] and collagen.[18] It is also useful for structural studies on galactose-containing polysaccharides.[19]

Preparation of the Apoenzyme.[14] To prepare the apoenzyme, treat the native protein with 10^{-3} *M* diethyldithiocarbamate and pass the reaction mixture through a column of Sephadex G-25 (Pharmacia). A column 1.5 × 10 cm is suitable for 3 mg of enzyme in 1 ml. The column should have a layer of (0.5 cm) Chelex 100 (Bio-Rad Laboratories) at the bottom to remove free copper. In this procedure only glassware rinsed with HNO_3 and quartz-distilled water should be employed. Follow the elution by measurement of absorbance at 280 mμ. To determine which fractions contain the included volume, test with a drop of 10^{-3} *M* $CuSO_4$, which will produce a yellow color with the diethyldithiocarbamate. The apoenzyme is reactivated by addition of copper salts but reactivation is not immediate.

Stability. The native enzyme and apoenzyme are stable, even when stored at room temperature. The enzyme is readily inactivated, especially in dilute solution, by shaking or if pigments have not been removed completely in the last step. The enzyme is very sensitive to H_2O_2, which is one of the reaction products; therefore, exposure to galactose in air should be avoided. It is for this reason that the cultures are grown on medium containing glucose, rather than galactose.

[17] B. W. Agranoff, N. Radin, and W. Suomi, *Biochim. Biophys. Acta* **57**, 194 (1962).
[18] O. O. Blumenfeld, M. A. Paz, P. M. Gallop, and S. Seifter, *J. Biol. Chem.* **238**, 3835 (1963).
[19] S. M. Rosen, M. J. Osborn, and B. L. Horecker, *J. Biol. Chem.* **239**, 3196 (1964).

[20a] Glucose Dehydrogenases—Particulate

I. *Pseudomonas* Species and *Bacterium anitratum*

By JENS G. HAUGE

$$\text{D-Glucose} \rightarrow \text{D-gluconolactone} + 2\,H^+ + 2\,e^-$$

A large number of bacteria have been reported to possess the ability to oxidize glucose and other aldoses to the corresponding lactone by a particulate fraction of the cell extracts. The reaction is followed by an enzymatic or nonenzymatic hydrolysis to yield the aldonic acids. Oxygen can function as the electron acceptor, and cytochromes of the particles become reduced by glucose. Alternatively, artificial acceptors such as 2,6-dichlorophenolindophenol and phenazine methosulfate may be used.

It is typical for this particulate glucose oxidation that no need for or reaction with NAD or NADP can be demonstrated.

The main source of the particle-bound glucose dehydrogenase are various *Pseudomonas* and *Acetobacter* species. Among these are *P. fluorescens*,[1] *P. pseudomallei*,[2] *P. quercito-pyrogallica*,[3] *Rhodopseudomonas spheroides*,[4] *P. fragi*,[5] and *P. graveolens*.[6] Twenty-one different strains of *Acetobacter* have been observed to possess a particulate glucose dehydrogenase.[7,8] Outside these two main groups, dehydrogenases of this general character have been reported present in *Xanthomonas phaseoli*,[9] and *Bacterium anitratum*.[10] This part describes the details of *P. fluorescens, P. quercito-pyrogallica, B. anitratum*, and *P. graveolens* (see [20b] for *A. suboxydans*) and gives a general comparison of all these five species.

Assay Methods

Principle. For the intact respiratory particles, the activity of the dehydrogenase can be followed as oxygen consumption with standard manometric techniques. Alternatively the reduction in optical density at 600 mμ of 2,6-dichlorophenolindophenol (DIP) is used to measure the enzyme activity. DIP, at least in one instance,[11] and probably in others is capable of reacting directly with the primary dehydrogenase.

Reagents

Glucose, 0.6 *M*
DIP solution, 1.2 micromole/ml in water
Phosphate, 0.1 *M*, pH 6.0 after 1:1 dilution
Enzyme solution

Procedure. Oxygen consumption or dye reduction is measured at 30° in a Warburg vessel or a 1-cm cuvette, respectively, containing 1.5 ml buffer, 0.1 ml glucose, and, for the dye assay, 0.1 ml DIP solution. Enough enzyme is added to attain a suitable rate of oxidation, and water

[1] W. A. Wood and R. F. Schwerdt, *J. Biol. Chem.* **201,** 501 (1953).
[2] J. H. Dowling and H. B. Levine, *J. Bacteriol.* **72,** 555 (1956).
[3] R. Bentley and L. Slechta, *J. Bacteriol.* **79,** 346 (1960).
[4] D. J. Niederpruem and M. Doudoroff, *J. Bacteriol.* **89,** 697 (1965).
[5] R. Weimberg, *Biochim. Biophys. Acta* **67,** 349 (1963).
[6] Y. Nishizuka and O. Hayaishi, *J. Biol. Chem.* **237,** 2721 (1962).
[7] T. E. King and V. H. Cheldelin, *J. Biol. Chem.* **224,** 579 (1957).
[8] A. H. Stouthamer. Thesis, University of Utrecht (1960).
[9] R. M. Hochster and H. Katznelson, *Can. J. Biochem. Physiol.* **36,** 669 (1958).
[10] J. G. Hauge, *Biochim. Biophys. Acta* **45,** 250 (1960).
[11] J. G. Hauge and P. A. Hallberg, *Biochim. Biophys. Acta* **81,** 251 (1964).

is added to a total volume of 3 ml. The particle preparation should be added to the assay mixture before the substrate in order to exhaust the endogenous reducing power or to obtain a correction for this. The reaction is started with mixing in the glucose, and the initial velocity recorded.

Definition of Unit and Specific Activity. One unit enzyme is defined as the amount of enzyme that oxidizes 1 micromole of glucose per minute under the above conditions. This corresponds in the DIP assay to a ΔE_{600}/minute of 5.0, in the manometric assay to 11.2 μl O_2 consumed per minute. Specific activity is expressed as units per milligram of protein.

Purification Procedures

Pseudomonas species *and B. anitratum.* The cell suspension may be broken by any of a number of ways, such as sonic vibration,[1,3,6] grinding with glass beads,[10] or rupture in a French press.[12] The present description follows the latter procedure.

Twenty grams of frozen or fresh bacterial paste is suspended in 20 ml of 0.1 *M* cold phosphate buffer, pH 6, and passed through a cold French pressure cell (American Instruments Company, Inc.) under 9 tons of pressure. The pressate is diluted with 35 ml of water, centrifuged twice, 5 minutes each, at 20,000 *g* to remove whole cells and large cell debris. A large part of the fragments containing the glucose-oxidizing system is now brought down by 90 minutes' centrifugation at 20,000 *g*. This precipitate is suspended in 0.05 *M* phosphate, pH 6, centrifuged again, and taken up in the same buffer so as to give a concentration of about 20 mg protein per milliliter. This may be stored at —20°.

Smaller glucose dehydrogenase-containing fragments may be brought down at larger gravitational forces[13] or with 50% ammonium sulfate.[1,10] These preparations are however, contaminated by ribonucleoproteins and, in the latter case, by soluble proteins as well.

No procedure has been reported for these organisms that leads to a separation of the glucose-oxidizing electron transport chain from the rest of the particles, which among other things carry NADH and gluconate oxidase.[1,10]

Properties

Catalytic Activity. The rate of reaction with washed but unfractionated particle preparations is shown in Table I. For three of the organisms the rate observed with oxygen as acceptor is two- to threefold

[12] This volume [21b].
[13] J. G. Hauge, *J. Bacteriol.* **82**, 609 (1961).

TABLE I
RATE OF GLUCOSE DEHYDROGENATION WITH UNFRACTIONATED PARTICLES

Species	Units per mg with:		
	O_2	DIP	Ferricyanide
A. suboxydans[a]	0.480	0.205	0.133
P. fluorescens[a]	0.013	0.005	0.002
B. anitratum[a]	0.300	0.107	0.002
P. graveolens[b]	0.080[d]	0.580[f]	—
P. quercito-pyrogallica[c]	0.170[e]	—	—

[a] J. G. Hauge, *J. Bacteriol.* **82,** 609 (1961).
[b] Y. Nishizuka and O. Hayaishi, *J. Biol. Chem.* **237,** 2721 (1962).
[c] R. Bentley and L. Schlechta, *J. Bacteriol.* **79,** 346 (1960).
[d] Calculated from observations at 35°, using $Q_{10} = 2$. Substrate, 10^{-2} *M* lactose.
[e] Tested with 0.01 *M* glucose.
[f] Calculated from observations at 23°, using $Q = 2$. Substrate, 3.3×10^{-3} *M* lactose

higher than the rate with the artificial acceptor, while the particles from *P. graveolens* appear to have a relatively inefficient respiratory chain.

Specificity. The range of aldoses attacked by these particulate dehydrogenases is fairly broad, and includes both mono- and disaccharides (Table II). No substrate is attacked more than 10% faster than D-

TABLE II
SUBSTRATE SPECIFICITY

Sugar	Per cent of rate with glucose			
	P. quercito-pyrogallica[a]	*P. graveolens*[b]	*B. anitratum*[c]	*A. suboxydans*[d]
D-Glucose	100	100	100	100
D-Galactose	45	85	—	High
D-Xylose	21	104	80	—
L-Arabinose	—	110	47	—
D-Mannose	12	35	—	Low
Lactose	9	20	—	0
Maltose	7	35	—	0
D-Alabinose	4	—	0.2	—
L-Xylose	—	<1	0	—

[a] Oxygen uptake data at 1×10^{-2} *M* substrate from R. Bentley and L. Slechta, *J. Bacteriol.* **79,** 346 (1960).
[b] DIP reduction data at 3.3×10^{-3} *M* substrate from Y. Nishizuka and O. Hayaishi, *J. Biol. Chem.* **237,** 2721 (1962).
[c] DIP reduction data at 4×10^{-4} *M* substrate from J. G. Hauge, *Biochim. Biophys. Acta* **45,** 263 (1960).
[d] Oxygen uptake data from T. E. King and V. H. Cheldelin, *J. Biol. Chem.* **224,** 579 (1957).

glucose, and the rate largely falls off as the configuration is removed from that of D-glucose. The designation glucose-dehydrogenase thus seems justified. The low activity with the 4-substituted D-glucose derivatives lactose and maltose may reflect a steric hindrance caused by the incorporation of the dehydrogenases in the particle, as the activities of the solubilized dehydrogenases of *P. graveolens* and *B. anitratum* with these two sugars are almost the same as with D-glucose itself.

K_m for 4-glucose for the *B. anitratum* particle is $1.5 \times 10^{-4}\,M$, and for *A. suboxydans* is $8.5 \times 10^{-2}\,M$. These values are not directly comparable, as the former was obtained with DIP and the latter with oxygen. The slower reoxidation of the primary dehydrogenase observed with DIP would have as effect a lower apparent Michaelis constant for the substrate.

In addition to DIP, ferricyanide is a relatively good acceptor for two of the organisms (Table I), while tetrazolium, NAD or NADP failed to accept electrons. With methylene blue, activity has sometimes been found,[1,6] other times not.[7,13] These variations for one acceptor between various organisms and for one and the same organism could reflect variations along the electron transport chain rather than true differences between the dehydrogenases.

Spectral Properties. The involvement of a cytochrome chain in the oxidation of aldoses by these particle preparations receive support from the spectral changes that take place upon addition of substrate. Table III gives the maxima of the difference spectra observed.

TABLE III
PEAKS IN DIFFERENCE SPECTRA PRODUCED WITH SUBSTRATE

Species	Deoxycholate	γ	β	α
A. suboxydans	–[a]	425	525	555
	+[b]	426–428	528–530	560
P. fluorescens	–[a]	425	525	558
	+[c]	425	528	558, 565
B. anitratum	–[a]	428	530	560
P. graveolens	+[d]	420	525	555

[a] J. G. Hauge, *J. Bacteriol.* **82,** 609 (1961).
[b] T. E. King and V. H. Cheldelin, *J. Biol. Chem.* **224,** 579 (1959).
[c] W. A. Wood and R. F. Schwerdt, *J. Biol. Chem.* **201,** 501 (1953).
[d] Y. Nishizuka and O. Hayaishi, *J. Biol. Chem.* **237,** 2721 (1962).

Inhibitors. At a concentration of $5 \times 10^{-4}\,M$, cyanide inhibits oxygen uptake for *A. suboxydans* particles 90%, for *P. fluorescens* particles, 30%; $10^{-2}\,M$ cyanide is required for particles from the latter organism and from *P. quercito-pyrogallica* to give 90 and 97% inhibition, respectively. A normal cytochrome oxidase thus appears not to be present in these two organisms. At 5×10^{-4} and 10^{-2} *M*, sodium azide gives 90% inhibition for *A. suboxydans* and *P. quercito-pyrogallica*, respectively.

Particles of *P. quercito-pyrogallica* lose their ability to use oxygen as acceptor when treated with EDTA followed by dialysis against water. The activity can be restored by adding various divalent cations. *P. fluorescens* particles do not release a cation with EDTA but the presence of 0.001 *M* EDTA during the assay nearly completely abolishes the oxygen uptake.[13] There is some indication that these observations reflect a structural requirement for the electron transport chain rather than a property of the dehydrogenase itself. *B. anitratum* and *A. suboxydans* particles are not affected by EDTA. *B. anitratum* particles are, however, 80% inhibited by 1.3×10^{-2} *M* atebrin.

Variation of Activity with pH. The pH optima in the aerobic assay are all within the range 5.5–6.5 (Table IV). A larger variation is found

TABLE IV
pH OPTIMA

Species	Acceptor	
	O_2	DIP
A. suboxydans	5.5	5
P. fluorescens	6.3	5.8
B. anitratum	5.9	4.3
P. graveolens	5.8	—
P. quercito-pyrogallica	6.5	—

with DIP as acceptor, but in every case the DIP optimum is lower than the O_2 optimum. The difference between the two sets of optima demonstrate that a different step is rate limiting for the two assay conditions.

Nature of the Primary Dehydrogenase. Further study of the primary dehydrogenase requires separation of the enzyme from the rest of the respiratory particle. This has been achieved for *P. graveolens*[6] and *B. anitratum*.[11]

From the particulate fraction of *P. graveolens*, the dehydrogenase is extracted and purified tenfold. The absorption spectrum gave evidence of flavins that were reducible by lactose, and of a heme component. It is probably premature, however, to conclude from this that the primary dehydrogenase is a flavoprotein.

The primary dehydrogenase has been extracted from the particulate fraction of *B. anitratum*,[11,14] and purified 3000-fold. After twentyfold purification, the preparation is dominated by flavoproteins. These are later eliminated, and the dehydrogenase has been demonstrated to be the same as the soluble glucose dehydrogenase of the same organism, which contains neither flavin, nicotinamide, nor heme. The prosthetic group is not yet identified. It is characterized, however, by a strong absorption band in the reduced state at 337 mμ.[12]

The absence of bound pyridine nucleotides is also indicated for the *A. suboxydans* dehydrogenase, through the failure to detect niacin microbiologically after acid digestion of the particles.[7] Suggestive of the wider occurrence of the *B. anitratum* prosthetic group are not only the many similarities between the particulate enzymes discussed above, but also the observation in *Rhodopseudomonas spheroides* of a factor with some of the same properties, here needed for phenazine methosulfate reduction by glucose in a particulate system.[4]

[14] J. G. Hauge and P. A. Hallberg, *Biochim. Biophys. Acta* **95**, 76 (1965).

[20b] Glucose Dehydrogenases—Particulate

II. *Acetobacter suboxydans*

By Tsoo E. King

$$\text{D-Glucose} \rightarrow \text{D-gluconolactone} + 2H^+ + 2e^-$$

For particulate glucose dehydrogenase from other microorganisms and general properties, see this volume [20a].

Assay Method

Principle. Glucose oxidation is determined manometrically following reaction (1).

$$\text{Glucose} + 1/2\ O_2 \rightarrow \text{gluconic acid} + \text{water} \quad (1)$$

Gluconolactone is a primary product, but there is no evidence to demonstrate the formation of H_2O_2 under the conditions tested.

Reagents

Glucose, 1.0 *M* in water
Phosphate buffer, 0.2 *M*, pH 5.8, Sørensen type
Albumin, 1% crystalline bovine serum albumin in water
Enzyme solution

Method. In the side arm of the Warburg flask is placed 0.2 ml glucose and in the main compartment are placed 1.2 ml phosphate buffer, 0.5 ml albumin, an appropriate amount of enzyme, and water to a total volume of 2.6 ml. The amount of enzyme is adjusted to give an oxygen uptake of 10–60 μl per 10 minutes. After a 6-minute equilibration at 29°, the glucose solution is tipped to the main compartment. The readings are taken at 10-minute intervals for 1 hour. The oxygen uptake is linear with time for at least 60 minutes using the particulate preparation of glucose dehydrogenase as described below; no carbon dioxide evolution or endogenous oxidation has been observed.

Protein is determined by the usual biuret method[1] with the following modification: 1 ml of sample is treated with 1.5 ml of 60% urea and 2.5 ml of the biuret reagent containing 0.3% $CuSO_4 \cdot 5\ H_2O$, 0.9% Rochelle salt, 0.5% KI, and 0.2 N NaOH. The reading is taken at 550 mμ after the mixture has stood at room temperature for 15 minutes. Crystalline bovine serum albumin is used as the standard.

Preparation Procedure

Preparation of Cells.[2] *Acetobacter suboxydans* (ATCC 621) cells are grown in 2 liters of a medium containing 5% glycerol, 1% Difco yeast extract, and 1% KH_2PO_4 in deionized or distilled water. The mixture is adjusted to pH 5.8–6.0, placed in a 5-liter round-bottom flask, and then sterilized at approximately 120° in an autoclave for 45 minutes. After cooling, the medium is inoculated with 50 ml of a 24- to 36-hour-old culture of the same basal medium.

The inoculated culture is incubated for about 36–44 hours at 30° with aeration through a very porous aeration bulb (such as "Gas Aerator Stone," No. 185-50, LaPine Scientific Company, Chicago). The air flow from the line is maintained at 1 liter per minute during the first 16 hours and then increased to 5–10 liters per minute. When a number of culture flasks are used, the air inlet can be connected in parallel. The air is sterilized through 50% sulfuric acid in the first bottle, 5% $KMnO_4$ in 0.01 N sulfuric acid in the second and third bottles, and water in the fourth, and then through a glass tubing of approximately 2 × 24 inches filled with cotton. The cotton plug, aeration bulb, and connections should be heat sterilized.

At the end of incubation, the cells are harvested in a Sharples centrifuge, washed four times in cold 0.05 M phosphate buffer, pH 6.0, and

[1] T. E. Weichselbaum, *Am. J. Clin. Pathol., Tech. Sect.* **10,** 16 (1946).

[2] T. E. King and V. H. Cheldelin, *J. Biol. Chem.* **198,** 127 (1952), and subsequent improvement.

suspended in about 50 ml of the same buffer. The mixture is shaken in air at room temperature for 30 minutes for the exhaustion of endogenous substrate. Finally it is centrifuged in a Servall centrifuge at 20,000 *g* and washed once with 50 ml distilled water. The processed cells may be used directly or dried *in vacuo* from the frozen state. The dried preparation can be stored for at least 10 months. Under the conditions described, the yield is about 0.8 g lyophilized cells per liter of the medium.

The scale of the culture may be increased. Commercial fermentors are convenient for this operation. The stock culture is maintained on agar slants with 5% glycerol, 0.5% phosphate, 0.5% glucose, and 1% Difco yeast extract. For active growth, the slant must be transferred every 3 or 4 days for at least three times. *Only actively growing cells give satisfactory results.*

Disintegration of Cells.[3] Disintegration of *A. suboxydans* cells is very difficult. However, two methods are found to be satisfactory. All manipulations hereafter are conducted at 0–4° unless otherwise specified. Five grams of lyophilized cells and 20 g alumina (Alcoa No. 301, the Aluminum Co. of America) are mixed well with 15 ml of 0.02 *M* phosphate buffer, pH 6.0, in a 15-cm porcelain mortar. The mortar is prechilled and kept in salt ice-water during the whole operation. The mixture is vigorously ground intermittently. Care must be taken to prevent freezing. During the intermission periods, the material adhering to the wall and the stem of the pestle is scraped off with a polyethylene spatula. The mixture is very tacky, but becomes less so after approximately the first 5 minutes of actual grinding time. At this time, a strong thiamine-like odor is detected. The grinding is continued for another 10 minutes, the mass is mixed with 50 ml of buffer and is allowed to stand for about 1 hour with occasional mixing. It is centrifuged at 10,000 *g* for 60 minutes. The residue is reextracted with 40 ml of buffer. The combined extracts are pooled.

An alternative procedure for disintegration is by sonication. Five grams of lyophilized cells are suspended in 50 ml of 0.02 *M* phosphate buffer with an aid of a Potter-Elvehjem homogenizer. The mixture is treated in a Raytheon 10 kc sonic vibrator (other sonication apparatus, such as MSC or Branson, is equally effective) equipped with a pump system so that ice water can be circulated through the jacket to keep the mixture below 8°. The machine is tuned up to the maximal energy output. The treatment requires 30 minutes net elapsed time. (A 1-minute

[3] T. E. King and V. H. Cheldelin, *Biochim. Biophys. Acta* **14**, 108 (1954); *J. Biol. Chem.* **220**, 177 (1956); and subsequent improvements.

interlude followed each 1.5-minute vibration.) The subsequent steps of separation of the debris, etc., by centrifugation are the same as above.

Preparation of Particulate Glucose Dehydrogenase.[4] The cell-free extract is centrifuged at 7500 rpm in a Spinco preparative centrifuge, rotor 30.2, for 15 minutes, and the residue is discarded. The supernatant is separated and further centrifuged for 60 minutes at 30,000 rpm. The clear yellow supernatant extract (nonparticulate) may be saved for studies of soluble enzymes. The upper gel-like red layer which consists usually of 0.7–0.9 of the total pellet over the white (or black if disintegration is done by sonication) precipitate is carefully removed and washed with about 30 ml of the buffer. Dispersion of the gel in this and all subsequent steps is carried out in a Potter-Elvehjem homogenizer. The second centrifugation at the same speed usually gives a very small amount of white material.

The pink gel is removed and washed as before. A third washing is performed in the same manner.[5] The pellet from the final centrifugation is suspended in 0.02 *M* glycylglycine buffer, pH 8.0, to a volume of 17 ml for each 5 g of lyophilized cells. At this point the preparation is free from soluble enzymes and inorganic material derived from disintegration of cells. It is also essentially free from the white, fine particles described above; however, contamination by the latter is not critical.

"Solubilization" of Particle.[4] The suspension obtained in the previous step is homogenized with an equal volume of 2% potassium deoxycholate (weight per volume) in 0.1 *M* glycylglycine buffer, pH 8.0–8.4, for 1 minute. The decrease of optical density at 550 mμ (due to turbidity) is immediate and usually amounts to more than 1.0 absorbance unit after a fivefold dilution. The mixture is then centrifuged for 30 minutes at approximately 100,000 *g*. The inactive precipitate is discarded. The supernatant fraction contains glucose dehydrogenase and cytochromes. The glucose dehydrogenase is not precipitated by further centrifugation at 140,000 *g* for 100 minutes. After dialysis, about 50% of the enzymatic activity is sedimented under the same conditions. However, in order to attain full activity, dialysis against 0.02 *M* glycylglycine or phosphate buffer for 24–48 hours is essential. The activity of the dialyzed extract shows at least 100 μatoms of oxygen consumption per milligram per hour in a 2.8 ml system at 29°.

The average yield of the soluble extract is about 280 mg protein; this represents 8% of the overall.

[4] T. E. King and V. H. Cheldelin, *J. Biol. Chem.* **224,** 579 (1957); and subsequent improvement.

[5] The red color of the gel usually becomes paler after washings.

Properties[4]

Nature and Specificity. A. suboxydans contains at least three enzyme systems for glucose oxidation: (1) oxidation after phosphorylation,[6] (2) TPN-linked direct oxidation,[7] and (3) the oxidation as described here. The first two occur in soluble fraction. The third is in particulate fraction and does not require exogenous factors, cations, or phosphate. Mammalian cytochrome *c*, DPN, TPN, FMN, or FAD added externally are not stimulatory.

The particle before solubilization can oxidize glucose, ethanol, propanol, mannitol, sorbitol, erythritol, glycerol, and butanol. All these activities with the exception of glucose oxidation disappear upon aging. The solubilized preparation oxidizes only glucose and galactose. The rate of mannose oxidation is very slow. The oxidation of glucose consumes a half mole of oxygen and gives one mole of gluconic acid and water as the products. A primary product is gluconolactone as indicated in the hydroxamate test, but the formation of H_2O_2 is uncertain because the soluble preparation catalyzes the decomposition of hydrogen peroxide. The K_m value for glucose under the assay condition is 85 m*M* and the activation energy is 11.4 kcal. Temperatures above 40° inactivate the enzyme within 10 minutes. Optimal pH is 5.5 in citrate buffer.

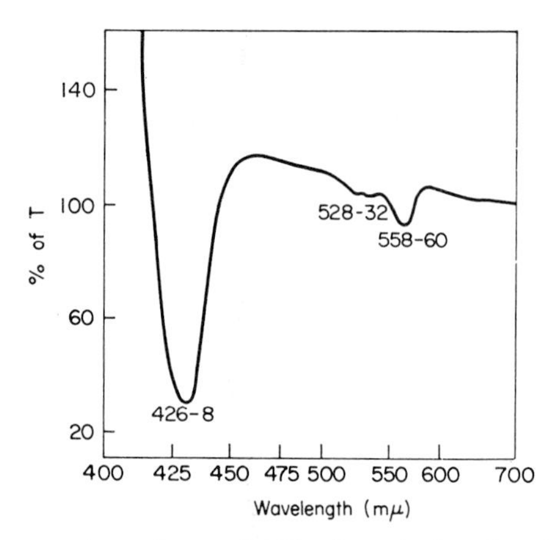

FIG. 1. Difference spectrum from solubilized particulate fraction of *A. suboxydans* as described in the text. Tracing from a Beckman DK-2 recording spectrograph performed at scanning rate = 10 minutes, sensitivity 40, photomultiplier IX. The protein concentration is about 6 mg/ml in 0.02 *M* buffer, pH 8.0. The reduced sample is formed by prior reaction with 10 mg of solid glucose.

[6] J. G. Hauge, T. E. King, and V. H. Cheldelin, *J. Biol. Chem.* **214,** 11 (1955).

[7] T. E. King and V. H. Cheldelin, *Biochem. J.* **68,** 31p (1958); and unpublished results.

The soluble preparation contains cytochromes as shown in Fig. 1. They are reduced by glucose. DPNH or TPNH can also reduce the cytochromes, but at a slower rate.

Acceptors. Oxygen is a good acceptor. Triphenyltetrazolium salt or methylene blue is inactive. However, 2,6-dichlorophenolindophenol is effective.

Inhibitors. The oxidation of glucose by oxygen is inhibited more than 90% by 0.5 m*M* cyanide (at pH 7.0) or azide (at pH 6.0 or 7.0). Antimycin A up to 10 μg per milliliter of assay mixture, 1 m*M* EDTA, or 1 m*M* 8-hydroxyquinoline did not show any effect. Deoxycholate at as low as 0.02% inhibits the oxidation.

[21a] Glucose Dehydrogenases—Soluble

I. *Bacillus cereus*

By H. L. SADOFF

$$\text{D-Glucose} + NAD^+ + H_2O \rightleftharpoons \text{gluconate} + NADH_2 + H^+$$

Sporulation in the Bacillaceae is a differentiation process which results in an extensive degradation of the vegetative bacterium and the resynthesis of a modified cell called the endospore. Glucose dehydrogenase is one of many spore-specific proteins and appears in *Bacillus cereus* cells after exponential growth is completed and sporulation has been initiated.[1,2] Other workers have reported a similar enzyme in *B. megaterium* cells and spores and in acetone-dried cells of *B. subtilis*.[3,4] The *B. cereus* protein is of considerable interest because its thermal stability can be modified over a millionfold range by the control of the pH and ionic strength of its suspending buffer.[5]

Assay Method

Principle. The spectrophotometric assay is based on the measurement of the rate of NAD^+ reduction at 340 mμ in the presence of the dehydrogenase and saturating levels of glucose. An $NADH_2$ oxidase in crude extracts of sporulating cells or spores interferes with the assay, but it

[1] R. Doi, H. Halvorson, and B. D. Church, *J. Bacteriol.* **77**, 43 (1959).

[2] J. A. Bach and H. L. Sadoff, *J. Bacteriol.* **83**, 699 (1962).

[3] R. Gavard and C. Combre, *Compt. Rend. Acad. Sci.* **249**, 2243 (1959).

[4] N. Kunita and T. Fukumaru, *Med. J. Osaka Univ.* **6**, 955 (1956).

[5] H. L. Sadoff, J. A. Bach, and J. W. Kools, *in* "Spores" (L. L. Campbell and H. O. Halvorson, eds.), Vol. III, p. 97. Am. Soc. Microbiology, Ann Arbor, Michigan, 1965.

can be removed if the extracts are adjusted to pH 6.5 and heated at 60° for 5 minutes.

Reagents

Tris buffer, *M*, pH 8
D-Glucose, *M*
NAD^+, 0.02 *M*
$MnSO_4$, 0.0001 *M*
Enzyme, Disrupted sporulating cells or spores in acetate buffer, 0.05 *M*, pH 5.0

Procedure. To 0.6 ml of Tris buffer in a 1-ml cuvette, 1-cm light path, add 0.1 ml glucose, 0.1 ml NAD^+, 0.1 ml $MnSO_4$, 0.1 ml of enzyme, and mix by inversion. The blank contains no glucose. Observe the increase in optical absorbance at 15-second intervals for 1–1.5 minutes. The course of the reaction is zero order. The use of an "optical density" converter and recorder facilitates the assay of this enzyme, but an automatic cuvette positioner is not necessary because of the short duration of the individual assays.

Definition of Unit Activity. One unit of enzyme catalyzes the oxidation of 1 micromole of substrate per minute.[6] Specific activity is expressed as units of enzyme per milligram of protein.

Purification Procedure

Growth Medium and Spore Production. B. cereus sporulates at 30° with aeration in the semisynthetic G medium,[7] which contains per liter: glucose, 4.0 g; yeast extract, 2.0 g; $(NH_4)_2SO_4$, 4.0 g; K_2HPO_4, 1.0 g; $MgSO_4$, 0.8 g; $MnSO_4 \cdot H_2O$, 0.1 g; $ZnSO_4$, 0.01 g; $CuSO_4$, 0.01 g; $CaCl_2$, 0.1 g; and $FeSO_4 \cdot 7\ H_2O$, 0.001 g. Dow Corning Antifoam B, 1 ml/liter, is included in the medium to control foaming. An active inoculum (5% of the final culture volume) is developed by serial transfers of the organism in its exponential growth phase (10% by volume) at 2-hour intervals through shake flasks and, where appropriate, intermediate size fermentors. The aeration rate is one volume of air per volume of culture medium per minute. Sporulation of the culture and the release of the spores into the medium is usually completed in 24 hours from the time of inoculation, and the spores are harvested in a Sharples centrifuge. The yield of wet spores is approximately 5 g/liter and volumes up to 100 liters

[6] The unit of activity was formerly designated as the amount of enzyme catalyzing a 0.001 change in absorbance at 340 mμ per minute at 25° in a reaction volume of 1 ml (cf. footnotes 2 and 5).

[7] B. T. Stewart and H. O. Halvorson, *J. Bacteriol.* **65**, 160 (1953).

have been utilized by appropriate scale-up procedures. Spores are stored at —20° as a frozen paste.

Step 1. Crude Extracts. Bacterial spores can be ruptured by mixing or shaking them at very high speeds with washed No. 110 pavement marking beads (Minnesota Mining and Manufacturing Co.). A 50-ml capacity Servall Omnimixer has been used for breaking 15 g (wet weight) of spores using 45 g of beads and sufficient buffer to fill the stainless steel cup. Since a considerable amount of heat is generated in the breaking process, the mixing device must be adequately cooled. The following procedure describes a purification process based on 500 g (wet weight) of spores, but the scale is easily modified. Although the enzyme is heat stable at pH 6.5, all purification procedures at pH 5.0 are carried out at 4°. The spores are suspended in 1200 ml 0.05 *M* acetate buffer (pH 5.0) with 800 g No. 110 beads in a refrigerated Eppenbach colloid mill Model MV-6-3[8] (Gifford-Wood Co., Hudson, New York), which is maintained at 10° during the course of spore rupture. After 20–30 minutes at top speed at setting of 0.028 inch between rotor and stator, the extract is removed from the instrument and the beads are permitted to settle. The washings from the beads (250 ml) and the extract are combined and are centrifuged at 15,000 *g* for 30 minutes. The supernatant extract (1600 ml) contains the soluble enzyme; the sediment, when frozen and reextracted in buffer, yields enzyme equivalent in amount to that in the original supernatant.

Step 2. Ammonium Sulfate Fractionation. The crude enzyme precipitates in the 0.5–0.8 saturation range on the addition of solid $(NH_4)_2SO_4$ and is resuspended in 150 ml of cold 0.05 *M* acetate buffer, pH 5.0. Upon refractionation with $(NH_4)_2SO_4$, the enzyme precipitates in the 0.35–0.6 saturation range. The precipitate is suspended in 15 ml of 0.01 *M* acetate buffer, pH 5, and dialyzed vs 1.5 liter of the same buffer. Globulinlike proteins which precipitate in the dialysis bag are removed by centrifugation at 30,000 *g* for 30 minutes.

Step 3. Gel Filtration. The enzyme from the preceding step is loaded on a Sephadex G-100 column, 4 × 45 cm, which has been equilibrated with 0.05 *M* acetate buffer, pH 5.0, and is then eluted with the same buffer. Five-milliliter fractions are collected. The elution volume for the enzyme lies between those for two protein peaks; one peak is excluded from the gel, and the other trails the enzyme. Under the conditions stated, the distribution coefficient, K_D, is 0.25. The recovery of enzyme from gel columns is 100%. The 10 fractions of highest specific activity are pooled and dialyzed vs 1 liter of 0.005 *M* acetate buffer, pH 5.0. The protein is

[8] J. C. Garver and R. L. Epstein, *Appl. Microbiol.* 7, 318 (1959).

lyophilized, reconstituted in 5 ml of water to effect a tenfold concentration, and rechromatographed on a Sephadex G-200 gel column, 2.5 × 25 cm, which has been equilibrated with 0.05 *M* acetate buffer, pH 5.0. The fractions of highest specific activity are pooled and yield enzyme of 80% purity. The colorless protein is stable indefinitely in the frozen state.

The purification procedure is summarized in the table below.

PURIFICATION PROCEDURE FOR *Bacillus cereus* GLUCOSE DEHYDROGENASE

Fraction	Total activity (units)	Protein (mg)	Specific activity (units/mg protein)	Purification ratio	Yield (%)
Crude extract	810	53,000	0.015	1	100
Ammonium sulfate					
0.5–0.8	820	4,200	0.195	13	101
0.35–0.6	690	650	1.06	71	85
Gel filtration I	430	20	21.5	1430	53
Gel filtration II	150	2.4	62.5	4160	19

Properties

Specificity. Glucose dehydrogenase purified from spores of *B. cereus* oxidizes only D-glucose with either NAD^+ or $NADP^+$. The pH range for activity of the two cofactors is slightly different, but generally the reaction rate with $NADP^+$ is one-third that with NAD^+. Only glucose or glucose 6-phosphate will bind to the apoenzyme and protect the protein from inactivation by 5-dimethylamino-naphthalene-sulfonylchloride. NAD^+ or $NADP^+$ will not protect the enzyme in the absence of substrate.

Kinetic Properties. The optimal pH for glucose oxidation is 8.7 with NAD^+ and 9 with $NADP^+$. These values differ from those first reported for this enzyme and from the corresponding glucose dehydrogenase in *B. megaterium.*[1,3] The enzyme is unstable under alkaline conditions, and thus is routinely assayed at pH 8. The Michaelis constants at pH 8 and 25° are 2×10^{-2} *M* for glucose and 3.5×10^{-4} *M* for NAD^+. These correspond well with 1.2×10^{-2} *M* and 0.7×10^{-4} *M* for glucose and NAD^+, respectively, reported for *B. megaterium.*[3] At pH 8, the K_m for glucose increases with temperature, a ninefold increase occurring between 35° and 55°, and is 1.8×10^{-1} at 60°. At pH 6.5, the K_m for glucose is 1×10^{-2} *M* and is constant over the temperature range 20–70°.

Activators and Inhibitors. The purified enzyme requires 10^{-5} *M* manganous ion for maximal activity. Iron, cobalt, magnesium, zinc, or copper have no effect on the activity. The enzyme is not inhibited by *p*-chloromercuribenzoate or iodoacetate over the pH range 5–8.5 and is very stable to photooxidation. High concentrations of ammonium ion inactivate the purified enzyme, and therefore the use of $(NH_4)_2SO_4$ fractionation should be restricted to treatment of crude extracts. Glucose dehydrogenase is inactivated by sulfonyl chlorides, chloro- and fluorobenzenes, or acetylating agents, but can be protected by the presence of substrate. Furthermore, neutral hydroxylamine, 0.16 *M*, partially reverses the inactivation due to acetic anhydride and fully reverses the inactivation produced by acetyl imidazole. The latter observation provides presumptive evidence for the presence of tyrosine at the active center of the enzyme.

Heat Resistance. The maximal thermal stability of glucose dehydrogenase occurs at pH 6.5. The half-life of the protein at 65° in 0.05 imidazole buffer, pH 6.5, is 3.5 minutes.[5] Over the pH range 8 to 6.5, the enzyme acquires two protons from the medium and undergoes a reversible dimer-to-monomer conversion. The thermal resistance of glucose dehydrogenase at pH 6.5 increases as a second order function of the concentration of the group 1 A cations, particularly Na^+ and K^+. The effect is reversible. The half-life of the enzyme at 85° in 0.5 *M* NaCl is 1 minute, whereas that in 5 *M* NaCl is 200 minutes. By comparison, in 0.05 *M* imidazole buffer alone, the calculated half-life of glucose dehydrogenase is 0.04 minute.

[21b] Glucose Dehydrogenases—Soluble

II. *Bacterium anitratum*

By Jens G. Hauge

$$\text{D-Glucose} \rightarrow \text{D-gluconolactone} + 2H^+ + 2e^-$$

Assay Method

Principle. The reduction in optical density at 600 mμ of 2,6-dichlorophenolindophenol (DIP) is used to measure the enzyme activity.

Reagents

Glucose, 0.6 *M*
DIP solution, 1.4 micromole/ml in water
Potassium phosphate, 0.1 *M*, pH 6.0 after 1:1 dilution
Enzyme, 0.4–40 units/ml

Procedure. The assay mixture ingredients, 1.5 ml of buffer, 0.1 ml of the glucose solution, 0.1 ml of the DIP solution, and a volume of water equal to 1.3 ml minus the volume of enzyme to be added, are mixed in a 1-cm cuvette and equilibrated at 25°. For routine purposes a larger portion of this assay mixture may be kept in a 25° waterbath ready for immediate use. A cuvette filled with water serves as the blank. The reaction is started with stirring in 1–50 μl of enzyme solution, depending on the activity. The optical density is recorded manually at 15-second intervals. The enzyme activity may be obtained more accurately from an automatic recording of the transmission of the solution, and this also expands the range of measurable concentrations to 400 units/ml. For crude extracts, a correction should be applied for the endogenous reducing power.

Definition of Unit and Specific Activity. One unit of enzyme reduces 1 μmole of DIP per minute, at an optical density of 0.6. This corresponds to a ΔE_{600}/minute of 5.0. ΔE_{600}/minute at this optical density is obtained either directly or alternatively from the slope of the tangent to the transmission plot in the inflection point (T, 0.375) through multiplication by the empirical factor 1.56.[1] Specific activity is expressed as units per milligram of protein. The protein concentration is determined by the biuret method.[2]

Purification Procedure

Medium. B. anitratum is grown with good aeration in a medium containing per liter of tap water 15 g of sodium succinate·6 H_2O, 4 g of NH_4Cl, 2 g of K_2HPO_4, and 0.06 g of $MgSO_4 \cdot 7\ H_2O$. The pH is kept between 6.6 and 7.6, and the culture is harvested when the yield reaches 1.0–1.5 g dry weight per liter. The cell paste is frozen and stored at —20°.

Step 1. A 137-g sample of frozen cell paste is thawed and suspended in 130 ml of cold 0.1 M phosphate, pH 6. All subsequent steps are performed at 2–6°. The suspension is passed in portions through a French pressure cell (American Instrument Company, Inc.) under 9 tons of pressure. The pressate is diluted with 45 ml of the above buffer and 200 ml of water, and whole cells and large fragments are removed by two centrifugations, 5 minutes each, at 20,000 g. The sediment is washed in 80 ml of 0.01 M phosphate, pH 6, and the second supernatant fraction is added to the first. This constitutes the crude extract.

Step 2. The crude extract is centrifuged for 30 minutes at 20,000 g to remove smaller cell fragments, and then is made 1% with respect to

[1] J. G. Hauge, *J. Biol. Chem.* **239,** 3630 (1964).
[2] A. G. Gornall, C. J. Bardawill, and M. M. David, *J. Biol. Chem.* **177,** **751** (1949).

protamine sulfate. After 30 minutes of stirring, the precipitate formed is removed by centrifugation. Solid ammonium sulfate is now added in steps to create 45, 55, 58, and 70% saturation, the pH being kept at 6.1–6.3. The first two steps remove a large portion of the enzyme units in the form of particulate glucose dehydrogenase,[3,4] whereas the soluble form, to be purified here, precipitates only when the saturation exceeds 58%. The yellow precipitate formed between 58 and 70% saturation is dissolved in 0.005 *M* phosphate, pH 7, and dialyzed with internal stirring against the same buffer for 5 hours. The material that becomes insoluble during the dialysis is removed. The preparation is routinely frozen at this point and may be stored for shorter or longer periods with only minor losses.

Step 3. The dialyzate is passed through a 2.5-g DEAE-cellulose column (0.7 meq/g) that has been equilibrated with 0.005 *M* phosphate, pH 7.

Step 4. The DEAE-cellulose effluent is adjusted to pH 6 and passed through a 1-g CM-cellulose column (0.8 meq/g) equilibrated with 0.01 *M* phosphate of pH 6. The column is developed with 60 ml of a linear gradient from 0.025 *M* phosphate, pH 6.3 to 0.025 *M* phosphate, 0.1 *M* NaCl, pH 7.0. Before storage, the enzyme of the peak-fractions is routinely concentrated by adsorption on 30 mg CM-cellulose and elution with a small volume 0.1 *M* phosphate, pH 7.

This purification procedure (see table) has been used a great many

PURIFICATION PROCEDURE FOR *Bacterium anitratum* GLUCOSE DEHYDROGENASE

Fraction	Volume (ml)	Total units	Total protein (mg)	Specific activity (units/mg)
1. Crude extract	505	6200	8070	0.8
2. Ammonium sulfate precipitate	25	2510	185	13.6
3. DEAE-cellulose effluent	29	2300	34	67.6
4. CM-cellulose peak fractions	3.3	820	1.4	570

times by the present author and found to be dependable. The specific activity of the final product may vary somewhat, mainly reflecting the specific activity of the crude extract. The specific activity reached in the purification described above was judged to be about 90% pure.

Alternatively the enzyme may be purified from the membrane frac-

[3] J. G. Hauge, *Biochim. et Biophys. Acta,* **45**, 250 (1960).

[4] This volume [20a, 20b].

tion. An initial step with deoxycholate treatment is here included, otherwise the procedure is the same.[5]

Properties

Stability. A variable tendency to lose some activity during frozen storage and also during the last stages of purification has been noted. This may be an expression of loss of the prosthetic group (see below).

Specificity. The enzyme acts on a number of aldoses with the same maximal velocity but with different Michaelis constants.[6] The Michaelis constant for D-glucose, with the electron acceptor in excess, is $5.3 \times 10^{-3} M$, that of D-xylose fiftyfold larger. L-Xylose is not attacked. Maltose, lactose, and cellobiose are attacked almost as readily as D-glucose, whereas melibiose is dehydrogenated very sluggishly. β-D-Glucose is preferred to α-D-glucose, but this specificity is not absolute.

In addition to 2,6-dichlorophenolindophenol ($K_m = 1.6 \times 10^{-4} M$) phenazine methosulfate is an efficient acceptor. Methylene blue accepts electrons at 2% of the rate of indophenol, and ferricyanide at 0.3%. Flavin or pyridine nucleotides are not measurably reduced, nor cytochrome *c* or triphenyltetrazolium chloride. The natural acceptor is not known. The soluble cytochrome *b* which accompanies the dehydrogenase during most of the purification, is however reduced by glucose via the dehydrogenase, either directly or through an intermediary carrier.

Inhibitors. High concentrations of substrate and acceptor inhibit the enzyme, apparently by mutual competition. Atebrin inhibits through competition with the acceptor ($K_i = 4 \times 10^{-3} M$). *p*-Chloromercuribenzoate, arsenite, cyanide, or *o*-phenanthroline do not inhibit.

pH Optimum. The pH optimum under standard assay conditions is 5.5. With lowered substrate and increased acceptor concentrations, the pH optimum is transposed to higher values.

Absorption Spectrum. The oxidized state of the enzyme is characterized by a broad absorption band in the region of 320–390 mμ, peak at 347 mμ, and by a ratio, $E_{280}:E_{260}$, of 1.65. On addition of glucose a sharper band appears in the near ultraviolet, with a maximum at 337 mμ. At the same time the absorption below 300 mμ is reduced, maximally at about 260 mμ, so that the ratio, $E_{280}:E_{260}$, is now 1.90. The 337-mμ band also appears with dithionite or borohydride. Through titration of the prosthetic group of the intact enzyme with glucose, the following molar extinction coefficients were found: ϵ_{350}(oxidized) = 15,600; ϵ_{337} (reduced) = 38,900; ϵ_{337}(difference) = 24,400; ϵ_{259}(difference) = 15,500.

[5] J. G. Hauge and P. A. Hallberg, *Biochim. Biophys. Acta* **81**, 251 (1964).

[6] J. G. Hauge, *Biochim. Biophys. Acta* **45**, 263 (1960).

Molecular Weight and Turnover Number. Glucose titration data on an average gave an equivalent weight for the enzyme of 86,000. With a sedimentation constant $S_{20,w}$ of 6.2, established in the analytical ultracentrifuge, this equivalent weight probably gives the molecular weight as well.

Extrapolation of the velocities observed of substrate and acceptor concentrations which permit linear Lineweaver-Burk plots to infinite concentrations give a turnover number of 320,000 min^{-1}. From this value and the K_m for the acceptor, the rate constant for the reaction between reduced enzyme and acceptor was calculated to be 3.3×10^7 $sec^{-1}M^{-1}$. This calculation assumes the reoxidation reaction not to be limiting at infinite acceptor concentration.

Dissociation and Reactivation of the Enzyme. The apoenzyme has most dependably been prepared by gel filtration on Sephadex G-25 at pH 2–2.5. The apoenzyme is inactive as such and does not show the 337-mμ absorption band upon chemical reduction or addition of glucose.

Apoenzyme preparations could be reactivated by addition of boiled extract or a neutralized perchloric acid extract of purified enzyme. A few microliters of apoenzyme were preincubated with a similar volume of boiled juice or perchlorate extract. Half-maximal activation was observed within 2 to 20 minutes, depending on the concentration of the reactants. With excess extract, the concentration of reactivable sites could be determined; and with excess apoenzyme, the concentration of the dissociated prosthetic group could be estimated similarly.

Prosthetic Group. The prosthetic group has been purified and concentrated by chromatography on DEAE-Sephadex, and some of its spectral properties have been investigated.[1] These studies indicate that it is not identical to any of the prosthetic groups or cofactors whose structure is known today. The high extinction coefficient for the bound and the free reduced group is particularly noteworthy.

The cytochrome chain-bound glucose dehydrogenase of *B. anitratum* has been demonstrated to carry the same prosthetic group,[5] and glucose oxidizing particles from various *Acetobacter* and *Pseudomonas* species may well have it.[4] Especially interesting is the case of *Rhodopseudomonas spheroides,* which has been observed to require an unknown factor present in the boiled soluble fraction for the reduction of phenazine methosulfate.[7] This factor was adsorbed by anion exchangers and charcoal, as is the *B. anitratum* factor. The *Rhodopseudomonas* enzyme, furthermore, could be reactivated by a boiled extract of *B. anitratum.*

[7] D. J. Niederpruem and M. Doudoroff, *J. Bacteriol.* **89**, 697 (1965).

[22] β-D-Galactose Dehydrogenase from *Pseudomonas saccharophila*

By K. WALLENFELS and G. KURZ

$$\beta\text{-D-Galactose} + DPN^+ \rightleftharpoons \gamma\text{-D-galactonolactone} + DPNH + H^+$$

Assay Method

Principle. According to Doudoroff,[1,2] the assay is based on the spectrophotometric determination of DPNH formed with galactose as substrate at high pH conditions.

Reagents

Tris-HCl buffer, 3.3×10^{-2} *M*, pH 8.6
DPN^+, 1.0×10^{-2} *M*, in Tris-HCl buffer, pH 8.6
Galactose, 7.5×10^{-1} *M*, in Tris-HCl buffer, pH 8.6

Enzyme. The enzyme is diluted with 3.3×10^{-2} *M* potassium sodium phosphate buffer (pH 6.8), to give a solution with an activity between 1.0×10^{-1} units/ml and 5.0×10^{-1} units/ml. (See definition of unit below.)

Procedure. A mixture of 0.1 ml of DPN^+ solution, 0.1 ml of galactose solution, and 2.7 ml of Tris-HCl buffer is incubated at 30° for 10 minutes in a spectrophotometer cell with a 1.0-cm light path. The reaction is started by the addition of 0.1 ml of the enzyme solution prewarmed at 30° for a minimum of 10 minutes.

Definition of Unit and Specific Activity. One unit of enzyme is defined as that amount which will catalyze the reduction of 1 micromole of DPN^+ per minute at 30° under the conditions of the test. The specific activity is defined as units per milligram of protein.[3]

Purification Procedure

Cultures of *Pseudomonas saccharophila* are grown with continuous aeration with a sparger aerator stone (sintered gloss) at 30° in 20-liter glass bottles in a medium as described by Doudoroff[1,2,4] and harvested

[1] M. Doudoroff, C. R. Contopoulou, and S. Burns, *Proc. Intern. Symp. Enzyme Chem. Tokyo-Kyoto, 1957,* p. 313. Academic Press, New York, 1958.

[2] M. Doudoroff, Vol. V [40].

[3] Enzyme Nomenclature, Recommendations (1964) of the International Union of Biochemistry on the Nomenclature and Classification of Enzymes, *in* "Comprehensive Biochemistry" (M. Florkin, and E. H. Stotz, eds.), 2nd ed., Vol. 13, p. 7. Elsevier, Amsterdam, 1965.

[4] K. Wallenfels and G. Kurz, *Biochem. Z.* **335,** 559 (1962).

by centrifugation with a Sharples T1P centrifuge in the early stationary phase of growth. The centrifuged cells are washed with cold $3.3 \times 10^{-2} M$ potassium sodium phosphate buffer (pH 6.8) and lyophilized. The dried bacteria are stable at $-15°$ for several months.

Step 1. Preparation of Cell-Free Extract. Lyophilized bacteria (100 g) are suspended in $3.3 \times 10^{-2} M$ potassium sodium phosphate buffer (pH 6.8) containing $1.0 \times 10^{-2} M$ 2-mercaptoethanol and $1.0 \times 10^{-3} M$ EDTA; 10 ml of buffer is used per gram cells. The suspension is stirred for several hours at 4°. To prevent foaming *n*-octanol is added. The mixture is centrifuged at 35,000 *g* and 4° for 90 minutes. A reddish, turbid, viscous solution (ca. 600 ml) is obtained.

Step 2. Precipitation with Protamine Sulfate. To the crude extract a 1.5% protamine sulfate solution of pH 6.0 (ca. 600 ml) is added dropwise at room temperature until the ratio of the absorbancy at 280 mμ to the absorbancy at 260 mμ of a centrifuged aliquot increases to 0.65–0.70. The pH of the mixture is then adjusted to 6.8. The required amount of protamine sulfate may be different for each purification procedure. Attention must be called to the ratio of the absorbancies 280 mμ to 260 mμ, for the yield of enzymatic activities decreases in the following steps if the value is lower than 0.6. After allowing the solution to stand at 4° for a few hours, the precipitate is removed by centrifugation at 10,000 *g* and 4° and discarded.

Step 3. Heat Treatment. The clear supernatant from the protamine sulfate step (1100 ml) is placed in a 58° water bath and vigorously stirred. After reaching the temperature of 57–58° in 1 minute, the solution is maintained at this temperature for 3–5 minutes and immediately chilled in a cold bath. The coagulated protein is removed by filtration or centrifugation at 5000 *g* and 4°. This step gives better results when performed in fractions of 200–250 ml.

Step 4. Precipitation with Ammonium Sulfate. The supernatant solution (1050 ml) is brought to 38% saturation by dropwise addition of a solution of ammonium sulfate saturated at 4° and adjusted to a pH of 6.8–7.0. After standing for 30 minutes at 4°, the precipitate is collected by centrifugation at 10,000 *g* and 4° and dissolved in a small volume (ca. 8 ml) of $3.3 \times 10^{-2} M$ potassium sodium buffer (pH 6.8) containing $1.0 \times 10^{-2} M$ 2-mercaptoethanol and $1.0 \times 10^{-3} M$ EDTA. Any insoluble material is removed by centrifugation. Salt fractionation at a pH lower than 6.5 causes a loss of enzyme activity.

Step 5. Acid Precipitation. The solution is brought to a protein concentration of 10 mg/ml (ca. 16 ml). Then $1.0 \times 10^{-1} M$ acetic acid is added dropwise at 4° under continuous stirring until the pH drops to 4.65 as indicated by a pH meter. After centrifugation at 10,000 *g* and 4° the precipitate is treated with $3.3 \times 10^{-2} M$ potassium sodium phosphate buffer (pH 6.8) containing 1.0×10^{-2} *M* 2-mercaptoethanol and 1.0×10^{-3} *M* EDTA. The pH is adjusted to 6.8 with a solution of 5.0×10^{-1} *M* Na_2HPO_4. After centrifugation the supernatant fluid is retained (2.5 ml) and the precipitate is extracted once more as above. Both supernatants are combined (5.0 ml).

Step 6. Precipitation with Phosphates. To the combined supernatants a solution of 3.9 *M* phosphate buffer is added at 25°. This buffer, containing K_2HPO_4 and NaH_2PO_4 in the ratio 4:1, is added until the concentration of phosphate in the mixture is 5.0×10^{-1} *M*. After it has been stirred for 20 minutes at 25°, the precipitate is centrifuged at the same temperature and dissolved in a small volume of 3.3×10^{-2} *M* potassium sodium phosphate buffer (pH 6.8) containing 1.0×10^{-2} *M* 2-mercaptoethanol and 1.0×10^{-3} *M* EDTA to a final concentration of 15–20 mg protein per milliliter (ca. 1.0 ml). The resulting preparation is stored at —15°, and the supernatant of the phosphate step is discarded.

A summary of a typical purification procedure is given in the table.

Properties

Homogeneity and Constitution. The purified enzyme appears to be homogeneous in the ultracentrifuge and in gel electrophoresis. From the sedimentation coefficient ($S^0_{20,w} = 6.23 \times 10^{-13}$ seconds) and the diffusion coefficient ($D^0_{20,w} = 6.01 \times 10^{-7}$ cm^2 sec^{-1}) the molecular weight was calculated to 101,000, assuming a partial specific volume of 0.75 ml g^{-1}. Treatment with guanidine-HCl causes dissociation into subunits.[5]

Stability. The purified enzyme may be kept for several weeks at —20° with gradual loss of activity. The decrease of the enzymatic activity caused by low temperatures can be reversed by removing to 30°. The reactivation depends somewhat on the nature of the enzyme sample.[6]

For a further description of properties see M. Doudoroff.[2]

Uses of the Enzyme as an Analytical Tool

At a high pH value the formed galactonolactone is hydrolyzed. Thus the reaction becomes irreversible and proceeds to completion. The enzyme is therefore a valuable analytical tool for the determination of D-galactose[4,7] by a simple test and also for the determination of derivatives of D-galactose like galactosides[4] and galactose -1-phosphate[7] in combined optical tests.

[5] G. Kurz, H. Sund, K. Wallenfels, and K. Weber, in preparation.
[6] K. Wallenfels and G. Kurz, in preparation.
[7] K. Wallenfels, and G. Kurz, unpublished.

PURIFICATION OF β-D-GALACTOSE DEHYDROGENASE FROM *Pseudomonas saccharophila*

Step	Volume (ml)	Activity (units/ml)	Total activity (units)	Protein (mg/ml)	Total protein (mg)	Specific activity (units/mg)	Yield (%)	Purification (fold)
1. Crude extract	600	4.5	2700	25.0	15000	0.18	100	—
2. Protamine supernatant	1100	2.2	2420	7.5	8250	0.29	89	1.6
3. Heat treatment supernatant	1050	1.9	2000	3.9	4100	0.48	74	2.6
4. $(NH_4)_2SO_4$ precipitation	8	200	1600	22.4	180	8.95	59	50
5. Acid precipitation	5	225	1125	10.5	52.5	21.4	41	120
6. Phosphate precipitation	1	1030	1030	12.7	12.7	81.0	38	450

Combined optical assays can also be used for measuring the activities of certain other enzymes like aldose 1-epimerase (mutarotase)[8,9] and β-galactosidase.[6]

[8] K. Wallenfels, K. Herrmann, and G. Kurz, *Abstr. VIth Intern. Congr. Biochem., New York, 1964,* p. 342 (Abstracts IV–192).

[9] K. Wallenfels, and K. Herrmann, this volume [109].

[23] Glucose 6-Phosphate Dehydrogenase (Crystalline) from Brewers' Yeast

By Stephen A. Kuby and Ernst A. Noltmann

D-Glucose 6-phosphate + $NADP^+ \rightleftharpoons$ 6-phosphoglucono-δ-lactone + NADPH + H^+

Determination of Enzymatic Activity

Glucose 6-phosphate dehydrogenase (D-glucose 6-phosphate: NADP oxidoreductase, EC 1.1.1.4.9) is most conveniently measured spectrophotometrically by the rate of formation of NADPH, as determined from its absorbance at 340 mμ. Historically, it was one of the first enzymes to be assayed by Warburg's "optical test."[1]

For measurement of the enzymatic activity, a thermostated (30°) recording spectrophotometer is employed, and the following reaction mixture components[2] are pipetted into a cuvette of 1-cm light path: 2.5 ml of 0.1 *M* glycylglycine buffer, pH 8.0, 0.1 ml of 0.03 *M* glucose 6-phosphate, 0.1 ml of 0.01 *M* NADP, 0.2 ml of 0.15 *M* magnesium sulfate. The reaction is initiated by the addition of 0.1 ml of properly diluted enzyme. All dilutions are made with ice-cold 0.05 *M* EDTA, pH 8.0; for specific activities greater than 2000 units per milligram, in addition to the EDTA, 1 mg of crystalline bovine plasma albumin (Armour Laboratories) per milliliter is necessary to stabilize the highly diluted enzyme.

One arbitrary unit of glucose 6-phosphate dehydrogenase is defined as that amount of enzyme, per 1 ml of reaction mixture, which catalyzes the reaction between NADP and glucose 6-phosphate and (under the conditions described above) requires a time of 1 minute to cause an increase in absorbance of 0.1.

Specific activity is expressed in terms of units per milligram of protein. To convert to micromoles of NADPH formed per milligram of

[1] O. Warburg, "Wasserstoffübertragende Fermente," p. 31ff. Cantor GmbH., Freiburg i. Br., 1949.

[2] E. A. Noltmann, C. J. Gubler, and S. A. Kuby, *J. Biol. Chem.* **236**, 1225 (1961).

protein per minute, the above-defined enzyme units per milligram of protein are multiplied by the factor 0.0161 (derived from the molar absorbance index[3] of 6.22×10^3 cm^{-1} M^{-1} and the above-defined unit which is in terms of 1 ml of reaction mixture).

Protein is determined by the colorimetric biuret procedure of Gornall *et al.*[4] A factor of 32.0 mg/10 ml reaction mixture per unit of absorbance at 540 mμ (1-cm light path) is used as an average biuret value for the purpose of determining the protein concentration throughout the purification procedure. In those cases in which interfering material is present, the protein is first precipitated with 10% trichloroacetic acid, and then the biuret procedure is carried through.

Materials

"Dried Brewers' Yeast for Enzyme Work" is purchased from Anheuser-Busch, Inc., St. Louis, and is the Anheuser-Busch strain of *Saccharomyces carlsbergensis.*

The crystalline barium salt of D-glucose 6-phosphate and "98% pure" NADP may be obtained commercially; the former is converted to the sodium salt after removal of the barium as $BaSO_4$, and both reagents are neutralized and assayed enzymatically.

DEAE-cellulose should have an exchange capacity of approximately 0.6 meq/mg since all washing and eluting volumes, which in the course of the procedure are given in milliliters per gram of dry DEAE-cellulose, are based on *that* exchange capacity. Before use, the DEAE-cellulose is cycled through H_2O–0.5 N NaOH–H_2O–0.5 N HCl–H_2O–0.5 N NaOH–H_2O. Finally, it is freed from as much liquid as possible on a large coarse sintered glass funnel and stored as a moist filter cake (approximately 30% dry weight) in a tightly stoppered bottle at 0°. Required amounts are calculated in terms of dry DEAE-cellulose.

Bentonite, U.S.P. (Powder), is obtained from the Fisher Scientific Company and used without further treatment.

Diisopropyl fluorophosphate should be of high purity (e.g., Merck and Company, Inc.). A 0.1 M solution of this extremely poisonous reagent is made with dried redistilled isopropanol and stored in sealed ampuls at 0°. For each preparation, a fresh ampul should be used to minimize the possibility of hydrolytic decomposition.

All other reagents, including ammonium sulfate and the disodium salt of EDTA are of analytical grade.

[3] B. L. Horecker and A. Kornberg, *J. Biol. Chem.* **175,** 385 (1948).

[4] A. G. Gornall, C. J. Bardawill, and M. M. David, *J. Biol. Chem.* **177,** 751 (1949).

Isolation Procedure

The following procedure is essentially that previously described,[2] except that it has now been slightly amended to permit the isolation of the enzyme protein from initially 10 kg of dried yeast compared to the 3 kg previously employed. In order to facilitate the operations at this increased scale some of the steps have been simplified and a second use of DEAE cellulose has been included (see below).

Unless otherwise stated, all steps are carried out in a cold room at 2–4° or in an ice bath. Required amounts of ammonium sulfate are calculated by the formula[5]

$$w = \frac{0.515 \times V_1 \times (S_2 - S_1)}{1 - (0.272 \times S_2)}$$

in which w equals the weight of ammonium sulfate in grams, V_1 is the volume of the solution in milliliters at fraction saturation S_1, and S_2 represents the fraction saturation desired at 0°.

Ammonium sulfate concentrations (S'') of dissolved ammonium sulfate precipitated pellets (of final volume V'') are estimated by the formula:

$$S'' = \frac{\Delta\nu}{V''} (S')$$

assuming that the volume increase due to the dissolved pellet ($\Delta\nu$) is of the same ammonium sulfate saturation as the previous fraction (S'). Obviously, application of the formula depends upon quantitative recovery of the liquid volumes.

All solvent additions (alcohol, acetone, saturated ammonium sulfate solutions) are made under the assumption that the volumes are additive, and are calculated by the formula

$$\nu = \frac{V_1(C_2 - C_1)}{1 - C_2}$$

in which ν equals the volume of solvent required in milliliters, V_1 is the volume of the solution at concentration C_1 (or fraction saturation S_1), and C_2 represents the concentration (or fraction saturation S_2) desired.

Fraction I. Dried brewers' yeast, 10 kg, is suspended in 100 liters of 0.2 M ammonium sulfate, pH 8.9, at room temperature [2643 g of $(NH_4)_2SO_4$ + 667 ml of concentrated ammonium hydroxide made up to 100 liters with deionized water]. The suspension is stirred for 40 minutes with a heavy duty mechanical stirrer, during which time the pH drops

[5] L. Noda and S. A. Kuby, *J. Biol. Chem.* **226**, 541 (1957).

to 8.7. The mixture is then incubated without stirring for 12 hours (overnight) at $26 \pm 2°$.

Fraction II. After this period of autolysis (the pH is approximately 8.5), the mixture is transferred to the cold room and brought to 0.48 saturation with ammonium sulfate, assuming the saturation to be 0.05 initially. The pH decreases further to about 8.3 and is adjusted to 5.5 with ice-cold 2 N H_2SO_4 which is also 1.87 M (0.48 saturation) with respect to ammonium sulfate. The sulfuric acid, of which about 4500 ml are required, is slowly added while the reaction mixture is stirred vigorously. The precipitated protein together with the insoluble cell debris is removed by centrifugation at 1300 g for 30 minutes in large capacity refrigerated centrifuges.[6]

The volume recovery of the opalescent, yellowish supernatant liquid (fraction II) is approximately 100 liters.

Fraction III. The solution is made 5×10^{-3} M with respect to silver nitrate by dissolving the calculated amount of crystalline $AgNO_3$ in 500 ml of cold distilled water and adding this solution to the vigorously stirred fraction II. The pH, about 5.4, is adjusted to 4.2 ± 0.05 with ice-cold 2 N H_2SO_4–1.87 M $(NH_4)_2SO_4$ (approximately 1500 ml is required) in the same manner as described for fraction II, and the suspension is allowed to stand in the cold room for 3 hours with gentle stirring. The precipitate is then collected with refrigerated Sharples super centrifuges running at 40,000 rpm while the suspension is siphoned in at a rate of approximately 200 ml per minute for each centrifuge.

The pellet is suspended in 0.05 M EDTA, pH 8.0 (one-eighth the volume of fraction II), with a Waring blendor at a low speed regulated with a variable autotransformer (Powerstat). Initially, the pellet dissolves, but with time a considerable amount of material, probably denatured, comes out of solution. The mixture is stirred gently overnight and shielded against light.

The next day, the precipitated material is centrifuged off at 4000 g for 30 minutes.[7] The supernatant liquid (yellow and slightly opalescent) is retained, and the pellet is extracted a second time with 0.05 M EDTA, pH 8.0 (1/48 the volume of fraction II). The very thick suspension this time is centrifuged[8] for 20 minutes at 15,000 g or 40 minutes at 4000 g. Both extracts are combined (fraction III), yielding a volume of about one-seventh that of fraction II.

[6] Either International refrigerated model 13L, or SR-3, or PR-2's with No. 276 rotors have been found adequate for this purpose.

[7] For example, International PR-2's, with No. 959 rotors.

[8] For example, Servall RC-2 with GSA rotor or International PR-2 with No. 850 rotor, respectively.

Fraction IV. The ammonium sulfate saturation of fraction III is estimated as indicated above (averaging approximately 0.06) and increased to 0.48. The small precipitate which appears is removed by centrifugation[8] for 20 minutes at 15,000 *g* or for 40 minutes at 4000 *g*. The supernatant liquid (yellow and opalescent) is made 0.67-saturated with ammonium sulfate, and the heavy precipitate is collected by centrifugation with refrigerated Sharples Super centrifuges, as before, but at a flow rate of approximately 75–100 ml per minute for each centrifuge. The pellet is dissolved in freshly prepared 0.05 *M* EDTA—0.01 *M* cysteine, pH 8.0 (1/20 the volume of fraction III). The resulting solution, slightly turbid and deeply yellow in color, is dialyzed for 10 hours against a total of 100 liters of 0.005 *M* magnesium sulfate by flow dialysis.[2] The small precipitate, which forms during dialysis, is removed by centrifugation for 15 minutes at 15,000 *g*.

The protein concentration of the clear, golden yellow supernatant fluid (fraction IV) is determined.

Fraction V. The initial conditions for the ethanol fractionation are: 10 mg of protein per milliliter, 0.005 *M* magnesium sulfate, 0.1 *M* magnesium acetate. A calculated volume of 0.005 *M* $MgSO_4$ is first added to fraction IV, followed by an aliquot (0.1 of required final volume) of a solution, which is 1.0 *M* with respect to magnesium acetate and 0.005 *M* with respect to magnesium sulfate,[9] to yield the desired final concentrations. The pH of this mixture (about 6.5) is adjusted to 5.5 with ice-cold 1 *N* H_2SO_4 (approximately 0.02 the volume of fraction IV, after dilution to 10 mg/ml, is required).

After the solution is chilled to —1° in an —8° bath, 0.25 volume of 95% ethanol (chilled to —10°) is added with mechanical stirring, at a rate such that the temperature does not exceed 0°. Stirring is continued, with care to avoid formation of foam, until the temperature has dropped to —5°. The milky precipitate is then collected by centrifugation with a refrigerated Sharples Super centrifuge running at 40,000 rpm. The temperature[10] of the outflow liquid from the centrifuge bowl should lie between 0 and 1°, which corresponds to a flow rate of approximately 200 ml per minute.

The small cream-colored precipitate is suspended in a volume of ice-cold 0.1 *M* magnesium acetate, which is equal to 1/12 the volume of fraction IV after dilution to 10 mg/ml. Extraction is allowed to take

[9] In the original publication,[2] due to a typographical error, this figure had read "0.05 *M* with respect to magnesium sulfate," although the final conditions given were correct.

[10] If the temperature of the outflow rises, incomplete recovery of the precipitate may result.

place by stirring in an ice bath until the suspension is homogeneous. The insoluble material is removed by centrifugation for 15 minutes at 15,000 g, and the clear supernatant liquid (light greenish yellow) is designated as fraction V.

Fraction VI. After Fraction V is analyzed for protein, it is diluted to 5 mg/ml, if necessary, and bentonite is added with stirring, at a ratio of 2.7 g of bentonite per gram of protein. Gentle stirring is continued for 15 minutes, and the adsorbent is removed by centrifugation for 40 minutes at 15,000 g. The slightly colored supernatant liquid (fraction VI) is retained.

Fraction VII. Fraction VI is made 0.66-saturated with ammonium sulfate, and the resultant semicrystalline-like material is centrifuged off at 15,000 g for 20 minutes and discarded. The ammonium sulfate saturation of the supernatant liquid is increased to 0.86; this second precipitate is collected by centrifugation at 15,000 g for 30 minutes and dissolved in 0.05 M EDTA, pH 8.0 (1/10 the volume of fraction VI).

The protein solution is made 10^{-4} M with respect to diisopropyl fluorophosphate by the addition of an ice-cold 0.1 M solution, and dialyzed for 8 hours against 40 liters of 0.01 M sodium succinate, pH 5.6, by flow dialysis. A trace of turbidity, which may appear after the dialysis, is removed by centrifugation at 15,000 g for 15 minutes. The clear yellow supernatant liquid, fraction VII, has about 1/6 the volume of fraction VI.

Fraction VIII. Initial conditions for the acetone fractionation are: 5 mg of protein per milliliter, 0.001 M manganese (-ous) sulfate, 0.02 M manganese (-ous) acetate. A calculated volume of 0.01 M succinate, pH 5.6, is added to fraction VII, followed by an aliquot (0.02 of required final volume) of a solution, which is 1.0 M with respect to manganese acetate and 0.05 M with respect to manganese sulfate, to yield the desired final concentrations. The pH should now be 5.6 ± 0.1.

The solution is chilled to $-1°$ in a $-8°$ bath and acetone (at $-10°$) is slowly added to give a final concentration of 15% (v/v); the temperature should not exceed 0°. The mixture is kept for 10 minutes at 0°, and the precipitate is collected by centrifugation at 15,000 g for 20 minutes. The supernatant liquid, which is very faintly yellowish brown in color, is discarded, and the gelatinous yellow pellet is dissolved with some difficulty in 0.05 M ammonium citrate, pH 8.0 (one-tenth the volume of fraction VII after dilution to 5 mg/ml). The protein concentration of the golden yellow solution (fraction VIII) is determined.

Fraction IX. Fraction VIII is diluted to 10 mg of protein per milliliter with 0.05 M ammonium citrate, pH 8.0, and made 0.60 saturated with ammonium sulfate. After removal of a yellow precipitate by centrifugation (20 minutes at 25,000 *g*), the ammonium sulfate saturation of the supernatant liquid is brought to 0.75, and a very light-colored precipitate is collected by centrifugation as above. It is dissolved in 0.05 *M* ammonium citrate, pH 8.0 (one-sixth the volume of fraction VIII after dilution to 10 mg/ml) and dialyzed overnight against 20 liters of 0.002 *M* dibasic ammonium phosphate (pH 7.8) by flow dialysis. The light yellow dialyzate is fraction IX.

Fraction X. After the total protein of fraction IX is determined, a 35-fold amount of DEAE-cellulose (prepared as described above and calculated as dry material) is suspended in cold distilled water and transferred to a coarse sintered glass funnel (9 cm in diameter), inserted tightly in a suction flask. The DEAE-cellulose is stirred to give a thick homogeneous slurry, which is finally packed to a wet cake of about 1 mm height per gram of dry cellulose.

Fraction IX is allowed to pass into the bed of DEAE-cellulose which is then washed with 0.025 *M* $(NH_4)_2HPO_4$ (100 ml per gram of dry DEAE-cellulose) with use of gentle suction. Elution of the enzyme by 0.15 *M* $(NH_4)_2HPO_4$, pH 7.8, follows with a total of 14 ml per gram of dry DEAE-cellulose. The first two volumes (2 ml per gram of DEAE-cellulose) may be discarded. The remaining 12 volumes are divided into three portions of 6, 4, and 2 volumes, respectively. Suction is applied to the flask but with care to avoid formation of foam, and after each of these three eluting portions the cake is pressed out with a flat glass rod. The eluates are combined and made 0.95-saturated with ammonium sulfate. Because the protein concentration is relatively low (approximately 1 mg/ml), at least 1 hour is allowed for precipitation; after this time the precipitate is collected by centrifugation at 15,000 *g* for 1 hour. The pellet is dissolved in 0.001 *M* EDTA, pH 7.0 (1 ml per gram of the dry DEAE-cellulose used) to give fraction X.

Fraction XI. The protein concentration of fraction X is determined and diluted if necessary to 10 mg/ml with 0.001 *M* EDTA, pH 7.0. The ammonium sulfate saturation is estimated as described before and increased to 0.64. After 1 hour, a small but significant turbidity is removed by centrifugation at 25,000 *g*. The ammonium sulfate saturation of the supernatant liquid is brought to 0.78, and the mixture is kept for at least ½ hour before centrifugation[11] at 25,000 *g* for 30 minutes.

The precipitate is dissolved in 0.05 *M* EDTA, pH 7.0 (1/3 the volume of fraction X after dilution to 10 mg/ml). It is dialyzed overnight against 20 liters of 0.002 *M* phosphate (NH_4^+), pH 6.1, to yield fraction XI (clear and yellow), the protein concentration of which is determined.

[11] For example, Servall RC-2 with SS-34 rotor.

Fraction XII. DEAE-cellulose (20 *g* per gram of protein) is equilibrated by suspending the cellulose in 0.2 *M* phosphate (NH_4^+), pH 5.8 (30 ml per gram of dry weight). The slurry is transferred to a coarse sintered glass funnel (6-cm diameter) and washed with 0.005 *M* phosphate (NH_4^+), pH 6.1 (300 ml per gram of dry weight). A filter cake is then prepared as described above, and fraction XI is allowed to pass into it. The cellulose is washed with 0.02 *M* phosphate (NH_4^+), pH 6.0 (100 ml per gram of dry DEAE-cellulose). The enzyme is then eluted, in a similar manner as described for fraction X, with 0.15 *M* phosphate (NH_4^+), pH 5.8. A total of 26 ml per gram of dry DEAE-cellulose are divided into portions of 2, 10, 6, 4, and 4 ml per gram of dry weight, of which the first 2 volumes may be discarded and the remaining 24 volumes are combined. The combined eluates are brought to 0.95 saturation with ammonium sulfate. After 1 hour to allow for complete precipitation, the suspension is centrifuged for at least 1 hour at 15,000 *g*.

Crystallization. The pellet is dissolved in 0.01 *M* NADP, pH 7.3 (about 1/15 to 1/10 the volume of fraction X after dilution to 10 mg of protein per milliliter). The ammonium sulfate saturation is estimated, and saturated (at 0°) ammonium sulfate solution of pH 7.3 is slowly added with gentle swirling to bring the saturation to 0.58. The final protein concentration should be approximately 25 mg/ml at this point. *Immediately* after the 0.58 ammonium sulfate saturation is reached, the enzyme solution is centrifuged for 5 minutes at 25,000 *g* to remove any traces of insoluble fibrous material. The clear light yellow liquid is then transferred into a round-bottom tube, which allows smooth swirling without causing formation of foam, and is kept at 0°. Within several hours a considerable increase in viscosity occurs, and, simultaneously, on cautious swirling, a very impressive silkiness can be observed with indirect light.

In the course of the next 3–4 days, the ammonium sulfate saturation is gradually increased to 0.62 (about 0.01 per day) by the addition of saturated ammonium sulfate solution, pH 7.3. Within 36–48 hours after the last increase, 80% or more of the active protein will have crystallized.

Recrystallizations. The crystalline suspension is centrifuged for 1 hour at 25,000 *g*, and the mother liquor is removed cautiously with a transfer pipette. (The specific activity of the yellow mother liquor is less than 1/3 that of the crystalline material.) After the crystalline pellet is dissolved in 0.01 *M* NADP, pH 7.3 (one-half the volume used for the first crystallization), the ammonium sulfate saturation is estimated and brought to 0.58 as described for the first crystallization; the second crystals usually appear within 10–60 minutes. Over a period of the next 2–3 days the ammonium sulfate saturation is increased to 0.62;

Fractionation of Glucose 6-Phosphate Dehydrogenase (Preparation No. 14)[a]

Fraction (Initially, 3 kg of dried brewers' yeast)	Volume (ml)	Protein (mg/ml)	Total activity (units $\times 10^{-5}$)	Specific activity (units/mg protein)	Purification		Recovery (%)	
					Overall	Over preceding step	Overall	From preceding step
I. Autolyzate	26,880[b]	15.3	68.3	16.6	—	—	(100)	(100)
II. $(NH_4)_2SO_4$ supernatant, 0.48 saturation	30,000	8.5	61.2	23.9	1.4	1.4	90	90
III. Extracted $AgNO_3$ precipitate	4,360	22.7	48.0	48.5	2.9	2.0	70	78
IV. $(NH_4)_2SO_4$, 0.48–0.67 saturation	1,470	39.7	44.5	76.3	4.6	1.6	65	93
V. Alcohol precipitate	605	4.6	38.4	1,380	83	18.1	56	86
VI. Bentonite supernatant	585	2.4	31.2	2,250	136	1.6	46	81
VII. $(NH_4)_2SO_4$, 0.66–0.86 saturation	95	8.8	28.5	3,400	205	1.5	42	91
VIII. Acetone precipitate	18.8	17.3	23.3	7,160	430	2.1	34	82
IX. $(NH_4)_2SO_4$, 0.60–0.73 saturation[c]	18.5	9.5	18.7	10,670	640	1.5	27	81
X. Concentrated DEAE-SF eluate	7.15	13.1	15.8	16,890	1,020	1.6	23	84
XI. $(NH_4)_2SO_4$, 0.64–0.78 saturation	2.4	27.0	14.2	21,960	1,320	1.3	21	90
Crystallizations[d]								
1. Crystals	1.44	24.3	11.4	32,550	1,960	1.5	17	80
2. Crystals	1.09	22.1	9.2	38,000	2,290	1.2	13	80
3. Crystals	0.95	20.5	8.1	41,900	2,520	1.1	12	89
4. Crystals	0.93	17.6	6.9	42,100[e]	2,535	1.006	10	85
4. Mother liquor	0.86	3.0	1.1	41,500	—	—	—	13

[a] Reproduced from E. A. Noltmann, C. J. Gubler, and S. A. Kuby, *J. Biol. Chem.* **236,** 1225 (1961).

[b] After subtraction of the volume included by the cell debris (16%).

[c] See text for change to 0.75 saturation.

[d] Fraction XII (see text) has a recovery of ca. 80% and a 1.3-fold purification over fraction XI.

[e] Corresponds to 678 micromoles of NADPH formed per minute per milligram of protein at 30°.

again, more than 80% of the activity of the first crystals can be obtained in crystalline form. The second mother liquor is only very light yellow and additional liquors are essentially colorless. Further recrystallizations are carried out in the same manner.

Usually, the third or fourth crystals yield a specific activity of about 650 micromoles of NADPH formed per milligram of protein per minute, under the conditions described above for analysis. A summary of the data for a typical preparation (No. 14), at 3-kg scale, is given in the table, taken from the original publication.[2] The reproducibility of the procedure has been found to be satisfactory in the hands of the authors in two different laboratories.

Properties

The sedimentation behavior of the ultracentrifugally homogeneous protein is significantly influenced by the presence of NADP.[12,13] The available evidence indicates an association phenomenon in the presence of the coenzyme. Preliminary values for $S_{20,w}$ (0.01 M phosphate, 0.15 M KCl, pH 6.8), extrapolated to zero protein concentration, were found to be 9.6 Svedberg units for the NADP saturated enzyme and 6.3 Svedberg units for the NADP-free enzyme.[14]

Following an observation made by Colowick and Goldberg[15] with a partially purified preparation, a four times crystallized sample has also been found[16] to catalyze a very slow reaction between NADP and β-D-glucose, the latter at relatively high concentrations.

For a compilation of miscellaneous kinetic properties of glucose 6-phosphate dehydrogenase from various sources, refer to Table I of Noltmann and Kuby.[13]

Acknowledgment

The experimental work on which the article is based was supported in part by grants from the National Institutes of Health and the National Science Foundation.

The authors wish to acknowledge the permission of the Journal of Biological Chemistry to make quotations from their original publication.[2]

[12] E. A. Noltmann and S. A. Kuby, unpublished experiments (1961); reported in part at the 45th Annual Meeting of the Federation of American Societies for Experimental Biology, Atlantic City, New Jersey, April 10–14, 1961.

[13] E. A. Noltmann and S. A. Kuby, *in* "The Enzymes" (P. D. Boyer, H. Lardy, and K. Myrbäck, eds.), 2nd ed., Vol. 7, p. 223. Academic Press, New York, 1963.

[14] The molecular weight of the NADP-free enzyme has been recently determined. R. H. Yue, R. N. Roy, E. A. Noltmann, and S. A. Kuby, *Federation Proc. Abstr.* **25**, 712 (1966).

[15] S. P. Colowick and E. B. Goldberg, *Bull. Res. Council Israel* **11A4**, 373 (1963).

[16] S. A. Kuby, unpublished experiments (1963).

[24] Glucose 6-Phosphate Dehydrogenase from Erythrocytes

By Robert G. Langdon

Glucose 6-phosphate + $TPN^+ \rightleftharpoons$ 6-phosphogluconate + TPNH + H^+

Assay Method

Principle. The reduction of TPN^+ to TPNH results in the appearance of an absorption band at 340 mμ. In the presence of saturating concentrations of G-6-P and TPN, the rate of change of absorbancy at this wavelength is proportional to the enzyme concentration.

Reagents

Tris-chloride buffer, 1 *M*, pH 7.5

Glucose 6-phosphate, 2.5×10^{-2} *M*. Dissolve 97.5 mg of the dipotassium salt of glucose 6-phosphate trihydrate in 10 ml of water.

TPN, 2×10^{-3} *M*. Dissolve a quantity of TPN equivalent to 16.7 mg of the free acid in 9 ml water. Adjust the pH to 7.5. Adjust the volume to 10 ml.

$MgCl_2$, 0.2 *M*. Dissolve 4.1 g $MgCl_2 \cdot 6\ H_2O$ in sufficient water to give a final volume of 100 ml

All buffers and reagents should be prepared in water distilled from a hard glass or quartz still.

Procedure. The assay is carried out at room temperature (25°).

To a 3-ml quartz cell having a 1-cm path length is added sufficient glass-distilled water to yield a final volume of 3 ml after additions of all reagents and enzymes have been made; 0.1 ml buffer, 0.1 ml G-6-P, 0.1 ml TPN, and 0.1 ml $MgCl_2$ are then added. These are mixed and the cell is placed in a spectrophotometer adapted for recording of absorbancy as a function of time. A volume of enzyme solution containing approximately 0.1–0.2 enzyme units is then quickly added and mixed, and the rate of change of absorbancy at 340 mμ is recorded. The velocity of the reaction during the first 10 seconds following mixing is taken to represent the initial velocity.

Definition of Enzyme Unit and Specific Activity. One unit of enzyme activity is defined as that quantity which catalyzes the reduction of 1 micromole of TPN per minute under the above assay conditions. This corresponds to an absorbancy change of 2.07 per minute. Protein is measured by the optical method of Warburg and Christian[1] or by the

[1] O. Warburg and W. Christian, *Biochem. Z.* **310**, 384 (1941). See also E. Layne, Vol. III, p. 451.

colormetric method of Lowry.[2] Specific activity is defined as the units of enzyme per milligram of protein.

Purification Procedure

The purification procedure given here is essentially that described by Chung and Langdon.[3] A similar procedure yielding a slightly less pure product has been described by Kirkman.[4]

During the purification it is important to use reagent grade chemicals and distilled water which has been redistilled from a hard-glass still or passed through a mixed-bed ion exchange column.

Preparation of Adsorbents

Commerial ion exchange cellulose is prepared for chromatography as follows: Each 100 g of the material is suspended in 2 liters of water and titrated with 1 *M* KH_2PO_4 or K_2HPO_4 to the desired pH value. After the suspension has settled for a few minutes, any fine particles are removed by decantation and the remainder is filtered through fine-mesh nylon in a Büchner funnel. The damp cake is washed twice in a similar manner with water and 3 times in the desired buffer. It is then stored in buffer at 0° until used.

Calcium phosphate gel is prepared as described by Keilin and Hartree.[5]

Preparation of Erythrocytes

The erythrocytes from 18 pints of human blood, collected in standard ACD solution by a blood bank and stored at 4° for 3–5 weeks, are allowed to settle and the supernatant plasma is removed by suction. The thick suspension is then centrifuged at low speed, and the erythrocytes are washed four times with 0.15 *M* KCl in 5×10^{-3} *M* potassium phosphate buffer, pH 7.0, containing 10^{-4} *M* EDTA. The erythrocytes may be stored at 4° for a short time prior to use.

Step 1. Preparation of Hemolyzate. The erythrocytes are mixed with an equal volume of distilled water, frozen solidly by immersing the container in a dry ice bath, and then thawed in a water bath at 10°. Most subsequent operations are carried out in a cold room at 4°.

Step 2. First DEAE-Cellulose Treatment. The hemolyzate from each 6 pints of blood is diluted with water to 2 liters and mixed with 150 g of DEAE-cellulose suspended in 2 liters of 5×10^{-3} *M* potassium phosphate,

[2] O. H. Lowry, N. J. Rosebrough, A. L. Farr, and R. J. Randall, *J. Biol. Chem.* **193**, 265 (1951). See also E. Layne, Vol. III, p. 448.

[3] A. E. Chung and R. G. Langdon, *J. Biol. Chem.* **238**, 2309 (1963).

[4] H. N. Kirkman, *J. Biol. Chem.* **237**, 2364 (1962).

[5] D. Keilin and E. F. Hartree, *Proc. Roy. Soc.* **B124**, 397 (1938).

pH 7.0, containing 10^{-4} M EDTA. After the suspension has been gently stirred for 30 minutes, it is poured into an 8×60 cm column. The packed cellulose is washed with an additional 4 liters of the buffer solution, which elutes most of the hemoglobin. Then 0.3 M KCl in the same buffer is passed through the column; fractions are collected and assayed for enzyme activity. Those fractions containing enzyme are pooled.

Step 3. Ammonium Sulfate Precipitation. The enzyme is precipitated by the addition of 351 g of solid ammonium sulfate per liter of solution obtained in the preceding step. It is collected by centrifugation, dissolved in 5×10^{-3} M potassium phosphate, pH 7.0, containing 10^{-4} M EDTA, and the ammonium sulfate precipitation is repeated. After solution in 5×10^{-3} M potassium phosphate, pH 6.0, containing 10^{-4} M EDTA, the solution is passed through a Sephadex G-25 column[6] equilibrated with the same buffer in order to free it from ammonium sulfate.

Step 4. Calcium Phosphate Gel Adsorption. Calcium phosphate gel is added to the enzyme solution in a ratio of 2 mg gel per milligram protein; after the gel has stood for several minutes, it is collected by centrifugation. The supernatant solution is assayed for enzyme activity, and successive additions of gel are made until the enzyme has been almost completely adsorbed. The supernatant fluid is discarded, and the enzyme is eluted from the gel by suspending it in an equal volume of 0.12 M potassium phosphate, pH 7.0; after 30 minutes the suspension is centrifuged and the supernatant fluid is removed and saved. The gel phase is treated twice more in an identical fashion with 0.12 M phosphate buffer, and the supernatant fractions are pooled.

Step 5. Ammonium Sulfate Precipitation. The enzyme is precipitated by the addition of 390 g of solid ammonium sulfate per liter of solution. It is collected by centrifugation and redissolved in a few milliliters of 5×10^{-3} M potassium phosphate, pH 6.0, containing 10^{-4} M EDTA, and the ammonium sulfate is removed by passing the enzyme solution through a Sephadex G-25 column equilibrated with the same buffer.

Step 6. CM-Cellulose Chromatography. The protein concentration in the Sephadex eluate is adjusted to 2–5 mg per milliliter by addition of more 0.005 M phosphate buffer; sufficient TPN is then added to give a final concentration of 2×10^{-6} M. A 4.5×40 cm column containing 60 g (dry weight) of carboxymethyl cellulose equilibrated with 5×10^{-3} M potassium phosphate, pH 6.0, 10^{-4} M EDTA is prepared. The enzyme is applied to the column and an additional 1000 ml of the buffer solution is allowed to pass through the column; this eluate is discarded. A linear

[6] J. Porath and P. Flodin, *Nature* **183**, 1657 (1959).

ionic strength gradient is begun with 2000 ml of buffer in the mixing chamber and 2000 ml of 1 M KCl in buffer in the other container. Fractions of 12 ml volume are collected and each is assayed for protein and enzyme. G-6-P dehydrogenase is eluted between 0.05 and 0.15 M KCl concentration. Fractions of higher specific activity are pooled.

Step 7. Calcium Phosphate Gel Adsorption. To the pooled fractions from the preceding step calcium phosphate gel is added at a gel to protein ratio of 6:1. The adsorbed enzyme is eluted from the gel by the addition of small increments of 0.12 M potassium phosphate pH 7.0 until no additional enzyme appears in the eluate.

Step 8. CM-Cellulose Chromatography. The ionic composition of the solution is changed to 5×10^{-3} M potassium phosphate, pH 6.0, 10^{-4} M EDTA by passing it through a Sephadex G-25 column equilibrated with this buffer; sufficient TPN is added to the eluate to yield a 2×10^{-6} M solution of this nucleotide. This solution is then added to a 1.2×22 cm column of CM-cellulose previously equilibrated with buffer. The column is washed with 50 ml of buffer and then is developed with a linear ionic strength gradient using 100 ml of buffer and 100 ml of 0.5 M KCl in buffer. Fractions of 2 ml volume are collected and assayed for both protein and glucose 6-phosphate dehydrogenase. The center fractions of the peak are combined.

Step 9. Calcium Phosphate Gel Adsorption and DEAE-Cellulose Chromatography. Enzyme in the pooled fractions is concentrated by adsorption to and elution from calcium phosphate gel as in step 7. The potassium phosphate concentration is then reduced to 0.05 M by addition of 1.4 volumes of water; the solution is made 10^{-4} M in EDTA and 2×10^{-6} M in TPN by addition of concentrated solutions of these reagents. The enzyme solution is then added to a 1.2×22 cm column of DEAE-cellulose which has been prewashed by passing through it 100 ml of 0.05 M potassium phosphate, pH 7.0, containing 10^{-4} M EDTA and 2×10^{-6} M TPN. After the enzyme is adsorbed, the column is washed with an additional 50 ml of this solution. The enzyme is eluted by a linear ionic strength gradient using 100 ml of buffer and 100 ml of 0.5 M KCl in buffer. Fractions of 2 ml volume are collected.

Step 10. Final Calcium Phosphate Gel Adsorption. The enzymatically active fractions from step 9 are pooled and passed through a Sephadex G-25 column previously equilibrated with 5×10^{-3} M potassium phosphate pH 6.0, 10^{-4} M EDTA. The enzyme is adsorbed to and eluted from calcium phosphate gel as in step 7.

The final product may be stored at 0° for several weeks or for prolonged periods at —20°.

PURIFICATION OF GLUCOSE 6-PHOSPHATE DEHYDROGENASE

Step[a]	Enzyme units	Enzyme specific activity	Cumulative purification	Cumulative yield (%)
1. Hemolyzate	5170	0.0026	0	100
2. DEAE chromatography	1810	0.123	47.2	35
3. $(NH_4)_2SO_4$ precipitation	1450	0.21	81	28
4. Gel adsorption	1230	0.31	119	25
5. $(NH_2)_2SO_4$ precipitation	970	0.60	230	19
6. CM-cellulose chromatography	955	7.55	2,900	18
7. Gel adsorption	538	11.1	4,270	10
8. CM-cellulose chromatography	571	23.1	8,900	11
9. Gel adsorption and DEAE-cellulose chromatography	434	93	35,800	8.4
10. Gel adsorption	390	113	43,500	7.5

[a] The number of each step corresponds to that in the text.

Properties of Erythrocyte G-6-P Dehydrogenase

Purity. Enzyme having a specific activity of 113 units per milligram appears homogeneous on free and starch gel electrophoresis. In the analytical ultracentrifuge, one major component comprising approximately 80% of the protein is observed. One minor peak, which is believed to represent the monomeric form, is also present.[3]

Physical Properties. The sedimentation constant determined by the moving boundary method is 7.1 S. The diffusion coefficient as determined by the porous diaphram method[7] is 3.4×10^{-7} cm^2 sec^{-1}. From these values the calculated molecular weight is 190,000 and the frictional ratio is 1.6.[3] Kirkman,[4] using different methods, has concluded that the molecular weight of the enzyme is 105,000.

Bound Coenzyme. The enzyme as prepared contains 2 moles of TPN which are very tightly bound.[8,9] These are reduced to enzyme-bound TPNH when the enzyme is treated with glucose 6-phosphate. The bound nucleotide may be removed from the enzyme by treatment with acid ammonium sulfate. The apoenzyme dissociates into catalytically inactive subunits having approximately one-half the molecular weight of the holoenzyme. The enzyme-bound TPN may also be destroyed by treatment of the enzyme with snake (*Agkistroden piscivorus*) venom which contains an active triphosphopyridine nucleotidase. Under appropriate

[7] C. E. Mize, T. E. Thompson, and R. G. Langdon, *J. Biol. Chem.* **237,** 1596 (1962).
[8] A. E. Chung and R. G. Langdon, *J. Biol. Chem.* **238,** 2317 (1963).
[9] H. N. Kirkman and E. M. Hendrickson, *J. Biol. Chem.* **237,** 2371 (1962).

circumstances,[8,9] the apoenzyme and coenzyme recombine with restoration of the physical and catalytic properties of the native enzyme.

Enzyme Structure. Two NH_2-terminal amino acids, alanine and tyrosine, have been detected.[3] Therefore, the enzyme has at least two nonidentical polypeptide subunits.

Stability. The purified native enzyme is stable in solution for several weeks at 0°, and for prolonged periods at −20°. It may also be kept in solution for several hours at 40° without appreciable loss of activity. In the absence of added TPN, the apoenzyme is very unstable at 40°; within 1 hour at this temperature most of the catalytic properties of the protein are irreversibly lost. Presumably the instability[10] of glucose 6-phosphate dehydrogenase in crude hemolyzates is due to a nucleotidase-catalyzed destruction of enzyme-bound TPN with production of the thermally unstable apoenzyme.

Kinetic Properties. At pH 7.6 the K_m for TPN has been reported to be $4.2 \times 10^{-6} M$ and for G-6-P to be $3.5 \times 10^{-5} M$.[4] Assuming enzyme which has a specific activity of 113 units mg^{-1} to be 80% pure and that the molecular weight of the functional unit is 190,000, it may be calculated that the turnover number of the enzyme is approximately 21,500 min^{-1}. The enzyme is inhibited by *p*-chloromercuribenzoate; it is protected against this agent by TPN but not by DPN, 2′-AMP, or several closely related compounds. It has also been reported that the enzyme is inhibited by pregnenolone and other steroids[11] in a noncompetitive manner.

[10] P. A. Marks, *Cold Spring Harbor Symp. Quant. Biol.* **26**, 343 (1961).
[11] P. A. Marks and J. Banks, *Proc. Natl. Acad. Sci. U.S.* **46**, 1483 (1960).

[25] Glucose 6-Phosphate Dehydrogenase—Clinical Aspects

By PAUL A. MARKS

$$\text{Glucose 6-phosphate} + \text{TPN}^+ \rightleftharpoons \text{6 phosphogluconolactone} + \text{TPNH} + \text{H}^+$$

Clinical Significance

A deficiency in glucose 6-phosphate dehydrogenase occurs as a genetically determined trait associated with an increased susceptibility to hemolytic anemia following ingestion of a large variety of drugs and other chemical agents, including sulfonamides, nitrofurantoins, primaquine, salicylates, naphthaline, and methylene blue; the fava bean; during certain viral and bacterial infections; and in association with metabolic disorders, such as diabetic acidosis and uremia.[1] There is considerable heterogenity in the expression of this trait which is manifest in the severity of the deficiency in activity of the enzyme, the tissues involved, the properties of the glucose 6-phosphate dehydrogenase, and the susceptibility of affected subjects to hemolytic anemia.

Assay Method

Principle. This enzyme catalyzes a reaction involving the reduction of TPN. TPNH has an absorption band with a maximum at 340 mμ, a wavelength at which TPN does not absorb. In the presence of excess glucose 6-phosphate and TPN, the rate of TPNH appearance is proportional to the concentration of enzyme. The spectrophotometric method described below is a modification of the procedure of Kornberg and Horecker.[2] For many clinical studies, a more rapid procedure not requiring a spectrophotometer is both desirable and adequate. Such an assay method is also described.

Reagents for Preparation of Erythrocytes for Assay of Glucose 6-Phosphate Dehydrogenase

Dried heparin
Acid–citrate–dextrose solution. Dissolve 0.44 g of anhydrous citric acid, 1.32 g of sodium citrate, and 1.47 g of dextrose in 100 ml of H_2O.
NaCl, 0.15 *M*
NaCl, 1.5 *M*
$MgCl_2$, 0.0015 *M*, in 0.001 *M* Tris buffer, pH 7.5

Additional Reagent Required for Separation of Erythrocytes, Leukocytes, and Platelets from Whole Blood

Dextran,[3] 3%, (mol. wt. about 225,000) and 3% glucose in 0.15 *M* NaCl

Reagents for Assay of Glucose 6-Phosphate Dehydrogenase

Glucose 6-phosphate,[4] 0.01 *M*
TPN,[4] 0.0025 *M*
$MgCl_2$, 0.1 *M*
Glycylglycine buffer, 0.25 *M*, pH 7.5

[1] P. A. Marks, *in* "The Red Blood Cell" (C. Bishop and D. M. Surgenor, eds.), p. 211. Academic Press, New York, 1964.
[2] A. Kornberg and B. L. Horecker, Vol. I, p. 323.
[3] Larco, grade HH, Henley & Co., New York, New York.
[4] Sigma Chemical Company, St. Louis, Missouri.

Enzyme. Aliquot of crude cell lysate sufficient to obtain a change of 0.010 to 0.050 optical density units per minute. If lysates of cells are prepared by the methods indicated below, the usual aliquot for hemolyzates prepared by freezing and thawing cells is 0.02–0.03 ml; for hemolyzates prepared by lysis of cells with water, 0.1 ml; for leukocyte lysates, 0.02 ml; and for platelet lysates, 0.1 ml.

Preparation of Crude Lysates of Blood Cells

ERYTHROCYTES. Centrifuge venous blood, anticoagulated with dried heparin (1.0 ml acid-citrate-dextrose (ACD) solution per 4 ml blood is also satisfactory as an anticoagulant) at 1500 *g* for 10 minutes and remove the plasma and buffy coat. Wash the erythrocytes twice with 4 volumes of 0.15 *M* NaCl and resuspend in this solution to yield a hematocrit of approximately 25%. This entire procedure is performed at 4°. Lyse the red cells by freezing and thawing twice, employing a dry ice-acetone bath for freezing. The concentration of hemoglobin in the hemolyzate is determined. In this procedure, the resuspended erythrocytes are generally contaminated with fewer than 0.05% leukocytes and 0.5% reticulocytes. Red cells may be prepared with a smaller degree of white cell contamination by employing a method for selective erythrocyte sedimentation, such as is described below for the preparation of leukocytes and platelets. Alternatively, red cells may be selectively hemolyzed by a short exposure to a hypotonic solution.[5,6] In this procedure, 4 volumes of a solution of 0.0015 *M* $MgCl_3$ in 0.001 *M* Tris buffer, pH 7.5 is added to the washed cells. After 1 minute the mixture is restored to isotonicity by addition of 0.1 volume of 1.5 *M* NaCl.

LEUKOCYTES AND PLATELETS. For studies in which enzyme activity is to be determined in erythrocytes, leukocytes, and/or platelets, draw venous blood into siliconized[7] syringes and transfer to siliconized test tubes. Various methods are available for separating the cellular elements of blood.[8] A convenient method is the following. The anticoagulated whole blood is added to an equal volume of a solution of 3% dextran and 3% glucose in 0.15 *M* NaCl and sedimentation is allowed to proceed for 30–45 minutes. This sedimentation and the ensuing steps in the

[5] H. J. Fallon, E. Frei, III, J. D. Davidson, J. S. Trier, and D. Burk, *J. Lab. Clin. Med.* **59,** 779 (1963).

[6] R. H. DeBellis, N. Gluck, and P. A. Marks, *J. Clin. Invest.* **43,** 1329 (1964).

[7] Equipment may be siliconized with Siliclad, a water-soluble concentrate, Clay-Adams, Inc., New York, New York.

[8] S. P. Martin, *in* "Methods in Medical Research," Vol. 7, p. 128. Year Book Publ., Chicago, Illinois, 1958.

separation of these cells are carried out at 4°. Remove the supernatant plasma and treat further as indicated below to recover platelets. Resuspend the sedimented white cells in a volume of 0.15 M NaCl to yield a concentration of approximately 1×10^7 white cells per milliliter. Disrupt the leukocytes by freezing and thawing 4 times. Perform leukocyte counts before and after this procedure to ensure that enzyme assays are done only on samples in which more than 95% of the white cells are destroyed. This procedure yields a leukocyte preparation which is generally contaminated with fewer than 5 red cells and less than 1 platelet for each white cell, a degree of contamination that has no detectable affect on the assay of glucose 6-phosphate dehydrogenase in white cells.

Platelets may be recovered from the plasma from which leukocytes were sedimented by recentrifuging at 3000 g for 45 minutes. Disrupt the platelets by freezing and thawing 4 times. Perform platelet counts before and after this procedure to ensure that assays are done only on samples in which more than 95% of the platelets are disrupted. The platelets prepared by this procedure are generally not contaminated with detectable leukocytes or erythrocytes.

After sedimentation in dextran, the red cells may be recovered by centrifuging at 1000 g for 10 minutes. Remove the supernatant solution, wash the red cells twice with 0.15 M NaCl, and resuspend in this solution. The erythrocyte samples prepared in this manner are contaminated with less than 0.03% leukocytes, 0.5% platelets, and 0.5% reticulocytes.

Spectrophotometric Assay Procedure:

To a 3.0-ml quartz or Corex cuvette having a light path of 1 cm, add 0.1 ml TPN, 0.5 ml $MgCl_2$, 0.5 ml glycylglycine buffer, 0.5 ml glucose 6-phosphate and enzyme solution. Make up to total volume of 2.5 ml with water. Take readings at 340 mμ at 1-minute intervals, starting immediately after addition of enzyme solution.

Definition of Unit of Enzyme Activity. A unit of enzyme activity in erythrocytes is defined as that amount in 1 ml which causes an initial change in 1.0 optical density unit per minute (room temperature 23–25°). The change in optical density (OD) units per minute may be converted to micromoles TPNH formed per minute by the equation:

$$\text{micromoles TPNH/min} = (\Delta\text{OD/min} \times 2.5)/6.22 \qquad (1)$$

This calculation is based on the use of an extinction coefficient for TPNH of 6.22×10^6 cm^2 mole.[9] Enzyme activity in red cells is referred to

[9] B. L. Horecker and A. Kornberg, *J. Biol. Chem.* **175**, 385 (1948).

hemoglobin concentration rather than cell count or cell volume because this parameter was found to provide most reproducible results.[10] Enzyme activity in leukocytes or in platelets is referred to cell count.

Application of Assay Method. In measuring glucose 6-phosphate dehydrogenase activity in crude cell lysates in the presence of an excess of glucose 6-phosphate and TPN, the rate of TPNH formation exceeds the rate of glucose 6-phosphate oxidation. This is attributable to the fact that the reaction catalyzed by 6-phosphogluconate dehydrogenase proceeds as 6-phosphogluconate is formed and results in the reduction of a second mole of TPN. The rate of TPN reduction attributable to 6-phosphogluconate dehydrogenase activity in crude lysates may be assayed directly by the procedure detailed in [27]. The difference in the rate of TPN reduction as determined in the presence of an excess of both glucose 6-phosphate and 6-phosphogluconate and that determined in the presence of 6-phosphogluconate alone has been taken as a measure of glucose 6-phosphate dehydrogenase activity.[11,12] This type of assay for glucose 6-phosphate dehydrogenase is performed by the procedure indicated above, except that two cuvettes are prepared for each assay, one with 0.5 ml of 6-phosphogluconate and another with 0.5 ml of glucose 6-phosphate and 0.5 ml of 6-phosphogluconate. However, the value for glucose 6-phosphate dehydrogenase activity in crude red cell lysates obtained by this "difference" method is on the average $71\% \pm 12\%$ of that obtained when this enzyme was assayed with only added glucose 6-phosphate. This probably reflects the fact that, in the assay with only glucose 6-phosphate added, the activity of 6-phosphogluconate dehydrogenase is not being measured under optimal conditions of substrate concentration. For purposes of clinical study, e.g., determining the possible presence of erythrocyte glucose 6-phosphate dehydrogenase deficiency, assay in the presence of the single substrate is satisfactory.

Dye Reduction Screening Assay Procedure[13]

Principle. The reduction of certain dyes may be coupled with the reoxidation of TPNH. In the presence of an excess of glucose 6-phosphate and TPN, the rate at which dye reduction occurs is proportional to the rate of TPNH formation, which, in turn, is proportional to the concentration of enzyme.

[10] P. A. Marks, *Science* **127,** 1339 (1958).

[11] G. E. Glock and P. McLean, *Biochem. J.* **55,** 400 (1953).

[12] W. H. Zinkham and R. E. Lenhard, *J. Pediat.* **55,** 319 (1959).

[13] This assay procedure is a modification of a method developed by Dr. Arno Motulsky. See A. G. Motulsky, *Human Biol.* **32,** 1, (1960).

Reagents

Color reagent.[14] Mix together (a) 40 ml tris(hydroxymethyl)aminomethane (Tris) buffer obtained by dissolving 8.95 g Tris in 100 ml water and adjust to pH 7.5 with concentrated (37%) HCl; (b) 0.165 g of glucose 6-phosphate, sodium salt; (c) 10 mg of TPN; (d) 16 mg of brilliant cresyl blue; and (e) 70 ml of H_2O. This reagent stored at 4° is stable for at least 2 weeks.

Mineral oil

Procedure. To 1 ml distilled H_2O in a 9 × 75 mm test tube, add 0.02 ml blood (hemoglobin pipette is convenient for this purpose). Blood may be taken directly from a finger tip puncture or a venous blood sample anticoagulated with heparin or ACD solution. Add 0.5 ml color reagent to tube. Mix thoroughly and cover surface with 1 ml mineral oil to prevent reoxidation of reduced dye by contact with air. Allow to stand in a 37° incubator or water bath and note time required for each tube to turn from blue to red, which is the background color due to the hemoglobin in solution.

Expression of Results. The results are expressed as the time required for decolorization. Blood samples of subjects with normal erythrocyte glucose 6-phosphate dehydrogenase activity decolorize within 30 to 60 minutes. The blood of affected male subjects requires more than 90 minutes to decolorize.

Application of Assay Method. This assay is particularly useful in screening large population groups for erythrocyte glucose 6-phosphate dehydrogenase deficiency. In employing the assay, it is desirable to determine a range of normal values for each new batch of reagents. Since the reaction time is temperature dependent, variations in incubation temperature will cause variations in the time required for decolorization. For example, if the assay is run at 25°, blood samples of subjects with normal erythrocyte glucose 6-phosphate dehydrogenase activity may require 60–120 minutes to decolorize and that of affected subjects more than 150 minutes. Bloods with known normal glucose 6-phosphate dehydrogenase activity should be assayed with each group of unknowns tested. The assay is most reliable in the detection of males with inherited glucose 6-phosphate dehydrogenase deficiency, whose red cells generally have very low levels of this enzyme.[1]

In the presence of anemia, the time required for decolorization is

[14] The color reagent is available from Sigma Chemical Company, St. Louis, Missouri, in powder form, containing all the components ready to be used upon addition of water.

prolonged, since a given volume of blood containing fewer red cells will have less enzyme. A further limitation to the use of this assay is its relative unreliability in the detection of moderate decreases in enzymatic activity, as may be encountered in most females with genetically determined glucose 6-phosphate dehydrogenase deficiency. The difficulty in detecting moderate decreases in red cell glucose 6-phosphate dehydrogenase by this dye reduction assay is attributable to the fact that, in the intermediate range of enzymatic deficiency, a small prolongation in decolorization time may reflect a relatively large difference in enzymatic activity. A number of females with a moderate decrease in glucose 6-phosphate dehydrogenase activity as determined by the spectrophotometric assay may give decolorization times within the range of normal values.

In addition to the above-described dye reduction screening assay, there are several other simple methods that have been developed to detect erythrocyte glucose 6-phosphate dehydrogenase deficiency. These other methods are all based on the principle of linking TPNH generation in the reaction catalyzed by glucose 6-phosphate dehydrogenase to the reduction of another compound, such as, dichloroindophenol,[15] 3(4,5-dimethylthylthiazolyl-1,2)2,5-diphenyltetrazolium bromide,[16] and methemoglobin.[17]

[15] H. E. Ells and H. N. Kirkman, *Proc. Soc. Exptl. Biol. Med.* **106**, 607 (1961).
[16] V. F. Fairbanks and E. Beutler, *Blood* **20**, 591 (1962).
[17] G. J. Brewer, A. R. Tarlov, and A. S. Alving, *Bull. World Health Organ.* **22**, 633 (1960).

[26] 6-Phosphogluconate Dehydrogenase—Crystalline

By SANDRO PONTREMOLI and ENRICO GRAZI

D-Gluconate 6-phosphate + TPN $\leftrightarrows$ D-ribulose 5-phosphate + CO_2 + TPNH + H^+

Assay Method

Principle. The method of determination of enzymatic activity is based on the spectrophotometric measurement of TPNH + H^+ formed during oxidation and decarboxylation of D-gluconate 6-phosphate to D-ribulose 5-phosphate and CO_2.

Reagents

Glycyglycine, 0.05 *M* buffer, pH 7.6
TPN, 0.01 *M*

$MgCl_2$, 0.1 *M*
D-Gluconate 6-phosphate, sodium salt, 0.15 *M*

Procedure. The incubation mixture (1.0 ml) contained: 0.26 ml of water, 0.5 ml of glycylglycine buffer, 0.2 ml of $MgCl_2$ 30 μl of TPN, 10 μl of D-gluconate 6-phosphate and a quantity of enzyme (usually between 1 and 5 μl) to produce an optical density change of 0.010–0.025 per minute at 340 mμ. A unit of enzyme is defined as that quantity in the cuvette which, under these conditions, would result in an optical density change of 1.0 per minute. Specific activity is the number of enzyme units per milligram of protein.

Starting Material. Candida utilis dried at low temperature was kindly provided by the Lake States Yeast Corporation, Rhinelander, Wisconsin, U.S.A.

Purification Procedure

Autolysis and First Ammonium Sulfate Fractionation. All operations are carried out at 2° except when otherwise indicated. The dried yeast preparation (50 g) is suspended in 150 ml of 0.1 *N* $NaHCO_3$. After 5 hours at 37° the autolyzed suspension is centrifuged for 20 minutes at 20,000 *g*. The extract (90 ml) is kept frozen overnight without loss of activity and is diluted with 117 ml of cold water for the subsequent procedures (yeast autolyzate, 207 ml, see the table).

The clear extract is adjusted to pH 6.2 with 5 *N* CH_3COOH and treated with 420 mg of protamine sulfate (Salmine Sigma) dissolved in 42 ml of water. The heavy precipitate is removed by centrifugation. The supernatant solution is adjusted to pH 4.8 with 5 *N* CH_3COOH, and the precipitate is discarded. The clear supernatant solution (230 ml) is treated with 59.3 g of ammonium sulfate, and the resulting precipitate is centrifuged and discarded. The supernatant solution is treated with 9.4 g of ammonium sulfate and the precipitate is collected by centrifugation and dissolved in 10 ml of water and brought to pH 6.3 with 1.0 *N* KOH. A final fraction is obtained by addition of 18.3 g of ammonium sulfate. The precipitate is collected by centrifugation, suspended in 10 ml of water and immediately adjusted to pH 6.3 with 1.0 *N* KOH. The last two fractions are tested separately and combined when necessary (ammonium sulfate fraction I, 28 ml).

Charcoal Treatment and Heat Denaturation. This step requires an amount of charcoal which must be determined by pilot trials with small volumes. The procedure should result in removal of about 65% of the protein and 50–70% of the D-glucose 6-phosphate dehydrogenase activity with essentially no loss of D-gluconate 6-phosphate dehydrogenase activity. For this purpose 0.1-ml aliquots of ammonium sulfate fraction I are diluted to 0.8 ml with water and quantities of acid-washed charcoal are added[1] that range from 0.75 to 1.50 mg per milligram of protein. The suspensions are heated in a water bath at 48° for 1 hour with occasional stirring, chilled to 0° and centrifuged at 25,000 *g* for 10 minutes. The residue is washed with 0.2 ml of cold water and discarded. The supernatant solutions and washings are combined and analyzed for protein, D-glucose 6-phosphate dehydrogenase and D-gluconate 6-phosphate dehydrogenase activity.

Usually a quantity of charcoal equivalent to 1.3 mg per milligram of protein is adequate. For the full scale procedure the ammonium sulfate fraction (26 ml) is diluted to 182 ml with cold water and treated with 3.24 g of charcoal. The mixture is heated in a water bath at 48° for 1 hour with occasional stirring, chilled to 0° and centrifuged at 25,000 *g* for 10 minutes. The residue is washed with 18 ml of cold water and discarded. The supernatant solution and washings are combined (charcoal fraction, 175 ml).

Gel Adsorption and Second Ammonium Sulfate Fractionation. The amount of gel to be used is determined in each case by pilot trials with small quantities of enzyme. Specifically 0.2-ml aliquots of the charcoal fraction are diluted with 1.0 ml of water and treated with quantities of calcium phosphate gel[2] (dry weight, 8 mg/ml), equivalent to 1.0–2.0 mg dry weight per milligram of protein. The suspensions are kept at 0° for 10 minutes and centrifuged at 25,000 *g*. The supernatants are tested for proteins and activity with D-glucose 6-phosphate and D-gluconate 6-phosphate. The quantity of gel selected (1.1 mg per milligram of protein) is sufficient to remove 90% of the D-glucose 6-phosphate dehydrogenase activity. This procedure is then carried out with the bulk of the enzyme preparation. The charcoal fraction (174 ml) is diluted with 617 ml of cold water, and 79 ml of calcium phosphate gel is added. The suspension is centrifuged and the residue discarded. The supernatant solution (850 ml) is again treated with calcium phosphate gel.

Pilot runs are made as before with 0.8–1.6 mg of gel per milligram of protein. D-Glucose 6-phosphate dehydrogenase activity should be removed completely. For the main batch the first calcium phosphate gel supernatant solution (847 ml) is treated with 28 ml of calcium phosphate gel (224 mg), and the supernatant solution is collected by centrifugation. The dilute solution (860 ml) is concentrated under re-

[1] Activated charcoal (Antichromos) Ditta Faravelli, Via Medardo Rosso, Milan, Italy; boiled for 10 minutes with 1 *N* HCl, filtered with suction and washed with deionized water until washings are neutral. The washed charcoal is dried in air.

[2] For the preparation see Vol. I [11].

duced pressure (bath temperature 30–32°) to a volume of 26 ml. This solution is treated with 16 g of ammonium sulfate, and the precipitate collected and dissolved in just enough glycylglycine buffer (0.25 *M*, pH 7.6) to bring the ammonium sulfate saturation to 50%, as measured with the Barnstead purity meter. About 3.0 ml is required. The solution is centrifuged at once (ammonium sulfate fraction II, 9.4 ml).

Crystallization. The 50% saturated solution is brought to pH 6.2 with 5 *N* CH_3COOH and kept at 0°. Crystal formation begins in a few minutes and is complete after several hours. The suspension may be stored overnight at 0°. The crystals are collected by centrifugation and dissolved in 2.0 ml of glycylglycine buffer, 0.25 *M* pH 7.6 (first crystals, 2.0 ml).

For recrystallization the solution is treated with an excess (1.75 ml) of cold saturated ammonium sulfate and diluted with 0.25 *M* glycylglycine buffer until only a faint turbidity remains. The pH is again brought to 6.2, and the solution is placed in the cold room for crystallization.

The crystal suspension is stored at 2°. For assay, aliquots are centrifuged and the precipitate is dissolved in 0.5 volume of glycylglycine buffer as above (second crystals).

The purification procedure is summarized in the table.

PURIFICATION OF D-GLUCONATE 6-PHOSPHATE DEHYDROGENASE

Step	D-Gluconate 6-phosphate dehydrogenase: Total units (1)	D-Gluconate 6-phosphate dehydrogenase: Specific activity	D-Glucose 6-phosphate dehydrogenase (total units) (2)	Ratio (1:2)
Yeast autolyzate	18.600	1.2	30.800	0.6
Ammonium sulfate fraction I	11.300	4.3	11.200	1.0
Charcoal fraction	6.850	11.8	5.400[a]	1.3
Ammonium sulfate fraction II	3.670	18.8	0[b]	∞
First crystals	2.760	119	—	—
Second crystals	2.020	173[c]	—	—

[a] Usually the activity of D-glucose 6-phosphate dehydrogenase at this point was less than in this particular preparation.

[b] Not detectable.

[c] Following a third recrystallization the specific activity varied from 180 to 220 units per milligram.

Properties

Specificity. The crystalline enzyme catalyzes the reversible decarboxylation of D-gluconate 6-phosphate to D-ribulose 5-phosphate and

CO_2. No evidence has been found for the existence of separate enzymatic activities catalyzing the oxidative and the decarboxylative step.[3]

The crystalline enzyme preparation contains no detectable D-glucose 6-phosphate dehydrogenase activity and is also completely free of D-ribose 5-phosphate isomerase and D-xylulose 5-phosphate 3-epimerase.

Stability and pH Optimum. The suspension of crystals in ammonium sulfate can be stored at 2° for many months without loss of activity. The pH optimum for the enzyme activity occurs between 7.4 and 7.6.

Affinity Constants. The Michaelis constants were calculated to be $K_{m(\text{TPN})} = 2.6 \times 10^{-5}\ M$ and $K_{m(\text{6-PG})} = 1.6 \times 10^{-4}\ M$. Kinetic data seem to indicate that the presence of one substrate does not interfere with the binding to the enzyme of the other.[4]

Activators and Inhibitors. The enzyme is activated by $MgCl_2$ at a concentration of 0.03 *M* but inhibited at high concentrations. $MgCl_2$ can be completely replaced by 0.2 *M* NaCl or KCl.

The enzyme contains 8 titrable SH groups. The carboxymethylation of one cysteine residue per mole of enzyme, is followed by an almost complete loss of activity.[5]

[3] S. Pontremoli, A. De Flora, E. Grazi, G. Mangiarotti, A. Bonsignore, and B. L Horecker, *J. Biol. Chem.* **236**, 2975 (1961).

[4] S. Pontremoli, E. Grazi, A. De Flora, and G. Mangiarotti, *Arch. Sci. Biol.* **46**, 83 (1962).

[5] E. Grazi, M. Rippa, and S. Pontremoli, *J. Biol. Chem.* **240**, 234 (1965).

[27] 6-Phosphogluconate Dehydrogenase—Clinical Aspects

By PAUL A. MARKS

$$\text{6-Phosphogluconate} + \text{TPN}^+ \leftrightarrows \text{ribulose-5-phosphate} + CO_2 + \text{TPNH} + H^+$$

Clinical Significance

A deficiency of 6-phosphogluconate dehydrogenase occurs as a genetically determined trait which is not associated with any, as yet, recognized physical or hematologic abnormalities.[1]

Assay Method

Principle. This enzyme catalyzes a reaction involving the reduction of TPN. TPNH has an absorption band with a maximum at 340 mμ, a wavelength at which TPN does not absorb. In the presence of excess 6-phosphogluconate and TPN, the rate of TPNH appearance is propor-

[1] G. J. Brewer and R. J. Dern, *Am. J. Human Genet.* **16**, 472 (1964).

tional to the concentration of the enzyme. The following assay method is a modification of the procedure of Horecker and Smyrniotis.[2]

Assay. For the preparation of erythrocytes, leukocytes and/or platelets for the assay of 6-phosphogluconate dehydrogenase, the procedure as detailed in [25] is employed.

Reagents

6-phosphogluconate, sodium salt,[3] 0.01 *M*
TPN,[3] 0.0025 *M*
$MgCl_2$, 0.1 *M*
Glycylglycine buffer, 0.25 *M*, pH 7.5
Enzyme, aliquot of crude cell lysate sufficient to obtain a change of 0.010–0.050 optical density units per minute. If lysates of cells are prepared by the methods indicated in [25], the aliquots are the same as those indicated for the assay of glucose 6-phosphate dehydrogenase [25].

Procedure. To a 3.0 ml quartz or Corex cuvette having a light path of 1 cm, add 0.1 ml TPN, 0.5 ml $MgCl_2$, 0.5 ml glycylglycine buffer, 0.5 ml 6-phosphogluconate and enzyme solution. Make up to total volume of 2.5 ml with water. Take readings at 340 mμ at 1-minute intervals, starting immediately after addition of enzyme solution.

Definition of Unit of Enzyme Activity. A unit of enzyme activity in erythrocytes is defined as that amount which causes an initial change in 1.0 optical density per minute. The change in optical density units per minute may be converted to micromoles TPNH formed per minute by the equation indicated under definition of unit of enzyme activity for glucose 6-phosphate dehydrogenase [25]. Enzyme activity in red blood cells is referred to hemoglobin concentration.

[2] B. L. Horecker and P. Z. Smyrniotis, Vol. I, p. 323.
[3] Sigma Chemical Company, St. Louis, Missouri.

[28] Mannitol Dehydrogenase (Crystalline) from *Lactobacillus brevis*[1]

By B. L. Horecker

D-Mannitol + DPN$^+$ $\rightleftharpoons$ D-fructose + DPNH + H$^+$

Assay Method

Principle. A spectrophotometric assay, based on the oxidation of DPNH by fructose is employed. The oxidation of DPNH is followed by the change in absorbance at 340 mμ.

Reagents

Fructose stock solution, 1 *M*. Dissolve 1.8 g of fructose (Pfanstiehl) in 10 ml of water and dilute to 10.0 ml. Store in the refrigerator.

DPNH, 10^{-2} *M* in 0.001 *N* NaOH

Buffer solutions, 0.05 *M* sodium acetate buffer, pH 5.35, and 0.05 *M* sodium acetate buffer, pH 6.0. The latter solution is made up to contain 2×10^{-4} *M* mercaptoethanol and 1% bovine serum albumin.

Procedure. To 0.5 ml of water and 0.4 ml of 0.05 *M* acetate buffer, pH 5.35, in a 1.0 ml quartz absorption cell add 0.01 ml of DPNH solution and sufficient diluted enzyme solution to produce an absorbance change of about 0.020 per minute. Start the reaction by addition of 0.1 ml of substrate and read every 30 seconds for 3 minutes. Make the enzyme dilutions in the 0.05 *M* acetate buffer, pH 6.0, containing albumin and mercaptoethanol. The unit of enzyme activity is defined as the amount required to oxidize 1 micromole of DPNH per minute in the spectrophotometric test carried out at 25°.

Purification Procedure

Growth of Cells. Obtain a lyophilized culture of *Lactobacillus brevis* (ATCC 367) from the American Type Culture Collection. Prepare growth medium of the following composition[2]: 0.5% yeast extract, 1.0% nutrient broth, 1.0% sodium acetate·3 H_2O, 0.001% NaCl, 0.001% $FeSO_4$·7 H_2O, 0.02% $MgSO_4$·7 H_2O, 0.001% $MnSO_4$·4 H_2O, and 1.0% fructose. Maintain on 2% agar stabs of this composition and transfer every 15 days.

[1] G. Martinez, H. A. Barker, and B. L. Horecker, *J. Biol. Chem.* **238**, 5 (1963).

[2] H. Gest and J. O. Lampen, *J. Biol. Chem.* **194**, 555 (1952).

Grow the cultures in liquid medium[3] by incubation at 32° without aeration. For preparation of large amounts of cells for enzyme purification, inoculate 10 ml of a 24-hour liquid culture into 1 liter of fresh medium. This is used 24 hours later as inoculum for 16 liters of medium in a 20-liter bottle. Incubate at 32° for about 17 hours, at which time nearly all the fructose will have been consumed, and collect the cells by centrifugation in a Sharples supercentrifuge. Wash the harvested cells twice with 150-ml portions of 0.05 *M* sodium phosphate buffer (pH 6.5) and store the packed cell paste at —16°. About 2.0 g of cells (wet weight) are obtained per liter of culture.

Preparation of Extracts. Perform all operations at 0° unless otherwise stated. Take about one-third of the cells produced in 16 liters of culture for each preparation. Suspend three 3.5-g portions of frozen cell paste in 7.0 ml of 0.005 *M* sodium phosphate buffer (pH 6.5) containing 2×10^{-4} *M* mercaptoethanol and shake with 8 g of glass beads in the Nossal tissue disintegrator[4] for four 20-second periods, cooling the cartridges in ice between each treatment. Centrifuge the suspensions for 10 minutes at 10,000 *g*. Wash the residues twice with a total of 20 ml of the same buffer and combine the washings with the first supernatant solution.

Protamine Fraction. Treat the clear extract with 6.0 ml of 2% protamine sulfate solution and centrifuge after 20 minutes at 0°. Discard the precipitate.

Ammonium Sulfate I. To the supernatant solution add 3.0 ml of 0.5 *M* sodium phosphate buffer (pH 6.5) and 12.85 g of ammonium sulfate. After 10 minutes remove the precipitate by centrifugation and discard it. Treat the clear solution (42 ml) with 2.86 g of ammonium sulfate; collect the precipitate by centrifugation after 10 minutes, and dissolve it in 7.5 ml of 0.05 *M* sodium phosphate buffer, pH 6.5.

Heated Fraction. To the solution add 0.7 ml of 0.5 *M* sodium phosphate buffer (pH 6.5) containing 2×10^{-4} *M* mercaptoethanol. Heat the diluted solution in a water bath at 63–65° for 5 minutes, centrifuge for 10 minutes at 10,000 *g*, and discard the precipitate.

Ammonium Sulfate II. Determine the amonium sulfate concentration in the supernatant solution (heated fraction, 7.3 ml) with a Barnstead PR2 conductivity meter and adjust to 70% saturation by addition of solid ammonium sulfate. After 10 minutes collect the precipitate by centrifugation at 10,000 *g* for 5 minutes. Extract this precipitate with 4.0

[3] The substitution of glucose for fructose in the growth media does not alter the yield of cells but reduces the enzymatic activity in the crude extract to about 5% of that found in the fructose medium.

[4] P. M. Nossal, *Australian J. Exptl. Biol. Med. Sci.* 31, 583 (1953).

ml of 65% saturated ammonium sulfate (pH 7.0), and after 10 minutes again collect the precipitate by centrifugation. Extract successively, as before, with 3.0 ml and 2.0 ml of 60% saturated ammonium sulfate. Combine the last two extracts.

Crystallization. Adjust the ammonium sulfate II solution to pH 6.0 with 0.1 N acetic acid and add a solution of saturated ammonium sulfate until faint turbidity is observed. Allow to stand overnight at 3° and collect the crystals by centrifugation. Dissolve them in 1.6 ml of 0.1 M sodium acetate buffer (pH 6.0) containing 2×10^{-4} M mercaptoethanol (first crystals). Repeat this procedure (second crystals), and, after centrifugation to separate, repeat again (third crystals). Finally add mercaptoethanol (2×10^{-4} M) and ammonium sulfate until crystallization is begun and store the crystal suspension at 3°. The crystals obtained in this manner are very transparent; for the purpose of visualization carry out another crystallization in the presence of 10^{-6} M methylene blue.

Specificity. The enzyme is specific for D-fructose and D-mannitol. It does not catalyze oxidation of the following polyols: glycerol, erythritol, xylitol, D-sorbitol, D-arabitol, L-arabitol, dulcitol, inositol, ribitol, melibiitol, L-glycero-D-glucoheptitol. The enzyme will not catalyze reduction of the following sugars: D-sorbose, D-xylulose, D-ribulose, sedoheptulose, fructose 6-phosphate, D-glucose. TPNH will replace DPNH in the reduction of fructose; with the standard assay conditions the reaction rate is about one-half as fast with this coenzyme. In the reaction between oxidized coenzyme and mannitol at pH 8.6 the rate with TPN is only 2.5% of that with DPN.

Distribution. An enzyme with this specificity has thus far been reported only in *L-brevis,* but it probably occurs in other species of *Lactobacillus* that accumulate mannitol.

Use for Analytical Purposes. The enzyme can be employed for the determination of D-mannitol and D-fructose. For fructose, prepare a reaction mixture containing 0.1 M sodium acetate buffer, pH 6.0, 0.1 mM DPNH, 0.02 mg of mannitol dehydrogenase, and 0.01–0.05 micromole of D-fructose in a total volume of 1.0 ml. Read the absorbance at 340 mμ before and after addition of D-fructose, and calculate the amount present in the assay cuvette from the change in absorbance.[5] For mannitol, prepare a reaction mixture (0.30 ml) containing 0.02 M Tris buffer, pH 8.0, 7.5 mM pyruvate, 2.0 mM DPN, 0.01 mg of lactic dehydrogenase,[6] and 0.2 mg mannitol dehydrogenase. Add 0.5–1.0 micro-

[5] B. L. Horecker and A. Kornberg, *J. Biol. Chem.* **175,** 385 (1948).

[6] Vol. I [67].

mole of mannitol, incubate at room temperature for 3 hours, and measure fructose colorimetrically.[7] D-Mannose can be determined with this procedure after reduction with sodium borohydride.

PURIFICATION PROCEDURE

Fraction	Total volume (ml)	Units/ml	Total units	Specific activity (units/mg)	Recovery (%)
Extract	34	21.3	725	0.75	—
Protamine fraction	35	20.3	690	2.7	95
Ammonium sulfate I	7.5	59.3	430	4.9	59
Heated fraction	7.3	41.8	305	8.6	42
Ammonium sulfate II	5.1	49.0	250	16.7	34
First crystals	1.6	90.7	145	23.7	20
Second crystals	1.6	65.7	105	28.9	14
Third crystals	1.6	50.0	80	27.3	11

Properties

Stability. The crude extracts show a rapid loss of activity at 2°, but this loss can be partially restored by incubation with 10^{-3} M mercaptoethanol at 37°. The purified enzyme is most stable in solution at pH 6.0 in a 0.02 M sodium acetate buffer containing 10^{-3} M mercaptoethanol. Such solutions can be stored for weeks at —16°. The enzyme solutions lose no activity when dialyzed at 3° against oxygen-free water containing 2×10^{-4} M mercaptoethanol for a period of 18 hours. Suspensions of the crystalline enzyme preparation kept in ammonium sulfate at 3° lose about 20% of their activity in 2 months. This loss is largely reversed by the addition of mercaptoethanol to 2×10^{-4} M followed by warming to 37° for 1 hour.

Effect of pH on Reaction Rate. The optimal pH for the conversion of fructose to mannitol, the reaction utilized in the enzyme assay, is 5.35. The optimal pH for the reverse reaction, the conversion of mannitol to fructose, is 8.6.

[7] J. H. Roe, *J. Biol. Chem.* **107,** 15 (1934).

[29] Mannitol (Cytochrome) Dehydrogenase

By N. L. EDSON and D. R. D. SHAW

D-Mannitol + oxidized cytochrome = D-Fructose + reduced cytochrome
Reduced cytochrome + 0.5 O_2 = oxidized cytochrome + H_2O
Sum: D-Mannitol + 0.5 O_2 = D-Fructose + H_2O

D-Mannitol: cytochrome oxidoreductase (EC 1.1.2.2) is extracted from *Acetobacter suboxydans* as a particulate preparation that contains several other enzymatic activities.[1]

Assay Methods

Principles. Mannitol dehydrogenase can be assayed by measuring the rate of oxygen uptake at the optimum pH (5.0). Alternatively, it can be assayed anaerobically by linking the dehydrogenation of mannitol to phenol blue and measuring the rate of reduction of the dye (decrease in absorbance at 655 mμ) at pH 5.4.

Manometric Assay (Warburg Technique)[2]

Reagents

Mannitol, 0.375 *M*. Dissolve 6.83 g of mannitol in water to make 100 ml of solution.
Sodium acetate buffer, 0.1 *M*, pH 5.0
Calcium chloride, 0.25 *M*
Magnesium chloride, 0.25 *M*
KOH solution, 10% (w/v)
Enzyme. The preparation should contain 1 unit to 6 units/ml. Dilute with water as required.

Procedure. Each Warburg cup contains 1.9 ml of acetate buffer, 0.1 ml of $CaCl_2$ solution, 0.1 ml of $MgCl_2$ solution, 0.5 ml of enzyme, and 0.4 ml of mannitol solution (added from the side bulb after equilibration). The center well contains 0.2 ml of KOH solution. Oxygen uptake at 30° is measured at 10-minute intervals for 30 minutes. There is no endogenous consumption of oxygen. A linear relationship between rate of reaction and enzyme concentration holds over the range 0.5–3 units per cup.

Definition of Unit. One unit of enzyme is the amount that causes an

[1] A. C. Arcus and N. L. Edson, *Biochem. J.* 64, 385 (1956).
[2] J. W. Lock, M.Sc. Thesis, University of Otago (1964).

uptake of 1 micromole of oxygen in 10 minutes. The plot of oxygen uptake is linear.

Spectrophotometric Assay[2]

Reagents

Sodium acetate buffer, 0.1 *M*, pH 5.4
Solutions of mannitol, $CaCl_2$ and $MgCl_2$ as above
Phenol blue (4′-dimethylaminophenyl-quinoneimine) obtainable from the British Drug Houses Ltd.
Phenol blue (22.6 mg) is dissolved in 1 ml of 1 *N* acetic acid and diluted with 9 ml of 0.02 *M* sodium acetate buffer, pH 5.4. The pH is adjusted to 5.4 with 0.02 *M* sodium acetate, and the solution is made up to 100 ml with sodium acetate buffer. The solution is stored at 0–1° and discarded after 3 days.
Enzyme. The enzyme preparation is diluted with water to give 0.013–0.26 spectrophotometric unit per milliliter.

Procedure. Each cuvette (1 cm light path) contains 1.6 ml of 0.1 *M* acetate buffer, 0.1 ml of $CaCl_2$ solution, 0.1 ml of $MgCl_2$ solution, 0.1 ml of phenol blue solution, and 0.5 ml of enzyme. The reaction is started by adding 0.4 ml of mannitol solution, and absorbance at 655 mμ is measured at 15-second intervals for 4 minutes. Temperature, 25°. Phenol blue is not reduced in the absence of substrate.

Definition of Unit. One unit of enzyme is defined as the amount that causes a decrease in absorbance of 1.0/minute (initial rate) under the conditions of the assay.

Preparation of the Enzyme

Cultivation of the Bacteria. Acetobacter suboxydans (American Type Culture Collection, No. 621) is grown at 30° for 3 days in a medium containing 0.5% of Difco yeast extract, 0.025 *M* potassium phosphate buffer, pH 5.5, and 2.5% of glycerol (or mannitol or sorbitol). The cultures (100 ml of medium in 500-ml flasks) are aerated by continuous shaking. The cells are harvested by centrifuging and washed three times with water.

Cell-Free Extract. Washed cells (10 g) are suspended in 20 ml of 0.01 *M* sodium phosphate and treated for 30 minutes in a Raytheon 9-kc sonic disintegrator. The cup assembly is maintained at 0–1° by circulating ice water. Unbroken cells and debris are removed by centrifuging at 2° and 20,000 *g* for 15 minutes. The supernatant is red-brown and clear but opalescent in reflected light.

Fractionation. The clarified extract is centrifuged for 2 hours at 4°

and 60,000 *g*. It separates into a red gel and a straw-colored supernatant which is discarded. The gel is washed twice by suspending in 0.02 *M* sodium phosphate buffer, pH 6.9, and recentrifuging. It is suspended in 20 ml of the same buffer and dialyzed against distilled water for 12 hours at 3°. This dispersion consists of ultramicroscopic particles.

Properties

Other Activities. In addition to oxidizing polyols, the preparation catalyzes oxygen consumption in the presence of D-gluconate, D-galactonate, D-xylonate, D-glucose, D-galactose, glycerol, and *n*-, *sec*- and isobutanol.[1,3-5] The nature of any requisite prosthetic group or cofactor other than the cytochromes is not known.

Substrate Specificity. The enzyme oxidizes a wide range of acyclic polyols containing the D-*erythro* configuration (Bertrand-Hudson rule):

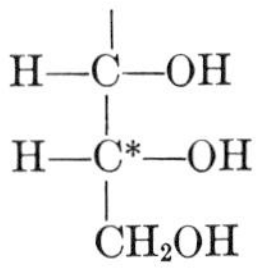

```
        |
    H—C—OH
        |
    H—C*—OH
        |
      CH2OH
```

* Site of oxidation.

The following are oxidized: Erythritol, ribitol, D-arabitol, allitol, D-mannitol, L-gulitol (sorbitol), 6-deoxy-L-gulitol, D-talitol, D-*glycero*-D-*manno*-heptitol, D-*glycero*-D-*galacto*-heptitol, *glycero*-*gulo*-heptitol, *glycero*-*allo*-heptitol, D-*glycero*-D-*gluco*-heptitol, D-*glycero*-D-*altro*-heptitol, and 6-deoxy-L-galactitol (L-fucitol).[1,2,5] The oxidation of L-fucitol is regarded as a special extension of the Bertrand-Hudson rule.[6] Cyclitols are not oxidized.

Enzyme–Substrate Affinity. K_m values determined from Lineweaver-Burk plots under the conditions of the phenol blue assay are: D-mannitol, 3.1×10^{-2} *M*; sorbitol, 2.0×10^{-2} *M*; ribitol, 4.4×10^{-2} *M*; D-arabitol, 1.7×10^{-2} *M*.

Anaerobic Reduction of Hydrogen Acceptors. In the presence of mannitol the enzyme will reduce phenol blue, ferricyanide, phenolindo-2,6-dichlorophenol, *o*-chlorophenolindo-2,6-dichlorophenol, and the endogenous cytochromes.

Stability. The enzyme is stable for several weeks at 0°. If activity decreases during this time, it can be restored by adding Ca^{++} and Mg^{++} in the concentrations employed in the assay.

[3] C. Widmer, T. E. King, and V. H. Cheldelin, *J. Bacteriol.* **71,** 737 (1956).

[4] J. De Ley and A. H. Stouthamer, *Biochim. Biophys. Acta* **34,** 171 (1959).

[5] K. Kersters, W. A. Wood, and J. DeLey, *J. Biol. Chem.* **240,** 965 (1965).

[6] N. K. Richtmeyer, L. C. Stewart, and C. S. Hudson, *J. Am. Chem. Soc.* **72,** 4934 (1950).

[30] D-Mannitol 1-Phosphate Dehydrogenase and D-Sorbitol 6-Phosphate Dehydrogenase from *Aerobacter aerogenes*[1]

By SUSAN B. HORWITZ

$$\text{D-Mannitol 1-P} + \text{NAD} \rightleftarrows \text{D-fructose 6-P} + \text{NADH}$$
$$\text{D-Sorbitol 6-P} + \text{NAD} \rightleftarrows \text{D-fructose 6-P} + \text{NADH}$$

Assay Method

Principle. The method depends on the increase in light absorbance at 340 mμ when NAD is reduced in the presence of either D-mannitol 1-phosphate or D-sorbitol 6-phosphate. The two enzymes, D-mannitol 1-phosphate dehydrogenase and D-sorbitol 6-phosphate dehydrogenase, are completely different enzymes as determined by their physical, chemical, and catalytic properties.

Reagents

D-Mannitol 1-P 0.05 *M* and D-sorbitol 6-P, 0.05 *M*. The barium salts of these two compounds are prepared by the potassium borohydride reduction of D-mannose 6-P and D-glucose 6-P, respectively.[2] The barium salts are converted to the free compound by mixing with Dowex 50 (H^+) until the pH drops to below 2.0. After filtration, the solution is immediately neutralized to pH 7.0. The exact concentration of the hexitol phosphate is determined by measuring inorganic phosphate after acid hydrolysis.[3] The stock solutions of hexitol phosphates are kept frozen in 5-ml aliquots.

NAD, 0.015 *M*

$NaHCO_3$ buffer, 0.1 *M*, pH 9.0 (or 0.1 *M* Tris-HCl, pH 9.0)

Enzyme. The enzyme is diluted with 0.05 *M* bicarbonate buffer, pH 8.0.

Procedure. To a 1-ml cell containing 0.7 ml buffer, 5 micromoles D-mannitol 1-P (or 10 micromoles D-sorbitol 6-P), 1.5 micromoles NAD, and suitably diluted enzyme are added. The final volume is 1.0 ml. The increase in optical density at 340 mμ during the 30- to 90-second interval is used to measure the reaction rate.

Definition of Enzyme Unit and Specific Activity. The enzyme unit is that amount which will produce a change of 0.01 in optical density at

[1] M. Liss, S. B. Horwitz, and N. O. Kaplan, *J. Biol. Chem.* **237**, 1342 (1962).

[2] J. B. Wolff and N. O. Kaplan, *J. Bacteriol.* **71**, 557 (1956).

[3] C. H. Fiske and Y. SubbaRow, *J. Biol. Chem.* **66**, 375 (1925).

340 mμ during the 30- to 90-second interval. Protein is determined by the biuret phenol method.[4] Specific activity is expressed as units per milligram of protein.

Purification Procedure for D-Mannitol 1-P Dehydrogenase

Growth. Aerobacter aerogenes ATCC 8724 was grown for approximately 20 hours with vigorous aeration at 30° in a synthetic medium that contained 0.54% KH_2PO_4, 0.12% $(NH_4)_2SO_4$, 0.04% $MgSO_4 \cdot 7\ H_2O$, and 1.5% D-mannitol. The cells were grown in 25-liter batches.

Step 1. Cells from 150 liters are harvested in Sharples centrifuge, washed with 0.85% saline solution, and suspended in 3 volumes of 0.05 *M* $NaHCO_3$, pH 8.0. The cell suspension is sonicated in 60-ml aliquots for 30 minutes in a 250-watt, 10-kc Raytheon oscillator. All subsequent steps are performed at 4°. The sonic extract is centrifuged for 30 minutes at 25,000 *g*, and the sediment is discarded.

Step 2. Streptomycin sulfate is added to the supernatant solution to bring the concentration to 1%, and the mixture is stirred for 1 hour before centrifugation at 25,000 *g* for 20 minutes.

Step 3. The supernatant solution is adjusted to pH 5.0 with a dilute acetic acid solution, and four volumes of calcium phosphate gel (23 mg of solids per milliliter) prepared according to Swingle and Tiselius[5] are added. The pH of the mixture is readjusted to pH 5.0 with additional acetic acid. The gel mixture is stirred for 30 minutes and then centrifuged for 30 minutes at 5000 *g*. The supernatant is discarded and the packed gel is then eluted several times with 400-ml aliquots of a 0.05 *M* $NaHCO_3$ solution of pH 8.0. In the first elution it is necessary to add alkali to maintain the pH at 8.0.

Step 4. All eluates are combined and the enzyme solution is brought to 45% saturation with $(NH_4)_2SO_4$ and allowed to stand overnight. The material is then centrifuged at 26,000 *g* for 20 minutes, and the supernatant is brought to 65% ammonium sulfate saturation and allowed to stand for 12 hours. The precipitate, recovered by centrifugation at 26,000 *g* for 20 minutes, is suspended in 40 ml of 0.05 *M* $NaHCO_3$, pH 8.0, containing 5×10^{-3} *M* 2-mercaptoethanol. The enzyme loses activity if a sulfhydryl protecting agent is not present in all subsequent purification steps. The material is centrifuged and the supernatant is dialyzed against a large volume of a 0.05 *M* $NaHCO_3$ solution, pH 8.0, containing 5×10^{-3} *M* 2-mercaptoethanol.

[4] O. H. Lowry, N. J. Rosebrough, A. L. Farr, and R. J. Randall, *J. Biol. Chem.* **193,** 265 (1951).

[5] S. M. Swingle and A. Tiselius, *Biochem. J.* **48,** 171 (1951).

Step 5. The dialyzed enzyme solution is placed on a 2.5 × 27-cm DEAE-cellulose column previously equilibrated with 0.05 *M* $NaHCO_3$, pH 8.0. Before and after application of the enzyme, the column is washed with 0.05 *M* $NaHCO_3$, pH 8.0, containing 5×10^{-3} *M* 2-mercaptoethanol. The enzyme is eluted by a linear gradient technique in which the mixing vessel contains 1 liter of 0.05 *M* $NaHCO_3$, pH 8.0, containing 5×10^{-3} *M* 2-mercaptoethanol, and the reservoir vessel contains 1 liter of 0.05 *M* $NaHCO_3$, pH 8.0, containing 0.3 *M* NaCl and 5×10^{-3} *M* 2-mercaptoethanol. Aliquots (15 ml) are collected; tubes 40 through 50, which contain the highest specific activity, are combined.

Step 6. The enzyme solution is dialyzed overnight against 3 liters of a solution containing 0.05 *M* $NaHCO_3$ pH 8.0, $(NH_4)_2SO_4$ 70% of saturation, and 5×10^{-3} *M* 2-mercaptoethanol. The dialyzed material is centrifuged for 20 minutes at 26,000 *g*. The sediment is taken up in 2 ml of 0.05 *M* Tris-HCl buffer, pH 7.6, containing 0.05 *M* NaCl and 5×10^{-3} *M* 2-mercaptoethanol and is centrifuged at 26,000 *g* for 20 minutes. The supernatant solution containing the enzyme is dialyzed overnight against 0.05 *M* Tris-HCl buffer, pH 7.6, containing 0.05 *M* NaCl and 5×10^{-3} *M* 2-mercaptoethanol. This dialyzed enzyme represents an 80-fold purification of D-mannitol 1-P dehydrogenase. This preparation is stable for at least 4 weeks if kept at 4°, but loses activity in several days in the absence of 2-mercaptoethanol or in dilute solution.

The purification procedure is summarized in Table I.

TABLE I
PURIFICATION PROCEDURE FOR D-MANNITOL 1-P DEHYDROGENASE

Step	Total units	Specific activity
1. Sonic extraction	6.3×10^7	3,000
2. Streptomycin treatment	4.5×10^7	5,400
3. Calcium phosphate gel	2.5×10^7	105,000
4. Dialysis, 45–65% $(NH_4)_2SO_4$	1.3×10^7	147,000
5. DEAE-cellulose eluate	8.3×10^6	200,000
6. Dialysis, 0–70% $(NH_4)_2SO_4$	2.7×10^6	240,000

Properties

Stimulation of Enzyme Activity. There is a wide variation in the amount of D-mannitol 1-P dehydrogenase activity present in cell extracts. The activity is dependent upon the carbon source used for growth. The largest amount of activity is present when the cells are grown on D-mannitol. Cells grown on D-sorbitol have one-fifteenth the amount of activity and cells grown on D-glucose have one twenty-fifth the amount of activity as compared to D-mannitol grown cells.

Substrate Specificity. The enzyme is specific for D-mannitol 1-P and NAD.

pH Optimum and Buffer Effects. The pH optimum of D-mannitol 1-P dehydrogenase in 0.07 *M* Tris-HCl buffer is 9.5. The rate observed is twice that observed at this pH with bicarbonate buffer.

Reactivity with NAD Analogs. D-Mannitol 1-P dehydrogenase does not react well with analogs of NAD which have been altered in the nicotinamide portion of the molecule. Alterations in the 6-amino position of the adenine moiety of NAD do not reduce the activity of the enzyme more than 25%.

Sedimentation Constant. $S_{20,w} = 3.06 \times 10^{-13}$.

Purification Procedure for D-Sorbitol 6-P Dehydrogenase

Growth. Cells are grown as described for D-mannitol 1-P dehydrogenase, except that the medium contains 1.5% D-sorbitol as a carbon source for growth.

Step 1. One hundred grams, wet weight, of D-sorbitol grown *A. aerogenes* are suspended in 200 ml of 0.05 *M* $NaHCO_3$, pH 8.0, and sonicated in 60-ml aliquots for 30 minutes in a 250-watt, 10-kc Raytheon oscillator. All subsequent steps are performed at 4°. The sonic extract is centrifuged for 30 minutes at 25,000 *g*, and the supernatant solution is recovered.

Step 2. Two grams of streptomycin sulfate is added, and the solution is stirred for 45 minutes before centrifugation at 25,000 *g* for 20 minutes. The supernatant solution is dialyzed against 6 liters of a 0.05 *M* $NaHCO_3$, pH 8.0, for 12 hours.

Step 3. The dialyzed preparation is chromatographed on two 3.5 × 27-cm DEAE-cellulose columns which have been equilibrated with 0.05 *M* $NaHCO_3$, pH 8.0. Each column is eluted with a linear gradient in which the mixing vessel contains 1 liter of 0.05 *M* $NaHCO_3$, pH 8.0, and the reservoir vessel contains 1 liter of 0.05 *M* $NaHCO_3$, pH 8.0, which is 0.3 *M* with respect to NaCl. Fifteen-milliliter fractions are collected. The major portion of D-sorbitol 6-P dehydrogenase is present in tubes 96 through 106.

Step 4. Tubes 96 through 106 are combined, and 10-ml aliquots are pipetted into 25-ml erlenmeyer flasks which are incubated in a 58° constant temperature water bath and then immediately cooled in ice. This heating step removes all the D-mannitol 1-P dehydrogenase activity.

The turbidity in the heated samples is removed by centrifuging for 40 minutes at 30,000 rpm in a Spinco ultracentrifuge with a No. 30 rotor.

Step 5. The enzyme solution is brought to 35% saturation with $(NH_4)_2SO_4$ and allowed to stand overnight; the precipitate is then removed by centrifugation at 25,000 *g* for 40 minutes. The supernatant solution is adjusted to 50% saturation with $(NH_4)_2SO_4$; after it has stood for 4 hours, the precipitate is collected by centrifugation at 25,000 *g* for 30 minutes.

Step 6. The sediment is taken up in 30 ml of 0.05 *M* $NaHCO_3$, pH 8.0. The enzyme solution is dialyzed overnight against 2 liters of 0.05 *M* $NaHCO_3$, pH 8.0, which is 57% saturated with $(NH_4)_2SO_4$ and contains 5×10^{-3} *M* 2-mercaptoethanol. The volume of the dialyzed material is reduced to one-half by this procedure, and this results in the formation of a flocculent precipitate which is collected by centrifugation at 25,000 *g* for 20 minutes. The precipitate is taken up in 2 ml of 0.1 *M* $NaHCO_3$, pH 8.0, containing 5×10^{-3} *M* 2-mercaptoethanol and is dialyzed against 2 liters of 0.05 *M* Tris-HCl buffer of pH 7.5 containing 0.05 *M* NaCl and 5×10^{-3} *M* 2-mercaptoethanol. This dialyzed enzyme represents a 100-fold purification of D-sorbitol 6-P dehydrogenase. The enzyme when stored frozen in buffer is stable over a period of several weeks.

The purification procedure is summerized in Table II.

TABLE II
PURIFICATION PROCEDURE FOR D-SORBITOL 6-P DEHYDROGENASE

Step	Total units	Specific activity
1. Sonic extraction	1.3×10^6	127
2. Streptomycin treatment	1.2×10^6	208
3. DEAE-cellulose elution	4.9×10^5	1,640
4. Incubation, 58°	4.2×10^5	2,900
5. $(NH_4)_2SO_4$, 35–50%	6.3×10^5 [a]	15,730[a]
6. Dialysis, 0–57% $(NH_4)_2SO_4$	8.5×10^4	13,280

[a] Reaction rate enhanced by $(NH_4)_2SO_4$.

Properties

Stimulation of Enzyme Activity. D-Sorbitol 6-P dehydrogenase activity is present only in cells in which D-sorbitol is used as the carbon source for growth.

Substrate Specificity. The enzyme is specific for D-sorbitol 6-P and NAD. It has very slight activity with D-dulcitol 6-phosphate and NAD.

pH Optimum and Buffer Effects. The pH optimum of D-sorbitol 6-P dehydrogenase in 0.06 *M* Tris-HCl buffer is pH 9.0. The rate observed is thirty times greater than the rate with $NaHCO_3$ buffer at this pH.

Reactivity with NAD analogs. D-Sorbitol 6-phosphate dehydrogenase activity is strongly inhibited by alterations of the 6-amino position of the adenine moiety of NAD.

Sedimentation Constant. $S_{20,w} = 6.0 \times 10^{-13}$.

[31] D-Mannitol 1-Phosphate Dehydrogenase and D-Sorbitol Dehydrogenase from *Bacillus subtilis*[1]

By SUSAN B. HORWITZ

$$\text{D-Mannitol 1-P} + \text{NAD} \rightleftarrows \text{D-fructose 6-P} + \text{NADH}$$
$$\text{D-Sorbitol} + \text{NAD} \rightleftarrows \text{D-fructose} + \text{NAD}$$

Assay Method

Principle. The method depends on the increase in light absorbance at 340 mμ when NAD is reduced in the presence of either D-mannitol 1-P or D-sorbitol. The two enzymes, D-mannitol 1-P and D-sorbitol dehydrogenase are completely distinct enzymes, which have different physical, chemical, and catalytic properties.

Reagents

D-Mannitol 1-P, 0.05 *M* and 0.5 *M* purified D-sorbitol. The barium salt of D-mannitol 1-P is prepared by the potassium borohydride reduction of D-mannose 6-P.[2] The barium salt is converted to the free compound by mixing with Dowex 50 (H^+) until the pH drops to below 2.0. After filtration, the solution is immediately neutralized to pH 7.0. The exact concentration of the hexitol phosphate is determined by measuring inorganic phosphate after acid hydrolysis.[3] The stock solution of hexitol phosphate is kept frozen in 5-ml aliquots.

NAD, 0.015 *M*

Tris-HCl buffer, 0.1 *M*, pH 9.0

Enzyme

Procedure. To a 1-ml cell containing 0.7 ml buffer, 5 micromoles D-mannitol 1-P or 50 micromoles D-sorbitol, 1.5 micromoles NAD, and

[1] S. B. Horwitz, and N. O. Kaplan, *J. Biol. Chem.* **239,** 830 (1964).
[2] J. B. Wolff, and N. O. Kaplan, *J. Bacteriol.* **71,** 557 (1956).
[3] C. H. Fiske, and Y. SubbaRow, *J. Biol. Chem.* **66,** 375 (1925).

suitably diluted enzyme are added. The final volume is 1.0 ml. The increase in optical density at 340 mμ during the 30- to 90-second interval is used to measure the rate of reaction.

Definition of Enzyme Unit and Specific Activivty. One enzyme unit is that amount which will produce a change of 0.01 in optical density at 340 mμ during the 30- to 90-second interval. Protein is determined by the biuret phenol method.[4] Specific activity is expressed as units per milligram of protein.

Purification Procedure for D-Mannitol 1-P Dehydrogenase

Growth. The amount of D-mannitol 1-P dehydrogenase present in *B. subtilis* depends upon the carbon source used for growth. Because of interfering impurities in commercial preparations of hexitols, D-mannitol is crystallized twice from water and D-sorbitol is purified by a modification of the method of Strain.[5] Forty-five grams of D-sorbitol is successively crystallized from 1000 ml, 330 ml, 260 ml, and 200 ml of pyridine. The white crystals, consisting of a mole per mole pyridine-sorbitol complex, are dried over H_2SO_4 at less than 1 mm of pressure for 24 hours. The dried white crystals are kept at 80° in a subliming apparatus at less than 1 mm of pressure for 24 hours. The removal of pyridine is gauged by the loss of weight of the sample and by the absorbancy of the material indicating the presence of pyridine. Thirty per cent of the weight of the sample is lost. The D-sorbitol is then recrystallized from absolute alcohol and dried over H_2SO_4. The yield is 37% of the starting material. D-Mannitol is by far the best carbon source for stimulating the formation of D-mannitol 1-P dehydrogenase. D-Sorbitol, recrystallized to remove any D-mannitol, is still capable of stimulating the synthesis of a significant amount of the enzyme. Relative to the amount of enzyme present in D-glucose grown cells, there is a 70-fold increase in D-mannitol grown cells and a 20-fold increase in D-sorbitol grown cells.

For purification of D-mannitol 1-P dehydrogenase, *Bacillus subtilis* W168[6] is grown in 25-liter batches with aeration for approximately 12 hours at 37°. One liter of the synthetic medium contains 1.5 g KCl, 5.0 g NaCl, 2.5 g NH_4Cl, 4.0 g K_2HPO_4, 1.0 g KH_2PO_4, 50 mg Na_2SO_4, 10 mg $MgSO_4 \cdot 7\ H_2O$, and 1 g of D-mannitol. The medium is adjusted to pH 7.0 with 5 *N* NaOH.

Step 1. Cells from 100 liters of medium are used for enzyme purification. The bacteria are harvested in a Sharples centrifuge and washed

[4] O. H. Lowry, N. J. Rosebrough, A. L. Farr, and R. J. Randall, *J. Biol. Chem.* **193**, 265 (1951).

[5] H. H. Strain, *J. Am. Chem. Soc.* **56**, 1746 (1934).

[6] Culture was obtained from Dr. Julius Marmur, Albert Einstein College of Medicine, Yeshiva University, New York, New York.

with cold sodium chloride–phosphate solution consisting of 0.85% NaCl and 0.02 *M* K_2HPO_4 adjusted to pH 7.0. The bacteria are suspended in 300 ml of 0.05 *M* Tris HCl buffer, pH 8.0. Aliquots (50 ml) are sonicated for 15 minutes in a 250-watt, 10-kc Raytheon oscillator. The sonic extract is centrifuged for 20 minutes at 25,000 *g*, and the precipitate is discarded.

Step 2. Streptomycin sulfate is added to the supernatant fluid to make a final concentration of 1%. After gentle stirring for 10 minutes, the solution remains at 4° for 30 minutes before centrifugation at 25,000 *g* for 20 minutes. The supernatant is dialyzed against 10 liters of 0.05 *M* $NaHCO_3$, pH 8.0, and 1.0 m*M* 2-mercaptoethanol for 2 hours.

Step 3. The dialyzed solution is chromatographed on a 3.5 × 27 cm DEAE-cellulose column that has been equilibrated with 0.05 *M* $NaHCO_3$, pH 8.0. The enzyme is eluted with a linear gradient in which the mixing vessel contains 1 liter of a 0.05 *M* $NaHCO_3$ solution, pH 8.0, and the reservoir vessel contains 1 liter of a 0.05 *M* $NaHCO_3$ solution, pH 8.0, that is 0.3 *M* with respect to NaCl. Twelve-milliliter fractions are collected; the major portion of the enzymatic activity is present in tubes 67 through 78, which are combined.

Step 4. The combined eluate is brought to 50% saturation with $(NH_4)_2SO_4$. After it has stood for 1 hour, the material is centrifuged at 25,000 *g* for 20 minutes. The precipitate is taken up in 20 ml of 0.05 *M* $NaHCO_3$, pH 8.0, and the fraction is dialyzed against 2 liters of 0.05 *M* $NaHCO_3$, pH 8.0, and 1.0 m*M* 2-mercaptoethanol for 1 hour.

Step 5. The dialyzed solution is chromatographed on a 2.1 × 11 cm DEAE-cellulose column that has been equilibrated with 0.05 *M* $NaHCO_3$, pH 8.0. The column is eluted with a linear gradient in which the mixing vessel contains 500 ml of 0.05 *M* $NaHCO_3$, pH 8.0, and the reservoir vessel contains 500 ml of 0.05 *M* $NaHCO_3$, pH 8.0, that is 0.3 *M* with respect to NaCl. Twelve-milliliter fractions of eluate are collected. Tube 26 contains the major portion of the enzyme. This purified enzyme is somewhat unstable losing 20% of its activity in 2 days at 4°.

The purification procedure is summarized in Table I.

TABLE I
PURIFICATION PROCEDURE FOR D-MANNITOL 1-P DEHYDROGENASE

Step	Total units	Specific activity
1. Sonic extraction	1.0×10^7	2,094
2. Streptomycin treatment	6.1×10^6	2,136
3. DEAE-cellulose eluate I	2.6×10^6	5,357
4. Dialysis 50–65% $(NH_4)_2SO_4$	1.5×10^6	30,200
5. DEAE-cellulose eluate II	3.8×10^5	124,194

Properties

Substrate Specificity. The enzyme is highly specific for D-mannitol 1-P and NAD.

Reaction with NAD Analogs. Any change in the nicotinamide moiety of NAD results in essentially no activity with D-mannitol 1-phosphate dehydrogenase. The enzyme is much less sensitive to alterations in the adenine moiety.

pH Optimum. The oxidation of D-mannitol 1-P is optimal at pH 10.0 in both 0.1 M Tris-HCl buffer and 0.1 M $NaHCO_3$ buffer.

Purification Procedure for D-Sorbitol Dehydrogenase

Growth. D-Sorbitol is the only compound capable of stimulating the production of D-sorbitol dehydrogenase in *B. subtilis. B. subtilis* W168 is grown as described for D-mannitol 1-P dehydrogenase, except that the medium contains 0.1% purified D-sorbitol instead of D-mannitol. To obtain the highest yield of enzyme, cells are harvested in the early log phase when the cultures are at an optical density of 0.15 at 600 mμ (18-mm I.D. tube) on the Coleman spectrophotometer.

Step 1. Cells from 2 liters of medium are harvested in a Sharples centrifuge and washed with cold sodium–phosphate solution consisting of 0.85% NaCl and 0.02 M K_2HPO_4 adjusted to pH 7.0. The cells are sonicated for 15 minutes in 20 ml of 0.1 M $NaHCO_3$, pH 8.0, in the presence of 1.0 mM 2-mercaptoethanol. The material is centrifuged at 4° at 25,000 g for 20 minutes, and the precipitate is discarded.

Step 2. The supernatant is divided in half, and each of the two fractions is placed in a 125-ml erlenmeyer flask. The two flasks are put in a 54° water bath and gently swirled for 6 minutes. The fractions are then combined and frozen overnight. Heating at 54° for 6 minutes inactivates the D-mannitol 1-P dehydrogenase but has no effect on D-sorbitol dehydrogenase.

Step 3. The heated extract is thawed and 10.4 ml of cold saturated $(NH_4)_2SO_4$, pH 7.0, is added dropwise (52% of saturation). After gentle mixing, the solution is allowed to stand in the cold for 1 hour before centrifugation for 20 minutes at 25,000 g. The supernatant is then dialyzed for 1 hour against 3 liters of 0.05 M $NaHCO_3$, pH 8.0, containing 1.0 mM 2-mercaptoethanol.

Step 4. The dialyzed preparation is adjusted to 65% saturation by adding 11.5 ml of saturated $(NH_4)_2SO_4$ solution. This material remains overnight in the cold and is then centrifuged at 25,000 g for 20 minutes. The precipitate is taken up in 2 ml of 0.05 M $NaHCO_3$, pH 8.0, and is dialyzed for 1 hour against 0.05 M $NaHCO_3$, pH 8.0, containing 1.0 mM 2-mercaptoethanol. This dialyzed enzyme is unstable and is used the same day it is prepared.

The purification procedure is summarized in Table II.

TABLE II
PURIFICATION PROCEDURE FOR D-SORBITOL DEHYDROGENASE

Step	Total units	Specific activity
1. Sonic extraction	12,300	123
2. Heated, 54°	12,300	123
3. 52% $(NH_4)_2SO_4$ supernatant	8,000	573
4. 52–65% $(NH_4)_2SO_4$ precipitate	4,800	1,200

Properties

Substrate Specificity. D-Sorbitol dehydrogenase in the presence of NAD will oxidize D-sorbitol, L-iditol, xylitol, allitol, and adonitol. D-sorbitol is oxidized to D-fructose.

Reaction with NAD Analogs. D-Sorbitol dehydrogenase reacts well with D-sorbitol and 3-acetylpyridine adenine dinucleotide, 3-thionicotinamide adenine dinucleotide, and nicotinamide hypoxanthine dinucleotide.

pH Optimum. The oxidation of D-sorbitol and xylitol is optimal at pH 10 in both 0.1 M Tris-HCl buffer and 0.1 M $NaHCO_3$ buffer.

[32] Sorbitol Dehydrogenase from Spermatozoa

By TSOO E. KING and T. MANN

Assay Method

Principle. The activity of sorbitol dehydrogenase is determined from the forward reaction of Eq. (1).

$$\text{Sorbitol} + \text{DPN}^+ \rightleftharpoons \text{fructose} + \text{DPNH} + \text{H}^+ \tag{1}$$

In the presence of excess sorbitol, the plot of the initial velocity measured by increase of the absorbancy at 340 mμ due to the formation of DPNH is linear with the amount of the enzyme present in the system. However, when ΔA_{340} is greater than 0.25 per minute, deviation of the linearity occurs.

Reagents

DPN, 3.3 m*M*
Tris-HCl buffer, 0.33 *M*, pH 8.6
Sorbitol, 0.5 *M*
Enzyme, diluted with 0.035 *M* phosphate buffer, pH 7.0, so that the sample being assayed contains an amount of enzyme in 0.05 ml which causes ΔA_{340} of 0.05–0.25 in the assay procedure described below.

Procedure. The reaction mixture in both the experimental and blank cuvettes (1-cm light path) contains 0.3 ml DPN, 0.3 ml Tris buffer, 0.3 ml sorbitol, and water to a final volume of 2.95 ml. The assay is run at room temperature. The reaction is started by addition of 0.05 ml enzyme to the experimental cuvette. The increase in absorbance at 340 mμ is observed at 30-second intervals for 3 minutes.

Purification Procedure

1. Collection of Ram Spermatozoa. Ram semen is used as the starting material for the extraction of the enzyme. There are several accepted methods for the collection of ram semen. The one to be recommended depends on the use of the so-called artificial vagina.[1] The artificial vagina (which may be purchased from Holborn Surgical Instrument Company, Ltd., London, England) consists of two cylinders; one the outer casing, about 20 cm × 5 cm, made of brass or heavy rubber hose; the other, an inner rubber sleeve, about 28 cm × 4 cm, made of fine-gauge latex rubber, which is folded over both ends of the outer casing. Warm water (42°) is introduced through the filling tap, under slight pressure (equivalent to approximately 50 cm mercury) into the space between the outer and inner cylinder. The inside of the inner sleeve is lightly lubricated with white vaseline. A test tube for collecting the semen is attached at one end of the artificial vagina. The other end is left open, and it is through that opening that the ram (specially trained for that purpose) is allowed to ejaculate into the artificial vagina.

A single ram-ejaculate obtained by that method of collection has a volume of 0.5–2 ml, and contains on the average 3,000,000 spermatozoa per microliter. A convenient quantity of ram semen for the preparation of the enzyme is 10 ml. For that purpose, ejaculates have to be collected from several rams and pooled.

2. Separation of the Spermatozoa from the Seminal Plasma. Ram

[1] J. P. Maule, "The Semen of Animals and Artificial Insemination," p. 221. Commonwealth Agricultural Bureaux: Farnham Royal, Buckinghamshire, England, 1962.

semen (about 10 ml) is diluted with 4 volumes of Ringer solution at room temperature, centrifuged at about 700 *g* for about 15 minutes. The supernatant liquid, which is the diluted seminal plasma, is discarded. The sedimented spermatozoa are washed three times, each time with approximately 10 ml of fresh Ringer solution; the last centrifugation has to be sharp (usually 15 minutes at 15,000 *g* in a Servall centrifuge) in order to obtain firmly packed spermatozoa and to keep the solvent retained by the pellet at a minimum. The Ringer solution used for that purpose is calcium free and made by mixing 100 ml 0.9% NaCl, 4 ml 1.15% KCl, 1 ml 2.11% KH_2PO_4, 1 ml 3.82% $MgSO_4 \cdot 7\ H_2O$, and 2 ml of a 1.3% solution of $NaHCO_3$ which has been saturated with CO_2.

3. Disintegration of Spermatozoa.[2] The mass from the last centrifugation is scooped out, transferred to a precooled porcelain mortar standing in an ice-water bath, and ground with levigeted aluminum oxide such as Alumina No. 301, manufactured by the Aluminum Company of America. Aluminium oxide powder, 5 g, is added to the sample of spermatozoa derived from about 10 ml of ram semen; 0.035 *M*-phosphate buffer, pH 7, is then added dropwise in a quantity sufficient to produce, on grinding, a thick, doughy mass. Vigorous grinding is continued for 1.5 minutes, and 1 minute is allowed for scraping off the mass from the sides of the pestle and mortar. This treatment is repeated seven or eight times and is completed in about 20 minutes.

4. Extraction and Partial Purification of Sorbitol Dehydrogenase. All subsequent manipulations are conducted at 0–4° unless otherwise specified. When the grinding is over, more 0.035 *M* phosphate buffer is added, so that altogether the volume of phosphate used for the sample derived from 10 ml semen is about 30 ml. The mixture is left in the ice bath for about an hour, with occasional stirring, and then centrifuged for 20 minutes at 2000 *g*. The opalescent extract thus obtained is centrifuged for a further 45 minutes in a Spinco preparative centrifuge (Model L) at 40,000 rpm, yielding a water-clear and completely transparent solution. The above procedure has been found equally suitable for samples ranging from 4 to 40 ml semen.

The clear solution thus obtained is treated with a minimum amount of 1% neutralized protamine sulfate solution.[3] The amount of protamine

[2] This as well as the extraction of sorbitol dehydrogenase is based on T. E. King and T. Mann, *Proc. Roy. Soc.* **B151,** 226 (1959), and unpublished observations.

[3] This and subsequent steps are usually performed from a pool of several batches of samples. The step of the protamine precipitation is advisedly first carried out with a small aliquot in order to find the minimal amount required. Alternatively it may be done by serial additions of protamine solution followed by centrifugation until practically no more precipitate is formed. However, even a slight excess of protamine must be avoided.

required varies from batch to batch. After it has stood for 15 minutes in ice bath, the voluminous precipitate is removed by centrifugation. The supernatant fraction is mixed with solid ammonium sulfate to 0.25 saturation. The precipitate is removed by centrifugation and discarded. The supernatant liquid is further treated with ammonium sulfate to 0.60 saturation. The colorless precipitate harvested by centrifugation is dissolved in a minimum amount of 0.035 *M* phosphate buffer, pH 7.0. The good preparations up to this stage show an activity of approximately 5 micromole sorbitol oxidized per minute per milligram of protein.

Properties of the Enzyme

Stability. The soluble enzyme is stable for at least 3 days at 4° at pH 6.5–7.0 in phosphate buffer. At pH 8.2 and 8.6 in Tris buffer, a decrease of about 25% after storage for 24 hours has been observed.

Specificity. In the oxidation of sugar alcohols the sorbitol dehydrogenase from ram spermatozoa shows a specificity similar to the enzyme obtained from rat liver,[4,5] guinea pig liver,[6] and guinea pig seminal vesicle.[7] Data relating to the specificity of the sperm enzyme are listed in the table along with those for the dehydrogenase from seminal vesicle.

SPECIFICITY OF SORBITOL DEHYDROGENASE IN THE OXIDATION OF SUGAR ALCOHOLS

Substrate	Source of enzyme	
	Ram spermatozoa[a]	Guinea pig seminal vesicle[b]
D-Sorbital	100	100
L-Iditol	90	101
D-Xylitol	79	51
D-Ribitol	36	76
D-Mannitol	17	5
D-Arabitol	4	0
meso-Inositol	0	0
meso-Erythritol	0	—

[a] Substrate concentration 0.033 *M* in the assay system of pH 8.6. See T. E. King and T. Mann, *Proc. Roy. Soc.* **B151**, 226 (1959).

[b] Substrate concentration 0.013 *M* (except for mannitol, 0.033 *M*) in the assay system of pH 8.9. See H. G. Williams-Ashman, J. Banks, and S. K. Wolfson, *Arch. Biochem. Biophys.* **73,** 485 (1957).

[4] R. L. Blakley, *Biochem. J.* **49,** 257 (1951).

[5] H. G. Williams-Ashman and J. Banks, *Arch. Biochem. Biophys.* **50,** 513 (1954).

[6] O. Touster, V. H. Reynolds, and R. M. Hutcheson, *J. Biol. Chem.* **221,** 697 (1956).

[7] H. G. Williams-Ashman, J. Banks, and S. K. Wolfson, *Arch. Biochem. Biophys.* **73,** 485 (1957).

The relative specificity for the reduction of various sugars to sugar alcohols, i.e., the reverse reaction of Eq. (1), for the sperm enzyme has been found as follows: D-fructose, 100; D-sorbose, 38; L-xylose, 2; and D-lyxose, 2; whereas D-ribose, 2-deoxy-D-glucose, D-mannose, and D-glucose are inactive.

Michaelis constants. It has been found that $K_m^{\text{sorbitol}} = 9.8$ mM and $K_m^{\text{idotol}} = 9.0$ mM.

[33a] Polyol Dehydrogenases of *Candida utilis*[1]

I. DPN-Linked Dehydrogenase

By M. CHAKRAVORTY and B. L. HORECKER

$$\begin{array}{c} H_2COH \\ | \\ CHOH \\ | \\ (CHOH)_n \\ | \\ H_2COH \end{array} + DPN \rightleftharpoons \begin{array}{c} H_2COH \\ | \\ C{=}O \\ | \\ (CHOH)_n \\ | \\ H_2COH \end{array} + DPNH + H^+$$

Assay Method

Principle. A spectrophotometric assay, based on the reduction of DPN by xylitol, is employed. The reaction is followed by the change in absorbance at 340 mμ.

Reagents

Xylitol solution (0.5 M). Dissolve 1.5 g of xylitol in water and dilute to 100 ml. Store at −15°.

DPN (0.034 M). Commercial DPN is dissolved in water.

Glycine buffer, 0.1 M, pH 8.6. This is prepared each day to contain 10^{-2} M mercaptoethanol.

Procedure. Place 0.9 ml of glycine–mercaptoethanol buffer, 0.01 ml of DPN solution, and 0.1 ml of xylitol solution in a cuvette and add water and enzyme to give a final volume of 1.0 ml and a change in absorbance of less than 0.01 per minute. Start the reaction by addition of enzyme and take readings at 340 mμ at 1-minute intervals. Make the enzyme dilution in the glycine-mercaptoethanol buffer.

Definition of Unit. One unit of enzyme is defined as the quantity re-

[1] M. Chakravorty, L. A. Veiga, M. Bacila, and B. L. Horecker, *J. Biol. Chem.* **237**, 1014 (1962).

quired to reduce 1 micromole of DPN per minute at 25° under the conditions of the assay.

Purification Procedure

Preparation of Extract. Perform all operations at 0° unless otherwise stated. Suspend 24 g of dry *C. utilis*[2] in 72 ml of 0.1 *M* $NaHCO_3$ containing 10^{-3} *M* mercaptoethanol and allow to autolyze at 25° for 15 hours. Collect the autolyzate by centrifugation and reextract the residue with 72 ml of 0.1 *M* $NaHCO_3$ containing mercaptoethanol. Combine the two extracts (crude extract).

Ammonium Sulfate Fractionation. Treat the crude extract (114 ml) with 11.4 ml of 1 *M* $MnSO_4$, allow to stand in the cold for 30 minutes, and centrifuge. Discard the precipitate. To the supernatant solution (120 ml) slowly add 23.3 g of solid ammonium sulfate with stirring. Keep the solution at 0° for 10 minutes, centrifuge, and discard the precipitate. To the supernatant solution (126 ml) add 14.9 g of solid ammonium sulfate. Collect the precipitate immediately by centrifugation and dissolve it in 11.4 ml of glycylglycine buffer (0.1 *M*, pH 7.4) containing 10^{-3} *M* mercaptoethanol. Dialyze this fraction against 3 liters of 0.1 *M* sodium acetate for 1 hour (dialyzed ammonium sulfate fraction I).

Acetone Fractionation. This step should be carried out on a small scale, using acetone cooled to —15°. To 9 ml of the dialyzed ammonium sulfate fraction I add 4.64 ml of acetone over a period of 4–5 minutes, while cooling the solution in an alcohol bath at —15° so that the final temperature is about —7°. Keep the mixture for 2 minutes and centrifuge briefly at —10°. Discard the precipitate. Treat the supernatant solution (11 ml) in a similar way with 1.1 ml of acetone. Collect the precipitate by centrifugation and suspend it in 2 ml of glycylglycine buffer (0.1 *M*, pH 7.3) containing 10^{-3} *M* mercaptoethanol. Remove the insoluble residue by centrifugation (acetone fraction).

Second Ammonium Sulfate Fractionation. To the acetone fraction (2.1 ml) add 822 mg of ammonium sulfate and remove by centrifugation the precipitate that forms. To the supernatant solution add 122 mg of ammonium sulfate and collect the precipitate. Dissolve it in 0.5 ml of glycylglycine buffer (0.1 *M*, pH 7.4) containing 10^{-3} *M* mercaptoethanol. Dialyze this solution against 2 liters of 0.1 *M* sodium acetate for 30 minutes (ammonium sulfate fraction II).

The purification procedure is summarized in the table.

[2] Low temperature-dried *C. utilis* can be obtained from the Lake States Yeast Corp., Rhinelander, Wisconsin.

PURIFICATION PROCEDURE FOR DPN-LINKED DEHYDROGENASE

Fraction	Total volume (ml)	Units/ml	Total units	Protein (mg/ml)	Specific activity (units/mg)	Recovery (%)
Crude extract	116	89	1040	2.5	0.36	100
Dialyzed ammonium sulfate fraction I	21	40	850	3.8	1.05	82
Acetone fraction	4.9[a]	100	490	18.9	5.3	47
Ammonium sulfate fraction II	1.9[a]	163	310	10	16.3	30

[a] Corrected to represent the total quantity obtained in a preparation with 24 g of dried *Candida*.

Properties

Specificity. The enzyme is most active with xylitol, but a number of other polyols are good substrates. The relative activities (xylitol = 100) were as follows: D-sorbitol, 87; L-iditol, 41; D-altritol, 32; D-*glycero*-D-*gluco*-heptitol, 27; L-*glycero*-D-*gluco*-heptitol, 27; D-mannitol, 23; ribitol, 23; D-arabitol, 7; L-glucitol, 5. Polyols with 3 or 4 carbon atoms (glycerol and erythritol) do not serve as substrates. No activity was observed with L-mannitol, D-iditol, dulcitol, and a number of heptitols other than those listed. The relative activity with xylitol and D-sorbitol remains unchanged during the purification procedure, but the activity toward D-mannitol and ribitol falls off by about one-half. However, competition experiments with the latter substrates suggest that a single enzyme is responsible for activity with all four of these substrates, as well as with the others mentioned.

The enzyme catalyzes the oxidation of DPNH by a number of ketoses, including D-xylulose, D-fructose, and D-ribulose. Neither D- nor L-sorbose reacts, and it has been demonstrated that the reaction product with both D-mannitol and D-sorbitol is D-fructose. The most rapidly reduced substrate is D-xylulose. The purified enzyme preparations show no activity with TPN or TPNH.

Stability of the Enzyme Preparations. The dialyzed ammonium sulfate fraction I can be kept frozen for months with little loss in activity. The final dialyzed preparation loses activity gradually on repeated freezing and thawing but can be stored for long periods at —15° without loss of activity. The addition of mercaptoethanol is essential to preserve activity.

Activators and Inhibitors. The enzyme is inactivated when it is incubated with EDTA at concentrations as low as 10^{-3} *M*. Inactivation is progressive and depends on the concentration of EDTA at the time of incubation. DPNH, but not DPN, protects against this inactivation by EDTA. Metal ions, such as Mg^{++}, Mn^{++}, or Ca^{++} are without effect.

Effect of pH. The rate of reaction with xylitol as substrate is maximal at pH 9.0. At pH 7.0 the rate is about one-third of that observed at 9.0.

Distribution. DPN polyol dehydrogenases have been described in rat liver[3] and other mammalian sources.[4-8] A DPN-xylitol dehydrogenase has been described in guinea pig liver.[9] In microorganisms the DPN-linked enzyme has been reported in *Acetobacter suboxydans* and *C. utilis*,[10] in *Penicillium chrysogenum*,[11,12] and in xylose-grown *C. albicans*.[13] The enzyme originally discovered by Arcus and Edson[10] is probably identical to that described here.

[3] R. L. Blakley, *Biochem. J.* **49**, 257 (1951).
[4] J. McCorkindale and N. L. Edson, *Biochem. J.* **57**, 518 (1954).
[5] H. G. Williams-Ashman and J. Banks, *Arch. Biochem. Biophys.* **50**, 513 (1954).
[6] T. E. King and T. Mann, *Proc. Roy. Soc. London,* **B151**, 226 (1959).
[7] J. Hickman and G. Ashwell, *J. Biol. Chem.* **234**, 758 (1959).
[8] H. G. Hers, *Biochim. Biophys. Acta* **37**, 120 (1960).
[9] S. Hollmann and O. Touster, *J. Am. Chem. Soc.* **78**, 3544 (1956).
[10] A. C. Arcus and N. L. Edson, *Biochem. J.* **64**, 385 (1956).
[11] C. Chiang and S. G. Knight, *Nature* **188**, 79 (1960).
[12] C. Chiang and S. G. Knight, *Biochim. Biophys. Acta* **46**, 271 (1961).
[13] L. A. Veiga, M. Bacila, and B. L. Horecker, *Biochem. Biophys. Res. Commun.* **2**, 440 (1960).

[33b] Polyol Dehydrogenases of *Candida utilis*[1]

II. TPN-Linked Dehydrogenase[2]

By B. M. Scher and B. L. Horecker

$$\begin{array}{c} HC{=}O \\ | \\ HCOH \\ | \\ (CHOH)_n \\ | \\ H_2COH \end{array} + TPNH + H^+ \rightleftharpoons \begin{array}{c} H_2COH \\ | \\ HCOH \\ | \\ (CHOH)_n \\ | \\ H_2COH \end{array} + TPN$$

[1] M. Chakravorty, L. A. Veiga, M. Bacila, and B. L. Horecker, *J. Biol. Chem.* **237**, 1014 (1962).
[2] This material was taken from a thesis presented in partial fulfillment for the degree of Doctor of Philosophy at New York University, 1965. See also this volume [37].

Assay Method

Principle. A spectrophotometric assay, based on the oxidation of TPNH by D-xylose, is employed. The reaction is followed by the change in absorbance at 340 mμ.

Reagents

D-Xylose stock solution (1 *M*). Dissolve 1.5 g of D-xylose in 8 ml of water and dilute to 10 ml. Store at —15°.

TPNH, 10^{-2} *M*. Prepare fresh daily in 10^{-3} *M* NaOH.

Buffer solutions: 0.25 *M* glycylglycine-KOH, pH 7.4; 0.05 *M* potassium phosphate, pH 6.2

Procedure. To 0.8 ml of water and 0.1 ml of 0.25 *M* glycylglycine buffer, pH 7.4, in a 1.0-ml quartz absorption cell add 0.01 ml of TPNH solution and sufficient diluted enzyme solution to produce an absorbance change of about 0.01–0.03 per minute. Begin the reaction by the addition of 0.1 ml of substrate and read the absorbance every minute for 6 minutes. Make the enzyme dilutions in 0.05 *M* potassium phosphate buffer, pH 6.2.

Definition of Unit. A unit of enzyme activity is defined as the quantity required to oxidize 1 micromole of TPNH per minute at 25° under the conditions of assay.

Purification Procedure

Preparation of Extracts. Autolyze 1.0 kg of frozen cells of *Candida utilis* (Lake States Yeast Corporation, Rhinelander, Wisconsin) with 50 ml of 1.0 *N* sodium acetate, pH 4.6, at 37° for 5 hours. Centrifuge the autolyzate at 30,000 *g* for 10 minutes. Bring the clear supernatant solution to pH 6.2 by the addition of 53 ml of 1.0 *N* $NaHCO_3$ (extract).

Acetone Fraction. To the extract (480 ml) add 58.5 ml of 4 *N* sodium acetate buffer, pH 6, to bring the acetate concentration to 0.53 *M*. To this solution, in a dry ice-ethanol bath at about —10°, add, with constant mechanical stirring, 335 ml of acetone which has been chilled to —70°. During the addition maintain the solution above the freezing point; the final temperature should be —6°. Keep the solution for 10 minutes at this temperature and centrifuge at 14,000 *g* in a centrifuge cooled to —10°. Discard the precipitate. Treat the supernatant solution as before with 93 ml of cold acetone, so that the final temperature of the solution is —12° to —14°. Collect the precipitate by centrifugation for 3 minutes at —10° at 14,000 *g*, evacuate the centrifuge bottles containing the precipitate for about 1 minute to remove excess acetone and suspend the precipitate in 62 ml of glass-distilled water at 0°. To the supernatant solution from the acetone fractionation again add 92 ml of cold acetone and after 10 minutes collect the third precipitate and suspend it in 62 ml of cold glass-distilled water. Assay the second and third acetone fractions; if they are of comparable specific activity, combine them. To the combined fractions (124 ml) add 124 ml of saturated ammonium sulfate solution (saturated at 0°, the pH adjusted to 7.0 with NH_4OH). Allow the suspension to stand in ice for 15 minutes and centrifuge at 30,000 *g*. Discard the precipitate. To the clear supernatant solution (238 ml) add 46.1 g of solid ammonium sulfate. Allow the suspension to stand in ice for 15 minutes, collect the precipitate by centrifugation and dissolve it in 56 ml of 0.05 *M* potassium phosphate buffer, pH 6.2, containing 5×10^{-3} *M* 2-mercaptoethanol (acetone fraction).

Fractionation on DEAE Columns. Equilibrate a Sephadex G-50 (coarse grade beads) column (2.5 cm × 49 cm) with the potassium phosphate (0.05 *M*)–mercaptoethanol (5×10^{-3} *M*) buffer (pH 6.2). Layer one-half of the acetone fraction (33 ml) onto the column and elute with 250 ml of the same buffer. Collect 15-ml fractions at a flow rate of 4 ml per minute. Assay the fractions for enzyme activity and determine their ammonium sulfate concentration with a Barnstead PR-2 conductivity meter. Pool the active fractions which contain less than 0.006 *M* ammonium sulfate. Layer the pooled G-50 fractions onto a DEAE-Sephadex A-50 column (2.5 × 39 cm) which had previously been equilibrated with the potassium phosphate-mercaptoethanol buffer. Wash the column with 400 ml of 0.05 *M* KCl in the same buffer. Elute the enzyme activity with 800 ml of 0.14 *M* KCl in this buffer. During the elution collect fractions of 11.5 ml volume at a flow rate of about 0.3 ml per minute. Pool the active fractions (470 ml) and add 153 g of solid ammonium sulfate. After 15 minutes, centrifuge the solution at 30,000 *g* for 10 minutes and discard the precipitate. Treat the supernatant solution (525 ml) with 488 g of solid ammonium sulfate. After 15 minutes, collect the precipitate by centrifugation at 30,000 *g* for 10 minutes and dissolve it in 13 ml of potassium phosphate–mercaptoethanol buffer (DEAE fraction).

ECTEOLA-Cellulose Chromatography. Layer 2 ml of the DEAE fraction onto a column of Sephadex G-50 (coarse beads, 0.8 × 30 cm) which has been equilibrated with potassium phosphate–mercaptoethanol buffer, pH 6.2. Elute the activity with 15 ml of the same buffer. Combine the active fractions having an ammonium sulfate concentration of 0.006 *M* or less, as before. Layer the combined fractions (4.0 ml) onto an ECTEOLA-cellulose column (1.0 cm × 29 cm). Wash the column with

90 ml of potassium phosphate-mercaptoethanol buffer, pH 6.2, until the absorbance of the eluate at 280 mμ is 0.04 or less. Elute the activity with 90 ml of 0.06 *M* KCl in the same buffer. Collect fractions of 6.2 ml each at a flow rate of about 0.3 ml per minute. Pool the active fractions (18.6 ml) and treat them with 5.25 g of ammonium sulfate. After 15 minutes in ice, centrifuge the solution, discard any precipitate which has formed, and add to the supernatant solution (19.3 ml) 4.44 g of ammonium sulfate. After 15 minutes collect the precipitate by centrifugation and dissolve it in 1 ml of potassium phosphate-mercaptoethanol buffer, pH 6.2 (ECTEOLA fraction).

The purification procedure is summarized in the table.

PURIFICATION PROCEDURE FOR TPN-LINKED DEHYDROGENASE

Fraction	Total volume (ml)	Units/ml	Total units	Specific activity (units/mg)	Recovery (%)
Extract	480	116	55,600	1.2	100
Acetone fraction	67	680	45,600	3.0	82
DEAE fraction	13.5	1200	16,200[a]	19.2	58
ECTEOLA fraction	1.1	1380	1,518[b]	36.8	38

[a] This value represents the amount obtained from one-half of the total starting material.

[b] This value represents the amount obtained from about one-fourteenth of the total starting material.

Properties

Stability. The crude extracts are stable for several days if they are quick-frozen and stored at —14°. The acetone fraction and DEAE fraction may be stored for months under these conditions. The most highly purified enzyme preparations are stable for at least several weeks at —15°.

Effect of pH on the Reaction Rate. The optimal pH range for the conversion of D-xylose to xylitol is 5.5–7.6.

Specificity. The enzyme catalyzes the reduction of a number of aldoses, including the following (D-xylose = 100): D-glyceraldehyde, 200; D-erythrose, 180; L-arabinose, 136; D-ribose, 42.5; D-galactose, 45; D-glucose, 13; D-glycero-D-galactoheptose, 12. D-Mannose, 2-deoxy-D-glucose, and 2-deoxy-D-ribose were poor substrates, showing less than 5% activity relative to D-xylose. The enzyme showed no activity with the following aldoses: L-glyceraldehyde, L-xylose, D-arabinose, D-lyxose, D-fructose, D-glucosamine, D-galactosamine, and D-ribose 5-phosphate.

The following polyols were found to be oxidized by TPN in the presence of the enzyme (xylitol = 100): galactitol, 187; L-arabitol, 167; sorbitol, 146; ribitol, 57.5; erythritol, 40; glycerol, 36.

The enzyme is specific for TPN or TPNH and shows no activity with DPN or DPNH.

Distribution. Enzymes with similar specificities have been described in *Candida albicans*,[3] *Penicillium chrysogenum*,[4] *Penicillium* sp., *Aspergillus* sp., *Rhodotorula* sp.,[5] sheep seminal vesicle and placenta,[6] rat lens,[7] and calf lens.[8]

[3] L. A. Veiga, M. Bacila, and B. L. Horecker, *Biochem. Biophys. Res. Commun.* **2**, 440 (1960).
[4] C. Chiang, C. J. Sih, and S. G. Knight, *Biochim. Biophys. Acta* **29**, 664 (1958).
[5] C. Chiang and S. G. Knight, *Nature* **188**, 79 (1960).
[6] H. G. Hers, "Le Métabolisme du Fructose." Editions Arscia, Brussels, 1957.
[7] R. Van Heyningen, *Biochem. J.* **73**, 197 (1959).
[8] S. Hayman and J. H. Kinoshita, *J. Biol. Chem.* **240**, 877 (1965).

[34] Polyol Dehydrogenases of *Gluconobacter*

By K. KERSTERS and J. DE LEY

According to the rule of Bertrand-Hudson, growing cultures of acetic acid bacteria oxidize only L-*erythro* polyols (I) to the corresponding ketoses (II).[1-3]

```
    CH2OH            CH2OH
      |                |
HO—C—H              C=O
      |       →        |
HO—C—H          HO—C—H
      |                |
    (I)³             (II)
```

These oxidations are catalyzed by a particulate L-*erythro* dehydrogenase, located on the cytoplasmic membrane of the cells.[4] Moreover four different soluble NAD- or NADP-linked polyol dehydrogenases occur in the cytoplasm of *Gluconobacter oxydans* (suboxydans) and oxidize several polyols which lack the Bertrand-Hudson (L-*erythro*) configuration.[4] This section describes the purification and properties of the soluble

[1] G. Bertrand, *Ann. Chim. Phys.* **3**, 181 (1904).
[2] R. M. Hann, E. B. Tilden, and C. S. Hudson, *J. Am. Chem. Soc.* **60**, 1201 (1938).
[3] The oxidized secondary OH group is always considered as the C_2-OH group.
[4] K. Kersters, W. A. Wood, and J. De Ley, *J. Biol. Chem.* **240**, 965 (1965).

and particulate polyol dehydrogenases from *Gluconobacter oxydans* (suboxydans), strain SU.

I. Soluble Polyol Dehydrogenases[4]

$$R\text{—CHOH—CH}_2\text{OH} + NAD^+ \rightleftharpoons R\text{—CO—CH}_2\text{OH} + NADH + H^+$$
$$R\text{—CHOH—CH}_2\text{OH} + NADP^+ \rightleftharpoons R\text{—CO—CH}_2\text{OH} + NADPH + H^+$$

where *R* represents a carbohydrate moiety

Assay Method

Principle. The assay is based on the rate of reduction of NAD or NADP at 340 mμ in the presence of polyol.

Reagents

Tris(hydroxymethyl)aminomethane (Tris) buffer, 0.1 *M*, pH 8.8
$MgCl_2$, 0.05 *M*
NAD or NADP, 0.005 *M*
Polyol, 0.2 *M*. The activity of NADP-xylitol-, NAD-D-*glycero*-, NAD-D-*xylo*-, and NADP-D-*lyxo*-dehydrogenases is routinely determined with *meso*-xylitol, *meso*-ribitol, D-glucitol, and D-mannitol as substrates, respectively
Enzyme. The enzyme is diluted in 0.01 *M* phosphate buffer at pH 7.0, to give a solution with an activity not greater than 0.3 unit/ml (see definition of unit below).

Procedure. Six-tenths milliliter of Tris buffer, 0.1 ml of $MgCl_2$, 0.1 ml of NAD or NADP, and 0.1 ml of suitably diluted enzyme are mixed in a spectrophotometer cell with a 1-cm light path. The reaction is initiated by the addition of 0.1 ml of polyol solution. The extinction at 340 mμ is measured at 1-minute intervals.

Definition of Unit and Specific Activity. One unit of polyol dehydrogenase is defined as that amount which causes the reduction of 1 micromole NAD or NADP per minute under the above conditions. Specific activity is expressed in terms of units of enzyme per milligram of protein, the latter determined spectrophotometrically.[5]

Purification Procedure

All operations were performed at 0–5°, unless otherwise stated. Precipitates are separated by centrifugation at 15,000 *g* for 15 minutes and dissolved in 0.01 *M* KH_2PO_4–Na_2HPO_4 buffer at pH 6.5. Sodium chloride and ammonium sulfate are removed from solutions by gel

[5] O. Warburg and W. Christian, *Biochem. Z.* **310,** 384 (1941).

filtration on Sephadex G-25 columns, equilibrated with the same buffer.

Step 1. Growth of Culture and Preparation of Particle-Free Extract. *Gluconobacter oxydans* (suboxydans), strain SU,[6] is grown on a rotary shaker at 30° in a medium containing 0.04 *M* KH_2PO_4-Na_2HPO_4 buffer at pH 6.0, 5% D-glucitol (sorbitol), and 0.5% yeast extract. The flasks are inoculated with 2% of a 20-hour culture, grown on the same medium. The cells are harvested by centrifugation after 2 days of growth and washed three times with 0.01 *M* phosphate buffer pH 6.5. The cell-free extract is prepared by sonic disintegration of a 30% (w/v) suspension of the cells in 0.01 *M* phosphate buffer at pH 6.5 in a 10-kc, 250-watt Raytheon magnetostrictor for 20 minutes at 4° under hydrogen atmosphere. Unbroken cells are removed by centrifugation at 20,000 *g* for 20 minutes. Centrifugation of the supernatant in a preparative Model L Spinco ultracentrifuge for 2 hours at 4° and 105,000 *g* results in the sedimentation of red particles which oxidize L-*erythro* polyols (II).

Step 2. Removal of Nucleic Acids. Two milliliters of 1.0 *M* $MnCl_2$ is added to 40 ml of supernatant solution. After 30 minutes the precipitate is removed by centrifugation and discarded. $MnCl_2$ is removed from the resulting supernatant solution by gel filtration on Sephadex G-25.

Step 3. Ammonium Sulfate Fractionation. Solid ammonium sulfate is added to 25% of saturation. The precipitate is removed by centrifugation and discarded. Ammonium sulfate is then added to 50% of saturation, and the precipitate is collected by centrifugation and dissolved in 12 ml 0.01 *M* phosphate buffer at pH 6.5. Ammonium sulfate is removed by gel filtration. The volume at this stage is 15 ml [$(NH_4)_2SO_4$ — 1].

An additional amount of solid ammonium sulfate is added to the supernatant solution to 70% of saturation. The precipitate is dissolved and dialyzed as above [$(NH_4)_2SO_4$ — 2].

Step 4. Chromatography on DEAE-Cellulose. The DEAE-cellulose columns are prepared as follows: 50-g batches of the exchanger (obtained from Serva Entwicklungslabor, Heidelberg, Germany; capacity: 0.68) are washed repeatedly with several liters of water. After 3 hours the turbid supernatants are decanted to remove the fines. Each batch is treated with 2 liters of 1 *N* NaOH. After 1 hour, the suspension is filtered on a large Büchner funnel and washed with water until the filtrate is neutral. The DEAE-cellulose is equilibrated 3 times for at least 3 hours with 1 liter 0.1 *M* KH_2PO_4–Na_2HPO_4 buffer at pH 6.5. A slurry of this suspension is poured into a 4.9 $cm^2 \times 30$ cm column and allowed to settle

[6] Available from the National Collection of Industrial Bacteria (NCIB 9108) Torry Research Station, Aberdeen, Scotland.

to 20 cm. Approximately 100 ml of 0.01 *M* phosphate buffer at pH 6.5 is passed through the column.

The dialyzed $(NH_4)_2SO_4$-1 fraction (15 ml) is applied to the column and washed on with 30 ml of 0.01 *M* phosphate buffer, pH 6.5. Proteins are eluted by an increasing NaCl-gradient in the same buffer. The mixing chamber contains 450 ml of 0.01 *M* KH_2PO_4–Na_2HPO_4 buffer at pH 6.5, and the reservoir contains 800 ml 0.3 *M* NaCl in the same buffer. The flow rate is maintained at 50–60 ml/hour, and fractions of 10 ml each are collected. Figure 1 represents the results of an experiment: the NADP-xylitol-, NAD-D-*glycero*-, and NAD-D-*xylo*-dehydrogenases are

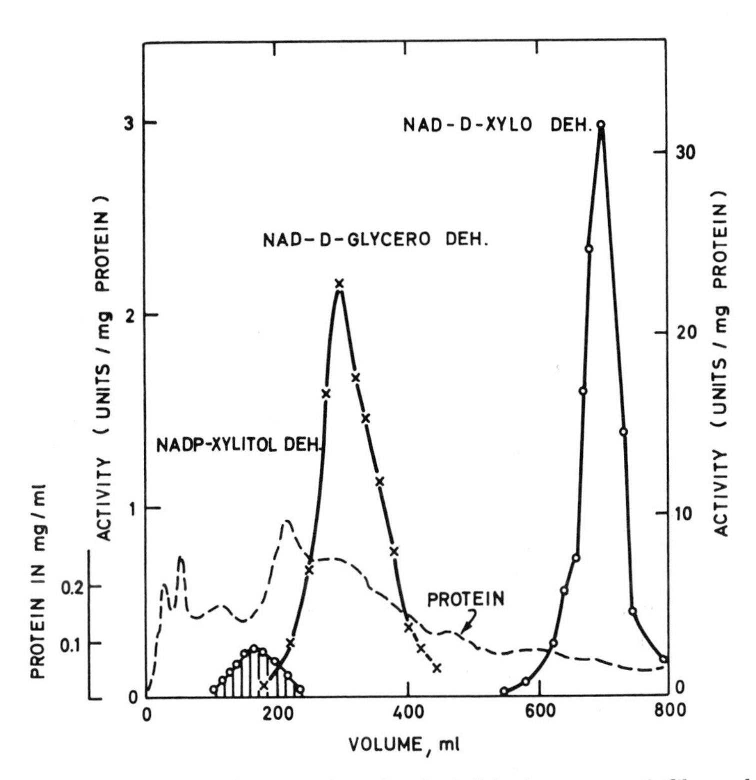

FIG. 1. Chromatographic separation of polyol dehydrogenases of *Gluconobacter oxydans* (suboxydans) on DEAE-cellulose. Conditions described in text. Ordinates at the left express protein and activity of NADP-xylitol dehydrogenase (*NADP-xylitol deh.*) (xylitol as substrate) and NAD-D-*glycero*-dehydrogenase (*NAD-D-glycero deh.*) (ribitol as substrate). The ordinate at the right expresses activity of NAD-D-*xylo*-dehydrogenase (*NAD-D-xylo deh.*) (D-glucitol as substrate).

eluted after 100, 200, and 600 ml of buffer, respectively, passed through the column. The major portion of each peak is contained in five to eight tubes. These fractions are pooled (DEAE-1), dialyzed, and again passed through a small DEAE-cellulose column (1.0 $cm^2 \times 2$ cm). The enzymes are eluted with 12 ml of 1 *M* NaCl, buffered with 0.01 *M* KH_2PO_4–Na_2HPO_4 buffer at pH 6.5 and 3 ml fractions are collected and dialyzed. The fractions which possess the highest specific activity (DEAE-2) are used for the specificity studies. The purified enzymes are stored at —15°.

The NADP-D-*lyxo*-dehydrogenase from the $(NH_4)_2SO_4$-2 fraction is further purified in an exactly similar manner, as described above for the other three dehydrogenases. The bulk of the enzyme is eluted in the 280–350 ml fraction. The enzyme is separated by chromatography from NADPH-2-ketogluconoreductase and from NADPH-5-ketogluconoreductase.[4]

A summary of the purification procedure is given in Table I.

Properties

Stability. Purified preparations of the NAD-D-*glycero*-, NAD-D-*xylo*-, and NADP-D-*lyxo*-dehydrogenases are stable at —15° for at least 2 months. The purified NADP-xylitol dehydrogenase is more labile. After 2 or 3 weeks at —15°, the enzyme loses 60–90% of its initial activity.

Specificity. The specificity of the four purified polyol dehydrogenases can be summarized as follows (see Table II, p. 176):

1. The NADP-xylitol dehydrogenase oxidizes the pentitol *meso*-xylitol and the corresponding acid D-xylonate and is NADP specific.[4] There is no detectable activity on other polyols.

2. The NAD-D-*glycero*-dehydrogenase does not oxidize tetritols or heptitols. Pentitols are oxidized at C-2 when the secondary OH-group at C_2 possesses a D-configuration. L-Arabitol has the required configuration both at C-2 and C-4 and is thus oxidized for 77% to L-xylulose and for 23% to L-ribulose.[4] Hexitols are oxidized when the OH-groups at C-2 *and* C-3 possess a D-configuration, i.e., when they have the D-*erythro* configuration at the top of the molecule. NADP is not a hydrogen acceptor.

3. The NAD-D-*xylo*-dehydrogenase oxidizes pentitols, hexitols, and heptitols with the D-*xylo* configuration. The following polyols are oxidized in decreasing order: D-glucitol, L-*glycero*-D-*galacto*-heptitol, *meso*-xylitol, D-*glycero*-D-*gluco*-heptitol, L-iditol, *meso*-*glycero*-*ido*-heptitol, and L-*glycero*-L-*ido*-heptitol. The weak oxidation of D-mannitol is an exception. Rechromatography on the DEAE-cellulose does not separate the activities for D-glucitol and D-mannitol. The ratio of the two activities remains

TABLE I

PURIFICATION OF FOUR SOLUBLE NAD- AND NADP-LINKED POLYOL DEHYDROGENASES[a]

Fraction	Volume (ml)	Protein (mg/ml)	NADP-xylitol dehydrogenase		NAD-D-*glycero*-dehydrogenase		NAD-D-*xylo*-dehydrogenase		NADP-D-*lyxo*-dehydrogenase	
			Total units	Specific activity (units/mg protein)	Total units	Specific activity (units/mg protein)	Total units	Specific activity (units/mg protein)	Total units	Specific activity (units/mg protein)
Particle-free extract	40	13.1	11.7	0.022	75	0.14	284	0.54	1040	2.0
$(NH_4)_2SO_4$-1 (25–50%)	15	19.0	9.1	0.032	62	0.22	235	0.83	146	0.5
$(NH_4)_2SO_4$-2 (50–70%)	15	5.7	0	0	1.3	0.015	11	0.13	460	5.4
DEAE-1 peak tubes	—	—	2.2	0.24	30	1.52	131	21.7	248	52
DEAE-2 peak tubes										
NADP-xylitol[b]	3	1.5	1.43	0.31	—	—	—	—	—	—
NAD-D-*glycero*[b]	6	0.49	—	—	6.7	2.3	—	—	—	—
NAD-D-*xylo*[b]	3	0.38	—	—	—	—	40	35	—	—
NADP-D-*lyxo*[b]	3	0.62	—	—	—	—	—	—	130	70

[a] The activity of NADP-xylitol-, NAD-D-*glycero*-, NAD-D-*xylo*- and NADP-D-*lyxo*-dehydrogenases is measured with respectively *meso*-xylitol, *meso*-ribitol, D-glucitol and D-mannitol as substrates.

[b] Represents the polyol dehydrogenases from the most active eluates of the second DEAE-cellulose column.

TABLE II
SPECIFICITY[a] AND OXIDATION PRODUCTS[b] OF THE NADP-XYLITOL-, NAD-D-*glycero*-, NAD-D-*xylo*-, AND NADP-D-*lyxo* DEHYDROGENASES

Substrate	NADP-xylitol dehydrogenase	NAD-D-*glycero*-dehydrogenase	NAD-D-*xylo*-dehydrogenase	NADP-D-*lyxo*-dehydrogenase
meso-Erythritol	0	0	0	0
meso-Xylitol	100[a] L-xylulose	23 D-xylulose	21 D-xylulose	0
meso-Ribitol	0	100[a] D-ribulose	0	0
D-Arabitol	0	0	0	3 D-xylulose
L-Arabitol	0	50 L-xylulose + L-ribulose	0	0
D-Mannitol	0	0	2 D-fructose	100[a] D-fructose
L-Mannitol	—	15 L-fructose	0	—
D-Glucitol	0	0	100[a] D-fructose	16 L-sorbose
L-Glucitol	0	2 D-sorbose	0	0
meso-Allitol	0	5 allulose	0	0
meso-Galactitol	0	0	0	0
L-Iditol	0	0	15 L-sorbose	0
L-Rhamnitol	0	15 rhamnulose	0	0
meso-glycero-allo-Heptitol	—	0	0	—
D-*glycero*-D-*manno*-Heptitol	—	0	0	0
D-*glycero*-D-*gluco*-Heptitol	—	0	20 D-altroheptulose	—
D-*glycero*-D-*galacto*-Heptitol	0	0	0	0
meso-glycero-gulo-Heptitol	0	0	0	0
meso-glycero-ido-Heptitol	—	—	9 D-idoheptulose	—
L-*glycero*-D-*galacto*-Heptitol	—	—	37 L-galactoheptulose	—
L-*glycero*-L-*ido*-Heptitol	—	—	3 L-glucoheptulose	—

[a] The relative oxidation rates are expressed against *meso*-xylitol, *meso*-ribitol, D-glucitol, and D-mannitol arbitrarily put at 100.
[b] The oxidation products were identified by chromatography, chemical, and enzymological methods.

constant over the whole peak. D-Gluconate is not oxidized; NAD is specifically required.

4. The NADP-D-*lyxo*-dehydrogenase oxidizes pentitols and hexitols with the D-*lyxo* configuration (D-mannitol, D-glucitol, and D-arabitol) and uses only NADP. Heptitols and D-gluconate are not oxidized.

Effect of pH. Dehydrogenation is favored in alkaline conditions. The pH optima of the NADP-xylitol-, NAD-D-*glycero-*, NAD-D-*xylo-*, and NADP-D-*lyxo*-dehydrogenases are, respectively, 9.5, 10.5, 9.8, and 8.5.

Inhibitors and Activators. Both NADP-xylitol- and NADP-D-*lyxo*-dehydrogenases are completely inhibited by 10^{-4} *M* *p*-chloromercuribenzoate, whereas this reagent is not inhibitory at 10^{-3} *M* for the NAD-D-*glycero-* and NAD-D-*xylo*-dehydrogenases. EDTA (2×10^{-4} *M*) inhibits the NAD-D-*xylo* dehydrogenase completely, but has no effect on the other three polyol dehydrogenases. The NAD-D-*xylo*-dehydrogenase is stimulated for 80% by 10^{-3} *M* $MgCl_2$.

Effect of Substrate Concentration. For the NAD-D-*xylo*-dehydrogenase a Lineweaver-Burk plot yields a K_m of 7×10^{-3} *M* for D-glucitol and 2×10^{-2} *M* for *meso*-xylitol (carbonate–bicarbonate buffer at pH 9.8 and 0.75×10^{-3} *M* NAD).

Distribution.[7] The NAD-D-*glycero-* and NAD-D-*xylo*-dehydrogenases are present in *Gluconobacter oxydans* (suboxydans), strains SU and 26, but absent in *Acetobacter aceti* (mesoxydans), strain NCIB 8622. The NADP-D-*lyxo* dehydrogenase was demonstrated in the three strains mentioned. The NADP-xylitol dehydrogenase however was observed only in *G. oxydans* (suboxydans), strain SU. The specificity of the latter enzyme resembles that of the NADP-xylitol (L-xylulose) dehydrogenase from guinea pig liver mitochondria.[8] The NAD-D-*xylo* dehydrogenase from *G. oxydans* (suboxydans) SU does not have the same substrate specificity as the NAD-linked L-iditol dehydrogenase isolated from rat liver,[9] sheep liver,[10] and *Azotobacter agilis*.[11] Both D-*ribo* and D-*xylo* polyols are oxidized by these L-iditol dehydrogenases. The substrate specificity of the here described NADP-D-*lyxo*-dehydrogenase resembles that of the D-arabitol dehydrogenase of *Aerobacter aerogenes*.[12] The cofactor required by the latter enzyme is, however, NAD.

[7] If not otherwise noted, these are unpublished data from K. Kersters and J. De Ley.

[8] S. Hollmann and O. Touster, *J. Biol. Chem.* **225,** 87 (1957). See also O. Touster and G. Montesi, Vol. V [34].

[9] J. McCorkindale and N. L. Edson, *Biochem. J.* **57,** 518 (1954).

[10] M. G. Smith, *Biochem. J.* **83,** 135 (1962).

[11] L. Marcus and A. G. Marr, *J. Bacteriol.* **82,** 224 (1961).

[12] E. C. C. Lin, *J. Biol. Chem.* **236,** 31 (1961).

II. Particulate Polyol Dehydrogenase

$$R—CHOH—CH_2OH \rightarrow R—CO—CH_2OH + 2\,H^+ + 2\,e$$

Assay Method

Principle. Since the particles contain the complete electron transport chain, the oxidation of polyols by the particulate enzyme(s) is followed by measurement of the rate of oxygen uptake in the Warburg respirometer.

Reagents

Polyol solution, 0.2 *M*

Enzyme. Particles are suspended in 0.02 *M* phosphate buffer at pH 6.2 to give a turbidity of 200 units in the Klett colorimeter with filter 66, corresponding to about 0.4 mg protein per milliliter.

Procedure. To the main compartment of a Warburg flask are added 1.7 ml of the particle suspension. Two-tenths milliliter of polyol solution is placed in the side arm and 0.1 ml 20% KOH in the center well. The total volume is 2 ml. Under these conditions the enzyme is saturated with substrate. Readings are taken at 5-minute intervals during 30 minutes. When the end products of the oxidation have to be isolated and identified, readings are taken until oxygen uptake ceases (at 0.5 mole oxygen per mole substrate). Oxygen uptake is corrected for endogenous respiration, which is usually very small (about 0.02 unit).

Definition of Unit. One unit of enzyme is defined as that amount which oxidizes 1 micromole of polyol to the corresponding ketose in 1 hour; it is calculated from the amount of oxygen uptake between the fifth and twentieth minutes.

Partial Purification Procedure

Cells of *Gluconobacter oxydans* (suboxydans), strain SU, are grown, harvested, washed, and sonicated as described for the soluble enzymes. The cell-free extract is centrifuged for 2 hours at 4° and 105,000 *g* in a preparative model L Spinco ultracentrifuge. The red precipitate is suspended in 0.01 *M* KH_2PO_4–Na_2HPO_4 buffer at pH 6.5 and again centrifuged at 105,000 *g* for 1 hour. These washed particles are used for activity determinations. Attempts to purify the particulate polyol dehydrogenase(s) were unsuccessful.[13]

[13] Most of the protein is released from the particles by treatment with 0.5% Triton-X-100 (Rohm & Haas Co., Philadelphia, Pennsylvania). This detergent solubilizes 75% primary alcohol dehydrogenase (see this volume [64]), but inactivates the

Properties

No appreciable loss of activity occurs when the particles are kept frozen at −15° for 6 months. The optimal pH for the polyol oxidation is about 6.0.

Specificity. The particulate polyol dehydrogenase(s) oxidize tetritols, pentitols, hexitols, and heptitols with the L-*erythro* (Bertrand-Hudson) configuration. In decreasing order of activity D-arabitol is oxidized to D-xylulose, *meso*-erythritol to L-erythrulose, D-glucitol to L-sorbose, D-mannitol to D-fructose, *meso*-ribitol to L-ribulose, *meso-glycero-gulo*-heptitol to L-glucoheptulose, *meso-glycero-allo*-heptitol to L-alloheptulose, D-*glycero*-D-*gluco*-heptitol to L-guloheptulose, D-*glycero*-D-*manno*-heptitol to a mixture of D-mannoheptulose and D-altroheptulose, D-*glycero*-D-*galacto*-heptitol to L-galactoheptulose, and *meso*-allitol to allulose. There is no measurable activity with the following sugar alcohols; *meso*-xylitol, L-arabitol, L-mannitol, L-glucitol, *meso*-galactitol, L-iditol, *meso-glycero-ido*-heptitol, L-*glycero*-D-*galacto*-heptitol, L-*glycero*-L-*ido*-heptitol, L-fucitol, and L-rhamnitol. L-Threitol is slowly oxidized to L-erythrulose.

Inhibitors. Under the assay conditions described above, the polyol dehydrogenase is completely inhibited by 10^{-3} *M* *p*-chloromercuribenzoate and 10^{-3} *M* EDTA and is inhibited to a considerable extent (60–70%) by 10^{-3} *M* semicarbazide, 10^{-3} *M* cyanide, and 10^{-3} *M* azide. Cyanide, azide, and EDTA are not inhibitory when 2,6-dichlorophenolindophenol is used as electron acceptor. This indicates that these substances inhibit the electron transport chain (probably cytochrome oxidase), but not the polyol dehydrogenase as such.

Other Properties. It is not known whether one or several polyol dehydrogenases are responsible for these particle-linked oxidations. Particles oxidize also glycols and primary and secondary alcohols.[14] It was found that the particulate primary and secondary alcohol dehydrogenases are different from the above-described polyol dehydrogenase(s).[15] This indicates that there is no connection between the oxidation of aliphatic glycols and the rule of Bertrand-Hudson for the polyols.

---

polyol dehydrogenase almost completely. The particulate polyol dehydrogenase could not be solubilized in an active state by pancreatin, trypsin, phospholipase D, acetone, Duponol, sodium deoxycholate, and Tween 80.

[14] K. Kersters and J. De Ley, *Biochim. Biophys. Acta* **71,** 311 (1963).

[15] K. Kersters and J. De Ley, this volume [64].

[35] Pentitol Dehydrogenases of *Aerobacter aerogenes*

By D. D. Fossitt and W. A. Wood

I. Ribitol Dehydrogenase

$$\text{Ribitol} + \text{DPN}^+ \rightleftarrows \text{D-ribulose} + \text{DPNH} + \text{H}^+$$
$$\text{Xylitol} + \text{DPN}^+ \rightleftarrows \text{D-xylulose} + \text{DPNH} + \text{H}^+$$
$$\text{L-Arabitol} + \text{DPN}^+ \rightleftarrows \text{L-xylulose} + \text{DPNH} + \text{H}^+$$

Assay Method[1,2]

Principle. Either the oxidation of ribitol or the reduction of D-ribulose with the accumulation or utilization of DPNH, respectively, may be used as an assay. Since the equilibrium lies far in the direction of ribitol formation, the reduction of D-ribulose by DPNH was routinely followed spectrophotometrically at 340 mμ. In the procedure described below, a longer linear reaction was realized by the use of elevated DPNH levels (i.e., corresponding to an initial absorbance of 2.0–2.5). Accurate recording of changes in this absorbance region was facilitated by the use of a Gilford log converter attached to a Beckman DU monochrometer,[3] and simultaneous assays of 2–4 cuvettes were accomplished with an automatic cuvette positioner.[4]

Reagents

DPNH (0.1 *M*). Dissolve 7 mg of DPNH in 1 ml water; determine the exact concentration from the absorbance at 340 mμ and the extinction coefficient of 6.25×10^{-3}.

D-Ribulose (0.1 *M*). Prepared from D-ribulose-*o*-nitrophenyl hydrazone, or as described by Mortlock and Wood [8].

Potassium phosphate or sodium citrate buffer, 0.1 *M*, pH 6, containing 18 mg glutathione per milliliter and 0.05 *M* with respect to $MgCl_2$.

Enzyme. Dilute enzyme solution to obtain 50–150 units of enzyme per milliliter.

Procedure. In a microcuvette of 1 cm light path and 0.5 ml volume, place 0.05 ml of buffer, 0.01 ml DPNH solution, 0.02 ml D-ribulose solution, diluted enzyme, and water to a volume of 0.2 ml. After the mixture

[1] R. P. Mortlock, D. D. Fossitt, and W. A. Wood, *Proc. Natl. Acad. Sci. U.S.* **54,** 572 (1965).
[2] W. A. Wood, M. J. McDonough, and L. B. Jacobs, *J. Biol. Chem.* **236,** 2190 (1961).
[3] W. A. Wood and S. R. Gilford, *Anal. Biochem.* **2,** 589 (1961).
[4] W. A. Wood and S. R. Gilford, *Anal. Biochem.* **2,** 601 (1961).

has been stirred by inversion or with a plastic toothpick, the rate of change of optical density at 340 mμ is measured. Microliter amounts of DPNH and enzyme are usually dispensed from a 10 μl microsyringe (Hamilton Co. No. 701).

Definition of Unit and Specific Activity. One unit of enzyme is defined as that amount which causes a change in absorbance at 340 mμ of 0.001 per minute under the conditions specified above. This corresponds to the reduction of 3.3×10^{-5} micromoles of D-ribulose per minute per 0.2 ml of reaction mixture. The specific activity is expressed as units of enzyme per milligram of protein. Protein is determined from the absorbancies at 280 mμ and 260 mμ as originally described by Warburg and Christian.[5]

Source of Enzyme. Ribitol dehydrogenase is induced by growth of *Aerobacter aerogenes,* PRL-R3, on ribitol, and maximum growth on ribitol-mineral medium (see below) occurs in 1 day. However, about 10-fold more enzyme is present in cells grown on xylitol or L-arabitol, although as many as 4 days may be required for full growth. During growth on the latter pentitols which are not inducers, mutants are selected which are constitutive for high levels of the dehydrogenase. After growth on xylitol or L-arabitol, the culture may be used as such or the mutant may be isolated using the plating procedure of Hulley *et al.*[6] Isolates thus obtained may then be cultured on peptone and still possess high ribitol dehydrogenase activity. The specific activivties of crude extracts obtained after inoculation of the wild type with media containing the following pentitols were: ribitol, 50–150; xylitol, 270–800; L-arabitol, 50–200, and for the constitutive mutant grown on peptone, 700–1000.

In the procedure to be described, the inducible dehydrogenase was isolated, but purification of the dehydrogenase from constitutive cells involves the same procedure.

Medium. Aerobacter aerogenes, PRL-R3, was grown on a medium described by Anderson and Wood[7] consisting of 1.35% $Na_2HPO_4 \cdot 7\ H_2O$, 0.15% KH_2PO_4, 0.3% $(NH_4)_2SO_4$, 0.02% $MgSO_4 \cdot 7\ H_2O$, 0.0005% $FeSO_4 \cdot 7\ H_2O$, and 0.5% ribitol (autoclaved separately). Forty-liter cultures were grown aerobically for 8–10 hours from a 5% to 10% inoculum. The cells were then collected in a Sharples centrifuge, washed once with cold water, and suspended in 3 volumes of 0.01 phosphate buffer, pH 7.2. The cells were disrupted either in a Raytheon sonic oscillator (10 kc,

[5] O. Warburg and W. Christian, *Biochem. Z.* **310,** 384 (1941). See also Vol. III [73].
[6] S. B. Hulley, S. B. Jorgensen, and E. C. C. Lin, *Biochim. Biophys. Acta* **67,** 219 (1963).
[7] R. L. Anderson and W. A. Wood, *J. Biol. Chem.* **237,** 296 (1962).

200 watt) for 20 minutes, or in a French pressure cell. The cellular debris was removed by centrifugation at 20,000 *g* for 15 minutes in a refrigerated centrifuge[8] to give the crude extract.

Purification Procedure

Step 1. The crude extract was diluted with 0.01 *M* phosphate buffer, pH 7.2, containing 10^{-4} *M* EDTA[9] to a final protein concentration of 10 mg/ml. Solid ammonium sulfate was added to a final concentration of 0.1 *M* and 0.1 volume (v/v) of 4% protamine sulfate was added slowly with stirring. The precipitate containing nucleic acids was removed by centrifugation and discarded.

Step 2. Solid ammonium sulfate was added to a final concentration of 1.5 *M* and the precipitate removed by centrifugation. Additional ammonium sulfate was added to raise the concentration to 2.1 *M*. The precipitate containing the enzyme was collected by centrifugation and dissolved in a small amount of 0.01 phosphate buffer. The solution was then dialyzed against 0.001 *M* potassium phosphate buffer, pH 7.2, containing 10^{-4} *M* EDTA for 18–24 hours.

Step 3. The dialyzed preparation was absorbed on a DEAE-cellulose column (4 cm^2 × 26 cm) which had been equilibrated with 0.001 *M* phosphate buffer, pH 7.2. Elution of the enzyme was accomplished with a linear gradient prepared by mixing 300 ml of 0.001 *M* with 300 ml 0.2 *M* potassium phosphate buffer, pH 7.2. The peak containing the enzyme activity was precipitated with ammonium sulfate (1.8–2.1 *M*) and rechromatographed on a second DEAE-cellulose column (1.2 cm^2 × 12 cm). Elution from the second column was accomplished with a gradient prepared from equal quantities of 0.001 *M* and 0.1 *M* buffer, pH 7.2.

Step 4. The enzyme fraction from the second DEAE-cellulose column was precipitated by addition of ammonium sulfate to a final concentration of 2.5 *M*. Following centrifugation, the precipitate was dissolved in 10–20 ml of water and the salt concentration adjusted to 0.03 *M* by dilution with distilled water.[10] Alumina Cγ (Sigma Chemical Co.) (1 mg per milligram of protein) was added; the mixture was stirred with a glass rod for 5 minutes, and then centrifuged. The supernatant solution containing the enzyme was then concentrated by precipitation with

[8] Unless otherwise stated, all centrifugations were performed with a refrigerated Lourdes high-speed centrifuge.

[9] All solutions used subsequently contained 10^{-4} *M* EDTA unless otherwise stated and all operations were performed at 0–4°C.

[10] The salt concentration was determined on 1:20,000 dilutions using a precalibrated Barnstead Purity Meter.

$(NH_4)_2SO_4$ (2.5 *M*), taken up in 5–10 ml of H_2O (about 6–10 mg protein/ml), and chromatographed on Sephadex G-200 (3.5 $cm^2 \times 100$ cm) equilibrated against 0.01 *M* potassium phosphate buffer, pH 7.2. The elution rate was 0.3–0.5 ml/min.

Step 5. The protein in the effluent fraction containing the enzyme was precipitated with ammonium sulfate and the precipitate was back-extracted with 2–5 ml portions each of 2.2 *M*, 2.0 *M*, and 1.8 *M* ammonium sulfate solutions. Following each extraction, the precipitate was recovered by centrifugation. These extracts were then allowed to stand in the cold. Usually crystals formed within 1 to 5 hours. The crystals are very small and nearly spherical in appearance. This crystallization step is not completely reproducible.

The purification procedure is summarized in Table I.

TABLE I
PURIFICATION OF RIBITOL DEHYDROGENASE FROM EXTRACTS OF *Aerobacter aerogenes*

Fraction	Total activity (units)	Yield (%)	Specific activity (units/mg protein)	Purification (fold)
Extract	525,000	—	180	1
Protamine sulfate	500,000	95	180	1
Ammonium sulfate, 1.3–2.0 *M* saturation	360,000	68	250	1.4
1st DEAE	250,000	48	2,450	10.4
2nd DEAE	232,000	44	9,500	53
Ammonium sulfate, 1.7–2.0 *M* saturation	149,000	28	19,600	109
Sephadex G-200	94,000	18	32,000	178
Crystallization	—	20–50	32,000	178

Properties

Stability. The enzyme is stable at pH 7.0 when frozen for a period of 6 months. Addition of glutathione increases the stability.

Specificity. The enzyme will oxidize ribitol to D-ribulose, xylitol to D-xylulose, and L-arabitol to L-xylulose using DPN^+ as coenzyme.[11] The rate of oxidation of xylitol and L-arabitol is about 10% the rate of oxidation of ribitol. The enzyme does not oxidize D-arabitol or hexitols. TPN does not serve as a cofactor for the reaction.

Effect of pH and Activators. The optimum pH for oxidation of the pentitols is 8.5 to 9.0 and oxidation is facilitated by the use of semicarbazide buffer which presumably reacts with the pentulose formed in the

[11] D. D. Fossitt, R. P. Mortlock, and W. A. Wood, *Bacteriol. Proc.* p. 82 (1965).

reaction. The optimum pH for reduction of pentuloses is 6.0–6.5, and the reaction proceeds nearly to completion.

For optimum activity, both 10^{-2} M Mg^{++} and a sulfhydryl compound such as glutathione are required.

Physical constants. The K_m values for ribitol, xylitol, and L-arabitol are 2.6×10^{-3} M, 2.9×10^{-1} M, and 2.9×10^{-1} M, respectively. The molecular weight, as estimated from the sedimentation coefficient, is 102,000–110,000.[12]

II. D-Arabitol Dehydrogenase

D-Arabitol + DPN^+ ⇄ D-xylulose + DPNH + H^+

Assay Method

The assay procedure and reagents were the same as for ribitol dehydrogenase except that 2.0 micromoles of D-xylulose were substituted for ribitol.

Purification Procedure

This procedure is generally similar to that described by Mortlock and Wood[13] but differs materially from the original one of Wood *et al.*[2]

Preparation of Extract. Aerobacter aerogenes PRL-R3 was grown as for ribitol dehydrogenase in the preceding section except that 0.5% D-arabitol was substituted for ribitol as an energy source. The cells were collected by centrifugation at 18,000 g and washed once with cold distilled water. Extracts were prepared by treating a 20% cell suspension in a French Press or in a 10 kc, 200-watt oscillator for 15 minutes. The cell debris was removed by centrifugation.

Protamine Sulfate Treatment. Solid ammonium sulfate was added to the extract to a final concentration of 0.1 M. One-tenth volume of 4% protamine sulfate (pH 5.0) was added dropwise with stirring over a period of 15 minutes, and the precipitate was removed by centrifugation.

Ammonium Sulfate Fractionation. Solid ammonium sulfate was added to give fractions of 0–0.3, 0.3–0.42, and 0.42–0.50 of saturation. The precipitates recovered by centrifugation were dissolved in distilled water. The 0.3–0.42 fraction contained almost all the activity.

Chromatography on DEAE-Cellulose. The ammonium sulfate fraction was diluted with water to 0.01 M concentration as determined with a precalibrated conductivity cell (Barnstead Purity Meter). The pH was

[12] R. P. Mortlock, D. D. Fossitt, D. H. Petering, and W. A. Wood, *J. Bacteriol.* **89**, 129 (1965).

[13] R. P. Mortlock and W. A. Wood, *J. Bacteriol.* **88**, 838 (1964).

adjusted to 7.5 with dilute ammonium hydroxide and the fraction then applied to a DEAE-cellulose column ($2 \text{ cm}^2 \times 50$ cm) which had been previously equilibrated with 0.01 M phosphate buffer, pH 7.5. Fractions were eluted from column with 100 ml portions of 0.04 M, 0.05 M, and 0.08 M phosphate buffer, pH 7.5. Most of the activity was eluted with 0.08 M buffer.

Ammonium Sulfate Fractionation. The pooled fractions from the DEAE-cellulose column were fractionated with solid ammonium sulfate to give precipitates at 0–0.35, 0.35–0.40, and 0.40–0.50 of saturation. The fraction obtained between 0.35 and 0.40 saturation contained most of the activity.

Treatment with Alumina Cγ. The fraction was diluted with water to 0.03 M ammonium sulfate concentration. Alumina Cγ (Sigma Chemical Co.) was added slowly with stirring at the rate of 2 mg of alumina Cγ per milligram of protein. The alumina was recovered by centrifugation, and the supernatant solution saved.

Ammonium Sulfate Fractionation. Solid ammonium sulfate was added to obtain a fraction between 0.35 and 0.40 of saturation. The precipitate was recovered by centrifugation and dissolved in distilled water.

In the purification cited in Table II, which is typical, about a 50-fold

TABLE II
PURIFICATION OF D-ARABITOL (→D-XYLULOSE) DEHYDROGENASE FROM *Aerobacter aerogenes*

Fraction	Total activity (units)	Yield (%)	Specific activity (units/mg protein)	Purification (fold)
Extract	46,100	—	76	1
Protamine sulfate	59,500	100	80	1
Ammonium sulfate, 0.3–0.42 saturation	58,500	98	334	4.4
DEAE-cellulose	40,000	67	1,800	24
Ammonium sulfate, 0.35–0.4 saturation	23,250	39	2,350	31
Al Cγ (negative)	18,000	30	2,715	36
Ammonium sulfate, 0.35–0.40 saturation	12,500	21	3,700	49

purification was achieved with a 20% recovery of activity. Specific activity values of > 6000 were obtained across the peak of D-arabitol dehydrogenase in sucrose density centrifugation runs. Therefore, it is likely that the final preparation is approximately 50% pure. The preparation could be stored at pH 7 and $-15°$ for at least one year without loss in activity.

Properties

D-Arabitol dehydrogenase is induced in *A. aerogenes* primarily by growth on D-arabitol but is present also in cells grown on D-xylose, D-lyxose, and xylitol.[14] It is likely that both D-arabitol and D-xylulose are inducers. The dehydrogenase reduces only D-xylulose of the ketopentoses as well as D-fructose at 2.5% the rate of D-xylulose. Aldopentoses, L-sorbose, and D-sedoheptulose were not reduced. Of the four pentitols, only D-arabitol was oxidized. TPNH did not serve as a hydrogen donor for D-xylulose reduction. The K_m values for D-xylulose and DPNH are $2 \times 10^{-3}\ M$ and $3.6 \times 10^{-5}\ M$, respectively. Approximately 8 micromoles per milliltier of D-xylulose and 0.25 micromole of DPNH per milliliter are needed for maximum activity.[2] The $S_{20,w}$ value estimated from sucrose density gradients was 4.5–4.8.[12] The dehydrogenase thus differs markedly from the inducible and constitutive ribitol dehydrogenase.[1] The pH optimum was 7.0, and higher activity was observed with potassium phosphate buffer than with glycine HCl.[2]

[14] R. P. Mortlock and W. A. Wood, *J. Bacteriol.,* 88, 845 (1964).

[36] Pentitol Dehydrogenases of *Saccharomyces rouxii*

By JORDAN M. INGRAM and W. A. WOOD

D-Ribulose + NADPH + H^+ → D-arabitol + NADP
D-Xylulose + NADH + H^+ → xylitol + NAD

Assay Method

Principle. The method is based on the absorption of the reduced nicotinic acid adenine dinucleotide at 340 mμ. With an excess of each substrate, the rate of reduced nicotinic acid adenine dinucleotide oxidation is proportional to the enzyme concentration.

Reagents

D-ribulose or D-xylulose (enzymatically assayed), 0.5 *M*
NADPH·H^+ or NADH·H^+, 0.01 *M*
Sodium cacodylate buffer, 0.5 *M*, pH 6.0
Enzyme. Dilute the preparation to give an approximate rate of 0.1 O.D. per minute or less.

Procedure. The reaction mixture contains: 0.1 micromole of NADPH·H^+ or NADH·H^+, 5.0 micromoles of D-ribulose or D-xylulose, 25 micromoles of cacodylate buffer and water to a total volume of 0.15 ml. The reactions are performed in 0.5 ml total volume silica glass cuvettes (light path, 1 cm). The assays were routinely followed in a Gilford modified Beckman DU spectrophotometer.[1]

Purification Procedure[2]

Preparation of Crude Extract. Saccharomyes rouxii, P_3a, was obtained from the Prairie Regional Laboratory, National Research Council, Saskatoon, Saskatchewan, Canada. The organism was grown on a medium composed of: 20% glucose, 1.0% yeast extract, and 0.23% urea. Growth proceeded in shake flasks under highly aerobic conditions at 30°C for 96 hours. The cells were recovered from the fermentation liquor by centrifugation at 30,000 *g* for 15 minutes, washed with water, and recentrifuged. This procedure was repeated once more. The cells were resuspended in one-twentieth of the original volume of water and disrupted by treatment in a 10 kc sonic oscillator for 25 minutes. Cellular debris was removed by centrifugation at 30,000 *g* for 15 minutes. The protein concentration of the crude extract was determined by the 28:260 ratio method.[3]

D-*Arabitol* (→ D-*Ribulose*) *Dehydrogenase.* The crude extract was made 0.1 *M* with respect to ammonium sulfate, and one-fifth the volume of 2% protamine sulfate was added with stirring at 4°. The precipitate was removed by centrifugation. The supernatant was brought to 50% of saturation with solid ammonium sulfate, and the precipitate was discarded. The supernatant was adjusted to 75% saturation of ammonium sulfate, and the precipitate was recovered by centrifugation and dissolved in one-tenth volume of cold water. Calcium phosphate gel was added in increments of 1 mg of gel per milligram of protein until the activity was absorbed. The gel fractions containing the activity were eluted with 20 ml of 0.2 *M*, pH 7.0, phosphate buffer. The procedure yielded a preparation 12- to 25-fold purified over the crude.

Xylitol (→ D-*Xylulose*) *Dehydrogenase.* This procedure was similar to that described above except that the activity was absorbed with 3.4 mg of gel per milligram of protein. After elution with 20 ml of 0.2 *M*, pH 7.0, phosphate buffer, the activity was fractionated between 60 and 75% of saturation ammonium sulfate. A preparation 33-fold purified over the crude extract was obtained.

[1] W. A. Wood and S. R. Gilford, *Anal. Chem.* **2,** 599 (1961).

[2] J. M. Ingram and W. A. Wood, *J. Bacteriol.* **89,** 1186 (1965).

[3] O. Warburg and W. Christian, *Biochem. Z.* **310,** 384 (1942).

Properties

Specificity. With use of the appropriate nicotinic acid adenine dinucleotide, each enzyme is specific for the respective substrate. Each enzyme preparation will oxidize the respective pentitol with difficulty. The preparations of each enzyme are contaminated with the other enzyme.

Kinetic Properties. The D-arabitol (→D-ribulose) dehydrogenase has a sharp optimum at pH 6.0 and a K_m value for D-ribulose of 2.0×10^{-2} *M* at this pH. The xylitol (→D-xylulose) dehydrogenase has a broad optimum at pH 6.0 and a K_m value for D-xylulose of 1.2×10^{-2} *M*.

[37] D-Xylose Reductase and D-Xylitol Dehydrogenase from *Penicillium chrysogenum*

By C. CHIANG and S. G. KNIGHT

$$\text{D-Xylose} + \text{TPNH} + \text{H}^+ \rightarrow \text{D-xylitol} + \text{TPN}^+$$
$$\text{D-Xylitol} + \text{DPN}^+ \rightleftharpoons \text{D-xylulose} + \text{DPNH} + \text{H}^+$$
$$\textit{Sum:}\ \text{D-Xylose} + \text{TPNH} + \text{DPN}^+ \rightleftharpoons \text{D-xylulose} + \text{TPN}^+ + \text{DPNH}$$

I. Xylose Reductase

Assay Method[1]

Principle. The enzymatic activity is determined by measurement of the decrease in optical density at 340 mμ caused by the oxidation of TPNH in the presence of D-xylose.[2,3]

Reagents

D-Xylose, 0.1 *M*
TPNH, 1.15×10^{-3} *M*, in 0.1 *M* Tris-HCl buffer, pH 7.5
Tris-HCl buffer or phosphate buffer, 0.1 *M*, pH 7.2
Enzyme. The enzyme is diluted with 0.01 *M* phosphate buffer, pH 7.2, to obtain a solution with an activity between 10 and 60 units per milliliter

Procedure. To a 1-ml quartz cell with a 1-cm light path, 0.1 ml of TPNH, 0.6 ml of Tris or phosphate buffer, 0.1 ml of distilled water or

[1] See also this volume [33b].
[2] C. Chiang, C. J. Sih, and S. G. Knight, *Biochim. Biophys. Acta* **29**, 664 (1958).
[3] C. Chiang and S. G. Knight, *Biochim. Biophys. Acta* **35**, 454 (1959).

solution to be tested, and 0.1 ml of the enzyme preparation are added. After equilibration for 1 minute at 25°, 0.1 ml of D-xylose solution is added and quickly mixed in; readings are taken immediately at 15-second intervals. A quartz cell containing all the components except TPNH serves as a blank. This procedure also is used to assay the crude cellular extracts.

Definition of Unit and Specific Activity. One unit of enzyme is defined as that amount which will cause a decrease in optical density of 0.01 per minute during the first minute of the reaction under the above conditions. Specific activity is expressed in terms of units of enzyme per milligram of protein.

Partial Purification Procedure

The purification procedure described herein is essentially the same as that previously reported.[3] The steps in the procedure are carried out at 0°–4°, unless otherwise indicated. All precipitations are separated by centrifugation at 10,000 *g* for 10 minutes in a RC-2 Servall refrigerated centrifuge.

Growth of Mycelial Cells. A little soil from the soil stock of *Penicillium chrysogenum* NRRL 1951-25 is sprinkled over the slanted agar surface on a D-xylose-mineral salts medium in a 6-ounce prescription bottle; the medium consists of 1% D-xylose; 0.5% NH_4NO_3; 0.05% $MgSO_4 \cdot 7\ H_2O$; 0.01% $FeSO_4 \cdot 7\ H_2O$; 0.001% each $MnSO_4 \cdot 4\ H_2O$, $CaCl_2$, and $ZnCl_2$; and 1.5% agar. The final pH is adjusted to 7.2 by adding 0.2% K_2HPO_4. After 4-days' incubation at 25°, heavy growth and sporulation is obtained. Fifteen milliliters of sterile distilled water is added to the growth and the bottle is shaken to obtain a uniform spore suspension. A 1-ml aliquot of the spore suspension is inoculated into a 250-ml Erlenmeyer flask containing 50 ml of the mineral salts medium with 2% D-xylose. The flask is incubated at 30° for 24 hours on a rotary shaker, and then is used for the final inoculum. A 5-ml aliquot of the growing cells is inoculated into a 500-ml Erlenmeyer flask containing 100 ml of 1% D-xylose-mineral salts medium; and the medium is incubated on the shaker for 48 hours. The mycelia are harvested on filter paper in a suction funnel and washed thoroughly with cold distilled water or 10^{-2} *M* phosphate buffer, pH 7.2. Approximately 2 g of damp mycelia are obtained per 100 ml of culture medium.

Preparation of Crude Extracts. Cell-free extracts usually are prepared by grinding the damp mycelia with an equal weight of acid-washed sand in a chilled mortar for 20–25 minutes; 3–4 volumes of 0.1 *M* phosphate buffer, pH 7.2, then is added. However, for a larger quantity, equal amounts of damp mycelia and acid-washed sand plus two volumes of buffer are ground for 10–15 minutes at the maximum speed in an Omni-mixer, the chamber of which is immersed into an ice-water bath. After centrifugation for 10 minutes at 2600 g in a refrigerated centrifuge, the supernatant fluid is collected as the crude extract, and its protein content is adjusted to about 10 mg/ml before proceeding with the following steps.

First Calcium Phosphate Gel Adsorption. One hundred milliliters of the crude extract is adjusted to pH 6.2 by adding 1 M acetic acid; then 50 ml of calcium phosphate gel (2.9 mg/ml and aged for 6 months) is added to it in a protein to gel ratio of 1:4 and the mixture is stirred for 5 minutes. The gel then is removed, washed once with 10 ml of 0.1 M phosphate buffer, pH 6.2, and discarded. The supernatant and wash are combined. This treatment removes the enzymes which endogenously reduce TPN.

First Ammonium Sulfate Fractionation. Thirty-seven grams of finely powdered $(NH_4)_2SO_4$ is added slowly to the combined supernatant and wash while stirring. After the mixture has stood for 20 minutes, the precipitate is removed and discarded. The supernatant is treated with an additional 26 g of $(NH_4)_2SO_4$ and allowed to settle for 30 minutes. The precipitate is recovered and taken into 0.02 M phosphate buffer, pH 7.2; the solution then is centrifuged to remove any insoluble substances. The enzyme preparation now is devoid of the ability to oxidize TPNH endogenously.

Second Calcium Phosphate Gel Adsorption. The enzyme solution is adjusted to pH 5.5 with acetic acid, and immediately an amount of calcium phosphate gel in a protein to gel ratio of 1:2 is added. After 2–3 minutes of stirring, the gel is removed and washed twice with 0.5-ml portions of 0.1 M acetate buffer, pH 5.5, and then discarded. The supernatant and washes are combined.

Second Ammonium Sulfate Fractionation. To the combined enzyme preparation obtained from the previous step, saturated $(NH_4)_2SO_4$ solution at 4°C is added to give 0.4 saturation. After the preparation has stood for 15 minutes, the precipitate is removed and discarded; the supernatant again is treated with $(NH_4)_2SO_4$ to 0.5 saturation. The precipitate is recovered by centrifugation after settling for 15 minutes, and then is taken into 0.1 M phosphate buffer, pH 7.2. As summarized in the table, this preparation represents a 20-fold purification.

Properties

Specificity and Kinetic Properties. The partially purified enzyme preparation also reduces L-arabinose and D-ribose at a rate about one-fourth that on D-xylose. It does not react with D-glucose, D-galactose,

Purification Procedure

Fraction	Total volume (ml)	Total units $\times 10^{-3}$	Total protein (mg)	Specific activity
Crude extract	100	26	720	36
1st calcium phosphate gel supernatant	143	25.8	172	150
1st ammonium sulfate precipitate	8	17.9	66.4	270
2nd calcium phosphate gel supernatant	10	14.4	40	360
2nd ammonium sulfate precipitate	5.4	12.9	17.3	750

D-mannose, D-rhamnose, D-glucuronic acid, D-galacturonic acid, D-arabinose, or L-xylose. DPNH cannot replace TPNH. The K_m values for D-xylose and for TPNH were 9×10^{-2} M and 2.2×10^{-5} M, respectively.

Stability. The enzyme preparations experienced no loss in activity during storage at 4° for 3 months, but activity was destroyed completely by freezing overnight. The maximum stability was found when the preparations were stored at pH 7.0–7.2, and declined gradually on both sides of this range.

Effect of pH and Buffers. The reaction rate was found to be maximal at pH 7.2. There was no difference in activity when the reactions were carried out in either Tris or phosphate buffer.

Activators and Inhibitors. No stimulation of the enzyme activity was observed with 1×10^{-3} M glutathione or 1×10^{-2} M cysteine; however, these compounds restored activity after the enzyme had been inactivated by *p*-chloromercuribenzoate, which gives 50% inhibition at 5×10^{-4} M. Arsenite, cyanite, and iodoacetate at 1×10^{-2} M do not affect the activity, but fluoride gives 50% inhibition at 5×10^{-2} M. Among the metal ions, Zn^{++} inhibits activity 80% at 1×10^{-3} M, but Mg^{++}, Ca^{++}, or Mn^{++} at 1×10^{-2} M have no effect. Similarly, there is no inhibition with 1×10^{-2} M EDTA or 1×10^{-3} M dipicolinic acid.

Equilibrium. At pH 7.0–7.2, D-xylose reductase favors the complete reduction of D-xylose to D-xylitol in the presence of TPNH. However, there is a detectable rate of reverse reaction when the enzyme is incubated at pH 9.0 with concentrations of TPN and D-xylitol at 1×10^{-3} M and 1 M, respectively.

II. Xylitol Dehydrogenase

Assay Method

Principle. The enzyme activity can be determined by (1) measuring the increase in optical density at 340 mμ caused by the reduction of

DPN in the presence of xylitol, or (2) measuring the formation of xylulose from xylitol.[3]

Reaction Mixtures and Procedures

(1) DPN-linked oxidation of xylitol
D-Xylitol, 40 micromoles
DPN, 0.9 micromole
$MgCl_2$, 10.0 micromoles
Tris buffer, pH 7.5, 200 micromoles
Crude cellular extract, 0.5 ml

Total volume of 2.5 ml, incubation at room temperature; reaction started by the addition of xylitol.

(2) Formation of xylulose from xylitol
D-Xylitol, 20 micromoles
DPN, 25 μmoles
$MgCl_2$, 10 μmoles
Tris buffer, pH 7.5, 320 micromoles
Crude cellular extract, 2.5 ml, in a total volume of 6.5 ml

Incubate at 30°. Remove 1-ml aliquots at 30-minute intervals, and precipitate with 2 ml 5% $ZnSO_4$ and 2 ml 0.3 N $Ba(OH)_2$, both at pH 7.2. Treat the supernatant with charcoal and analyze for D-xylulose by the method of Dische and Borenfreund[4] with a 2-hour incubation at 37° and D-xylulose monoacetate as the standard.

Partial Purification Procedure

The cells are grown and a crude extract is prepared as for the demonstration of D-xylose reductase. The enzyme seems to be associated with a particulate cellular component, and activity varies with the amount of cell and particle breakage. Purification beyond the crude cellular extract has not been achieved.

Properties

Specificity and Kinetic Properties. There is a rapid decrease in adsorption with a DPNH-D-xylulose mixture and no decrease with DPNH-L-xylulose. Hence the oxidation product of D-xylitol is D-xylulose and the equilibrium favors the reduction of D-xylulose. L-Arabitol also is oxidized to L-xylulose and L-ribulose by two independent DPN-linked dehydrogenases.[5] The K_m values have not been determined.

[4] Z. Dische and E. Borenfreund, *J. Biol. Chem.* **192,** 583 (1951).
[5] C. Chiang and S. G. Knight, *Biochim. Biophys. Acta* **46,** 271 (1961).

Stability. Because of the apparent lack of enzyme stability the cellular preparations must be used immediately. Extensive studies on stability have not been done.

Effect of pH. The activity of D-xylitol dehydrogenase is maximal at pH 7.5; L-arabitol dehydrogenases apparently are maximal at pH 8.5.

Activators and Equilibrium. Mg^{++} is necessary for the reaction; other studies have not been made.

Equilibrium. In the complete system containing Tris buffer at pH 7.5, Mg^{++}, and a xylitol to DPN ratio of 1:1.25, the yield of the ketopentose was 15% after 2 hours' incubation. An increase to 20% ketopentose was obtained by coupling the reaction with a lactic dehydrogenase–pyruvate system to remove the DPNH. Both D-xylitol and L-arabitol oxidation proceeds at a much faster rate at pH 8.5.

[38] 2,3-*cis*-Polyol Dehydrogenase[1]

By WILLIAM B. JAKOBY

$$\begin{array}{c} CH_2OH \\ | \\ H\text{—}C\text{—}OH \\ | \\ H\text{—}C\text{—}OH \\ | \\ CH_2OH \end{array} + DPN^+ \rightleftharpoons \begin{array}{c} CH_2OH \\ | \\ C\text{=}O \\ | \\ H\text{—}C\text{—}OH \\ | \\ CH_2OH \end{array} + DPNH + H$$

The enzyme, originally named erythritol dehydrogenase,[1] catalyzes the reversible, DPN-linked oxidation of polyols containing 3, 4, or 5 carbons with the hydroxyl groups at carbons 2 and 3 *cis* to each other.

Assay

Principle. Enzyme is assayed by determining the increase in absorption at 340 mμ due to the formation of DPNH.

Reagents

Tris-chloride, *M*, at pH 8.3

Mercaptoethanol, 0.1 *M*, prepared just before use by adding 0.07 ml of the compound to 10 ml of water

DPN, 20 m*M*

$MnCl_2$, 1 m*M*

Erythritol, 50 m*M*

[1] W. B. Jakoby and J. Fredericks, *Biochim. Biophys. Acta* **48**, 26 (1961).

Procedure. The reagents are added in the following order and brought to a final volume of 1.0 ml with water: 50 μl Tris, 50 μl mercaptoethanol, 50 μl DPN, 50 μl $MnCl_2$, an appropriate dilution of enzyme and 100 μl of erythritol. The initial rate of the reaction as measured at 340 mμ is linear with respect to time and protein concentration over a 3-minute period when absorbancy values of less than 0.300 are encountered.

Definition of Unit. A unit of activity is defined as that amount of enzyme required to form 1 micromole of DPNH per minute. Specific activity is defined as the number of units per milligram of protein; protein is measured by the method of Lowry *et al.*[2] with bovine serum albumin as a standard.

Purification Procedure

Growth. The enzyme is obtained from a strain of *Aerobacter aerogenes*[3] which had been isolated from an enrichment culture with erythritol as sole carbon source. The medium for its growth contains the following in grams per liter: K_2HPO_4, 0.48; KH_2PO_4, 1.1; $MgSO_4 \cdot 7\ H_2O$, 0.25; NH_4NO_3, 1.0; erythritol, 2.0. Washings from a 24-hour agar slant culture are used to inoculate 20 ml of liquid medium which, in turn, serves as inoculum for 1 liter of medium in a 6-liter flask. The liter of culture is used as inoculum for 10 liters of medium in a 5-gallon carboy equipped for vigorous aeration. At this stage the medium is supplemented with 0.001% Dow-Corning AF antifoam emulsion and 0.004% yeast extract.

Cells are harvested with a Sharples centrifuge after 24 hours of growth at room temperature, washed with 5 volumes of saline, and stored at —15°. Since storage under these conditions results in drastic loss of enzyme activity within 1 week, it is advisable that the cells be used shortly after harvesting.

Step 1. Extract. Cells suspended in 5 volumes of 0.1 *M* potassium phosphate at pH 7.0 containing 5 m*M* mercaptoethanol and 3 m*M* ethylenediaminetetraacetic acid, are treated in a 10-kc Raytheon sonic oscillator for 10 minutes. The residue after centrifugation is discarded. This and all subsequent steps are to be carried out at approximately 2°.

Step 2. Protamine. A 1% solution of protamine sulfate is added so that 10 ml of extract is treated with 1 ml of protamine. The resultant precipitate is removed by centrifugation.

Step 3. Ammonium Sulfate. To 100 ml of supernatant fluid are added 35 g of ammonium sulfate and the precipitate is discarded. An additional

[2] O. H. Lowry, N. J. Rosebrough, A. L. Farr, and R. J. Randall, *J. Biol. Chem.* **193,** 265 (1951).

[3] Available from American Type Culture Collection as no. 13797.

14 g of ammonium sulfate are added and the resultant precipitate is dissolved in 0.1 *M* potassium phosphate at pH 7.0 containing 5 m*M* mercaptoethanol and 3 m*M* EDTA.

Step 4. DEAE-Cellulose. The enzyme preparation, in 10-ml batches, is used to charge a 3 × 8 cm column of DEAE-cellulose which has been equilibrated with 0.03 *M* Tris at pH 7.3. The same buffer is used to elute the enzyme which is recovered between 40 and 60 ml of eluent. The active fractions are charged onto a second column differing only in that the adsorbant has been washed with 0.03 *M* Tris at pH 8.1. The column is attached to a mixing flask containing 300 ml of 0.03 *M* Tris at pH 8.0 which is, in turn, connected to a reservoir containing 0.5 *M* sodium chloride in the same buffer. Enzyme activity is found in the fractions eluted between 25 and 50 ml. The eluates containing the activity are pooled and treated with an equal volume of alkaline,[4] saturated ammonium sulfate solution. The resulting precipitate is collected by centrifugation and dissolved in 0.05 *M* Tris at pH 7.1 containing 5 m*M* mercaptoethanol and 3 m*M* ethylenediaminetetraacetic acid.

Properties of the Enzyme

Specificity. Polyols of 3 to 5 carbons having *cis* hydroxyl groups at C-2 and C-3 are active. The following relative rates were obtained when the polyols were tested at 20 m*M*: erythritol, 100; ribitol, 27; L-arabitol (L-lyxitol), 16; D-arabitol (D-lyxitol), 10; glycerol, 6. The following were active in the reverse direction using DPNH although limiting supplies of substrates did not allow rate determinations at saturating concentrations: L-erythrulose, D-ribulose, L-ribulose, D-xylulose, and dihydroxyacetone.

The optimal concentrations of erythritol and DPN were 50 m*M* and 1.5 m*M*, respectively; TPN did not serve as substrate.

Metal. Either manganous ion or very much higher concentrations of

Purification Procedure

Step	Volume (ml)	Total activity (units)	Total protein (mg)	Specific activity (units/mg)
1. Extraction	77	157	477	0.3
2. Protamine treatment	81	200	478	0.4
3. Salt precipitation	27	61	243	0.3
4. Elution and precipitation	2	5	1.6	3.1

[4] Freshly prepared by the addition of 5 ml of concentrated commercial ammonium hydroxide to 100 ml of a saturated ammonium sulfate solution.

magnesium ions are required. Optimum concentrations were 2.5×10^{-5} *M* $MnCl_2$ and 5×10^{-3} *M* for $MgSO_4$. The following do not satisfy the metal requirement: $ZnCl_2$, $CnCl_2$, $CaCl_2$, $NiCl_2$, $CoCl_2$, and $FeSO_4$.

pH Optima. With erythrytol and DPN, activity was optimal at pH 8.6. A sharp optimum was obtained at pH 6.7 with L-erythrulose and DPNH.

[39] 2-Ketogluconic Acid Reductase

By J. De Ley

$$\text{2-Ketogluconate} + \text{NADPH} + \text{H}^+ \rightleftharpoons \text{gluconate} + \text{NADP}^+ \qquad (1)$$

Assay Method

Principle. The reaction may be measured in both directions: either in neutral or slightly acid medium by the disappearance of NADPH in the presence of 2-ketogluconate, or in alkaline medium by the formation of NADPH from $NADP^+$ and an excess of gluconate. The oxidation or reduction of the coenzyme is followed spectrophotometrically at 340 mμ.

Reagents

NADPH, 0.001 *M*, in 0.02 *M* phosphate buffer, pH 7
$NADP^+$, 0.001 *M*, in 0.02 *M* glycine-NaOH buffer, pH 10
Sodium 2-ketogluconate, 0.02 *M*
Sodium gluconate, 0.1 *M*
$MgCl_2$, 0.05 *M*
Phosphate buffer, 0.04 *M*, pH 7
Glycine-NaOH buffer, 0.04 *M*, pH 10
Enzyme in 0.005 *M* phosphate buffer pH 7

Procedure for the Measurement of the Reductase Activity. The reaction mixture in quartz Beckman cuvettes ($l = 1$) contains 0.35 micromole NADPH, 0.2–0.8 ml of enzyme, 5 micromoles of $MgCl_2$, 1.5 ml of phosphate buffer pH 7, and water to a final volume of 2.9 ml. The optical density is measured at 340 mμ. Two micromoles of sodium 2-ketogluconate in 0.1 ml is mixed in quickly, and the decrease in optical density is recorded or measured every 30 seconds for 10 minutes.

Procedure for the Measurement of the Dehydrogenase Activity. The reaction mixture in quartz Beckman cuvettes ($l = 1$) contains 0.35 micromole $NADP^+$, 0.2–0.8 ml of enzyme, 5 micromoles of $MgCl_2$, 1.5 ml 0.04 *M* glycine-NaOH buffer pH 10, 50 micromoles of sodium gluconate in 0.5 ml and water for a final volume of 3.0 ml. The increase in optical density is followed at 1-minute intervals for some 10 minutes.

Definition of Unit. One unit of enzyme activity is defined as that amount which causes the oxidation of 1 micromole of NADPH per minute under the assay conditions described. Specific activity (units per milligram of protein) is based on the spectrophotometric protein estimation.[1]

Source of Enzyme

The enzyme has been detected in several bacteria: *Brevibacterium helvolum*[2] (ATCC 19239), *Aerobacter cloacae*,[3] spores of *Bacillus cereus*.[4] It is particularly active in the acetic acid bacteria of *Gluconobacter*[5] and the mesoxydans group of *Acetobacter*. It has been most extensively studied with *B. helvolum*[2] and *Gluconobacter oxydans* (suboxydans).[6] It is also present in the yeast *Debaryomyces hansenii* (20-fold less activity) and the mold *Aspergillus nidulans* (30-fold less activity).[2]

Purification Procedure

Preparation from B. helvolum[2]

Step 1. Growth of Culture. The bacteria are grown in a medium containing 0.1% Difco yeast extract, 0.5% KH_2PO_4, 0.5% NaCl, 0.025% $MgSO_4 \cdot 7\ H_2O$, 0.025% $FeSO_4 \cdot 7\ H_2O$, and 0.5% sodium 2-ketogluconate (the latter sterilized cold and separately). The pH is 7.2. The culture is grown on a reciprocal shaker at 30° for 24 hours. The cells are centrifuged, washed, and suspended in 0.01 *M* phosphate buffer pH 7. The yield of wet cells is about 8 g per liter of medium.

Step 2. Preparation of Crude Enzyme and Ammonium Sulfate Fractionation. A suspension (270 mg wet living cells per milliliter 0.01 *M* phosphate buffer) is disrupted for 45 minutes at 4° in a 250-watt, 10-kc Raytheon sonic oscillator. The suspension is centrifuged for 2 hours at 100,000 *g* at 4°. The yellowish supernatant is dialyzed overnight at 4° against 0.01 *M* phosphate buffer, pH 7.0, to eliminate endogenous sub-

[1] O. Warburg and W. Christian, *Biochem. Z.* **310**, 384 (1941).

[2] J. De Ley and J. Defloor, *Biochim. Biophys. Acta* **33**, 47 (1959). Culture is available from The National Collection of Industrial Bacteria (NCIB 9792) Torry Research Station, Aberdeen, Scotland.

[3] J. De Ley and S. Verhofstede, *Enzymologia* **18**, 47 (1957).

[4] R. Doi, H. Halvorson, and B. Church, *J. Bacteriol.* **77**, 43 (1959).

[5] J. De Ley and A. H. Stouthamer, *Biochim. Biophys. Acta* **34**, 171 (1959).

[6] K. Kersters, W. A. Wood, and J. De Ley, *J. Biol. Chem.* **240**, 965 (1965). NCIB 9108.

strates. The enzyme is present mainly in the 0.35–0.63 saturated ammonium sulfate fraction.

Preparation of Purified Enzyme from Gluconobacter oxydans (suboxydans)[6]

Steps 1 and 2. The growth of the culture, preparation of the particle-free extract, and removal of the nucleic acids is the same as described in another contribution.[7]

Step 3. Ammonium Sulfate Fractionation. Solid ammonium sulfate is added to 50% saturation and the precipitate discarded. Ammonium sulfate it then added to 70% saturation, and the precipitate is collected by centrifugation in 10 ml phosphate 0.01 *M* buffer, pH 6.5, and dialyzed by gel-filtration.

Step 4. Chromatography on DEAE-Cellulose. The DEAE-cellulose columns are prepared as described elsewhere.[7] The dialyzed fraction (15 ml) is applied to the 4.9 $cm^2 \times 20$ cm column and washed on with 30 ml of 0.01 *M* phosphate buffer pH 6.5. Proteins are eluted by an increasing NaCl-gradient in the same buffer. The mixing chamber contains 450 ml of 0.01 *M* KH_2PO_4–Na_2HPO_4 buffer at pH 6.5, and the reservoir contains 0.3 *M* NaCl in the same buffer. The flow rate is maintained at 50–60 ml/hour and fractions of 10 ml each are collected. The bulk of the 2-ketogluconic acid reductase is eluted in the 170–250 ml fraction. It is completely separated from the 5-ketogluconic reductase and contains only traces of the NADP-D-*lyxo*-dehydrogenase. The purification, compared to the original crude extract, is nine-fold.

Properties

Coenzyme Specificity. With the enzyme from *B. helvolum* the reaction with NADH proceeds at 6% the velocity of that with NADPH.[2] The purified enzyme from *G. oxydans* (suboxydans) does not react with NADH.[6]

Substrate Specificity. The specificity is slightly different for the *B. helvolum* and the *G. oxydans* enzyme. The results are compiled in the table. When gluconic acid is oxidized only 2-ketogluconate accumulates, as shown by paper chromatography; 5-ketogluconate is not formed.[2] Galactonate is oxidized to 2-ketogalactonate.[5] The reduction of 2-ketogluconate yields only gluconate.[2] Since the enzyme from *G. oxydans* oxidizes likewise D-galactonate, D-xylonate, and 5-ketogluconate, this indicates that the substrate needs an L-*threo*-configuration to be oxidized.[6]

[7] K. Kersters and J. De Ley, this volume [34].

SPECIFICITY OF 2-KETOGLUCONIC ACID REDUCTASE: REDUCTION OF NADP[a]

Substrate	*B. helvolum* enzyme[2]	*G. oxydans* (suboxydans) enzyme[5,6]
D-Gluconate	100	100
D-Mannonate	7	—
D-Gulonate	2	—
D-Galactonate	1	37
L-Galactonate	—	0
2-Keto-3-deoxygalactonate	3	—
D-Lactobionate	4	—
Saccharate	3	—
D-Xylonate	7	46
D-Ribonate	0	4
D-Arabonate	0	0
Glucuronate	2	—
Mannuronate	0	—
Galacturonate	1	—
6-Phosphogluconate	25	—
5-Ketogluconate	—	195
2-Ketogluconate	0	0
Glycerol	—	0
meso-Erythritol	—	0
meso-Ribitol	—	0
D-Mannitol	—	0
D-Glucitol	—	0

[a] The relative reaction rates are expressed against gluconate, arbitrarily put at 100.

pH Optimum. The *B. helvolum* enzyme is weakly active at pH 5 and 11. The pH optimum for the reduction is 7–7.3 and for the oxidation it is about 10. The *G. oxydans* enzyme has a pH optimum for the reduction at 7.0–7.5 and for the oxidation at 11.3.

Physical Constants. The equilibrium constant of reaction (1), in which the pH is included, is $K_c = 0.52\ (\pm 0.19)\ 10^{-10}\ M$. The standard free energy change is $\Delta G°\ (\Delta F°) = 13{,}800$ cal $\pm$ 200 cal. $\Delta G°$ for reaction (2) is 8580 cal.

$$\text{Gluconate}_{(aq)} \rightarrow \text{5-ketogluconate}_{(aq)} + H_{2(gas)} \qquad (2)$$

The equilibrium constant of this reaction is $3.99\ 10^{-7}\ M$. At pH 7 and 20° the E'_0 of the system gluconate$^-$/2-ketogluconate$^-$ is —0.222 volt.

Physiological Function. In nature 2-ketogluconate is mainly produced by pseudomonads, either with an enzyme of the type described here or with a particulate gluconic acid dehydrogenase. Other microorganisms can grow at the expense of 2-ketogluconate by reducing it to gluconate, probably followed by gluconokinase and 6-phosphogluconate dehydrogenase. The latter enzyme would form the NADPH required for 2-ketogluconate reduction. Both enzymes form an internal oxidation-reduction system.

2-Ketogluconic acid reductase is different from the particulate gluconate dehydrogenase in *G. oxydans*,[5] *Pseudomonas aeruginosa*,[8] and *P. fluorescens*,[9] which is probably released from the cytoplasmic membrane.

[8] T. Ramakrishnan and J. J. R. Campbell, *Biochim. Biophys. Acta* **17**, 122 (1955).
[9] W. A. Wood and R. F. Schwerdt, *J. Biol. Chem.* **201**, 501 (1953).

[40] 5-Ketogluconic Acid Reductase

By J. De Ley

$$\text{5-Ketogluconate} + \text{NADPH} + \text{H}^+ \rightleftharpoons \text{gluconate} + \text{NADP}^+ \quad (1)$$

Assay Method

Principle. The reaction may be measured in two directions: either in neutral or slightly acid medium by the disappearance of NADPH in the presence of 5-ketogluconate, or in alkaline medium by the formation of NADPH from $NADP^+$ and an excess of gluconate. The oxidation or reduction of the coenzyme is followed spectrophotometrically at 340 mμ.

Reagents

NADPH, 0.001 *M*, in 0.2 *M* Tris-HCl buffer pH 7.5
$NADP^+$, 0.001 *M*, in 0.02 *M* glycine-NaOH buffer pH 10
Sodium 5-ketogluconate, 0.02 *M*
Sodium gluconate, 0.1 *M*
$MgCl_2$, 0.05 *M*
Tris-HCl buffer, 0.04 *M*, pH 7.5
Glycine-NaOH buffer, 0.04 *M*, pH 10
Enzyme in 0.005 *M* phosphate buffer, pH 7.0

Procedure for the Measurement of the Reductase Activity. The reaction mixture in quartz Beckman cuvettes ($l = 1$) contains 0.35 micromole NADPH, 5 micromoles $MgCl_2$, 0.025 ml enzyme, 1.5 ml Tris-HCl buffer pH 7.5, and water for a final volume of 2.9 ml. The optical density is measured at 340 mμ. Two micromoles of sodium 5-ketogluconate in 0.1 ml is mixed in quickly, and the decrease in optical density is recorded or measured every 30 seconds for 10 minutes.

Procedure for the Measurement of the Dehydrogenase Activity. The reaction mixture in quartz Beckman cuvettes ($l = 1$) contains 0.35 micromole $NADP^+$, 5 micromoles of $MgCl_2$, 0.1 ml of enzyme, 1.5 ml 0.04 *M* glycine-NaOH buffer pH 10, 50 micromoles of sodium gluconate in 0.5 ml, and water for a final volume of 3.0 ml. The increase in optical density is followed at 1-minute intervals for some 10 minutes.

Definition of Unit. One unit of enzyme activity is defined as that amount which causes the oxidation of 1 micromole of NADPH per minute under the assay conditions described. Specific activity (units per milligram of protein) is based on the spectrophotometric protein estimation.[1]

Source of Enzyme

The enzyme has been detected in bacteria. It is inducible in *Klebsiella* and *Escherichia*.[2] It is particularly active in the acetic acid bacteria of *Gluconobacter* and the mesoxydans group of *Acetobacter*. The detection of the enzyme in *Gluconobacter oxydans* (suboxydans)[3] was later confirmed[4,5] in other strains of the same variety.

Purification Procedure

It is preferable to select a strain of acetic acid bacteria which produce copious amounts of the insoluble calcium 5-ketogluconate, e.g., *Gluconobacter oxydans* (suboxydans).

Steps 1, 2, and 3. These are the same as described in another contribution.[6]

Step 4. Chromatography on DEAE-Cellulose. The DEAE-cellulose columns are prepared as described in another contribution.[7] The dialyzed fraction is applied to the 4.9 $cm^2 \times 20$ cm column and washed on with 30 ml of 0.01 *M* phosphate buffer pH 6.5. Proteins are eluted by an increasing NaCl gradient in the same buffer. The mixing chamber contains 450 ml of 0.01 *M* KH_2PO_4–Na_2HPO_4 buffer at pH 6.5 and the reservoir contains 0.3 *M* NaCl in the same buffer. The flow rate is maintained at 50–60 ml per hour, and fractions of 10 ml each are collected. The bulk of the enzyme is eluted in the 600–700 ml fractions. It is completely separated from the 2-ketogluconic reductase and NADP-D-*lyxo*-dehydro-

[1] O. Warburg and W. Christian, *Biochem. Z.* **310**, 384 (1941).
[2] J. De Ley, *Biochim. Biophys. Acta* **27**, 652 (1958).
[3] J. De Ley and A. H. Stouthamer, *Biochim. Biophys. Acta* **34**, 171 (1959).
[4] E. Galante, P. Scalaffa, and G. A. Lanzani, *Enzymologia* **27**, 176 (1964).
[5] K. Okamoto, *J. Biochem.* (*Tokyo*) **53**, 448 (1963).
[6] J. De Ley, this volume [39].
[7] K. Kersters and J. De Ley, this volume [34].

genase. The purification, compared to the original crude extract, is ninefold.

A partial-purification procedure by treatment of the crude extract with acrinol, charcoal and ammonium sulfate fractionation, has been described.[5]

Properties

Coenzyme Specificity. With the enzyme from *Klebsiella* and *Escherichia* the reaction proceeds twice as fast with NADH as with NADPH.[2] The purified enzyme from *G. oxydans* (suboxydans) reacts only with NADPH.[8]

Substrate Specificity. The enzyme from *Klebsiella* and *Escherichia* did not react with 2-ketogluconate, fructose or L-sorbose.[2] The results with two strains of *G. oxydans* are summarized in the table. When

SPECIFICITY OF 5-KETOGLUCONIC ACID REDUCTASE FROM *Gluconobacter oxydans* (SUBOXYDANS): REDUCTION OF NADP[a]

Substrate	Strain SU: De Ley and Stouthamer[b]	Strain SU: Kersters *et al.*[c]	Strain IFO 3432, Okamoto[d]
D-Gluconate	—	100	100
D-Mannonate	—	—	28
L-Galactonate	0	—	—
D-Ribonate	—	0	—
L-Idonate	—	—	12
D-Arabonate	—	0	—
L-Arabonate	—	—	0
D-Xylonate	—	0	0
5-Keto-D-gluconate	—	0	—
2-Keto-D-gluconate	—	0	—
Glucose	—	0	—
Glycerol	—	0	—
meso-Erythritol	—	0	—
meso-Ribitol	—	0	—
D-Mannitol	—	0	—
D-Glucitol	—	0	—
Pyruvate	0	—	—
D-Glucuronate	0	—	0
D-Galacturonate	0	—	—

[a] The relative reaction rates are expressed against gluconate, arbitrarily put at 100.
[b] J. De Ley and A. H. Stouthamer, *Biochim. Biophys. Acta* **34,** 171 (1959).
[c] K. Kersters, W. A. Wood, and J. De Ley, *J. Biol. Chem.* **240,** 965 (1965).
[d] K. Okamoto, *J. Biochem.* (*Tokyo*) **53,** 448 (1963).

[8] K. Kersters, W. A. Wood, and J. De Ley, *J. Biol. Chem.* **240,** 965 (1965).

gluconate is used as a substrate, only 5-ketogluconate is produced, as shown by paper chromatography.[2,5] In the reverse reaction, only gluconate accumulates, not idonate.[2,5] The enzyme is thus stereospecific. This enzyme is different from the isomeric reduction carried out by the 5-ketoglucono-*idono*-reductase present in *Fusarium*.[9]

pH Optimum. The enzyme from *Klebsiella* and *Escherichia* is active between pH 4 and 11. At pH 6–7 the reduction of 5-ketogluconate is heavily favored, whereas at pH 10–11 the reverse reaction occurs.[2] The enzyme from *G. oxydans* is active within the same limits; the pH optimum for the reduction is at 7.3–7.5, and for the oxidation at 9.3.[5,8]

Physical Constants. The equilibrium constant for reaction (1) is $3.5 \times 10^{-12}\ M$.[5] Starting with this value, we can calculate the change in standard free energy $\Delta G°$ ($\Delta F°$) to be 15,360 cal. Reaction (1) can be split into its two parts:

$$\text{Gluconate}_{(aq)} \rightarrow \text{5-ketogluconate}_{(aq)} + H_{2(g)} \quad (2)$$

$$NADP^+ + H_2 \rightarrow NADPH + H^+ \quad (3)$$

$\Delta G°$ for reaction (3) is known to be 5440 cal.[10] Therefore $\Delta G°$ of reaction (2) would be 9920 cal. The equilibrium constant of this reaction would be $4.0 \times 10^{-8}\ M$. At pH 7 and 20° the E'_0 of the system gluconate$^-$/5-ketogluconate$^-$ would be —0.191 volt.

[9] Y. Takagi, *Agr. Biol. Chem.* **26**, 719 (1962).
[10] K. Burton and T. H. Wilson, *Biochem. J.* **54**, 86 (1953).

[41] 3-Deoxy-D-*glycero*-2,5-hexodiulosonic Acid Reductase[1]

By JACK PREISS

$$H^+ + \text{3-deoxy-D-}glycero\text{-2,3-hexodiulosonic acid} + \begin{matrix}DPNH\\ \text{or}\\ TPNH\end{matrix} \rightleftharpoons \text{2-keto-3-deoxygluconic acid} + \begin{matrix}DPN^+\\ \text{or}\\ TPN^+\end{matrix}$$

Assay Method

Principle. The reduction of 3-deoxy-D-*glycero*-2,3-hexodiulosonic acid to 2-keto–3-deoxygluconic acid by extracts of polygalacturonate-grown cells is followed by measuring the oxidation of reduced pyridine nucleotide spectrophotometrically.

[1] J. Preiss and G. Ashwell, *J. Biol. Chem.* **238**, 1577 (1963). Also see this volume [107].

Reagents

Potassium phosphate, 1 *M*, pH 7.0

3-deoxy-D-*glycero*-2,5-hexodiulosonic acid, 0.01 *M*. This compound is prepared by the isomerization of 4-deoxy-L-*threo*-5-hexosulose uronic acid.[1]

TPNH, 0.01 *M*

Reductase

Procedure. The reaction mixture consisted of 0.05 ml of phosphate buffer, 0.01 ml of the hexodiulosonic acid, 0.01 ml of TPNH, and 0.002–0.008 unit of enzyme in a total volume of 1.0 ml. The reaction was initiated by the addition of enzyme, and the oxidation of TPNH at 340 mμ was recorded between 1 and 5 minutes. In the absence of the hexodiulosonic acid, little or no TPNH oxidation was observed, even in the crude extract, whereas with DPNH an appreciable blank value was noted.

Definition of Unit and Specific Activity. A unit is defined as that amount of enzyme required to catalyze the oxidation of 1 micromole of TPNH per minute under the above conditions. Specific activity is defined as units per milligram of protein. Protein concentrations were determined by the method of Lowry *et al.*[2]

Purification Procedure

Cultivation of Bacteria. The organism, ATCC 14968, used in this study, which is classified as a pseudomonad, was originally isolated by Dr. J. D. Smiley on the basis of its ability to utilize alginic acid as the sole carbon source. It was grown aerobically on a minimal medium consisting of ammonium nitrate 1%, dibasic potassium phosphate 1.5%, monobasic sodium phosphate 0.5%, magnesium sulfate heptahydrate 0.1%, and sodium polygalacturonate 0.11%. The organism was grown and harvested as described in this volume [107].

Step 1. Crude Extract. The frozen bacterial paste was suspended in 2 volumes of 0.1 *M* potassium buffer, pH 7.5, containing 0.1% cysteine-HCl (or 0.01 *M* GSH), and was disrupted by sonic vibration for 20 minutes in a 10-kc Raytheon oscillator. The broken-cell mixture was centrifuged at 10,000 rpm for 15 minutes, and the supernatant solution was used as the starting material for purification of the enzyme. All ensuing operations were carried out at 0–3°.

Step 2. Ammonium Sulfate Fractionation. To 20 ml of the crude

[2] O. H. Lowry, N. J. Rosebrough, A. L. Farr, and R. J. Randall, *J. Biol. Chem.* **193**, 265 (1951).

extract, 7.5 ml of a 5% solution of streptomycin sulfate was added slowly and with continuous stirring. After 10 minutes, the suspension was centrifuged at 10,000 rpm, and the precipitate was discarded. To 24 ml of the supernatant fluid was added 20 ml of a saturated solution of ammonium sulfate. After 10 minutes, the precipitate was removed by centrifugation, and 13.5 g of solid ammonium sulfate was added. The suspension was stirred continuously for 10 minutes and centrifuged, and the precipitate was dissolved in 10 ml of 0.05 *M* Tris-HCl buffer, pH 7.5, containing 0.02 *M* mercaptoethanol, 0.005 *M* EDTA. The resulting solution was dialyzed overnight against 500 ml of the same buffer solution.

Step 3. Calcium Phosphate Gel Adsorption. To 10 ml of the ammonium sulfate fraction was added 50 ml of $CaPO_4$ gel suspension (10.8 mg of solids per milliliter). After 15 minutes, the gel was centrifuged and the residue was discarded. The supernatant fluid contained 80% of the total activity and about 20% of the protein. To this was added another 30 ml of calcium gel suspension, and after 15 minutes the mixture was centrifuged. The gel was then washed with 20 ml of 0.05 *M* Tris, pH 7.4, containing 0.02 *M* mercaptoethanol and was eluted three times with 8 ml of 0.01 *M* phosphate, pH 7.5, containing 0.02 *M* mercaptoethanol and 0.001 *M* EDTA. The eluates were pooled, and DPN was added to a concentration of 0.001 *M*.

Step 4. DEAE-Cellulose Chromatography. To a DEAE-cellulose column, 1 cm wide × 7.6 cm, that had been equilibrated with 0.03 *M* Tris buffer, pH 7.4, was added 21 ml of the calcium phosphate gel eluate. The column was washed with 10 ml of 0.01 *M* phosphate, pH 7.4, containing 0.001 *M* EDTA and 0.02 *M* mercaptoethanol and then with 20 ml of 0.1 *M* phosphate, pH 7.4, containing similar amounts of EDTA and mercaptoethanol. The enzyme was eluted from the column with the use of a linear gradient that contained 100 ml of 0.1 *M* phosphate, pH 7.4, in the mixing chamber and 100 ml of 0.4 *M* phosphate, pH 7.4, in the reservoir, both solutions containing 0.02 *M* mercaptoethanol and 0.001 *M* EDTA. After collection of 60 ml, the enzyme appeared in the effluent, and 48% of the absorbed activity was collected in the next 52 ml. The enzyme was pooled as three separate fractions designated DEAE-cellulose fractions I, II, and III, respectively, and made 0.001 *M* with respect to DPN. Results of the overall purification procedure are summarized in the table.

Properties

Stability. The final DEAE-cellulose fractions were markedly unstable in the absence of DPN or TPN and lost all their activity overnight

PURIFICATION OF REDUCTASE

Step and fraction	Volume (ml)	Activity (units/ml)	Protein (mg/ml)	Specific activity (units/mg)
1. Crude extract	20	79	53	1.5
2. Dialyzed ammonium sulfate	13.0	83	11.0	7.5
3. $CaPO_4$ gel eluate	24	17	0.32	53
4. DEAE-cellulose fractions				
I	15	3.2	0.04	80
II	25	3.9	0.02	195
III	12	2.4	0.03	80

at 0–3°. In the presence of 0.001 *M* DPN or TPN, the enzyme retained 50–67% of its initial activity after storage at 0–3° for 5 weeks.

Effect of pH. Maximal activity was obtained at pH 7.0–8.0, whether phosphate or Tris-HCl was the buffer or whether DPN or TPN was the reducing agent.

Specificity. The purified dehydrogenase exhibited activity with both TPNH and DPNH as electron donors. Neither 4-deoxy-5-L-*erythro*-hexosulose uronic acid nor 4-deoxy-5-L-*threo*-hexosulose uronic acid can be reduced by the reductase.

Occurrence of Enzyme. The reductase is not found in extracts of glucose-grown pseudomonad cells.[1] It appears to be involved in the metabolism of one of the intermediates of polygalacturonate degradation.[1,3]

[3] J. Preiss and G. Ashwell, *J. Biol. Chem.* **238,** 1571 (1963).

[42] 4-Deoxy-L-*erythro*-5-hexosulose Uronic Acid Reductase

By JACK PREISS

$$\text{4-Deoxy-L-}\mathit{erythro}\text{-5-hexosulose uronic acid} + \text{TPNH} + \text{H}^+ \rightleftharpoons \text{2-keto-3-deoxygluconic acid} + \text{TPN}^+$$

Assay Method

Principle. The reduction of the keto-deoxy-uronic acid, a product of bacterial degradation of alginic acid[1,2] is followed by measuring the oxidation of TPNH spectrophotometrically.

[1] J. Preiss and G. Ashwell, *J. Biol. Chem.* **237,** 309 (1962). Also see Vol. VIII [110].
[2] J. Preiss and G. Ashwell, *J. Biol. Chem.* **237,** 317 (1962).

Reagents

4-Deoxy-L-*erythro*-5-hexosulose uronic acid, 0.004 *M*. This compound can be prepared by enzymatic degradation of alginic acid by alginate lyase.[1]

Potassium phosphate buffer, 0.5 *M* pH 7.0

TPNH, 0.01 *M*

Reductase

Procedure. The phosphate buffer 0.1 ml, TPNH 0.01 ml, keto-deoxyuronic acid 0.1 ml, and reductase (0.002–0.008 unit) are mixed in a total volume of 1.0 ml. The reaction was initiated by the addition of enzyme, and the oxidation of TPNH was recorded at 340 mμ between 1 and 5 minutes. In the absence of substrate, no TPNH oxidation was observed.

Definition of Unit and Specific Activity. One unit is defined as that amount of enzyme required to catalyze the oxidation of 1 micromole of TPNH per minute under the above conditions. Specific activity is defined as units of enzyme per milligram of protein. Protein concentration was determined by the method of Lowry *et al.*[3]

Purification Procedure

The organism, classified as a pseudomonad (ATCC 14968), was grown aerobically on a minimal salt medium composed of NH_4NO_3 1%, K_2HPO_4 1.5%, NaH_2PO_4 0.5%, $MgSO_4 \cdot 7\ H_2O$ 0.1%, and sodium alginate[4] 0.11%. After 5 days of growth at room temperature, the bacteria were harvested in a Sharples centrifuge. The bacterial paste, bright orange in color, was stored frozen at —15°. Approximately 1 g of cells (wet weight) was obtained per liter of culture fluid. The organism is maintained in the laboratory at room temperature in the media described above and is transferred to fresh media monthly.

Step 1. Crude Extract. The frozen bacterial paste was suspended in 4 volumes of 0.1 *M* potassium phosphate buffer, pH 7.5, containing 0.01 *M* glutathione, and disrupted by sonic vibration for 20 minutes in a 10-kc Raytheon oscillator. The broken-cell mixture was centrifuged at 10,000 rpm for 15 minutes, and the supernatant solution was used as the starting material for purification of the enzyme. All ensuing operations were carried out at 0–3°.

Step 2. Streptomycin Sulfate Fractionation. To 28 ml of the crude extract, 18 ml of a 5% solution of streptomycin sulfate was added slowly

[3] O. H. Lowry, N. J. Rosebrough, A. L. Farr, and R. J. Randall, *J. Biol. Chem.* **193,** 265 (1951).

[4] Obtained from the Kelco Company, Los Angeles, California.

and with continuous stirring. After 15 minutes the suspension was centrifuged at 10,000 rpm and the precipitate was discarded.

Step 3. Ammonium Sulfate Fractionation. To 42 ml of the above streptomycin fraction was added 33 ml of a saturated solution of ammonium sulfate. After 10 minutes the precipitate was removed by centrifugation and an additional 99 ml of the saturated ammonium sulfate solution was added. The suspension was stirred continuously for 20 minutes and centrifuged; the precipitate was dissolved in 14 ml of 0.05 *M* Tris buffer, pH 7.5, containing 0.02 *M* β-mercaptoethanol, 0.001 *M* EDTA, and 0.0005 *M* TPN. The resulting solution was dialyzed overnight against 500 ml of the same buffer solution.

Step 4. Calcium Phosphate Gel. To 15 ml of the above fraction was added 13 ml of a calcium phosphate gel suspension (10.8 mg of solids per milliliter). After 15 minutes, the gel was centrifuged and to the resulting supernatant 0.015 ml of 0.1 *M* TPN, 0.13 ml of *M* β-mercaptoethanol, and 0.25 ml of 0.1 *M* EDTA were added.

Step 5. Alumina Gel Cγ Adsorption. To 25 ml of the gel fraction, 3 ml of alumina gel Cγ suspension (9.5 mg of solids per milliliter) was added. After standing for 15 minutes, the suspension was centrifuged and the supernatant solution, containing about 20% of the total activity, was discarded. The gel was then eluted twice with 10-ml portions of 0.01 *M* potassium phosphate, pH 7.5, containing 0.05 *M* β-mercaptoethanol and 0.001 *M* EDTA. The two eluates were combined and made 0.001 *M* with respect to TPN.

Step 6. DEAE-Cellulose Chromatography. To a DEAE-cellulose column (1 × 3.5 cm) which had been equilibrated with 0.03 *M* Tris buffer, pH 7.4, were added 18 ml of the alumina gel Cγ fraction. The column was washed successively with 5 ml of 0.01 *M* potassium phosphate, pH 7.4, containing 0.05 *M* β-mercaptoethanol and 0.001 *M* EDTA and then with 5 ml of 0.05 *M* potassium phosphate, pH 7.4. Elution was begun with a 0.2 *M* phosphate buffer, pH 7.4, containing 0.05 *M* β-mercaptoethanol. Approximately 84% of the adsorbed activity was recovered in the eluate designated as DEAE-cellulose fractions I, II, and III. Each of these fractions was made 0.001 *M* with respect to TPN and EDTA and stored at 0–3°. Results of the overall purification procedure are summarized in the table.

Properties

Stability. The final DEAE-cellulose fractions are markedly unstable in the absence of TPN and lose approximately 50% of their activity upon standing overnight at 0–3°. In the presence of 0.001 *M* TPN, the enzyme retained 90% of its initial activity after storage at 0–3° for 3 weeks.

PURIFICATION OF 4-DEOXY-L-*erythro*-5-HEXOSULOSE URONIC ACID REDUCTASE

Step	Volume (ml)	Activity (units/ml)	Protein (mg/ml)	Specific activity	Yield (%)
1. Crude extract	28	7.9	16.9	0.46	100
2. Streptomycin supernatant fluid	44	5.3	7.5	0.71	105
3. Dialyzed ammonium sulfate	17.5	12.4	4.7	2.6	102
4. $CaPO_4$ gel supernatant fluid	28	6.2	1.6	3.9	97
5. Aluminum gel $C\gamma$ eluate	20	4.6	0.67	6.9	59
6. DEAE-cellulose fractions:					
I	6.0	9.0	1.00	9.0	49
II	6.0	6.1	0.40	15.5	
III	6.0	2.7	0.17	15.9	

Effect of pH. Maximal activity was obtained at pH 7.0–7.5 with potassium phosphate buffer.

Inhibtiors. No significant effect upon the rate of the reaction was observed with β-mercaptoethanol, 10^{-2} *M*; bovine serum albumin, 100 g/ml; EDTA, 10^{-2} *M*; or sodium arsenite, 10^{-3} *M*. In the presence of *p*-hydroxymercuribenzoate, 10^{-4} *M*, the reaction rate was inhibited 26%; at 10^{-3} *M*, the inhibition amounted to 91%. Tris-HCl buffer, pH 7.5, 0.05 *M*, inhibited the reaction rate by 67%.

Effect of Substrate Concentration. At a 10^{-4} *M* concentration of TPNH, the K_m value for 4-deoxy-5-keto-uronic acid was 3.6×10^{-4} *M* and the V_{max} was 30.3 micromoles per milligram per minute. With 4-deoxy-5-keto-uronic acid at a concentration of 1.5×10^{-3} *M*, the K_m of TPNH was 3.2×10^{-5} *M*.

Specificity. D-Mannuronic acid, L-guluronic acid, D-glucuronic acid. D-galacturonic acid, L-iduronic acid, and 4-deoxy-L-*threo*-5-hexosulose uronic acid were inactive with the reductase. At 10^{-4} *M*, a concentration which is saturating for TPNH, DPNH oxidation proceeded at 1/14 the rate of TPNH oxidation.

Occurrence of Enzyme. The enzyme is not present in extracts of glucose-grown or of polygalacturonate-grown cells.[5]

[5] J. Preiss and G. Ashwell, *J. Biol. Chem.* **238,** 1577 (1963).

[43] Glyceraldehyde 3-Phosphate Dehydrogenase—Crystalline

By William S. Allison

D-Glyceraldehyde 3-phosphate + NAD^+ + AsO_4 →
3-phosphoglycerate + NADH + H^+ + AsO_4^{2-}

I. Human Heart Muscle[1]

Assay Method

The enzyme is assayed spectrophotometrically as described below for the lobster enzyme. Since human heart muscle contains α-glycerophosphate dehydrogenase and triosephosphate isomerase, crude extracts cannot be assayed by this method. However, it is not necessary to assay the crude extract.

Purification Procedure

General Considerations. Unless otherwise indicated, all steps are carried out at 0–5°. Saturated ammonium sulfate solutions are prepared at 23° from the reagent grade salt and contain 10^{-3} *M* EDTA.

During purification, protein is estimated spectrophotometrically by the method of Warburg and Christian.[2]

Preparation of the Extract. Eight to ten human hearts which have been frozen at autopsy are thawed, trimmed free of fat, and passed through a commercial meat grinder. The mince is stirred with 10^{-3} *M* EDTA, pH 7.0, which contains 10^{-3} *M* β-mercaptoethanol (βMSH) (2 liters per kilogram of muscle) for 2 hours. The suspension is then centrifuged at 1200 *g* in an International PR II refrigerated centrifuge. The clear, red supernatant is strained through a double layer of guaze to remove floating debris.

Ammonium Sulfate Fractionation. To each liter of extract, 472 g of solid ammonium sulfate is added over a 1-hour period. The resulting protein precipitate is removed by gravity filtration on large, fluted filter papers. Then approximately 200 g of solid ammonium sulfate is added to each liter of supernatant to saturate it with ammonium sulfate. The resulting protein precipitate is allowed to equilibrate overnight and is collected on large fluted filter papers.

Carboxymethyl Cellulose Chromatography. The 70–100% ammonium sulfate precipitate is scraped off the filter papers and dissolved in a mini-

[1] W. S. Allison and N. O. Kaplan, *J. Biol. Chem.* **239,** 2140 (1964).
[2] O. Warburg and W. Christian, *Biochem. Z.* **310,** 384 (1941).

mal volume of 0.02 M sodium phosphate, pH 6.5, which contains 10^{-3} M EDTA and 10^{-3} M βMSH. The protein solution is then dialyzed against 30 volumes of the same buffer with two changes at 10-hour intervals. The dialyzed protein solution is clarified by centrifugation at 20,000 g and then applied to a 3.5×50 cm column of CM-cellulose, which is equilibrated with 0.02 M sodium phosphate, pH 6.5, which contains 10^{-3} M EDTA and 10^{-3} M. βMSH. The CM-cellulose (type 20, purchased from Brown Company, Berlin, New Hampshire) is washed and equilibrated as described by Peterson and Sober.[3] The column is packed by gravity in the cold. After application of the protein solution, the enzyme is eluted with 0.02 M sodium phosphate, pH 6.5, which contains 10^{-3} M EDTA and 10^{-3} M βMSH. Under these conditions the TPD activity is washed through the column while hemoglobin is absorbed. The 10-ml effluent fractions containing enzyme activity are pooled and dialyzed against 30 volumes of saturated ammonium sulfate containing 10^{-3} M EDTA.

Crystallization. The precipitated, chromatographed protein is dissolved in enough 10^{-3} M EDTA, pH 7.0, to give a final protein concentration of 8–10 mg/ml. Then 2 volumes of ammonium sulfate solution, saturated at 23° and adjusted to pH 8.3 with 15% ammonium hydroxide, are added slowly to the protein solution which is swirled in an ice bath. Any precipitated protein is removed immediately by centrifugation at 20,000 g. The clear solution is then placed in a refrigerator at 5°. Within a few hours fine crystalline needles are observed under the microscope. The enzyme is recrystallized twice as described above and is stored as a crystalline suspension at 5° in 70% saturated ammonium sulfate, pH 8.0, which contains 10^{-3} M EDTA.

The third crystals have a specific activity of 40,000 units per milligram of protein per minute when assayed under the previously described conditions when it is assumed that the $E^{1\%}_{1\text{ cm}} = 1.0$ at 280 mμ. The third crystals are homogeneous as assessed by electrophoretic and ultracentrifugal analyses.

Properties

Stability of the Enzyme. The enzyme can be stored for several months at 5° in the ammonium sulfate mother liquor in the presence of EDTA without any loss in activity.

Comparison with Rabbit Muscle TPD. Like the rabbit muscle enzyme,[4,5] the human TPD crystallizes with bound DPN and the en-

[3] E. A. Peterson and H. A. Sober, Vol. V [1].

[4] G. T. Cori, M. W. Slein, and C. F. Cori, *J. Biol. Chem.* **173**, 605 (1948).

[5] I. Krimsky and E. Racker, *J. Biol. Chem.* **198**, 721 (1952).

zyme-coenzyme complex exhibits the familiar 360-mμ absorption band when the enzyme is in its active state.[1] The 280 mμ:260 mμ absorption ratio of the pure enzyme is 1·1.

The amino acid composition and catalytic properties of the human TPD are very similar to those of the rabbit muscle enzyme; however, slight differences in both these properties have been detected. The electrophoretic properties and sedimentation characteristics of the two enzymes are identical. The $S^0_{20,w}$ value for both enzymes is 7.45 ± 0.15 S.[1]

TABLE I
PURIFICATION PROCEDURE: HUMAN HEART MUSCLE ENZYME

Fraction	Volume (ml)	Protein (g)	Specific activity
1. Extract	7600	150	—
2. 70% saturated $(NH_4)_2SO_4$ supernatant	7900	50	800
3. 70–100% saturated $(NH_4)_2SO_4$ fraction	85	9.3	2,900
4. Pooled active fractions after CM-cellulose chromatography	180	0.720	32,000
5. First crystals	180	0.590	38,000
6. Third crystals	120	0.500	40,000

II. Lobster Tail Muscle[1]

Assay Method

Reagents

Sodium pyrophosphate buffer, 0.05 *M*, adjusted to pH 8.5 with HCl
DPN, 0.015 *M*
D-Glyceraldehyde 3-phosphate as an equimolar mixture of triosephosphate esters prepared by the method of Beisenherz *et al.*[6]
Disodium arsenate, 0.3 *M*

Procedure. Activity measurements are carried out spectrophotometrically essentially as described by Velick.[7] The assay reaction mixtures are prepared at room temperature and contain 2.70 ml of pyrophosphate buffer; 0.05 ml of DPN; 0.05 ml of D-glyceraldehyde 3-phosphate; 0.1 ml of diluted enzyme solution; and 0.05 ml of disodium arsenate. All reagents with the exception of the arsenate are mixed and the zero time

[6] G. Beisenherz, T. Bücher, and K.-H. Garbade, Vol. I, p. 391.
[7] S. F. Velick, Vol. I, p. 401.

optical density is read at 340 mμ. Then the arsenate is added with rapid mixing and the increase in optical density is recorded at 15-second intervals for 2 minutes.

Definition of Enzyme Units and Specific Activity. One unit of enzyme is defined as the amount of enzyme which produces an increase of optical density of 0.001 per minute at 340 mμ. Since the rate of reaction decreases with time, enzyme units are calculated by multiplying the increment in optical density (OD) between 15 and 45 seconds after initiating the reaction by 2000. Specific activity is expressed as units per milligram of protein:

$$\text{Sp. Act.} = \frac{\Delta OD_{15-45} \times 2000}{\text{mg protein}} = \frac{\text{units}}{\text{mg protein}}$$

When measured in this way, enzyme units are linearly proportional to the concentration of pure enzyme for the range of 3–10 μg of enzyme.

Since lobster muscle contains a potent α-glycerophosphate dehydrogenase and triosephosphate isomerase, crude extracts cannot be assayed by this method. However, an assay of crude extracts is not necessary.

Purification Procedure

General Considerations. Unless otherwise indicated all steps are carried out at 0–5°. Saturated ammonium sulfate solutions are prepared at 23° with the reagent grade salt and contain 10^{-3} *M* EDTA. Dilutions of these solutions are made by volume with 10^{-3} *M* EDTA, pH 7.0.

During the purification, protein is estimated spectrophotometrically by the method of Warburg and Christian.[2]

Preparation of the Extract. The shell is most easily removed from deep-frozen tail muscle of the North American lobster (*Homarus americanus*) with tin snips. Two kilograms of shelled tail muscle from approximately 20 one-pound lobsters is passed through a commercial meat grinder. The mince is stirred with 6 liters of 10^{-3} *M* EDTA, pH 7.0 which contains 10^{-3} *M* β-mercaptoethanol (βMSH) for 3 hours. The suspension is then centrifuged in an International PR II refrigerated centrifuge at 1200 *g*, and the supernatant is filtered through a double layer of gauze to remove floating debris.

Fractionation with Ammonium Sulfate. To each liter of extract, 472 g of solid ammonium sulfate is added over a 1-hour period. The protein precipitate is immediately removed by gravity filtration on large, fluted filter papers. The supernatant is then saturated with ammonium sulfate by the addition of approximately 200 g of solid ammonium sulfate to each liter. The resulting precipitate is equilibrated for at least 5 hours and is then collected on large fluted filter papers.

Crystallization: The 70–100% saturated ammonium sulfate precipitate is scraped off the filter papers with a large spatula and is dissolved in a minimal volume of 10^{-3} *M* EDTA, pH 7.0, which contains 10^{-3} *M* βMSH and is filtered through glass wool to remove filter paper scraps. The protein solution is then dialyzed against 50 volumes of 10^{-3} *M* EDTA, pH 7.0 which contains 10^{-3} *M* βMSH to remove occluded ammonium sulfate. After dialysis the protein solution is centrifuged and is brought to 50% saturation by the addition of solid ammonium sulfate (31.3 g/100 ml). Then an equal volume of saturated ammonium sulfate (prepared at 23° in 10^{-3} *M* EDTA, pH 7.0) is added to this solution slowly with swirling in an ice bath. The resulting precipitated is removed immediately by centrifugation at 20,000 *g*. The supernatant is filtered through glass wool. Then saturated ammonium sulfate is added dropwise with swirling to the clear supernatant in an ice bath until it becomes very slightly turbid. The solution is then placed in a refrigerator at 5°. Within a few hours after the first appearance of turbidity, small polyhedral crystals are observed under the microscope.

Recrystallization. The first crystals are allowed to grow for a few days and are then harvested by centrifugation at 20,000 *g*. They are dissolved in a volume of 50% saturated ammonium sulfate containing 10^{-3} *M* βMSH, to give a final protein concentration of 8–10 mg/ml. Then saturated ammonium sulfate prepared at 23° is added slowly to the protein solution, which is swirled in an ice bath, until slight turbidity appears. Turbidity first appears when approximately an equal volume of saturated ammonium sulfate is added. Crystals again form within a few hours at 5°. After a second recrystallization the enzyme is stored at 5° in 80% saturated ammonium sulfate which contains 10^{-3} *M* EDTA, pH 7.0 and 10^{-3} *M* βMSH.

The second crystals have a specific activity of 40,000 units per milligram of enzyme per minute which is not increased by further recrystallization. The second crystals are homogeneous as determined by ultracentrifugal, electrophoretic, and immunological methods.[1]

Yield. The overall yield of crystalline enzyme is not increased significantly by reextraction. However, yields are increased by carrying out all steps in the presence of 10^{-3} *M* βMSH. It is also important to carry out the first ammonium sulfate fractionation as soon as possible after having made the extract in order to obtain high yields. Since hemocyanin is removed in the first ammonium sulfate precipitate, it is possible that its removal prevents copper from being liberated into solution.

Properties

Stability of the Enzyme. When the crystalline enzyme is stored at 5° in 80% saturated ammonium sulfate in the presence of 10^{-3} *M* EDTA, but in the absence of thiols, maximal activity is maintained from 6 to 12 weeks.[8] Then losses in activity are observed which can be partially restored by thiols. Maximal activity is observed for a longer period if the crystals are stored in 80% saturated ammonium sulfate in the presence of 10^{-3} M EDTA and 10^{-3} M βMSH, especially if they are resuspended in fresh βMSH-EDTA-ammonium sulfate solution monthly.

Physical and Chemical Properties. Like the rabbit muscle enzyme,[5] lobster triosephosphate dehydrogenase crystallizes with firmly bound DPN which cannot be removed by recrystallization.[1] The 280 mμ:260 mμ absorption ratio is 1.1. The active enzyme-DPN complex exhibits a broad absorption band with a maximum at 360 mμ, similar to the one described by Racker for the rabbit muscle enzyme. The $E^{1\%}_{1\ \text{cm}} = 9.6$ at 280 mμ in 0.05 M sodium pyrophosphate, pH 8.5.

The amino acid composition of lobster TPD differs significantly from that of the rabbit muscle enzyme. However, when the enzyme is digested with trypsin after specific inhibition with 2-^{14}C-iodoacetate at 0° at pH 7.0, a single radioactive peptide is obtained which contains 2-^{14}C carboxymethylcysteine. The amino acid sequence around the radioactive carboxymethylcysteine residue is identical to the "active sulfhydryl" sequence described by Harris *et al.*[9,10] for the rabbit, pig, and yeast TPD's over a range of 12 amino acid residues.[11]

Chemical and physical evidence indicates that the enzyme is composed of four identical subunits of molecular weight 35,000.[11] There is no evidence of isoenzymes of TPD in lobster muscle.[1]

Watson and Banaszak have found that crystals of the native enzyme and an isomorphous PCMB derivative are suitable for extensive X-ray diffraction analysis.[12]

TABLE II
PURIFICATION PROCEDURE: LOBSTER TAIL MUSCLE ENZYME

Fraction	Volume (ml)	Protein (g)	Specific activity
1. Extract	5800	80	—
2. 70% saturated $(NH_4)_2SO_4$ supernatant	5950	40	2,200
3. 70–100% saturated $(NH_4)_2SO_4$ fraction	180	12.2	6,700
4. First crystals	400	2.10	36,000
5. Third crystals	200	1.750	40,000

[8] W. S. Allison and N. O. Kaplan, *Biochemistry* **3,** 1792 (1964).
[9] J. I. Harris, B. P. Meriwether, and J. H. Park, *Nature* **197,** 154 (1963).
[10] R. N. Perham and J. I. Harris, *J. Mol. Biol.* **7,** 316 (1963).
[11] W. S. Allison and J. I. Harris, *J. Mol. Biol.* in press (1966).
[12] H. C. Watson and L. J. Banaszak, *Nature* **204,** 918 (1965).

[44] D-3-Phosphoglycerate Dehydrogenase

By H. J. Sallach

D-3-Phosphoglycerate + DPN^+ ↔ phosphohydroxypyruvate + DPNH + H^+

Assay Method[1]

Principle. Due to the greater stability and availability of D-3-phosphoglycerate (3-PGA), as compared to phosphohydroxypyruvate, the most convenient assay for the enzyme is a spectrophotometric measurement of the increase in optical density at 340 mμ, the absorption maximum of DPNH. The equilibrium of the reaction strongly favors the reduction of phosphohydroxypyruvate. However, since the oxidative reaction results in the liberation of a proton and the production of a carbonyl compound, it is shifted to the right by carrying out the reaction at pH 9 and in the presence of a carbonyl-trapping reagent, hydrazine.

Reagents

3-PGA, potassium salt, 0.1 *M*, pH 7.0
DPN, 0.015 *M*
Glutathione, 0.05 *M*, adjusted to pH 6 with 1 *N* KOH
Stock buffer solution. Mix 5 volumes of 1.0 *M* Tris-chloride, pH 9.0, 4 volumes of 1.0 *M* hydrazine acetate, pH 9.0, and 1 volume of 0.25 *M* EDTA, pH 9.0, on day of use.
Enzyme. Prior to assay, dilute in 0.01 *M* phosphate buffer, pH 6.0, so that 0.1 ml will contain approximately 20 units as defined below.

Procedure. Place 1.0 ml of buffer solution, 0.1 ml of glutathione, 0.25 ml of 3-PGA (omit from control and blank), 0.1 ml of enzyme, and water to a final volume of 2.15 ml in each of three quartz cells with 1.0-cm light paths. The reaction is initiated by the addition of 0.1 ml of DPN (replaced by water in blank). Take readings at 340 mμ at 1 minute intervals for 5 minutes against the blank.

Definition of Unit and Specific Activity. One unit of enzyme is defined as that amount which causes an initial rate of change in absorbance (ΔA_{340}) of 0.001 per minute under the above conditions. Specific activity is expressed as units per milligram of protein as determined by the

[1] J. E. Willis and H. J. Sallach, *Biochim. Biophys. Acta* **81**, 39 (1964).

method of Lowry *et al.*[2] or by the optical density at 280 mμ after correcting for nucleic acid absorption at 260 mμ.[3]

Application of Assay Method to Crude Tissue Extracts. With crude enzyme preparations there is an appreciable change in absorbance observed with the control due to the nonspecific reduction of DPN by contaminating substrates. A correction for this ΔA_{340} must be made in calculating enzyme units.

Purification Procedure[4]

The following general conditions are used in the purification unless otherwise specifically stated. All operations are conducted at 4°. Fractionations with ammonium sulfate are made by the slow addition, with stirring, of the calculated amount of solid ammonium sulfate. The resulting suspensions are equilibrated for 30 minutes before centrifugation at 14,000 *g* for 25 minutes. In the chromatographic procedures employing cellulose derivatives, protein in the eluates from the columns is detected by the absorbance at 280 mμ.[5] All buffers are prepared from the potassium salts. A typical enzyme preparation is described below.

Step 1. Preparation of Crude Extract. Chicken liver is obtained fresh from the slaughter house and freed of connective tissue and fat.[6] The liver (2.13 kg) is homogenized in a Waring blendor for 1 minute with 2 volumes of 0.15 *M* acetate buffer, pH 4.6, containing 0.003 *M* EDTA and 0.001 *M* β-mercaptoethanol. The homogenate is centrifuged, and the residue is discarded.

Step 2. Ammonium Sulfate Fractionations. To the above supernatant solution is added 24 millimoles of EDTA, pH 6.0, and 3.5 millimoles of β-mercaptoethanol to give a final volume of 3.53 liters. The solution is brought to 43% saturation with 954 g of ammonium sulfate. The precipitate obtained on centrifugation is resuspended in 353 ml of 0.005 *M* EDTA containing 0.002 *M* β-mercaptoethanol (0.1 of original volume). The residue obtained on centrifugation is discarded and the supernatant fluid is adjusted to pH 7.2 by the addition of 465 ml of 0.5 *M* phosphate buffer, pH 7.8 (0.0005 *M* β-mercaptoethanol and 0.003 *M* EDTA) and is brought to 21% saturation with 111.6 g of ammonium sulfate. The precipitate is removed by centrifugation, and the supernatant fluid is

[2] O. H. Lowry, N. J. Rosebrough, A. L. Farr, and R. J. Randall, *J. Biol. Chem.* **193**, 265 (1951).

[3] O. Warburg and W. Christian, *Biochem. Z.* **310**, 384 (1942). See also Vol. III [73].

[4] D. A. Walsh and H. J. Sallach, *Biochemistry* **4**, 1076 (1965).

[5] A recording ultraviolet analyzer for flow streams greatly facilitates the monitoring of the eluates.

[6] Livers may be frozen and used later with slight loss in activity.

brought to 41% saturation with 111.6 g of ammonium sulfate. The precipitate is recovered by centrifugation and may be stored at —15° for several days.

Step 3. Chromatography on DEAE- and CM-Celluloses. (All buffers used in the remaining steps of the purification contain 0.0005 *M* β-mercaptoethanol and 0.003 *M* EDTA except where otherwise stated.) The above precipitate is dissolved in 200 ml of 0.01 *M* phosphate buffer, pH 6.0, and the solution dialyzed against the same buffer with regular changes until the dialyzate gives a negative reaction with Nessler's reagent (approximately 6 hours[7]). The enzyme solution is adjusted to pH 7.6 with 0.3 *M* ammonium hydroxide and applied to a DEAE-cellulose column (14 × 6 cm) equilibrated with 0.01 *M* phosphate buffer, pH 7.6. The column is washed with the same buffer (flow rate, 440 ml/hour). The first 680 ml of solution, after the hold-up volume, are collected and adjusted to pH 5.6 with 1 *N* acetic acid. The enzyme solution is applied to a CM-cellulose column (40 × 6 cm) equilibrated with 0.01 *M* phosphate buffer, pH 6.0. The column is washed with the same buffer (flow rate, 900 ml/hour) until the eluate is free of protein. The column is eluted with 2.5 liters of 0.01 *M* phosphate buffer, pH 6.9.[8] The protein in this fraction is discarded. The enzyme is eluted with 0.025 *M* phosphate buffer, pH 7.4 in the next protein peak.[9] To the 500 ml of enzyme solution are added 250 ml of 0.1 *M* phosphate buffer, pH 6.0. The solution is adjusted to pH 6.0 with 1 *N* acetic acid and applied to a CM-cellulose column (14 × 4.5 cm) equilibrated with 0.01 *M* phosphate buffer, pH 6.0. Under these conditions, the enzyme is not retained and is washed through with 0.05 *M* phosphate buffer, pH 6.0; 800 ml of enzyme solution is collected and is brought to 62% saturation with 320 g of ammonium sulfate. The precipitate is recovered by centrifugation and may be stored at —15° for several days.

Step 4. Chromatography on Phosphocellulose. The above precipitate is dissolved in 0.01 *M* phosphate buffer, pH 6.0 (final volume, 35 ml) and the ammonium sulfate is removed by gel filtration on a G-25 Sephadex column (54 × 4.6 cm) equilibrated with the same buffer. The enzyme solution (110 ml) is applied to a phosphocellulose column (15.5 × 4.7 cm) equilibrated with 0.01 *M* phosphate buffer, pH 6.0. The column is washed with an additional 50 ml of the same buffer and eluted with a gradient of phosphate concentration. The constant-volume mixing vessel contains 350 ml of 0.01 *M* phosphate buffer, pH 6.0, and the reservoir

[7] Extended dialysis (e.g., overnight) results in significant losses in activity.
[8] The greatest percentage of lactate dehydrogenase is removed at this step.
[9] It may be followed grossly by the appearance and movement of a red band.

contains 0.3 *M* phosphate buffer, pH 6.0. The enzyme is eluted at a molarity of 0.12 *M* phosphate in the fractions of the first protein peak. The eluate containing the enzyme (88 ml) is brought to 62% saturation by the addition of 35.2 g of ammonium sulfate. The precipitate is recovered by centrifugation and may be stored at −15° for several weeks with little loss in activity. This fraction usually contains small amounts of L-malate and L-lactate dehydrogenases. These may be removed by gel filtration as outlined below.

Step 5. Gel Filtration on Bio-Gel P-200. An aliquot of the above precipitate (400,000 units of enzyme) is dissolved in 2 ml of 0.2 *M* phosphate buffer, pH 7.2 (0.2 *M* NaCl, 0.003 *M* EDTA, 0.0003 *M* dithiothreitol[10]) and centrifuged. The supernatant fluid is applied to a column of Bio-Gel P-200 equilibrated with 0.1 *M* phosphate buffer, pH 6.1 (0.003 *M* EDTA, 0.0003 *M* dithiothreitol). The column is washed with the same buffer. The enzyme is eluted in the first protein peak (54 ml) before a number of lower molecular weight impurities. The enzyme solution is brought to 62% saturation with 21.6 g of ammonium sulfate and the precipitate is collected by centrifugation. Rechromatography of the enzyme on P-200 under identical conditions yields a single symmetrical enzyme and protein peak.

A summary of the purification data is given in the table.

PURIFICATION DATA

Fraction	Specific activity	Total units[a]	% Recovery
1. Initial extraction	88	2.70	100
2. Ammonium sulfate ppt.	480	2.50	92
3. 2nd CM-cellulose eluate	5,595	1.39	51
4. Phosphocellulose eluate	15,600	1.10	40
5. Bio-Gel P-200 eluate	22,000	0.86	32

[a] Millions per kilogram of liver.

Properties of the Enzyme[4]

Stability. The enzyme is very susceptible to inactivation, particularly at high dilutions. Dithiothreitol, the new protective reagent for thiol groups introduced recently by Cleland,[10] has been found to be extremely useful in stabilizing and reactivating the enzyme.

Specificity. TPN can only be substituted for DPN at abnormally high substrate concentrations. The following analogs of DPN react to

[10] W. W. Cleland, *Biochemistry* 3, 480 (1964).

a varying extent: 3-acetylpyridine-, 3-acetylpyridine-deamino-, 3-pyridinealdehyde-, and deamino-DPN. The enzyme is specific for 3-PGA as substrate. A number of other compounds, including D-glycerate, hydroxypyruvate, glycolate, glyoxylate, formate, DL-phosphoglyceraldehyde, DL-glyceraldehyde, glycolaldehyde, L-malate, and L-lactate, have been shown to be inactive as substrates.

Kinetic Properties. The kinetic constants for phosphoglycerate dehydrogenase have been determined at pH 8 and 25°. The K_m values are: DPN = 6.0×10^{-5} *M*; 3-PGA = 2.5×10^{-4} *M*; DPNH = 5.0×10^{-6} *M*; phosphohydroxypyruvate (PHPA) = 1.0×10^{-5} *M*. The dissociation constants for DPN and DPNH are 7.0×10^{-4} *M* and 5.0×10^{-6} *M*, respectively. The equilibrium constant (K_{eq} = [PHPA][DPNH][H^+]/[3-PGA][DPN]) has been calculated to be 1×10^{-12} *M*.

Inhibitors and Activators. The enzyme is inhibited by sulfhydryl reagents. Using the standard assay system, 6.1×10^{-3} mg of the enzyme (from step 4) is inhibited 50% by: 2.2×10^{-7} *M* *p*-mercuriphenylsulfonate or *p*-mercuriphenylbenzoate; 2.9×10^{-4} *M* 3-bromopyruvate; 5.9×10^{-4} *M* *N*-ethylmaleimide. Preincubation of the enzyme with DPN or DPNH, but not 3-PGA, partially protects the enzyme against inhibition by the latter two reagents. No activation or inhibition by a number of closely related metabolites, or by purine nucleotides, has been noted thus far. Activation by anions has been observed but has not been investigated systematically.

Other Sources

The enzyme is widely distributed in animal tissues[1] and in higher plants.[11,12] Among the former, avian liver (chicken, pigeon) is the richest source, but the enzyme is also found in liver, kidney, and brain of other animals. Of the plant tissues investigated,[12] wheat germ and seeds (pea, soybean, mungbean) have relatively high concentrations of the enzyme while little or no activity is found in the green leaves of a variety of plants. The enzyme is present in *Salmonella typhimurium*[13] and *Escherichia coli*.[14]

[11] J. Hanford and D. D. Davies, *Nature* **182**, 532 (1958).
[12] H. J. Sallach, unpublished results.
[13] H. E. Umbarger and M. A. Umbarger, *Biochim. Biophys. Acta* **62**, 193 (1962).
[14] L. Pizer, *J. Biol. Chem.* **238**, 3934 (1963).

[45] D-Glycerate Dehydrogenases of Liver and Spinach

By H. J. Sallach

D-Glycerate + $DPN^+ \leftrightarrow$ Hydroxypyruvate + DPNH + H^+

I. Calf Liver

Assay Method[1]

Principle. Hydroxypyruvate is reduced to L-glycerate in the presence of mammalian L-lactate dehydrogenases and reduced pyridine nucleotides.[2,3] Therefore, in the assay for D-glycerate dehydrogenase the reaction is measured in the direction of D-glycerate (D-GA) oxidation. The increase in optical density at 340 mμ, due to the formation of DPNH, is utilized to determine the rate of reaction under standardized conditions. Although the equilibrium of the reaction strongly favors the reduction of hydroxypyruvate, the oxidative reaction is shifted to the right by carrying out the reaction at pH 9 and in the presence of a carbonyl-trapping reagent.

Reagents

D-GA, potassium salt, 0.1 M, pH 7.0[4]
DPN, 0.015 M
Glutathione, 0.05 M, adjusted to pH 6 with 1 N KOH
Stock buffer solution. Mix 5 volumes of 1.0 M Tris-chloride, pH 9.0, 4 volumes of 1.0 M hydrazine acetate, pH 9.0, 1 volume of 0.25 M EDTA, pH 9.0, and 1 volume of 1 M NaCl on the day of use.
Enzyme. Prior to assay, dilute in 0.01 M phosphate buffer, pH 7.0, so that 0.1 ml will contain approximately 20 units as defined below.

Procedure. Place 1.1 ml of buffer solution, 0.1 ml of glutathione, 0.25 ml of D-GA (omit from control and blank), 0.1 ml of enzyme, and water to a final volume of 2.15 ml in each of three quartz cells with a light path of 1.0 cm. The reaction is initiated by the addition of 0.1 ml of DPN (replaced by water in blank). Take readings at 340 mμ at 1-minute intervals for 5 minutes against the blank.

[1] J. E. Willis and H. J. Sallach, *J. Biol. Chem.* **237**, 910 (1962).
[2] A. Meister, *J. Biol. Chem.* **197**, 309 (1952).
[3] S. R. Anderson, J. R. Florini, and C. S. Vestling, *J. Biol. Chem.* **239**, 2991 (1964).
[4] E. Baer, J. M. Grosheintz, and H. O. L. Fischer, *J. Am. Chem. Soc.* **61**, 2607 (1939).

Definition of Unit and Specific Activity. One unit of enzyme is defined as that amount which causes an initial rate of change of absorbance (ΔA_{340}) of 0.001 per minute under the above conditions. Specific activity is expressed as units per milligram of protein as determined by the method of Lowry *et al.*[5] or by the optical density at 280 mμ after correcting for nucleic acid absorption at 260 mμ.[6]

Application of Assay Method to Crude Tissue Extracts. With crude enzyme preparations there is an appreciable change in absorbance observed with the control due to the nonspecific reduction of DPN by contaminating substrates. A correction for this ΔA_{340} must be made in calculating enzyme units.

Purification Procedure[7]

The following general conditions are used in the purification of the enzyme unless otherwise specifically indicated. All operations are conducted at 4°. Fractionations with ammonium sulfate are made by the slow addition, with stirring, of the calculated amount of solid ammonium sulfate. The resulting suspensions are equilibrated for 30 minutes before centrifugation at 14,000 *g* for 25 minutes. In the chromatographic procedures employing phosphocellulose, protein in the eluates from the column is detected by absorbance at 280 mμ. A typical enzyme preparation is described below.

Step 1. Preparation of Crude Extract. A 730-g sample of fresh calf liver is homogenized with 2.34 volumes of 0.154 *M* KCl for 1 minute in a Waring blendor. The homogenate is centrifuged and the residue discarded.

Step 2. Treatment with Cetyl-Trimethyl Ammonium Bromide (CTAB). To the above supernatant solution (1582 ml) is added an equal volume of 0.5% CTAB.[8] The suspension is stirred for 30 minutes and centrifuged; the residue is discarded.

Step 3. Ammonium Sulfate Fractionation. The supernatant fluid from the above step (3000 ml) is brought to 30% saturation with 540 g of ammonium sulfate. The precipitate is removed by centrifugation and the supernatant fluid is brought to 55% saturation by the further addition of 510 g of ammonium sulfate. The precipitate is recovered by centrifuga-

[5] O. H. Lowry, N. J. Rosebrough, A. L. Farr, and R. J. Randall, *J. Biol. Chem.* **193**, 265 (1951).

[6] O. Warburg and W. Christian, *Biochem. Z.* **310**, 384 (1941). See also Vol. III [73].

[7] D. H. Antkowiak and H. J. Sallach, to be published.

[8] The solution of CTAB is at room temperature because this detergent is not sufficiently soluble in the cold at this concentration (obtained from Eastman Organic Chemical Co).

tion and may be stored at —15° for several weeks with little loss in activity.

Step 4. Dialysis. (All buffers in the remaining steps contain 0.001 *M* EDTA and 0.0005 *M* β-mercaptoethanol.) The 30–55% ammonium sulfate precipitate is dissolved in 79.1 ml of 0.01 *M* phosphate buffer, pH 8.0 (0.05 of original KCl supernatant volume) and is dialyzed against 6 liters of the same buffer with six changes for 12–14 hours.[9] Precipitated protein is removed by centrifugation. The supernatant solution is adjusted to pH 6.5 with 1 *N* acetic acid, diluted with an equal volume of 0.01 *M* phosphate buffer, pH 6.5, and allowed to stand in the cold for 1 hour. The small amount of precipitate that forms during this time interval is removed by centrifugation and discarded.

Step 5. Chromatography on Phosphocellulose. The supernatant solution from the above step is applied to a phosphocellulose column (19 × 4.7 cm) that is equilibrated with 0.01 *M* phosphate buffer, pH 6.5, at a flow rate of 180 ml per hour. The column is eluted successively with 100 ml of 0.01 *M* phosphate buffer, pH 6.5, 150 ml of 0.05 *M* phosphate buffer, pH 6.5, and 1400 ml of 0.05 *M* phosphate buffer, pH 7.4; the protein in these fractions is discarded. The enzyme is eluted from the column with 0.1 *M* phosphate buffer, pH 7.4, in the next protein peak (after approximately 250 ml). The enzyme solution (155 ml) is brought to 62% saturation with 62 g of ammonium sulfate, and the precipitate is recovered by centrifugation. It may be stored at —15° for several weeks with slight loss in activity.

Step 6. Removal of Inactive Protein with Calcium Phosphate Gel. The above precipitate is dissolved in 0.05 *M* phosphate buffer, pH 6.0 (final volume, 7 ml), and the ammonium sulfate is removed by gel filtration on a G-25 Sephadex column (30 × 2 cm) equilibrated with the same buffer. The enzyme solution (30 ml) is mixed with a volume of calcium phosphate gel[10] (25 mg/ml) to give a gel-protein ratio of 1:1. After standing for 15 minutes with occasional stirring, the gel is removed by centrifugation and discarded. The supernatant solution[11] is

[9] The first five changes are made at hourly intervals after the start of the dialysis. The final change is made the following morning, and the dialysis is continued for 1 hour.

[10] D. Keilin and E. F. Hartree, *Proc. Roy. Soc.* **B124**, 397 (1938). See also Vol. I [11].

[11] The solution is routinely assayed for D-glycerate and L-lactate dehydrogenases; the traces of the latter activity that are usually present in the fraction from the phosphocellulose step are generally removed by the first gel treatment. Due to the variability in calcium phosphate gels and in the amount of protein in this fraction, it is advisable to carry out a pilot experiment on a small aliquot before subjecting all the enzyme solution to the second gel treatment.

again treated with the same amount of calcium phosphate gel. After removal of the gel the supernatant solution (36 ml) is brought to 62% saturation with 14.4 g of ammonium sulfate. The precipitate is recovered by centrifugation and may be stored at —15° for several weeks with slight loss in activity. A summary of the purification data is given in Table I.

TABLE I
PURIFICATION DATA FOR CALF LIVER ENZYME

Step	Total volume (ml)	Total units[a]	Protein (mg/ml)	Specific activity	Recovery (%)
1. KCl supernatant	1582	171.4	28.0	4	100
2. CTAB supernatant	3000	174.0	7.5	8	101
3. $(NH_4)_2SO_4$, 30–55%	139	175.1	73.5	17	102
4. Diluted dialyzed solution	312	229.6[b]	24.0	32	133
5. $(NH_4)_2SO_4$, 0–62% on eluate from phosphocellulose	7	217.0	16.5	1885	127
6. First calcium phosphate supernatant	31.5	153.4	1.5	3240	89
Second calcium phosphate supernatant	36	108.7	0.5	6040	63

[a] Thousands per 730 g of calf liver.
[b] Activation was observed rather routinely at this step.

Properties of the Enzyme

Specificity.[1,12] DPN and TPN are almost equally effective as coenzymes. The ratio of activities, DPN:TPN, is 1.1 and remains relatively constant during the purification procedure. The following compounds are inactive as substrates with the enzyme preparation under the usual conditions of assay: L-glycerate, L-lactate, D-3-, or D-2-phosphoglycerates, glycolate, and L-malate. Glyoxylate is reduced by the enzyme preparation in the presence of either DPNH or TPNH, but at a slower rate than that observed with hydroxypyruvate. The enzyme from rat liver has essentially the same substrate specificity.[13]

Kinetic Properties. The following apparent Michaelis constants, K_m, have been determined at pH 9 for the oxidative reaction: DPN $= 3.7 \times 10^{-5}$ M and TPN $= 1.0 \times 10^{-5}$ M (with 0.02 M D-GA); D-GA $= 1.2 \times 10^{-3}$ M (with 0.001 M DPN) and 1.5×10^{-3} M (with 0.001 M TPN).[14]

[12] F. Heinz, K. Bartelsen, and W. Lamprecht, *Z. Physiol. Chem.* **329,** 222 (1962).
[13] P. D. Dawkins and F. Dickens, *Biochem. J.* **94,** 353 (1965).
[14] J. E. Willis, Ph.D. Thesis, Univ. of Wisconsin, 1963.

For the reverse reaction the reported K_m values are: hydroxypyruvate $= 4.5 \times 10^{-5} M$ (with DPNH) and $2.0 \times 10^{-5} M$ (with TPNH); DPNH and TPNH both $= 1.0 \times 10^{-5} M$; glyoxylate $= 1.4 \times 10^{-4} M$ and $2.5 \times 10^{-4} M$ with DPNH and TPNH respectively.[12]

Activators and Inhibitors. The enzyme is activated by anions (halides > NO_3^- > SO_4^{2-}). The activation is pH dependent and requires rather high concentrations (0.05–0.08 *M*) for maximal effect.[12,15] The enzyme has a broad pH-activity range for the reduction of hydroxypyruvate with a maximum between 6.8 and 7.0.[12,14] Sulfhydryl reagents are inhibitory; *p*-chloromercuribenzoate ($1.9 \times 10^{-5} M$) inhibits hydroxypyruvate reduction with DPNH, 40%, and with TPNH, 78%.[12]

II. Spinach Leaves

This enzyme was first described by Stafford *et al.*[16] and is present in a variety of plant tissues, the highest activity being found in green leaves. The methods and purification procedure described here are those of Holzer and Holldorf[17] with minor modifications.

Assay Method

Principle. The oxidation of DPNH is observed at 340 mμ under standardized conditions and in the absence of interfering reactions is a direct measure of the reduction of hydroxypyruvate to D-GA. Since the equilibrium of the reaction greatly favors the reduction of hydroxypyruvate, this is the most convenient method of assay.

Reagents

Lithium hydroxypyruvate, 0.008 *M*, pH 6.0[18]
DPNH, 0.001 *M*
Ammonium sulfate solution, 0.4 *M*
Potassium phosphate buffer, 0.6 *M*, pH 6.0
Enzyme. Dilute in 0.01 *M* phosphate buffer, pH 6.0, so that 0.1 ml will contain approximately 20 units as defined below.

Procedure. 1.0 ml of buffer, 0.5 ml of ammonium sulfate solution, 0.25 ml of hydroxypyruvate (omit from blank and control), and 0.25 ml of DPNH (omit from blank) are added to water to give a final volume of 2.9 ml in each of three quartz cells with a light path of 1.0 cm. After an initial determination of the optical density at 340 mμ, 0.1

[15] J. E. Willis and H. J. Sallach, *Biochim. Biophys. Acta* **62**, 443 (1962).
[16] H. A. Stafford, A. Magaldi, and B. Vennesland, *J. Biol. Chem.* **207**, 621 (1954).
[17] H. Holzer and A. Holldorf, *Biochem. Z.* **329**, 292 (1957).
[18] F. Dickens and D. H. Williamson, *Biochem. J.* **68**, 74 (1958). See also Vol. III [44].

ml of enzyme is added and the optical density is recorded at 30-second intervals for 3 minutes.

Definition of Unit and Specific Activity. Same as for liver enzyme above.

Application of Assay Method to Crude Tissue Extracts. The amount of enzyme required for the assay is of the order of 30 μg of protein; at this dilution, crude leaf extracts oxidize DPNH very slowly without added substrates. This activity is usually not great, compared to that in the presence of hydroxypyruvate, and a suitable correction for the endogenous activity can be applied as determined with the control.

Purification Procedure

The same general conditions are used as described above for the liver enzyme unless otherwise indicated.

Step 1. Preparation of Crude Extract. Fresh spinach leaves (1 kg) are homogenized with 1 liter of 0.001 *M* EDTA in a Waring blendor for 3 minutes. The homogenate is centrifuged, and the supernatant fluid is filtered by gravity through a pad of glass wool.

Step 2. First Ammonium Sulfate Fractionation. To each 100 ml of crude extract is added 13.3 g of ammonium sulfate. The precipitate is removed by centrifugation and discarded. An additional 15 g of ammonium sulfate per 100 ml of crude extract is added to the supernatant fluid. The precipitate is recovered by centrifugation and may be stored at —15° for several days.

Step 3. Acid Precipitation. The above precipitate is dissolved in 300 ml of 0.002 *M* phosphate buffer, pH 6.0. The dark green solution is adjusted to pH 4.85 with 1 *N* acetic acid and allowed to stand in the cold for 10–12 hours. The precipitated protein is removed by centrifugation and discarded. The supernatant solution is clear and light in color.

Step 4. Second Ammonium Sulfate Fractionation. To each 100 ml of the above solution is added 12.1 g of ammonium sulfate. The precipitate is removed by centrifugation and an additional 7.9 g of ammonium sulfate per 100 ml of original solution is added to the supernatant fluid. The precipitate is recovered by centrifugation and the supernatant fluid discarded.

Step 5. Adsorption on Alumina $C\gamma$ and Elution. The above precipitate is dissolved in 100 ml of 0.001 *M* EDTA containing 0.0001 *M* β-mercaptoethanol and is dialyzed against 5 liters of the same solution with regular hourly changes until the dialyzate gives a negative reaction with Nessler's reagent (approximately 5 hours). Precipitated protein is removed by centrifugation at 30,000 *g* for 10 minutes and discarded. The supernatant solution is mixed with 0.5 volume of alumina Cγ[19] (16.6 mg, dry weight, per milliliter) and is stirred slowly for 15 minutes. The gel is collected by centrifugation and the supernatant fluid, which should contain only traces of enzyme, is discarded. The enzyme is eluted from the gel by admixing with a volume of 0.15 *M* phosphate buffer,[20] pH 6.0, equal to that of the supernatant fluid from the gel centrifugation, and by stirring for 30 minutes before centrifugation.

Step 6. Third Ammonium Sulfate Fractionation. Sixteen grams of ammonium sulfate is added per 100 ml of gel eluate. If any precipitate forms, it is removed by centrifugation[21] and an additional 5 g of ammonium sulfate per 100 ml of gel eluate is added to the supernatant solution. The precipitate is collected by centrifugation at 30,000 *g* for 15 minutes, and the supernatant solution is discarded. A summary of the purification data is given in Table II.

TABLE II
PURIFICATION DATA FOR SPINACH ENZYME

Step	Total volume (ml)	Total units[a]	Protein (mg/ml)	Specific activity	Recovery (%)
1. Crude extract	1575	38.9	8.3	2,900	(100)
2. $(NH_4)_2SO_4$, 23–46%	331	33.1	22.5	4,400	85
3. Acid precipitation	300	20.8	5.4	12,800	54
4. $(NH_4)_2SO_4$, 22–33%	101	14.0	4.8	33,000	36
5. Alumina Cγ eluate	152	10.8	1.3	57,000	28
6. $(NH_4)_2SO_4$, 28–35%	51	8.9	1.0	175,000	23

[a] Millions per kilogram of spinach.

Properties of the Enzyme

Stability. A suspension of the purified enzyme in 20% ammonium sulfate decreases about 15% in activity in 10 days at 3–4°, but is stable for a month at −16° to −18°.[17]

Specificity. The enzyme is specific for DPNH. No reaction is observed

[19] R. Willstätter and H. Kraut, *Ber.* **56**, 1117 (1923). See also Vol. I [11].

[20] A pilot experiment on a small aliquot is required to determine the correct concentration of buffer to use to elute the enzyme from the particular gel preparation used. In the original article,[17] Holzer and Holldorf report that with one gel preparation 0.1 *M* buffer was adequate, but with a second, newer gel, 0.2 *M* buffer was required for the elution of the enzyme.

[21] Under the conditions described here, no precipitate was obtained in five separate preparations of the enzyme; however, the original method[17] indicates that inactive protein is precipitated at this point.

with TPNH with either crude or purified preparations.[17] The only substrate reactive with the enzyme, other than hydroxypyruvate, is glyoxylate. The maximal rate with hydroxypyruvate is four to five times that observed with glyoxylate; however, the ratio of the two activities remains relatively constant throughout the purification steps.[17] The DPNH-linked glyoxylate reductase from tobacco leaves[22] utilizes hydroxypyruvate as well as glyoxylate; however, in this case, the maximal rate with glyoxylate is three times that observed with hydroxypyruvate. The question whether one or more enzymes are involved has not been completely resolved (cf. footnotes 17, 22, and 23). A TPNH-linked glyoxylate reductase from spinach has been reported.[24] The rate of reduction of glyoxylate with this enzyme is 17 times that observed with hydroxypyruvate.

Kinetic Properties.[17] The reported Michaelis constants for hydroxypyruvate (HPA) and glyoxylate at pH 6 are $1.2 \times 10^{-4}\,M$ and $1.4 \times 10^{-2}\,M$, respectively. The equilibrium constant for the reaction (K_{eq} = [D-GA] [DPN]/[HPA] [DPNH] [H^+]) is $3.02 \pm 0.62 \times 10^{-12}\,M$.

Activators and Inhibitors. Activation of D-glycerate dehydrogenase by various anions ($Br^- > SO_4^{2-} > Cl^- = I^- > F^-$) has been demonstrated with hydroxypyruvate, but not with glyoxylate, as the substrate. The activation is greatest at lower pH values and is maximal with 0.06 M sulfate.[17, 23]

A number of carbonyl-binding and sulfhydryl reagents inhibit the enzyme. Under the conditions of the assay, $5 \times 10^{-6}\,M$ *p*-chloromercuribenzoate inhibits 55%, $1 \times 10^{-4}\,M$ iodoacetate, 37%, $1 \times 10^{-3}\,M$ semicarbazide, 36%, and $1 \times 10^{-2}\,M$ cyanide, 42%.[17] Metal binding reagents, e.g., *o*-phenanthroline and α,α'-dipyridyl at $1 \times 10^{-3}\,M$ are without effect.[17] The enzyme exhibits a broad pH range with a maximum at pH 6.2 under the conditions of the assay.[17]

[22] I. Zelitch, *J. Biol. Chem.* **216**, 553 (1955).
[23] G. Laudahn, *Biochem. Z.* **337**, 449 (1963).
[24] I. Zelitch and A. M. Gotto, *Biochem. J.* **84**, 541 (1962).

[46] Hydroxypyruvate Reductase (D-Glycerate Dehydrogenase; Crystalline) *Pseudomonas*

By Leonard D. Kohn and William B. Jakoby

$$\begin{array}{c} COOH \\ | \\ C{=}O \\ | \\ H{-}C{-}OH \\ | \\ H \end{array} + \begin{array}{c} DPNH \\ \text{or} \\ TPNH \end{array} + H^+ \rightleftarrows \begin{array}{c} COOH \\ | \\ H{-}C{-}OH \\ | \\ H{-}C{-}OH \\ | \\ H \end{array} + \begin{array}{c} DPN^+ \\ \text{or} \\ TPN^+ \end{array}$$

Assay Method

Principle. The reaction may be measured spectrophotometrically by utilizing the decrease in absorption at 340 mμ due to the conversion of DPNH to DPN^+ in the presence of hydroxypyruvate.

Reagents

Potassium phosphate, *M*, pH 6.3
Lithium hydroxypyruvate, 0.01 *M*, freshly prepared each hour by dissolving 12.8 mg in 10 ml of water
DPNH, 2 m*M*, or TPNH, 1 m*M*, adjusted to pH 8.0 with KOH

Procedure. The reagents are added in the order indicated and brought to a final volume of 1.0 ml: phosphate buffer, 100 μl; hydroxypyruvate, 400 μl; appropriate amounts of enzyme; and reduced pyridine nucleotide, 100 μl. The reaction is conveniently measured at 340 mμ with a Gilford log converter and cuvette changer attached to a Beckman DU monochromator. The reaction is linear at 25° with respect to time and protein concentration when absorbancy changes of less than 0.3 are recorded. Since limiting quantities of reduced pyridine nucleotide are used, care must be exercised to include the specified concentration of that reagent in order to maintain reproducibility of the assay.

Definition of a Unit. A unit of activity is defined as that amount of enzyme required to convert 1 micromole of DPNH to DPN^+ per minute using the above-noted procedure. Specific activity is defined as the number of units per milligram of protein. Protein is measured by the method of Lowry *et al.*,[1] using bovine serum albumin as a standard.

Purification Procedure

Growth. The enzyme is obtained from a strain of *Pseudomonas acidovorans* (ATCC 17455), originally isolated as *Pseudomonas* A by

[1] O. H. Lowry, N. J. Rosebrough, A. L. Farr, and R. J. Randall, *J. Biol. Chem.* **193**, 265 (1951).

U. Bachrach[2] from an enrichment culture containing uric acid. The organism grows well with L-(+)-tartrate as the sole carbon source.[3] Growth medium contains the following per liter: K_2HPO_4, 1.1 g; KH_2PO_4, 0.48 g; NH_4NO_3, 1 g; potassium L-(+)-tartrate, 1 g; and $MgSO_4 \cdot 7\ H_2O$, 0.25 g; the magnesium salt is added last, just prior to autoclaving. A fresh agar slant culture is used to inoculate 20 ml of the medium which, after 24 hours of growth, serves as an inoculum for 1 liter of medium in a 6-liter flask. After 24 hours of vigorous aeration on a reciprocating shaker, the contents of the flask are transferred to a 5-gallon carboy containing 5 liters of medium which is vigorously aerated for 18 hours at room temperature. Cells are harvested in the cold with a Sharples centrifuge, washed once with 0.02 *M* potassium phosphate at pH 7.2, and stored at —15°. Enzyme activity remains constant for at least 3 months under these conditions.

Purification Steps 1–6. The first six steps in the purification procedure are identical to those used in the purification of tartaric acid dehydrogenase, whch is described elsewhere in this volume.[4] The fractionation procedure, the results of which are summarized in the table, includes sonication, precipitation of inactive protein with acid, treatment with protamine sulfate, precipitation with ammonium sulfate, adsorption and elution from an aluminum hydroxide gel,[5] and chromatography on DEAE-cellulose. Under the conditions used,[4] hydroxypyruvate reductase is eluted from DEAE-cellulose in fractions 100–135, thereby resulting in separation from tartaric acid dehydrogenase. Except where specifically indicated, all procedures are carried out at approximately 0° and in the presence of 3 m*M* mercaptoethanol.

Step 7. The DEAE-cellulose fractions containing hydroxypyruvate reductase activity are pooled and treated with solid ammonium sulfate to 75% of saturation (52.5 g/100 ml). After centrifugation the precipitate is suspended (1 ml/10 mg) in 0.03 *M* potassium phosphate, pH 7.2, 50% saturated with respect to ammonium sulfate.[6] After it has been stirred for 15 minutes, the suspension is centrifuged and the residue is suspended in an equal volume of the same buffer except that the ammonium sulfate concentration is reduced to 45% saturation. The procedure is repeated in sequence with ammonium sulfate concentrations of 40%, 35%, 30%, and 20% of saturation.

[2] U. Bachrach, *J. Gen. Microbiol.* **17**, 1 (1957).

[3] S. Dagley and P. W. Trudgill, *Biochem. J.* **89**, 22 (1963).

[4] This volume [48].

[5] 800 mg of gel per milliliter was used instead of the smaller amount necessary for tartaric dehydrogenase.

[6] These solutions are prepared by appropriate dilution of stock solutions of ammonium sulfate saturated at 0°.

Those fractions containing substantial enzyme activity, usually the 45%, 40%, and 35% extracts,[7] are pooled and dialyzed overnight against 0.04 *M* Tris chloride buffer at pH 7.4.

Step 8. A column of hydroxyapatite, 2.5 × 16.5 cm, is washed overnight with 3 to 5 bed-volumes of 0.04 *M* Tris buffer at pH 7.4. The enzyme preparation from step 7 is applied and washed with a linear gradient of 500 ml of 0.1 *M* potassium phosphate at pH 7.5 into 500 ml of the Tris buffer at pH 7.4. When 4-ml volumes are collected, the enzyme is eluated in fractions 50–70. The eluates are pooled and concentrated by ultrafiltration with a Schleicher and Schuell apparatus to approximately 5 ml.

Step 9. A Biogel P-200 column, 1.5 × 75 cm, is prepared in the cold using 0.04 *M* Tris buffer at pH 7.4. The pooled concentrate from step 8 is applied to the column and is eluted with the same Tris buffer. The enzyme is collected between 145 ml and 170 ml and is concentrated by ultrafiltration to 20% of its volume.

Step 10. The protein is salted out by the addition of ammonium sulfate to 75% of saturation and the protein is extracted as described in step 7 with decreasing concentrations of ammonium sulfate. The resulting suspensions are stored at room temperature for 1 day and then transferred to 2° until crystal formation is judged to be complete. After 1–3 days the crystals are harvested; the active material obtained for the summary in the table was from the 35% and 40% fractions.

Step 11. Recrystallization is a repetition of step 10.

FRACTIONATION DATA

Fraction	Volume (ml)	Total activity (units)	Total protein (mg)	Specific activity (units/mg)
1. Extract	285	334	8800	0.04
2. Acid treatment	365	356	7740	0.05
3. Protamine sulfate	350	378	6650	0.06
4. Ammonium sulfate	185	390	3400	0.12
5. Gel	240	412	896	0.45
6. DEAE-cellulose	270	254	81	3.14
7. Ammonium sulfate	30	231	41	5.64
8. Hydroxyapatite	6	203	23	8.82
9. P-200	25	163	12	13.5
10. Crystals	3	115	8	14.5
11. Recrystallization	2	94	6.4	14.6

[7] These solutions, if left at room temperature, will form crystals of hydroxypyruvate reductase. However, a higher specific activity and greater total yield are obtained if the subsequent purification steps are followed.

Properties

Although the reduction of hydroxypyruvate by the enzyme appears to be analogous to that catalyzed by lactic dehydrogenase and glyoxylic acid reductase, the hydroxypyruvate reductase differs in its specificity for hydroxypyruvate. Pyruvic acid, glyoxylic acid, and dihydroxyfumaric acid do not serve as substrate. Preparations of tartronic semialdehyde have some activity as substrate although it is stressed that such preparations contain hydroxypyruvate. At a DPNH concentration of 0.15 mM, a K_m of 2.1×10^{-3} M is obtained for hydroxypyruvate. At a TPNH concentration of 0.15 mM, a K_m of 1.3×10^{-4} M is obtained for hydroxypyruvate. At a hydroxypyruvate concentration of 5 mM, the K_m for DPNH and TPNH are 2.4×10^{-4} M and 1.5×10^{-5} M, respectively. The appropriate kinetic constants obtained by the method of Alberty[8] are presented in Eqs. (2) and (3).

$$\frac{V_M}{v} = 1 + \frac{6.04 \times 10^{-3}\,M}{[\text{hydroxypyruvate}]} + \frac{1.5 \times 10^{-3}\,M}{[\text{DPNH}]} + \frac{4.2 \times 10^{-6}\,M}{[\text{hydroxypyruvate}][\text{DPNH}]} \quad (2)$$

$$\frac{V_M}{v} = 1 + \frac{7.4 \times 10^{-5}\,M}{[\text{hydroxypyruvate}]} + \frac{5.5 \times 10^{-6}\,M}{[\text{TPNH}]} + \frac{3.3 \times 10^{-9}\,M}{[\text{hydroxypyruvate}][\text{TPNH}]} \quad (3)$$

The protein at this stage of purification yields one band on disc gel electrophoresis and is homogeneous in the ultracentrifuge with an $S_{25,\text{w}}$ of 4.81 after extrapolation to zero protein concentration. The diffusion constant is 0.6167×10^{-6} and the partial specific volume is 0.734; the latter is an estimate based on amino acid analysis. From these data the molecular weight and frictional coefficient may be calculated as 73,000 and 1.27, respectively.

The optimal pH for hydroxypyruvate reduction is at 6.5 with half of maximal activity at pH 5.3 and 6.9. The reaction is experimentally reversible in the presence of D-glycerate, hydrazine, and either DPN^+ or TPN^+.

The enzyme has remained stable for 3 months when stored as a crystalline suspension in ammonium sulfate. Dilute solutions of protein (0.1 mg/ml) in 0.1 M potassium phosphate buffer at pH 7.5 appear to be stable for several days.

[8] R. A. Alberty, *J. Am. Chem. Soc.* **75,** 1932 (1953).

[47] L-Threonic Acid Dehydrogenase[1]

By ANITA J. ASPEN and WILLIAM B. JAKOBY

$$\begin{array}{c} COOH \\ | \\ H—C—OH \\ | \\ HO—C—H \\ | \\ CH_2OH \end{array} + DPN^+ \rightarrow \begin{array}{c} COOH \\ | \\ H—C—OH \\ | \\ C{=}O \\ | \\ CH_2OH \end{array} + DPNH + H^+$$

Assay Method

Principle. The reaction may be followed spectrophotometrically by utilizing the increase in absorption at 340 mμ due to the formation of DPNH.

Reagents

Tris-HCl, *M*, at pH 9.0

Mercaptoethanol prepared immediately before use by diluting 0.07 ml of the compound to 10 ml

Potassium L-threonate, 0.1 *M*. L-Threonic acid is prepared as the brucine salt.[2] The free acid may be obtained by passage of the salt (2 g/100 ml) through a Dowex 50 (H^+ form) column. The acid is adjusted to pH 7.0 with 0.2 *M* KOH.

DPN, 0.02 *M*, adjusted to pH 7.0

Procedure. The following volumes of the above reagents are added in the order indicated to a final volume of 1.0 ml: Tris, 100 μl; mercaptoethanol, 50 μl; potassium threonate, 50 μl; DPN, 100 μl; appropriate amounts of enzyme. The reaction is conveniently followed at 340 mμ with a Gilford log converter and cuvette changer attached to a Beckman DU spectrophotometer. The reaction is linear at 25° with respect to time and protein concentration over a 3- to 4-minute period when absorbancy changes of less than 0.3 are recorded.

Definition of a Unit. A unit of activity is defined as that amount of enzyme required to form 1 micromole of DPNH per minute. Specific activity is defined as the number of units per milligram of protein; protein is measured by the method of Lowry *et al.*[3] with bovine serum albumin as a standard.

[1] A. J. Aspen and W. B. Jakoby, *J. Biol. Chem.* **239,** 710 (1963).

[2] R. W. Herbert, E. L. Hirst, E. G. V. Percival, R. J. W. Reynolds, and F. Smith, *J. Chem. Soc.* p. 1270 (1933).

[3] O. H. Lowry, N. J. Rosebrough, A. L. Farr, and R. J. Randall, *J. Biol. Chem.* **193,** 265 (1951).

Purification Procedure

Growth. The enzyme is obtained from a species of *Pseudomonas* which had been isolated from an enrichment culture with L-threonic acid as sole carbon source. Growth medium contains the following per liter: K_2HPO_4, 0.48 g; KH_2PO_4, 1.1 g; $MgSO_4 \cdot 7\ H_2O$, 0.25 g; NH_4NO_3, 1 g; potassium L-threonate,[4] 1 g. A 24-hour-old agar slant culture is used to inoculate 20 ml of medium which, after 24 hours of growth, serves as inoculum for a liter of medium in a 6-liter flask. After 24 hours of vigorous aeration on a reciprocating shaker, the contents of the flask are transferred to a 5-gallon carboy containing 11 liters of medium and aerated overnight at room temperature. Cells are harvested in the cold with a Sharples centrifuge, washed with saline, and stored at −15°; enzyme activity remains constant for several months.

Step 1. Extract. Frozen cells are suspended in twice their volume of *M* potassium phosphate buffer, pH 7.2, containing 5 m*M* mercaptoethanol and are disrupted by sonication for 20 minutes in a 10-kc Raytheon sonic oscillator. The extract is centrifuged at 16,000 *g* for 30 minutes and the precipitate is discarded.

All subsequent operations are conducted at 0°–2° and in the presence of 5 m*M* mercaptoethanol. The outlined procedure is based on 100 ml of extract.

Step 2. Heating. The extract is heated in a water bath for 10 minutes at 55° with continuous stirring and then supplemented with sufficient mercaptoethanol to produce a 5 m*M* solution. The precipitate is discarded after centrifugation.

Step 3. Ammonium Sulfate. Solid ammonium sulfate is added to 25% of saturation, and the precipitate is removed by centrifugation and dissolved in approximately 10 ml of 5 m*M* potassium phosphate at pH 7.2. Subsequent fractions are obtained by the addition of 5% increments of ammonium sulfate. Each fraction obtained by centrifugation is dissolved in the same buffer. The highest enzyme activity is usually found in the 30–35% fraction.

Step 4. Gel Eluate. A 2-ml volume of the active fraction is placed on a column of Sephadex 50, 1×12 cm, that has been equilibrated with 0.03 *M* Tris buffer, pH 7.2, containing 5 m*M* mercaptoethanol. The column is eluted with the same buffer; the several 1-ml fractions containing activity are combined and mixed with 1.1 g of calcium phosphate

[4] For large-scale growth, synthesis of L-threonic acid may be carried out as described[2] through the step requiring precipitation of manganese hydroxide. Potassium phosphate buffer, 1 *M* at pH 7.2, is then added until there is no further precipitation. This solution may be added directly to the liquid medium.

gel, allowed to stand for 5 minutes, and centrifuged for 10 minutes. The supernatant fraction is mixed with a second 1.1 g batch of gel and the procedure is repeated. Each of the two gel samples is eluted with 2 ml of 0.05 M potassium phosphate buffer, pH 6.5.

Step 5. DEAE-Cellulose. The first 2-ml eluate from the calcium phosphate gel is placed on the Sephadex column, eluted as described above, and used to charge a column of DEAE-cellulose, 1×18 cm, that has been equilibrated with 0.03 M Tris buffer, pH 7.5. The column is eluted with buffer from a mixing flask equipped for magnetic stirring and containing 150 ml. The mixing flask is in turn connected for gradient elution to a container holding 450 ml of the same buffer supplemented with 0.5 M sodium chloride. Fractions of 4 ml are collected, the enzyme usually emerging in fractions 20–25. A summary of the results of one purification is presented in the table.

PURIFICATION OF L-THREONIC ACID DEHYDROGENASE

Fraction	Volume (ml)	Total activity (units)	Total protein (mg)	Specific activity (units/mg)
1. Sonicate	91	1150	955	1.20
2. Heating	82	1000	439	2.28
3. Ammonium sulfate (30–35%)	10	436	58.4	7.47
4. Gel eluate	2	134	4.20	31.9
5. DEAE-cellulose (fraction 21)	4	20.8	0.168	124

The dilute enzyme as obtained on elution is labile, losing most of its activity after overnight storage at −15°. After concentration by filtration under vacuum through dialysis tubing at 3°, a relatively stable preparation can be obtained. One fraction, subjected to intermittent thawing and refreezing, retained 44% of its activity after storage for 8 weeks at −15°.

Properties

The enzyme is specific for L-threonic acid and DPN in the forward reaction. TPN was inactive as substrate as were the following compounds: L-erythronic acid; L-arabonic acid; L-xylonic acid; L-gulonic acid; α-ketothreonic acid; DL-α-hydroxybutyric acid; L-β-hydroxybutyric acid; D-β-hydroxybutyric acid; γ-hydroxybutyric acid; dihydroxyacetone; L-glyceraldehyde. An analysis of the kinetics of the reaction is consistent with the formation of a ternary complex involving enzyme, threonic acid, and DPN. The velocity of the reaction may be expressed in terms of Eq. (1) where V_M is the maximum velocity at a given enzyme concentration.

$$\frac{V_M}{v} = 1 + \frac{1.1 \times 10^{-3}}{[\text{threonate}]} + \frac{7.6 \times 10^{-4}}{[\text{DPN}^+]} + \frac{4.9 \times 10^{-7}}{[\text{threonate}][\text{DPN}^+]} \quad (1)$$

An equilibrium constant of 4×10^{-11} has been calculated for the reaction according to Eq. (2) although this value is

$$K_{eq} = \frac{[\text{DPNH}][\text{H}^+][\beta\text{-ketothreonate}]}{[\text{DPN}][\text{threonate}]} \quad (2)$$

subject to errors due to the ease of decarboxylation of β-ketothreonate.

The reaction has a sharp pH optimum at pH 10.8. Half the maximal activity is attained at pH 8.9 and 11.5.

[48] L- and Mesotartaric Acid Dehydrogenase (Crystalline)

By LEONARD D. KOHN and WILLIAM B. JAKOBY

$$\begin{array}{c} \text{COOH} \\ | \\ \text{H—C—OH} \\ | \\ \text{H—C—OH} \\ | \\ \text{COOH} \end{array} + \text{DPN}^+ \rightleftarrows \begin{array}{c} \text{COOH} \\ | \\ \text{C—OH} \\ \| \\ \text{C—OH} \\ | \\ \text{COOH} \end{array} + \text{DPNH} + \text{H}^+$$

Assay Method

Principle. The reaction can be measured spectrophotometrically by following the increase in absorption at 340 mμ caused by the formation of $DPNH^+$.

Reagents

Tris-HCl, *M*, at pH 8.4
Manganous chloride, 2 m*M*
Mercaptoethanol 0.01 *M*, prepared immediately before use by diluting 0.07 ml of the compound to 100 ml
Ammonium sulfate, 0.2 *M*
Mesotartaric acid, 0.1 *M*, adjusted to pH 7.0 with KOH
DPN, 20 mg/ml, adjusted to pH 6.5

Procedure. Reagents are added in the order indicated and brought to a final volume of 1.0 ml: Tris, 100 μl; manganous chloride, 200 μl; mercaptoethanol, 100 μl; ammonium sulfate, 100 μl; *meso*-tartrate, 100 μl; appropriate amounts of enzyme; DPN, 100 μl. The reaction is conveniently followed at 340 mμ with a Gilford log converter and cuvette changer attached to a Beckman DU spectrophotometer. The reaction is linear with respect to time and protein concentration over a 10-minute period when absorbancy changes of less than 1.0 are recorded.

Definition of a Unit. A unit of activity is defined as that amount of enzyme required to form 1 micromole of DPNH per minute. Specific activity is equal to the number of units per milligram of protein. Protein is measured by the method of Lowry *et al.*,[1] using bovine serum albumin as a standard.

Purification Procedure

Growth. The enzyme is obtained from a strain of *Pseudomonas putida* which is grown on a medium containing the following per liter: KH_2PO_4, 0.48 g; K_2HPO_4, 1.1 g; NH_4NO_3, 1 g; and $MgSO_4 \cdot 7\ H_2O$, 0.25 g. Meso-tartaric acid, 2 g/liter, is neutralized with KOH, sterilized separately by filtration, and added to the medium immediately before inoculation. Growth from a 24-hour-old agar slant is used to inoculate 20 ml of the medium. After 24 hours of growth, the liquid medium serves as inoculum for another 1 liter of medium in a 6-liter flask. After 24 hours of vigorous aeration on a reciprocating shaker, the contents of this flask are transferred to a 5-gallon carboy containing 15 liters of medium which is aerated overnight at room temperature. Cells are harvested in the cold with a Sharples centrifuge, washed with 0.02 *M* potassium phosphate at pH 7.2, and stored at —15°. Enzyme activity remains constant for several months.

Step 1. Extract. Frozen cells are suspended in 4 times their volume of 0.02 *M* potassium phosphate at pH 7.2, 3 m*M* in mercaptoethanol, and are disrupted by sonication for 10 minutes in a 10-kc Raytheon sonic oscillator at 2°. The extract is centrifuged for 30 minutes at 18,000 *g*, and the precipitate is discarded.

All subsequent operations are conducted at approximately 0° in the presence of 5 m*M* mercaptoethanol.

Step 2. Acid Treatment. The extract is brought to pH 5.6 by the slow addition of 0.1 *N* HCl with constant stirring. After centrifugation for 20 minutes at 18,000 *g*, the precipitate is discarded and the supernatant fluid is neutralized to pH 7.0 with 0.2 *N* KOH.

Step 3. Protamine Sulfate. Solid protamine sulfate, 2 mg/ml is slowly added, and, after the mixture has been stirred for 15 minutes, the precipitate is centrifuged at 18,000 *g* for 20 minutes and discarded.

[1] O. H. Lowry, N. J. Rosebrough, A. L. Farr, and R. J. Randall, *J. Biol. Chem.* **193**, 265 (1951).

Step 4. Ammonium Sulfate. Solid ammonium sulfate is slowly added to the supernatant fluid until 31% of saturation (22 g/100 ml) is reached. The precipitate is discarded after centrifugation at 18,000 *g* for 10 minutes. Additional ammonium sulfate is added to 60% of saturation (20 g/100 ml), and the resulting precipitate is suspended in 0.04 *M* Tris-chloride at pH 7.4 which is 5 m*M* in mercaptoethanol.

Step 5. Gel Eluate. The active fraction is dialyzed overnight against 200 volumes of 0.04 *M* Tris-chloride, pH 7.4, 5 m*M* with respect to mercaptoethanol. Aluminum hydroxide gel[2] is added to the dialyzate in an amount determined by a pilot experiment to adsorb all enzyme activity; usually 400 mg of gel per milliliter of dialyzate is required. The suspension is allowed to stand for 10 minutes and then is centrifuged at 5000 *g* for 10 minutes. The gel is eluted twice with 0.1 *M* Tris-chloride, pH 7.4, and twice with 0.1 *M* potassium phosphate at pH 7.5; both buffers contain 5 m*M* mercaptoethanol. The phosphate eluates, containing the enzyme activity, are pooled and dialyzed overnight against 0.04 *M* Tris-chloride, pH 7.4, containing 5 m*M* mercaptoethanol.

Step 6. DEAE-Cellulose. The dialyzate serves to charge a DEAE-cellulose column, 5×30 cm, equilibrated with the same Tris-chloride buffer. The column is eluted with a linear gradient arranged so that the reservoir contains 1500 ml of 0.04 *M* Tris-chloride at pH 7.4, 5 m*M* mercaptoethanol, and 0.5 *M* sodium chloride; the mixing flask contains 1500 ml of the same buffer without NaCl. Fractions of 10 ml are collected at a rate of 1–2 ml per minute, the active material being eluted in tubes 270–340.

Step 7. Crystallization. The pooled enzyme fractions are brought to 7.5 m*M* with respect to mercaptoethanol and treated with solid ammonium sulfate to 75% of saturation (52.5 g/100 ml). After centrifugation the precipitate is suspended in 10 ml of 0.03 *M* potassium phosphate, pH 7.2, containing 5 m*M* mercaptoethanol/liter and 40% saturated with respect to ammonium sulfate.[3] The suspension is stirred for 15 minutes, then centrifuged; the residue is suspended in 10 ml of the same buffer except that the ammonium sulfate concentration is reduced to 35% of saturation. The procedure is repeated in sequence with ammonium sulfate concentrations of 30% and 20% of saturation. All fractions are stored under vacuum at room temperature until crystal formation is judged to

[2] Amphogel, a commercial preparation of aluminum hydroxide gel, is used after washing the material with distilled water to a point where the peppermint flavor is removed.

[3] These solutions are prepared by appropriate dilution of stock solutions of ammonium sulfate saturated at 0°.

be complete. After 1–3 days the rectangular prisms are harvested by centrifugation at room temperature and may be recrystallized by repetition of step 7.

A summary of the purification procedure is presented in the table.

FRACTIONATION DATA

Fraction	Volume (ml)	Total activity (units)	Total protein (mg)	Specific activity (10^3)
1. Extract	365	97.6	16,400	6
2. Acid treatment	500	94.1	12,300	8
3. Protamine	215	95.1	12,200	8
4. Ammonium sulfate	250	78.7	3,600	22
5. Gel eluate	700	68.3	1,100	60
6. DEAE-Cellulose	3	50.2	108	465
7. Crystallization[a]	10	2.6	2.7	956
Supernatant fluid	—	7.0	20	354

[a] Results of crystallization are presented for the 35% ammonium sulfate extract although each of the other extracts also gave rise to crystals. The total yield of crystalline protein from all fractions represents 10% of the activity present in the extract.

Properties

Both meso- and L-(+)-tartaric acid serve as substrate whereas D-(—)-tartaric acid is inactive. At saturating concentrations of DPN, the K_m values for meso- and L-(+)-tartrate are $4.5 \times 10^{-4}\ M$ and $7.2 \times 10^{-3}\ M$, respectively. At a saturating concentration of *meso*-tartrate, two K_m values for DPN, $6.0 \times 10^{-5}\ M$ and $1.5 \times 10^{-4}\ M$, were obtained; with L-(+)-tartrate the same two K_m values were calculated. The pH optimum with either isomer is at 8.5, with 50% of optimal activity attained at approximately pH 7 and 9. Both manganese and a mercaptan are required. Manganous chloride, 0.5 mM, and mercaptoethanol, 0.1 mM, are optimal at the indicated concentration. Ammonium sulfate, as well as other salts, at 20 mM, is stimulatory for activity with *meso*-tartrate as substrate; this same concentration of salt results in 50% inhibition with L-(+)-tartrate as substrate.

Dihydroxyfumaric acid has been identified as the product of the reaction. With dihydroxyfumarate and DPNH, activity is optimal at pH 6.5.

Sources. The presence of a DPN- and Mg-requiring tartaric acid oxidation system has been found in the mitochondrial fraction of several organs of both the rat and cow.[4] Extracts of higher plants contain a DPN-dependent system similar to that described.[5]

The formation of glycerate from L-(+)-tartrate by a DPN-linked system from a species of *Pseudomonas* (*Pseudomonas* A[6]) has been reported[7]; this pseudomonad has been found to have two DPN-linked tartaric dehydrogenases,[8] one of which is specific for the *meso* form and the other of which can utilize both meso- and L-(+)-tartrate. The dihydroxyfumarate thus formed, in both *Pseudomonas* A and the strain of *Pseudomonas putida* described here, can be converted to glycerate by a DPNH-linked reductase.[8]

[4] E. Kun and M. G. Hernandez, *J. Biol. Chem.* **218,** 201 (1956).
[5] H. A. Stafford, *Plant Physiol.* **32,** 338 (1957).
[6] U. Bachrach, *J. Gen. Microbiol.* **17,** 1 (1957).
[7] S. Dagley and P. W. Trudgill, *Biochem. J.* **89,** 22 (1963).
[8] L. D. Kohn and W. B. Jakoby, *Federation Proc.* **25,** 711 (1966).

[49] Tartronic Semialdehyde Reductase (Crystalline)

By H. L. KORNBERG and A. M. GOTTO

Tartronic semialdehyde + NAD(P)H + $H^+ \rightleftharpoons$ glyceric acid + NAD(P)

Tartronic semialdehyde reductase, which catalyzes the reduction of tartronic semialdehyde to glycerate with concomitant oxidation of reduced NAD or NADP, plays an important role in the growth of microorganisms (other than *Micrococcus denitrificans*[1]) on glycolate or on other precursors of glyoxylate.[2] The enzyme may be readily obtained in a crystalline state from extracts of glycolate-grown *Pseudomonas* sp.[3]

Assay Method

Principle. The enzyme can be assayed either by determining the rate of oxidation of reduced NAD or NADP when tartronic semialdehye is enzymatically transformed to glycerate (method A); or by coupling the oxidation of glycerate to the reduction of a redox dye such as 2,6-dichlorophenolindophenol (method B). Both methods have inherent difficulties: method A necessitates the availability of tartronic semialdehyde, which is not commercially available and must either be synthesized chemically or be prepared enzymatically each day; and method B is

[1] H. L. Kornberg and J. G. Morris, *Biochem. J.* **95,** 577 (1965).
[2] H. L. Kornberg and A. M. Gotto, *Biochem. J.* **78,** 69 (1961).
[3] A. M. Gotto and H. L. Kornberg, *Biochem. J.* **81,** 273 (1961).

subject to interference by other glycerate-oxidizing enzymes which may be present in the biological extracts used. However, both methods give reproducible results with the purified enzyme; with crude extracts, method A is recommended.

Method A

Reagents

Disodium hydrogen phosphate, 1 *M*, pH 8.5
NADH or NADPH, 0.01 *M*
Tartronic semialdehyde (abbreviated TSA), 0.01 *M*

Procedure. Into a silica cell (1-cm light path, approximately 1.5 ml volume), pipette 0.1 ml of phosphate buffer, 0.02 ml of NADH or NADPH, an appropriate amount of enzyme, and water to 0.95 ml. A "blank" cell receives the same reagents with the exception of the reduced NAD or NADP. Record the extinction at 340 mμ for 1–2 minutes, then add the TSA and continue to record the decrease in extinction at 340 mμ. For optimal assay, the amount of enzyme added should suffice to catalyze a decrease in extinction of not less than 0.02 and not more than 0.1 unit per minute. Although the enzymatic reduction of TSA at pH 8.5 occurs more rapidly with NADH than with NADPH (see below), the presence of NADH-oxidase in crude extracts and the relative absence of similar enzymes catalyzing the removal of NADPH, indicates NADPH to be preferable to NADH when crude extracts are assayed.

Preparation of TSA. (a) TSA may be prepared chemically by the method of Fischer *et al.*[4] or, more easily, by the procedure described by Jaenicke and Koch.[5] However, both methods are time consuming and neither gives high yields of TSA.

(b) TSA may be prepared enzymatically by treatment of glyoxylate with glyoxylate carboligase[6] as follows:

Reagents

KH_2PO_4/Na_2HPO_4 buffer, 0.5 *M*, pH 6.5
Thiamine pyrophosphate, 0.01 *M*
$MgCl_2$, 0.1 *M*
Glyoxylate carboligase
Sodium glyoxylate, 0.4 *M*

[4] H. O. L. Fischer, E. Baer, and H. Nidecker, *Helv. Chim. Acta* **20,** 1226 (1937).
[5] L. Jaenicke and J. Koch, *Biochem. Z.* **336,** 432 (1962).
[6] See also this volume [124].

$HClO_4$, 20%
KOH, 2 *N*

Procedure. Into the main compartment of a double-armed Warburg cup place 0.2 ml of phosphate buffer, pH 6.5, 0.05 ml of thiamine pyrophosphate solution, 0.05 ml of $MgCl_2$ solution, 0.05 ml of purified glyoxylate carboligase (see below).[6] Into sidearm 1, pipette 0.2 ml of the sodium glyoxylate solution; into sidearm 2, 0.2 ml of perchloric acid. Attach the flask to its manometer, gas with nitrogen and, after equilibration for 15 minutes at 30°, add the glyoxylate from the first sidearm. When evolution of CO_2 ceases (approximately 10 minutes), add the contents of the second sidearm, detach the manometer cup, and centrifuge its contents at 20,000 *g* and 0° for 2 minutes. (Continued shaking of the flask, after addition of the $HClO_4$, was found to result in considerable decarboxylation of the TSA.) Pour off the supernatant solution and adjust its pH to 7.0–7.5 by dropwise addition of KOH; remove the $KClO_4$ by centrifugation for 15 minutes at 0°.

The concentration of TSA in the resultant solution can be calculated from the total change in extinction at 340 mμ when samples are incubated with purified TSA reductase (see later) and an excess of NADH. The yields obtained vary between 30 and 60% of those expected from the measured evolution of CO_2 from glyoxylate; since the TSA is rather unstable, solutions such as this are best prepared freshly each day.

Method B

Reagents

Na_2HPO_4 buffer, 1 *M*, pH 8.5
2,6-dichlorophenolindophenol, 0.01%
KCN, 0.1 *M*
NADH-dehydrogenase (see below)
NAD^+, 0.01 *M*
Potassium D-glycerate, 0.07 *M*

Procedure. Into a silica cell (1-cm light path, approximately 1.5 ml volume) place 0.4 ml of the 2,6-dichlorophenolindophenol, 0.1 ml of phosphate buffer, 0.05 ml of KCN, 0.02 ml of NADH-dehydrogenase, 0.1 ml of NAD, an appropriate amount of enzyme, and water to 0.95 ml. Record the extinction at 600 mμ for 1–2 minutes, and start the reaction by adding 0.05 ml of the glycerate solution.

Definition of Unit and Specific Activity. One unit of TSA reductase is defined as that quantity of protein which catalyzes the oxidation of

1 micromole of NADH per minute when incubated with TSA; under the conditions of method A, 1 unit of enzyme catalyzes a $\Delta E_{340\,m\mu}$ of —6.28 units per minute. Under the conditions of method B, this quantity of enzyme catalyzes a $\Delta E_{600\,m\mu}$ of —2.56 units per minute. Specific activity is expressed as units of TSA reductase per milligram of protein.

Purification Procedure

Pseudomonas ovalis Chester (NCIB 8296; ATCC 8209), grown on glycolate as sole carbon source in the medium described by Dixon *et al.*,[7] are harvested and washed in 5 m*M* sodium–potassium phosphate buffer, pH 7.0 (prepared by mixing 61.1 volumes of 5 m*M* Na_2HPO_4 with 38.9 volumes of 5 m*M* KH_2PO_4); the washed packed cells may be stored frozen at —12° for several months without loss of activity.

Step 1. Preparation of Ultrasonic Extracts. Frozen cells (80 g wet weight) are thawed and diluted to 200 ml with the 5 m*M* phosphate buffer, pH 7.0, and are disintegrated by subjecting the suspension (in batches of approximately 20 ml) to the output of a magnetostrictor oscillator. The sonic extracts are combined and centrifuged for 1 hour at 35,000 *g*; this, and all subsequent operations, are carried out at 0–2°. The precipitated material is discarded.

Steps 2 and 3. Treatment with Protamine Sulfate and Alumina Cγ gel. To the supernatant solution, add 2% (w/v) protamine sulfate (1.5 mg of protamine sulfate for each 10 mg of protein), and remove the precipitate by centrifugation for 30 minutes at 35,000 *g*. To each 100 ml of the resultant supernatant solution, add 20 ml of alumina Cγ gel (30 mg dry weight per milliliter, prepared as described by Colowick[8]). Remove and discard the precipitate.

Step 4. Fractionation with Ammonium Sulfate. To each 100 ml of the supernatant solution, add 24.3 g of solid $(NH_4)_2SO_4$ (40% saturation) and discard the precipitate. Bring the supernatant solution to 75% saturation by addition of 24.5 g solid $(NH_4)_2SO_4$ to each 100 ml, collect the precipitate by centrifuging, and dissolve it in 15–20 ml of the 5 m*M* phosphate buffer, pH 7.0, to which mercaptoethanol to, 1 m*M* has been added. Dialyze the suspension against the 5 m*M* phosphate–1 m*M* mercaptoethanol buffer mixture overnight; remove and discard any material precipitated during this dialysis procedure.

Step 5. Column Chromatography. Suspend a batch of 7 g of diethylaminoethyl cellulose (such as Whatman DE 50) in 500 ml of the 5 m*M* phosphate buffer, and adjust the pH of the mixture to 7.0 with 2 *N* HCl.

[7] G. H. Dixon, H. L. Kornberg, and P. Lund, *Biochim. Biophys. Acta* **41,** 217 (1960).
[8] S. P. Colowick, Vol. I [11].

Wash the cellulose repeatedly with 5 mM phosphate buffer, pH 7.0, and remove by decantation any particles that do not sediment. Pour the slurry into a chromatography column (2×30 cm), the lower end of which is closed with glass wool. Equilibrate the column contents by allowing at least 1 liter of 5 mM phosphate buffer, pH 7.0, to run through it at 2°. Now apply the dialyzed material to the top of the column, at a rate just sufficient to keep the top moist, and apply a linear gradient of KCl to the column. This may be achieved by allowing 400 ml of a solution containing 5 mM phosphate buffer, 1 mM mercaptoethanol, and 450 mM KCl to flow, with constant stirring, into 400 ml of a solution containing 5 mM phosphate buffer, pH 7.0, 1 mM mercaptoethanol, and 50 mM KCl; the resultant mixtures are allowed to flow through the column at 40–50 ml per hour, the effluent being collected continuously at approximately 8 ml per fraction.

a. NADH-DEHYDROGENASE. The hold-up volume, i.e., the volume required to pass the material which is not adsorbed on the column, is rich in NADH-dehydrogenase activity but devoid of TSA reductase; it is precipitated with $(NH_4)_2SO_4$ to 75% saturation and the precipitate is dissolved in 5 mM phosphate buffer, pH 7.0. This solution, which can be stored at 2° for several weeks without loss of activity, may, when suitably diluted, be used as the source of NADH-dehydrogenase for assay method B.

b. TSA REDUCTASE. The fractions eluted from the column between 110 and 180 mM KCl contain most of the TSA reductase activity, the peak appearing at approximately 130 mM KCl. Those fractions containing the enzyme at a specific activity greater than 30 are combined, precipitated with $(NH_4)_2SO_4$ to 75% saturation, and dissolved in 3–4 ml of 5 mM phosphate buffer, pH 7.0.

Steps 6 and 7. Crystallization. Bring the solution to first turbidity (approximately 40% saturation) by the slow addition of solid $(NH_4)_2SO_4$; now allow to warm to approximately 20°; at this point crystallization begins. After 1 hour at room temperature, keep for a further 4–5 hours at 0° and collect the crystals by centrifugation. This material may be redissolved in a minimal volume of 5 mM phosphate buffer, pH 7.0, and recrystallized with $(NH_4)_2SO_4$ as described above: three successive recrystallizations yield enzyme of constant specific activity. The crystals appear to have the shape of two pyramids fused at their bases and have been found[3] to sediment in the ultracentrifuge as a single symmetrical peak.

The simple and reproducible purification procedure is summarized in the table.

PURIFICATION PROCEDURE[a]

Step	Volume of solution (ml)	Units	Concentration of protein[b] (mg/ml)	Specific activity (units/mg protein)	Recovery (%)
1. Sonic extract	180	10,550	65	0.90	100
2. Supernatant from protamine sulfate	230	9,380	21.5	1.90	89
3. Supernatant from Cγ-gel adsorption	290	9,380	13.5	2.40	89
4. Ammonium sulfate precipitate, 40–75% saturated	20	5,860	100	2.93	56
5. Pooled selected fractions from DEAE-cellulose column	100	2,730	1.82	15.0	26
6. Crystals, first	1	840	5.5	153	8.0
7. Crystals, after 3 crystallizations	2	800	2.5	160	7.6

[a] The enzymatic activity was assayed throughout by method B.

[b] The protein content of the material obtained in steps 1–4 was estimated by the procedure of A. G. Gornall, C. J. Bardawill, and M. M. David [*J. Biol. Chem.* **177,** 751 (1949)]; that of steps 5–7 by the method of O. Warburg and W. Christian [*Biochem. Z.* **310,** 384 (1941)].

C. GLYOXYLATE CARBOLIGASE. The material eluted from the DEAE-cellulose column between 200 and 250 m*M* KCl contains the glyoxylate carboligase activity of the original extract; it is precipitated with $(NH_4)_2SO_4$ to 75% saturation; when stored in the presence of crystalline bovine serum albumin in a solution of 5 m*M* phosphate buffer, pH 7.0, 1 m*M* $MgCl_2$, and 0.1 m*M* thiamine pyrophosphate, at 2°, it retains full enzymatic activity for several months. This material may be used for the preparation of TSA (method A), though even relatively crude cell-free extracts of *P. ovalis* Chester also serve this purpose.

Properties of TSA Reductase

Specificity. Crystalline TSA reductase does not catalyze the oxidation of NADH in the presence of glyoxylate, oxalacetate, pyruvate, glycolaldehyde, DL-glyceraldehyde, glyoxal, reductone, formaldehyde, or mesoxalate. Some reaction occurs with dihydroxyfumarate, but, as this reaction is not accompanied by any decrease in extinction at 290 mμ (as would be expected from the utilization of dihydroxyfumarate), it is likely that this reaction is due to TSA formed through decomposition of the unstable dihydroxyfumarate.[9] At saturating substrate concentrations, the rate of reduction of hydroxypyruvate was less than 17% of that noted with TSA,[3] and, at concentrations sufficient to saturate the enzyme with respect to TSA, the rate of reaction with hydroxypyruvate was less than 3%. (Note: the D-glyceric dehydrogenase of plants catalyzes the rapid reduction of hydroxypyruvate, but does not act on TSA.) Similarly, malonic semialdehyde and mesoxalic semialdehyde are reduced by NADH in the presence of crystalline TSA reductase, but at less than 15% of the rate observed with TSA. These findings imply that the enzyme is capable of catalyzing the reduction of C_3-compounds of general structure $CHO \cdot R \cdot CO_2H$; on the other hand, the enzyme has been found to catalyze the oxidation only of glyceric and hydroxypyruvic acids. In both directions, NADP or NADPH react, though (at pH 8.5) not as readily as do NAD or NADH.

Kinetic Properties. At pH 8.5, 20°, the K_m for TSA was measured to be 2×10^{-4} *M*, that for NADH was 2×10^{-5} *M*, and that for NADPH was 5×10^{-5} *M*. Under the conditions of method A, and at saturating concentrations of reactants, the crystalline enzyme oxidized 160 micromoles of NADH per minute per milligram of protein with concomitant reduction of TSA; on the basis of the measured[10] weight-average molecular weight M_w of 91,000, this corresponds to a turnover number of 14,600 moles of NADH oxidized per minute per mole of enzyme. For the oxidation of glycerate ($K_m = 4 \times 10^{-4}$ *M*) with concomitant reduction of NAD, the equilibrium constant K at 23° was 2×10^{-6} at pH 7.5 and 1.6×10^{-5} at pH 8.5; K_H was calculated to be 5.1×10^{-14} *M*. From these data, $\Delta G'$[11] at pH 7, 25°, was determined as 8.6 kcal/mole and $\Delta G°$ as 18 kcal/mole. Similarly, the oxidation-reduction potential for the reaction.

$$\text{Glycerate}^- \rightleftharpoons \text{TSA}^- + 2\,\text{H} + 2\,e$$

was found to be $E'_0 = -0.092$ volt at 25° and pH 7.

Activators and Inhibitors. No activators for TSA reductase have been described. The enzyme is not inhibited by 1 m*M* iodoacetate, 10 m*M* glycolaldehyde, DL-glyceraldehyde, sodium fluoride, 3-phosphoglycerate, or 4 m*M* phosphoenolpyruvate. Slight inhibition is produced by 10 m*M* pyruvate, L-malate, formaldehyde, oxalate, glyoxal, or tartonate; and strong inhibition by 10 m*M* glyoxylate, fluoroacetate, and glycolate.

Stability and pH Optimum. The crystalline enzyme, stored as a

[9] C. T. Chow and B. Vennesland, *J. Biol. Chem.* **233**, 997 (1958).
[10] A. Rodgers, *Biochem. J.* **81**, 285 (1961).
[11] $\Delta G'$ = standard free energy change at pH specified.

suspension in ammonium sulfate, pH 7.5, at 2°, loses less than 10% of its activity over a month. The enzyme catalyzes the oxidation of NADH by TSA over a wide range of H^+-ion concentration but with maximal velocity between pH 6.2 and pH 8.7. However, even at pH 5, the rate is approximately twice as rapid with NADH as with NADPH; but at pH 6.5 both nucleotides give similar rates.

[50] Purification and Resolution of the Pyruvate Dehydrogenase Complex (*Escherichia coli*)

By LESTER J. REED and CHARLES R. WILLMS

$$\text{Pyruvate} + \text{CoA} + \text{DPN}^+ \rightarrow \text{acetyl CoA} + \text{CO}_2 + \text{DPNH} + \text{H}^+ \quad (1)$$

Enzyme systems which catalyze reaction (1) have been isolated from pigeon breast muscle,[1] *Escherichia coli*,[2] pig heart muscle,[3] and beef kidney mitochondria[4] as multienzyme complexes of high molecular weight. The *E. coli* pyruvate dehydrogenase complex has been separated into three enzymes, pyruvate decarboxylase (E_1), lipoyl reductase-transacetylase (E_2), and dihydrolipoyl dehydrogenase (E_3), and has been reconstituted from the isolated enzymes.[5] The available evidence indicates that the oxidative decarboxylation of pyruvate represented by reaction (1) proceeds via the sequence shown in reactions (2–6). In its functional form the lipoyl moiety [$LipS_2$] is bound in amide linkage to

$$\text{CH}_3\text{COCO}_2\text{H} + \text{TPP-E}_1 \rightarrow [\text{CH}_3\text{CH(OH)-TPP}]\text{-E}_1 + \text{CO}_2 \quad (2)$$

$$[\text{CH}_3\text{CH(OH)-TPP}]\text{-E}_1 + [\text{LipS}_2]\text{-E}_2 \rightarrow [\text{CH}_3\text{CO-SLipSH}]\text{-E}_2 + \text{TPP-E}_1 \quad (3)$$

$$[\text{CH}_3\text{CO-SLipSH}]\text{-E}_2 + \text{HSCoA} \rightarrow [\text{Lip(SH)}_2]\text{-E}_2 + \text{CH}_3\text{CO-SCoA} \quad (4)$$

$$[\text{Lip(SH)}_2]\text{-E}_2 + \text{E}_3\text{-FAD} \rightarrow [\text{LipS}_2]\text{-E}_2 + \text{reduced E}_3\text{-FAD} \quad (5)$$

$$\text{Reduced E}_3\text{-FAD} + \text{DPN}^+ \rightarrow \text{E}_3\text{-FAD} + \text{DPNH} + \text{H}^+ \quad (6)$$

the ϵ-amino group of a lysine residue.[6] The purification and properties of the *E. coli* pyruvate dehydrogenase complex and the separation of the complex into its constituent enzymes are described below.

[1] V. Jagannathan and R. S. Schweet, *J. Biol. Chem.* **196,** 551 (1952). See also R. S. Schweet, B. Katchman, R. M. Bock, and V. Jagannathan, *J. Biol. Chem.* **196,** 563 (1952).

[2] M. Koike, L. J. Reed, and W. R. Carroll, *J. Biol. Chem.* **235,** 1924 (1960).

[3] T. Hayakawa, H. Muta, M. Hirashima, S. Ide, K. Okabe, and M. Koike, *Biochem. Biophys. Research Commun.* **17,** 51 (1964).

[4] E. Ishikawa and L. J. Reed, unpublished experiments (1965).

[5] M. Koike, L. J. Reed, and W. R. Carroll, *J. Biol. Chem.* **238,** 30 (1963).

[6] H. Nawa, W. T. Brady, M. Koike, and L. J. Reed, *J. Am. Chem. Soc.* **82,** 896 (1960).

I. Purification and Properties of the Pyruvate Dehydrogenase Complex

Assay Method

Principle. The dismutation assay[7] may be used at all levels of purity of the enzyme complex. It is based on measurement of acetyl phosphate [Eq. (9)] by means of the hydroxamic acid method.[8] Reaction (1) is coupled with the phosphotransacetylase- and lactate dehydrogenase-catalyzed reactions [Eqs. (7) and (8)] to give the pyruvate dismutation reaction [Eq. (9)].

$$\text{Pyruvate} + \text{CoA} + \text{DPN}^+ \rightarrow \text{acetyl CoA} + CO_2 + \text{DPNH} + H^+ \quad (1)$$
$$\text{Acetyl CoA} + P_i \rightarrow \text{acetyl-P} + \text{CoA} \quad (7)$$
$$\text{Pyruvate} + \text{DPNH} + H^+ \rightarrow \text{lactate} + \text{DPN}^+ \quad (8)$$
$$\textit{Sum: } 2\ \text{Pyruvate} + P_i \rightarrow \text{acetyl-P} + CO_2 + \text{lactate} \quad (9)$$

Reagents

Potassium phosphate buffer, 1 *M*, pH 7.0
Potassium pyruvate, 0.5 *M*
Magnesium sulfate, 0.2 *M*
Thiamine pyrophosphate, 0.01 *M*
CoA, 0.001 *M*, containing 0.064 *M* cysteine, prepared before use
DPN, 0.0023 *M*
Potassium citrate buffer, 0.1 *M*, pH 5.4
Hydroxylamine solution, 2 *M*, pH 6.4. This solution is freshly prepared by adjusting 4 *M* hydroxylamine hydrochloride to pH 6.4 with 3.5 *M* sodium hydroxide and diluting to a final concentration of 2 *M*.
Ferric chloride solution. Equal volumes of 3 *N* hydrochloric acid, 12% trichloroacetic acid, and 5% $FeCl_3$ in 0.1 *N* hydrochloric acid are mixed.
Phosphotransacetylase. A cell-free extract prepared from *Clostridium kluyverii* dried cells[9] is suitable.
Crystalline lactate dehydrogenase
Enzyme. The enzyme is diluted with 0.05 *M* phosphate buffer, pH 7.0, and the aliquot assayed is such that 0.5–2.0 micromoles of acetyl phosphate will be formed during incubation.

Procedure. To a 13 × 100 mm Pyrex test tube are added 0.1 ml of phosphate buffer, 0.02 ml of thiamine pyrophosphate, 0.02 ml of mag-

[7] S. Korkes, A. del Campillo, I. C. Gunsalus, and S. Ochoa, *J. Biol. Chem.* **193**, 721 (1951). See also L. J. Reed, F. R. Leach, and M. Koike, *ibid.* **232**, 123 (1958).
[8] F. Lipmann and L. C. Tuttle, *J. Biol. Chem.* **159**, 21 (1945). See also E. R. Stadtman, Vol. III [39].
[9] E. R. Stadtman, Vol. I [84].

nesium sulfate, 0.1 ml of CoA-cysteine solution, 0.1 ml of DPN, 0.1 ml of potassium pyruvate, 2000 units (optical units)[10] of lactate dehydrogenase, 10 units[11] of phosphotransacetylase, enzyme fraction to be assayed, and water to make a final volume of 1.0 ml. The control contains all reagents except enzyme fraction. The mixture is incubated for 30 minutes at 30°. One milliliter of citrate buffer and 1 ml of hydroxylamine solution are added, and the mixture is allowed to stand for 10 minutes at room temperature. Three milliliters of ferric chloride solution is added and the mixture is centrifuged for 7 minutes in a clinical centrifuge. The optical density of the supernatant fluid is determined at 540 mμ against the control in a suitable spectrophotometer. A standard curve is prepared with synthetic acethydroxamic acid.[8,12]

Definition of Unit and Specific Activity. One unit is the amount of enzyme required to produce 1 micromole of acetyl phosphate per hour under the conditions specified. Specific activity is expressed as units per milligram of protein. Protein is determined by the method of Lowry *et al.*[13] using crystalline bovine serum albumin as standard.

Other Methods of Assay. A spectrophotometric assay based on measurement of the rate of formation of DPNH [Eq. (1)] can be applied after a certain purification of the enzyme complex, i.e., with preparations free of DPNH oxidase and lactate dehydrogenase. The assay is carried out essentially as described for the α-ketoglutarate dehydrogenase complex.[14]

Purification Procedure

Reagents

Protamine solution, 2%, pH 6.2, prepared before use and kept at room temperature. Protamine sulfate (Nutritional Biochemicals Corporation) is suspended in water, the pH is adjusted to 6.2 with 10% KOH, and the mixture is centrifuged to remove insoluble material.

Potassium phosphate buffer, 0.02 *M*, pH 7.0

Potassium phosphate buffer, 0.05 *M*, pH 7.0

Potassium phosphate buffer, 0.1 *M*, pH 7.0

[10] A. Kornberg, Vol. I [67]. See also A. H. Mehler, A. Kornberg, S. Grisolia, and S. Ochoa, *J. Biol. Chem.* **174**, 961 (1948).

[11] E. R. Stadtman, Vol. I [98].

[12] W. P. Jencks, *J. Am. Chem. Soc.* **80**, 4585 (1958).

[13] O. H. Lowry, N. J. Rosebrough, A. L. Farr, and R. J. Randall, *J. Biol. Chem.* **193**, 265 (1951).

[14] B. B. Mukherjee, J. Matthews, D. L. Horney, and L. J. Reed, *J. Biol. Chem.* **240**, PC2268 (1965).

Acetic acid, 1% (v/v)

Yeast ribonucleic acid solution, 1%, pH 6.2. Yeast ribonucleic acid (Nutritional Biochemicals Corporation) is suspended in water, the pH is adjusted to 6.2 with 10% KOH, and the mixture is centrifuged to remove insoluble material. Soluble ribonucleic acid from yeast (General Biochemicals) is also suitable.

Preparation of Cell-Free Extract. E. coli, Crookes strain, is grown according to the procedure of Hager[15] (see also Korkes[16]), harvested with a refrigerated Sharples centrifuge, and stored at —20° until needed. The cells can be grown more conveniently in 45-liter batches of medium in a Biogen continuous culture apparatus. The conditions are 3.5 hours of growth at 35°, a chamber speed of 308 rpm, and an air pressure of 10 psi. The yield of cell paste is about 10 g per liter of medium. All subsequent operations are carried out at 0–5° unless stated otherwise. The thawed cells are suspended in 0.02 *M* phosphate buffer, pH 7.0, by means of a Waring blendor, at a concentration of 20 g of cell paste per 50 ml of buffer. Seventy-milliliter portions of the suspension are treated for 7 minutes in a 10-kc sonic oscillator (Raytheon). Cell debris is removed by centrifugation at 53,700 *g* (20,000 rpm) for 30 minutes in the No. 21 rotor of a Spinco Model L ultracentrifuge.

Fractionation with Protamine. The cell-free extract is diluted with 0.02 *M* phosphate buffer, pH 7.0, to a protein concentration of 20 mg/ml, as estimated by the method of Lowry *et al.*[13] A pilot run is advisable to determine how an individual batch may fractionate. A representative purification is described below. Twelve hundred milliliters of diluted extract are adjusted to pH 6.15 with 1% acetic acid, and 288 ml (0.24 volume) of 2% protamine solution, pH 6.2, is added dropwise with stirring. The mixture is stirred for an additional 15 minutes and then centrifuged at maximum speed for 30 minutes in the No. 845 head of an International Model PR-2 centrifuge. The gelatinous, white precipitate is discarded, and 36 ml (0.03 volume) of 2% protamine solution is added with stirring to the supernatant fluid. The mixture is stirred for 15 minutes and the precipitate is collected by centrifugation. Two 36-ml portions (0.03 volume) of 2% protamine solution are added successively to the supernatant fluid as described above, and the precipitates are collected by centrifugation. Aliquots of the supernatant fluids from the four successive additions of 2% protamine solution are assayed for pyruvate dismutation activity. Usually, the third and/or fourth addition of 2% protamine solution results in precipitation of over 90% of the pyruvate dehydrogenase complex. The grayish yellow precipitate is sus-

[15] L. P. Hager, Ph.D. Dissertation, University of Illinois, 1953.
[16] S. Korkes, Vol. I [77].

pended in 120 ml of 0.1 *M* phosphate buffer, pH 7.0, by means of a large glass homogenizer equipped with a motor-driven Teflon pestle. The suspension is stirred for 45 minutes and then centrifuged for 20 minutes at 47,000 *g* (20,000 rpm) in the No. 30 rotor of a Spinco Model L ultracentrifuge. The precipitate is discarded, and the yellow solution is allowed to stand overnight in an ice bath. If the solution becomes cloudy it is centrifuged for 20 minutes at 47,000 *g*, and the precipitate is discarded.

Ultracentrifugation. The clear, yellow solution from the previous step, designated protamine eluate, is centrifuged for 3 hours at 144,000 *g* (40,000 rpm) in the No. 40 rotor of a Spinco Model L ultracentrifuge. The supernatant fluid is removed carefully from the gelatinous yellow pellet and/or viscous yellow layer and is discarded. The yellow pellets and/or viscous yellow layers are dissolved in a total volume of 60 ml of 0.05 *M* phosphate buffer, pH 7.0, with the aid of a motor-driven Teflon pestle. The solution is centrifuged for 20 minutes at 36,000 *g* (20,000 rpm) to remove a small amount of insoluble material, and the supernatant fluid is centrifuged for 2½ hours at 144,000 *g*. The yellow pellets are dissolved in a total volume of about 12 ml of 0.05 *M* phosphate buffer, pH 7.0. If the solution is cloudy it is centrifuged for 20 minutes at 36,000 *g*, and the precipitate is discarded. The supernatant fluid contains the pyruvate and α-ketoglutarate dehydrogenase complexes and variable, but small amounts of lower molecular weight contaminant(s).

Isoelectric Precipitation. The clear, yellow solution from the previous step is diluted with 0.05 *M* phosphate buffer, pH 7.0, to a protein concentration of 5 mg/ml. For every 10 mg of protein, 0.04 ml of 1% yeast ribonucleic acid solution, pH 6.2, is added. The pH of the solution is carefully adjusted to 6.6 by dropwise addition, with stirring, of 1% acetic acid. The mixture is stirred for an additional 10 minutes, centrifuged for 10 minutes at 14,000 *g* (13,000 rpm) in an International Model HR-1 centrifuge, and the gray precipitate is discarded. The pH is lowered to 5.5 as described above. The mixture is stirred for an additional 5 minutes, and the yellow precipitate is collected by centrifugation. This precipitate contains the α-ketoglutarate dehydrogenase complex (approximately 115 mg of protein). The pH of the supernatant fluid from the previous step is lowered to 4.9. The mixture is stirred for an additional 5 minutes, and the yellow precipitate is collected by centrifugation. This precipitate contains the pyruvate dehydrogenase complex. The two precipitates are dissolved separately in 2–3 ml of 0.02 *M* phosphate buffer, pH 7.0. The solutions are clarified, if necessary, by centrifugation for 20 minutes at 36,000 *g*. The preparations are stored in the frozen state, preferably at −90°, and retain full activity for several months. A summary of the purification procedure is given in the table.

PURIFICATION PROCEDURE[a]

Fraction	Volume (ml)	Protein (mg)	Specific activity	Recovery (%)
Diluted extract	1200	24,000	8	(100)
Protamine eluate	120	853	180	80
Ultracentrifuge pellet	12	300	480	75
Precipitate, pH 5.5–4.9	2.4	115	1000	60[b]

[a] Three hundred twenty grams of cell paste.
[b] The overall recovery varied between 45% and 70% in about 50 purifications.

Comments. In the procedure described above both the pyruvate and α-ketoglutarate dehydrogenase complexes are purified together, and are separated in the final step by acid fractionation. Usually, the preparations obtained contain less than 5% of impurity, as revealed by ultracentrifugal and electrophoretic analyses. If larger amounts of impurity are present the isoelectric precipitation step is repeated. Alternatively, the preparations are subjected to fractionation with solid ammonium sulfate. The pyruvate dehydrogenase complex precipitates between 0.40 and 0.48 saturation, and the α-ketoglutarate dehydrogenase complex precipitates between 0.25 and 0.32 saturation. α-Ketoglutarate dehydrogenase activity is determined by measuring the rate of α-ketoglutarate oxidation with ferricyanide as electron acceptor, essentially as described below for pyruvate decarboxylase.

Ribonucleic acid is added prior to the isoelectric precipitation to remove protamine. Otherwise, a sharp separation of the two complexes may not be achieved.

Properties

Specificity. The enzyme complex is specific for DPN. It will oxidatively decarboxylate α-ketobutyrate [Eq. (1)], but is inactive toward α-ketoglutarate.

Physical Constants.[2] The electrophoretic mobility of different preparations of the pyruvate dehydrogenase complex varied from -7.4 to -6.7×10^{-5} cm^2 volt^{-1} sec^{-1} in 0.05 M potassium phosphate buffer, pH 6.9. This difference may be due to variable amounts of bound protamine. The enzyme complex exhibits a sedimentation coefficient ($S^0_{20,w}$) of 64 S, a diffusion coefficient ($D_{20,w}$) of 1.20×10^{-7} cm^2 sec^{-1} (protein concentration, 1.5 mg/ml), and a partial specific volume ($\bar{v}$) of 0.735 ml/g.[17] Based on these data the calculated molecular weight is 4.8×10^6. The frictional ratio (f/f_0) is 1.6.

[17] D. W. Kupke, J. Senter, C. R. Willms, and L. J. Reed, unpublished experiments.

Other Properties. In the electron microscope[18] the complex appears as a polyhedron with a diameter about 300 Å and a height about 200 Å.

II. Resolution of the Pyruvate Dehydrogenase Complex[5]

Currently available data indicate that the *E. coli* pyruvate dehydrogenase complex is an organized mosaic consisting of about 12 molecules of pyruvate decarboxylase (molecular weight 183,000), 6 molecules of dihydrolipoyl dehydrogenase (molecular weight 112,000) and 24 subunits (molecular weight 70,000) comprising the lipoyl reductase- transacetylase aggregate. The flavoprotein, dihydrolipoyl dehydrogenase, is selectively dissociated from the complex by fractionation, in the presence of 4 *M* urea, on a column of calcium phosphate gel suspended on cellulose. This procedure yields free flavoprotein and a complex of the decarboxylase and lipoyl reductase-transacetylase. The latter complex is separated into its two component enzymes by fractionation on gel-cellulose at pH 9.5. The decarboxylase can be selectively dissociated from the pyruvate dehydrogenase complex by fractionation on gel-cellulose at pH 9.5. This procedure yields free decarboxylase and a complex of lipoyl reductase-transacetylase and flavoprotein. The latter complex can be separated into its two component enzymes by fractionation on gel-cellulose in the presence of 4 *M* urea. Lipoyl reductase-transacetylase appears to possess specific binding sites for the decarboxylase and for the flavoprotein. The former enzyme combines spontaneously with the latter two enzymes to produce a large unit resembling the native pyruvate dehydrogenase complex.[5,18]

A purification procedure described earlier by Hager[15] and Gunsalus[19] (see also Korkes[16]) yielded two enzyme fractions, designated A and B, from *E. coli* extracts. It is apparent that fraction A contains the partial complex consisting of pyruvate decarboxylase and lipoyl reductase-transacetylase and that fraction B contains dihydrolipoyl dehydrogenase.

A. Dihydrolipoyl Dehydrogenase

$$\text{Lipoamide} + \text{DPNH} + \text{H}^+ \rightleftharpoons \text{dihydrolipoamide} + \text{DPN}^+ \qquad (10)$$

Assay Method

Principle. The assay is based on spectrophotometric determination of the rate of DPNH oxidation (at 340 mμ) in the presence of the dehydrogenase and lipoamide.

[18] H. Fernández-Morán, L. J. Reed, M. Koike, and C. R. Willms, *Science* **145**, 930 (1964).

[19] I. C. Gunsalus, *in* "The Mechanism of Enzyme Action" (W. D. McElroy and B. Glass, eds.), p. 545. Johns Hopkins Press, Baltimore, Maryland, 1954.

Reagents

Potassium phosphate buffer, 0.3 *M*, pH 8.1
DPNH, 0.001 *M*
DPN, 0.001 *M*
(±)-Lipoamide,[20] 0.1 *M*, in 95% ethanol, prepared before use and kept at room temperature

Procedure. To a cuvette with a 1-cm light path are added 1.5 ml of phosphate buffer, 0.3 ml of DPNH, 0.9 ml of DPN, 0.09 ml of lipoamide (shake during addition to prevent local precipitation), water, and enzyme to make a total volume of 3.0 ml. The reaction is initiated at room temperature by the addition of enzyme. The blank cell contains 1.5 ml of phosphate buffer, 1.41 ml of water, and 0.09 ml of lipoamide. The decrease in absorbance at 340 mμ is followed with a recording spectrophotometer. A decrease in absorbance of 0.05–0.2 during the initial phase of the reaction (about 1 minute) is a linear function of protein concentration.

Definition of Unit and Specific Activity. One unit is defined as the amount of enzyme that catalyzes an initial rate of oxidation of 1 micromole of DPNH per minute. Specific activity is expressed as units of enzyme per milligram of protein.

Other Methods of Assay. Reaction (10), from right to left, can be coupled with the lactate dehydrogenase-catalyzed reaction [Eq. (8)], and the disappearance of -SH groups can be measured.[15,21] The activity of the flavoprotein can be determined also in the pyruvate dismutation assay by adding a suitable excess (usually twofold) of pyruvate decarboxylase and lipoyl reductase-transacetylase or a complex of the latter two enzymes.[5,15,21,22] This latter assay is applicable to each of the three enzymes comprising the pyruvate dehydrogenase complex.[5,23] Its application to lipoyl reductase-transacetylase is described below.

Resolution Procedure

Reagents

Urea, 4 *M*, and 2% ammonium sulfate in 0.1 *M* potassium phosphate buffer, final pH, 7.5. Reagent-grade urea is recrystallized from 95% ethanol.

[20] Prepared as described in L. J. Reed, M. Koike, M. E. Levitch, and F. R. Leach, *J. Biol. Chem.* **232,** 143 (1958), or obtained commercially.
[21] L. P. Hager and I. C. Gunsalus, *J. Am. Chem. Soc.* **75,** 5767 (1953).
[22] M. Koike, P. C. Shah, and L. J. Reed, *J. Biol. Chem.* **235,** 1939 (1960).
[23] U. Henning, C. Herz, and K. Szolyvay, *Z. Vererbungslehre* **95,** 236 (1964).

Urea, 4 *M*, and 1% ammonium sulfate in 0.1 *M* potassium phosphate buffer, final pH 7.5

Ammonium sulfate, 1%, in 0.1 *M* potassium phosphate buffer, final pH 7.5

Ammonium sulfate, 10%, in 0.1 *M* potassium phosphate buffer, final pH 7.5

Calcium Phosphate Gel Suspended on Cellulose.[24] To 31 g of cellulose powder (Whatman standard grade) suspended in 200 ml of distilled water is added 50 ml of a solution containing 0.1 mole of $CaCl_2$, 0.067 mole of KH_2PO_4, and 0.033 mole of HCl. The mixture is stirred rapidly for 2 minutes, 50 ml of 8 *N* NH_4OH is added, and stirring is continued as the mixture thickens. The mixture is allowed to stand overnight at 5°. The supernatant fluid is decanted, and the gel-cellulose is washed by decantation with 3-liter volumes of distilled water until the supernatant fluid is negative to Nessler's reagent. During the washing procedure, which should be completed in 1 day, care is taken to remove fine particles. The gel-cellulose is collected by low-speed centrifugation and resuspended in 600 ml of 0.02 *M* phosphate buffer, pH 6.0. The mixture is stored at 5° for at least 1 week before use. Prior to packing columns of gel-cellulose at room temperature, it is advisable to remove dissolved gases by warming the suspension to room temperature and then evacuating the mixture with a water aspirator for about 10 minutes.

Procedure. To a 2.8 $\times$ 23 cm glass chromatographic tube, fitted with a detachable sintered glass plate covered with a circle of filter paper, is added 0.02 *M* phosphate buffer, pH 6.0. A slurry of gel-cellulose is added in portions to the tube. The gel-cellulose is allowed to pack by gravity to give a column approximately 3.9 cm in height, comprising about 24 ml of gel-cellulose. The column is transferred to a cold room to equilibrate. Alternatively, a jacketed column may be used in conjunction with cold, circulating water to maintain the temperature of the column below 10°. A flow rate of 30–40 ml per hour is maintained throughout the resolution procedure by applying slight pressure to the column.

A solution of the pyruvate dehydrogenase complex containing 50 mg of protein in about 1.5 ml of 0.02 *M* phosphate buffer, pH 7.0, is applied carefully to the column. After this solution has passed into the column, approximately 1.0 ml of 0.02 *M* phosphate buffer, pH 7.0, is used to wash

[24] This procedure is a modification of that described in V. E. Price and R. E. Greenfield, *J. Biol. Chem.* **209,** 363 (1954). Gel-cellulose prepared as described by V. Massey [*Biochim. Biophys. Acta* **37,** 310 (1960)] has been used to resolve the pig heart α-ketoglutarate dehydrogenase complex [V. Massey, *Biochim. Biophys. Acta* **38,** 447 (1960)].

down the wall of the glass tube near the top of the gel bed. A 3-ml portion of a solution containing 4 *M* urea and 2% ammonium sulfate in 0.1 *M* phosphate buffer, pH 7.5, is passed into the column. This is followed by 3 ml of a solution containing 4 *M* urea and 1% ammonium sulfate in 0.1 *M* phosphate buffer, pH 7.5. The column is then washed with about 60 ml of a solution of 1% ammonium sulfate in 0.1 *M* phosphate buffer, pH 7.5. A colorless protein fraction is eluted, leaving a broad yellow, fluorescent band on the column. Emergence of the colorless fraction can be detected usually by a slight turbidity, Alternatively, the protein content of the fractions is estimated visually after adding 0.1-ml aliquots to 1-ml portions of 12% trichloroacetic acid. Immediately after elution the colorless fraction (about 10 ml) is dialyzed with stirring against 0.05 *M* phosphate buffer, pH 7.0, to remove the urea. The yellow, fluorescent fraction is eluted from the column with a solution of 10% ammonium sulfate in 0.1 *M* phosphate buffer, pH 7.5, and dialyzed overnight against 0.05 *M* phosphate buffer, pH 7.0.

The dialyzed colorless fraction is brought to 0.4 ammonium sulfate saturation by adding slowly with stirring 0.28 g of the salt per mililiter. The mixture is stirred for an additional 15 minutes, centrifuged for 10 minutes at 14,000 *g*, and the precipitate is discarded. The supernatant fluid is brought to 0.44 ammonium sulfate saturation by addition of 0.03 g of the salt per milliliter. The precipitate is collected by centrifugation and dissolved in about 1 ml of 0.05 *M* phosphate buffer, pH 7.0. The solution is dialyzed overnight against the same buffer. Approximately 35 mg of protein, which is a complex of pyruvate decarboxylase and lipoyl reductase-transacetylase, is recovered in the dialyzed solution.

The dialyzed yellow fraction is brought to 0.5 ammonium sulfate saturation by slow addition of 0.36 g of the salt per milliliter. Stirring is continued for an additional 15 minutes, and the suspension is centrifuged for 10 minutes at 14,000 *g*. The light yellow precipitate contains unresolved pyruvate dehydrogenase complex and/or a complex of lipoyl reductase-transacetylase and flavoprotein. The supernatant fluid is brought to 0.8 ammonium sulfate saturation by addition of 0.21 g of the salt per milliliter. The bright yellow precipitate is collected by centrifugation and dissolved in 0.05 *M* phosphate buffer, pH 7.0. The solution is dialyzed overnight against the same buffer. The recovery of dihydrolipoyl dehydrogenase in this fraction is approximately 5 mg; specific activity is 140 units/mg. The preparations are stored in the frozen state (—20°) and retain full activity for at least several months.

Comments. Pyruvate decarboxylase is somewhat labile to urea. Therefore, it is essential that the period between application of the urea solution to the column and dialysis of the colorless fraction be kept to a minimum (usually 1 hour). The urea lability of the decarboxylase precludes use of larger columns. However, larger amounts of the pyruvate dehydrogenase complex can be processed by running two or more columns simultaneously. The resolution procedure described is more effective with preparations of the pyruvate dehydrogenase complex which have been frozen and thawed once than with preparations which have not been frozen.

Properties

Physical Constants. At a protein concentration of 8.0 mg/ml, dihydrolipoyl dehydrogenase exhibits a sedimentation coefficient ($S_{20,w}$) of 6.24 S and a diffusion coefficient ($D_{20,w}$) of 5.01×10^{-7} cm^2 sec^{-1}. There is little concentration dependence of these coefficients. From these data and an assumed partial specific volume ($\bar{v}$) of 0.73 ml/g, the molecular weight is calculated to be 112,000.

Specificity. The natural substrate of dihydrolipoyl dehydrogenase is the dihydrolipoyl moiety which is covalently bound to the ε-amino group of a lysine residue in lipoyl reductase-transacetylase. However, the enzyme also catalyzes reaction (10) with lipoic acid and derivatives thereof. Both optical isomers of lipoic acid undergo reaction.[21,22] The oxidation of DPNH by lipoic acid and derivatives thereof requires the presence of DPN.[25,26] The DPN apparently prevents conversion of the flavoprotein to an inactive form in which the flavin is fully reduced. The enzyme is inactive with TPN but it does catalyze the reduction of DPN analogs by dihydrolipoamide.[26]

Inhibitors. The enzyme is inhibited by arsenite or cadmium ion in the presence of DPNH.[26] The inhibition is reversed by dithiols and less effectively by monothiols, suggesting that *E. coli* dihydrolipoyl dehydrogenase, like the pig heart enzyme (Straub's diaphorase),[27] contains a reactive disulfide group, presumably that of a cystine residue, which is reduced to a dithiol in the catalytic cycle.

Other Properties. The flavoprotein contains 2 molecules of FAD per molecule of enzyme. It exhibits absorption maxima at 273, 359, and 456 mμ with a shoulder at 480 mμ. It is extremely fluorescent for a flavoprotein.

[25] V. Massey and C. Veeger, *Biochim. Biophys. Acta* **48**, 33 (1961).

[26] J. Matthews, Ph.D. Dissertation, The University of Texas, 1961.

[27] V. Massey, *in* "The Enzymes" (P. D. Boyer, H. Lardy, and K. Myrbäck, eds.), p. 275. Academic Press, New York, 1963.

B. Pyruvate Decarboxylase

$$\text{Pyruvate} + 2\ Fe(CN)_6^{-3} + H_2O \rightarrow \text{acetate} + CO_2 + 2\ Fe(CN)_6^{-4} + 2\ H^+ \quad (11)$$

Assay Method

Principle. The assay is a modification of that described by Hager.[15] It is based on colorimetric determination of ferrocyanide (as prussian blue) produced by oxidative decarboxylation of pyruvate with ferricyanide as electron acceptor [Eq. (11)].

Reagents

Potassium phosphate buffer, 1.0 *M*, pH 6.0
Thiamine pyrophosphate, 0.002 *M*
Magnesium sulfate, 0.003 *M*
Potassium pyruvate, 0.5 *M*
Potassium ferricyanide, 0.25 *M*
Trichloroacetic acid, 10%
Duponol (sodium lauryl sulfate), 4%
Ferric ammonium sulfate-Duponol reagent. Prepared by dissolving 1.7 g of ferric ammonium sulfate in 10 ml of distilled water, filtering, and adding the filtrate to 20 ml of a solution containing 1.5 g of Duponol in 20 ml of water. To this solution is added 27 ml of 85% phosphoric acid, and the mixture is diluted with water to a final volume of 140 ml.
Enzyme. The enzyme is diluted with 0.02 *M* potassium phosphate buffer, and the aliquot assayed is such that 1.8 to 5.4 micromoles of ferrocyanide will be formed during incubation.

Procedure. To a 13 × 100 mm Pyrex test tube are added 0.15 ml of phosphate buffer, 0.1 ml of thiamine pyrophosphate, 0.1 ml of magnesium chloride, 0.1 ml of potassium pyruvate, 0.1 ml of potassium ferricyanide, enzyme fraction to be assayed, and water to make a final volume of 1.4 ml. The control contains all reagents except enzyme. The mixture is incubated for 30 minutes at 30°. The reaction is terminated by the addition of 1 ml of 10% trichloroacetic acid, and the mixture is centrifuged for 5 minutes in a clinical centrifuge to remove denatured protein. To a 0.1- or 0.2-ml aliquot of the supernatant fluid are added 1 ml of 10% trichloroacetic acid, 0.1 ml of potassium ferricyanide, and water to make a final volume of 2.4 ml. To the latter solution are added 1 ml of 4% Duponol and 0.5 ml of ferric ammonium sulfate-Duponol reagent. The mixture is allowed to stand at room temperature for 30 minutes. The optical density is determined at 540 mμ against the control. A standard curve is prepared with potassium ferrocyanide.

Definition of Unit and Specific Activity. One unit is the amount of enzyme required to produce 2 micromoles of ferrocyanide per hour under the conditions described. Specific activity is expressed as units of enzyme per milligram of protein.

Other Methods of Assay. The rate of CO_2 evolution [Reaction (11)] can be used to estimate pyruvate decarboxylase activity.[15] A $^{14}CO_2$-pyruvate exchange assay for the decarboxylase has been described.[28] The activity of the decarboxylase can be determined in the pyruvate dismutation assay by adding a suitable excess (usually twofold) of lipoyl reductase-transacetylase and dihydrolipoyl dehydrogenase or a complex of the latter two enzymes.[5] The dismutation assay is the most reliable for determining uncomplexed pyruvate decarboxylase in crude extracts of *E. coli* mutants.[23]

Resolution Procedure

Reagents

Ethanolamine, 0.16 *M*

Ethanolamine-phosphate buffer, pH 9.5; 0.5 ml of redistilled ethanolamine is dissolved in 160 ml of 0.05 *M* potassium phosphate buffer, pH 7.0, the pH is adjusted to 9.5 with 1 *M* potassium phosphate buffer, pH 6.0, and the solution is diluted with water to a final volume of 400 ml. The final solution is approximately 0.02 *M* with respect to ethanolamine and 0.025 *M* with respect to phosphate.

Procedure. A column (2.8 × 3.2 cm) of calcium phosphate gel suspended on cellulose (aged 2–3 weeks) is prepared as described above. The column is washed with ethanolamine-phosphate buffer until the pH of the effluent is 9.5. The column is transferred to a cold room to equilibrate. To a solution of 70 mg of the colorless fraction (complex of pyruvate decarboxylase and lipoyl reductase-transacetylase obtained from urea resolution of the pyruvate dehydrogenase complex)[29] in 2 ml of 0.05 *M* phosphate buffer, pH 7.0, are added 1.5 ml of water and 0.5 ml of 0.16 *M* ethanolamine. The pH is adjusted quickly to 9.5 with 1 *M* phosphate

[28] A. D. Gounaris and L. P. Hager, *J. Biol. Chem.* **236**, 1013 (1961).

[29] Preparations of the colorless fraction which have been stored at −20° for several days and then thawed are more easily resolved into pyruvate decarboxylase and lipoyl reductase-transacetylase than are preparations which have not been frozen.

buffer, pH 6.0. This solution is applied to the gel-cellulose column. Pressure is applied to the column as needed to achieve a flow rate of 30–40 milliliters per hour. When the enzyme solution has passed into the gel-cellulose bed, washing with ethanolamine-phosphate buffer, pH 9.5, is begun. Emergence of protein (pyruvate decarboxylase) is detected by adding 0.1-ml aliquots of the effluent to 1-ml portions of 12% trichloroacetic acid. Immediately after elution the pyruvate decarboxylase fraction is dialyzed with stirring against 0.05 *M* phosphate buffer, pH 7.0. Washing of the column with ethanolamine-phosphate buffer is continued until virtually no protein can be detected in the effluent. Approximately 80 ml of buffer are used. A second protein fraction (lipoyl reductase-transacetylase) is eluted from the column with a solution of 1% ammonium sulfate in 0.1 *M* phosphate buffer, pH 7.5. Emergence of this fraction is detected turbidimetrically with trichloroacetic acid as described above. It is dialyzed overnight with stirring against 0.01 *M* phosphate buffer, pH 7.0. The dialyzed solution is concentrated to a volume of 2–3 ml by ultrafiltration. The recovery of protein is approximately 20 mg; specific activity is 800–900 units/mg (see transacetylase assay below).

The dialyzed pyruvate decarboxylase fraction is brought to 0.5 ammonium sulfate saturation by slow addition, with stirring, of 0.36 g of the salt per milliliter. The mixture is stirred for 15 minutes and centrifuged for 10 minutes at 14,000 *g*; the precipitate is discarded. The supernatant fluid is brought to 0.6 ammonium sulfate saturation by addition of 0.07 g of the salt per milliliter. The precipitate is collected by centrifugation and dissolved in 0.05 *M* phosphate buffer, pH 7.0; the solution is dialyzed overnight against the same buffer. Approximately 25 mg of protein is recovered in the dialyzed solution; specific activity is 25–30 units/mg. The pyruvate decarboxylase preparation may be stored in the frozen state (—20°) for several months without significant loss of activity.

Alternative Procedure for Preparing Pyruvate Decarboxylase.[5] At pH 9.5 the pyruvate dehydrogenase complex dissociates into pyruvate decarboxylase and a complex consisting of lipoyl reductase-transacetylase and dihydrolipoyl dehydrogenase. The two components can be separated effectively by fractionation on a column of gel-cellulose at pH 9.5, essentially as described above. The complex of lipoyl reductase-transacetylase and dihydrolipoyl dehydrogenase can be detected on the column as a yellow, fluorescent band, and is eluted with a solution of 4% ammonium sulfate in 0.1 *M* phosphate buffer, pH 7.5. In a representative experiment 26 mg of pyruvate decarboxylase (0.5–0.6 ammonium sulfate fraction) and 22 mg of the lipoyl reductase-transacetylase–dihydrolipoyl dehydrogenase complex were recovered from 63 mg of the pyruvate dehydrogenase complex.

Properties

Specificity. The enzyme will oxidatively decarboxylate [Eq. (11)] α-ketobutyrate, but is inactive toward α-ketoglutarate. In its physiological role pyruvate decarboxylase utilizes the lipoyl moiety covalently bound to lipoyl reductase-transacetylase as the oxidant for oxidation of "active acetaldehyde," i.e., 2-(α-hydroxyethyl)thiamine pyrophosphate [Eq. (3)]. The lipoyl moiety is reductively acetylated in this reaction. Ferricyanide can replace the protein-bound lipoyl moiety, leading to formation of acetate [Eq. (11)] rather than a thioester. Free lipoate or lipoamide can also replace the protein-bound lipoyl moiety. In the latter reaction, which is a model of the physiological reaction [Eq. (3)], *S*-acetyldihydrolipoate or *S*-acetyldihydrolipoamide are produced.[30] However, the rate of oxidative decarboxylation of pyruvate with ferricyanide, lipoate, or lipoamide as oxidant is considerably slower than with the lipoyl moiety bound covalently to lipoyl reductase-transacetylase.

Physical Constants. At a protein concentration of 5.9 mg/ml, pyruvate decarboxylase exhibits a sedimentation coefficient ($S_{20,w}$) of 9.24 S and a diffusion coefficient ($D_{20,w}$) of 4.54×10^{-7} cm² sec⁻¹. There is little concentration dependence of the sedimentation coefficient. From these data and an assumed partial specific volume of 0.73 ml/g, the molecular weight is calculated to be 183,000.

Cofactors. The purified enzyme shows complete dependence on added thiamine pyrophosphate in the ferricyanide reduction assay [Eq. (11)] and in the pyruvate dismutation assay (in the presence of excess lipoyl reductase- transacetylase and dihydrolipoyl dehydrogenase).[5]

Unusual Properties. At pH 9.5 pyruvate decarboxylase undergoes a marked conformational change accompanied by dimer formation, as indicated by a doubling of the molecular weight and a drop in sedimentation coefficient ($S_{20,w}$) to 6.2 S (protein concentration, 4.6 mg/ml).[5,31] These changes, which can be reversed by lowering the pH to 7.0, are apparently responsible for dissociation of the decarboxylase at pH 9.5 from the pyruvate dehydrogenase complex and from the partial complex consisting of decarboxylase and lipoyl reductase-transacetylase.

[30] E. Thompson, M. Koike, and L. J. Reed, unpublished observations (1962).

[31] C. R. Willms and L. J. Reed, unpublished observations (1963).

C. Lipoyl Reductase-Transacetylase

$$\text{Acetyl CoA} + \text{Lip(SH)}_2 \rightarrow \text{acetyl-SLipSH} + \text{CoA} \qquad (12)$$

The partial purification and properties of lipoic transacetylase from *E. coli* have been presented in an earlier volume of this series.[32] In the present section the preparation and properties of lipoyl reductase-transacetylase from the *E. coli* pyruvate dehydrogenase complex are described. The two enzymes are apparently identical. However, the lipoyl reductase-transacetylase has been obtained in a more purified state, free of decarboxylase and flavoprotein.

Assay Method

Principle. The assay is a modification of that described by Hager[15] (see also Knight and Gunsalus[32]). It is based on colorimetric determination of *S*-acetyldihydrolipoamide as the ferric acethydroxamate complex.[8] Acetyl CoA is generated from acetyl phosphate in the presence of catalytic amounts of CoA and phosphotransacetylase [Eq. (7)], and the acetyl group is transferred to dihydrolipoamide in the presence of lipoyl reductase-transacetylase [Eq. (12)].

$$\text{Acetyl-P} + \text{CoA} \rightleftharpoons \text{acetyl CoA} + \text{P}_i \qquad (7)$$
$$\text{Acetyl CoA} + \text{Lip(SH)}_2 \rightarrow \text{acetyl-SLipSH} + \text{CoA} \qquad (12)$$
$$\textit{Sum: } \text{Acetyl-P} + \text{Lip(SH)}_2 \rightarrow \text{acetyl-SLipSH} + \text{P}_i \qquad (13)$$

The heat stability of the thioester, *S*-acetyldihydrolipoamide, at acid pH permits it to be distinguished from the heat-labile acetyl phosphate.[33] The activity of the lipoyl reductase-transacetylase also can be determined in the pyruvate dismutation assay by adding a suitable excess (usually twofold) of pyruvate decarboxylase and dihydrolipoyl dehydrogenase.

Reagents

Tris buffer, 1 *M*, pH 7.0
CoA, 0.001 *M*
Acetyl phosphate, 0.1 *M* (Worthington Biochemical Corporation)
(±)-Dihydrolipoamide,[20] 0.2 *M*, in 95% ethanol, prepared before use
Phosphotransacetylase, a purified preparation of C. F. Boehringer and Sons

[32] E. Knight, Jr., and I. C. Gunsalus, Vol. V [90].
[33] E. R. Stadtman, *J. Biol. Chem.* **196**, 535 (1952).

Enzyme .The enzyme is diluted with 0.01 *M* potassium phosphate buffer, pH 7.0, containing 5 mg of crystalline bovine serum albumin per milliliter; the aliquot assayed is such that 0.5–2 micromoles of *S*-acetyldihydrolipoamide will be formed during incubation.

Procedure. Dihydrolipoamide Transacetylase Assay. To a 13×100 mm Pyrex test tube are added 0.1 ml of Tris buffer, 0.1 ml of acetyl phosphate, 5 units[11] of purified phosphotransacetylase, 0.1 ml of CoA, enzyme fraction to be assayed, and water to make a final volume of 0.95 ml. A 0.05-ml aliquot of 0.2 *M* dihydrolipoamide is added with shaking to prevent local precipitation. The blank contains all reagents except enzyme fraction. The tubes are incubated for 30 minutes at 30°. The reaction is stopped by the addition of 0.1 ml of 1 *N* hydrochloric acid, and the mixture is heated in a boiling-water bath for 5 minutes to destroy unreacted acetyl phosphate. The mixture is cooled to room temperature and assayed for hydroxylamine-reactive acetyl groups as described above (see Procedure, pages 248, 249).

Modified Pyruvate Dismutation Assay. A mixture of 1 μg of lipoyl reductase-transacetylase, 3 μg of pyruvate decarboxylase, and 1 μg of dihydrolipoyl dehydrogenase in a total volume of 0.2 ml of 0.02 *M* phosphate buffer, pH 7.0, is allowed to stand for 10 minutes at 0°. The other components required in the pyruvate dismutation assay are added to give a final volume of 1 ml, and the assay is carried out as described above (see Procedure, pages 248, 249).

Definition of Unit and Specific Activity. One unit is the amount of enzyme required to produce 1 micromole of *S*-acetyldihydrolipoamide per hour (or 1 micromole/hour of acetyl phosphate in the modified pyruvate dismutation assay) under the conditions specified. One unit of activity in the transacetylase assay is equivalent to about 0.3 unit in the modified pyruvate dismutation assay. Specific activity is expressed as units per milligram of protein.

Enzyme Preparation

Lipoyl reductase-transacetylase is obtained by resolution of its complex with pyruvate decarboxylase on gel-cellulose at pH 9.5 as described above (see Procedure, page 259). It can be obtained also by resolution of the complex consisting of lipoyl reductase-transacetylase and dihydrolipoyl dehydrogenase (see Alternate Procedure, page 260) on gel-cellulose in the presence of 4 *M* urea.[5] However, a complete separation is difficult to achieve by the latter procedure.

Properties

Stability. Preparations of lipoyl reductase-transacetylase can be stored in the frozen state (—20°) for several weeks without significant loss of activity if the protein concentration is at least 5 mg/ml. Prolonged storage or repeated freezing and thawing results in aggregation, as indicated by formation of an insoluble, gelatinous precipitate.

Physical Constants. The purified enzyme exhibits a sedimentation coefficient ($S^0_{20,w}$) of 23 S. It comprises approximately one-third of the protein of the pyruvate dehydrogenase complex, which indicates that its "molecular weight" is about 1.6×10^6.

Bound Prosthetic Group. The purified enzyme contains approximately 1 mole of covalently-bound lipoyl moiety per 35,000 g of protein. The lipoyl moiety is bound in amide linkage to the ε-amino group of a lysine residue.[6] The lipoyl moiety is released, concomitant with loss of activity in the pyruvate dismutation assay, by incubation with the enzyme, lipoamidase.[34] The lipoyl moiety can be reincorporated, concomitant with restoration of activity, by incubation with lipoic acid, ATP and a lipoic acid-activating enzyme system.[35]

Specificity. In its physiological role lipoyl reductase-transacetylase catalyzes a transfer of the acetyl group from the *S*-acetyldihydrolipoyl moiety (bound covalently to the enzyme) to CoA [Eq. (4)]. However, the enzyme also catalyzes the transfer of the acetyl group from acetyl CoA to dihydrolipoate (6,8-dithioloctanoate) and to dihydrolipoamide [Eq. (12)]. Apparently, the sulfhydryl group attached to C-6 of the carbon skeleton is acetylated.[36] This nonphysiological reaction does not involve the covalently bound lipoyl moiety.[35] In contrast to earlier reports[15,19] indicating absolute optical specificity of lipoic transacetylase [for (—)-dihydrolipoate] in reaction (13), we have observed, with lipoyl reductase-transacetylase and with the pyruvate dehydrogenase complex, that more than one-half of the (±)-dihydrolipoate and (±)-dihydrolipoamide is acetylated.[30,37] The latter observation indicates that both optical isomers of dihydrolipoate and dihydrolipoamide are active. Pre-

[34] K. Suzuki and L. J. Reed, *J. Biol. Chem.* **238,** 4021 (1963).

[35] M. Koike and L. J. Reed, *J. Biol. Chem.* **235,** 1931 (1960).

[36] I. C. Gunsalus, L. Barton, and W. Gruber, *J. Am. Chem. Soc.* **78,** 1763 (1956).

[37] In a representative experiment, a reaction mixture containing (in micromoles) Tris buffer (100), pH 7.5, acetyl phosphate (10), CoA (0.1), (±)-dihydrolipoate or (±)-dihydrolipoamide (5), 2 units of phosphotransacetylase, and 400 μg of pyruvate dehydrogenase complex in a final volume of 1 ml was incubated for 1 hour at 30° under an atmosphere of nitrogen. The yield of heat-stable thioester obtained from (±)-dihydrolipoate as 3.5 micromoles, and from (±)-dihydrolipoamide, 3.9 micromoles.

sumably, the rate with the natural (—)-isomer is considerably faster than with the (+)-isomer.

Dissociation and Reconstitution of Lipoyl Reductase-Transacetylase.[38] In 0.83 *M* acetic acid, pH 2.6, the enzyme dissociates into disorganized, inactive subunits with a molecular weight of approximately 70,000. Removal of the dissociating agent by rapid dilution into appropriate buffers results in restoration of enzymatic activity and reconstitution of the unique structure of the lipoyl reductase-transacetylase. Electron microscope studies indicate that in both the native and reconstituted lipoyl reductase-transacetylase the subunits are arranged into four stacks, comprising a tetrad.[18, 38]

[38] C. R. Willms, R. M. Oliver, and L. J. Reed, unpublished observations.

[51] Pyruvic (Cytochrome b_1) Dehydrogenase (Crystalline)

By F. ROBERT WILLIAMS and LOWELL P. HAGER

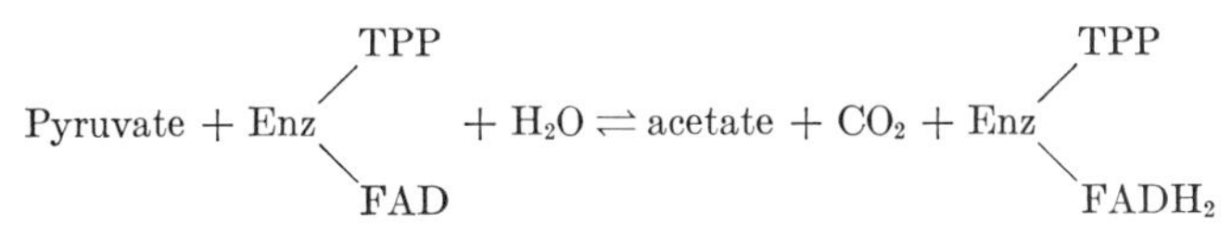

In addition to the lipoic acid, coenzyme A linked pyruvate dehydrogenase complex,[1] *Escherichia coli* cells oxidize pyruvate via a flavoprotein pyruvate dehydrogenase enzyme which links to membrane-bound cytochrome b_1. This flavoprotein, which contains both thiamine pyrophosphate and flavin adenine dinucleotide as prosthetic groups, has been isolated in crystalline form.[2]

Assay Method

Principle. The assay is based on the measurement of CO_2 release from pyruvate during the oxidative decarboxylation reaction. Ferricyanide serves as the terminal electron acceptor so that the flavoprotein serves in a catalytic fashion.

Reagents

Potassium phosphate buffer, 1.0 *M*, pH 6.0
$MgCl_2$, 0.1 *M*
TPP, 0.01 *M*

[1] See Vol. I [77].
[2] F. R. Williams and L. P. Hager, *J. Biol. Chem.* **236**, PC36 (1961).

Potassium pyruvate, 0.5 *M*
$K_3Fe(CN)_6$, 0.25 *M*
Sodium lauryl sulfate (Duponol), 0.01 *M*
Ovalbumin, 5 mg/ml
Enzyme. Use an amount of enzyme sufficient to give a rate of CO_2 evolution of 0.3–3 micromoles per 5 minutes.

Procedure. Each Warburg cup contains 100 micromoles of phosphate buffer, 10 micromoles of $MgCl_2$, 0.1 micromole of TPP, 50 micromoles of pyruvate, 1.0 micromole of sodium lauryl sulfate, 0.5 mg of ovalbumin, and enzyme in the main compartment of the flask plus 25 micromoles of ferricyanide in the side arm of the flask. After temperature equilibration at 30°, the ferricyanide is tipped into the main compartment of the Warburg flask and CO_2 evolution is measured at 5-minute intervals.

Definition of Unit and Specific Activity. One unit of enzymatic activity is defined as the evolution of 1 micromole of CO_2 per 30 minutes under the above conditions. The rate of CO_2 evolution was estimated from the first 10 minutes of the reaction. Specific activity is defined as units per milligram of protein.

Specificity of Assay. The inclusion of sodium lauryl sulfate in the incubation medium makes this assay quite specific for pyruvic (cytochrome b_1) dehydrogenase. The carboxylase component of the pyruvate dehydrogenase complex which normally would have activity in the ferricyanide-linked assay is not active in the presence of this concentration of sodium lauryl sulfate.

Preparation of Enzyme

Growth Medium. The cells are grown on a medium containing 0.2% NH_4Cl; 0.4% glucose; 0.25% sodium glutamate; 0.0005% yeast extract; 0.15% KH_2PO_4; 1.35% Na_2HPO_4; 0.02% $MgSO_4 \cdot 7\ H_2O$; 0.001% $CaCl_2$, and 0.00005% $FeSO_4 \cdot 7\ H_2O$. The phosphate, glutamate, and glucose are sterilized in separate containers and added to the sterile salt solution. When the acetate-requiring mutant is used as the source of enzyme, the medium is further supplemented with 20 micromoles of potassium acetate per milliliter. The cells are grown under forced aeration, either in 10-liter batches in a New Brunswick fermentor or in 100-liter batches in a 50-gallon fermentor manufactured by Stainless and Steel Products Co.

Inoculum. Escherichia coli, strain B, Crookes, W, or the acetate-requiring mutant 191-6[3] derived from strain W, is grown on a complex medium containing 1% tryptone, 1% yeast extract, 0.5% K_2HPO_4, and

[3] This mutant was obtained from Professor B. Davis, Bacteriology Department, Harvard Medical School, Boston, Massachusetts.

0.3% glucose. Actively growing cells (log phase) are inoculated at a 0.5% level into the growth medium.

Growth. Cells are grown to the stationary phase (15–20 hours) and are then harvested on a Sharples centrifuge. The yield is approximately 10 g of cell paste per liter of culture medium.

Purification Procedure

Step 1. Preparation of Cell Extract. Six hundred grams of cell paste is suspended in 500 ml of 0.02 *M* potassium phosphate buffer, pH 7, and mixed thoroughly in a Servall omnimixer. Six hundred milliliter of washed glass beads[4] (120 μ in diameter) are added to the cell suspension. The cell-glass bead mixture is then ground in an Eppenbach colloid mill[5] (rotor-stator setting of 0.030) for 30 minutes at 15–20° with continuous recycling.[6] The resulting slurry is centrifuged at 15,000 *g* for 10 minutes to remove the glass beads.[7] The supernatant fluid (supernatant 1) is saved. The precipitate (together with the glass beads) is washed by resuspension in 1000 ml of phosphate buffer and is centrifuged at 15,000 *g* for 10 minutes. The supernatant fluid (supernatant 2) is again saved. The precipitate (consisting mostly of glass beads) is again resuspended in 1000 ml of phosphate buffer and allowed to sediment for 1 hour. The supernatant fluid (supernatant 3) is decanted and saved. Supernatants 1 and 2 are combined and centrifuged at 15,000 *g* for 1 hour. Three layers are formed during the centrifugation: a clear liquid phase; a dense, viscous particulate phase; and a precipitate of cell debris. The liquid phase and the particulate phase are removed together and saved (supernatant 4). The precipitate of cell debris is washed by resuspension in supernatant 3 and is centrifuged at 15,000 *g* for 1 hour. The precipitate is discarded. The supernatant (supernatant 5) is removed and combined with supernatant 4 to yield a crude extract.

Step 2. First Ammonium Sulfate Precipitation. The crude extract is brought to 0.25 ammonium sulfate saturation by the addition of 14.4 g of ammonium sulfate per 100 ml of crude extract. The suspension is stirred for 4 hours. The precipitate is removed by centrifugation at

[4] This was obtained from Minneapolis Mining and Manufacturing Company, Minneapolis, Minnesota.

[5] This mill was obtained from Gifford-Wood Co., Hudson, New York.

[6] J. C. Garver and R. L. Epstein, *Appl. Microbiol.* **1**, 318 (1959). Other methods routinely used to prepare bacterial cell-free extracts, such as sonic oscillation, treatment in a French pressure cell, and alumina grinding, may be used to prepare the cell-free extract. Grinding with glass beads in the colloid mill is especially useful for handling large volumes of cell paste.

[7] All operations on the enzyme preparation were carried out at 3–4° unless otherwise stated.

15,000 g for 75 minutes. This precipitate is saved and is used for the preparation of crystalline cytochrome b_1. The supernatant fluid is brought to 0.75 ammonium sulfate saturation by the addition of 34.4 g of ammonium sulfate per 100 ml of supernatant fluid. The precipitate is removed by centrifugation at 15,000 g for 30 minutes and redissolved in a volume of buffer equal to one-half the volume of the initial crude extract.

Step 3. Removal of Nucleic Acid. The 0.25–0.75 saturated ammonium sulfate fraction is adjusted to pH 6 by the addition of 1 N acetic acid. Nucleic acid and inactive protein is precipitated by the slow addition of an equal volume of 2% protamine sulfate, pH 6. The resulting mixture is dialyzed overnight against a total of 20 volumes of phosphate buffer. The insoluble protamine nucleate, which precipitates during dialysis, is removed by centrifugation at 15,000 g for 20 minutes and discarded. At this stage the ratio of UV absorption at 280 mμ and 260 mμ (280:260) of the supernatant solution should be above 1, although successful purifications have been carried out with fractions having a ratio of 0.8.

Step 4. Second Ammonium Sulfate Precipitation. The supernatant fluid from the protamine sulfate treatment is brought to 0.36 ammonium sulfate saturation by the addition of 25.2 g of ammonium sulfate per 100 ml of supernatant fluid. After stirring for 2–3 hours, the resulting precipitate is removed by centrifugation at 15,000 g for 30 minutes and discarded. The supernatant fluid obtained from the 0.36 saturated ammonium sulfate fraction is brought to 0.55 ammonium sulfate saturation by the addition of 12.1 g of ammonium sulfate per 100 ml of supernatant fluid and stirred for 2–3 hours. The precipitate is removed by centrifugation at 15,000 g for 30 minutes and redissolved in a volume of buffer equal to 0.06 volume of the crude extract.

Step 5. Heat Precipitation. The 0.36–0.55 saturated ammonium sulfate fraction is rapidly warmed to 60° in a 70–80° bath with vigorous stirring and then held at 60° in a 60° bath for 5 minutes. The heat-treated fraction is rapidly cooled in an ice bath and the precipitate of denaturated protein is removed by centrifugation at 35,000 g for 1 hour. The precipitate of denaturated protein is washed twice with phosphate buffer (1 ml of buffer per 100 ml of 0.36–0.48 saturated ammonium sulfate fraction) and the washes are combined with the original supernatant fluid. At this stage, the enzyme is stored in 0.70 saturated ammonium sulfate in batches of approximately 500,000 units and stored at −20°. From this point on the enzyme fractions are never allowed to freeze since 20–50% of their activity is lost upon freezing and thawing.

Step 6. Chromatography on DEAE-Sephadex. Forty grams of DEAE-

Sephadex A-50 is swelled in an excess of water. Fines are then removed by decantation 15–20 times. The Sephadex is then washed in a sintered-glass filter with 2 liters of 0.5 *N* of HCl, followed quickly by several washes with water. The Sephadex is then washed with 2 liters of 0.5 *N* NaOH, again followed by large wash volumes of water. At this point, fines are again removed by decantation. The material is then allowed to settle overnight to remove bubbles which formed during the washing periods. The washed material is adjusted to pH 5.7 by the addition of 1 *N* phosphoric acid. When the pH remains steady at 5.7, the DEAE-Sephadex is washed with 0.02 *M* phosphate buffer, pH 5.7, and then suspended in phosphate buffer of the same concentration.

A 40 mm × 400 mm column is packed at room temperature to a height of 300 mm with the washed Sephadex under gravity flow. Only that material which settles on the column bed within 5 minutes is used for the packing. This is accomplished by pouring the washed Sephadex suspension back and forth between two beakers. In packing, the liquid head is allowed to stand only 5 minutes and then poured back into the beaker. A thin layer of glass wool is used to cover the top of the column.

The DEAE-Sephadex column is charged with enzyme in the following manner. A batch of 500,000 units of enzyme in 0.70 saturated ammonium sulfate is centrifuged for 30 minutes at 35,000 *g*. The resulting precipitate is redissolved in 100–200 ml of 0.02 *M* potassium phosphate buffer, pH 5.7, and dialyzed 18 hours against a total volume of 6 liters of 0.02 *M* phosphate buffer, pH 5.7, in three successive steps, using 2 liters of dialysis fluid for each step. The precipitate formed during dialysis is removed by centrifugation at 35,000 *g* for 10 minutes and discarded. The supernatant fluid containing the flavoprotein is now ready for charging the column. The column is washed with 200 ml of 0.02 *M* potassium phosphate buffer, pH 5.7, and then cooled to 3–4°. The flow rate for the column should be at least 70–80 ml per hour at 3–4°. The dialyzed supernatant fluid containing the flavoprotein is applied to the DEAE-Sephadex column, and the column is washed with a volume of 0.02 *M* phosphate buffer, pH 5.7, equal to one-half the volume of the dialyzed supernatant fluid. The enzyme is eluted from the column using a linear gradient as described by Bock and Ling.[8] The mixing flask contains 1100 ml of 0.02 *M* potassium phosphate buffer, pH 5.7, and the reservoir contains 1100 ml of 0.3 *M* potassium phosphate buffer, pH 5.3. Five milliliter fractions are collected. The enzyme usually is eluted in tubes 200–260. The enzyme is easily located by its yellow-green color and is usually the first of two closely associated colored peaks. The enzyme is assayed and

[8] R. M. Bock and N. Ling, *Anal. Chem.* **26**, 1543 (1954).

the peak tubes are pooled so as to give an average specific activity of 1100 or more. The DEAE-Sephadex column fractionation should be completed in 18 hours or less since longer periods of time lead to severe losses of enzyme activity.

Step 7. Protamine Sulfate Fractionation. The pooled DEAE-Sephadex column fractions are brought to 0.70 ammonium sulfate saturation by the addition of 47.2 g of ammonium sulfate per 100 ml of enzyme solution. The precipitate is removed by centrifugation at 35,000 *g* for 20 minutes and redissolved in a volume of 0.02 *M* potassium phosphate buffer, pH 5.7, such as to give a final concentration of approximately 20,000–30,000 units of enzyme per milliliter. The redissolved precipitate is then rapidly dialyzed against 50 volumes of 0.02 *M* potassium phosphate buffer, pH 5.7, in five successive steps using 10 volumes of dialysis fluid in each step. The precipitate formed during dialysis is removed by centrifugation at 35,000 *g* for 10 minutes and discarded.[9] The supernatant fluid from the dialysis is treated with 1 volume of 2% protamine sulfate, pH 8, to precipitate the flavoprotein (if no precipitate forms immediately, the solution is dialyzed against distilled water to cause precipitation). The precipitate is removed by centrifugation at 35,000 *g* for 10 minutes and suspended in approximately 1 ml of 0.2 *M* potassium phosphate buffer, pH 5.7, for each 10 ml of the above supernatant fluid. The insoluble precipitate is removed by centrifugation at 35,000 *g* for 10 minutes and discarded. If the soluble flavoprotein remains in the supernatant after protamine treatment and dialysis, it can be precipitated by a second protamine treatment. All the supernatant protamine eluate fractions are now pooled in preparation for the crystallization step.

Step 8. Crystallization. The combined supernatant fluid from the protamine fractionation step is slowly diluted with distilled water until a very slight cloudiness appears. The solution is then left to stand at 3° for 48 hours. At this point, impurities often precipitate with a small amount of the enzyme. The precipitate is removed by centrifugation at 35,000 *g* for 10 minutes and, if necessary, the precipitate is eluted with 0.2 *M* potassium phosphate buffer, pH 5.7, to recover flavoprotein (which is added back to the mother liquor). The enzyme solution is then further diluted until a slight cloudiness appears (final buffer concentration at this point is 0.05–0.1 *M*). This solution is left to crystallize at 3° for a week or until the supernatant fluid is visibly clear of the yellow color. The enzyme preparation is then recrystallized in the same manner

[9] If this precipitate is yellow, it is redissolved in 1 ml of 0.2 *M* potassium phosphate buffer, pH 5.7, and the insoluble material is removed by centrifugation at 35,000 *g* for 10 minutes. The supernatant solution is added to the combined supernatant fractions from the protamine step.

until constant specific activity crystals are obtained and the specific activity of the mother liquor equals that of the crystals. The results of a purification run are shown in the table.

PURIFICATION OF FLAVIN PYRUVATE OXIDASE[a]

Fraction	Volume (ml)	Protein (mg)	280:260 mμ	Enzyme activity: Total units (1 × 10^{-3})	Enzyme activity: Specific activity (units/mg protein)	Enzyme activity: Recovery (%)
Crude extract	3,800	236,000	0.6	1,698	7.2	100
Ammonium sulfate, 0.25–0.75, saturation	2,100	130,600	0.6	1,700	13.0	100
Protamine sulfate supernatant	3,480	33,200	1.2	1,592	26.2	94
Ammonium sulfate, 0.36–0.55, saturation	520	26,500	1.6	1,200	48.0	70
Supernatant fraction from heat treatment	598	10,000	1.5	1,130	113.0	67
DEAE-Sephadex fractions	520	582	1.5	754	1,300.0	45
Eluate fraction from protamine sulfate	15	294	1.3	587	2,000.0	35
First crystallization	5	134	1.3	469	3,500.0	28
Second crystallization	5	69	1.2	418	6,000.0	25
Third crystallization	4	50	1.2	300	6,000.0	
Mother liquor of third crystallization	12	19	1.2	118	6,000.0	

[a] Starting from 700 g of *Escherichia coli* paste.

Properties

The crystalline flavoprotein has a molecular weight of approximately 265,000 as determined by the Archibald method.[2] The sedimentation constant ($S^0_{20,w}$) of the enzyme in 0.2 M potassium phosphate buffer, pH 5.7, is 11.5 S. Flavin analysis performed on the enzyme indicate that flavin adenine dinucleotide is the sole flavin component and is present in a ratio of 4 moles of FAD per mole of enyzme. The crystalline enzyme does not contain thiamine pyrophosphate but does have an absolute thiamine pyrophosphate requirement for the formation of reduced enzyme from pyruvate.

The reduced flavoprotein substrate complex is not autoxidizable. However, it is air oxidizable in the presence of cytochrome b_1-containing membrane fragments obtained from *E. coli,* and under certain conditions it will react with soluble cytochrome b_1 preparations derived from these membrane fragments.[10] Artificial electron acceptors such as 2,6-dichlorophenolindophenol and ferricyanide will serve as electron acceptors for the reduced enzyme. Untreated crystalline preparations of the enzyme have a turnover number of 1800 when ferricyanide serves as the electron acceptor. Maximal enzymatic activity with a turnover number of approximately 53,000 can be achieved in the ferricyanide reaction by exposure of the flavoprotein to partial trypsin or chymotrypsin digestion[2] or to certain ionic surface active agents, such as sodium lauryl sulfate. The mechanism of activation in these instances is not known.

The flavoprotein has absorption peaks in the visible spectrum at 370, 438, and 460 mμ with shoulders at 355 and 415 mμ and minima at 390 and 455 mμ. The calculated millimolar extinction coefficients at 355, 370, 415, 438, and 460 mμ are 10.7, 11.3, 10.5, 14.6, and 12.7, respectively. The ultraviolet spectrum of the flavoprotein shows a ratio of optical density at 280 mμ to that at 260 mμ (280:260) of 1.2.

[10] S. S. Deeb and L. P. Hager, *J. Biol. Chem.* **239**, 1024 (1964).

[52] Lipoyl Dehydrogenase from Pig Heart

By Vincent Massey

$$\text{Dihydrolipoyl derivative} + DPN^+ \rightleftarrows \text{oxidized lipoyl derivative} + DPNH + H^+$$

Assay Methods

General Considerations. While the physiological function is that given in the reaction above, the substrate specificity of lipoyl dehydrogenase is fairly broad. Thus, suitable hydrogen donors are dihydrolipoyl derivatives, DPNH,[1] and other reduced pyridine nucleotide analogs; suitable acceptors are oxidized lipoyl derivatives, DPN, and oxidized pyridine nucleotide analogs, $K_3Fe(CN)_6$, methylene blue, 2,6-dichlorophenolindophenol, menadione, and to a limited extent, O_2. Thus a large number of catalytic reactions are possible between pairs of hydrogen donors and acceptors. Three suitable assays are described below.

[1] The following abbreviations are used: DPN, diphosphopyridine nucleotide; DPNH, reduced diphosphopyridine nucleotide; TPN, triphosphopyridine nucleotide; TPNH, reduced triphosphopyridine nucleotide; FAD, flavin adenine dinucleotide; EDTA, ethylenediaminetetraacetic acid adjusted to pH 7.0 with *N* NAOH.

Reagents

Sodium citrate, 1 *M*, pH 5.65
Bovine serum albumin (crystalline), 2% (w/v), in 3×10^{-2} *M* EDTA
DL-Lipoic acid, 2×10^{-2} M
DPNH, 10^{-2} *M*
DPN, 10^{-2} *M*
Potassium phosphate, 0.3 *M*, pH 7.2
Dichlorophenolindophenol, 10^{-3} *M*
Sodium acetate, 1 *M*, pH 4.8
$K_3Fe(CN)_6$, 10^{-2} *M*

DPNH–Lipoyl Reductase Activity. Lipoic acid + DPNH + $H^+ \rightarrow$ dihydrolipoic acid + DPN^+. The activity measured is dependent both on ionic strength[2] and on the presence of DPN.[3] Activity is measured spectrophotometrically at 340 mμ in the following reaction mixture at 25°; 2.5 ml of citrate buffer; 0.1 ml of bovine serum albumin in EDTA; 0.1 ml of DL-lipoic acid; 0.03 ml of DPNH, 0.03 ml of DPN; water and finally enzyme to a total volume of 3.0 ml. A blank without lipoic acid must be run to correct for the DPNH oxidase activity (about 10% of the lipoyl reductase activity). After correction for the DPNH-O_2 blank, the pure enzyme has a specific activity of 24–25 micromoles DPNH oxidized per minute per milligram protein. This assay can be conducted with tissue suspensions with fair accuracy.

DPNH–Dichlorophenolindophenol Reductase (Diaphorase) Activity. This activity is measured spectrophotometrically at 600 mμ at 25° by following the reduction of the dye. Each cuvette contains 0.5 ml of phosphate buffer; 0.1 ml of bovine serum albumin in EDTA; 0.12 ml of dichlorophenolindophenol; 0.06 ml of DPNH; water and finally enzyme to a total volume of 3.0 ml. With the native enzyme this activity is very small (0.4–0.5 micromole DPNH oxidized per minute per milligram protein). However, with the Cu^{++}-modified enzyme (see later) the activity may be as high as 30 times this value. This activity may be followed in crude suspensions, but is not very meaningful owing to the presence of other enzymes whch catalyze the dye reduction.

DPNH–Ferricyanide Reductase Activity. This activity is measured spectrophotometrically at 420 mμ at 25° by following the reduction of $K_3Fe(CN)_6$. Each cuvette contains 0.5 ml of acetate buffer; 0.1 ml of bovine serum albumin in EDTA; 0.2 ml of $K_3Fe(CN)_6$, 0.06 ml of

[2] C. Veeger and V. Massey, *Biochim. Biophys. Acta* **64,** 83 (1962).
[3] V. Massey and C. Veeger, *Biochim. Biophys. Acta* **48,** 33 (1961).

DPNH; water and finally enzyme to a volume of 3.0 ml. The specific activity of pure enzyme is 210 micromoles DPNH oxidized per minute per milligram protein. Owing to the high activity and the low pH optimum, this assay can be conducted readily with crude suspensions and is a fairly meaningful assay of lipoyl dehydrogenase activity under these conditions.

Other Assays. Details of other assays are given in footnote 2.

Isolation Procedure

Owing to the sensitivity of the enzyme to modification by Cu^{++} it is essential to carry out the purification employing glass-distilled water and including EDTA in all steps after the initial extraction. Unless otherwise specified operations are carried out at 0–5°.

Step 1. Preparation of Pig Heart Muscle Particles.[4] Prior to extraction of the lipoyl dehydrogenase from the muscle, much purification is obtained by washing away most of the readily soluble enzymes. Pig hearts (12 is a convenient number) are collected in ice at the slaughter house, freed of fat and connective tissues, and passed through a mincer with approximately ⅛-inch diameter holes. The mince is then suspended in about 20 liters cold deionized H_2O and stirred vigorously with a mechanical stirrer for 5 minutes. The mince is allowed to settle for a few minutes, and the supernatant solution is siphoned off. This procedure is repeated 4 or 5 times until the supernatant is only lightly colored. The mince is then squeezed as dry as possible by hand in several layers of cheesecloth. The squeezed mince is then homogenized in a Waring blendor at top speed for 1 minute with 5 times (w/v) cold 0.02 *M* Na_2HPO_4, and the homogenate is centrifuged for 15 minutes at 1400 *g*. The cloudy supernatant is then decanted carefully from the rather loose pack of cell debris, adjusted to pH 5.4 by the addition of 1 *N* acetic acid, and centrifuged at 1400 *g* for 20 minutes. The supernatant is discarded, and the precipitate is washed with about 2 liters cold glass-distilled H_2O and centrifuged again. The first pH 5.4 supernatant is a rich source of fumarase[5]; the precipitate (washed heart muscle preparation) can be used for the extraction of DPNH-cytochrome *c* reductase, succinic dehydrogenase, α-ketoglutarate dehydrogenase, or lipoyl dehydrogenase.[6]

Step 2. Extraction of Lipoyl Dehydrogenase.[7] The washed heart muscle preparation is blended briefly (15 seconds at half speed) with a solution of 1% (w/v) $(NH_4)_2SO_4$ containing 3×10^{-4} *M* EDTA (1 liter

[4] V. Massey, *Biochim. Biophys. Acta* **37**, 310 (1960).
[5] V. Massey, *Biochem. J.* **51**, 490 (1952).
[6] V. Massey, *Biochim. Biophys. Acta* **37**, 314 (1960).
[7] V. Massey, Q. H. Gibson, and C. Veeger, *Biochem. J.* **77**, 341 (1960).

per kilogram unwashed mince) and extracted overnight. The suspension is then centrifuged for 30 minutes at 10,000 *g*, and the precipitate is discarded. The clear reddish solution contains lipoyl dehydrogenase, both free and in the α-ketoglutarate dehydrogenase complex.

Step 3. Dissociation of Enzyme from α-Ketoglutarate Dehydrogenase Complex. To the supernatant from step 2 is added 1/500 volume 3×10^{-2} *M* EDTA and 3% (v/v) ethanol. The mixture is heated at 45° for 15 minutes with continuous stirring. This procedure serves to dissociate lipoyl dehydrogenase bound in the α-ketoglutarate dehydrogenase complex, which would otherwise be discarded in the subsequent fractionations. After cooling to 10°, the turbid solution is centrifuged for 20 minutes at 10,000 *g* and the precipitate is discarded.

Step 4. Fractionation with $(NH_4)_2SO_4$. To the supernatant from step 3, 1/30 volume 0.3 *M* Na_2HPO_4 is added, and the solution then fractionated by the addition of solid $(NH_4)_2SO_4$. Protein precipitating between 0.4 and 0.80 saturation is collected. The enzyme is dissolved in a minimum volume of H_2O and dialyzed overnight against 5 liters of 0.001 *M* phosphate, pH 6.3, containing 3×10^{-4} *M* EDTA. [Owing to the sensitivity of the enzyme to Cu^{++}, it is essential to soak dialysis tubing in EDTA (approximately 10^{-3} *M*) before use.] The dialyzed enzyme is heated at 55° for 5 minutes and cooled to 10°; the precipitate formed is removed by centrifugation (40,000 *g* for 20 minutes or longer if necessary).

Step 5. Preparation of Calcium Phosphate Gel-Cellulose Column. To a suspension of 200 ml of 10% (w/v) cellulose powder (Whatman Standard Grade, CF11) is added, with stirring, 100 ml calcium phosphate gel (30 mg/ml) prepared by the method of Swingle and Tiselius.[8] After removal of dissolved air by evacuation on a water pump the gel-cellulose suspension is packed by gravity in a column above a plug of glass wool and a 1–2 cm bed of the CF11 cellulose. This gives a packed column 25 cm high and 2.2 cm in diameter. The column is then equilibrated at 0° with 0.1 *M* phosphate, pH $7.6 + 3 \times 10^{-4}$ *M* EDTA and should have a flow rate of 20–60 ml per hour.

Step 6. Calcium Phosphate Gel-Cellulose Column Chromatography. The supernatant from step 4 is made 0.1 *M* with respect to phosphate by the addition of 0.5 volume 0.3 *M* phosphate pH 7.6 and applied carefully to the calcium phosphate gel-cellulose column. The lipoyl dehydrogenase is adsorbed strongly at the top of the column as a bright yellow and strongly fluorescent band; other colored components are not adsorbed strongly and are eluted with 0.1 *M* phosphate. Elution with

[8] S. M. Swingle and A. Tiselius, *Biochem. J.* **48,** 171 (1951).

the buffer is continued until the D_{280} of the eluate is less than 0.05; conveniently this is done overnight with a constant head device. The lipoyl dehydrogenase is now eluted with 0.1 *M* phosphate, pH 7.6, +4% (w/v) $(NH_4)_2SO_4$ $+3 \times 10^{-4}$ *M* EDTA.

Step 7. Fractionation with $(NH_4)_2SO_4$. The eluate from step 6 is fractionated by the addition of solid $(NH_4)_2SO_4$; the enzyme is located in the fraction precipitating between 0.55 and 0.70 saturation.

Step 8. Heat Precipitation of Impurities. The 0.55–0.70 $(NH_4)_2SO_4$ fraction is dissolved to a final concentration of 2–6 mg/ml in a solution of 0.06 *M* phosphate, pH 7.0, containing 0.1 *M* $(NH_4)_2SO_4$ and 10^{-4} *M* EDTA. The solution is then heated at 72–73° for 6 minutes with careful stirring to avoid local overheating. The white precipitate is centrifuged off, and the supernatant is fractionated between 0.50 and 0.70 saturation with solid $(NH_4)_2SO_4$. At this stage the enzyme is generally about 90% pure; it may be dissolved and stored as described below, or further purified by a second chromatography on calcium phosphate. For this process the precipitate is dissolved in and dialyzed against 0.1 *M* phosphate buffer, pH 7.6, containing 3×10^{-4} *M* EDTA and applied directly to another calcium phosphate gel-cellulose column. Elution is repeated as previously, and the bright yellow enzyme fraction is precipitated by addition of solid ammonium sulfate to 80%. The precipitate is then dissolved in a convenient buffer such as 0.03 *M* phosphate pH 7.0 containing 3×10^{-4} *M* EDTA. The enzyme is stable almost indefinitely if stored frozen under these conditions. Recovery and specific activity data of a typical preparation are given in the table.

ACTIVITY DATA FROM A TYPICAL PREPARATION OF LIPOYL DEHYDROGENASE

Procedure	Protein (mg)	DPNH-lipoyl reductase units (μmoles/min)	Specific activity (μmoles/min/mg)
1% $(NH_4)_2SO_4$ extract from 1400 g washed heart muscle (12 pig hearts)	5500	2830	0.515
0.4–0.7 $(NH_4)_2SO_4$ fraction (end of step 4)	1200	2300	1.91
Eluate from calcium phosphate–cellulose column	130	2000	15.4
Pure enzyme at end of step 8	72	1800	25

Properties

Physical. Lipoyl dehydrogenase is a flavoprotein containing FAD as prosthetic group. It is identical with the enzyme originally isolated by Straub[9] and known as Straub diaphorase. The oxidized enzyme has absorption peaks at 455 mμ, 370–355 mμ, and 274 mμ. The ratio of optical densities $D_{280}:D_{455}$ characteristic of the pure enzyme is 5.35. There is a distinct shoulder between 465 and 485 mμ, a phenomenon shown by many purified flavoproteins. A distinctive feature of lipoyl dehydrogenase is its very pronounced fluorescence, which is greater than that of free FAD.[10] The molecular weight, determined by the Archibald approach to equilibrium procedure, is 102,000.[11] The sedimentation constant, $S^0_{20,w}$, is 5.7 Svedbergs; the diffusion constant determined at a concentration of 10.6 mg/ml is 4.63×10^{-7} cm^2 sec^{-1}. The unit molecular weight, per molecule of enzyme-bound FAD is approximately 50,000. The enzyme thus contains two molecules of FAD per molecule of protein. From studies of the enzyme denatured by substrate in the presence of 6.5 *M* urea, and from peptide mapping, it has been determined that the enzyme is composed of two identical peptide chains linked by two disulfide bonds, which form parts of the two active centers.[11]

Effect of Cu^{++}. Lipoyl dehydrogenase provided one of the first clear-cut examples of profound modification of catalytic properties of an enzyme as a result of very mild treatment. Incubation of the enzyme at 0° with less than stoichiometric amounts of Cu^{++} leads to the complete loss of the catalytic reactions of the enzyme involving oxidized or reduced lipoyl derivatives; at the same time the DPNH-dichlorophenolindophenol reductase activity increases by some thirtyfold.[2] This conversion is accompanied by oxidation of apparently essential sulfhydryl groups.[12]

Nature of the Active Center and the Catalytic Mechanism. A wide variety of evidence has been obtained that shows that, as well as the FAD, a protein disulfide group is involved in catalysis. The reactive intermediate in catalysis, characterized by a long wavelength absorption band centered at 530 mμ,[7] appears to be a compound in which both the flavin and the disulfide are half reduced. The 4-electron, fully reduced form of the enzyme appears to be catalytically inert in the reactions involving lipoyl derivatives. The conversion of the disulfide to a dithiol in catalysis is the basis of the sensitivity to inhibiton by arsenite. The evidence for the involvement of an active center disulfide in catalysis is reviewed by Massey.[10]

Kinetic Constants. In view of the complex kinetics of lipoyl dehydrogenase[7] (and indeed of all flavoproteins) very few of the kinetically

[9] F. B. Straub, *Biochem. J.* **33**, 787 (1939).

[10] V. Massey, *in* "The Enzymes" (P. D. Boyer, H. Lardy, and K. Myrbäck, eds.), 2nd ed., Vol. VII, p. 275. Academic Press, New York, 1963.

[11] V. Massey, T. Hofmann, and G. Palmer, *J. Biol. Chem.* **237**, 3820 (1962).

[12] L. Casola and V. Massey, unpublished results.

significant constants (obtained by extrapolation to infinite concentration of both hydrogen donor and acceptor) have been determined, especially since DPNH is an inhibitor when added in excess. In general the K_m values for lipoyl derivatives are fairly high (about 1 mM) and those for DPN and DPNH somewhat lower (10^{-5} to 10^{-4} M) (Massey[10] should be consulated for details). The DPNH-lipoyl reductase activity varies markedly with the nature of the lipoyl derivative; turnover numbers under comparable conditions range from 1000 moles DPNH oxidized per minute per mole enzyme-bound FAD with lipoic acid as substrate to $\geq$ 80,000 with lipoamide as substrate.[6] The reactivity also varies markedly with the nature of the pyridine nucleotide. In the dihydrolipoamide-pyridine nucleotide reductase assay, DPN and thionicotinamide-DPN have essentially identical activity, acetylpyridine DPN and pyridine aldehyde DPN react at less than one-tenth the rate, deamino DPN at less than one-hundredth the rate, and deamino acetyl pyridine DPN and deamino pyridine aldehyde DPN at less than one-thousandth the rate of the reaction with DPN.[3] TPN and TPNH have essentially no activity in catalytic tests; however TPNH will slowly reduce the enzyme, at a rate of the order $1/10^6$ to $1/10^7$ times the corresponding rate with DPNH.[13]

pH Optima. The optimum pH for catalysis of reactions by lipoyl dehydrogenase varies somewhat depending on the nature of the lipoyl derivative employed. In the DPNH-lipoyl reductase assay the pH optimum varies from pH 5.65 with lipoic acid to pH 6.5 with lipoamide. In the dihydrolipoyl-DPN reductase assay the pH optimum is 7.0 for dihydrolipoyl glycine and pH 7.9 for dihydrolipoamide.[6]

[13] V. Massey and G. Palmer, *J. Biol. Chem.* **237**, 2347 (1962).

[53] Lactic Dehydrogenases (Crystalline)

By Francis Stolzenbach

$$\text{L-Lactate} + \text{NAD}^+ \rightleftharpoons \text{pyruvate} + \text{NADH} + \text{H}^+$$

Assay Procedure

In a total volume of 3 ml the following are contained:
- Potassium phosphate, 300 micromoles, pH 7.0
- Sodium pyruvate, 2 micromoles
- NADH, 0.4 micromole

Lactic dehydrogenase to give a change in optical density at 340 mμ of between 0.1 and 0.2 in 1 minute

One unit of enzyme is that amount which will give a change in optical density of 1.0 in 1 minute under the conditions of the assay.

General Comments on the Isolation Procedures

In general, the present methods developed in our laboratories for the purification of lactic dehydrogenases involve several basic steps: fractionation by solubility, separation by charge, and separation by molecular size.

The mainstay of all the procedures is fractionation by solubility, using ammonium sulfate.[1] Gross precipitations for the purpose of concentrating as well as for purification are used in most of our procedures. Repeated ammonium sulfate fractionations using finer cuts as the purification progresses are utilized in most procedures.

Separation by charge using cation and anion exchange celluloses (celluloses washed as described by Pesce *et al.*[2]) is the other main purification step. During the course of electrophoretic investigation, it was noted that some proteins had their mobility changed drastically while others were unaffected by changes in pH. This helped greatly in determining the exchanger and pH to use in our procedures. Proteins which chromatograph as one symmetrical peak may be separated on a subsequent column simply by changing the pH conditions for equilibration and running of the column. Whenever practical, a negative absorption step using ion exchange celluloses is utilized to facilitate subsequent purification steps.

Sephadex molecular sieves and sucrose density gradient centrifugations are used for separation by molecular size. If necessary these steps are usually used at a very late stage in the procedure, since the volume would tend to limit the amount of enzyme which could be processed by these methods.

It was found that some lactic dehydrogenases are stabilized by the presence of β-mercaptoethanol during the course of the procedures as stated. As the enzymes are purified, the dilutions for assay might tend to become less stable. This is overcome by diluting just prior to assay, using cold buffer containing 1% bovine serum albumin or 10^{-3} *M* β-mercaptoethanol, or sometimes both, as protecting agents.

[1] Percentage of saturation was based on Table I, p. 76 of Vol. I, even though the enzymatic solutions were kept at 4°.

[2] A. Pesce, R. H. McKay, F. Stolzenbach, R. D. Cahn, and N. O. Kaplan, *J. Biol. Chem.* **239**, 1753 (1964).

M_4 Lactic Dehydrogenases from Bird Muscle

Pheasant, duck, turkey, rhea, and ostrich M_4 lactic dehydrogenases were prepared following the same basic procedure with small variations as indicated.

Crude Extract. Muscle was ground in a commercial meat grinder and extracted with two volumes of cold 10^{-1} *M*, pH 7 Tris in 10^{-3} *M* EDTA for 1 hour and was filtered through cheesecloth. The mixture was then clarified by centrifuging in a Sharples continuous flow centrifuge.

First Ammonium Sulfate. The clear pink extract was brought to 70% saturation with solid ammonium sulfate and left at 0° for 1 hour. The suspension was then filtered overnight on fluted filter papers.

Second Ammonium Sulfate. The precipitate was scraped off and suspended in an equal volume of cold distilled water. The ammonium sulfate concentration was estimated at 35% saturation. After 1 hour at 0° the suspension was centrifuged at 13,000 *g* for 1 hour, and the precipitate was discarded. (In some cases the suspension was too viscous, and an equal volume of 35% saturated ammonium sulfate was added to facilitate the centrifugation.) To the clear supernatant solid ammonium sulfate was added to 70% saturation. After 1 hour at 4°, the suspension was centrifuged for 1 hour at 13,000 *g*, and the precipitate was dissolved in a minimal amount of cold distilled water.

Negative Absorption on DEAE-Cellulose. The enzyme solution was dialyzed for 16 hours against two changes of 5×10^{-3} *M* Tris, pH 7.6, with 20-liter volumes used for each change. The dialyzate was then centrifuged for 10 minutes at 13,000 *g*, and the precipitate was discarded. After centrifugation the enzyme solution was passed through a DEAE cellulose column, equilibrated against 5×10^{-3} *M* Tris, pH 7.6, and the column was flushed with 5×10^{-3} *M*, pH 7.6 Tris until all the enzyme activity was eluted.

Third Ammonium Sulfate Fractionation. To the combined DEAE fractions, ammonium sulfate was added to 60% saturation. After it had stood for 1 hour at 0°, the suspension was centrifuged 1 hour at 13,000 *g*, and the precipitate was dissolved in a minimal amount of cold distilled water.

Carboxymethyl Cellulose. The dissolved precipitate was dialyzed against 20 liters of 10^{-2} *M* potassium phosphate, pH 6.5, for 16 hours. After dialysis the enzyme was absorbed onto a CM-cellulose column that had been equilibrated to pH 6.5 with potassium phosphate. The enzyme was then eluted with a gradient of NaCl as described for beef M_4 lactic dehydrogenase (see below). All the fractions with a specific activity of

600 or better were combined and precipitated with solid ammonium sulfate at 60% saturation.

Crystallization. After standing overnight at 4°, the enzyme was centrifuged and dissolved in enough distilled water to give a protein concentration of about 10 mg/ml. The insoluble material was centrifuged at 13,000 *g* and discarded. Solid ammonium sulfate was added to the supernatant until a faint turbidity appeared. After 30 minutes the enzyme began to crystallize. The crystals were allowed to form for several days at 4° and were then harvested by centrifugation at 25,000 *g*. The crystals were dissolved and recrystallized as before. After two recrystallizations there was no increase in specific activity. The enzyme was stored at 4° as a crystalline suspension. All the muscle enzymes from birds crystallized between 45% and 55% saturation.

TABLE I
TURKEY M_4 LACTIC DEHYDROGENASE

Step	Total activity	Specific activity (units/mg protein)
I. Crude extract	5.1×10^6	1.5×10^1
II. First ammonium sulfate fractionation	3.3×10^6	4.3×10^1
III. DEAE chromatography	1.9×10^6	5.9×10^1
IV. Second ammonium sulfate fractionation	1.1×10^6	5.7×10^2
V. CMC chromatography	1.0×10^6	1.1×10^3
VI. Crystallization	1.0×10^6	1.1×10^3

Preparation of Chick H_4 Lactic Dehydrogenase

The same basic procedure is applicable to turkey, pheasant, ostrich and duck H_4 lactic dehydrogenase with slight variations in the crystallization. The ammonium sulfate fractionations, acetone fractionation and chromatography on DEAE are interchangeable in these bird heart lactic dehydrogenase preparations.

Crude Extract. Fresh chicken hearts (25 pounds) were ground in a commercial meat grinder and extracted in 14 liters of cold distilled water for 1 hour with occasional stirring. The mixture was then strained through a double layer of cheesecloth, and the filtrate was spun in a Sharples continuous flow centrifuge. The supernatant was poured through a funnel containing a layer of glass wool between layers of cheesecloth. This removed a major part of the remaining fat.

First Ammonium Sulfate Precipitation. Solid ammonium sulfate was added to 70% saturation, and the mixture was left at 4° for 1 hour. The resulting precipitate was then collected on fluted filter papers by an overnight filtration. The precipitate was scraped off and dissolved in 1 liter of cold distilled water and then dialyzed overnight against 20 liters of cold distilled water.

Second Ammonium Sulfate Precipitation. The dialyzate was spun to clarify it, then was made to 25% saturation with solid ammonium sulfate. The precipitate formed was spun down and discarded. To the clear red supernatant, solid ammonium sulfate was added to 60% saturation. After 1 hour at 4°, this precipitate was collected by centrifugation at 18,000 g in a Servall refrigerated centrifuge and dissolved in 300 ml of cold $10^{-2} M$ Tris buffer, pH 7.6. The resultant solution was then dialyzed overnight against $10^{-2} M$, pH 7.6 Tris buffer.

Acetone Fractionation. Acetone, precooled to —15°, was then added to the dialyzate to yield a precipitate at 25% v/v. After 10 minutes at 0°, the precipitate was spun out and discarded. To the 25% supernatant, —15° acetone was added to a final concentration of 50%, and this was left at 0° for 20 minutes. This precipitate was collected by centrifugation at 18,000 g for 30 minutes and extracted with 200 ml of cold $10^{-2} M$, pH 7.6 Tris. The insolubles were then spun out and discarded.

DEAE-Cellulose. The clarified extract, which was still red, was then passed through a 4.8 cm × 45 cm DEAE-cellulose column equilibrated with $10^{-2} M$ Tris, pH 7.6, and flushed with $10^{-2} M$, pH 7.6 Tris, until all the enzyme was eluted. Most of the red and brown color remained on the column while the enzyme passed through.

Crystallization. To the pooled active eluates ammonium sulfate was added to 40% saturation. The enzyme began to crystallize immediately. After allowing additional crystallization overnight, the crystals were

TABLE II
CHICK H_4 LACTIC DEHYDROGENASE

Step	Total activity	Specific activity (units/mg protein)
I. Crude extract	1.4×10^6	2.5×10^1
II. First ammonium sulfate fractionation	1.35×10^6	3.6×10^1
III. Second ammonium sulfate fractionation	1.25×10^6	6.5×10^1
IV. Acetone fractionation	8.75×10^5	1.3×10^2
V. DEAE chromatography	8.0×10^5	4.5×10^2
VI. Crystallization	7.2×10^5	9.6×10^2

harvested by centrifugation at 18,000 *g* and then washed three times with 40% saturated ammonium sulfate to rid them of any remaining reddish color.

Recrystallization. The washed crystals were then dissolved in a minimal amount of cold distilled water and spun at 25,000 *g* for 10 minutes. This final spinning clears out any remaining color. Slow additions of ammonium sulfate, eventually reaching 35% saturation, start the crystallization. This may be stored at 4° for several months with no appreciable loss of activity.

Isolation of Dogfish M_4 Lactic Dehydrogenase

Crude Extract. Approximately 10 kg of frozen dogfish muscle was ground in a commercial meat grinder and extracted in 11 liters of cold distilled water. The suspension was stirred for about 1 hour at 4°, strained through two layers of cheesecloth, and then centrifuged in a Sharples continuous flow centrifuge. The supernatant was collected through glass wool. The last step removed much of the remaining fat in the preparation.

Fractionation by Ammonium Sulfate. Ammonium sulfate to 45% saturation was added, slowly and with stirring, to the supernatant from the previous step. The suspension was allowed to stand at 4° for 2 hours and then was filtered overnight through fluted filter paper.

Ammonium sulfate was added to the filtrate from the previous step to 70% saturation in the manner described for M_4 lactic dehydrogenase from bird muscle. The suspension was allowed to stand at 4° for 2 hours and then filtered overnight as before. The precipitate was scraped off the filter paper and dissolved in 1.5 liters cold distilled water.

The solution was spun at 18,000 *g* for 30 minutes and filtered through glass wool. This eliminated most of the remaining fat. The solution was then dialyzed overnight *vs* two changes of 12 liters of 5×10^{-3} *M* Tris, pH 7.6, at 4°.

DEAE-Cellulose Chromatography. The dialyzed solution from the previous step was centrifuged to remove the insoluble material. It was loaded onto a DEAE-cellulose column of 80 cm × 10 cm which had been equilibrated against 5×10^{-3} *M* Tris, pH 7.6, and kept at 4°. The fractions were eluted with the same buffer. The fractions in which the activity appeared were pooled.

The column also removed two small components which migrated differently from the pure M_4 type on a starch gel.

Final Purification with Ammonium Sulfate. Ammonium sulfate was added as before to the combined fractions from the column until 50% saturation was reached and the suspension was allowed to stand for 1 hour at 4°. The suspension was then centrifuged at 18,000 *g* for 30 minutes.

Ammonium sulfate was added to the supernatant from the preceding step until 70% saturation was reached and then was allowed to stand overnight at 4°. The precipitate was collected by centrifugation at 18,000 *g*.

Crystallization of the Enzyme. The precipitate from the previous step was dissolved in 100 ml cold distilled water, and ammonium sulfate was added to 30% saturation. The enzyme began to crystallize almost immediately. The suspension was allowed to stand overnight. The crystals were then harvested by centrifugation at 33,000 *g* for 30 minutes and washed three times with a solution of 30% saturated ammonium sulfate. The enzyme dissolved in the washings, which were collected and kept cold. The remaining color was in the undissolved precipitate.

TABLE III
DOGFISH M_4 LACTIC DEHYDROGENASE

Step	Total activity	Specific activity (units/mg protein)
I. Crude extract	4.1×10^6	1.8×10^1
II. First ammonium sulfate fractionation	3.9×10^6	5.7×10^1
III. DEAE chromatography	3.3×10^6	2.8×10^2
IV. First crystallization	2.0×10^6	6.9×10^2
V. Second crystallization	2.0×10^6	8.3×10^2

Recrystallization of the Enzyme. Ammonium sulfate was added slowly to the washings from the preceding step until 55% saturation was reached. The suspension was allowed to stand overnight at 4° and kept as stock for further experiments. Under the microscope, the suspension showed large crystals with a minimum of extraneous material.

Beef M_4 Lactic Dehydrogenase

Crude Extraction. Twelve pounds of round steak were ground in a commercial meat grinder and extracted with 17 liters of cold distilled water for 1 hour with occasional stirring. The mixture was then filtered through a double layer of cheesecloth, and the filtrate was then passed through a Sharples continuous flow centrifuge to remove as much of the finer debris as possible.

First Ammonium Sulfate Step. To reasonably clear supernatant from the centrifugation, solid ammonium sulfate was added to give a saturation of 70%. The pH was maintained at 7 with dilute ammonium hydroxide during the addition of the salt. After being allowed to stand for several hours, the suspension was filtered overnight on fluted filter papers. The precipitate was then scraped off the filter papers and suspended in about 1 liter of cold distilled water. The insolubles were then spun out and discarded, and virtually all the starting activity was present in the supernatant.

Second Ammonium Sulfate Step. The enzyme solution was then dialyzed against two 20-liter changes of 5×10^{-3} *M*, pH 7, Tris, for a total of 12 hours. Ammonium sulfate was then added to yield a fraction at 25% saturation. This fraction was then centrifuged and the precipitate was discarded. To the supernatant, ammonium sulfate was added to 50% saturation. After standing for 1 hour the suspension was centrifuged and the precipitate containing the enzyme was dissolved in 500 ml of 5×10^{-3} *M*, pH 7, Tris.

Chromatography on Carboxymethyl Cellulose. The enzyme solution was dialyzed against two changes of 20 liters of 5×10^{-3} *M*, pH 7 Tris for 8 hours each. After dialysis, the protein solution was centrifuged to clarify and then placed on a 4.8 cm × 45 cm column of carboxymethyl cellulose equilibrated against 5×10^{-3} *M* Tris pH 7. The enzyme was eluted from the column with a gradient of 2 liters of 2×10^{-1} *M* NaCl in 5×10^{-3} *M* Tris, pH 7, running into 2 liters of 5×10^{-3} *M*, pH 7 Tris.

Crystallization. The major enzyme peak from the carboxymethyl cellulose column was then dialyzed overnight against saturated ammonium sulfate. The precipitated protein was then centrifuged and dissolved in a minimal amount of cold distilled water and immediately centrifuged to clarify. Upon standing in ice, the enzyme began to crystallize. After several days the enzyme was harvested, dissolved in cold distilled water, and recrystallized by slow addition of solid ammonium sulfate up to about 30% saturation.

TABLE IV
BEEF M_4 LACTIC DEHYDROGENASE

Step	Total activity	Specific activity (units/mg protein)
I. Crude extract	4.1×10^6	4.2×10^1
II. First ammonium sulfate fractionation	2.6×10^6	6.0×10^1
III. Second ammonium sulfate fractionation	1.6×10^6	2.0×10^2
IV. CMC chromatography	9.5×10^5	8.5×10^2
V. Crystallization	9.0×10^5	1.4×10^3

Separation of Hybrids from the Pure M_4 Lactic Dehydrogenase. In some cases the hybrids of lactic dehydrogenase are not separated by the first carboxymethyl cellulose column. In these instances, a similar chromatography was run on the crystalline enzyme.

Beef Heart Lactic Dehydrogenase

Crude Extraction. Eight beef hearts were trimmed of fat and connective tissue and ground in a commercial meat grinder. The ground hearts (7.5 kg) were then stirred with 30 liters of cold distilled water. After 15 minutes the suspension was filtered through two layers of cheesecloth to remove the residue and then through glass wool to remove most of the remaining fat.

Calcium Phosphate Gel Step. The clear red filtrate was titrated with calcium phosphate gel[3] until 90% of the activity was absorbed onto the gel. The amount needed varied with the gel preparation and the efficiency of the extractions of the mince. The gel was then washed with 3-liter portions of 0.2 *M* potassium phosphate buffer, pH 7.2, until no more activity was eluted. The enzyme was concentrated by making the pooled active eluates 60% saturated with ammonium sulfate. After standing for several hours, the precipitate was centrifuged and dissolved in enough 0.1 *M* potassium phosphate buffer, pH 7.2, to bring the protein concentration to 30 mg/ml.

First Acetone Fractionation. To the dissolved precipitate, acetone, precooled to —15°, was slowly added at a ratio of 580 ml of acetone for each 960 ml of enzyme solution. The temperature was then raised to 13° and maintained there for 10 minutes. After the suspension had been centrifuged at 10,000 *g* for 10 minutes, the supernatant was discarded and the precipitate was thoroughly suspended in 400 ml of 30% saturated ammonium sulfate.

Ammonium Sulfate Fractionation. The suspension was left at 4° for 1 hour and centrifuged at 13,000 *g* for 15 minutes. The precipitate was discarded and the supernatant was raised to 50% saturation with solid ammonium sulfate and left 6–18 hours at 4°. The active precipitate was then centrifuged at 13,000 *g* for 20 minutes and dissolved in enough cold distilled water to yield a protein concentration of 30–35 mg/ml.

Second Acetone Fractionation. Deep freeze acetone was then added to the enzyme solution at a ratio of 66 ml of acetone for each 110 ml of solution. The temperature was then raised to 18° for 10 minutes. The precipitate was centrifuged at 10,000 *g* for 15 minutes, suspended in 50 ml of 30% saturated ammonium sulfate, and dialyzed overnight against one liter of 30% saturated ammonium sulfate.

[3] A. Meister, *Biochem. Prep.* **2,** 18 (1952).

Crystallization. The dialyzed suspension was centrifuged at 13,000 *g* for 20 minutes and the precipitate was discarded. To the supernatant, 2.6 g of ammonium sulfate was added for each 44 ml of solution, and the suspension was immediately centrifuged for 5 minutes at 7500 *g*. The precipitate was discarded and an additional 2.6 g of ammonium sulfate was added slowly over a 2 to 4 hour period. The enzyme began to crystallize almost at once on the first few additions of ammonium sulfate. The crystals were left overnight at 4°, harvested by centrifugation at 13,000 *g* for 30 minutes, and finally washed several times with 50% saturated ammonium sulfate. The crystals were stored as a crystalline suspension in ammonium sulfate at 4°. This method is approximately the same as the Neilands preparation (see Vol. I [69]), and is preferable to modified preparations, since it also yields some hybrids that can be isolated in pure form as described in this article.

TABLE V
BEEF HEART LACTIC DEHYDROGENASE

Step	Total activity	Specific activity (units/mg protein)
I. Crude extract	5.3×10^6	4.0×10^1
II. Calcium phosphate gel eluate	2.5×10^6	6.6×10^1
III. First acetone fractionation	1.5×10^6	3.65×10^2
IV. Ammonium sulfate fractionation	1.4×10^6	3.88×10^2
V. Second acetone fractionation	1.2×10^6	8.0×10^2
VI. Crystallization	1×10^6	1.2×10^3

Preparation and Separation of Hybrids of Beef Lactic Dehydrogenases

Preparation of the hybrids from pure beef M_4 and pure beef H_4 is based on the "quick-freeze" and "slow-thaw" method of Chilson *et al.*[4] One hundred milligrams of beef H_4 and 150 mg of beef M_4 were thoroughly dialyzed against 0.1 *M* sodium phosphate, pH 7.0. Concentrations were then adjusted to give a solution of 0.1 *M* sodium phosphate, 0.5 *M* sodium chloride, 10^{-3} *M* β-mercaptoethanol, pH 7.0, containing a total protein concentration of 1 mg/ml. The solution was poured into a flask and was rapidly frozen over a mixture of dry ice and methanol. The mixture was then allowed to thaw slowly, undisturbed, at room temperature. Freezing and thawing are repeated. About 85–90% of the initial enzyme activity remained. The solution was thoroughly dialyzed against 10^{-2} *M* Tris, pH 7.3, containing 10^{-3} *M* β-mercaptoethanol. The dialyzed solution was then placed on a column of DEAE-cellulose (2.5 cm × 60 cm) that had been previously equilibrated in the same

[4] O. P. Chilson, L. Costello, and N. O. Kaplan, *Biochemistry* 4, 271 (1965).

buffer. The column was washed with 10^{-2} *M* Tris until no more protein appeared. The unbound enzyme is M_4. Then a gradient was applied. A 500-ml amount of 0.22 *M* sodium chloride in 10^{-2} *M*, pH 7.3 Tris was allowed to flow from overhead into 500 ml 10^{-2} *M* Tris, pH 7.3, in the mixing flask in order to establish an exponential gradient. Four-milliliter fractions were collected. The enzymes were eluted from the column in the following order: M_3H, H_2M_2, H_3M, and H_4. Fractions shown by starch gel electrophoresis[5] to be homogeneous were combined. The hybrids H_2M_2 (75 mg) and H_3M (25 mg) were both concentrated by dialysis against saturated ammonium sulfate and stored as ammonium sulfate suspensions. Unlike the other four enzymes, M_3H is unstable in concentrated solutions of ammonium sulfate. The combined fractions of M_3H (60 mg) were placed in dialysis tubing (2.5 cm in diameter), and surrounded by powdered aquacide II (obtained from Calbiochem). The tubing and powder were wrapped in a sheet of parafilm, and the solution was concentrated at 4° to give a final protein concentration of 15 mg/ml. The enzyme was then thoroughly dialyzed against 10^{-1} *M* potassium phosphate, 10^{-1} *M* ammonium sulfate, 10^{-3} *M* β-mercaptoethanol, pH 7.2, and stored at 4°. Although the enzyme was stable for several months in this buffer, all experiments with M_3H were performed within a few days after its preparation.

Measurement of "low" to "high" pyruvate ratios[2] for each of these enzymes yields the following values when 10^{-4} *M* NADH serves as coenzyme: H_4, 3.0; H_3M, 2.38; H_2M_2, 2.00; M_3H, 1.60; and M_4, 0.93. Complement fixation studies have demonstrated that these hybrids are immunologically identical with the natural hybrids.[6,7]

[5] I. H. Fine and L. Costello, Vol. VI [127].
[6] O. P. Chilson, personal communication.
[7] The technique of hybrid separation was developed by S. Anderson, (personal communication).

[54] L-Lactic Dehydrogenase: Heart (H_4)

By W. J. Reeves, Jr., and Grace M. Fimognari

Crystalline lactic dehydrogenase prepared from pig or beef heart muscle according to Straub[1] contains up to four enzymatically active proteins as determined by starch gel electrophoresis.[2] These lactic de-

[1] F. B. Straub, *Biochem. J.* **34,** 483 (1940). See Vol. I [69].
[2] C. L. Markert and F. Møller, *Proc. Natl. Acad. Sci. U.S.* **45,** 753 (1959).

hydrogenase "isozymes"[3] are members of a family of structurally and functionally interrelated L-lactic dehydrogenases. The five isozymes commonly described each contain four subunits. The major isozyme of skeletal muscle and liver, M_4, has four M (muscle) subunits; while H_4, the principal isozyme of heart muscle in most species, contains four H (heart) subunits. Isozymes HM_3, H_2M_2, and H_3M contain both types of subunits.[4]

Assay Method

Principle. Lactic dehydrogenase activity is assayed at 25° and pH 7.4 by measuring the rate of decrease in absorbance of NADH at 340 mμ. This reaction, of course, measures the sum of all lactic dehydrogenase isozyme activities and is not specific for H_4.

Reagents

NADH, 0.005 *M*[5]
Sodium pyruvate, 0.0227 *M*
Potassium phosphate buffer, pH 7.4, 1.00 *M*, 1% in bovine serum albumin, fraction V.
Bovine serum albumin, fraction V, 1%, in 0.15 *M* NaCl with 0.05 *M* potassium phosphate, pH 7.4, for diluting enzyme for assay
Sodium chloride, 0.9%

Procedure. The assay mixture contains: NADH, 0.1 ml; pyruvate, 0.1 ml; phosphate buffer, 0.3 ml; and 2.4 ml distilled water. In the above proportions, it is convenient to make enough assay mixture for 10–30 assays. This should be used within a few hours. To 2.9 ml of assay mixture at 26° in a 3-ml cuvette, add 0.1 ml of ice cold diluted enzyme containing 0.02–0.05 enzyme units (final temperature, 25°; ΔOD/min 0.040–0.100). Mix quickly by inversion and immediately begin readings at 340 mμ in a thermostatted spectrophotometer. The reaction velocity for the first minute is used to compute enzyme activity. Linearity should be retained for at least 1 minute.

Best accuracy is obtained if enough NADH is added to the blank to make the assay cuvette read about 0.45 absorbancy unit (reading against water, approximately 1.0 unit). When scanning column eluate for major enzyme peaks, assay mixture showing nil activity can be used for one or

[3] The term "isozyme" was proposed by Markert and Møller.[2]

[4] Discussions on the physiochemical nature, differences in catalytic and immunologic properties, and distribution of lactic dehydrogenase isozymes are available.[7,8]

[5] NADH solutions develop an inhibitor on standing. A fresh solution should be made each day just prior to use. The other reagents can be stored for several days at 4°.

two additional enzyme samples as great precision is unnecessary at this stage.

Definition of Unit and Specific Activity. One unit of activity is the amount of enzyme oxidizing 1 micromole of NADH per minute under the above conditions. Specific activity is stated as units of activity per milligram of protein.

Protein concentration is determined by measuring A_{215}. Little protein is required, and the determination is quick, easy, and accurate. Sodium chloride 0.9% is used as blank and diluent. Protein concentration is equal to $0.0663 \times A_{215}$ mg/ml ($E_{215}^{1\%} = 151$) for A_{215} measurements of 0.000–1.000. Cuvettes must be scrupulously clean, and a blank correction should be determined just before use.

Purification Procedure

Several procedures are available for the preparation of crystalline lactic dehydrogenase, H_4, from various sources.[6-8] The M and H subunits of lactic dehydrogenase isozymes have characteristic amino acid compositions and consequently differ in charge. This feature is made use of in removing contaminating isozymes. The procedure used below is for the preparation of lactic dehydrogenase, H_4, from hog heart and employs an initial DEAE-cellulose column chromatography step to separate the desired isozyme.[9] It has been used successfully by both experienced investigators and students and results in recovery of 40–60% of the original total activity in crystalline form.

Step 1. Preparation of DEAE-Cellulose Column. DEAE-cellulose is prepared for use by washing successively with 0.1 *N* NaOH, 0.1 *N* HCl, distilled water, 95% ethanol, 0.1 *N* NaOH, and distilled water, and finally suspending in a suitable volume of 0.5 *N* NaCl. After decanting of fines, a column (60×4.7 cm) containing 120–150 g of DEAE-cellulose is poured (sufficient to process 1200–1500 ml of water extract from 750 g of heart). The bed is equilibrated initially with 1.0 *M* potassium phosphate buffer, pH 7.0, and finally with potassium phosphate buffer, pH 7.0, ionic strength 0.06. Equilibration with the latter solution is tested by measuring either the buffer capacity or electrical conductivity of the effluent. The column is then placed in a cold room at 4° and all subsequent steps are carried out at 0–4° unless otherwise indicated.

The column may be regenerated after use by successive washings with

[6] J. S. Nisselbaum and O. Bodansky, *J. Biol. Chem.* **236**, 323 (1961).

[7] A. Pesce, R. H. McKay, F. Stolzenbach, R. D. Cahn, and N. O. Kaplan, *J. Biol. Chem.* **239**, 1753 (1964).

[8] C. L. Markert and E. Appella, *Ann. N.Y. Acad. Sci.* **94**, 678 (1961).

[9] W. J. Reeves, Jr., and G. M. Fimognari, *J. Biol. Chem.* **238**, 3853 (1963).

0.1 *N* NaOH, distilled water, 0.1 *N* HCl and 0.1 *N* NaOH and then equilibrated with potassium phosphate buffer.

Step 2. Preparation of Water Extract. Hog hearts, obtained as soon as possible after slaughter and stored on ice until used, are freed of fat, vessels, and atrial fibers, cut into 1–2 cm pieces, weighed and mixed with 2 volumes of ice-cold distilled water and ice chips. The mixture is homogenized for 1 minute in a Waring blendor and then centrifuged at 3200 *g* for 20–30 minutes. Traces of fat are removed by filtering the clear red supernatant liquid through glass wool and the filtrate pH adjusted to 7.0 with a few milliliters of 1.0 *M* K_2HPO_4.

Step 3. Column Chromatography on DEAE-Cellulose. From 1200 to 1500 ml of the water extract is loaded on the column at a rate of 4 ml per minute. During application of the water extract enzymatically active protein is bound, while two-thirds of the total protein (including the red-colored material of the extract) passes through the column under the conditions employed. Enzyme activity bound to the column is stable for several days at 4°.

A pH and salt gradient[10] is used to obtain maximum resolution of proteins bound to the column. Initially the mixing chamber contains 500 ml of 0.06 ionic strength phosphate buffer, pH 7.0. The reservoir is filled first with 1 liter of 0.06 *M* KH_2PO_4, then successively with 1 liter 0.06 *M*, 0.09 *M*, 0.14 *M*, and 0.20 *M* in both KH_2PO_4 and KCl. Elution rate is 2–4 ml per minute and eluate fractions of 20–25 ml are collected. A typical elution pattern is shown in Fig. 1.

For the isolation of isozyme H_4, generally only those eluate fractions having a pH of 6.0 or lower need be assayed for enzyme activity. Isozyme H_4 is eluted in the above program when the pH of the eluate drops to between 5.9 and 5.7. To locate the fractions containing enzyme activity, 1-ml aliquots are taken from every 3rd or 4th tube; 4 or 5 of these are mixed, and the mixture is assayed. When the major peak of activity is located, every second fraction may be assayed to define the limits of the peak. Fractions of 90 U/ml (ΔOD 190/min/ml) and with a specific activity greater than 135 U/mg (ΔOD 280/min/mg) are pooled for further purification. The preparation is stable for several weeks at this level of purification, if stored at 4°.

Step 4. Acetone Fractionation. With the pooled eluate kept at 2° in an ice-salt bath, 0.4 volume of redistilled acetone, at −15°, is added slowly with constant stirring. The suspension is brought to 5°, allowed to remain at this temperature 10–30 minutes and then is centrifuged at 10,000 *g* for 10 minutes. The yellow precipitate is discarded and the

[10] See Vol. V [1].

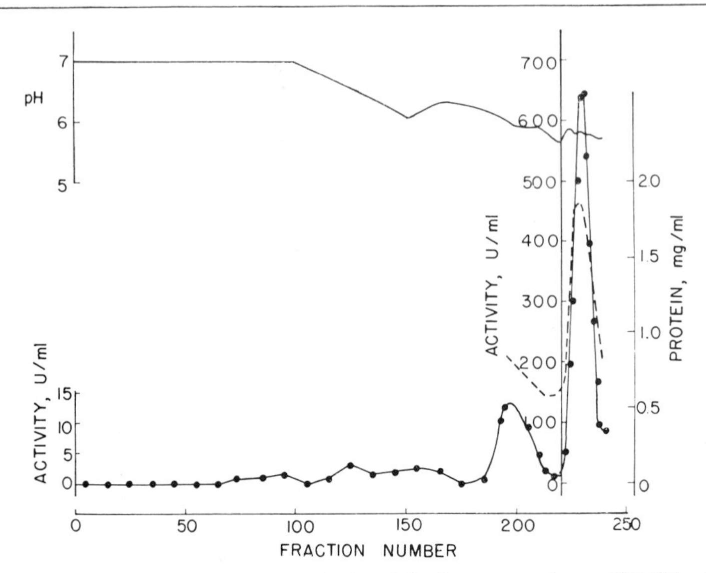

FIG. 1. Elution diagram of the lactic dehydrogenase from DEAE-cellulose. Experimental conditions are given in the text. The reservoir initially contained 1 liter of 0.06 M KH_2PO_4; at fraction 53 the reservoir was changed to a solution 0.06 M in KH_2PO_4 and 0.06 M in KCl; at fraction 113, to 0.09 M in KH_2PO_4 and 0.09 M in KCl; at fraction 172, to 0.14 M in KH_2PO_4 and 0.14 M in KCl; and at fraction 227, to a solution 0.20 M in KH_2PO_4 and 0.20 M in KCl. Activity is indicated by ●, and protein by - - -.

supernatant liquid is treated by adding redistilled acetone, at —15° as above, to a total of 0.75 volume of acetone per volume of original pooled eluate. The suspension is allowed to stand at 5° and centrifuged as above. All but a few milliliters (1–2 ml per 30 ml of supernatant liquid) of the supernatant liquid are carefully removed, taking care not to dry the precipitate.[11] The enzyme is now dissolved in a volume of 0.1 M potassium phosphate buffer, pH 7.0, at least 10 times that of the remaining supernatant solution.

Step 5. Ammonium Sulfate Fractionation and Crystallization. After slowly adding solid ammonium sulfate to 40% saturation (24 g per 100 ml of solution), the suspension is allowed to stand at 5° for 10–30 minutes and is then centrifuged at 10,000 g for 10 minutes. At least 80% of the activity which is recovered from the acetone step should remain in solution. The yellow precipitate is discarded, and ammonium sulfate is slowly added to the supernatant liquid to 55% saturation (10.0 g per 100 ml of solution). After several minutes a silky flow

[11] Contact of the precipitate surface with air leads to rapid loss of enzyme activity.

birefringence typical of crystalline lactic dehydrogenase is observed. The crystals are collected, after standing overnight, by centrifuging the suspension for 10 minutes at 10,000 *g*. The supernatant liquid is decanted, with caution to avoid air contact. The crystals are then dissolved in 10–20 ml of 0.1 *M* potassium phosphate buffer, pH 7.0 and stored at 4°. Recrystallization is done by repetition of the ammonium sulfate fractionation step but does not result in further purification. A typical purification is summarized in the table.

PURIFICATION PROCEDURE FOR H_4[a]

Step	Volume (ml)	Total activity (units × 10^{-3})	Total protein (g)	Specific activity (units/mg)	Enrichment (-fold)	Recovery (%)
Crude water extract	1200	122.0	15.7	7.8	—	100
Total column eluate	—	104.0	—	—	—	85
Fractions 224–237	250	85.0	0.397	214	27.4	69
0.4 Volume acetone supernatant liquid	345	84.2	—	—	—	68
0.4–0.75 Volume acetone precipitate	112	75.8	0.265	286	36.7	62
40% saturation ammonium sulfate supernatant liquid	121	72.0	0.217	332	42.6	59
Crystals	20	71.6	0.208	344	44.1	58

[a] Values shown are for 750 g of heart tissue.

Notes on Purification

Preparation of DEAE-Cellulose Column. In the procedure described, DEAE-cellulose purchased from Eastman Kodak was used. If highly purified commercial DEAE-cellulose is used, it is suspended in 0.5 NaCl, fines are decanted, and the column is poured.

Column Chromatography on DEAE-Cellulose. Although with any particular type of DEAE-cellulose the elution pattern is reproducible, use of various types of DEAE-cellulose (e.g., obtained from different sources or with different capacities) can result in differences in the elution pattern. Both the pH at which the major activity peak is eluted and the sharpness and concentration of enzymes in eluate fractions at the peak can be affected by the type of DEAE-cellulose employed. Therefore, it is recommended that on first using this procedure the entire eluate be scanned to locate the position of enzyme activity peaks.

Acetone Fractionation. In cases where broad and less concentrated activity peaks occur, or where the principal enzyme peak is eluted at a substantially different pH and salt concentration, somewhat higher concentrations of acetone may be required to precipitate the enzyme. Consequently, a preliminary experiment to determine the limits of acetone solubility is recommended. This is done by measuring the residual activity in the supernatant liquid after sequential additions of 0.1 volume of acetone to 10.0 ml of pooled eluate from the major peak. Once determined, the limits of acetone solubility for pooled eluate obtained under fixed conditions are constant. With high acetone concentrations it may be necessary to dissolve the acetone precipitate in 0.1 *M* potassium phosphate buffer, pH 7.0, and precipitate it with ammonium sulfate at 0.55–0.60 saturation prior to the usual ammonium sulfate fractionation and crystallization step.

Quantity of Enzyme. The above procedure may be scaled up or down to obtain as much as 1 g or as little as a few milligrams of crystalline lactic dehydrogenase, H_4. For reproducible results, it is necessary to maintain the proportionality between loading, weight of DEAE-cellulose, mixing chamber volume, reservoir volume, and flow rate.

Properties

Purity. Only one component is apparent in the crystalline preparation by both starch gel electrophoresis and free boundary electrophoresis. However, polyacrylamide gel electrophoresis as well as sedimentation velocity analysis indicate two protein components. The major component has a sedimentation constant of 7.1 S at a protein concentration of 6.5 g per liter, and the minor component a constant of 19 S. The high molecular weight component comprises approximately 1% of the protein in the crystalline preparation as determined by analysis of the sedimentation diagram.

Stability. The concentrated crystalline preparation is stable in either 0.1 *M* potassium phosphate, pH 7.0, or 40% ammonium sulfate for several months at 4°.

[55] Lactic Dehydrogenase (Clinical Aspects)

By MORTON K. SCHWARTZ and OSCAR BODANSKY

Introduction

Lactic dehydrogenase catalyzes the reaction: pyruvate + DPNH $\rightleftharpoons$ lactate + DPN. It is one of the group of ubiquitous enzymes whose activity in serum is elevated in a considerable variety of disease states such as myocardial infarction, anemia, cancer, including leukemias and lymphomas, and liver disease.[1,2] As will be noted in detail in a later section, the extent of elevation is dependent upon the particular disease, and very often on the stage of the disease.

Lactic dehydrogenase has been shown to exist in five molecular forms or isozymes. Kaplan[3] and Markert[4] demonstrated that these forms represent the five possible combinations in tetramer form of two different subunits. These subunits have been designated as A and B by Markert and as M and H by Kaplan. The five varieties would therefore be termed A_4B_0, A_3B_1, A_2B_2, A_1B_3, and A_0B_4 according to Markert and, according to Kaplan, as MMMM, MMMH, MMHH, MHHH, and HHHH. These teramers have also been designated as lactic dehydrogenases 1 to 5; LDH-1, LDH-2, LDH-3, LDH-4, and LDH-5. The tetramer, LDH-1 (A_4B_0 or MMMM), is the most cathodic, whereas LDH-5 (A_0B_4 or HHHH) is the most anodic. The kinetic and immunochemical characteristics of these various tetramers have been considered in great detail.[2] Attempts have been made to utilize diagnostically the properties of isozymes in various diseases. Kaplan and his associates[5] have proposed the use of a coenzyme analog ratio method in which the enzyme activity obtained with a reaction mixture containing reduced nicotinamide hypoxanthine dinucleotide and low pyruvate concentration is compared to that making use of DPNH and high pyruvate concentration. The greater heat lability of the LDH-1 isozyme than of the other isozymes has been utilized in a method whereby this component was found to be elevated in the serum in patients with hepatic disease, whereas the heat-stable lactic dehydrogenase, or LDH-5, was elevated in the serum of patients with myocardial infarction.[6]

Lactic dehydrogenase, like many other enzymes, is present in the urine, and its daily excretion has been reported to be increased in renal disease,[7] particularly in cases of carcinoma of the kidney and bladder.[8] It has been reported[9] that 15% of a series of 38 patients with malignant neoplasms of the urinary tract had normal excretions and conversely, that approximately 70% of 63 patients with benign conditions had in-

[1] R. J. Erickson and D. R. Morales, *New Engl. J. Med.* **265**, 478 (1961).
[2] O. Bodansky, *Adv. Cancer Res.* **6**, 1 (1961).
[3] N. O. Kaplan, *Bacteriol. Rev.* **27**, 155 (1963).
[4] C. L. Markert, *Harvey Lectures Ser.* **59**, 187 (1965).
[5] R. D. Goldman, N. O. Kaplan, and T. C. Hall, *Cancer Res.* **24**, 389 (1964).
[6] F. Wroblewski and K. F. Gregory, *Ann. N.Y. Acad. Sci.* **94**, 912 (1961).
[7] S. B. Rosalki and J. H. Wilkinson, *Lancet* **ii**, 327 (1959).
[8] W. E. C. Wacker and L. E. Dorfman, *J. Am. Med. Assoc.* **181**, 972 (1962).
[9] R. S. Riggins and W. S. Kiser, *J. Urol.* **90**, 594 (1962).

creased excretions of LDH. It has recently been found[10] that urines, free of leukocytes or erythrocytes, from patients with bladder carcinoma have isozyme patterns (high LDH-1 and low LDH-5) that are similar to those obtained from normal urine. Urines which contained either leukocytes or erythrocytes or both whether from patients with cystitis or from patients with bladder carcinoma showed increased urinary lactic dehydrogenase activity with a relatively high LDH-5.

The lactic dehydrogenase activity in pleural and peritoneal effusions containing or in contact with malignant cells and/or tissues has been reported to be higher than the serum lactic dehydrogenase of the same patients, whereas the lactic dehydrogenase activity in benign effusions was found to be lower.[11,12] This characteristic was proposed as a possible aid in the diagnosis of neoplastic disease but has not been confirmed by subsequent investigators.[13,14]

Assay Procedure

Principle. The assay most commonly used is based on the principle described by Kubowitz and Ott in 1943.[15] This principle involves the reduction of pyruvate to lactate and the concomitant oxidation of DPNH to DPN as observed at 340 mμ. For their determinations, Kubowitz and Ott used a final volume of 3 ml in which the final concentrations of reactants were: pyruvate, 1 m*M*; DPNH, 0.23 m*M*; phosphate buffer, 33 m*M* (pH 7.3). Subsequent investigators,[16-22] whether they utilized this principle for assay of crystalline preparations of tissue lactic dehydrogenase activity, or introduced it for assay of lactic dehydrogenase in other body fluids, have varied these concentrations. The temperature has ranged from presumably room temperature (24–25°) to 38°. In the method used in our laboratory the final concentrations of pyruvate and

[10] A. H. Gelderman, H. V. Gelboin, and A. C. Peacock, *J. Lab. Clin. Med.* **65**, 132 (1965).
[11] F. Wroblewski, *Am. J. Med. Sci.* **234**, 301 (1957).
[12] F. Wroblewski and R. Wroblewski, *Ann. Internal Med.* **48**, 813 (1958).
[13] J. E. Horrocks, J. King, A. P. B. Waind, and J. Ward, *J. Clin. Pathol.* **15**, 57 (1962).
[14] M. J. Brower, M. West, and H. J. Zimmerman, *Cancer* **16**, 533 (1963).
[15] F. Kubowitz and P. Ott, *Biochem. Z.* **314**, 94 (1943).
[16] A. Kornberg, Vol. I, p. 441.
[17] M. K. Schwartz and O. Bodansky, *Methods Med. Res.* **9**, 5 (1961).
[18] R. J. Henry, N. Chiamori, O. J. Golub, and S. Berkman, *Am. J. Clin. Pathol.* **34**, 381 (1960).
[19] L. P. White, *New Engl. J. Med.* **255**, 984 (1956).
[20] F. Wroblewski and J. S. LaDue, *Proc. Soc. Exptl. Biol. Med.* **90**, 210 (1955).
[21] B. R. Hill and C. Levi, *Cancer Res.* **14**, 513 (1954).
[22] A. Meister, *J. Natl. Cancer Inst.* **10**, 1263 (1949–1950).

DPNH are 0.33 mM and 0.1 mM, respectively, and the temperature is 37°.

Reagents

Phosphate buffer, 0.067 M pH 7.40. Dissolve 7.61 g of Na_2HPO_4 (anhydrous) and 1.81 g of $NaH_2PO_4 \cdot H_2O$ in about 800 ml of water. Adjust the pH electrometrically to 7.40 by addition of 1 N NaOH or 1 N HCl. Transfer quantitatively to a volumetric 1-liter flask. Bring final volume to 1000 ml with water.

Sodium pyruvate, 0.01 M. Dissolve 0.055 g sodium pyruvate in about 40 ml of water. Adjust to pH 7.40 electrometrically and bring final volume to 50 ml with water. Store in refrigerator.

NaCl, 0.85%. Dissolve 8.5 g NaCl in 1000 ml of distilled water.

DPNH, 3 mM. Dissolve 22.9 mg of $Na_2DPNH \cdot 3\ H_2O$ in 10 ml of water, or correspondingly larger amounts if needed. The solution should be made up fresh each day. It has been noted that thawing and freezing of stock solutions results in the development of an inhibitor.[23]

Enzyme. The serum is diluted 1:10 with 0.85% NaCl or, in the case of greater lactic dehydrogenase activity, 1:20 or 1:50 with 0.85% NaCl. After collection of blood and separation of serum, lactic dehydrogenase is stable for at least 8 hours at room temperature, for 4 days at refrigerator temperature and for about 20 days in the deepfreeze.[24]

Procedure. Into a small test tube pipette 2.3 ml of the phosphate buffer, 0.1 ml of the DPNH solution and 0.5 ml of the appropriately diluted serum. Mix well and place tube into water bath at 37° for about 20 minutes. At the end of that period, the enzyme reaction is initiated by adding 0.1 ml of the sodium pyruvate solution. Mix, dry the outside of the tube, and transfer the contents as rapidly as possible into a Beckman cuvette with a 10-mm light path. The cuvette is placed immediately into a Beckman DU spectrophotometer equipped with thermospacers maintained at 37°. The decrease in absorbance at 340 mμ is observed at 1 minute after the addition of the sodium pyruvate and thereafter at 1-minute intervals for 5 minutes or longer, if necessary, to obtain a reliable reaction velocity in the zero order portion of the reaction. Use a blank cuvette containing 0.85% NaCl to adjust the dark current and sensitivity. Plot the readings of absorbance against time. Draw a straight

[23] C. P. Fawcett, M. M. Ciotti, and N. O. Kaplan, *Biochim. Biophys. Acta* **54**, 210 (1961).

[24] O. Bodansky, *Methods Med. Res.* **9**, 1 (1961).

line through the initial zero order portion of the reaction, and determine the change in absorbance per minute. If the absorbance change per minute is 0.060 or greater, repeat the determination with a more dilute sample of tissue or serum.

Definition of Unit. Most of the clinical literature on the activity of serum lactic dehydrogenase, and indeed of other serum enzymes involving the conversion DPNH ⇌ DPN, is based on the absorbance change at 340 mμ instead of the officially designated unit[25]—that is, the amount of enzyme that transforms 1 micromole of substrate per minute. Since it might prove confusing at this time to interpret the values in the clinical literature by the latter definition, our unit is defined as 1000 times the change per minute in absorbance produced by 0.5 ml of a 1:10 dilution of serum in the 3 ml reaction mixture, or by 0.0167 ml serum per milliliter reaction mixture. The units used in our assay may be converted to International Units by multiplying by 0.16.[17] If dilutions greater than 1:10 are used, the values obtained must be adjusted accordingly.

Alternate Methods

As has been noted earlier, there are several modifications of the method described in this section. These modifications involve other concentrations of DPNH, pyruvate, or serum and reaction temperatures other than 37°. Three methods deserve more detailed consideration: an automated method based on the same principle as described above[26]; a colorimetric method based on the decrease of pyruvate, measured by hydrazone formation in the interaction of DPNH and pyruvate[27,28]; and a spectrophotometric method based on the interaction of lactate and DPN to form pyruvate and DPNH.[29]

In the automated method,[26] a continuous flow instrumental system of analysis, the AutoAnalyzer (Technicon Instruments Corporation) equipped with a phototube colorimeter is used. Through a system of plastic and glass tubing, a constant flow pump aspirates the reagents from various bottles. The enzyme sample is aspirated at the same time from plastic cups on a constant speed turntable set so that a sample from each particular cup is drawn for precisely 60 seconds. At the end of this interval the aspirating tube is automatically withdrawn from the sample, and air is aspirated during the succeeding 30 seconds, a period that permits the segregation of the samples and the cleaning of

[25] R. H. S. Thompson, *Science* **137**, 405 (1962).
[26] M. K. Schwartz, G. Kessler, and O. Bodansky, *J. Biol. Chem.* **236**, 1207 (1961).
[27] L. Berger and D. Broida, *Sigma Technical Bull.* **500**, St. Louis, Missouri, 1957.
[28] P. G. Cabaud and F. Wroblewski, *Am. J. Clin. Pathol.* **30**, 234 (1958).
[29] E. Amador, L. E. Dorfman, and W. E. C. Wacker, *Clin. Chem.* **9**, 391 (1963).

the system. The flow rates are arranged so that the concentrations in the reaction mixture correspond to those in the manual method described above. The complete reaction mixture passes through an incubation bath at 37° for a precisely measured interval of about 6 minutes. At the end of this period, the reaction mixture flows into a cuvette in the colorimeter, and the absorbance of the reduced DPNH appears as a deflection on a moving chart on a suitable recorder. During the 30-second period that the aspirating tube is out of the sample cup, the deflection returns to a baseline representing the absorbance of the unreacted DPNH and other reagents. Obviously, the reaction of DPN and lactate to be presently described may be automated in the same manner.

A second method[27,28] utilizes the well known interaction of phenylhydrazine with ketones which had been previously applied to the determination of pyruvic acid,[30] ascorbic acid,[31] and aldolase activity.[32] Hydrazone formation is measured 30 minutes after the interaction of pyruvate with DPNH has been initiated by the addition of the serum lactic dehydrogenase. The authors state that the amount of pyruvate remaining after the incubation is inversely proportional to the amount of lactic dehydrogenase present. In general, methods depending on the formation and measurement of hydrazones are laborious and nonspecific.[33] Unless a laboratory does not possess a spectrophotometer that is capable of measuring absorbance at 340 mμ, it would seem advisable to use a method based on the reduction of DPN or oxidation of DPNH.

A third type of method involves the interaction of lactate and DPN to form DPNH and pyruvate, the so-called "forward" reaction. Amador and his associates[29] have proposed the following procedure. Sodium pyrophosphate, 6.2 g, is dissolved in 250 ml of hot distilled water. The solution is cooled and lactic acid, 2.0 ml, is added; the pH is adjusted to 8.8 with 1 N NaOH. DPN, 1.10 g, is dissolved in this solution, and the volume is adjusted to 280 ml, after which 2.8-ml aliquots are stored at —20°. The concentrations of the reactants at this stage are stated to be 50 mM sodium pyrophosphate, 775 mM lactate, and 5.25 mM DPN. When the assay is to be done, the aliquot of 2.8 ml is thawed and brought to 25°. Serum, 0.2 ml, is added at zero time, mixed with the reagents, and transferred to a 3-ml cuvette. The increase in absorbance at 340 mμ is then followed at 25°. One unit of serum activity is defined as the increase in absorbance of 0.001 per minute per milliliter of serum.

[30] I. S. MacLean, *Biochem. J.* **7,** 611 (1913).
[31] J. H. Roe and C. A. Kuether, *J. Biol. Chem.* **147,** 399 (1943).
[32] J. A. Sibley and A. L. Lehninger, *J. Biol. Chem.* **177,** 859 (1949).
[33] W. J. P. Neish, *Methods Biochem. Anal.* **5,** 107 (1957).

TABLE I
NORMAL VALUES FOR SERUM LACTIC DEHYDROGENASE ACTIVITY

Method	Mean ± SD ($A \times 1000$ per min)	Concentration of serum upon which units are based (ml per ml reaction mixture)	Concentration of pyruvate (mM)	Concentration of DPNH (mM)	Actual concentration of serum (ml/ml reaction mixture)	Temperature (°C)	Equalization of units on basis of 0.33 ml serum per 1 ml reaction mixture per minute and 37°[a] (units)
Present[b]	19 ± 3.3	0.0167	0.33	0.10	0.0167	37	380
Present, children 5–10 years[c]	29 ± 6	—	—	—	—	—	—
W + L[d]	470 ± 130	0.33	0.75	0.13	0.037	24–27	—[g]
W[e]	90[f]	0.0032	0.33	0.14	0.0032	38	290
H *et al.*[a]							
Males	321 ± 61	0.33	0.60	0.19	0.033	32	445
Females	297 ± 64	0.33	0.60	0.19	0.033	32	420

[a] Temperature correction based on data of R. J. Henry, N. Chiamori, O. J. Golub, and S. Berkman, *Am. J. Clin. Pathol.* **34,** 381 (1960).

[b] M. K. Schwartz and O. Bodansky, *Methods Med. Res.* **9,** 5 (1961).

[c] O. Bodansky, S. Krugman, R. Ward, M. K. Schwartz, J. P. Giles, and A. M. Jacobs, *A.M.A. J. Diseases Children* **98,** 166 (1959).

[d] F. Wroblewski and J. S. LaDue, *Proc. Soc. Exptl. Biol. Med.* **90,** 210 (1955).

[e] L. P. White, *New Engl. J. Med.* **255,** 984 (1956).

[f] These units are based on a reaction time of 30 minutes. In the other methods, the units are expressed on the basis of a reaction time of 1 minute.

[g] The authors present a normal value of 470 units ± 130 units at a room temperature of 24–27°. However, it has been our experience that, without a thermospacer, the temperature in the cuvette compartment may be 5–8° higher. Henry *et al.*[a] have stated that it is about 5° higher than room temperature for a Beckman Model DU spectrophotometer. A literal rendition of a value of 470 units at a room temperature of 25° would yield a value of 1120 units at 37°, in obvious disagreement with all other values.

Interpretation of the Assay

Using the method described in this paper, we have obtained a mean value of 19 units ±3.3 for a series of normal adults. Since several clinical studies in the literature have utilized modifications of the method and of the units described in this article, it may be of value to tabulate these also (Table I). The various modifications give essentially the same values, about 400 units, when the activities are corrected to a concentration of 0.33 ml serum per milliliter reaction mixture and a temperature of 37°.

The frequency and extent of elevations in various acute and chronic diseases are illustrated by several representative studies, listed in Table II. For example, in 60 patients studied sequentially after myocardial infarction, 70–80% showed elevations above normal from day 1 to day

TABLE II
ELEVATION OF SERUM LACTIC DEHYDROGENASE IN VARIOUS DISEASES

Disease	Percentage of cases having values in					
	Normal range	1.1–2.0 × normal range	2.1–3.0 × normal range	3.1–4.0 × normal range	4.1–5.0 × normal range	5.1–8.0 × normal range
Myocardial infarction[a]						
Day 1	28	25	22	14	11	—
Day 2	23	30	16	23	7	—
Day 3	23	36	25	13	2	—
Day 4	32	35	26	2	4	—
Day 6	40	46	14	—	—	—
Day 10	91	9	—	—	—	—
Infectious hepatitis in children[b]	20	60	20	—	—	—
Extensive hepatic metastasis[c]	8	58	33	—	—	—
Myopathies[d]	9	23	14	18	18	18
Lymphocytic leukemia[e] untreated						
<15 years	13	47	20	20	—	—
>15 years	10	45	22	13	10	—
Granulocytic leukemia[e] untreated						
<15 years	0	40	20	40	—	—
>15 years	0	20	27	20	20	13

[a] A. Kontinnen and P. I. Halonen, *Am. J. Cardiol.* **10**, 525 (1962).
[b] O. Bodansky, S. Krugman, R. Ward, M. K. Schwartz, J. P. Giles, and A. M. Jacobs, *A.M.A. J. Diseases Children* **98**, 166 (1959).
[c] M. West and H. J. Zimmerman, *A.M.A. Arch. Intern. Med.* **102**, 103 (1958).
[d] G. Schapira and J. Dreyfus, *Compt. Rend. Soc. Biol.* **151**, 22 (1957).
[e] H. R. Bierman, B. R. Hill, L. Reinhardt and E. Emory, *Cancer Res.* **17**, 660 (1957).

4 after the attack; and this incidence decreased to 9% by day 10. The highest elevations, three- to fivefold the upper limit of normal, were seen most frequently on day 1 and day 2.[34] In a study of 22 cases with various myopathies, 90% had elevations of serum lactic dehydrogenase, and in 54% of these cases the elevations were more than three times the upper limit of normal.[35]

In a survey of the literature of inherited disorders, Harris[36] found at least twenty different conditions in which a specific enzyme deficiency had been demonstrated by *in vitro* studies. The existence of an enzyme in multiple molecular forms, as in the isozymes of lactic dehydrogenase, raises the possibility that the subunit may be genetically altered, and that other molecular forms of the enzyme may result. In the course of examining the electrophoretic patterns of red cell hemolyzates of 300 American Negroes, 50 white persons, 100 Papuans, and 200 Nigerians, a lactic dehydrogenase variant was found in a 25-year-old Nigerian male.[37] The variant was due to a modification of the subunit B, leading to 5 bands in LDH-1 position, 4 bands in LDH-2 position, 3 bands in LDH-3, and 2 bands in LDH-4. A variant leading to a doubling of the LDH-2 and the LDH-3 bands[38] and explained by a mutant subunit A has been reported in four members of a Brazilian family. Additional instances of polymorphism of human lactic dehydrogenase isozymes have been considered more recently.[39,40] There appear to be few studies concerning the activity or kinetic characteristics of the lactic dehydrogenase variants which have so far been described.

[34] A. Kontinnen and P. I. Halonen, *Am. J. Cardiol.* **10,** 525 (1962).
[35] G. Schapira and J. Dreyfus, *Compt. Rend. Soc. Biol.* **151,** 22 (1957).
[36] H. Harris, "The Inborn Errors Today. Supplement to Garrod's Inborn Errors of Metabolism." Oxford Univ. Press, London and New York, 1963.
[37] S. H. Boyer, D. C. Fainer, and E. J. Watson-Williams, *Science* **141,** 642 (1963).
[38] V. E. Nance, A. Claflin, and O. Smithies, *Science* **142,** 1075 (1963).
[39] A. P. Kraus and C. L. Neely, Jr., *Science* **145,** 595 (1964).
[40] E. S. Vessell, *Science* **148,** 1103 (1965).

[56] D-(—)-Lactate Cytochrome *c* Reductase

By THOMAS P. SINGER and TERENZIO CREMONA

D-(−)-Lactate + 2 cytochrome $c^{3+} \rightarrow$ 2 cytochrome c^{++} + 2 H^+ + pyruvate

Assay Method

Principle. The oxidation of D-(—)-lactate may be followed spectrophotometrically either with phenazine methosulfate (PMS) as mediator and 2,6-dichlorophenolindophenol (DCIP) as terminal electron acceptor at 600 mμ or with cytochrome *c* as the oxidant.[1] The advantages of the PMS–DCIP assay over the cytochrome *c* assay are threefold. First, the activity with PMS is 8 times higher than with the most active species of cytochrome *c* tested. Second, while the activity is greatly influenced by the quality of the cytochrome *c* and by its biological source,[2] the PMS–DCIP assay is free from this complication. Third, since the inhibition of yeast cytochrome oxidase by cyanide or azide is incomplete at concentrations which do not inactivate the reductase, in particulate preparations containing active cytochrome oxidase reliable activity measurements with cytochrome *c* as the acceptor are difficult to achieve, while the PMS–DCIP assay is satisfactory under these conditions.

The ratio of activities at V_{max} with respect to PMS and cytochrome *c*, respectively, is 8.0 at 30° in favor of the former. At the fixed PMS and cytochrome *c* concentrations recommended below the ratio is 2.91.

Phenazine Methosulfate Assay

Reagents

Imidazole buffer, 0.1 *M*, pH 7.5 at 30°
D-(—)-Lactate, 0.2 *M*, adjusted to pH 7.5
DCIP in imidazole buffer, pH 7.5, 2.1×10^{-3} *M*
Phenazine methosulfate in water, 1% (w/v)

Procedure. Although satisfactory results may be obtained with any recording spectrophotometer which is equipped for absorbance read-out, because of their stability and scale expansion features the instruments manufactured by the Gilford Instruments Company, Oberlin, Ohio, in conjunction with a recorder capable of instantaneous change of chart speed in the range of 1–12 inches per minute are most convenient.

When work is done at fixed PMS concentration, each cuvette receives 1.5 ml of imidazole buffer, 0.1 ml of DCIP, 0.2 ml of PMS, and water to a final volume of 3 ml, allowing for the volume of enzyme to be added. The cuvette is incubated for 3 minutes at 30° and transferred to the sample chamber of the spectrophotometer which is thermostated at 30°; the reaction is started by the addition of sufficient enzyme in a volume of 1–25 μl to allow an absorbance change of the order of 0.3 absorbance units in a 10-mm light path at 600 mμ per minute. The absorbance scale used is 1.0 or 0.5 unit, and a chart speed in the range of 3–12 inches per

[1] C. Gregolin and T. P. Singer, *Biochim. Biophys. Acta* **67**, 201 (1963).
[2] C. Gregolin and T. P. Singer, *Nature* **193**, 659 (1962).

minute is recommended. Activity is calculated from the relation ϵ_{mM} at 600 mμ = 19.1.

In order to prevent reoxidation of reduced DCIP by cytochrome oxidase, preparations up to the point of digestion with bacterial proteinase are assayed in the presence of 1 to 5×10^{-4} M neutral cyanide, which is added to the cuvette immediately before the enzyme.

Since the apparent K_m for PMS is relatively constant,[1] routine assays may be performed at the fixed PMS concentration mentioned and the factor 3.12 may be used to correct activity to V_{max} with respect to PMS. (Concentrations of DCIP higher than 7×10^{-5} M, as used in the procedure above, do not influence the activity.) For kinetic studies and accurate activity determinations each sample is assayed at a series of 5 or 6 PMS concentrations in the range of 0.03–0.2 ml of 1% PMS per 3 ml reaction mixture and the results are extrapolated to V_{max} with respect to PMS (Fig. 1).

Definition of Unit and Specific Activity. One unit of enzyme is that quantity which catalyzes the oxidation of 1 μmole of D-(—)-lactate per

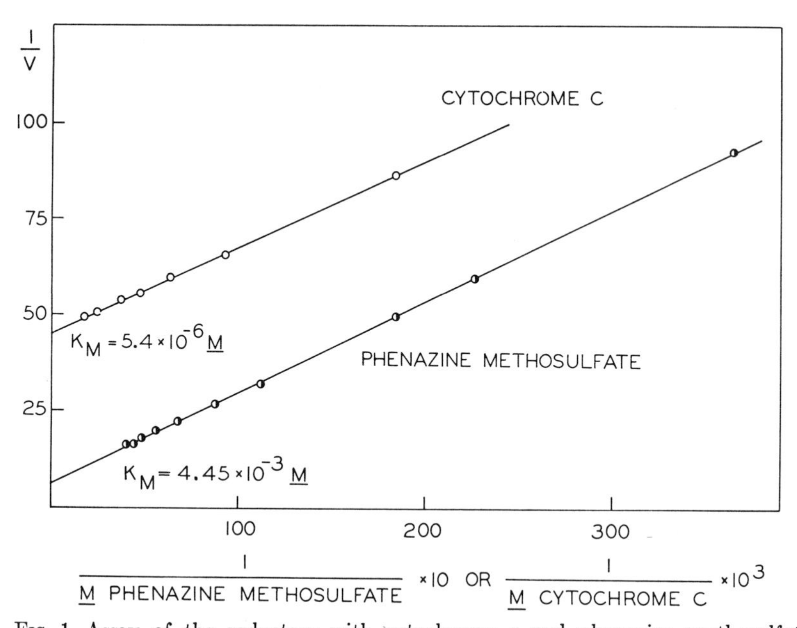

FIG. 1. Assay of the reductase with cytochrome *c* and phenazine methosulfate as electron acceptors. Abscissa, reciprocal concentration of acceptor; ordinate, reciprocal activity in arbitrary units. Test conditions as described in the text. Two micrograms of purified enzyme was employed, and the reaction period was 15 seconds. Reproduced from Gregolin and Singer[1] by permission of the publisher.

minute at V_{max} with respect to PMS in the assay described. Specific activity is stated as units per milligram protein. The latter is determined by the biuret method with the coefficients 0.095 per milligram protein in 3 ml volume at 540 mμ through the ammonium sulfate step and 0.111 for the succeeding steps.

Cytochrome c Assay

Reagents

Imidazole buffer, 0.1 *M*, pH 7.5 at 30°
D-(—)-Lactate, 0.2 *M*, adjusted to pH 7.5
Horse heart cytochrome *c*, free from modified cytochrome, 2% (w/v)

Procedure. The spectrophotometric equipment used is the same as in the PMS assay, except that in the determination of V_{max} with respect to cytochrome *c*, the availability of scale expansion and absorbance features are highly desirable, since at low cytochrome concentrations the rate falls off rapidly and scale expansion to 0.25 absorbance unit in conjunction with high speeds of recording (6–12 inches per minute) permits the measurement of linear initial rates.

Since the curve relating reciprocal activity to reciprocal cytochrome *c* concentration is quite flat (i.e., the K_m is low), for routine work the fixed cytochrome *c* concentration recommended below is satisfactory. The V_{max} with respect to cytochrome *c* and the K_m for cytochrome *c* must be determined for each batch of cytochrome, however, in order to ascertain the absence of polymeric cytochrome *c* or of other inhibitory material (cf. below).

The experimental procedure is exactly as described for PMS except that 0.15 μmole of cytochrome *c* (5×10^{-5} *M* final concentration) is substituted for PMS and DCIP and the reaction is followed at 550 mμ. In the determination of V_{max} the cytochrome concentration is varied in the range of 5×10^{-5} to 5×10^{-6} *M*. At concentrations below 2×10^{-5} *M* scale expansion to 0.5 or 0.25 absorbance unit, full recorder scale and high recording speeds are recommended. The inclusion of cyanide is essential at all stages prior to the DEAE chromatography step. Activity is calculated from the relation[3] ϵ_{mM} at 550 mμ = 19.1. A typical assay is reproduced in Fig. 1.

[3] The correct value[4] is 22.1, but the earlier figure (19.1) in the literature is retained here, since kinetic data and conversion factors from the cytochrome *c* to the phenazine methosulfate assay in the literature are based on the latter value.

[4] T. Flatmark, *Acta Chem. Scand.* **18,** 157 (1964).

Purification Procedure

Step 1. Preparation of Particles. The procedure given below is that described by Gregolin and Singer[1] (Procedure A) and modified by Lusty *et al.*[5] (Procedure B).

Procedure A

For large scale preparations 480 g of pressed Red Star bakers' yeast were suspended with 1% (w/v) NaCl (adjusted to pH 7.5) to 1.2 liter and mixed with 1 liter of precooled Ballotini beads (No. 12). The suspension was blended with an overhead blendor[6] operated at 18,000 rpm for 15 minutes in an 8-liter stainless steel beaker surrounded by an ice bath. For somewhat smaller scale preparations the Lourdes homogenizer with a Teflon-lined container may be substituted. The speed setting on the rheostat with this equipment is 62, the blending time 20 minutes, and the suspension consists of 120 g yeast diluted to 300 ml with 1% NaCl plus 250 ml Ballotini beads. With other blending equipment or different yeast the ratio of yeast: NaCl solution: Ballotini beads and the time of breakage for optimal yield of enzyme must be redetermined.

Following blending, the beads were allowed to settle and the supernatant suspension was decanted and diluted to 4 liters with 0.05 *M* phosphate, pH 7.5 (per 480 g yeast), using part of this buffer to wash the beads 2 or 3 times. The suspension was centrifuged for 4 minutes at 1300 *g* in an International Model SR-3 centrifuge; the turbid supernatant was collected, and the precipitate containing residual cells and cell walls was discarded. The supernatant solutions obtained from four 480 g batches were united and centrifuged in a refrigerated Sharples centrifuge at 50,000 rpm at a flow rate of 30 ml/min; the effluent appeared only slightly turbid and was discarded.

Step 2. Preparation of Acetone Powder. The thick paste obtained from 1920 g of yeast was homogenized and stirred for 5 minutes with 4 liters of acetone, precooled to —10°, in an 8-liter stainless steel beaker, using the overhead blendor operated at 8000 rpm. The suspension was then centrifuged for 5 minutes at 1400 *g* at —10°. The precipitate was homogenized and stirred with acetone as before, rapidly filtered by suction through a Büchner funnel, and the moist cake was washed with 500 ml of ether (—10°). Residual solvent was removed by spreading the cake on heavy paper in the cold room in front of a fan and then was dried in a high vacuum at room temperature. The resulting powder weighed 40–60 g per 1920 g of pressed yeast. The enzyme in this form is very

[5] C. J. Lusty, J. M. Machinist, and T. P. Singer, *J. Biol. Chem.* **240,** 1804 (1965).
[6] T. P. Singer, E. B. Kearney, and P. Bernath, *J. Biol. Chem.* **223,** 599 (1956).

stable, and the powder can be stored at least for several weeks without appreciable loss of activity.

Step 3. Extraction with Triton X-100. Treatment of the acetone powder suspension with a buffer of moderately high ionic strength extracts a certain amount of inert protein and thereby simplifies the subsequent purification of the reductase. Acetone powder (40 g) suspended by brief homogenization in 1 liter of 0.1 *M* sodium phosphate (pH 6.5) were stirred vigorously for 15 minutes at 0° and then centrifuged for 30 minutes at 18,000 rpm in the batch rotor of the Spinco Model L ultracentrifuge (43,400 *g*) or similar high capacity rotor. The yellowish supernatant, containing essentially no enzyme, was discarded and the precipitate was resuspended in 800 ml 0.0125 *M* sodium phosphate (pH 6.5). To this 200 ml of a 20% solution (v/v) of Triton X-100 in water was added and the suspension was slowly stirred for 15 minutes, particular care being taken to avoid excessive foaming. The suspension was then centrifuged 30 minutes at 20,000 rpm in a Spinco Model L ultracentrifuge (rotor No. 21), and the resulting clear, intensely brown-yellow supernatant (about 1000 ml) contained 90–100% of the enzyme present in the suspended powder.

Step 4. Purification with Calcium Phosphate Gel and Lyophilization. The protein dispersed by Triton X-100,[7] although not suitable for salt fractionation, can be partially purified by means of calcium phosphate gel. For best yield and purity it is advisable to determine for a given extract the amount of gel which gives 90–95% adsorption of the enzyme, the usual range being 0.4–0.6 mg gel per milligram protein.

In practice, the Triton extract was treated with 70 ml of a gel suspension containing 27 mg gel per milliliter (on a dry weight basis); after 10 minutes of slow stirring, it was centrifuged for 5 minutes at 1400 *g*. The supernatant was discarded and the gel was washed with 1 liter of 0.05 *M* phosphate buffer (pH 6.5). The wash was discarded and the enzyme was eluted from the gel with 400 ml of 0.25 *M* potassium phosphate buffer (pH 6.5) to which 4 ml of 20% Triton X-100 had been added. The clear, yellow-brown supernatant contained the enzyme in a yield of 75–85%. The salt concentration in the eluate was lowered by passing it through a column of Sephadex G-25 ($V_0 = 500$ ml), equilibrated with 5 m*M* phosphate (pH 6.5). The solution was then lyophilized. The resulting powder (350–450 mg protein) could be preserved for several weeks at —20° without appreciable loss of activity.

Step 5. Treatment with Phospholipase A and Bacterial Proteinase. Digestion of the preparation at this stage with phospholipase A and

[7] Prior to preparation of the solution traces of aldehyde are removed from the Triton X-100 by distillation in high vacuum below 40°

bacterial proteinase converts it to a soluble form, as evidenced by solubility in $(NH_4)_2SO_4$ solution and behavior on ion exchange columns. Solubilization is principally accomplished by the former enzyme, while the latter serves mainly to remove interfering proteins, such as cytochrome *c*.

Five batches of lyophilized powder were pooled and dissolved in cold distilled water to a concentration of 25 mg protein per milliliter. The dark brown, clear solution was dialyzed for 8–10 hours against 5 m*M* phosphate buffer (pH 6.5) to reduce the salt concentration. Following 5 minute temperature equilibration the preparation was incubated for 45 minutes at 30° in the presence of *Naja naja* venom (0.04 mg per milligram of protein) and crystalline bacterial proteinase (0.005 mg per milligram protein). Turbidity developed during incubation, but no inactivation of the enzyme occurred. The mixture was then rapidly cooled to 0° in a salt-ice mixture and centrifuged for 10 minutes at 144,000 *g*. The yellow-brown supernatant solution contained 85–90% of the enzyme.

Step 6. Chromatography on DEAE-Cellulose. This step yields fourfold purification and further serves to remove residual Triton and bacterial proteinase.

The soluble enzyme was adsrobed on a DEAE-cellulose column (2 × 15 cm) equilibrated with 5 m*M* potassium phosphate (pH 6.5). The column was washed with 0.05 *M* potassium phosphate (pH 6.5) until the A_{280} of the effluent was almost zero; the enzyme was then eluted with 0.2 *M* NaCl-0.1 *M* phosphate (pH 6.0). The yellow enzyme band was collected batchwise with a yield of 90–95%. From this point particular care was taken to protect the enzyme from light.

The enzyme was immediately concentrated by the addition of solid $(NH_4)_2SO_4$ to 0.75 saturation while the pH was maintained at 6 by the addition of 2 *N* NH_4OH. After 15 minutes stirring the precipitate was collected by 15 minutes centrifugation at 24,000 *g*.

Step 7. Salt Fractionation. The precipitate was dissolved in 0.1 *M* phosphate (pH 6.5) to a concentration of 15 mg protein per milliliter, and the resulting $(NH_4)_2SO_4$ concentration was calculated from the volume of the precipitate in the preceding step. Saturated $(NH_4)_2SO_4$ solution was added to 0.53 saturation, and the pH was maintained at 6.5. The resulting precipitate was centrifuged and discarded and the enzyme was precipitated from the supernatant solution by the further addition of saturated $(NH_4)_2SO_4$ solution to 0.68 saturation. The precipitate, dissolved in a minimal volume of 0.02 *M* phosphate (pH 6.5) containing 65% of the enzyme, was dialyzed for 6–8 hours against the same buffer in order to remove low-molecular-weight protein impurities resulting, presumably, from the digestion with bacterial proteinase which had not been removed in the preceding steps.

Step 8. Carboxymethylcellulose Chromatography. The dialyzed enzyme was equilibrated with 0.02 *M* sodium acetate (pH 5.2) on a column of Sephadex G-25; turbidity was removed by 5 minutes centrifugation, and the enzyme was then adsorbed on a CM-cellulose column (1×10 cm), equilibrated with the same acetate buffer. The column was washed with the equilibrating solution, and a first inactive fraction was removed. The fraction containing the enzyme was then eluted with a pH gradient obtained as follows: 75 ml of 0.01 *M* sodium succinate buffer (pH 5.2) in the mixer, 0.025 *M* sodium phosphate (pH 7) in the reservoir. Fractions of 2 ml were collected at a flow rate of about 0.5 ml per minute. The enzyme was distributed in 12 fractions, and the 6 central fractions containing about 60% of enzyme of highest specific activity were pooled, adjusted to pH 6.5 with 2 *N* NH_4OH, and concentrated by precipitation with solid $(NH_4)_2SO_4$ at 0.70 saturation.

At this stage the enzyme is 55–60% pure, as judged by sedimentation analysis. The remaining contaminant is a high-molecular-weight protein which sediments almost 3.5 times as fast as the reductase. While this contaminant may be removed by differential ultracentrifugation and thus an apparently close to homogeneous preparation is obtained, there is always an attendant inactivation, since from this point on the enzyme is relatively unstable and fairly readily loses its flavin component.

Step 9. Ultracentrifugation. The enzyme obtained as described above was dissolved to a concentration of about 10 mg/ml and equilibrated with 0.1 *M* NaCl–0.01 *M* potassium phosphate (pH 6.5) by passage through Sephadex G-25. It was ultracentrifuged in the SW 39L swinging-bucket rotor of the Spinco Model L ultracentrifuge (4 ml tubes, 173,000 *g*). After 6 hours' centrifugation, the content of the upper half of the tube was carefully removed. In this fraction the enzyme was present in nearly homogeneous form. Partial inactivation occurred during ultracentrifugation, since about 20% of the activity could not be accounted for; the activity observed in the final product was, therefore, corrected for the inactivation factor (total protein recovery)/(total activity recovered in ultracentrifugation).

The outline of a typical preparation is summarized in the table.

Procedure B

An alternative procedure[5] for the terminal two stages of the purification involves chromatography on Sephadex G-200. The product is of comparable specific activity but of somewhat lower purity, as judged by sedimentation analysis. Although in this method the enzyme is not exposed to low pH, the product is labile just as is that obtained by Procedure A. Hence the lability may be due to the removal of protective proteins in the terminal steps.

PURIFICATION PROCEDURE

Step	Total units	Specific activity
Procedure A		
Yeast particles (from 9000 g)	115,000	0.4
Triton extract of acetone powder	97,200	7
Eluate from Ca phosphate gel	84,000	32
After digestion with *Naja naja* phospholipase A and bacterial proteinase	69,600	75
Eluate from DEAE-cellulose	62,400	300
53–68% $(NH_4)_2SO_4$ fraction	40,000	440
Eluate from CM-cellulose	24,000	940
After preparative ultracentrifugation	8,000[a]	1670[a]
Procedure B		
Hydroxylapatite eluate	32,400	753
G-200 eluate, peak tubes	—	1450

[a] Corrected for inactivation during ultracentrifugation.

Step 8. Hydroxylapatite Chromatography. The salt-fractionated enzyme was equilibrated against 0.03 *M* phosphate, pH 6.5, on Sephadex G-25 and adsorbed on a column of hydroxylapatite (equilibrated with the same buffer). The adsorbent was washed with 0.2 *M* phosphate, pH 6.5, until the absorbance at 280 mμ decreased to 0.08 or less and the enzyme was then eluted in a very sharp band with 0.3 *M* phosphate, pH 6.5. The contents of the peak tubes were pooled and the enzyme was precipitated with 0.65 saturated $(NH_4)_2SO_4$ and equilibrated against 0.05 *M* phosphate, pH 6.5, on Sephadex G-25.

Step 9. A concentrated solution of the enzyme (3 ml) was placed on a 1.2 × 40 cm column of Sephadex G-200 equilibrated with 0.05 *M* phosphate, pH 6.5, and fractions of 2 ml were collected with the same buffer. The excluded volume was discarded; all colored fractions in the included volume were assayed, and those with the highest specific activity were pooled.

Molecular Properties

From the sedimentation constant ($S_{20,w} = 6.8$) and the flavin content, the molecular weight of about 100,000 has been suggested. There appear to be 2 moles of FAD and 4–6 g atoms of Zn^{++} per mole of enzyme.[1,8] Preparations containing 2 or 3 g atoms of Zn^{++} per mole of FAD with the same specific activity have been obtained. The FAD may be dissociated by acid-$(NH_4)_2SO_4$ without loss of Zn^{++} and the apoenzyme reactivated

[8] C. Gregolin and T. P. Singer, *Biochim. Biophys. Acta* **57**, 410 (1962).

by FAD but not FMN. The Zn^{++} is liberated in inorganic form on 10 minutes exposure to pH 2.5 at 0°, but the enzyme is denatured under these conditions and thus reversible resolution with respect to the metal has not yet been achieved.[9]

The enzyme is very stable in the cold up to the carboxymethylcellulose step but becomes increasingly labile thereafter. Inactivation is accompanied by the loss of flavin.[1] The pH of optimal stability at low ionic strength in the cold is 6.5. Certain metal chelating agents (oxalate, EDTA) inactivate the enzyme rapidly; others, such as *o*-phenanthroline, react relatively slowly.[9] Inactivation by *o*-phenanthroline is due to the formation of an inactive enzyme–Zn^{++}–*o*-phenanthrolinate complex, not to the dissociation of the metal. The inactive complex is fully reactivated by dialyzing away the chelator. Reactivation is also achieved by incubation at elevated temperatures: under these conditions the chelating agent is not dissociated, but an intramolecular transformation of the enzyme-chelate from an inactive to an active form occurs.[9,10] Inhibition by —SH reagents is not marked and, where it occurs, appears to be incomplete.[1]

Kinetic Properties

The only known substrates of the enzyme are D-lactate and D-α-hydroxybutyrate: the relative rates of oxidation of these substrates (at V_{max}) is 1:0.41 in favor of lactate, and the K_m values are 2.85×10^{-4} *M* and 1.4×10^{-3} *M* at 30°, respectively. Pyruvate does not inhibit the enzyme, nor can it serve as a substrate in the reverse reaction.[1]

Among electron acceptors tested only PMS and cytochrome *c* from various sources have been found to be active. Crystalline, monomeric cytochrome *c* preparations of various origins (beef, horse, and tuna heart, pigeon breast muscle, and yeast) react at different rates with the enzyme and yield very different K_m values.[2] Curiously, yeast cytochrome *c* is much less active and has a much lower affinity for the enzyme than mammalian cytochromes.

The pH of optimum activity at 30° is 7.5 in imidazole buffer in both the assays described but is at a more acid pH in phosphate buffer, particularly in the cytochrome *c* assay.[1] The reaction of the reduced enzyme with cytochrome *c* is competitively inhibited by salts at elevated ionic strengths; the inhibition is greater with polyvalent than with monovalent ions at equal ionic strength.[11]

[9] T. Cremona and T. P. Singer, *J. Biol. Chem.* **239,** 1466 (1962).
[10] T. Cremona and T. P. Singer, *Biochim. Biophys. Acta* **57,** 412 (1962).
[11] A. P. Nygaard, *J. Biol. Chem.* **236,** 2128 (1961).

On reduction of the enzyme with D-lactate the appearance of a pronounced absorption band at 530 mμ has been noted, which may suggest the occurrence of an unusually stable semiquinoid intermediate.[1]

Reactions in Intact Mitochrondria

In respiratory particles isolated with the Nossal shaker in 1% NaCl, D-lactic cytochrome *c* reductase is not linked to the respiratory chain, but in mitochondria disrupted in 0.25 *M* sucrose–0.02 *M* phosphate, pH 7.3,–0.001 *M* EDTA the enzyme is directly linked to the respiratory chain, its activity appears to be stimulated by Mg^{++}, and the reaction with cytochrome *c* is hindered and is measurable only under special conditions.[12–14] The reaction with PMS is, however, readily measurable also in such preparations. The assay conditions for such preparations have been recently defined by Gregolin *et al.*[14]

Alternative Preparation

An enzyme with similar properties has been isolated from yeast by Nygaard.[15] The specific activity reported by Nygaard is about 28% of the value attained by Gregolin and Singer[1]; the turnover number of the former preparation[15] is 15,000, whereas that of the latter[1] is 90,000 to 120,000. The preparation of Nygaard is said to be rather labile even at early stages of purification, possibly as a result of damage during the heating and acetone fractionation steps. Although in other catalytic properties the two preparations appear to be quite similar, the enzyme isolated by Nygaard is reported to react with ferricyanide and DCIP, whereas the preparation of Gregolin and Singer does not. The reactivity with ferricyanide is enhanced and that with cytochrome *c* is suppressed by polyvalent cations such as protamine.[16]

Determination of the Quality of Cytochrome *c* Preparations with the Aid of D-Lactic Cytochrome *c* Reductase

Although, contrary to statements in the literature, several enzymes utilizing cytochrome *c* as a reaction partner are sensitive to the presence of modified cytochrome *c* in cytochrome preparations,[1] the D-lactic cytochrome reductase test is, without doubt, the most sensitive enzymatic method available for the detection of modified cytochrome *c*. From an extensive survey of different lots of horse heart cytochrome *c* from all

[12] C. Gregolin and A. D'Alberton, *Biochem. Biophys. Res. Commun.* **14,** 103 (1964).
[13] B. R. Roy, *Nature* **201,** 80 (1964).
[14] C. Gregolin, P. Scalella, and A. D'Alberton, *Nature* **204,** 1302 (1964).
[15] A. P. Nygaard, *J. Biol. Chem.* **236,** 920 (1961).
[16] A. P. Nygaard, *Acta Chem. Scand.* **15,** 1627 (1961).

known commercial sources, Gregolin and Singer[2] concluded that (1) except for certain batches from one supplier all other preparations were markedly modified, showed low activity, irregular kinetic behavior, and often also a high K_m for the reductase; (2) the purity stated by the manufacturers bore no obvious relation to the quality of the preparations; in fact, even crystalline commercial samples may be grossly modified; (3) different lots from the same manufacturer may vary widely in regard to the presence of modified cytochrome. Some of these points are illustrated in Fig. 2, reproduced from these authors' paper.[2]

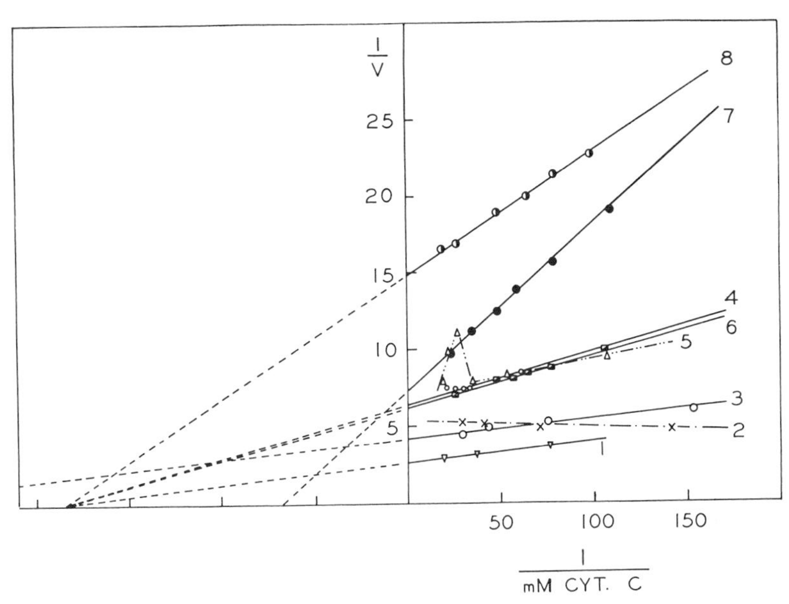

FIG. 2. Comparison of different commercial cytochrome *c* preparations from horse heart in the D-lactic cytochrome reductase test. *1*, Sigma type III, lot No. 120B706 (the standard preparation); *2* and *3* were the gifts of Mann Research Laboratories, Inc., identified as lot Nos. 1035 and 247254, respectively; *4* was a Keilin-Hartree preparation purchased from C. F. Boehringer and Soehne, lot No. 6111202; *5* was a crystalline preparation kindly supplied by the same firm; *6*, *7*, and *8* were type II (Keilin-Hartree procedure) preparations from horse heart from the Sigma Chemical Company, lot numbers 81B47-1, 120B47, 81B48, respectively. Reproduced from Gregolin and Singer[2] by permission of the publisher.

The presence of modified cytochrome *c* in a given preparation from horse heart may be readily ascertained by performing the assay at V_{max} with respect to cytochrome *c*, as described above. For practical purposes a solution of the lyophilized powder is the most convenient source of the enzyme, since it may be preserved for prolonged periods below 0° in the dry state. The quantity of enzyme is so chosen as to yield an absorbance change of approximately 0.3 absorbance units per minute. Since satisfactory commercial sources which may serve as a standard are not available at present, the V_{max} may be compared with the activity in the PMS assay and should be one-eighth of that value, while the K_m for cytochrome *c* should be no higher than 6×10^{-6} *M* for a satisfactory sample.

[57] L-Lactate (Cytochrome) Dehydrogenase (Crystalline, Yeast)

By R. H. Symons and L. A. Burgoyne

L-Lactate and other α-hydroxy acids → pyruvate and other α-oxoacids + $2H^+ + 2\epsilon$

(Known oxidants: oxygen, ferricytochrome *c*, ferricyanide, phenazine methosulfate, redox dyes, quinones)

This enzyme (cytochrome b_2) is the L-lactate:cytochrome *c* oxidoreductase of bakers' yeast (EC 1.1.2.3). In the preparation of cytochrome b_2 described here, this flavohemoprotein is obtained as type 1 crystals with associated DNA.[1-4] Removal of the DNA, which is probably an artifact of the crystallization procedure,[5,6] is described below as is an improved method for the preparation of oxidized enzyme free of substrate. The dehydrogenase catalyzes electron transport from L-lactate and a number of α-hydroxy acids to a wide range of oxidants. The function of this enzyme in the yeast cell is not known.

Assay Method

Principle. The activity of the enzyme can be followed by the reduction of cytochrome *c*, methylene blue, 2,6-dichlorophenolindophenol, or ferricyanide with L-lactate as substrate.[1] The ferricyanide assay is commonly used and is described here; reduction of ferricyanide is followed by measurement of the decrease of absorbancy at 420 mμ, at which wavelength ferrocyanide has negligible absorption.[1]

Reagents

Sodium DL-lactate, 5 *M*, pH 6.8; a stock solution used for preparing all lactate solutions. In order to hydrolyze the lactide present in

[1] C. A. Appleby and R. K. Morton, *Biochem. J.* **71**, 492 (1959).
[2] C. A. Appleby and R. K. Morton, *Biochem. J.* **73**, 539 (1959).
[3] C. A. Appleby and R. K. Morton, *Biochem. J.* **75**, 258 (1960).
[4] R. K. Morton and K. Shepley, *Biochem. J.* **89**, 257 (1963).
[5] R. H. Symons, *Biochim. Biophys. Acta* **103**, 298 (1965).
[6] L. A. Burgoyne and R. H. Symons, unpublished.

commercial lactic acid, the lactate solution is prepared by slowly adding 200 g of solid NaOH to a solution of 500 g DL-lactic acid (A.R.) in 200 ml of water, under these conditions the solution becomes very hot. After cooling, the alkaline solution is adjusted to pH 6.8 and diluted to 1 liter. Commercial DL-lactate often contains a large excess of L-lactate.[7]

Solution A: 0.3 M sodium DL-lactate, 0.05 M tetrasodium pyrophosphate, 0.1 mM EDTA, adjusted to pH 6.8 with HCl

Potassium ferricyanide, 0.01 M

n-Butanol

Assay Procedure.[1] This is carried out at 25°. 2.65 ml of Solution A and 0.25 ml of 0.01 M ferricyanide ($\epsilon_{mM} = 1.04$ at 420 mμ) are pipetted into a 1.0 cm cuvette and mixed. Enzyme (0.1 ml) is added, and the rate of decrease of absorbancy at 420 mμ determined.

Units of Activity. One unit is taken as that quantity of enzyme that will reduce 1 micromole of ferricyanide per minute at 25°.

Specific Activity. This is expressed as micromoles of ferricyanide reduced per minute per milligram enzyme under the above conditions. The concentration of reduced enzyme (in the presence of lactate) is determined from the absorbancy of the Soret peak at 423 mμ; a 0.1% solution has an absorbancy of 2.90 at this wavelength ($\epsilon_M = 2.32 \times 10^5$ at 423 mμ and molecular weight 80,000 per mole of protoheme).

Application of Assay to Whole Yeast. The above method can be used to estimate the activity of whole yeast (dried or undried) provided n-butanol is present to render the cell membrane permeable to the reagents. For example, to 1.0 g of yeast is added 1 ml of n-butanol, and the suspension is diluted to 10.0 ml with Solution A. After vigorous shaking to ensure thorough mixing, 0.1 ml is taken for assay as above. Assays are not as accurate as those carried out with purified or partly purified preparations of enzyme because of turbidity changes and of interference by endogenous substrates and enzymes, but usually a reasonable estimate can be made. The specific activity of these whole yeast samples is expressed as units per gram dry weight.

Purification Procedure

Principle. The procedure described is essentially that of Appleby and Morton[1] with a number of minor modifications.[4-6,8] The enzyme is extracted from dried yeast, which has been thoroughly powered to break the cell wall, by briefly stirring with a solution of butanol and lactate.

[7] H. A. Krebs, *Biochem. Prep.* 8, 75 (1961).

[8] R. G. Nicholls, M. R. Atkinson, L. A. Burgoyne, and R. H. Symons, *Biochim. Biophys. Acta* in press.

The aqueous extract is fractionated with acetone and the fraction precipitating between 25 and 35% acetone is dissolved and dialyzed against lactate solutions of low ionic strength at 4°. Crystals appear after 24–48 hours and are purified further by two recrystallisations.

Reagents

Solution A: As above

Solution B: The extracting medium; 0.1 *M* sodium DL-lactate, 1.0 m*M* $MgSO_4$, 0.1 m*M* EDTA, pH 6.8. For convenience, this is prepared at 10 times the final concentration and diluted as required.

Solution C: As for Solution B but also containing 25% (v/v) acetone.

Solution D: Dialysis solution; 0.025 *M* sodium DL-lactate, 0.02 *M* triethanolamine-HCl, 0.1 m*M* EDTA adjusted to pH 6.8 with HCl

Acetone: Redistilled commercial acetone is satisfactory.

n-Butanol: Untreated commercial butanol is satisfactory.

Commercial pressed bakers' yeast: This should have an activity, when assayed as above, of at least 40 units per gram dry weight.

Preparation of Yeast. Fresh commercial yeast is crumbled by hand, spread in a thin layer on plastic sheets, and dried in a draught of air in a draught cupboard. This usually takes 1–2 days depending on room temperature and humidity. When the yeast is thoroughly dry, it is stored in a sealed tin at —15° until required and is preferably used within 3 months.

A few days before an enzyme preparation, the dried yeast is ground in ball mills at —15° until as fine as talc powder in order to break the cell wall. This usually takes at least 48 hours; the efficiency of milling may be increased by further drying the yeast overnight in a freeze-drying apparatus prior to milling. The dry, powdered yeast is stored at —15° until used. Ball mills can vary considerably in their efficiency of grinding; satisfactory mills consist of earthenware jars 30 cm long × 19 cm in diameter of 4.5 liter capacity, half full of porcelain balls or cylinders 2–2.5 cm in diameter and each weighing approximately 25 g. The mills are rotated at 60–90 rpm. It is essential not to overfill the mills; for a mill of the above size, 900 g of dried yeast is satisfactory.

Fractionation Procedure

Step 1. Extraction of the Enzyme. One kilogram of powdered yeast and 3 liters of butanol are added to a stainless steel bucket and stirred by hand into a slurry. Ten liters of Solution B at room temperature are added, and the mixture is stirred vigorously for about 10 minutes with a top-drive stirrer until a smooth, lump-free suspension is obtained. This suspension is centrifuged in a large-capacity (4–6 liters) refrigerated centrifuge at 1600 *g* or higher for 20–30 minutes with the refrigeration set at —10° so that the temperature of the extract falls to approximately —5° by the end of the run. This centrifugation resolves the suspension into three layers, an upper lipid layer, a lower solid yeast layer and a middle aqueous layer (approximately 6 liters); the aqueous layer is drawn off into a Büchner flask with suction from a water pump.

Prolonged stirring of the suspension at room temperature can cause decreased yields of enzyme; it is therefore desirable to centrifuge the suspension as soon as possible. Also, maximum extraction of the enzyme occurs as soon as the butanol and extracting medium are mixed together.

Step 2. First Acetone Fractionation. The aqueous extract from step 1 is added to a stainless steel bucket and cooled to —5° in a —20° alcohol bath, at which point the solution begins to freeze. The extract is stirred by hand, and acetone at —15° is added slowly to give a 15% (v/v) solution (177 ml per liter of extract). The precipitate is removed by centrifugation at 1600 *g* for 20 minutes with the refrigeration set at —5°; the supernatant is carefully decanted and kept at —5°.

Step 3. Second Acetone Fractionation. Acetone at —15° is added with stirring to the supernatant from step 2 to increase the concentration to 35% (v/v) (360 ml/1.18 liters of step 2 supernatant), and the resultant precipitate is sedimented for 5 minutes at 1600 *g* with refrigeration set at —5°.

Step 4. Third Acetone Fractionation. The gummy 35% acetone precipitate from step 3 is thoroughly dispersed in 800 ml of Solution C, and the suspension is stirred for about 20 minutes at +4° to reextract the cytochrome b_2. The suspension is centrifuged for 20 minutes at no less than 1600 *g* with the refrigeration set at +2°. The precipitate is discarded, but, if there is a large bulk of red-brown precipitate, which indicates incomplete extraction, another extraction with a reduced volume of Solution C can be profitable.

Step 5. Fourth Acetone Fractionation. To the brown supernatant from step 4, acetone at —15° is added to increase the concentration from 25% to 35% (v/v) (154 ml per liter of step 4 supernatant). The precipitate is sedimented at 1600 *g* for 5 minutes with refrigeration at 2°.

Step 6. First Crystallization. The final 35% acetone precipitate from step 5 is liquefied with a minimum volume of Solution D (5–10 ml) and the viscous, red-brown liquid is poured gently into a dialysis bag; the end of the bag tied, leaving as little air as possible in the bag. The dialysis bag is added to a Büchner flask containing 2 liters of Solution D at 4° (previously degassed with a water pump at room temperature). The air is flushed out with nitrogen, and the flask is sealed and left at 4°. A few drops of chloroform can be added to inhibit microbial growth. After 40–48 hours, the red crystals are harvested by centrifugation for 15 minutes at 5000 *g* and washed two or three times, by resuspension and centrifuging, in Solution D. Under the light microscope, the crystals appear as flat, square plates the sides of which are approximately 0.05 mm long.

Step 7. The Second and Third Crystallizations. The washed crystals from step 6 are dissolved in 5 ml of Solution A and the solution is centrifuged at 10,000 *g* for 15 minutes to remove insoluble material. The solution is dialyzed again as in step 6, and the crystals are harvested after the solution has stood overnight at 4°. This crystallization process is then repeated to give the third crystals. These are then washed in the dialysis solution, dissolved in Solution A, and centrifuged to remove a small amount of insoluble material. The yield of third crystals is usually 25–50 mg (as estimated from the absorbancy at 423 mμ) starting from dried yeast which has a specific activity of approximately 65 units per gram dry weight. The purified enzyme has a specific activity of approximately 140 units per milligram. The enzyme solution is stored in a sealed tube at —15° with the gas space filled with nitrogen.

Comments on the Purification Procedure. Using 1 kg of powdered yeast and two refrigerated centrifuges, it is possible to carry out steps 1–6 in approximately 4 hours; 2–3 kg can be readily processed in a normal working day. It is desirable to carry out the second and third crystallizations as soon as possible, as the first crystals can be inactivated rapidly on storage at —15°. A summary of the results obtained for a typical purification are given in the table.

Preparation of DNA-Free Enzyme (Type II)[5,6]

When type I cytochrome b_2 in Solution A (prepared as above) is passed through a DEAE-cellulose column equilibrated with Solution A, the type II enzyme passes straight through whereas the DNA component is adsorbed. On dialysis of the effluent against Solution D, crystals of type II enzyme appear overnight. The red crystals, which tend to adhere to the wall of the dialysis tubing and are visible to the naked eye, are hexagonal bipyramids 0.1–0.5 mm in diameter. Dilute solutions of type II enzyme (less than about 2 mg/ml) will not crystallize; these can be concentrated by precipitating the enzyme with 35% acetone at —5°. After the DEAE-cellulose column has been washed with 0.4 *M* NaCl, pH 7.0, the DNA component is eluted with 1.0 *M* NaCl, pH 7.0.

PURIFICATION PROCEDURE[a,b]

Step	Volume (ml)	Total units	Specific activity[c] (units/mg protein)	Enzyme yield (%)
1. Initial suspension	13,500	70,000	0.2	100
Butanol/lactate extract	6,000	17,000	0.5	25
2. 15% Acetone supernatant	7,050	17,000	0.9	25
3. 35% Acetone precipitate	—	—	—	—
4. 25% Acetone/lactate extract	800	12,000	3.0	17
5. 35% Acetone precipitate	—	—	—	—
6. Dissolved pellet of step 5	60	10,600	3.5	15
Dissolved first crystals	5	6,300	135	9
7. Dissolved second crystals	5	5,400	140	8
Dissolved third crystals	5	5,100	140	7

[a] Starting material, 1 kg of dried and powdered yeast.

[b] Some of the values given are likely to vary considerably from one preparation to another.

[c] Units per milligram protein was determined by the method of Lowry *et al.* with bovine serum albumin as standard; for solutions of crystals, protein concentration was taken as equal to enzyme concentration determined from the absorbancy at 423 mμ [O. H. Lowry, N. J. Rosebrough, A. L. Farr, and R. J. Randall, *J. Biol. Chem.* **193,** 265 (1951)].

Preparation of Oxidized Cytochrome b_2[6,9]

As the enzyme is autoxidizable, it can be oxidized by transferring it from buffers containing lactate to air-saturated, lactate-free buffers. This is readily carried out on columns of Sephadex G-75 (bead form) with lactate-free buffers containing glutathione as a protecting agent.[4] The oxidized enzyme is relatively unstable and can lose about 20% of its activity overnight at —15°; however, at the temperatures of dry ice, it can be stored for several weeks without appreciable deterioration. Ferricyanide can also be used to oxidize the enzyme; a small volume of ferricyanide solution is added to the Sephadex column and allowed to move down a short distance. The enzyme solution is then added to the column, and the cytochrome b_2 is oxidized as it passes through the ferricyanide band.

Properties of the Purified Enzyme

Enzyme purified by the above procedure gives 5 bands of enzymatic activity on electophoresis in starch gels whereas only one band is obtained on electophoresis of the partially purified enzyme at the end of

[9] K. Hiromi and J. M. Sturtevant, *J. Biol. Chem.* **240,** 4662 (1965).

step 4 of the purification procedure.[8] The changes responsible for this effect occur during the first crystallization or on storage of partially purified enzyme. These changes are, so far as has been determined, a necessary prerequisite of crystallization[8] and are accompanied by an increase in K_m for L-lactate.[8,10]

Physical Properties.[2,3,11,12] Type I cytochrome b_2 prepared as described above contains, in addition to protein, equimolecular amounts of FMN and protoheme and 5–6% by weight of DNA. From analyses of flavin and heme, the minimum molecular weight is 80,000, but the enzyme appears to exist as a dimer in solution. The $S^0_{20,w}$ values are: type I enzyme, 9.2 S; type II enzyme, 8.6 S; DNA component, 5.8 S. The molecular weight of the DNA component is probably about 120,000.

Crystals of type I enzyme are probably very flat tetragonal bipyramids, but in the absence of the DNA component, the type II enzyme crystallizes in a different crystal system as hexagonal bipyramids. The addition of the DNA component to solutions of type II enzyme produces, on dialysis, type I crystals, provided that the specific activity of the enzyme is high (not less than 120 units/mg) and the secondary structure of the DNA has not been disrupted.[5,6] The DNA and enzyme do not appear to be associated in solution but come together only at, or just prior to, crystallization.[5] In type I crystals, there is one mole of DNA to approximately 24 moles (heme) of enzyme. The DNA component is a breakdown product of high molecular weight yeast DNA.[6]

Spectral Properties.[13] Reduced type I cytochrome b_2 has absorption maxima at 557, 528, 423, 330, and 265 mμ; the ratio $A_{265\ m\mu}$: $A_{423\ m\mu}$ = 0.80–0.85. Reduced type II cytochrome b_2 (DNA-free) has absorption maxima at 557, 528, 423, 330, and 269 mμ; the ratio $A_{269\ m\mu}$: $A_{423\ m\mu}$ = 0.50–0.60. Oxidized type I cytochrome b_2 has absorption maxima at 530–560 (broad band), 413, and 269 mμ. Oxidized type II cytochrome b_2 has absorption maxima at 530–560 (broad band), 413, and 275 mμ.

Substrate Specificity.[13] Cytochrome b_2 catalyzes the reduction of ferricyanide by a number of α-hydroxymoncarboxylic acids, but, of the compounds tested, L-lactate is clearly the most effective substrate. The K_m of crystallized enzyme for L-lactate is 1.6 m*M* with ferricyanide as acceptor.[2,14]

[10] F. Labeyrie and P. P. Slonimski, *Bull. Soc. Chim. Biol.* **12**, 1293 (1964).
[11] J. McD. Armstrong, J. H. Coates, and R. K. Morton, *Biochem. J.* **86**, 136 (1963).
[12] J. McD. Armstrong, J. H. Coates, and R. K. Morton, *Biochem. J.* **88**, 266 (1963).
[13] R. K. Morton, J. McD. Armstrong, and C. A. Appleby, "Haematin Enzymes" (J. E. Falk, R. Lemberg, and R. K. Morton, eds.), p. 501. Pergamon Press, Oxford, 1961.
[14] R. K. Morton and J. M. Sturtevant, *J. Biol. Chem.* **239**, 1614 (1964).

Inhibitors.[13,15] Cytochrome b_2 is readily inactivated, a process usually accompanied by release of FMN and the appearance of fluorescence. The inactivation caused by copper ions can be prevented by EDTA. *N*-Ethylmaleimide and arsenite are effective inhibitors. *p*-Hydroxymercuribenzoate (10^{-2} to 10^{-4} *M*), but not iodoacetate (10^{-2} *M*), strongly inhibits lactate dehydrogenase activity.

pH Optimum.[2] For L-lactate as substrate and ferricyanide as acceptor, cytochrome b_2 has a broad pH optimum between 7 and 8.5. For cytochrome *c* as acceptor, there is a broad pH optimum between 5.5 and 9.0.

[15] R. K. Morton and K. Shepley, *Biochem. Z.* **338,** 122 (1963).

[58] DL-Lactate Dehydrogenases (NAD^+-Independent) from *Lactobacillus arabinosus*

By ALAN M. SNOSWELL

Lactobacillus arabinosus (*plantarum*) contains D- and L-lactate dehydrogenases which are NAD^+-dependent,[1] but in addition it also contains D- and L-lactate dehydrogenases which are NAD^+-independent,[2] and these latter enzymes are the ones referred to in this chapter. Similar NAD^+-independent lactate dehydrogenases occur in other bacteria.[3-7]

Assay Method

Principle. The NAD^+-independent lactate dehydrogenases are assayed by following the rate of reduction of 2,6-dichlorophenolindophenol at 600 mμ with either D- or L-lactate as a substrate. The combined lactate dehydrogenase activity may be measured using lithium DL-lactate as substrate.

Procedure. The following reagents are added to a cuvette of 1 cm light path; 0.3 ml 0.2 *M* Tris-maleate buffer (pH 6.0 for the L-lactate dehydrogenase and pH 6.7 for the D-enzyme), 0.1 ml 0.001 *M* 2,6-di-

[1] D. Dennis and N. O. Kaplan, *J. Biol. Chem.* **235,** 810 (1960).
[2] A. M. Snoswell, *Biochim. Biophys. Acta* **77,** 7 (1963).
[3] J. De Ley and J. Schel, *Biochim. Biophys. Acta* **35,** 154 (1959).
[4] R. Molinari and F. J. S. Lara, *Biochem. J.* **75,** 57 (1960).
[5] P. M. Wilson and E. Holdsworth, *Australian J. Sci.* **24,** 134 (1961).
[6] E. Kaufmann and S. Dikstein, *Nature* **190,** 346 (1961).
[7] K. Yamada, H. Yamada, Y. Takesue, and S. Tanaka, *J. Biochem.* (*Tokyo*) **50,** 72 (1961).

chlorophenolindophenol, 0.5 ml 0.2 *M* sodium D- or L-lactate,[8] and distilled water to give a volume of 2.9 ml. The reaction is started by the addition of 0.1 ml enzyme solution and the absorbancy at 600 mμ is measured at 30-second intervals for 3 minutes in a spectrophotometer with the cell compartment maintained at 28°. The rate of decrease in absorbancy is thus calculated with reference to a blank containing the incubation mixture without lactate.

Definition of Unit. One unit of enzyme activity is defined as that amount catalyzing the oxidation of 1 micromole of lactate per minute and specific activity as units per milligram protein, determined by the biuret method.[9]

The amount of lactate oxidized is determined from the amount of indophenol reduced (1 mole lactate is used to reduce 1 mole of indophenol). The molar extinction coefficient of oxidized indophenol at 600 mμ is taken as 13.8 and 19.4 cm^{-1} $mole^{-1} \times 10^{-3}$ at pH 6.0 and 6.7, respectively.[10]

Application of the Assay Procedure to Crude Extracts. This assay procedure may be applied to crude bacterial extracts, as it has been found that 10^{-3} *M* oxalate, a competitive inhibitor of both the D- and L- NAD^+-independent lactate dehydrogenases, completely abolishes the activity in crude extracts while the NAD^+-dependent enzymes are not affected. Also, the addition of *Neurospora* DPNase has no effect on the estimation of the enzymes by the above procedure.[5]

Purification Procedure

Large Scale Production of L. arabinosus. The medium used for the large scale preparation of the organism contains 20 g enzymatically hydrolyzed casein (Nutritional Biochemical Corporation), 20 g glucose (autoclaved separately), 15 g sodium citrate, 5 g sodium acetate, 1 g KH_2PO_4, 1 g $MgSO_4 \cdot 7\ H_2O$, 50 mg $MnCl_2$ and 80 ml yeast extract[11] per liter. The medium is dispensed into 10-liter Florence flasks which are subsequently inoculated with 250-ml subcultures of *L. arabinosus* (ATCC 8014 or NCIB 6376) which have been grown for 7 hours at 30°. The large flasks are incubated for 18 hours at 30°. The cells are then harvested by centrifuging in a Sharples centrifuge and washed twice with distilled water. The yield is about 2.8 g dry weight of cells per liter of medium.

[8] Prepared from Calbiochem calcium salts by treatment with cation-exchange resin.

[9] A. G. Gornall, C. J. Bardawill, and M. M. David, *J. Biol. Chem.* **177,** 751 (1949).

[10] J. M. McD. Armstrong, *Biochim. Biophys. Acta* **86,** 194 (1964).

[11] Prepared by boiling compressed bakers' yeast with an equal volume of distilled water and filtering.

Preparation of Cell-Free Extract. The cells are suspended in 0.1 *M* sodium phosphate buffer (pH 7.0) to give a concentration of approximately 100 mg dry weight of cells per milliliter, and 100-ml fractions of this suspension are then placed in a 300 ml stainless-steel vessel of a Lourdes high-speed homogenizer together with 60 g No. 14 Ballotini glass beads. The whole vessel is immersed in crushed ice and the cells are disrupted by homogenizing for 3 periods of 5 minutes at 17,000 rpm. The glass beads are removed by centrifugation at 100 *g*. The cell debris and unbroken cells are subsequently removed by centrifuging at 15,000 *g* for 20 minutes. The supernatant is referred to as the cell-free extract and contains approximately 17 mg protein per milliliter. The subsequent fractionation steps are carried out at 4°.

Step 1. Ammonium Sulfate Precipitation. The cell-free extract is first diluted with 0.1 *M* sodium phosphate buffer (pH 7.0) until the protein concentration is reduced to 10 mg/ml. Solid ammonium sulfate is added to give 50% saturation, and the precipitate is removed by centrifugation at 10,000 *g* for 20 minutes. The supernatant is then brought to 70% saturation by the further addition of solid ammonium sulfate, and the precipitate is collected as above. This precipitate is dissolved in 0.005 *M* Tris-maleate buffer (pH 7.6), containing 0.01 *M* sodium lactate, and dialyzed for 16 hours against 48 liters (six changes of 8 liters) of 0.005 *M* Tris-maleate buffer (pH 7.6), containing 0.002 *M* sodium lactate in a rocking dialyzer.[12]

Step 2. Protamine Sulfate Precipitation. The protein concentration of the solution is then reduced to 5 mg/ml by dilution with 0.005 *M* Tris-maleate (pH 7.6) containing 0.002 *M* sodium lactate, and the pH is lowered to 5.9 by the cautious addition of 0.25 *M* acetic acid. Protamine sulfate[13] (2%, pH 4.0) is added to give a final concentration of approximately 0.18 mg/mg protein, or sufficient to raise A_{280}:A_{260} ratio from 0.5–0.6 to greater than 1.0, thus indicating the extensive removal of nucleic acids. The precipitate is removed by centrifugation at 1000 *g* for 10 minutes.

Step 3. Absorption on DEAE-Cellulose. The pH of the supernatant is then readjusted to pH 7.6 by the addition of 1 *M* Tris solution, and the protein concentration is reduced to 1.75 mg/ml by the addition of 0.005 *M* Tris-maleate buffer (pH 7.6), containing 0.002 *M* sodium lactate. A suspension of DEAE-cellulose is added to the dilute protein solution (2 mg DEAE-cellulose per 1 mg protein) and stirred with a magnetic

[12] It is essential that the dialysis be as complete as possible to ensure the success of the subsequent fractionation step.

[13] Only protamine sulfate (ex herring) supplied by I. Light & Co., Colnbrook, Bucks, England, was found to give successful results.

stirrer for 15 minutes the DEAF-cellulose is removed, by centrifugation, washed twice with 0.005 *M* Tris-maleate buffer (pH 7.6), containing 0.002 *M* sodium lactate, and the protein is eluted with 0.18 *M* sodium lactate. The eluate contains some 90% of the total lactate dehydrogenase activity, but only 15% of the original protein.

Step 4. Chromatography on First DEAE-Cellulose Column. The eluate is dialyzed for 4 hours against 0.005 *M* Tris-maleate buffer (pH 7.6), containing 0.002 *M* sodium lactate, and then freeze dried. The freeze-dried material is dissolved in a small volume of distilled water and dialyzed for a further 4 hours against the above buffer. This concentrated fraction is then applied to a DEAE-cellulose column (23 × 4 cm, containing 35 g DEAE-cellulose) which has been previously equilibrated overnight against the dilute Tris-maleate buffer described above. The protein is then eluted with sodium lactate (pH 7.6), the concentration of the sodium lactate being increased in a linear gradient from 0.002 to 0.3 *M* in a volume of 400 ml. The fractions containing total lactate dehydrogenase activity are pooled, freeze-dried, and dialyzed as before.

Step 5. Chromatography on Second DEAE-Cellulose Column. The concentrated, dialyzed fraction from the previous step is then applied to a second DEAE-cellulose column (21 × 2.0 cm, containing approximately 15 g DEAE-cellulose) and the process is repeated as for the first column. If total lactate dehydrogenase activity is determined using DL-lithium lactate as substrate, the enzyme activity is found partially separated into two peaks.[2] The first peak eluted (designated peak 1) contains predominantly D-lactate dehydrogenase activity whereas the second peak eluted (designated peak 2) contains predominantly L-lactate dehydrogenase activity. The fractions contained in each peak are pooled separately and freeze dried.

Step 6. Chromatography of L-Lactate Dehydrogenase on CM-Cellulose. The freeze-dried material of the second peak (predominantly L-lactate dehydrogenase) is dissolved in distilled water and dialyzed for 4 hours against 8 liters of 0.005 *M* Tris-maleate buffer (pH 6.0), containing 0.002 *M* sodium lactate. This fraction is then applied to a CM-cellulose column (24 × 1 cm, containing 5 g CM-cellulose) which has been equilibrated overnight against 0.005 *M* Tris-maleate buffer (pH 6.0), containing 0.002 *M* sodium lactate. The protein is subsequently eluted with Tris-maleate buffer by increasing the concentration and pH in a linear gradient to 0.05 *M* and pH 7.5 in a volume of 50 ml. The L-lactate dehydrogenase is eluted as a sharp peak in about 10 ml and is completely separated from the small amount of contaminating D-lactate dehydrogenase, which is eluted after about 40 ml of buffer has passed through the column.[2] The yield of L-lactate dehydrogenase is approximately 18%

PURIFICATION PROCEDURE FOR L- AND D-LACTATE DEHYDROGENASES

		L-Lactate dehydrogenase						D-Lactate dehydrogenase					
Fractionation step	Treatment	Total volume (ml)	Units/ml	Total units	Protein (mg/ml)	Specific activity (units/mg protein)	Recovery (%)	Total volume (ml)	Units/ml	Total units	Protein (mg/ml)	Specific activity (units/mg protein)	Recovery (%)
	Cell-free extract	555	0.530	294	10.2	0.052	—	555	0.540	300	10.2	0.053	—
1.	50–70% $(NH_4)_2SO_4$ fraction	205	0.933	191	18.6	0.050	65	205	0.938	192	18.6	0.050	64
2.	Protamine sulfate supernatant	1760	0.166	292	1.85	0.063	99	1760	0.016	292	1.85	0.089	97
3.	DEAE-cellulose eluate	22.4	10.8	242	29.4	0.377	90	22.4	11.7	263	29.4	0.40	88
4.	Eluate from 1st DEAE-column	20.7	7.60	151	6.4	1.19	54	20.7	8.10	168	6.4	1.27	56
5.	Peak 1 eluted from 2nd DEAE-column	—	—	—	—	—	—	3.2	17.2	54.9	2.8	6.50	18
	Peak 2 eluted from 2nd DEAE-column	3.1	19.9	61.8	4.2	4.74	21	—	—	—	—	—	—
6.	Peak 2 after chromatography on CM-cellulose	2.2	24.7	54.3	4.3	5.74	18	—	—	—	—	—	—
7.	Peak 1 after chromatography on CM-cellulose	—	—	—	—	—	—	1.7	12.0	21.3	0.4	30	7

and represents a 110-fold purification (see the table). The L-lactate dehydrogenase may be further purified by column chromatography on TEAE- and Ecteola-cellulose columns.[2]

Step 7. Chromatography of D-*Lactate Dehydrogenase on CM-Cellulose.* The freeze-dried material of the first peak from the second DEAE-cellulose column (which contains predominantly D-lactate dehydrogenase) is treated in a similar manner to the first peak and applied to a CM-cellulose column as described above. The small amount of contaminating L-lactate dehydrogenase is eluted after about 10 ml of buffer has passed through the column whereas the D-lactate dehydrogenase comes off as a sharp peak after about 30 ml. The yield of D-lactate dehydrogenase is approximately 7% with a 560-fold purification.

Properties

Stability. The D-lactate dehydrogenase is considerably less stable than the L-lactate dehydrogenase during the fractionation procedure (see the table). The final D-lactate dehydrogenase preparation retains less than 10% of its original activity on storage at —15° for 2 weeks whereas the L-lactate dehydrogenase retains 30% of its original activity after 2 months' storage under similar conditions.

Optimum pH. Optimum activity of the L-lactate dehydrogenase is observed at pH 6.0 compared with 6.7 for the D-lactate dehydrogenase.

Substrate Specificity. Neither the D- nor the L-lactate dehydrogenase oxidize any 2-hydroxy acids other than lactic acid to any significant extent. The D-lactate dehydrogenase of *Leuconostoc mesenteroides* shows a similar specificity.[6]

K_m and K_i Values. The K_m value of the L-lactate dehydrogenase is $1.6 \times 10^{-2} M$ and of the D-lactate dehydrogenase is $2.3 \times 10^{-3} M$. Oxalate acts as a powerful competitive inhibitor, K_i for the L-enzyme is $2.0 \times 10^{-4} M$ ($K_{m,\ \text{L-lactate}}:K_{i,\ \text{oxalate}} = 80$) compared with $1.1 \times 10^{-5} M$ ($K_{m,\ \text{D-lactate}}:K_{i,\ \text{oxalate}} = 210$) for the D-enzyme. Oxamate acts as a non-competitive inhibitor of the L-enzyme; 50% inhibition is observed at $2 \times 10^{-3} M$. D-Lactate is a weak competitive inhibitor of the L-enzyme, ($K_{m,\ \text{L-lactate}}:K_{i,\ \text{D-lactate}} = 0.27$) and L-lactate is a weak competitive inhibitor of the D-enzyme ($K_{m,\ \text{D-lactate}}:K_{i,\ \text{L-lactate}} = 0.06$).

Electron Acceptors. Neither enzyme reduces NAD^+ (or $NADP^+$) in the presence of lactate nor oxidizes NADH (or NADPH) in the presence of pyruvate. Only oxidation reduction dyes with an E'_0 of $+0.1$ volt or greater are utilized as electron acceptors, and 2,6-dichlorophenolindophenol ($E'_0 = +0.217$ volt at pH 7.0) is the most efficient.

Prosthetic Group. Highly purified L-lactate dehydrogenase contains

FMN in the approximate ratio of 1 molecule of FMN per molecule of enzyme. Addition of sodium L-lactate reduces the absorbancy approximately 20% in the 360–460 mμ region, and the difference spectrum (oxidized minus reduced) reveals absorption peaks 375 and 455 mμ.

Electrophoretic Mobility. Electrophoresis for 16 hours on No. 3 MM paper at 10 volts cm in 0.05 *M* Tris-maleate buffer (pH 7.9) results in the D-enzyme moving approximately 4 cm toward the anode, and the L-enzyme approximately 12 cm[2].

[59] D-α-Hydroxy Acid Dehydrogenase

By Terenzio Cremona and Thomas P. Singer

$$\text{D-}\alpha\text{-Hydroxy acid} + 2\ Fe(CN)_6^{-3} \rightleftarrows \alpha\text{-Keto acid} + 2\ Fe(CN)_6^{-4} + 2\ H^+$$

Assay Method

Principle. The oxidation of the substrate [usually D-(—)-lactate] is followed spectrophotometrically with ferricyanide as electron acceptor at 420 mμ. Other oxidants, such as 2,6-dichlorophenolindophenol or, anaerobically, methylene blue may be substituted for ferricyanide. When menadione is used as the direct oxidant cytochrome *c* is usually added as the terminal electron acceptor, and the reaction is then followed at 550 mμ; cytochrome *c* itself does not react with the enzyme directly. The reactivity with ferricyanide, 2,6-dichlorophenolindophenol, methylene blue, and menadione changes during the purification procedure.[1] This may be due either to the presence of interfering enzymes in crude preparations or to the modification of the reaction sites of these oxidants in the course of purification.[1] Hence, in routine assays the use of ferricyanide is recommended.

Reagents

Tris buffer, 0.1 *M*, pH 8.0 at 30°
D-(—)-Lactate, 0.2 *M*, adjusted to pH 8.0
$K_3Fe(CN)_6$, 0.1 *M*

Procedure. The assay mixture contains 0.15 ml of 0.2 *M* D-(—)-lactate, 0.030 ml of 0.1 *M* $K_3Fe(CN)_6$, and sufficient Tris buffer to give a final volume of 3 ml. The reactants are incubated in a spectrophotometer cuvette for 3 minutes at 30°, and the reaction is started by the addition of the enzyme in a small volume. The amount of enzyme recommended

[1] T. Cremona, *J. Biol. Chem.* **239**, 1457 (1964).

is such as to give a change of 0.3–0.4 absorbance units at 420 mμ per minute in a 10-mm light path. Activity is calculated from the absorbance charge at 420 mμ with the millimolar absorbance index of ferricyanide taken as 1.0. Extrapolation to V_{max} with respect to the oxidant is not necessary, since the measured activity varies relatively little with ferricyanide concentration and since the apparent K_m for ferricyanide does not vary during purification.

Although the initial rate remains linear for at least a minute so that manual spectrophotometers may be used for assay, for optimal results a recording spectrophotometer is recommended.

Definition of Unit and Specific Activity. One unit of enzyme is that quantity which catalyzes the oxidation of 1 micromole of D-(—)-lactate per minute under the conditions described. Protein is determined by the biuret method[2] with the coefficient of 0.095 per milligram protein in 3 ml volume at 540 mμ. Specific activity is stated as units per milligram protein.

Purification Procedure

The purification and properties of the enzyme summarized below refer to preparations obtained from a strain of yeast originally isolated from commercial yeast used in alcohol production (Distillerie Italiane, Ferrara, Italy). When grown anaerobically in the complex medium to be described, which approximates that used in commercial production, the properties of the enzyme agree with those of preparations isolated from slurries supplied by the Ferrara plant of Distillerie Italiane. The same strain grown on other media yielded preparations which were not suitable for isolation because of lability (cf. below).

Growth of Cells. Subcultures were grown aerobically at 28° for 24 hours in a medium containing per 100 ml 25 g malt solids,[3] 2 g of glucose, 1 g of peptone, 0.1 g of KH_2PO_4 and 2.5 g of agar. From these slants subcultures were inoculated and grown in air for 24 hours at 28° in a solution containing 5 ml of molasses,[4] 3 g of malt solids, 0.2 g of peptone, and 0.3 g of KH_2PO_4 per 100 ml. In the next subculture growth was at 28° in a medium containing 20 ml of molasses, 0.6 g of malt solids, and 2.5 g of KH_2PO_4 per 100 ml, adjusted to pH 4.5. After 48 hours' growth the contents of each flask (1.2 liters) was used to inoculate 12 liters of a medium containing 200 ml of molasses, 6 g of malt solids, and 25 g of

[2] A. G. Gornall, C. J. Bardawill, and M. M. David, *J. Biol. Chem.* **177,** 751 (1949).

[3] The malt extract (wort) containing 0.4 g of solids per milliliter was obtained from the Rickel Co., Detroit, Michigan.

[4] Beet sugar molasses are obtainable from the Michigan Sugar Company, Carrolton, Michigan.

KH_2PO_4 per liter. The medium was adjusted to pH 4.5 and was placed in 5-gallon bottles fitted with a rubber stopper through which were passed an inlet ending in an efficient ceramic sparger which reached the bottom of the flask, a gas outlet, and a stirring propeller. Growth was continued for 48 hours at 28°. During the first hour N_2 was passed through the sparger; at later stages of growth anaerobic conditions were maintained by the CO_2 evolution resulting from fermentation. Mechanical stirring was applied during the first hours of growth and then for 1-hour periods after 8, 24, and 32 hours. Cells were harvested and washed with glass-distilled water by centrifugation in the cold until the supernatant solution was almost colorless. From the packed cells excess water was removed by vacuum filtration on Büchner funnels as completely as possible (about 2 hours at room temperature). The yield of filter cake was 4.3 to 6.1 g per liter. The packed cells were then preserved at —15° until used.

Extraction. A 40 g batch of frozen, pressed yeast was thawed in 100 ml distilled water at 0° and mixed with 240 g of precooled Ballotini beads (No. 12) in the 500 ml bowl of the VirTis 45 blendor. The blendor was operated at a rheostat setting of 90 for 10 minutes. During this operation the bowl was immersed in chipped ice. The beads were allowed to settle; the supernatant suspension was decanted and the beads and residue were washed 3 to 4 times with 25–30 ml portions of cold water. The combined supernatant suspensions were centrifuged for 20 minutes at 105,000 g_{max} (Spinco No. 30 rotor, 30,000 rpm). The clear, dark brown extract (pH 6.1 to 6.2 at 0°) was united with that resulting from the extraction of another 40 g batch of yeast.

Heat Inactivation. The extract was incubated in a water bath regulated at 45° for 30 minutes without stirring, cooled to 0°; the resulting precipitate was removed by centrifugation (30 minutes at 105,000 *g*) and discarded. This treatment causes no loss of activity; although it gives no significant purification, it does facilitate subsequent operations.

Adsorption on Alumina Gel. The enzyme was treated with 20% of its volume of alumina gel[1] (20 mg dry weight per milliliter) by slow addition and efficient stirring. The pH was maintained at 6.8–7.0. After 15 minutes' stirring at 0°, the gel was removed by centrifugation for 10 minutes at 2000 g_{max}. The supernatant solution contained all the activity.

Fractionation with $(NH_4)_2SO_4$. The enzyme was next precipitated by the addition of 410 g of solid $(NH_4)_2SO_4$ per liter, while the pH was maintained at 6.8–7.0. After 15 minutes' stirring at 0°, the precipitate was collected by centrifugation for 20 minutes at 47,000 g_{max} (Spinco No. 30 rotor, 20,000 rpm). The precipitate was resuspended by homogenization in a sufficient volume of a 1.96% (w/v) solution of $(NH_4)_2SO_4$ to yield exactly one-tenth the volume prevailing at the end of the alumina gel step. This corresponds to a protein concentration of about 40–50 mg/ml and 0.45 saturation with respect to $(NH_4)_2SO_4$. Undissolved protein was removed by centrifugation and discarded. For each ml of the clear solution 0.099 g of solid $(NH_4)_2SO_4$ was added, while the pH was maintained at 6.8 to 7.0. The suspension was stirred and centrifuged as before. The sedimented enzyme was redissolved in a minimal volume of 0.05 *M* phosphate–1 m*M* D-lactate, pH 6.8, and preserved in ice.

At this stage of purification, in the presence of D-lactate and the $(NH_4)_2SO_4$ carried over from the last precipitation, is stable for a few weeks at 0°. Prior to assay salts and lactate are readily removed by passage through a column of Sephadex G-25, equilibrated with 0.05 *M* phosphate, pH 6.8, at 0°.

Gel Exclusion. Further purification was achieved by dilution of the enzyme with the phosphate-lactate solution to a protein concentration of 25–30 mg/ml and passage at 0° through a column of Sephadex G-100, which had been equilibrated with the same phosphate–lactate solution. For 600 mg of protein in 20 ml solution a column of 3.2×15 cm was used. The excluded protein (42% of that placed on the column) contained all the activity. This fraction was placed directly on a column of Sephadex G-200 (3.4×17.6 cm), equilibrated as above. The enzyme was included on this gel and was eluted with 0.05 *M* phosphate–1 m*M* D-lactate, pH 6.8.

Fractionation with $(NH_4)_2SO_4$. The eluate from Sephadex G-200 was treated with 0.277 g of solid $(NH_4)_2SO_4$ per milliliter at pH 6.8–7.0. Precipitated impurities were sedimented by centrifugation and discarded. The enzyme was precipitated by the further addition of 0.099 g of $(NH_4)_2SO_4$ per milliliter of solution. The precipitate was collected, redissolved, and preserved as at the end of the first $(NH_4)_2SO_4$ fractionation cycle. At this stage of purification the enzyme is quite labile: 76% inactivation occurred in 48 hours at 0°, and essentially all activity was lost in 3 days.

A summary of the purification procedure is presented in the table.

Molecular Properties

The presence of flavin prosthetic groups is clearly indicated by the difference spectrum of the enzyme recorded after reduction with D-lactate.[1,5] The enzyme has not been sufficiently purified to permit analytical determination of the type of flavin present, but preparations resolved by

[5] E. Boeri, T. Cremona, and T. P. Singer, *Biochem. Biophys. Res. Commun.* **2**, 293 (1960).

PURIFICATION PROCEDURE

Step	Volume (ml)	Total activity units (μmoles lactate/min)	Total protein (mg)	Specific activity (units/mg)	Recovery of activity (%)
Extract	300	1770	4440	0.40	100[a]
Heat inactivation	295	1725	3570	0.48	98
After alumina gel	303	1725	2450	0.70	98
After $(NH_4)_2SO_4$ fractionation	10	1600	605	2.64	90
Sephadex G-100, excluded fraction	30	1550	255	6.06	87
Eluate from Sephadex G-200	40	1300	65	20.0	74
$(NH_4)_2SO_4$ precipitate	4	1240	44	28.0	70

[a] Activity of extract arbitrarily taken as 100%.

acid and $(NH_4)_2SO_4$ are reactivated by FAD but not by FMN.[6] The dehydrogenase is readily inactivated by EDTA, *o*-phenanthroline, and oxalate.[5,7-9] Inactivation is due to the formation of an enzyme-metal-chelator compound, not to dissociation of the metal.[9,10] The loss of activity is readily reversed by removing the chelator either with added metals,[5,9-11] by dialysis,[9,10] or by incubation at elevated temperatures.[10] The nature of the metal component of the flavoprotein is not established, but since preparations inactivated by $(NH_4)_2SO_4$ treatment at pH 5 are reactivable with added $Zn,^{++}$ it has been suggested that Zn^{++} may be naturally occurring metal in the dehydrogenase.[6]

The inactivation by chelators is prevented by substrates.[1,12] This suggests that the metal moiety may be the substrate-binding site.

The molecular weight, estimated by the sucrose gradient method and by Sephadex filtration, has been reported to be 105,000.[13]

Kinetic Properties

The dehydrogenase has a wide range of specificity toward hydroxy-acids and oxidizes D-lactate, D-α-hydroxybutyrate, D-malate, and DL-

[6] M. Iwatsubo and A. Curdel, *Biochem. Biophys. Res. Commun.* **6,** 385 (1961).
[7] A. Curdel, L. Naslin, and F. Labeyrie, *Compt. Rend. Acad. Sci.* **249,** 1959 (1959).
[8] E. Stachiewicz, F. Labeyrie, A. Curdel, and P. P. Slonimski, *Biochim. Biophys. Acta* **50,** 45 (1961).
[9] A. Ghiretti-Magaldi, T. Cremona, T. P. Singer, and P. Bernath, *Biochem. Biophys. Res. Commun.* **5,** 334 (1961).
[10] T. Cremona and T. P. Singer, *Nature* **194,** 836 (1962).
[11] A. Curdel and F. Labeyrie, *Biochem. Biophys. Res. Commun.* **4,** 175 (1961).
[12] F. Labeyrie and E. Stachiewicz, *Biochim. Biophys. Acta* **52,** 136 (1961).
[13] M. Iwatsubo and A. Curdel, *Compt. Rend. Acad. Sci.* **256,** 5224 (1963).

glycerate.[1,5] The action of the enzyme is readily reversible: the reduction of pyruvate by $FMNH_2$ may be followed spectrophotometrically at 450 mμ under anaerobic conditions.[1,5] Cyanide is a competitive inhibitor ($K_i = 2.1 \times 10^{-6} M$).[1]

Comments

Despite the large number of papers from different laboratories dealing with D-α-hydroxyacid dehydrogenase from yeast, the purification procedure detailed above is the only one which has been published.[1] When the same yeast is grown in a different medium (e.g., 10% w/v glucose–1% (w/v) yeast extract–1% (w/v) peptone, with or without DL-lactate), the properties of the enzyme appear to be different and the material is unsuitable for isolation.[1] Prominent among the differences is a great lability even at early stages of purification, particularly in the presence of $(NH_4)_2SO_4$, but, in addition, the relative rates of oxidation of various substrates and their K_m values are different from those of the enzyme grown in the molasses-malt medium.[1]

From Other Sources

A soluble preparation of an enzyme with similar properties has been obtained from liver and kidney mitochondria by Tubbs, who has also studied the kinetic properties of the enzyme and the effect of chelating agents in considerable detail.[14,15]

[14] P. K. Tubbs and G. D. Greville, *Biochem. J.* **81**, 104 (1961).
[15] P. K. Tubbs, *Biochem. J.* **82**, 36 (1962).

[60] Propanediol Dehydrogenase[1]

By WILLIAM G. ROBINSON

$$\text{D-Lactaldehyde} + \text{TPNH} + \text{H}^+ \rightleftharpoons \text{1,2-propanediol} + \text{TPN}^+$$

Assay Method

Principle. Since the equilibrium of the reaction catalyzed by propanediol dehydrogenase strongly favors the reduction of lactaldehyde, the assay of the enzyme is based on the spectrophotometric measurement at 340 mμ of the rate of TPNH oxidation in the presence of enzyme and excess lactaldehyde. The initial rate of TPNH oxidation is proportional to enzyme concentration, and the course of the reaction is zero order for

[1] N. K. Gupta and W. G. Robinson, *J. Biol. Chem.* **235**, 1609 (1960).

several minutes. An alternative but less convenient assay procedure is based on the colorimetric determination[2] of the propanediol formed.

Reagents

Potassium phosphate buffer, 0.1 *M*, pH 7.4
TPNH, 0.006 *M*
DL-Lactaldehyde,[3] 1 *M*
Enzyme. The enzyme is diluted with 0.01 *M* potassium phosphate buffer, pH 7.4, so that 0.1 ml when added to the assay system gives an absorbancy change of 0.020–0.040 per minute.

Procedure. The reagents are added in the following order to cuvettes having a 1-cm light path: 0.50 ml of potassium phosphate buffer, 2.25 ml of water (2.30 ml in the blank cell), 0.05 ml TPNH (omitted from the blank cell), 0.10 ml of enzyme, and 0.10 ml of lactaldehyde. The reaction is carried out at room temperature, 23–25°. Absorbancy readings at 340 mμ are taken at 30-second intervals.

Definition of Unit and Specific Activity. One unit of dehydrogenase is the amount of enzyme catalyzing the transformation of 1 micromole of substrate per minute in the standard assay. Specific activity is expressed as units per milligram of protein.

Application of Assay Method to Crude Tissue Preparation. No interfering reactions have been observed with crude pig kidney extracts. Pig kidney is a particularly suitable starting material for the preparation of the enzyme because it does not contain alcohol dehydrogenase which catalyzes the DPNH-dependent reduction of lactaldehyde to propanediol.[4,5] In contrast, enzyme preparations obtained from pig liver catalyze so many interfering reactions that it is not yet possible to state whether liver contains a specific propanediol dehydrogenase.

Purification Procedure

The purification procedure for the dehydrogenase is summarized in the table. All operations are carried out at 0–4° unless otherwise stated.

Step 1. Extraction. Frozen pig kidney (200 g) is minced and homogenized with 200 ml of 0.5 *M* potassium chloride in a Waring blendor for

[2] L. R. Jones and J. A. Riddick, *Anal. Chem.* **29**, 1214 (1957).

[3] Crystalline DL-lactaldehyde is prepared by reducing the dibutyl acetal of methylglyoxal with lithium aluminum hydride and subsequent hydrolysis of the hydroxy acetal [L. Hough and J. K. N. Jones, *J. Chem. Soc.* p. 4052 (1952)]. D-(or L-) Lactaldehyde is prepared from L- (or D-) threonine by the ninhydrin reaction [E. Huff and H. Rudney, *J. Biol. Chem.* **234**, 1641 (1959)].

[4] E. Huff, *Biochim. Biophys. Acta* **48**, 506 (1961).

[5] S.-M. Ting, O. Z. Sellinger, and O. N. Miller, *Biochim. Biophys. Acta* **89**, 217 (1964).

5 minutes. The homogenate is diluted with 200 ml of 0.5 M potassium chloride and 10 ml of 1 M potassium phosphate buffer, pH 7.4, and stirred mechanically for 30 minutes. The preparation is chilled in a bath at —5°, and 400 ml of alcohol solution (prepared by mixing equal volumes of absolute ethanol and 0.5 M potassium chloride), previously chilled to —20°, is added slowly with stirring. After the resulting mixture has been stirred for 30 minutes, it is centifuged at 23,000 g for 30 minutes, and the supernatant solution is dialyzed for 20 hours against 8 liters of 0.01 M phosphate buffer, pH 7.4.

Step 2. Ammonium Sulfate Fractionation. Solid ammonium sulfate is added slowly to the dialyzed alcohol extract to give a final salt concentration of 0.21 g/ml (0.40 saturation).[6] The mixture is stirred for 30 minutes; the precipitate obtained after centrifugation at 23,000 g is discarded. The supernatant solution is brought to a salt concentration of 0.32 g/ml (0.60 saturation), and the mixture is stirred for 30 minutes and centrifuged. The resulting precipitate is dissolved in a small volume of 0.01 M phosphate buffer and dialyzed overnight against 4 liters of 0.001 M phosphate buffer, pH 7.4.

Step 3. First Purification on DEAE-Cellulose. A sample (20 g) of DEAE-cellulose (capacity, 1.0 meq/g) is washed with 0.1 M phosphate buffer, pH 7.4, until the supernatant liquid resulting upon centrifugation is at pH 7.4. The cellulose is resuspended in 1 liter of 0.01 M phosphate buffer, pH 7.4, and packed in a column, 2.5 × 23 cm. The dialyzed ammonium sulfate fraction (protein concentration 45–60 mg/ml) is passed into the column, and the dehydrogenase is eluted with 0.01 M phosphate buffer, pH 7.4. The flow rate is maintained at 0.5 ml per minute by gravity or slight pressure, and 5-ml portions of eluate are collected. Protein elution is followed spectrophotometrically by determining the absorbancy at 280 mμ. Dehydrogenase activity is present in approximately 20 fractions collected immediately after the initial appearance of protein in the eluate. The fractions of highest specific activity are pooled. The enzyme solution at this stage of purification is usually light pink in color, although cloudy solutions of lower specific activity are

[6] The amount of ammonium sulfate is calculated by using the relationship: $X = 0.526\,Y\ [Z + (X/2)]$, where X = grams of ammonium sulfate, Y = fraction of saturation, and Z = volume of solution before the addition of any ammonium sulfate. The concentration of a saturated solution of ammonium sulfate is 0.526 g/ml. For every gram of ammonium sulfate that dissolves, the volume of the solution increases by approximately 0.5 ml. The loss in volume resulting from the removal of a protein precipitate is ignored in calculating the amount of ammonium sulfate required to raise the salt concentration.

obtained with preparations of DEAE-cellulose having a capacity less than 1.0 meq/ml.

Step 4. Treatment with Alumina Gel Cγ and Concentration with Ammonium Sulfate. A sample (40 ml) of alumina gel Cγ[7] (10.5 mg/ml) is centrifuged lightly, and the supernatant fluid is discarded. The enzyme solution from the above step is added to the packed gel, and the mixture is stirred mechanically for 30 minutes and centrifuged. The gel is discarded and the supernatant solution is brought to an ammonium sulfate concentration of 0.42 g/ml (0.80 saturation) by the slow addition of the solid salt. The mixture is stirred for 30 minutes and centrifuged at 15,000 *g* for 20 minutes, and the supernatant fluid is discarded. The precipitate is dissolved in a small quantity of 0.005 *M* Tris buffer, pH 7.4, and dialyzed overnight against 4 liters of the same buffer.

Step 5. Second Purification on DEAE-Cellulose. Approximately 1 g of DEAE-cellulose is washed as in step 3, suspended in 50 ml of 0.005 *M* Tris buffer, pH 7.4, and packed in a column, 1.0 × 15 cm. The dialyzed enzyme solution from the previous step is passed into the column at the rate of 0.5 ml per minute. After the column has been washed with approximately 30 ml of 0.005 *M* Tris buffer, pH 7.4, the enzyme is eluted with 0.015 *M* Tris buffer, pH 7.4, and 5-ml fractions collected. The dehydrogenase is present in the third or fourth tube after elution is started. A high percentage of the enzyme applied to the column is recovered in three fractions, but the fraction with the highest specific activity is used routinely. When stored in the frozen state, preparations of the purified enzyme undergo 10–15% loss in activity overnight but retain about 60% of the activity after several weeks. The dehydrogenase is even more unstable at high dilution.

Purification of Propanediol Dehydrogenase from Pig Kidney

Preparation	Volume (ml)	Protein (mg)	Units	Specific activity (× 10²)	Yield (%)
Extract	1000	12,300	240	2.0	100
$(NH_4)_2SO_4$ precipitate (0.4–0.6 saturation)	100	4,550	202	4.3	84
DEAE-cellulose column eluate	98	1,078	184	16.7	77
$(NH_4)_2SO_4$ precipitate (0–0.8 saturation) of alumina gel Cγ supernatant fraction	13.2	332	111	33.6	46
DEAE-cellulose column eluate	5.0	23.5	51.4	218	21

[7] See Vol. I pp. 97–98.

Properties

Specificity. The purified dehydrogenase catalyzes the reduction of the D-isomer[8] of lactaldehyde to 1,2-propanediol in the presence of TPNH. DPNH is not oxidized. TPNH oxidation does not occur in the presence of acetaldehyde, propionaldehyde, glycoaldehyde, acetol, glyoxylic acid, pyruvic acid, D-galactose, D-mannose, L-fucose, D-fructose, D-galacturonic acid, D-lyxose, D- or L-arabinose, L-rhamnose, D- or L-ribulose, or D-xylulose. Glucose 6-phosphate dehydrogenase and TPN-specific isocitric dehydrogenase are absent from the purified preparation. When the enzyme is incubated with methylglyoxal, glucuronic acid, glyceraldehyde, ribose, or xylose, TPNH is reduced, presumably by other enzymes present as impurities.

Inhibitors. The dehydrogenase is completely inhibited by 3.3×10^{-4} *M* *p*-chloromercuribenzoate and 50% inhibited by 3.3×10^{-3} *M* *N*-ethylmaleimide. 1,2-Dimercaptopropanol at low concentrations strongly inhibits the reduction of lactaldehyde whereas thioethanol, glutathione, and cysteine are inhibitory at higher concentrations. High concentrations of 1,10-phenanthroline, 8-hydroxyquinoline, and EDTA do not appreciably alter the reaction rate.

Other Properties. The K_m value for DL-lactaldehyde is 3.8×10^{-3} *M* with TPNH at 10^{-4} *M*. The affinity of the enzyme for TPNH is so great that the K_m cannot be determined by means of the optical assay. Lactaldehyde is reduced in the pH range 6–9 with maximal activity at about pH 7.4. The equilibrium constant for the dehydrogenase reaction is 2.4×10^{-13}.

[8] A D-lactaldehyde preparation used in earlier work (see footnote 1) was incorrectly identified as L-lactaldehyde. The L-isomer is not reduced.

[61] 1,2-Propanediol Phosphate (PDP) Dehydrogenase

By O. NEAL MILLER

$$\text{Propanediol phosphate} + NAD^+ \rightleftarrows \text{acetol phosphate} + NADH + H^+$$

Introduction

PDP dehydrogenase has not been isolated from tissues free of other dehydrogenase activities. It has been isolated as a part of "myogen A" complex of enzymes which is known to contain aldolase, α-glycero-P dehydrogenase, triose isomerase, and PDP dehydrogenase.[1] Also, it has

[1] O. N. Miller, C. G. Huggins, and K. Arai, *J. Biol. Chem.* **202,** 263 (1953).

been found in a preparation of glycero-P dehydrogenase as originally prepared by Beisenherz *et al.*[2] and previously reported.[3] The oxidation of PDP by a NAD-linked dehydrogenase of *Micrococcus ureae* grown on 2,3-butanediol has also been reported.[4] Evidence for the existence of a distinct dehydrogenase for PDP is adduced from the differential effect of dialysis, the resulting change in the ratios of the specific activities of glycero-P dehydrogenase and PDP dehydrogenase with dihydroxyacetone-P and acetol-P as respective substrates, the differential behavior of the 2 enzymatic activities upon storage and also the differences in susceptibility to inhibition by a number of anions.

Assay

This enzyme activity is preferably assayed by adding reduced coenzyme and acetol-P to the reaction mixture and measuring the oxidation of NADH in a spectrophotometer at 340 mμ.

Materials. One can use either enzymatically synthesized acetol-P prepared from an incubation mixture containing acetol kinase from rabbit kidney, acetol, and ATP (see the article on Acetol kinase, this volume [86] for details) or one can use acetol-P chemically synthesized by the method previously reported[5] and shown in detail in this volume [86]. The use of chemically synthesized DL-PDP is not advised owing to the slowness of the reaction going to the right; furthermore, the unnatural isomer of PDP may be inhibitory to the dehydrogenase reaction.

It is recommended that a reference source of PDP dehydrogenase be prepared in order to assure that inactive reaction mixtures are complete. Either fresh myogen A[6] or a fresh preparation of α-glycero-P dehydrogenase[2] may be used.

Procedure. The reaction mixture for assay of PDP-dehydrogenase should contain, 3 micromoles of acetol-P, 0.3 micromoles of NADH, 120 micromoles of Tris buffer, pH 7.7, and enzyme. The final volume is 3 ml. The reaction is followed by reading the absorbance at 340 mμ in a suitable spectrophotometer, and the reaction is initiated by the addition of the enzyme. The control contains all components, minus NADH. The reaction is allowed to run for 120 seconds.

$$\frac{\text{Decrease in absorbance} \times 0.48}{\text{Tissue equivalent (mg)} \times 2\ \text{(time)}} \times 1000 + \text{micromoles/g/minute}$$

[2] G. Beisenherz, H. J. Boltze, T. Bücher, R. Czok, K. H. Garbade, E. Meyer-Arendt, and G. Z. Pfleiderer, *Z. Naturforsch.* **8b**, 555 (1953).

[3] O. Z. Sellinger and O. N. Miller, *J. Biol. Chem.* **234**, 1641 (1959).

[4] E. Juni and G. A. Heym, *J. Bacteriol.* **74**, 757 (1957).

[5] O. Z. Sellinger and O. N. Miller, *Biochim. Biophys. Acta* **29**, 74 (1958).

[6] Y. T. Baranowski, *Z. Physiol. Chem.* **260**, 43 (1939).

Properties

Stimulation by Dialysis. PDP dehydrogenase not only is stable when dialyzed for 12–24 hours at 4° against 300–400 volumes of deionized water, but also there is an appreciable net increase in dehydrogenase observed. It was found that dialysis causes almost complete disappearance of NADH oxidase activities which make measurement of PDP dehydrogenase technically easier.

Inhibition by Anions. PDP dehydrogenase is inhibited by 70% with sulfate at an ionic strength of 0.57; 41% by sulfite and 28% by formate at the same ionic strength. Bromide and phosphate inhibit 45 and 32%, respectively, when added at an ionic strength of 0.19.

Product Inhibition. Acetol-P, the product of PDP dehydrogenase activity, was found to competitively inhibit α-glycero-P dehydrogenase activity. An inhibition constant (K_i) of 1.3×10^{-3} *M* has been obtained.

Kinetic Constants. The K_m of PDP dehydrogenase for NADH is 0.5×10^{-4} *M*, for acetol-P, 1.0×10^{-4} *M* and for DL-PDP, 4.4×10^{-3} *M*.

[62] Glycolate Oxidase (Ferredoxin-Containing Form)

By A. L. Baker and N. E. Tolbert

$$\begin{array}{l} COOH \\ | \\ CH_2OH \end{array} \xrightarrow{+O_2} \begin{array}{l} COOH \\ | \\ CHO \end{array} + H_2O_2$$

$$\begin{array}{l} COOH \\ | \\ CHO \end{array} \xrightarrow{+O_2} \begin{array}{l} COOH \\ | \\ COOH \end{array} + H_2O_2$$

Glycolate oxidase or glycolate:O_2 oxidoreductase (EC 1.1.3.1 was originally isolated by ammonium sulfate precipitation,[1] further purified by ethanol fractionation,[2] and crystallized from ammonium sulfate.[3] FMN (10^{-4} *M*) is a required cofactor.[2–4] The enzyme catalyzes the oxidation of both glycolate and glyoxylate.[5] However, if the H_2O_2 is not rapidly removed by a large excess of catalase, the nonenzymatic peroxide

[1] C. O. Clagett, N. E. Tolbert, and R. H. Burris, *J. Biol. Chem.* **178,** 977 (1949). See also N. E. Tolbert, C. O. Clagett, and R. H. Burris, *J. Biol. Chem.* **181,** 905 (1949).

[2] I. Zelitch and S. Ochoa, *J. Biol. Chem.* **201,** 707 (1953). See also I. Zelitch, Vol. I, p. 528.

[3] N. A. Figgerio and N. A. Harbury, *J. Biol. Chem.* **231,** 135 (1958).

[4] R. H. Kenton and P. J. G. Mann, *Biochem. J.* **52,** 130 (1952).

[5] K. E. Richardson and N. E. Tolbert, *J. Biol. Chem.* **236,** 1280 (1961).

oxidation of glyoxylate to CO_2 and formic acid predominates as a second reaction.[1-3]

Another form of glycolate oxidase exists in leaves,[6,7] and properties of this alternate form are described in this section. The alternate form can be isolated only if the enzyme is protected with its substrate and not precipitated from solution. Reactions catalyzed are similar to those of the FMN-requiring form of the oxidase. The alternate form does not require added FMN for activity although it contains FMN. When the alternate form is purified 500-fold it has the absorption spectra of ferredoxin, and added ferredoxin stimulates activity.[7]

Assay Method

Principle. Three methods are available: (a) manometric,[1] (b) measurement of reduction of 2,6-dichlorophenolindophenol at 620 mμ[2], or (c) measurement of glyoxylate phenylhydrazone formation at 324 mμ.[8,9] Most of the research on glycolate oxidase has been carried out with the manometric assay. The phenylhydrazone assay for glycolate oxidase as described below has not been used extensively.

Reagents

Sodium glycolate 0.04 *M*, 0.02 *M*, and 0.004 *M*, adjusted to about pH 8
Potassium phosphate buffer, 0.1 *M*, adjusted to pH 8.3
Cysteine·HCl 0.1 *M*, prepared fresh and adjusted to pH 6
Phenylhydrazine·HCl, 0.1 *M*, adjusted to pH 6

Procedure for Glyoxylate Phenylhydrazone. In a 1-cm cell are added 2 ml of phosphate buffer (200 micromoles), 0.1 ml of cysteine (10 micromoles), 0.1 ml of phenylhydrazine (10 micromoles), and 0.5 ml of sodium glycolate (20 micromoles). The solution is aerated with bubbling oxygen for 30 seconds, and then the reaction is immediately initiated by addition of 0.2 ml of solution containing about 0.01 unit of enzyme. The rate of increase in absorbancy is then followed spectrophotometrically at 324 mμ.

Manometric Procedure. In the main part of the Warburg vessel are placed 2.0 ml of phosphate buffer, 0.5 ml of 0.004 *M* glycolate, and 0.05–0.3 units of enzyme. The maintenance of substrate with the enzyme

[6] N. E. Tolbert and M. S. Cohan, *J. Biol. Chem.* **204,** 639 (1953). See also M. Kuczmak and N. E. Tolbert, *Plant Physiol.* **37,** 729 (1962).
[7] A. L. Baker and N. E. Tolbert, *Plant Physiol.* **40,** 57S (1965).
[8] G. H. Dixon and H. L. Kornberg, *Biochem. J.,* **72,** 3P (1959).
[9] R. K. Yamazaki, A. L. Baker, and N. E. Tolbert, unpublished (1965).

at all times necessitates addition of glycolate when temperature-equilibrating the enzyme in the vessel. After 5 minutes' equilibration at 30°, zero time is designated when the manometers are closed and an additional 0.5 ml of 0.02 *M* glycolate is added from the side arm. The center well contains NaOH.

Definition of Unit and Specific Activity. For the manometric assay one enzyme unit is defined as the amount required for the uptake of 1 micromole of oxygen per minute during a 25-minute period of incubation. Specific activity is expressed as units per milligram of protein.

For the glyoxylate phenylhydrazone assay, one unit of activity is the production of 1 micromole of glyoxylate per minute. Since ϵ for glyoxylate phenylhydrazone[8] is 1.7×10^4 liter $mole^{-1}$ cm^{-1}, a ΔOD of 5.67 per minute is equal to 1 unit of activity in 3 ml, or in the assay a ΔOD of 0.057 equals 0.01 of a unit.

Application of Assay Methods to Crude Tissue Preparations. Zelitch has discussed the limitation of the 2,6-dichlorophenolindophenol procedure.[2] The manometric procedure has generally been used for assaying crude extracts, but side reactions such as peroxide activity or inhibition by oxalate may exist. Because of its sensitivity, the glyoxylate phenylhydrazone assay can generally be used with sufficient dilutions of crude sap to allow spectrophotometric assay.

Purification Procedure

The procedure for purification of the original form of the enzyme has previously been described.[2,3] The following procedure applies only for purification of the alternate form of glycolate oxidase. Etiolated leaves are used because only the alternate form of the oxidase is present. Although both forms of the enzyme are present in young green leaves of tobacco and wheat, no extensive purification of the alternate form from green leaves has been tried. Throughout the procedure the enzyme must be maintained in the presence of excess glycolate (generally, 0.01 *M*) or otherwise all activity even at 2° is either lost, or there is a change to the original form of the enzyme which requires excess FMN.

Step 1. Preparation of Crude Extract. Etiolated Thatcher wheat is grown in moist Vermiculite at 20° in total darkness for 7–8 days. The leaves are washed in cold tap water and frozen at —20°. The frozen leaves may be stored for weeks before use. Four kilograms of frozen tissue is thawed at 2° and cut into inch segments. All subsequent steps are conducted at 2°. Four 1-kg weights of tissue, without added solvent, are each homogenized with 1 g of solid sodium glycolate in a large Waring blendor. The homogenates are squeezed through cheesecloth, combined, and centrifuged at 15,000 *g* for 30 minutes.

Step 2. Acid Treatment. Cold acetic acid is added to the crude extract until the pH is 5.3. After centrifugation at 15,000 *g* for 15 minutes, the supernatant fluid is adjusted to pH 6.5. Due to instability of the enzyme at pH 5.3, this procedure is run as rapidly as possible. At pH 6.5 the enzyme is stable at —20°.

Step 3. Fractionation with Alumina Cγ Gel. This gel (about 15 g) is added until the enzyme is completely adsorbed as determined by assaying the supernatant fluid. The gel is removed by centrifugation and washed 1 hour by resuspension in 4 liters of 0.1 *M* Tris-acetate buffer, pH 7.0, containing 0.001 *M* sodium glycolate. After recentrifugation, the gel is suspended for 90 minutes in 250 ml of 0.1 *M* potassium phosphate, pH 8.3, containing 0.01 *M* sodium glycolate. After centrifugation, the supernatant fluid which contains the enzyme is lyophilized with small loss of activity.

Step 4. Sephadex G-200 Chromatography. The lyophilized enzyme is dissolved in 30 ml of 0.02 *M* sodium glycolate and chromatographed on a Sephadex G-200 column (4 × 70 cm). The Sephadex is prepared 3 weeks in advance in 0.1 *M* potassium phosphate, pH 8.3, containing 0.01 *M* sodium glycolate. After packing the column and equilibration, elution of the enzyme at a flow rate of 60 ml per hour is achieved with the same phosphate buffer containing glycolate. Twelve-milliliter fractions are collected from the column and the enzyme fractions (about 16–33) are located by assay. Fractions with activity in excess of 0.13 micromole of O_2 per minute per milligram protein are pooled.

Step 5. TEAE-Cellulose Chromatography. The combined fractions from the Sephadex column are dialyzed 4 hours against 20 volumes of solution containing 0.01 *M* glycolate and 0.001 *M* cysteine. The enzyme is then passed through a TEAE-cellulose column (2.2 × 40 cm) which has been preequilibrated with 0.01 *M* potassium phosphate, pH 8.3, containing 0.01 *M* sodium glycolate. The activity is not adsorbed and emerges with the front.

A further purification of about twofold can be obtained by lyophilizing the TEAE-cellulose eluate and passing the concentrated enzyme through a second Sephadex G-200 column. However, a large loss of active protein generally occurs upon repeating the Sephadex step.

The purification procedure is summarized in the table.

Properties

Unlike the original glycolate oxidase, the alternate form is unstable to fractionation by ammonium sulfate and ethanol precipitation or dialysis against water.

Prosthetic Group. The purified alternate form of glycolate oxidase

PURIFICATION OF ENZYME

Fraction	Total protein (mg)	Total activity (units)	Units per mg protein	Per cent recovery
Extract	114,300	768	.0067	100
Acid treated	69,200	652	.0094	85
Alumina gel	4,200	261	.0622	34
Sephadex G-200	865	154	.178	20
TEAE-cellulose	74	115	1.56	15
Sephadex G-200	12	38	3.16	5

has an absorption spectrum with maxima at 420, 320, and 278 mμ. This resembles the absorption spectrum of spinach ferredoxin. The activity of the enzyme is enhanced by the addition of either *Clostridium* or plant ferredoxin. When the enzyme is boiled, FMN in the supernatant fluid can be detected by chromatography. Enzyme activity is only slightly stimulated by addition of FMN.

Physical Constants. The pH optimum is at 8.3. Since the enzyme is irreversibly destroyed in the absence of substrate, K_m values have not been determined.

Inhibitors. The use of Tris buffer results in a 27% inhibition of activity at pH 8.3 as compared to phosphate buffer. *p*-Hydroxymercuribenzoate (10^{-4} *M*) immediately causes a 50–100% loss of activity which cannot be reversed with excess cysteine. The original form of the oxidase is inhibited only after an hour of incubation with 5×10^{-2} *M* PCMB. Fifty per cent inhibition of the enzyme occurs with about 3×10^{-4} *M* hydroxymethyl sulfonates or α-hydroxy-2-pyridylmethane sulfonate. The enzyme is not inhibited by sodium azide, sodium cyanide, or EDTA.

[63] Glyoxylate Dehydrogenase

By J. R. QUAYLE

$$\text{Glyoxylate} + \text{CoA-SH} + \text{TPN}^+ \rightleftharpoons \text{oxalyl CoA} + \text{TPNH} + \text{H}^+$$

Glyoxylate dehydrogenase is an enzyme catalyzing the reversible oxidation and acylation of glyoxylate to oxalyl CoA, analogous to the acetaldehyde dehydrogenase discovered in *Clostridium kluyverii*.[1] It has been found in oxalate-grown *Pseudomonas oxalaticus* and *Pseudomonas*

[1] R. M. Burton and E. R. Stadtman, *J. Biol. Chem.* **202**, 873 (1953).

ODl,[2] and is probably present in shoots of *Oxalis pes caprae*.[3] In view of its probable metabolic function, viz. reduction of oxalyl CoA to glyoxylate, it may be termed oxalyl CoA reductase.

The purification from *Pseudomonas oxalaticus* and its properties which are described here have been published previously.[2,4]

Assay Method

Principle. The enzyme is most conveniently assayed in the direction of glyoxylate oxidation, the equilibrium of the reaction being pulled toward oxalyl CoA formation by working at pH 8.6. Under these conditions, the assay is performed by measuring the rate of increase of optical density at 340 mμ consequent on the reduction of TPN in the presence of glyoxylate and CoA. Cysteine is included in the reaction mixture to diminish oxidation of CoA; its presence enhances the linearity of the rate of TPN reduction at higher reaction rates.

Reagents

Sodium pyrophosphate buffer, 0.15 *M*, pH 8.6
Cysteine, 0.1 *M*
Sodium glyoxylate, 0.1 *M*
Coenzyme A, 0.01 *M*
TPN, 0.01 *M*

Procedure. The complete reaction system in a 1.5-ml quartz cell (light path 1 cm), consists of 0.2 ml of pyrophosphate buffer, 0.02 ml of cysteine, 0.05 ml of glyoxylate, 0.02 ml of CoA, 0.02 ml of TPN, enzyme extract, and water to total volume of 1 ml. CoA and TPN are omitted from the blank cell. The reaction is started by addition of TPN, whose reduction is followed at 340 mμ over about 3 minutes. The initial rate is proportional to the amount of enzyme used over the range 0.002–0.02 unit.

Definition of Unit and Specific Activity. One unit of enzyme is defined as the amount that catalyzes the reduction of 1 micromole of TPN per minute under the conditions of assay. Specific activity is expressed as units of enzyme per milligram of protein.

Growth Conditions. The source of *Pseudomonas oxalaticus,* its maintenance, and method of large-scale growth on oxalate is described elsewhere in this volume.[5]

[2] J. R. Quayle and G. A. Taylor, *Biochem. J.* **78**, 611 (1961).
[3] A. Millerd, R. K. Morton, and J. R. E. Wells, *Biochem. J.* **88**, 281 (1963).
[4] J. R. Quayle, *Biochem. J.* **87**, 368 (1963).
[5] See this volume [67].

Purification Procedure

Step 1. Preparation of Cell-Free Extract. Frozen bacteria (32 g wet weight) are crushed in a Hughes press at —25° and then mixed with 250 ml of 0.03 *M* phosphate buffer (pH 7.6) at 2°. Crystalline ribonuclease and deoxyribonuclease (L. Light & Co., Ltd. Colnbrook, Bucks, England), 2 mg each, are added. The resulting extract is centrifuged at 16,000 *g* for 20 minutes and the precipitate is discarded. All subsequent operations are performed at 2°.

Step 2. Treatment with Protamine Sulfate. Protamine sulfate is added to the extract in the proportion of one part to 10 parts of bacterial protein (w/w). The resulting suspension is centrifuged and the precipitate is discarded.

Step 3. Ammonium Sulfate Precipitation and Dialysis. Solid ammonium sulfate is added to the supernatant solution to give 40% of saturation. The precipitated protein is collected by centrifuging and discarded. To the supernatant solution is added ammonium sulfate to 60% of saturation. The pH of the solution is maintained at 7.0–7.2 throughout the precipitation procedure by addition of 0.01 *N* sodium hydroxide. The protein precipitating between 40 and 60% of ammonium sulfate saturation is collected by centrifugation and dissolved to a final volume of 7.6 ml in 0.03 *M* sodium pyrophosphate buffer (pH 7.8), it is dialyzed for 1.5 hours against 3 liters of the same buffer. The resulting solution (11.7 ml) is stored overnight at —15°.

Step 4. Alumina Cγ Gel Adsorption. The dialyzed fraction is diluted to 114 ml with water and the enzyme adsorbed on to Cγ-gel by treatment with 0.34 volume of gel suspension (30.6 mg/ml). The gel is separated by centrifugation, and the adsorbed enzyme is eluted by four successive extractions with 30 ml of 0.01 *M* sodium phosphate–potassium phosphate buffer (pH 7.5). The four eluates are combined (113 ml).

Step 5. Concentration of Enzyme by Ammonium Sulfate Precipitation, Followed by Dialysis. To the combined eluates solid ammonium sulfate is added to 80% of saturation. The precipitated protein is dissolved in 3 m*M* tris-HCl buffer (pH 7.3) (final volume 9.6 ml) and dialyzed against 3 liters of the same buffer for 2 hours.

Step 6. Ion Exchange Chromatography. A slurry of 5 g of diethylaminoethyl cellulose (DEAE-cellulose, Whatman DE 50) in 3 m*M* tris-HCl buffer (pH 7.3) is poured into a chromatographic column (2.5 cm × 25 cm), and this is washed with 1 liter of the same buffer. The dialyzed extract is diluted with an equal volume of water and is applied to the column, which is subsequently washed with 200 ml of the same buffer. The enzyme is not eluted under these conditions. The column is then eluted with a solution of Tris-HCl buffer–potassium chloride in which the potassium chloride forms a linear gradient of increasing concentration. This is formed by drawing the column eluent from 600 ml of 3 mM Tris-HCl buffer (pH 7.3) connected by siphon tube to 600 ml of molar potassium chloride. The levels of solution in both flasks should drop at the same rate throughout, and the flask containing the tris buffer should be stirred continuously. Fractions of eluate (approximately 6.5 ml) are collected at a flow rate of 40–50 ml per hour. Under these conditions the glyoxylate dehydrogenase is eluted mainly in five fractions around fraction number 50. These are combined, and the resulting solution (41 ml) is stored as separate 2-ml samples at −15°.

A summary of the purification procedure is given in the table.

SUMMARY OF PURIFICATION PROCEDURE

Step	Volume (ml)	Activity (units/ml)	Protein (mg/ml)	Specific activity (units/mg)	Yield (%)
1. Cell-free extract	210	1.87	6.4	0.29	100
2. Treatment with protamine sulfate	205	1.93	6.3	0.30	100
3. 40–60% Ammonium sulfate precipitate	9.2	40.5	66	0.61	94
4. Eluate from C_γ gel adsorption	121	2.28	1.37	1.7	69
5. Dialyzed ammonium sulfate precipitate	9.6	15.1	11	1.4	37
6. Pooled selected fractions after column chromatography	41	1.0	0.24	4.2[a]	10

[a] At the peak of activity eluted, the specific activity was 7.2.

Properties

Specificity. As far as has been tested, the enzyme is specific for glyoxylate: no activity is observed with formaldehyde, acetaldehyde, glycolaldehyde, pyruvate, oxalacetate, α-ketoglutarate, or hydroxypyruvate. The enzyme is specific for TPN. CoA may be replaced by pantetheine, N-acetylcysteamine reacts very slowly in place of CoA, cysteamine not at all. The preparation appears to be free of interfering enzymes.

Activators and Inhibitors. No metal-ion requirements have been found. The enzyme is inhibited in the presence of Tris buffer. Replacement of pyrophosphate buffer in the standard assay system by Tris buffer of the same strength and pH halves the initial rate of the reaction. This fact, together with the increased hydrolysis of oxalyl thiol compounds which takes place in amino buffers such as Tris or triethanolamine as compared with inorganic buffers such as phosphate,[6] makes Tris buffer unsuitable for use with the enzyme.

Equilibrium. Using a measured value for the equilibrium constant of glyoxylate dehydrogenase,[4] the $\Delta F'$ at pH 7 has been calculated to be —2.54 kcal. This compares with the value +0.9 kcal calculated previously.[2] Some of the free energy values used in the latter calculation are of doubtful accuracy.

The equilibrium of the reaction may be upset if other thiol compounds besides CoA are present as a result of nonenzymatic ester interchange between oxalyl CoA and the thiol compounds. This effect becomes apparent at pH values above 7; at pH >8 with excess cysteine present, the requirement for CoA becomes a catalytic one only. The reaction is readily reversible; at pH 6.5 in the presence of a fivefold excess of TPNH, oxalyl CoA may be reduced to glyoxylate within 1% of completion.[7]

Stability. The activity of glyoxylate dehydrogenase rapidly diminishes in crude cell-free extracts, and losses of up to 70% have been observed within 1 day at 0°. The enzyme becomes more stable on purification and may be stored at —15° for several weeks without loss of activity. Rather variable stability of the purified enzyme at 0° has been observed.

pH Optima. The pH optimum for oxidation of glyoxylate in pyrophosphate buffer is 8.6. The optimum for reduction of oxalyl CoA to glyoxylate in phosphate buffer is 6.7.

Kinetic Properties. The K_m values for glyoxylate and TPN, measured at pH 8.6 in pyrophosphate buffer at 25°, are $5.7 \times 10^{-4}\,M$ and $3.4 \times 10^{-5}\,M$, respectively.

[6] J. Koch and L. Jaenicke, *Ann. Chem.* **652**, 129 (1962).

[7] The preparation and properties of oxalyl CoA have been described in the reference cited in footnote 4.

[64] Primary and Secondary Alcohol Dehydrogenases from *Gluconobacter*

By K. Kersters and J. De Ley

Acetic acid bacteria oxidize primary and secondary alcohols and glycols with a primary alcohol function to the corresponding carboxylic or hydroxycarboxylic acids. Secondary alcohols and glycols with a secondary alcohol function are oxidized to the corresponding ketoses. These oxidations are catalyzed by two different enzyme systems: one is soluble, the other particulate.[1,2] The particulate enzymes are probably localized on the cell envelope (considered to be the cytoplasmic membrane). This section describes the purification and the properties of two soluble NAD-linked primary and secondary alcohol dehydrogenases and two particulate primary and secondary alcohol dehydrogenases from *Gluconobacter oxydans* (suboxydans), strain SU (NCIB 9108).

I. Soluble NAD-Linked Primary and Secondary Alcohol Dehydrogenases

$$R—CH_2OH + NAD^+ \rightleftharpoons R—CHO + NADH + H^+$$
$$R—CHOH—R' + NAD^+ \rightleftharpoons R—CO—R' + NADH + H^+$$

Assay Method

Principle. The assay is based on the rate of reduction of NAD at 340 mμ in the presence of an alcohol.

Reagents

Tris(hydroxymethyl)aminomethane (Tris) buffer, 0.1 *M*, pH 8.8
$MgCl_2$, 0.05 *M*
NAD, 0.005 *M*
Substrate: A 2 *M* solution is used for primary alcohols and glycols with a primary alcohol function. A 0.2 *M* solution is used for all other compounds. The primary and secondary alcohol dehydrogenases are routinely tested with respectively *n*-propanol and *meso*-2,3-butanediol as substrates.
Enzyme. The enzyme is diluted in 0.01 *M* phosphate buffer at pH 7.0, to give a solution with an activity not greater than 0.3 unit/ml.

Procedure. Six-tenths milliliter of Tris buffer, 0.1 ml of $MgCl_2$, 0.1 ml of NAD, and 0.1 ml of suitably diluted enzyme are mixed in a spectrophotometer cell with a 1-cm light path. The reaction is initiated by the addition of 0.1 ml of alcohol solution. The extinction at 340 mμ is measured at 1-minute intervals.

Definition of Unit and Specific Activity. One unit of alcohol dehydrogenase is defined as that amount which causes the reduction of 1 micromole of NAD per minute under the assay conditions described. Specific activity is expressed in terms of units of enzyme per milligram of protein, and protein is determined spectrophotometrically.[3]

[1] K. Kersters and J. De Ley, *Biochim. Biophys. Acta* **71**, 311 (1963).
[2] J. De Ley and K. Kersters, *Bacteriol. Rev.* **28**, 164 (1964).
[3] O. Warburg and W. Christian, *Biochem. Z.* **310**, 384 (1941).

Purification Procedure

All operations are performed at 0°–5°, unless otherwise stated. Precipitates are separated by centrifugation at 15,000 *g* for 15 minutes, and dissolved in 0.01 *M* KH_2PO_4-Na_2HPO_4 buffer at pH 6.5. Sodium chloride and ammonium sulfate are removed from solutions by gel filtration on Sephadex G-25 columns, equilibrated with the same buffer.

Step 1. Growth of Culture and Preparation of Particle-Free Extract. Gluconobacter oxydans (suboxydans), strain SU, is grown at 30° on a solid medium containing 2% D-mannitol, 3% $CaCO_3$, 0.5% yeast extract, and 2.5% agar. The Roux flasks are inoculated with a 20-hour slant culture, grown on the same medium. The cells are harvested by centrifugation after 2 days of growth, and washed three times with 0.01 *M* phosphate buffer, pH 6.5. A cell-free extract is prepared by sonic disintegration of a 30% (w/v) suspension of the cells in 0.01 *M* phosphate buffer at pH 6.5 in a 10 kc, 250-watt Raytheon magnetostriction oscillator for 20 minutes at 4° under hydrogen atmosphere. Unbroken cells are removed by centrifugation at 20,000 *g* for 20 minutes. Centrifugation of the supernatant in a preparative Model L Spinco ultracentrifuge for 2 hours at 4° and 105,000 *g* results in the sedimentation of red particles, which also oxidize primary and secondary alcohols (see Part II).

Step 2. Removal of Nucleic Acids. Two and half milliliters of 1.0 *M* $MnCl_2$ is added to 50 ml of supernatant solution. After 30 minutes, the precipitate is removed by centrifugation and discarded. $MnCl_2$ is eliminated from the resulting supernatant by gel filtration on Sephadex G-25.

Step 3. Ammonium Sulfate Fractionation. Solid ammonium sulfate is added to 25% of saturation. The precipitate is removed by centrifugation and discarded. Ammonium sulfate is then added to 55% of saturation and the precipitate is collected by centrifugation and dissolved in 5 ml of 0.01 *M* phosphate buffer at pH 6.5. The ammonium sulfate is removed by gel filtration on a Sephadex G-25 column, previously equilibrated with the same buffer.

Step 4. Chromatography on DEAE-Cellulose. The DEAE-cellulose column is prepared as described elsewhere in this volume.[4] The dialyzed ammonium sulfate fraction (7 ml) is applied to the column, and proteins are eluted with an increasing NaCl-gradient in the same buffer. The mixing chamber contains 450 ml of 0.01 *M* KH_2PO_4-Na_2HPO_4 buffer (pH 6.5), and the upper vessel contains 500 ml 0.15 *M* NaCl in the same buffer. The flow rate is maintained at 40–50 ml per hour and fractions of 4 ml are collected. The primary and secondary alcohol dehydrogenases

[4] K. Kersters and J. De Ley, this volume [34].

are eluted after 220 ml and 300 ml of buffer, respectively, passed through the column. Fractions 62–65, representing the peak of the primary alcohol dehydrogenase, and the fractions 84–87, representing the peak of the secondary alcohol dehydrogenase are collected and used to determine the enzyme specificity.

A summary of the purification procedure is given in Table I.

TABLE I
PURIFICATION OF NAD-LINKED PRIMARY AND SECONDARY ALCOHOL DEHYDROGENASES

			Primary alcohol dehydrogenase *n*-propanol		Secondary alcohol dehydrogenase *meso*-2,3-butanediol	
Fraction	Volume (ml)	Protein (mg/ml)	Total units	Specific activity (units/mg protein)	Total units	Specific activity (units/mg protein)
Particle-free extract	50	3.6	72	0.40	162	0.9
Ammonium sulfate (25–55%)	7	14	61	0.62	138	1.4
DEAE-cellulose						
Fractions 62–65	15	0.06	4.6	5.2	—	—
Fractions 84–87	15	0.11	—	—	18	11.0

Properties

The purified enzyme preparations are stable when kept frozen for at least 2 months. Both dehydrogenases show optimal activity between pH 8.8 and 9.2.

Specificity. (1) NAD-LINKED PRIMARY ALCOHOL DEHYDROGENASE. The enzyme is NAD-specific and catalyzes the oxidation of various primary alcohols and glycols with primary OH-groups at the following relative rates: *n*-propanol, 100; ethanol, 67; 1,7-heptanediol, 67; 1,6-hexanediol, 42; 1,5-pentanediol, 39; ethylene glycol, 28; 1,4-butanediol, 25; 1,3-propanediol, 22; ethylene glycol monomethyl ether, 17; *n*-butanol, 17; *n*-hexanol, 16; allyl alcohol, 8; isobutanol, 3; and DL-1,3-butanediol, 2. The following compounds were found to be inactive: methanol, diethylene glycol, triethylene glycol, 2-ethyl-2-nitro-1,3-propanediol, *sec*-butanol, cyclopentanol, cyclohexanol, *tert*-butanol, DL-1,2-propanediol, *meso*-2,3-butanediol, DL-2,3-butanediol, *meso*-3,4-hexanediol, glycerol, DL-lactate, and several polyols.

(2) NAD-LINKED SECONDARY ALCOHOL DEHYDROGENASE. The enzyme is NAD-specific and catalyzes the oxidation of various secondary alcohols

and glycols with a secondary OH-group at the following relative rates: DL-2,3-butanediol, 100; *meso*-2,3-butanediol, 50; (—)-3,4-hexanediol, 32; cyclooctanol, 15; cycloheptanol, 15; L-1,2-propanediol, 9; *sec*-butanol; 6; *meso*-3,4-hexanediol, 5; DL-1,2-propanediol, 5; cyclohexanol, 2; cyclopentanol, 1 and *sec*-propanol, 0.5. The following compounds were found to be inactive: primary alcohols and glycols with a primary OH-group, all polyols tested, glycerol, DL-1,3-butanediol, DL-lactate, DL-β-hydroxybutyrate, and phosphoglycerate.

II. Particulate Primary and Secondary Alcohol Dehydrogenases

$$R{-}CH_2OH \rightarrow R{-}CHO + 2\,H^+ + 2\,e$$
$$R{-}CHOH{-}R' \rightarrow R{-}CO{-}R' + 2\,H^+ + 2\,e$$

Assay Method

Principle. The particles contain the complete electron transport chain. The oxidation of alcohols and glycols by the particulate enzymes can thus be followed by measuring the rate of oxygen uptake in the Warburg respirometer. Once solubilized, the electron transport to oxygen is broken. Therefore the oxidation of primary and secondary alcohols is routinely determined spectrophotometrically by the reduction of 2,6-dichlorophenolindophenol. In spite of the disadvantages inherent to the use of this electron acceptor, the assay was found to be reproducible for routine determinations in following the purification of the enzymes.

Reagent

KH_2PO_4–Na_2HPO_4 buffer, 0.03 *M*, pH 5.8
2,6-Dichlorophenolindophenol, 10^{-3} *M*
Primary or secondary alcohol, 0.2 *M*. Ethanol and *meso*-2,3-butanediol are routinely used as substrates.
Enzyme. The enzyme is diluted in 0.03 *M* phosphate buffer at pH 5.8 to give a solution with an activity not greater than 0.2 unit/ml.

Procedure. One-tenth milliliter of suitably diluted enzyme, 2.9 ml of phosphate buffer, and 0.3 ml of 2,6-dichlorophenolindophenol are mixed in a colorimetric tube. The reaction is initiated by the addition of 0.2 ml of alcohol solution. The rate of decolorization is followed during 3 minutes at 660 mμ in a Beckman Model C colorimeter. The optical density change is calculated from the percentage transmittance recorded on a Varian recorder.

Definition of Unit and Specific Activity. One unit of primary and secondary alcohol dehydrogenase is defined as that amount which causes the oxidation of 1 micromole of alcohol, or the reduction of 1 micromole of 2,6-dichlorophenolindophenol per minute under the assay conditions described. In the Beckman C colorimeter, filter 66, with 12-mm tubes, 0.1 absorbance unit equals 0.06 micromole 2,6-dichlorophenolindophenol. Specific activity is expressed in terms of units of enzyme per milligram of protein, and protein is determined by the colorimetric method of Lowry *et al.*[5]

Purification Procedure

All operations are performed at 0–5° unless otherwise stated. Precipitates are separated by centrifugation at 15,000 *g* for 15 minutes, and dissolved in 0.01 *M* KH_2PO_4–Na_2HPO_4 buffer at pH 6.0. Sodium chloride, ammonium sulfate, and sucrose are removed from solutions by dialysis against 0.005 *M* phosphate buffer at pH 6.0.

Step 1. Growth of Culture and Preparation of Particles. Cells of *Gluconobacter oxydans* (suboxydans), strain SU, are grown, harvested, washed, and sonicated as described for the soluble enzymes. The cell-free extract is centrifuged for 2 hours at 4° and 105,000 *g* in a preparative Model L Spinco ultracentrifuge. The red precipitate (particles) is suspended in 0.01 *M* KH_2PO_4–Na_2HPO_4 buffer at pH 6.5 and again centrifuged at 105,000 *g* for 1 hour.

Step 2. Solubilization of Particulate Dehydrogenases. Washed particles (20 g, wet weight) are suspended in 60 ml 0.5% Triton-X-100 (Rohm & Haas Co, Philadelphia, Pennsylvania) in 0.025 *M* KH_2PO_4–Na_2HPO_4 buffer with 10^{-4} *M* EDTA at pH 7.6. The mixture is stirred during 1 hour at 4° and centrifuged for 90 minutes at 105,000 *g*.

Step 3. Removal of Nucleic Acids. Three milliliters of 1.0 *M* $MnCl_2$ is added to 60 ml of the red-pigmented supernatant solution. After 30 minutes a white precipitate is removed by centrifugation and discarded. $MnCl_2$ is eliminated from the supernatant solution by dialysis against phosphate buffer.

Step 4. First Ammonium Sulfate Fractionation. The dialyzed extract is fractionated with solid ammonium sulfate into the following fractions 0–30%, 30–40%, 40–45%, and 45–50%. The precipitate of each fraction is dissolved in 10 ml of 0.01 *M* phosphate buffer at pH 6.0 and dialyzed. Usually the 40–45% fraction, $(NH_4)_2SO_4$-1, contains a tenfold purified

[5] O. H. Lowry, N. J. Rosebrough, A. L. Farr, and R. J. Randall, *J. Biol. Chem.* **193**, 265 (1951).

ethanol dehydrogenase. It is, however, important to determine the activity in each fraction. The fraction with the highest specific activity possesses a pronounced red color.

Step 5. Second Ammonium Sulfate Fractionation. To the dialyzed 40–45% ammonium sulfate fraction, solid ammonium sulfate is added to 35% of saturation. The yellow precipitate is removed by centrifugation and discarded. Ammonium sulfate is then added to 80% of saturation, and the deep-red colored precipitate is dissolved in 3.5 ml of 0.01 *M* phosphate buffer (pH 6.0) and dialyzed [$(NH_4)_2SO_4$–2].

Step 6. Centrifugation in a Sucrose Gradient. The sucrose gradients are prepared in 35 ml lusteroid tubes by layering sucrose solutions of 25.7%, 18.7%, 11.8%, and 4.9% on top of each other. Linear sucrose gradients are obtained by leaving the tubes undisturbed overnight at 4°. Eight-tenths milliliter of the $(NH_4)_2SO_4$–2 fraction is layered on top of the gradient. The tubes are centrifuged in the Spinco SW-25 rotor during 8 hours at 25,000 rpm, the contents of the tubes are collected in several fractions. The peak of the alcohol dehydrogenase activity is located at approximately 12% sucrose. The fractions with the highest specific activity are pooled, dialyzed, and concentrated by saturating with ammonium sulfate. The red precipitate is dissolved in 3 ml 0.01 *M* phosphate buffer at pH 6.0 and dialyzed.

A summary of the purification procedure is given in Table II.

TABLE II
PURIFICATION OF THE PARTICULATE PRIMARY AND SECONDARY ALCOHOL DEHYDROGENASES[a]

			Primary alcohol dehydrogenase[a]		Secondary alcohol dehydrogenase[a]	
Fraction	Volume (ml)	Protein (mg/ml)	Total units	Specific activity (units/mg protein)	Total units	Specific activity (units/mg protein)
Washed particles	60	31	1210	0.65	750	0.40
Triton X-100 extract	60	11	935	1.4	52	0.08
$(NH_4)_2SO_4$-I (40–45%)	10	7.5	540	7.2	34	0.46
$(NH_4)_2SO_4$-2 (35–80%)	3.5	5	156	9	8.4	0.48
Sucrose-gradient centrifugation[b]	3	1.2	84	24	5.4	1.5

[a] The activity of primary and secondary alcohol dehydrogenases is measured with, respectively, ethanol and *meso*-2,3-butanediol as substrates.

[b] Represents ammonium sulfate precipitates of the most active fractions after sucrose-gradient centrifugation.

Properties

No appreciable loss of activity occurred when the washed particles were kept frozen at —15° during 6 months. The purified enzyme is stable for at least 1 month under the same conditions. The optimal pH for oxidation of primary and secondary alcohols is 5.8 with a sharp decrease in activity below this point.

Substrate Specificity. The purified enzyme has a very broad specificity. It catalyzes the oxidation of various primary and secondary alcohols, glycols, aldehydes, and polyols at the following relative rates.

PRIMARY ALCOHOLS. Methanol, 10; ethanol, 100; *n*-propanol, 110; *n*-butanol, 110; *n*-amyl alcohol; 100; *n*-hexanol, 110; *n*-octyl alcohol, 90; isobutanol, 60; allyl alcohol, 100.

GLYCOLS WITH PRIMARY ALCOHOL FUNCTION. 1,2-Ethylene glycol, 30; 1,3-propanediol, 70; 1,4-butanediol, 80; 1,5-pentanediol, 80; 1,6-hexanediol, 90; 1,7-heptanediol, 140; DL-1,3-butanediol, 60; ethylene glycol monomethyl ether, 24; diethylene glycol, 1; triethylene glycol, 18; DL-1,2-propanediol, 9; 1,2,6-hexanetriol, 16; 2-amino-2-ethyl-1,3-propanediol, 8; 2-nitro-2-ethyl-1,3-propanediol, 16.

AROMATIC ALCOHOLS. Cinnamyl alcohol, 34; Anisyl alcohol, 32; coniferyl alcohol, 18; benzyl alcohol, 10.

COMPOUNDS WITH SECONDARY ALCOHOL FUNCTION. sec-Propanol, 32; cyclopentanol, 21; cyclooctanol, 14; cyclohexanol, 13; cycloheptanol, 11; 2,5-hexanediol, 11; *meso*-3,4-hexanediol, 9; *meso*-2,3-butanediol, 6; *sec*-butanol, 5; DL-lactate, 4. Several polyols are only slowly oxidized.

It was shown that the purified enzyme contained at least three different enzymes for the oxidation of alcoholic functions: (1) one or more primary alcohol dehydrogenase, (2) one or more secondary alcohol dehydrogenase, (3) one or more polyol dehydrogenase. The primary and secondary alcohol dehydrogenase are different enzymes, since 75% of the former and only 7% of the latter enzyme are solubilized by treatment with Triton X-100 (see Table II). Eighty percent of the *meso*-2,3-butanediol dehydrogenase remains on the nonsolubilized fraction. The primary alcohol dehydrogenase is completely inhibited by 10^{-3} *M* *p*-chloromercuribenzoate, whereas the secondary alcohol dehydrogenase is not. The Triton X-100 treatment has thus released the enzymes from the bulk of the insoluble cell hull. Nevertheless we are led to believe that several apoenzymes are still linked together as a larger aggregate, as they could not be separated by column chromatography on several ion exchange resins (DEAE-, CM-, SE-cellulose, DEAE-Sephadex, and Amberlite XE-64).

Electron Acceptor Specificity. Foı the crude particles oxygen could act as electron acceptor. In the purified state 2,6-dichlorophenolindophenol, ferricyanide, phenazine methosulfate, thionine, and methylene blue are efficient electron acceptors.

Coenzyme. No evidence was found for the participation of NAD, NADP, FAD, FMN, the usual ubiquinones, and heavy-metal ions. All preparations display the spectrum of a reduced cytochrome-553 on addition of ethanol, *meso*-2,3-butanediol, or hydrosulfite. In all purification steps the ratio specific activity:specific hemoprotein content remained constant, indicating that enzyme activity is probably associated with this cytochrome-553. The purified enzyme in the oxidized form displays the following maxima: 530 mμ (diffuse), 412 mμ, and 279 mμ. In the reduced form the following maxima were observed: 553 mμ, 522 mμ, 418 mμ, 315 mμ, 279 mμ, and a shoulder at 345 mμ. The spectral properties of this enzyme are identical with the ethanol-cytochrome-553 reductase from *Acetobacter aceti* (*rancens*).[6]

Inhibitors. The primary alcohol dehydrogenase is completely inhibited by 10^{-3} *M* *p*-chloromercuribenzoate. The primary and secondary alcohol dehydrogenases are not inhibited by the following compounds (10^{-2} *M*): arsenite, semicarbazide, hydroxylamine, cyanide, azide, *o*-phenanthroline, 8-hydroxyquinoline, and EDTA.

[6] T. Nakayama, *J. Biochem.* (*Tokyo*) **49**, 240 (1961).

[65] Ethanolamine Oxidase

By STUART A. NARROD and WILLIAM B. JAKOBY

$$\text{Ethanolamine} + 1/2\ O_2 \rightarrow \text{glycolaldehyde} + NH_3$$

Assay Method[1]

Principle. The assay is based on measuring the rate of formation of glycolaldehyde. This is accomplished by formation of the bis-2,4-dinitrophenylhydrazone of glycolaldehyde followed by alkali treatment and measurement of the resultant purple color.

Reagents

Sodium borate, 0.05 *M*, adjusted to pH 8.0

Ethanolamine, 0.025 *M*, adjusted to pH 8.0 with HCl just prior to use. Commercial ethanolamine is redistilled and the fraction boiling between 171° and 173° is collected.

[1] S. A. Narrod and W. B. Jakoby, *J. Biol. Chem.* **239**, 2189 (1964).

2,4-Dinitrophenylhydrazine, 0.1%, in 2 N HCl
Alcoholic KOH, 1 N, prepared just prior to use by diluting 4 N aqueous KOH with 95% ethanol

Procedure. Appropriate dilutions of enzyme are incubated at 37° in a vessel containing 1 ml of ethanolamine and 1 ml of sodium borate in a total volume of 5 ml. At 1, 15, 30, and 60 minutes, 1-ml aliquots are removed and added to 1 ml of 2,4-dinitrophenylhydrazine solution. The mixture is boiled for 10 minutes and cooled; 5 ml of alcoholic KOH is added. The absorption of the resulting purple bis-2,4-dinitrophenylhydrazone is measured at 550 mμ ($\epsilon_{550} = 6.9 \times 10^4$). Under these conditions, the reaction is linear with respect to time and protein concentration when absorbancies of less than 1.0 are measured.

The formation of glycolaldehyde may also be determined with 3-methyl-2-benzothiozolone hydrazine.[2]

Definition of Unit. One unit is defined as the amount of enzyme which forms 1.0 millimicromole of glycolaldehyde per minute under the above conditions. Specific activity is in units per milligram of protein; protein was measured by the method of Lowry *et al.*[3] with bovine serum albumin as standard.

Purification Procedure

Ethanolamine oxidase was found in a bacterium[1] originally isolated by the enrichment culture technique. The organism is now available from the American Type Culture Collection as No. 13796. Large quantities of cells may be obtained by growth in 6-liter flasks containing 1 liter of medium on a reciprocating shaker at 25°. The medium contains 0.1% ethanolamine, 0.1% $MgSo_4 \cdot 7\ H_2O$, 0.2% KH_2PO_4, 0.05% yeast extract and is adjusted to pH 7.0. It was convenient to neutralize the ethanolamine in a 1:20 dilution prior to addition to the other medium components.

A 0.1% inoculum, grown for 24 hours in the same medium, is used. Cells are harvested after 18 hours, washed with 0.85% saline solution, and dried *in vacuo*. Approximately 1 g of dried cells per 4 liters of medium is usually obtained by this procedure.

All subsequent manipulations are conducted at approximately 2°.

1. Crude Extract. Cell-free extracts are prepared by suspending dried cells, 50 mg/ml, in 0.005 M Tris at pH 7.3 and passing the suspension

[2] E. Sawicki, T. R. Hauser, T. W. Stanley, and W. Elbert, *Anal. Chem.* **33**, 93 (1961).
[3] O. H. Lowry, N. J. Rosebrough, A. L Farr, and R J. Randall, *J. Biol. Chem.* **193**, 265 (1951).

twice through a French pressure cell. Cell debris is removed by centrifugation at 20,000 g, resulting in a yellow, opalescent solution.

2. Ammonium Sulfate. Thirty milliliters of 1 M $MnCl_2$ is slowly added with stirring to each 100 ml of crude extract, and the precipitate is removed by centrifugation. After the addition of 53 g of ammonium sulfate, the precipitate formed is collected by centrifugation and dissolved in 70 ml of 0.02 M Tris at pH 8.5. This solution is then dialyzed for 18 hours at 5° against distilled water. At this stage the absorbance ratio, A280:A260, is approximately 1.0.

3. Second Ammonium Sulfate. To each 100 ml of fraction 2 is added 17.4 g of ammonium sulfate; the precipitate is discarded. The precipitate formed after the further addition of 17.4 g of ammonium sulfate is suspended in 30 ml of 0.02 M Tris at pH 8.5 and dialyzed against distilled water for 5 hours.

4. DEAE-Cellulose. A DEAE-cellulose column (3 $cm^2 \times 14$ cm), exhaustively washed with 5 mM Tris at pH 7.2, is charged with 15 ml of fraction 3. Protein is eluted with a linear gradient of sodium chloride, achieved by mixing 500 ml of 5 mM Tris at pH 7.2 with 500 ml of 0.5 M sodium chloride in 5 mM Tris at pH 7.2. Activity is eluted in the fractions between 550 ml and 750 ml. The combined eluates of active material are concentrated by precipitation with ammonium sulfate (70 g/100 ml). The resulting precipitate is dissolved in 7 ml of 0.02 M Tris at pH 8.2 and dialyzed against distilled water for 5 hours.

The results of such a course of purification are summarized in the table.

SUMMARY OF PURIFICATION PROCEDURE

Fraction	Volume (ml)	Activity (units)	Protein (mg)	Specific activity (units/mg)
I. Cell-free extract	128	1150	1130	1.0
II. $MnCl_2 + (NH_4)SO_4$	65	2960	325	9.1
III. $(NH_4)_2SO_4$	20	2120	176	12.1
IV. DEAE-cellulose	10	435	7	62.1

Properties[1]

The enzyme is active with ethanolamine, 3-amino-1-propanol, or 1-amino-2-propanol as substrate at maximal relative velocities of 100, 56, and 18, respectively. Other amines such as ethylamine, n-propylamine, putrescine, choline, and phosphoethanolamine were inactive at a concentration of 0.1 M. The K_m for ethanolamine is 5×10^{-3} M at pH 8.0.

The enzyme is active at pH values between 7 and 10.5, with optimum activity at pH 8.0. Preparations of purified enzyme have been stored at $-5°$ for 6 months without any appreciable loss of activity.

Inhibition was obtained with the following reagents at the concentrations noted: 0.2 m*M* quinacrine, 48%; 1 m*M* hydrazine, 97%; 0.1 m*M* hydroxylamine, 100%; 10 m*M* iproniazid, 64%; 10 m*M* isoniazid, 8%; 0.1 m*M* semicarbazide, 96%; 20 m*M* sodium bisulfite, 68%; 10 m*M* cysteine, 80%; 10 m*M* glutathione, 25%; 10 m*M* mercaptoethanol, 68%.

[66] Formaldehyde Dehydrogenase

By Zelda B. Rose and Efraim Racker

$$\text{Formaldehyde} + \text{DPN}^+ \longrightarrow \text{formate} + \text{DPNH} + \text{H}^+$$

Assay Method

Principle. The assay of the enzyme in crude extracts is based on the formation of acid.[1] Alcohol dehydrogenase present in the extracts reacts with formaldehyde and DPNH to yield methanol and DPN. The overall reaction catalyzed by formaldehyde dehydrogenase and alcohol dehydrogenase is a dismutation of formaldehyde to methanol and formic acid. Catalytic amounts of DPN (NAD) and GSH are required. The reaction is followed by measuring manometrically the rate at which the acid produced releases carbon dioxide from bicarbonate buffer.

After acetone treatment (step 2), the enzyme can be assayed spectrophotometrically by observing the initial rate of increase of absorbancy at 340 mμ due to the reduction of DPN (NAD). Residual alcohol dehydrogenase is inhibited by hydroxylamine.[2]

Manometric Assay for Crude Extracts

Reagents

Formaldehyde, 0.12 *M*
Glutathione, 0.067 *M*
DPN (NAD), 0.012 *M*
$KHCO_3$, 2 *M* (freshly prepared)

Procedure. To the main compartment of a Warburg flask the following are added to a final volume of 2.67 ml: 0.10 ml of GSH, 0.10 ml of

[1] Z. B. Rose and E. Racker, *J. Biol. Chem.* **237**, 3279 (1962).
[2] N. O. Kaplan and M. M. Ciotti, *J. Biol. Chem.* **201**, 785 (1953).

DPN (NAD), 0.10 ml of $KHCO_3$, and enzyme. Into the sidearm 0.3 ml of formaldehyde and 0.03 ml of $KHCO_3$ are added. The flasks are placed in a 30° bath and flushed with 5% carbon dioxide 95% nitrogen for 10 minutes, after which the contents of the flask are mixed. The rate of acid release is linear during the period from 5 to 20 minutes after the addition of the substrate.

Spectrophotometric Assay

Reagents

Triethanolamine buffer, 1 *M*, pH 8.5
GSH, 0.067 *M*
DPN (NAD), 0.012 *M*
$NH_2OH \cdot HCl$, 0.02 *M*
HCHO, 0.02 *M*

Procedure. A cuvette with a 10-mm light path contains, in a 1-ml volume: 0.10 ml of triethanolamine buffer, pH 8.5, 0.05 ml of GSH, 0.03 ml of DPN (NAD), 0.04 ml of $NH_2OH \cdot HCl$, and enzyme. After 2 minutes of incubation, 0.05 ml of HCHO is added to start the reaction. The initial rate of increase of absorbancy at 340 mμ is measured.

Definition of Unit and Specific Activity. A unit of enzyme is the amount necessary to oxidize 1 micromole of formaldehyde per minute under the conditions of the assay. Specific activity is expressed as units per milligram of protein. Protein was determined by the biuret test[3] or the method of Warburg and Christian.[4]

Purification Procedure

Step 1. Extraction. Fleischmann's bakers' yeast is air-dried and ground in a ball mill. The yeast powder, 300 g, is extracted with 0.066 *M* disodium phosphate, 900 ml, for 2 hours at 37° and then for 3 hours at room temperature. The mixture is centrifuged at 4° for 80 minutes at 3000 *g*. The extract is heated for 15 minutes at 55°, cooled rapidly, and centrifuged at 4° for 10 minutes.

Step 2. Precipitation with Acetone. The enzyme solution, 410 ml, is maintained at 0° to —2° while cold acetone, 205 ml, is added. The precipitate that forms is removed by centrifugation at 4000 *g* for 15 minutes at 0°. Cold acetone, 205 ml, is added to the supernatant solution, and the precipitate is removed by centrifugation, dissolved in a minimal amount of cold water, and dialyzed overnight against distilled water at 4°.

Step 3. Precipitation with Ammonium Sulfate. To the dialyzed solu-

[3] A. G. Gornall, C. J. Bardawill, and M. M. David, *J. Biol. Chem.* **177,** 751 (1949).
[4] O. Warburg and W. Christian, *Biochem. Z.* **310,** 384 (1941).

tion from step 2, 39 g of solid ammonium sulfate is added per 100 ml of solution. The preparation is centrifuged at 0° for 15 minutes at 12,000 *g*, and the precipitate is discarded. To the supernatant fluid an additional 18 g of ammonium sulfate is added per 100 ml of solution. After centrifugation, the precipitate is dissolved in a minimal amount of cold water. This fraction can be stored for at least several months at —20° without loss of activity. At this stage, most of the alcohol dehydrogenase is removed, but residual DPNH oxidase, thiolesterase, and glyoxalase I and II activities interfere with further investigations of certain properties of the dehydrogenase.

Step 4. Calcium Phosphate Gel Treatment. The active ammonium sulfate fraction is dialyzed overnight at 4° against distilled water. After the pH is adjusted to 6, a calcium phosphate gel suspension[5] is added in small increments until approximately 80% of the enzyme is adsorbed. The dehydrogenase is eluted from the gel with five 10-ml portions of 0.1 *M* potassium phosphate buffer, pH 7.4. Eluates with high specific activities are combined. The enzyme is unstable at this point. It cannot be stored at 0° for more than 1 or 2 hours and loses considerable activity upon freezing. Preparations stable to freezing are obtained by adding solid ammonium sulfate to saturation, removing the supernatant liquid by centrifugation, and dissolving the precipitate in the minimal volume of a solution containing phosphate buffer (pH 7.5, 0.05 *M*), EDTA (0.01 *M*), and bovine serum albumin (4 mg/ml). DPNH (NADH) oxidase and glyoxalase II are no longer detected, but glyoxalase I is still present.

The results of the purification procedure are summarized in the table.

PURIFICATION OF FORMALDEHYDE DEHYDROGENASE FROM BAKERS' YEAST

Fraction	Volume (ml)	Units per ml	Protein (mg/ml)	Specific activity
Crude extract	450	0.38	44	0.0086
Heated extract	410	0.34	26	0.013
Acetone fraction	138	1.36	24	0.057
Ammonium sulfate precipitate	14	6.60	75	0.088
Combined eluates from calcium phosphate gel	54	0.64	1	0.64

Properties

Specificity. Glutathione is required for the reaction; no activity is obtained upon substitution of cysteine, 2,3-dimercaptopropanol, or thioglycolate. DPN (NAD) cannot be replaced by TPN (NADP). Methyl-

[5] D. Keilin and E. F. Hartree, *Proc. Roy. Soc.* **B124,** 397 (1938).

glyoxal and glyoxal, as well as formaldehyde causes DPN (NAD) to be reduced in the presence of the enzyme. Aldehydes found inactive as substrates are acetaldehyde, glycolaldehyde, glyoxylic acid, benzaldehyde, and glucosone.

Other Properties. The enzyme is active between pH 6 and 8.5, with an optimum at pH 8.0. The K_m for DPN is 6.8×10^{-5} *M*. HCHO inhibits the enzyme under some conditions. This inhibition can be overcome by the addition of adequate amounts of glutathione. These findings made it difficult to obtain meaningful K_m values for the substrates and favor the idea that the actual substrate of the reaction is the thiohemiacetal of glutathione and formaldehyde. The spontaneous formation of such addition compounds is well known.[6]

S-Formylglutathione,[7] a proposed intermediate of the reaction, causes a rapid oxidation of DPNH (NADH), but not of TPNH (NADPH) in the presence of the enzyme. Addition of *S*-acetylglutathione[7] or *S*-lactylglutathione[8] to the enzyme does not result in DPNH (NADH) oxidation, nor is either inhibitory.

Synthetic *S*-formylglutathione is hydrolyzed rapidly by the most highly purified enzyme preparations obtained; *S*-acetylglutathione is hydrolyzed very slowly, and *S*-lactylglutathione not at all.

[6] M. P. Schubert, *J. Biol. Chem.* **114,** 341 (1936).

[7] *S*-Formylglutathione is synthesized by the anhydride method used by T. Wieland and H. Köppe [*Ann. Chem.* **581,** 1 (1953)] for the synthesis of *S*-lactylglutathione. The acylating agent was formic acid (98–100%). *S*-Acetylglutathione was similarly prepared from glacial acetic acid.

[8] E. Racker, *J. Biol. Chem.* **190,** 685 (1951). See also I. A. Rose, *Biochim. Biophys. Acta* **25,** 214 (1957).

[67] Formate Dehydrogenase

By J. R. Quayle

$$\text{Formate}^- + \text{DPN}^+ + H_2O \rightarrow HCO_3^- + \text{DPNH} + H^+$$

Assay Method

Principle. This assay[1] depends on measurement of the rate of increase of optical density at 340 mμ consequent on the reduction of DPN in the presence of formate.

[1] P. A. Johnson, M. C. Jones-Mortimer, and J. R. Quayle, *Biochim. Biophys Acta* **89,** 351 (1964).

Reagents

Sodium phosphate buffer, 0.2 *M* pH 7.0
Sodium formate, 0.2 *M*
DPN, 0.01 *M*

Procedure. The complete reaction system in a 1.5 ml silica cell (1 cm light path) consists of 0.25 ml of phosphate buffer, 0.25 ml of sodium formate, 0.1 ml of DPN, enzyme extract, and water to a total volume of 1 ml. The blank cell lacks formate and DPN. The reaction is started by addition of enzyme to both cells, and the rate of increase of extinction at 340 mμ is measured. Alternatively, enzyme may be placed initially in both test and blank cell and the reaction be started by adding formate to the test cell. If, however, the reaction is started by the addition of DPN to the test cell, considerably lower rates of reaction are observed. This last procedure is not recommended. In crude extracts the rate of DPN reduction is linear for only a short time owing to DPNH oxidase activity. The initial rate of reaction is proportional to the quantity of enzyme present over the range 0.002–0.02 unit of activity.

Definition of Unit and Specific Activity. One unit of enzyme is defined as the amount of enzyme that catalyzes the reduction of 1 micromole of DPN per minute under the conditions of assay. Specific activity is expressed as units of enzyme per milligram of protein.

Growth Conditions

Cultures of *Pseudomonas oxalaticus* (strain OX 1) may be obtained from the National Collection of Industrial Bacteria, Torry Research Station, Aberdeen, Scotland (Culture No. 8642) and maintained on slopes consisting of nutrient agar (Oxo Ltd., London) to which 5 m*M* sodium or potassium oxalate is added. The organism is subcultured every month onto fresh slopes, grown at 30°, and stored at 2°.

After several months of subculture the organism sometimes becomes sluggish in growing up on oxalate or formate as sole carbon source and is best discarded. It is therefore advisable to make several freeze-dried cultures from the parent culture, for storage at 2°.

The organism is grown in 10-liter batches under forced aeration at 25–30° in a growth medium consisting of: 25 g of $(NH_4)_2SO_4$, 2 g of $MgSO_4 \cdot 7\ H_2O$, 0.01 g of $CaCl_2 \cdot 2\ H_2O$, 0.05 g of $FeSO_4 \cdot 7\ H_2O$, 0.025 g of $MnSO_4 \cdot 5\ H_2O$, 0.025 g of $Na_2MoO_4 \cdot 2\ H_2O$, 87 g of K_2HPO_4, 78 g of $NaH_2PO_4 \cdot 2\ H_2O$, and 67 g of disodium oxalate, dissolved in 10 liters. The organism may also be grown on formate as carbon source, but the amount of enzyme obtained is less than that from oxalate. The pH of the medium is maintained at 7.5–8.0 as growth proceeds by addition of oxalic acid. The bacteria are harvested by centrifuging while in the logarithmic phase of growth at a cell density of approximately 8 g wet weight of cell paste per 10 liters and stored at —15°. It is of crucial importance that the culture is not allowed to become more alkaline than pH 8.0 and that the cells are growing rapidly (mean generation time of 3–4 hours) when harvested. The activity of the enzyme greatly decreases at the end of the exponential growth phase, and only very poor purification of the enzyme from such cells is obtained.

Purification Procedure[1]

Step 1. Preparation of Cell-Free Extract. Bacteria (8 g wet weight) are either crushed in a Hughes press at —25° or suspended in 24 ml of 0.04 *M* phosphate buffer (pH 7.5) and passed twice through a French pressure cell at 2°. The crushed cells from the Hughes press treatment are suspended in 24 ml of 0.04 *M* phosphate buffer (pH 7.5). The crude extracts from either type of breakage method are then handled in the same way at 2°. Crystalline DNase and RNase (about 200 μg of each) (L. Light and Co., Ltd., Colnbrook, Bucks, England) are added, and the extract is centrifuged at 25,000 *g* for 20 minutes.

Step 2. Protamine Sulfate Treatment. To the supernatant is added protamine sulfate in the proportion of 1 part to 10 parts of bacterial protein (w/w). The precipitate is removed by centrifuging and discarded.

Step 3. First Ammonium Sulfate Precipitation. To the supernatant is added sufficient saturated ammonium sulfate solution to bring the ammonium sulfate concentration to 30% of saturation. After 10 minutes the resulting precipitate (if any) is removed by centrifuging. More ammonium sulfate solution is added to raise the final concentration to 40%. The resulting precipitate is collected after 10 minutes by centrifuging and dissolved in 3 ml of 0.04 *M* phosphate buffer (pH 7.5). This solution of enzyme is reasonably stable to storage at —15°. It is advisable to store any quantity of enzyme at this stage and purify it further as needed, as it becomes less stable on further purification.

Step 4. Second Ammonium Sulfate Precipitation. The solution of enzyme obtained from the first ammonium sulfate precipitation is adjusted to pH 6.0 by addition of 0.5 *M* KH_2PO_4. Solid ammonium sulfate is then added to give 30% of saturation; the resulting precipitate is collected by centrifuging and dissolved in 2 ml of 0.04 *M* phosphate buffer (pH 7.5) to give the required enzyme fraction. Step 4 should be performed quickly, and dilution of the enzyme solution should be avoided.

A summary of the purification procedure is given in the table.

PURIFICATION PROCEDURE

Step	Volume (ml)	Activity (units/ml)	Protein (mg/ml)	Specific activity (units/mg)	Yield (%)
1. Cell-free extract	21.5	2.7	14.2	0.19	100
2. Treatment with protamine sulfate	21.5	2.8	13.3	0.21	100
3. 30–40% $(NH_4)_2SO_4$ precipitate	3.6	6.1	13.5	0.45	38
4. 0–30% $(NH_4)_2SO_4$ precipitate	2.9	7.6	16.5	0.46	38

Properties

Specificity. As far as has been tested, the purified enzyme is specific for formate: no activity has been detected with methanol, ethanol, formaldehyde, oxalate, acetate, pyruvate, or malate. The preparation contains slight DPNH oxidase activity; otherwise it appears to be free of interfering enzymes.

Activators and Inhibitors. No metal-ion requirements have been found. After 30 minutes preincubation with the following inhibitors at 1 mM concentration, the resulting inhibition of the enzyme is given by the respective percentage figures: potassium fluoride, 80; sodium azide, 90; potassium cyanide, 100; sodium hypophosphite, 80.

Equilibrium. The equilibrium of the reaction strongly favors dehydrogenation of formate. Using listed thermodynamic data,[2] the $\Delta F'$ at pH 7 of the reaction is calculated to be —7.95 or —4.15 kcal/mole, depending upon which value for the free energy of formation of formate is used.

Stability. The enzyme obtained from step 4 loses half its activity after storage at —15° for 1 month. It is less stable if stored at 2°; suspension in 2.7 M ammonium sulfate does not stabilize it.

pH Optimum and Kinetic Properties. The pH optimum is at 7.6. The K_m for formate at pH 7.5 is 7.7×10^{-5} M; the K_m for DPN under these conditions is 6.4×10^{-4} M.

Use of the Enzyme for Determination of Formate

The enzyme may be used for the direct spectrophotometric assay of formate.[1]

For this assay the following complete mixture is made up in a 1.5-ml

[2] *In* "Biochemists' Handbook" (C. Long, ed.), p. 90. Spon, London, 1961.

quartz cuvette (1 cm light path): 50 micromoles of phosphate buffer (pH 7–7.5), 1 micromole of DPN, sample containing 0.05–0.15 micromole of formate, 0.1–0.2 unit of enzyme, water to 1 ml. The blank cell contains no formate. The reaction is started with enzyme and followed by change of absorbancy at 340 mμ; this should be complete in approximately 5 minutes. The enzyme preparation catalyzes a slow removal of DPNH as evidenced by a decrease in the absorbancy at 340 mμ at the end of the reaction. The effect of this on the estimation of the formate is corrected for by extrapolating the steady rate of decrease to the time of commencement of the reaction and taking the resultant value of the absorbancy as the correct one. The assay is specific for formate, and accuracy within $\pm 6\%$ should be obtained.

The only other DPN-linked formate dehydrogenase which has been purified is that from pea seeds,[3] but its affinity for formate is too low for it to be used in the assay of micromolar amounts of formate.

[3] M. B. Mathews and B. Vennesland, *J. Biol. Chem.* **185**, 667 (1950).

[68] Aldehyde Oxidase

By K. V. Rajagopalan and P. Handler

$$R{-}CHO + H_2O + O_2 \rightarrow R{-}COOH + H_2O_2 \quad (1)$$

$$N^1\text{Methylnicotinamide} + H_2O + O_2 \rightarrow N^1\text{-methyl-6-pyridone-3-carboxamide} + H_2O_2 \quad (2)$$

Assay Method

Principle. The reaction catalyzed by aldehyde oxidase is essentially a hydroxylation of the substrate from the elements of water. Hence the role of molecular oxygen is only that of an electron acceptor. Several artificial electron acceptors may be used instead of oxygen. The aerobic oxidation of N^1-methylnicotinamide provides a simple routine assay for the enzyme, since the formation of the pyridone can be measured spectrophotometrically as an increase in absorbance at 300 mμ. The use of ferricyanide as an artificial electron acceptor provides a more general spectrophotometric assay for the enzyme and can be used with a wide variety of substrates. Since the rate of ferricyanide reduction is the same aerobically or anaerobically, the ferricyanide assay can be carried out in cuvettes open to air.

Reagents

FOR ROUTINE ASSAY

Potassium phosphate buffer, 0.05 *M*, pH 7.8, containing 0.005% EDTA
Solution of N^1-methylnicotinamide chloride in water (NMN), 0.3 *M*
Enzyme, 6 units/ml, in buffer

FOR FERRICYANIDE ASSAY

Phosphate buffer, as above
Potassium ferricyanide solution, 0.01 *M*, in buffer
Enzyme, 10–12 units/ml, in buffer

Procedure. ROUTINE ASSAY. To 0.05 ml of substrate solution and 2.9 ml of buffer in a quartz cuvette, 0.05 ml of enzyme is added. After mixing, the increase in optical density at 300 mμ is measured spectrophotometrically. Readings are taken in a Zeiss spectrophotometer at 15-second intervals. The change in optical density in the first minute of reaction is taken as the rate of enzyme activity. If a recording spectrophotometer is used, only the initial linear portion of the tracing is taken into account.

FERRICYANIDE ASSAY. Since reduction of ferricyanide results in a decrease in absorbance at 420 mμ, measurement of small changes in absorbance is best carried out by the "blank reversal" technique. A typical assay system is as follows:

	Blank (ml)	Reference (ml)
NMN	0.05	0.05
Ferricyanide	0.30	0.30
Buffer	2.60	2.65
Enzyme, at 0 time	0.05	—

Enzyme is added to the blank cuvette at 0 time, and the blank cuvette is adjusted to 100% transmission at 420 mμ at precisely 30-second intervals and the absorbance of the reference cuvette measured each time. The decrease in absorbance in the blank cuvette thus appears as an increase in absorbance in the reference cuvette. A Zeiss or a Beckmann DU spectrophotometer can be used in this assay. When a recording spectrophotometer is used, the cuvette with enzyme is placed in the blank compartment and the cuvette without enzyme is used as the reference. When testing for possible substrates it will be necessary to use several concentrations of the substrate for setting up optimum conditions.

Definition of Unit and Specific Activity. A unit of enzyme activity is that amount of enzyme which produces an absorbancy change of 1 per minute per cm at 25° with NMN as substrate in the routine assay. Specific activity is defined in terms of units of activity per milligram of protein. Protein concentration may be determined by any of several standard methods.

Purification Procedure

The following is a modification of a previously published procedure.[1]

Frozen washed livers from adult rabbits are obtained from Pel-Freez Rabbit Biologicals, Rogers, Arkansas. The livers can be stored frozen for several months without loss of enzyme activity. Two pounds of frozen liver are allowed to thaw for 15–30 minutes, cut up into chunks, and then homogenized for 3 minutes in 5 volumes of distilled water at room temperature. The homogenate is rapidly heated, with continuous stirring, to 70° by placing it in a boiling water bath. After 2 minutes at 70°, the homogenate is cooled to 55° by addition of ice and then transferred to an ice bath and permitted to cool to 10–15° with stirring. The suspension is centrifuged at 8000 rpm for 20 minutes in a Servall refrigerated centrifuge, and the supernatant is then filtered through glass wool to yield a clear brown solution. Subsequent steps in purification are carried out at 0–4° in ice.

Solid ammonium sulfate is added to the solution to 50% saturation and the precipitate collected by centrifugation at 8000 rpm as above. The precipitate is dissolved in a small volume of 0.05 *M* potassium phosphate, pH 7.8, containing 0.005% EDTA, and the pH is carefully adjusted to 7.8–8.0 by the addition of 5 *N* NaOH. The solution is made up to 80 ml and stored at 4° overnight.

The solution obtained in the previous step is cooled in an ice bath, and acetone, previously chilled to 0°, is added at the rate of about 10 ml per minute, with stirring, to a final concentration of 43%. The bulky precipitate obtained at this stage is readily removed by centrifugation for 5 minutes at 3000 rpm in an International refrigerated centrifuge. Cold acetone is then added to the yellowish brown supernatant to a final concentration of 50%. The precipitate is collected by centrifugation at 3000 rpm as above and then dissolved in minimum volume of cold distilled water containing 0.005% EDTA. The solution is then dialyzed overnight against a large volume of cold distilled water containing 0.005% EDTA.

[1] K. V. Rajagopalan, I. Fridovich, and P. Handler, *J. Biol. Chem.* **237**, 922 (1962).

It is essential that only dialysis tubing treated with boiling EDTA solution (mM) be used for dialyzing the enzyme at any stage. Similarly, all solutions must contain 0.005% EDTA to prevent inactivation of enzyme by heavy metals present in buffer salts and perhaps in distilled water.

The enzyme solution is centrifuged to remove insoluble material, and calcium phosphate gel[2] is then added until all the activity is adsorbed onto the gel. The gel is collected by centrifugation and washed with cold distilled water by stirring for 20 minutes. After centrifugation the washing is repeated, once with distilled water and once with 0.003 M potassium phosphate buffer, pH 7.8. The enzyme is then extracted from the gel by stirring for 30 minutes with 0.02 M potassium phosphate, pH 7.8, followed by centrifugation. Under optimal conditions the clear brownish red supernatant may have as much as 75% of the activity originally present in the heated supernatant.

The enzyme is precipitated by addition of solid ammonium sulfate to 50% saturation to the solution from the previous step. The precipitate is collected by centrifugation at 18,000 rpm in a Servall refrigerated centrifuge, dissolved in a small volume of 0.005 M potassium phosphate, pH 7.8, and dialyzed overnight against the same buffer.

The dialyzed solution is then applied on a 2×40 cm column of DEAE-cellulose equilibrated with 0.005 M potassium phosphate, pH 7.8. Elution is effected with a linear gradient of the same buffer between 0.005 M and 0.3 M. The eluates containing enzyme with highest specific activity are pooled and treated with solid ammonium sulfate to 50% concentration. After centrifugation, the precipitate is dissolved in 0.05 M potassium phosphate, pH 7.8, and the pH of the solution adjusted to 7.8–8.0. The preparation thus obtained is ultracentrifugally homogeneous.

Properties

Rabbit liver aldehyde oxidase has a molecular weight of about 280,000 with $S_{20,w}$ of 11·4 S, and contains 2 moles of FAD, 2 g atoms of molybdenum and 8 g atoms of nonheme iron per mole of enzyme as its prosthetic groups. Significant amounts of coenzyme Q_{10} have also been detected in the purified enzyme.[1] Flavin and nonheme iron contribute to the absorption spectrum of the enzyme in the region 300–600 mμ.[3]

Specificity. The enzyme is capable of oxidizing aliphatic and aromatic aldehydes such as acetaldehyde, crotonaldehyde, benzaldehyde,[4] pyri-

[2] T. P. Singer and E. B. Kearney, *Arch. Biochem. Biophys.* **29**, 190 (1950).
[3] K. V. Rajagopalan and P. Handler, *J. Biol. Chem.* **239**, 1509 (1964).
[4] W. E. Knox, *J. Biol. Chem.* **163**, 699 (1946).

doxal,[5] and salicylaldehyde. A wide variety of aromatic heterocyclic compounds containing quaternary or tertiary nitrogen as the heteroatom are also oxidized by the enzyme.[4,6] Representative of this group of substrates are purine, quinoline, and N^1-methylnicotinamide. Electron acceptors include molecular oxygen, several dyes, ferricyanide, silicomolybdate, and organic nitrocompounds.[7] Reduction of cytochrome *c* by the enzyme is oxygen dependent.[1]

Inhibitors. Aldehyde oxidase is progressively and irreversibly inhibited by cyanide,[5] and resembles xanthine oxidase in this respect. However, unlike xanthine oxidase, aldehyde oxidase is susceptible to strong inhibition by a wide variety of reagents which include amytal, antimycin A, oligomycin, hormones like estradiol and progesterone, several quinones and nonionic detergents like Triton X 100.[7]

Comments

The name aldehyde oxidase was originally given to an enzyme from hog liver by Gordon *et al.*[8] to distinguish it from xanthine oxidase, which also can oxidize aldehydes but can in addition oxidize several purines. A similar enzyme was later isolated from rabbit liver by Knox.[4] Subsequent studies have shown[4,6] that the substrate specificity of aldehyde oxidase is even broader than that of xanthine oxidase. It would seem, therefore, that it is a euphemism to call this enzyme aldehyde oxidase.

[5] J. Hurwitz, *J. Biol. Chem.* **212,** 757 (1955).
[6] K. V. Rajagopalan and P. Handler, *J. Biol. Chem.* **239,** 2027 (1964).
[7] K. V. Rajagopalan and P. Handler, *J. Biol. Chem.* **239,** 2022 (1964).
[8] A. H. Gordon, D. E. Green, and V. Subrahmanyan, *Biochem. J.* **34,** 764 (1940).

Section IV

Kinases and Transphosphorylases

[69a] Hexokinase

I. Brain

By MIRA D. JOSHI and V. JAGANNATHAN

Hexose + ATP → hexose 6-phosphate + ADP

The preparation of a particulate form of brain hexokinase has been described.[1] A method for obtaining this enzyme in soluble form[2] and its purification are described here.

Assay Method

Principle. Glucose 6-phosphate formed by the hexokinase reaction is measured by adding glucose 6-phosphate dehydrogenase and NADP and following NADPH formation.[3] This method minimizes inhibition due to glucose 6-phosphate by oxidizing it to 6-phosphogluconic acid.

Reagents

Glucose, 0.15 *M*
$MgCl_2 \cdot 6\ H_2O$, 0.2 *M*
Tris-HCl buffer, 0.2 *M*, pH 7.6
NADP, 0.0013 *M*
Disodium ethylenediaminetetraacetate (EDTA), 0.0001 *M*
ATP, sodium salt, 0.3 *M*, pH 7.6
Glucose 6-phosphate dehydrogenase, 2 units/ml (for activity determination, see Vol. I [42]). It should have negligible glucose 6-phosphatase, NADPase, NADPH oxidase, hexokinase, glucose-NADP reductase, and 6-phosphogluconic dehydrogenase.[4]

Procedure. Mix 0.3 ml each of glucose, $MgCl_2$, Tris, EDTA, NADP, and glucose 6-P dehydrogenase and 1.0 ml of water in a cuvette (1-cm light path): add 0.1 ml of hexokinase and then 0.1 ml of ATP and note the rate of change in optical density at 340 mμ. The change in optical density should be between 0.005 and 0.020 per minute and is measured from the second to the tenth minute after adding ATP. Correct for glu-

[1] See Vol. I [33].
[2] V. Jagannathan, *Indian J. Chem.* **1**, 192 (1963).
[3] M. W. Slein, G. T. Cori, and C. F. Cori, *J. Biol. Chem.* **186**, 763 (1950).
[4] Glucose 6-phosphate dehydrogenase, Type V, Sigma Chemical Co., St. Louis, Missouri, has been routinely used.

cose-NADP reductase and hexokinase activities of the glucose 6-P dehydrogenase by measuring NADP reduction in controls without ATP and without brain hexokinase, respectively. (ΔO.D. of the blanks should not be more than 0.001 per minute.) Crude hexokinase is diluted with 0.1 *M* phosphate, pH 7.6, and the purified enzyme with buffer containing glucose (0.5 *M*), serum albumin (0.01%), and thioethanol (0.001 *M*).

Definition of Unit and Specific Activity. One unit is defined as that amount of enzyme which catalyzes the formation of 1 micromole of glucose 6-P (which corresponds to the reduction of 1 micromole of NADP) per minute at 30°. Specific activity is defined as the units of enzyme per milligram of protein. Protein is determined either by Warburg's ultraviolet absorption method or Lowry's colorimetric procedure (see Vol. III [73]). Dialysis is required to remove thioethanol (TE), phosphate (more than 0.1 *M*), glucose, and ammonium sulfate which interfere in Lowry's method. The Warburg method is used for routine determinations. A blank with the same concentration of TE and buffer as in the enzyme solution is used for correction for the UV absorption of TE.

Purification Procedure

All operations are carried out at 0–4° unless stated otherwise. All solutions are made with glass-distilled water, and phosphate buffers are prepared from potassium salts.

Step 1. Preparation of Particulate Brain Hexokinase. The particulate enzyme is prepared according to the method of Crane and Sols.[1] Wash fresh ox brains with water and scrape the cortex free from the white matter. Homogenize 400 g of cortex with 800 ml of 0.1 *M* phosphate, pH 6.8, for 3 minutes in a Waring blendor in two lots (maximum temperature 10–12°). Mix with 400 ml of the same buffer and centrifuge the homogenate at 800 *g* for 20 minutes. Mix the residue with 1200 ml of the same buffer and recentrifuge as before. Centrifuge the combined supernatants at 3500 *g* for 30 minutes. Suspend the sediment in 150 ml of 0.05 *M* phosphate, pH 6.2, and centrifuge at 6000 *g* for 25 minutes. Disperse the residue in 0.05 *M* phosphate, pH 7.2, to give a final volume of 90 ml.

Step 2. Elastase Treatment. Mix 90 ml of the particulate enzyme with 18 ml of 0.09 *M* sodium pyrophosphate-HCl buffer, pH 8.4, and 2.6 mg of twice crystallized pancreatic elastase and keep at 0° for 24 hours with occasional mixing. Freeze and thaw the enzyme 5 or 6 times by keeping at —20° for 24 hours and then thawing slowly at 5°. Centrifuge the enzyme, which becomes a thick jelly at this stage, at 14,000 *g* for 50 minutes to obtain soluble hexokinase. The enzyme is not sedimentable at 100,000 *g* in 2 hours. In all subsequent steps the enzyme is stored at —20° when not in use.

Step 3. Protamine Sulfate and Ammonium Sulfate Treatment. Mix the soluble enzyme with 0.1 volume of 1 *M* phosphate, pH 7.5, and then with 0.2 volume of 5% protamine sulfate (final pH about 7.2–7.3). Centrifuge at 14,000 *g* for 30 minutes. Add ammonium sulfate to the supernatant (22.5 g for 100 ml) and centrifuge as before. Precipitate the enzyme by further addition of 33.4 g of ammonium sulfate for every 100 ml of supernatant. Collect the precipitate by centrifugation, dissolve it in 0.04 *M* phosphate-0.001 *M* TE, pH 7.5, and dialyze with gentle stirring against three changes of the same buffer for 3–4 hours.

Step 4. Calcium Phosphate Gel Treatment. Centrifuge the dialyzed enzyme if it is turbid, and add calcium phosphate gel and phosphate buffer (final concentrations, 2 units of enzyme and 30 mg of gel per milliliter of 0.05 *M* phosphate, pH 7.5). Stir occasionally for 30 minutes, centrifuge, and wash the gel 3 or 4 times with 100-ml lots of 0.05 *M* phosphate, pH 7.5. Elute the gel successively with 90–100 ml and 30–40 ml of 0.2 *M* phosphate, pH 7.5. Add 3.6 g of glucose and 60.2 g of ammonium sulfate to every 100 ml of the combined eluates, and after 40 minutes centrifuge at 14,000 *g* for 60 minutes. Dissolve the precipitate in 5–10 ml of 0.005 *M* phosphate–0.001 *M* TE, pH 7.5, dialyze against three changes of the same buffer, and clarify by centrifugation if necessary.

Step 5. DEAE-Cellulose Chromatography. Wash DEAE-cellulose successively with 1 *N* NaOH, water, 1 *N* HCl, water, 1 *N* NaOH and finally with water until free from alkali. Equilibrate with 0.5 *M* phosphate, pH 7.5, wash 3 or 4 times with water and finally equilibrate with 0.005 *M* phosphate–0.001 *M* TE–0.2 *M* glucose, pH 7.5, and suspend in the same buffer. Prepare a 5 × 9-cm column containing about 25 g of cellulose with a layer of acid-washed glass wool at the bottom, and pass 400 ml of the same buffer through the column under gentle pressure. Subsequent operations are carried out without applying pressure. Add about 5–10 ml of enzyme (about 25 mg of total protein) to the column, rinse the sides twice with 5 ml and then wash with 400 ml of 0.01 *M* phosphate–0.001 *M* TE–0.2 *M* glucose, pH 7.5. Elute with 0.07 *M* phosphate–0.001 *M* TE–0.2 *M* glucose, pH 7.5, and collect 5-ml fractions. Assay each fraction for activity and protein (by the Warburg method), and pool fractions having a specific activity of more than 25. Precipitate the enzyme by adding 60.2 g of ammonium sulfate per 100 ml and after 1 hour centrifuge at 14,000 *g* for 1 hour. Dissolve the precipitate in 0.1 *M* phosphate-0.01 *M* TE–0.5 *M* glucose, pH 7.5, and centrifuge to remove any insoluble material. The enzyme can be dialyzed against buffer of the same composition to remove ammonium sulfate. It is stored at —20.° A summary of the purification procedure is given in the table.

PURIFICATION PROCEDURE FOR BRAIN HEXOKINASE

Step No.	Fraction	Volume (ml)	Total units	Specific activity (units/mg protein)
	Crude extract	2100	2100	0.05–0.08
1	Particulate preparation	90	610	0.3
2	Elastase treatment	76	296	0.8
3	Protamine and ammonium sulfate treatment	11	272	1.1
4	Calcium phosphate gel treatment	12	146	6
5	DEAE-cellulose	2	108	31

Properties

The enzyme obtained after gel elution can be kept at —20° (or at 0° in 0.5 saturated ammonium sulfate) for a few days, but more purified preparations rapidly lost activity even at —20°. Further work was possible only when it was found that high concentrations of glucose or sucrose (0.5–1.0 *M*) have a marked stabilizing effect on the enzyme. The purified enzyme (25 units/ml) is stable in 0.5 *M* glucose, 0.1 *M* phosphate-0.01 *M* TE, pH 7.5 for at least 2 weeks at —20°.

The properties of the soluble enzyme are in general very similar to those of the particulate preparation with regard to its activity with different substrates, K_m values for ATP, Mg^{++}, and hexose and inhibition by glucose 6-P and ADP (see Vol. I [33]). Fructose is phosphorylated at 1.4 times the rate with glucose, and the rate with ITP is about 12% that with ATP. GTP and UTP are inactive as phosphate donors. Iodoacetate (8×10^{-3} *M*) and *p*-chloromercuribenzoate (2.5×10^{-7} *M*) cause complete inhibition of the enzyme. Cysteine, insulin, and avidin show no significant effect on enzyme activity.

Notes on Purification Procedure

1. Reproducible results have been obtained with commercial samples of crystalline elastase and with enzyme prepared according to the procedure of Lewis *et al.*[5]

[5] U. J. Lewis, D. E. Williams, and N. G. Brink, *J. Biol. Chem.* **222**, 705 (1956).

2. Freezing and thawing are necessary in addition to elastase treatment for obtaining the soluble enzyme. About 40–60% of the activity is obtained in soluble form. The insoluble residue can be rehomogenized with phosphate and pyrophosphate buffer and treated with more elastase to obtain an additional 20–30% of the enzyme in soluble form, but it has not been used for further purification.

3. Enzyme with a specific activity of 5 or more is highly susceptible to inhibition or inactivation by contact with several commonly used materials (glassware washed with chromic acid, some varieties of rubber or plastic tubing, stopcock grease, sintered glass, and detergents). Columns are set up with only ground glass joints and stopcocks without grease. All glassware is successively washed with sodium carbonate solution, water, dilute nitric acid, and finally with glass-distilled water. Dialysis tubing is boiled with water for 10 minutes and rinsed with water.

4. Calcium phosphate gel is prepared according to Swingle and Tiselius.[6] A preliminary experiment with a small amount of enzyme should be carried out with calcium phosphate gel since the amount of gel required for adsorption and of buffer for elution varies slightly with different batches of gel. Protamine sulfate (ex-herring) (Koch-Light Laboratories Ltd., Colnbrook, Bucks, England) and salmine sulfate (British Drug Houses Ltd., Poole, England) have been routinely used. Other samples of protamine sulfate should be tested for their suitability.

5. DEAE-cellulose prepared according to Peterson and Sober[7] (100–200 mesh, 0.50 meq/g) and Cellex-D (0.62 meq/g) (Bio-Rad Laboratories, Richmond, California) have been used. A column of the size described in the text is used for a load of about 25 mg of total protein and several columns are run simultaneously for processing larger batches. DEAE-cellulose is not re-used.

6. Ammonium sulfate precipitations are carried out by adding small amounts of solid at a time without too rapid stirring. Care is necessary to avoid loss of enzyme during transfers, since the precipitate tends to float to the surface in concentrated ammonium sulfate.

7. The first four steps of the above procedure have been repeated about 30 times with reproducible results and ten DEAE-cellulose fractionations gave enzyme of specific activity varying from 25 to 51. Ultracentrifugally the preparation is about 85% pure.

[6] S. W. Swingle and A. Tiselius, *Biochem. J.* **48,** 171 (1951).
[7] E. A. Peterson and H. A. Sober, *J. Am. Chem. Soc.* **78,** 751 (1956).

[69b] Hexokinase

II. Bakers' Yeast (Modifications in Procedure)[1]

By I. T. Schulze, J. Gazith, and R. H. Gooding

Hexose + ATP → hexose 6-phosphate + ADP

Crystalline hexokinase prepared by the method described in Volume V [25] contains trace amounts of a proteolytic enzyme. During storage of such preparations at 4° in 50–70% saturated ammonium sulfate, or during recrystallization, the hexokinase is converted to more acidic chromatographic forms without loss of enzyme activity.[2] A DEAE-cellulose chromatographic step has now been introduced into the procedure in order to remove the protease and other contaminating proteins so that stable preparations with high specific activity can be obtained without repeated recrystallization. In addition, the preparation time before the DEAE-cellulose chromatography step has been shortened. A rapid test for detecting hexokinase in fractions from columns is included. Further, an alternative method is described for minimizing protease activity by treatment with the nontoxic protease inhibitor phenylmethanesulfonyl fluoride.[3] Finally, an alternative to the bentonite step is introduced.

Spot Test for Hexokinase

The use of a spot test for determining the presence of hexokinase in effluents from DEAE-cellulose and Sephadex gel filtration columns eliminates the time consuming process of doing spectrophotometric assays on fractions with little or no activity.

The reaction mixture used for the spot test is basically that described by Darrow and Colowick for use in the spectrophotometric assay (Vol. V [25]). The amount of ATP is increased to ensure sufficient enzymatic acid production to permit complete titration of the glycylglycine and cresol red. The reaction mixture is prepared from the reagents described in Vol. V [25] as follows: Add 3.0 ml of 0.1 *M* ATP to 14 ml of solution containing 0.006% cresol red and 1.6% $MgCl_2 \cdot 6\ H_2O$. Add 0.1 *M* NaOH until the indicator becomes reddish-purple (pH is approximately 9.0).

[1] Supported by USPHS Grant AM-03914 awarded to Dr. Sidney P. Colowick.

[2] I. T. Schulze, in preparation.

[3] D. E. Fahrney and A. M. Gold, *J. Am. Chem. Soc.* **85**, 997 (1963).

Add 3 ml of 0.1 M glycylglycine–NaOH and dilute with distilled water to 30 ml. Add 5.0 ml of 0.2 M glucose.

The reaction is carried out in spot test plates at room temperature. One drop of each fraction to be tested for hexokinase is mixed with 0.4 ml of the reaction mixture. Those wells containing high levels of hexokinase will become yellow immediately, whereas those with low levels (less than 1 unit per milliliter of the fraction tested) require longer incubation (1 hour or more).

Purification Procedure

Only those steps in the purification procedure which have been altered are described here. The enzyme yield and specific activity through step 3 are comparable to those originally described. These three steps now require 1½ working days; steps 1 and 2 are carried out on the first day.

Step 1. Preparation of Crude Extract. Prewarmed buffer is now used in the preparation of crude extract, so that 2 hours (rather than 3 hours) at 37° gives maximum enzyme yields. The autolyzate is cooled in ice to 10° to prevent foaming during centrifugation.

Step 2. Ammonium Sulfate Fractionation. This step in the procedure has not been altered except that a simple apparatus for continuous flow dialysis has been devised by which the ammonium sulfate can be removed from the redissolved enzyme by overnight dialysis.

The apparatus described here accommodates 4 batches of enzyme (see Vol. V [25]). It consists of a 56 cm $\times$ 8 cm glass column fitted at the bottom with a Teflon stopcock. The top of the column is closed with a one-holed rubber stopper into which metal hooks have been inserted so that the dialysis bag can be suspended in the column. Buffer from a reservoir enters the column through the rubber stopper and is drained from the bottom at a rate which permits approximately 16 liters of buffer to pass through the apparatus overnight. This method removes essentially all the ammonium sulfate. Small amounts of ammonium sulfate if present do not, however, interfere with adsorption to bentonite as previously thought (Vol. V [25]).

Step 3. Bentonite Adsorption. As noted by Darrow and Colowick (Vol. V [25]), excess bentonite leads to poor elution. Since recent lots of bentonite[4] vary in their ability to adsorb hexokinase, it is now necessary to determine the amount of each lot which is to be used. Bentonite is added on the basis of enzyme activity, rather than optical density at

[4] Purchased from Fisher Scientific Co., Fairlawn, New Jersey.

280 mμ, without adjusting the volume of the dialyzed enzyme. An amount of bentonite which adsorbs 70–90% of the hexokinase should be used. One gram of bentonite will adsorb 7000–20,000 units of hexokinase.[5]

After the enzyme is eluted from the bentonite (see step 3, Vol. V [25]), it is precipitated from the neutralized bentonite eluate with ammonium sulfate in order to reduce the volume in subsequent steps.

The insoluble gelatinous material in the redissolved enzyme is not removed by centrifugation as previously described since this material is removed during the subsequent DEAE-cellulose chromatography.

At this stage of purity the level of contaminating protease is the highest; it is therefore important that the enzyme be stored at —50° until subsequent steps can be carried out. Freezing is necessary to prevent conversion to the more acidic forms.

Trace amounts of ammonium sulfate will prevent hexokinase from binding to DEAE-cellulose. The enzyme is therefore equilibrated with pH 6.0 succinate buffer ($5 \times 10^{-3} M$) containing EDTA ($10^{-4} M$) by passing it through a Sephadex G-25 column of sufficient size to separate the hexokinase from the ammonium sulfate. The bed volume of the column should be about 6 times the volume of the sample. Spot tests are used to locate the hexokinase and $BaCl_2$ to determine the ammonium sulfate front. The resulting enzyme solution is put directly onto DEAE-cellulose columns or is refrozen at —50°.

Step 3a. Dialysis in Acetate Buffer (Alternative to step 3). Because of difficulties with the bentonite step, it has recently been replaced as follows. The material from the ammonium sulfate fractionation is dialyzed 24 hours against acetate buffer (0.1 ionic strength, pH 4.4) instead of phosphate buffer using the continuous flow apparatus described above. The large precipitate is removed and the supernatant adjusted to pH 6.0 with NaOH. The precipitate which forms is removed, the supernatant dialyzed against succinate buffer (pH 6.0), and another precipitate removed. The specific activity of the supernatant is 20–30 units per milligram. Although this activity is only about 20% of that obtained when the bentonite adsorption (step 3) is successfully applied, step 3a is more reproducible and the material obtained is suitable for chromatography as described in step 4.

Step 4. DEAE-Cellulose Chromatography. Succinate buffers ($5 \times 10^{-3} M$) with EDTA ($10^{-4} M$) are used throughout this step.[6] DEAE-

[5] Hexokinase from some batches of dried yeast is not adsorbed by bentonite. We have found that this hexokinase is usually protease-modified enzyme. The acidic form of hexokinase produced by the yeast protease or by trypsin treatment is not adsorbed by bentonite.

[6] See K. A. Trayser and S. P. Colowick, *Arch. Biochem. Biophys.* **94**, 177 (1961).

cellulose, washed with acid and alkali as described by Peterson and Sober (Vol. V [1]) is stored in the cold as a wet pack in pH 6 buffer. Six-gram (dry weight) DEAE-cellulose[7] columns (18 cm × 1.8 cm) are loaded with 30,000–50,000 units of hexokinase and are washed with pH 6 buffer at a flow rate of 90–100 ml per hour for approximately 4 hours to remove nonadsorbed protein. This flow rate is maintained throughout elution. The hexokinase is removed from the column by eluting with pH 5.4 buffer for 1 hour and then with pH 4.5 buffer for at least 6 hours (usually overnight). All fractions positive by spot test are pooled, and the eluate is adjusted to pH 6.0 with NaOH. Hexokinase eluted in this manner contains two chromatographic forms. The two forms can be isolated separately by prolonging the elution with pH 5.4 buffer until the first fraction is removed. The second fraction is then eluted by pH 4.5 buffer.[8] Each form is rechromatographed to ensure separation. Recovery through each column is 70–90%; this dilute material however, loses activity upon standing.

Step 5. Concentration Column. The enzyme in the eluate is too dilute to permit ammonium sulfate precipitation. It is therefore concentrated by adsorbing the enzyme to a DEAE-cellulose column (10,000–15,000 units per gram of DEAE-cellulose) and eluting it in a small volume with pH 4.8 buffer containing 30% (w/v) ammonium sulfate. Columns with height to diameter ratios from 6:1 to 10:1 are used. The flow rate during adsorption of the enzyme may be 400 ml per hour; however, during elution a low flow rate (approximately 40 ml per hour) and small fraction size (5 ml) facilitate concentration. Recovery from this column is 80–90%. The enzyme is precipitated from the eluate with ammonium sulfate and the precipitate is dissolved and crystallized as previously described (step 4, Vol. V [25]). The specific activity of precrystalline preparations containing both chromatographic forms is approximately 450 units per milligram. Crystals form in approximately 1 week without seed crystals and have a specific activity of 500–600 units per milligram.

[7] Selectacel from Brown Co., Berlin, New York, or from Carl Schleicher and Schuell Co., Keene, New Hampshire.

[8] When hexokinase has undergone proteolytic modification, it is not eluted by the procedure described here, but requires sodium chloride in addition for elution. This "sodium chloride form" formerly constituted about one-half of the crystalline product [K. A. Trayser and S. P. Colowick, *Arch. Biochem. Biophys.* **94**, 177 (1961)] and was thought to be a naturally occurring "isozyme" [Kaji *et al.*,[13] *Ann. N.Y. Acad. Sci.* **94**, 798 (1961)]. The fraction now eluted at pH 5.4 was not present at all in previous preparations of crystalline enzyme, because it disappears completely during proteolytic modification. Whether the fraction eluted with pH 4.5 buffer is naturally occurring or is derived from the 5.4-elutable fraction has not been established.

One recrystallization is usually sufficient to bring the hexokinase to maximum specific activity.

That form of the enzyme eluted with pH 5.4 buffer cannot be crystallized by the regular procedure and has a specific activity of approximately 200 units per milligram, although apparently as pure as the crystalline material by several criteria. This form of the enzyme is similar to that available commercially (Boehringer-Mannheim Corp.).

When enzyme is prepared by step 3a the form eluted with pH 4.5 buffer contains a yellow contaminant but has the usual specific activity. The yellow material may be removed by overloading a DEAE-cellulose column at pH 6.0 with 50,000–100,000 units per gram dry weight. Under these conditions the yellow material adsorbs to the DEAE-cellulose and 80–90% of the hexokinase is recovered.

Use of Phenylmethanesulfonyl Fluoride (PMSF) for Inhibition of Yeast Protease. If DEAE-cellulose chromatography cannot be carried out, or if delays in the preparative procedure are unavoidable (e.g., between steps 2 and 3, where freezing the enzyme results in poor adsorption to bentonite) PMSF can be used to inhibit the protease.

When added at a final concentration of 2×10^{-3} *M* in 5% ethanol, PMSF reduces the protease activity to a level below that detectable by hydrolysis of benzoyl-L-arginine ethyl ester.[9] It also prevents the formation of the "sodium chloride form."[8]

PMSF at 20 times the desired concentration is dissolved in 95% ethanol and is immediately added to the preparation. No reduction in hexokinase activity is observed if the preparation is adequately mixed during the addition. Protease activity cannot be detected after 40 minutes at 0°.

Sufficient PMSF (2×10^{-3} *M*) can be added to the crude extract to inhibit essentially all the protease present. However, when it is added during autolysis, the resulting crude extract contains about half the protease usually present and the hexokinase yields are reduced.

Properties of the Enzyme[10]

Hexokinase prepared through step 5 is free of detectable protease when assayed using casein (Vol. II [3]) or benzoyl-L-arginine ethyl ester[9] as substrates. (First crystalline suspensions prepared without DEAE-cellulose chromatography or PMSF treatment have 1.3–4.0 tryp-

[9] See G. W. Schwert and Y. Takenaka, *Biochim. Biophys. Acta* **16**, 570 (1955).
[10] See also Vol. I [32] and Vol. V [25].

sin units per milligram of protein when assayed by the method of Schwert and Takenaka.[11])

The chromatographic behavior and electrophoretic mobility of this hexokinase remains unchanged after prolonged storage at 4° as a crystalline suspension in ammonium sulfate. However, redissolved hexokinase, stored for 2–3 weeks at 4° is converted to the more acidic molecular forms, suggesting that the protease is not completely removed.

The enzyme can be frozen or lyophilized without loss of activity. No activity is lost when dilute solutions (approximately 50 μg/ml) are incubated overnight at 30° in pH 8 Tris buffer containing EDTA.[2] This is in contrast to previous observations of unstable preparations.[12]

The enzyme retains its ATPase activity.[13]

Hexokinase, whether modified or unmodified,[8] is a tetramer at pH 5 (mol. wt. 96,000) which dissociates into four inactive subunits (mol. wt. 24,000) in acid or alkali. The subunits reassociate and full activity returns upon neutralization if oxidation of —SH groups is prevented.[14] The modified enzyme exists largely in the dimer form in neutral solution, even at low ionic strength. The unmodified enzyme is largely in tetramer form at pH 8 at low ionic strength, but is converted to a dimer (mol. wt. 48,000) at pH 8 in high ionic strength or in glucose.[2,15] Under these conditions, incubation with trace amounts of trypsin results in formation of modified enzyme which resembles the acidic forms produced by the contaminating yeast protease. This does not occur when the enzyme is in the tetramer form during incubation with trypsin.[2]

[11] A change in absorbancy at 253 mμ of 0.003 per minute is defined as one trypsin unit. Two times crystallized trypsin measured under identical conditions has approximately 3500 units per milligram.

[12] A. Kaji, *Arch. Biochem. Biophys.* **112**, 54 (1965).

[13] A. Kaji, K. Trayser, and S. P. Colowick, *Ann. N.Y. Acad. Sci.* **94**, 798 (1961).

[14] U. W. Kenkare and S. P. Colowick, *J. Biol. Chem.* **240**, 4570 (1965).

[15] H. K. Schachman, *Brookhaven Symp. Biol.* **13**, 49 (1960).

[70a] Glucokinase

I. Liver

By D. G. WALKER and M. J. PARRY

ATP + glucose → ADP + glucose 6-phosphate + H^+

In addition to one or more hexokinases of low K_m (a crude preparation of which is described from rat liver in Vol. I [33]), liver tissue of many mammalian species, but not ungulates, contains a glucokinase[1-3] which has a narrower specificity for the phosphoryl acceptor than other mammalian hexokinases. This enzyme has a comparatively high K_m value for glucose, which makes it of unique significance in the regulation of the uptake of glucose by the liver when the blood glucose level increases above the fasting level.

Assay Method

Principle. A wide variety of methods are available for following the hexokinase reaction (see Vol. I [32] and [33] and Vol. V [25]). In crude systems, however, and for following this preparation the only suitable assay procedure is that of DiPietro and Weinhouse,[4] in which the product of the reaction, glucose 6-phosphate, is rapidly oxidized by $NADP^+$ in the presence of an excess of glucose 6-phosphate dehydrogenase; the formation of NADPH is followed spectrophotometrically at 340 mμ.

Reagents

Glycylglycine buffer, 0.25 *M* pH 7.5
Magnesium sulfate, 0.075 *M*
ATP (disodium salt), 0.075 *M*, neutralized to pH 7.0
$NADP^+$, 0.0075 *M*
Glucose, 0.75 *M*
Potassium chloride, 1.0 *M*
Glucose 6-phosphate dehydrogenase diluted to contain 40 units/ml

Procedure. A known volume of the enzyme preparation is added to a cuvette, having a light path of 1 cm, containing 0.3 ml of glycylglycine buffer, 0.15 ml of magnesium sulfate, 0.1 ml of ATP, 0.1 ml of $NADP^+$, 0.15 ml of potassium chloride, 0.2 ml of glucose, 0.4 unit of glucose 6-phosphate dehydrogenase, final volume 1.5 ml, and previously equilibrated at 28°. The increase in extinction at 340 mμ is followed in a recording spectrophotometer such as the Gilford model 2000. A ΔOD of 4.14/minute corresponds to glucokinase activity of 1 micromole of glucose phosphorylated per minute when only 1 mole of $NADP^+$ is reduced per mole of glucose phosphorylated. For crude tissue preparations and during early stages of the purification procedure the presence of 6-phosphogluconate dehydrogenase activity results in up to 2 moles of $NADP^+$ being reduced per mole of glucose phosphorylated. Crude tissue prepara-

[1] D. L. DiPietro, C. Sharma, and S. Weinhouse, *Biochemistry* **1,** 455 (1962).
[2] E. Viñuela, M. Salas, and A. Sols, *J. Biol. Chem.* **238,** PC 1175 (1963).
[3] D. G. Walker, *Biochim. Biophys. Acta* **77,** 209 (1963).
[4] D. L. DiPietro and S. Weinhouse, *J. Biol. Chem.* **235,** 2542 (1960).

tions also contain low-K_m hexokinase activity, and a correction for this is applied by subtracting the activity measured in a cuvette containing 0.2 ml of 0.00375 M glucose (i.e., a final glucose concentration of 0.0005 M) and the other reagents; all activities are corrected for the very low optical density increase observed in a cuvette containing all reagents except ATP.

Another useful procedure applicable to purified preparations (but not to crude systems) involves measuring ADP formation by coupling via pyruvic kinase (with added phosphoenolpyruvate) and lactic dehydrogenase (muscle) to the oxidation of NADH. This method is of particular value in certain kinetic studies.

Definition of Unit and Specific Activity. One unit of activity is defined as that amount of enzyme catalyzing the phosphorylation of 1 micromole of glucose per minute at 28°. Specific activity is expressed as units per milligram of protein. Protein is determined in the crude preparations by a biuret method[5] and in more purified preparations by the method of Lowry *et al.*[6] Protein in the fractions from columns is rapidly monitored by Warburg's ultraviolet absorption method (see Vol. III [73]).

Purification Procedure

The following procedure requires 7 working days with certain of the column fractionations continuing overnight. The livers from 10 rats (approximately 120 g) has proved to be a convenient quantity of starting material; larger quantities would require the use of rather large columns for effective fractionation to be achieved. With minor variations the method has been performed in our laboratory during the past two years over thirty times as far as step 6 and ten times to completion and has given consistent results.

Step 1. Preparation of Crude Extract. Ten normal well-fed adult male rats of the Wistar albino strain are killed, exsanguinated, and the livers removed and chilled in ice-cold homogenizing medium (see below). All subsequent operations are performed at 0–4° in a cold room. Homogenates (33%, w/v) of each liver are prepared in a Potter-Elvehjem-type homogenizer using 2 volumes of homogenizing medium (0.15 M potassium chloride containing 0.004 M magnesium sulfate, 0.004 M EDTA and 0.004 M N-acetyl cysteine, pH 7.0) and pooled. This crude homogenate is centrifuged for 10 minutes at 40,000 g (M.S.E. high-speed 18 cen-

[5] A. G. Gornall, C. J. Bardawill, and M. M. David, *J. Biol. Chem.* **177**, 751 (1949).

[6] O. H. Lowry, N. J. Rosebrough, A. L. Farr, and R. J. Randall, *J. Biol. Chem.* **193**, 265 (1951).

trifuge; 8 × 50 ml rotor) and the supernatant is further centrifuged at 100,000 *g* for 1 hour (M.S.E. super-speed 40 centrifuge; 8 × 50 ml rotor).

Step 2. Ammonium Sulfate Fractionation. The crude supernatant extract (186 ml) is brought to 0.45 saturation with respect to ammonium sulfate by the addition, per 100 ml of extract, of 82 ml of saturated (at 4°) ammonium sulfate solution in 0.004 *M* magnesium sulfate–0.004 *M* EDTA–0.004 *M* *N*-acetyl cysteine, previously adjusted to pH 7.0 with 1.0 *N* KOH. The precipitate is removed by centrifugation for 10 minutes at 40,000 *g* and discarded. The clear 0.45 saturated supernatant solution is brought to 0.65 saturation by the addition of 57 ml of the same ammonium sulfate solution per 100 ml of this supernatant solution. The precipitated protein is collected by centrifugation for 10 minutes at 40,000 *g*. This ammonium sulfate paste can be stored at —15° for several days without appreciable loss of activity.

Immediately before the next step the ammonium sulfate paste is dissolved in a minimum volume (about 40 ml) of 0.020 *M* Tris-HCl buffer containing 0.004 *M* magnesium sulfate, 0.004 *M* EDTA, and 0.004 *M* *N*-acetyl cysteine, pH 7.0 (buffer A) to which KCl has been added to 0.1 *M* concentration. The resulting solution is placed in a cellophane dialysis bag and dialyzed with gentle agitation against 800 ml of 0.1 *M* KCl in buffer A for 1 hour.

Step 3. Batch Chromatography on DEAE-Sephadex. Batch adsorption onto this material is achieved by adding about 40 ml (containing approximately 3 g of dry adsorbent) of a suspension of DEAE-Sephadex A-50 (Medium) (Pharmacia) which has been previously equilibrated with 0.1 *M* KCl in buffer A. After 15 minutes the mixture is centrifuged for 10 minutes at 40,000 *g*. The supernatant is removed, and an aliquot is assayed for glucokinase activity. This adsorption process is repeated with further lots of adsorbent (usually two or three times) until less than 5% of the original glucokinase activity remains in the supernatant. The several lots of adsorbent are pooled and transferred to a glass column, 5 cm in diameter, using a small amount of the 0.1 *M* KCl in buffer A. The height of the material in the column is now 6–8 cm. The column is washed with two bed volumes of 0.1 *M* KCl in buffer A. The glucokinase activity is eluted at a rate of 100 ml per hour with 0.3 *M* KCl in buffer A. Fractions (8 ml) are collected, and those containing the bulk of the activity are pooled.

Step 4. Gradient Chromatography on DEAE-Sephadex. The pooled fractions (103 ml) from step 3 are diluted by the addition of an equal volume of buffer A to bring the KCl concentration to 0.15 *M*. This diluted solution is applied to a column (3 cm in diameter × 17 cm in height) of DEAE-Sephadex A-50 (medium) previously equilibrated with 0.15 *M*

KCl in buffer A. After adsorption the column is washed with 1 bed volume of this same solution and then a linear gradient of KCl in buffer A from 0.15 *M* to 0.6 *M* is applied; the total volume of eluting medium is 500 ml. Activity is eluted between approximately 0.25 *M* and 0.35 *M* KCl. Fractions (3–4 ml) are collected at a flow rate of 10 ml per hour; this operation proceeds overnight. The active fractions are pooled (18 ml).

Step 5. Gradient Chromatography on DEAE-Cellulose. The pooled fractions from step 4 are diluted with 54 ml of buffer A to give a final KCl concentration of approximately 0.1 *M*. The diluted solution is applied to a column (3 cm × 31 cm) of DEAE-cellulose powder previously equilibrated with 0.1 *M* KCl in buffer A. Activity is now eluted (overnight) by a linear gradient of KCl in buffer A from 0.1 *M* to 0.6 *M* KCl, total volume 500 ml, at a flow rate of 20 ml per hour and collecting 3–4 ml fractions. Activity again appears in the region of 0.25 *M* to 0.35 *M* KCl. The active fractions are pooled (63 ml).

Step 6. Concentration Step. The pooled fractions are diluted with 63 ml of buffer A to give a final KCl concentration just below 0.2 *M*. The diluted solution is applied to a very small column (1.0 × 1.5 cm) of DEAE Sephadex A-50 (medium) previously equilibrated with 0.2 *M* KCl in buffer A. The activity is eluted with 0.4 *M* KCl in buffer A at a flow rate of 0.25 ml per minute. Small fractions of about 6 drops each are collected manually. By this means almost all the activity can be collected into a volume less than 3 ml.

Step 7. Gel Filtration. The concentrated fraction from step 6 is applied to a column (3.5 × 30 cm) of Bio-Gel P. 225 (Bio-Rad Laboratories) previously equilibrated with 0.4 *M* KCl in buffer A. (The P. 225 column is made by mixing equal wet volumes of the P. 150 and P. 300 gels, each having previously been allowed to swell for 3 days in water.) Activity is eluted with 0.4 *M* KCl in buffer A at a rate of 3–4 ml per hour (usually overnight) and 3–4 ml fractions are collected. The fractions containing activity are pooled (about 20 ml).

Step 8. Final Concentration Step. The pooled fractions are diluted with an equal volume of buffer A, and step 6 is repeated. The activity is thus concentrated into a volume of about 2 ml.

A summary of a typical preparation is given in the table. The purified preparations (in 0.4 *M* KCl in buffer) are kept at 0° and retain essentially full activity for over a month.

Notes on Purification Procedure

1. Glucokinase activity can be precipitated from the crude homogenate supernatant over a wide range of ammonium sulfate concentrations,

PURIFICATION OF GLUCOKINASE FROM RAT LIVER

Step and fraction	Volume (ml)	Total units	Total protein (mg)	Specific activity (units/mg)	Yield (%)
1. Crude homogenate supernatant	186	96[a]	9570	0.01	100
2. Dialyzed solution of ammonium sulfate paste	37	—[b]	2700	—[b]	—
3. First DEAE-Sephadex eluate	103	50	170	0.29	52
4. Second DEAE-Sephadex eluate	18	43	31	1.4	45
5. DEAE-cellulose eluate	63	32	12	2.8	33
6. Concentration column eluate	2.0	14.6	3.0	4.9[c]	15
7. Bio-Gel eluate	22	14.3	1.8	7.9[d]	15
8. Final concentration	2.2	9.6	1.1	8.7[d]	10

[a] This activity is based on 2 moles of $NADP^+$ being reduced per mole of glucose phosphorylated; other later activities are calculated on a 1:1 basis because 6-phosphogluconate dehydrogenase activity has been removed.

[b] Activities rather lower than those recorded for the following step are usually obtained here; the contribution of 6-phosphogluconate dehyrogenase activity at this stage is difficult to assess.

[c] The specific activity in the peak fraction is over 6 units/mg.

[d] The specific activity in the peak fraction is over 11 units/mg.

but the 0.45–0.65 saturation gives the best yield assessed against protein fractionation. Further ammonium sulfate fractionations result in large losses in activity and have been abandoned.

2. Glucokinase is very sensitive to —SH inactivation and the presence of a sulfhydryl reagent through out the preparation is essential. Mercaptoethanol was used in many early trials, but its volatility, odor, and UV absorption make it inconvenient. The use of *N*-acetylcysteine has proved to be a major factor in the success of the preparation.

3. Potassium ions also stabilize glucokinase. The enzyme is assayed in the presence of 0.1 *M* KCl and the concentration of KCl never falls below this level at any stage of the preparation. The enzyme also appears to be comparatively stable while adsorbed on the columns.

4. The trace of low-K_m hexokinase activity remaining at the end of step 2 is removed during step 3. 6-Phosphogluconate dehydrogenase activity is also removed completely during step 3.

5. In each of the fractionation steps involving elution from columns (steps 3, 4, and 5), the glucokinase activity is eluted just after the main protein peak but is not completely separated in the first two of these steps. A fairly detailed survey of both glucokinase activity and protein concentration in each of the column fractions is necessary to decide which of them to pool and pass to the next stage.

Properties

The preparation represents a 850-fold purification over the original supernatant and the highest specific activity for a mammalian hexokinase yet reported, but it is by no means pure. The following enzymes are absent: hexokinase, 6-phosphogluconate dehydrogenase, glucose 6-phosphate dehydrogenase, phosphoglucoseisomerase, phosphomannoseisomerase, phosphoglucomutase, and glucose 6-phosphatase. Insufficient material has yet been obtained for analytical studies, but several protein bands are evident after electrophoresis on a 7.5% polyacrylamide gel.

Substrate Specificity. The specificity of rat hepatic glucokinase is much narrower than for other mammalian hexokinases (see Vol. I [33]). The relative maximum rates[7] are glucose 1.0, mannose 0.9, and 2-deoxyglucose 0.5; fructose is also phosphorylated (see below); other analogs tested had rates less than 0.01. The K_m values[7] for those sugars phosphorylated are glucose $2 \times 10^{-2}\,M$, mannose $5 \times 10^{-2}\,M$, and 2-deoxyglucose $9.5 \times 10^{-2}\,M$. By extrapolation of Lineweaver-Burk plots through points obtained with fructose concentrations up to $1.2\,M$, fructose has an apparent K_m of about $2\,M$ with a very high V_{max} at saturation. The K_i values (measured against glucose as substrate) are rather lower than the K_m values: mannose $1.4 \times 10^{-2}\,M$, 2-deoxyglucose $1.6 \times 10^{-2}\,M$, and fructose $0.35\,M$. The K_i[7] values for sugars not phosphorylated are *N*-acetylglucosamine $5 \times 10^{-4}\,M$, glucosamine $1.0 \times 10^{-3}\,M$, galactose $0.67\,M$, lyxose $8.3 \times 10^{-2}\,M$, and xylose $0.12\,M$. The glucokinase from rabbit liver shows similar specificity characteristics.[8]

Half-maximal activity is obtained with $5 \times 10^{-4}\,M$ ATP with a Mg^{++}: ATP ratio of 2:1. Other natural analogs of ATP give negligible rates.

Product Inhibition. ADP^{3-} is a competitive inhibitor ($K_i = 2 \times 10^{-3}\,M$) (i.e., when [$Mg^{++}$] not greater than [ATP]) with respect to both $MgATP^{2-}$ and glucose. $MgADP^-$ is a noncompetitive inhibitor ($K_i = 2 \times 10^{-3}\,M$) with respect to glucose (in the presence of $0.005\,M$ $MgATP^{2-}$) but gives mixed inhibition with respect to $MgATP^{2-}$. Thus in the presence of $0.1\,M$ glucose, $MgADP^-$ is a competitive inhibitor ($K_i = 9 \times 10^{-4}\,M$) at $MgATP^{2-}$ concentrations above $0.001\,M$ but with $MgATP^{2-}$ concentrations below $0.001\,M$ the inhibition is noncompetitive ($K_i = 1.4 \times 10^{-3}\,M$). Similar results have been reported for rabbit liver glucokinase.[8]

Unlike most mammalian hexokinases, which are inhibited by glucose 6-phosphate in a manner considered to be of physiological significance

[7] M. J. Parry and D. G. Walker, *Biochem. J.* **99,** 266 (1966).

[8] J. Salas, M. Salas, E. Viñuela, and A. Sols, *J. Biol. Chem.* **240,** 1014 (1965).

(see Vol. I [33]), hepatic glucokinase is not subject to inhibition by this product at physiological concentrations.[2,7] At a $MgATP^{2-}$ concentration of 0.005 *M*, glucose 6-phosphate is a competitive inhibitor ($K_i = 6.5 \times 10^{-2} M$) with respect to glucose.[7]

Stability. The enzyme is very unstable in the absence of KCl (and NaCl and NH_4Cl are not very effective stabilizers) but is protected to a limited extent against inactivation by saturating concentrations of the substrates. The presence of 0.1 *M* KCl in the assay procedure does not affect either the V_{max} or the K_m values.[7,8] The enzyme is inhibited by reagents such as *p*-chloromercuribenzoate, and this inhibition is reversed by sulfhydryl reagents. The Q_{10} is 1.9.

[70b] Glucokinase

II. *Aerobacter aerogenes*

By R. L. Anderson and M. Y. Kamel

$$ATP + glucose \rightarrow ADP + glucose\ 6\text{-phosphate} + H^+$$

Assay Method

Principle. Many assay methods are available.[1] In the procedure described here, the reaction is measured spectrophotometrically by observing TPN reduction in a glucose 6-phosphate dehydrogenase-linked assay.

Reagents

Glycylglycine buffer, 0.2 *M*, pH 7.5
$MgCl_2$, 0.1 *M*
ATP, 0.05 *M*
TPN, 0.01 *M*
D-Glucose, 0.2 *M*
Purified glucose 6-phosphate dehydrogenase[2]

Procedure. The following are added to a microcuvette with a 1.0-cm light path: 0.05 ml of buffer, 0.01 ml of $MgCl_2$, 0.01 ml of ATP, 0.01 ml of TPN, 0.01 ml of D-glucose, excess glucose 6-phosphate dehydrogenase, D-glucokinase, and water to a volume of 0.15 ml. The reaction is initiated by the addition of D-glucokinase. The rate of absorbance in-

[1] See Vol. I [32] and [33].
[2] The preparation must be free from interfering enzymes such as 6-phosphogluconate dehydrogenase and hexokinase.

crease at 340 mμ may be conveniently measured with a Gilford multiple sample absorbance recorder. The cuvette compartment should be thermostated at 25°.

Evaluation of the Assay. When care is taken to see that glucose 6-phosphate dehydrogenase is in excess, the reaction rate is constant with time and proportional to the D-glucokinase concentration. The rate of D-glucose phosphorylation should be equivalent to the rate of TPN reduction. In crude enzyme preparations, however, the observed rate may be falsely high due to the presence of 6-phosphogluconate dehydrogenase. In the extract described below, the 6-phosphogluconate dehydrogenase activity was never more than 20% of the activity of D-glucokinase.

Definition of Unit and Specific Activity. A unit of D-glucokinase is defined as the amount that catalyzes the phosphorylation of 1 micromole of D-glucose per hour in the assay described. Specific activity (units per milligram of protein) is based on a spectrophotometric determination of protein.[3]

Purification Procedure[4]

Aerobacter aerogenes PRL-R3 was grown in 100-liter volumes in a New Brunswick Model 130 Fermacell fermentor at 30° with an aeration rate of 6–8 cubic feet per minute and an agitation speed of 300 rpm. The medium consisted of 1.35% $Na_2HPO_4 \cdot 7\ H_2O$, 0.15% KH_2PO_4, 0.3% $(NH_4)_2SO_4$, 0.02% $MgSO_4 \cdot 7\ H_2O$, 0.0005% $FeSO_4 \cdot 7\ H_2O$, 0.02% Dow Corning Antifoam B, and 0.5% D-glucose (autoclaved separately). The inoculum was 2.5 liters of an overnight culture in the same medium minus the antifoam. The cells were harvested with a Sharples AS-12 centrifuge 8–9 hours after inoculation. The yield was about 10 g (wet weight) of cells per liter.

Extracts were prepared by treating cell suspensions for 10 minutes in a Raytheon 10-kc sonic oscillator circulated with ice water. The broken-cell suspension was centrifuged at 13,200 *g*, and the resulting supernatant solution was used as the cell extract. The extract used in the purification described below was derived from 700 g (wet weight) of cells. All fractionation procedures were carried out at 0–4°.

Bentonite Fractionation. Bentonite, 223 g, was added with stirring to 3350 ml of cell extract containing 54 mg of protein per milliliter with a 280:260 mμ ratio of 0.69. The bentonite was then removed by centrifugation and discarded. The supernatant (2300 ml) contained 6.6 mg of protein per milliliter and had a 280:260 mμ ratio of 0.61.

[3] O. Warburg and W. Christian, *Biochem. Z.* **310**, 384 (1941).

[4] M. Y. Kamel, D. P. Allison, and R. L. Anderson, *J. Biol. Chem.* **241**, 690 (1966).

First Ammonium Sulfate Fractionation. Ammonium sulfate, 30.3 g, was dissolved in the supernatant from the bentonite step. Protamine sulfate (100 ml of a 7.6% solution at room temperature) was then added slowly with stirring, and the precipitate that formed was removed by centrifugation and discarded. To the supernatant (2375 ml) was added 1169 g of ammonium sulfate (80% of saturation). The precipitated protein was collected by centrifugation and dissolved in water. This fraction (142 ml) contained 22 mg of protein per milliliter and had a 280:260 ratio of 1.15.

Acid-Precipitation. The above fraction was diluted to 600 ml with water and the pH was lowered to 4.4 by the addition of acetic acid. The precipitated protein was removed by centrifugation and discarded, and the pH of the supernatant (which contained the D-glucokinase) was raised immediately to 7.0 with ammonium hydroxide. The protein concentration of this solution (600 ml) was 3.8 mg/ml and the 280:260 mμ ratio was 1.12.

Second Ammonium Sulfate Fractionation. To the above fraction was added 550 ml of saturated ammonium sulfate (pH 7.0). The precipitate of crystalline and amorphous protein which appeared was removed by centrifugation and discarded. To the supernatant was added 400 ml of saturated ammonium sulfate (pH 7.0). The precipitated protein was collected and dissolved in water to yield 16 ml of solution containing 28 mg of protein per milliliter with a 280:260 mμ ratio of 1.18.

Sephadex G-100 Chromatography. The above fraction was placed on a column (5×153 cm) of Sephadex G-100 equilibrated with 0.01 *M* sodium phosphate buffer (pH 6.5) and eluted with the same buffer. Twenty milliliter fractions were collected, and those which contained most of the activity were pooled. This solution (120 ml) contained 0.33 mg of protein per milliliter and had a 280:260 mμ ratio of 1.55.

DEAE-Cellulose Chromatography. DEAE-cellulose (Bio-Rad Cellex D, exchange capacity = 0.95 milliequivalents per gram) was pretreated as recommended by Peterson and Sober[5] and equilibrated with 0.01 *M* sodium phosphate buffer (pH 6.5) in a column 1.5×12 cm. The above fraction was added to the column and eluted with 500 ml (5-ml fractions) of the same buffer containing NaCl in a linear gradient from 0 to 0.8 *M*. The five fractions containing most of the activity (6.6% of the D-glucokinase activity of the crude extract) were 1530- to 1980-fold purified, contained 0.20–0.35 mg of protein per milliliter, and had 280:260 mμ ratios of 1.61–1.71. A summary of the purification is shown in the table.

[5] See Vol. V [1].

PURIFICATION OF D-GLUCOKINASE FROM *A. aerogenes*

Fraction	Total activity (units[a])	Recovery (%)	Specific activity (units/mg)
Cell extract	401,000	100	2.2
Bentonite	229,000	57	15.1
Ammonium sulfate I	146,000	37	46.7
Acid	158,000	39	69.2
Ammonium sulfate II	73,400	18	164
Sephadex G-100	49,100	12	1230
DEAE-cellulose, fraction 24	7,000		4000
fraction 25	7,000		4360
fraction 26	4,420	6.6	3360
fraction 27	4,310		4100
fraction 28	3,500		3500

[a] Micromoles of D-glucose phosphorylated per hour.

Properties[4]

Effect of Growth Substrate. D-Glucokinase is a constitutive enzyme in *Aerobacter aerogenes* PRL-R3, its level being about the same whether the cells are grown on D-glucose, glycerol, or peptone.

Substrate Specificity. Of 37 sugars and polyols tested, only D-glucose ($K_m = 8 \times 10^{-5} M$) and D-glucosamine were phosphorylated. Phosphorylation of the following compounds could not be detected: 2-deoxy-D-glucose, α-methyl-D-glucoside, D- and L-mannose, D- and L-fructose, D- and L-galactose, D-allose, D-altrose, L-glucose, D-gluconic acid, D-glucuronic acid, D-galacturonic acid, D- and L-fucose, L-rhamnose, L-sorbose, D- and L-arabinose, D- and L-xylose, D-ribose, D-lyxose, D- and L-xylulose, D- and L-ribulose, sucrose, D-sorbitol, D-mannitol, D- and L-arabitol, xylitol, and ribitol.

Phosphoryl Donor Specificity. At 3.3 mM concentrations of phosphoryl donor (in 6.6 mM $MgCl_2$), the relative rates of D-glucose phosphorylation were as follows: ATP, 100; ITP, 13; GTP, 3; and UTP, 3. Compounds that could not be shown to serve as phosphoryl donors were CTP, ADP, acetyl phosphate, carbamyl phosphate, and creatine phosphate. The K_m for ATP is $8 \times 10^{-4} M$.

Metal ion specificity. A metal ion is required for activity. At 6.6 mM concentrations of metal salts (twice the ATP concentration), the relative rates of D-glucose phosphorylation were as follows: Mg^{++}, 100; Mn^{++}, 43; Co^{++}, 24; Ni^{++}, 3; Ca^{++}, 0; and Zn^{++}, 0.

pH Optimum. Activity as a function of pH is optimal at pH 7.5 in glycylglycine buffer and at about pH 8.9 in glycine buffer.

Stability. Purified D-glucokinase has been kept at room temperature for several days or at 0° (unfrozen) for several weeks without a significant loss of activity. It is unstable to storage in the frozen state at —20°.

[71] Acyl Phosphate : Hexose Phosphotransferase (Hexose Phosphate : Hexose Phosphotransferase)

By R. L. ANDERSON and M. Y. KAMEL

Acetyl phosphate + D-glucose → D-glucose 6-phosphate + acetate
D-Glucose 6-phosphate + D-mannose ⇄ D-mannose 6-phosphate + D-glucose

Assay Method

Principle. The assay of choice involves the measurement of D-glucose 6-phosphate formed from D-glucose plus acetyl phosphate by coupling the reaction to glucose 6-phosphate dehydrogenase.[1] The increase in absorbance at 340 mμ associated with TPN reduction is measured with a spectrophotometer.

Reagents

Glycylglycine buffer, 0.2 *M*, pH 7.5
D-Glucose, 0.1 *M*
Acetyl phosphate, 0.1 *M*
TPN, 0.01 *M*
EDTA, 0.01 *M*
Purified glucose 6-phosphate dehydrogenase (free from 6-phosphogluconate dehydrogenase)

Procedure. The following are added to a microcuvette with a 1.0-cm light path: 0.05 ml of buffer, 0.01 ml of D-glucose, 0.01 ml of acetyl phosphate, 0.02 ml of TPN, 0.01 ml of EDTA, excess glucose 6-phosphate dehydrogenase, acyl phosphate:hexose phosphotransferase, and water to a volume of 0.15 ml. The reaction is conveniently measured with a Gilford multiple sample absorbance recorder. The cuvette compartment should be thermostated at 25°.

Evaluation of the Assay. The reaction rate is linear with time and phosphotransferase concentration. The rate of D-glucose phosphorylation should be equivalent to the rate of TPN reduction. In *Aerobacter aerogenes* PRL-R3, however, the phosphotransferase activity is low relative to 6-phosphogluconate dehydrogenase activity. This causes the observed rate to be twice the actual rate of phosphorylation except when

[1] M. Y. Kamel and R. L. Anderson, *J. Biol. Chem.* **239**, PC 3607 (1964).

the phosphotransferase has been purified through the heat step. The pH of the assay used in the purification described here is below the optimum because comparisons were being made on other substrates using coupling enzymes with low pH optima. The use of glycine buffer at pH 8.5 or 9 would give higher rates.

Definition of Unit and Specific Activity. A unit of acyl phosphate: hexose phosphotransferase is defined as the amount that catalyzes the phosphorylation of 1 micromole of D-glucose per hour in the assay described. Specific activity (units per milligram of protein) is based on a spectrophotometric determination of protein.[2]

Purification Procedure

Growth of Bacteria. Acyl phosphate:hexose phosphotransferase is a constitutive enzyme in *Aerobacter aerogenes* PRL-R3 and is present in cells grown under a variety of conditions. For the purification described here, *A. aerogenes* PRL-R3 was grown in 100-liter volumes in a New Brunswick Model 130 Fermacell fermentor. The temperature was 30°, the aeration rate was 6–8 cubic feet per minute, and the agitation rate was 300 rpm. The medium consisted of 1.35% $Na_2HPO_4 \cdot 7\ H_2O$, 0.15% KH_2PO_4, 0.3% $(NH_4)_2SO_4$, 0.02% $MgSO_4 \cdot 7\ H_2O$, 0.0005% $FeSO_4 \cdot 7\ H_2O$, 0.02% Dow Corning Antifoam B, and 0.5% D-glucose (autoclaved separately). The inoculum was 2.5 liters of an overnight culture in the same medium minus the antifoam. The cells were harvested with a Sharples AS-12 centrifuge 8–9 hours after inoculation. The yield was about 10 g (wet weight) of cells per liter.

Preparation of Extracts. Seventy grams (wet weight) of cells were suspended in 80 ml of water. Extracts were prepared by treating the cell suspension for 15 minutes in a Raytheon 10-kc sonic oscillator circulated with ice water. The broken-cell suspension was centrifuged at 13,200 *g*, and the resulting supernatant fluid was used as the cell extract. Unless stated otherwise, the fractionation procedures described below were performed at 0–4°.

Protamine Treatment. The cell extract was diluted with water to give 675 ml with a protein concentration of 21 mg/ml and a 280:260 mμ ratio of 0.73. Ammonium sulfate (8.91 g) was added to give a concentration of 0.1 *M*, followed by the slow addition of 136 ml of a 2% solution of protamine sulfate. The mixture was stirred for 10 minutes and the precipitate that formed was removed by centrifugation and discarded. The supernatant solution (780 ml) contained 9 mg of protein per milliliter and had a 280:260 mμ ratio of 0.90.

[2] O. Warburg and W. Christian, *Biochem. Z.* **310**, 384 (1941).

First Ammonium Sulfate Fractionation. Ammonium sulfate (210.8 g) was added to the protamine sulfate-treated extract, and the precipitate that formed was removed by centrifugation and discarded. To the supernatant solution was added 233.6 g of ammonium sulfate (50–90% of saturation), and the resulting precipitate was collected by centrifugation and dissolved in water. This solution (85 ml) contained 29 mg of protein per milliliter and had a 280:260 mμ ratio of 1.11.

Heat Treatment. The temperature of the above fraction was raised quickly to 50°, held for 5 minutes, and quickly cooled. The precipitated protein was removed by centrifugation and discarded. The supernatant solution contained 13 mg of protein per milliliter and had a 280:260 mμ ratio of 1.10.

Second Ammonium Sulfate Fractionation. Ammonium sulfate, 19.7 g (0–50% of saturation), was added to 70 ml of the above fraction, and the resulting precipitate was collected by centrifugation and dissolved in water. This solution (30 ml) contained 14 mg of protein per milliliter and had a 280:260 mμ ratio of 1.20.

Chromatography on Sephadex G-200. The above fraction was placed on a column (5×50 cm) of Sephadex G-200 and eluted with water at a flow rate of 80 ml per hour. Five-milliliter fractions were collected, and those which contained the highest specific activities (tubes 60–70) were pooled. Turbidity was removed by centrifugation. The supernatant solution (46 ml) contained 0.81 mg of protein per milliliter and had a 280:260 mμ ratio of 1.52.

Chromatography on DEAE-Cellulose. The above fraction was concentrated by lyophilization and redissolved in 5.5 ml of water. Turbidity was removed by centrifugation. The solution was kept on ice overnight, and a precipitate of crystalline and amorphous protein which appeared was removed by centrifugation and discarded. All the phosphotransferase activity remained in the supernatant. DEAE-cellulose (Bio-Rad Cellex D, exchange capacity = 0.95 meq/g) was treated with 0.2 *M* glycylglycine buffer, pH 6.5, and equilibrated with 0.02 *M* glycylglycine, pH 6.5. A column (1.5×10 cm) was prepared, and 4.4 ml of the lyophilized Sephadex fraction was added. The protein was eluted with 100 ml of 0.02 *M* glycylglycine, pH 6.5, containing NaCl in a linear gradient from 0 to 0.3 *M*. Four-milliliter fractions were collected, and those which contained the highest specific activity were pooled to yield 8 ml of 710-fold purified phosphotransferase. The protein concentration was 0.16 mg/ml, and the 280:260 mμ ratio was 1.7. Correcting for the portion of fractions not used for further purification, the yield was 8.5% of the activity in the crude extract. A summary of the purification procedure is given in the table.

PURIFICATION OF ACYL PHOSPHATE:HEXOSE PHOSPHOTRANSFERASE

Fraction	Total activity[a] (units)	Yield (%)	Specific activity (units/mg)
Cell extract	1570[b]	(76)	0.11
Protamine sulfate	2060[b]	100	0.29
Ammonium sulfate I	1240[b]	60	0.50
Heat	1250	61	1.1
Ammonium sulfate II	1300	63	2.5
Sephadex G-200	294	14	6.5
DEAE-cellulose	174	8.5	78

[a] Micromoles of D-glucose phosphorylated per hour; corrected for the portion of fractions not used for further purification.

[b] Corrected for the 6-phosphogluconate dehydrogenase contribution by dividing the observed rates by 2.

Properties[3]

Phosphoryl Donor Specificity. The relative rates at which various compounds (at 6.7 mM concentrations) served as phosphoryl donors for D-glucose (6.7 mM) were as follows: acetyl phosphate, 100; carbamyl phosphate, 100; D-mannose 6-phosphate, 71; D-fructose 6-phosphate, 25; D-ribose 5-phosphate, 18; D-fructose 1-phosphate, 15; D-glucose 1-phosphate, 5; 6-phosphogluconate, 2.5; and D-sorbitol 6-phosphate, 2.5. Compounds which could not be shown to serve as phosphoryl donors were: 3-phosphoglycerate, α-glycerol phosphate, D-mannitol 1-phosphate, D-mannose 1-phosphate, D-galactose 1-phosphate, phosphoenolpyruvate, creatine phosphate, adenosine triphosphate, adenosine diphosphate, inorganic pyrophosphate, and inorganic orthophosphate.

Substrate Specificity. Compounds that can be phosphorylated in addition to D-glucose are D-mannose, D-fructose, and D-mannitol.

pH Optimum. Phosphotransferase activity as a function of pH is maximal at pH 9 and half-maximal at pH 7.5.

Kinetic Constants. K_m values are as follows: acetyl phosphate, 4×10^{-4} M; carbamyl phosphate, 4×10^{-4} M; D-mannose 6-phosphate, 4×10^{-4} M; D-ribose 5-phosphate, 2×10^{-3} M; D-glucose, 1.6×10^{-4} M; D-mannose, 1.2×10^{-2} M; and mannitol, 6.7×10^{-2} M. With acetyl phosphate as the phosphoryl donor, the V_{max} values are about the same for D-glucose, D-mannose, and mannitol. With D-glucose as the phosphoryl acceptor, the relative V_{max} values are: acetyl phosphate, 100; carbamyl phosphate, 100; D-mannose 6-phosphate, 75; and D-ribose 5-phosphate, 31.

[3] M. Y. Kamel and R. L. Anderson, *Federation Proc.* **24,** 422 (1965).

Stability. The most highly purified fractions of phosphotransferase have been stored at —20° with repeated thawing and freezing with no detectable loss of activity.

[72] A Phosphoenolpyruvate-Hexose Phosphotransferase System from *Escherichia coli*

By WERNER KUNDIG and SAUL ROSEMAN

$$\text{(A)}\quad \text{Phosphoenolpyruvate} + \text{heat-stable protein (HPr)} \underset{\text{Mg}^{++}}{\overset{\text{enzyme I}}{\rightleftharpoons}} \text{phospho-HPr} + \text{pyruvate}$$

$$\text{(B)}\quad \text{Phospho-HPr} + \text{hexose} \xrightarrow[\text{Mg}^{++}]{\text{enzyme II}} \text{hexose-6-P} + \text{HPr}$$

$$\text{(A + B)}\quad \text{Phosphoenolpyruvate} + \text{hexose} \xrightarrow[\text{HPr, Mg}^{++}]{\text{enzymes I + II}} \text{hexose-6-P} + \text{pyruvate}$$

Assay Method

Principle. The phosphotransferase system was isolated from *Escherichia coli*,[1] and has been detected in *Aerobacter*, *Lactobacillus*, and *Bacillus subtilis*.[2] The system catalyzes the transfer of phosphate from phosphoenol pyruvate (PEP) to a variety of sugars and sugar derivatives, and requires three protein fractions, enzymes I and II, and a heat-stable protein, called HPr. The latter acts as a "phosphate carrier" in the overall reaction; in phospho-HPr, the phosphate group is covalently linked to protein-bound histidine.[1] The complete system involves two discrete reactions: enzyme I catalyzes the transfer of phosphate from PEP to HPr; enzyme II catalyzes the transfer of phosphate from phospho-HPr to suitable sugar acceptors. The specificity toward the sugars is confined to enzyme II. At least three different enzymes II have been demonstrated; they are located in the membrane fraction obtained from lysed *E. coli* spheroplasts, and *B. subtilis* protoplasts.[2] Enzyme I and HPr are found primarily in the supernatant fractions of the lysed preparations.

Although the rates of the individual reactions can be determined with purified preparations, a more convenient procedure involves determination of the rate of the complete system where one of the three protein fractions is used in rate-limiting concentration. With purified HPr as the

[1] W. Kundig, S. Ghosh, and S. Roseman, *Proc. Natl. Acad. Sci. U.S.* **52**, 1073 (1964).
[2] W. Kundig, F. Dodyk Kundig, B. Anderson, and S. Roseman, *Federation Proc.* **24**, 658 (1965).

rate-limiting protein, using the conditions described below, product formation is proportional to time of incubation in the range 5–15 μg. The assay mixtures contain either ^{14}C-labeled hexose and unlabeled PEP, or ^{32}P-labeled PEP and unlabeled hexose. Labeled products are separated from excess substrate by high voltage paper electrophoresis and determined by counting in the Packard Tri-Carb liquid scintillation spectrometer. A useful acceptor with crude extracts is ^{14}C-acetyl labeled *N*-acetylmannosamine,[1] since the product *N*-acetylmannosamine-6-P is not further metabolized with preparations obtained from glucose- or glycerol-grown cells. With more purified preparations, a number of sugars can be used as substrates.

Reagents

Phosphate buffers (K^+), 1 *M*, pH 7.6 and 6.5
Tris-HCl buffers, 1 *M*, pH 7.6 and 7.4
Sodium glycinate buffer, 1 *M*, pH 8.0
EDTA, 0.1 *M*, pH 7.6
$MgCl_2$, 0.1 *M*
Hexose, 0.1 *M*; hexosamine, 0.1 *M*; or *N*-acylhexosamine, 0.1 *M*
Phosphoenolpyruvate, sodium salt, 0.1 *M*
^{14}C-labeled hexose or hexosamine, 0.1 *M*. The preferred substrate for crude extracts is *N*-acetyl-(^{14}C)-D-mannosamine (specific activity, 5×10^5 cpm per micromole) (Vol. VIII [26])
^{32}P-labeled PEP, 0.1 *M*, prepared by incubation of chicken liver mitochondria with $^{32}P_i$, in the presence of K malate (specific activity, 5×10^5 cpm per micromole)[3]
Tergitol-4, 1%, in water (purchased from Union Carbide, New York)

Procedure

The rate of phosphorylation is determined with either ^{14}C-labeled sugar (such as *N*-acetylmannosamine) and unlabeled PEP, or with unlabeled sugar and ^{32}P-labeled PEP. The incubation mixtures contain the following components in final volumes of 0.18 ml: 0.020 ml hexose, hexosamine or *N*-acetylmannosamine; 0.020 ml PEP; 0.010 ml $MgCl_2$; 0.010 ml of either Tris-HCl or the glycine buffer (pH 7.4 and 8.0, respectively); purified HPr, 20 μg; 0.010 ml Tergitol-4; one unit of enzymes I and II. To estimate the concentration of enzyme I, an excess of enzyme II (4 units) is added, while the reverse is true for estimating the concentration of enzyme II. The concentration of HPr given above represents

[3] J. Mendicino and M. F. Utter, *J. Biol. Chem.* **237**, 1716 (1962).

an excess for the complete system. To assay for HPr, 4 units of enzymes I and II are employed, with the equivalent of 5–15 μg of purified HPr. After 30 minutes of incubation at 37°, the reaction is stopped by heating for 2 minutes at 100°, and an aliquot is subjected to electrophoresis on Whatman 3 MM paper saturated with 0.05 *M* pyridinium-acetate buffer, pH 6.5 (50 volts/cm). When the labeled substrate is the hexose, electrophoresis is conducted for 15 minutes, and 30 minutes is employed when the substrate is ^{32}P-labeled PEP; these conditions will separate the product (a sugar-P), P_i, and PEP. The control incubation mixtures contain heat-inactivated in place of active enzyme, or are incubated 0 minute, or lack hexose in the case of the ^{32}P-labeled PEP. In addition, standard samples are electrophoresed at the same time as the incubation mixtures. The determination of $^{32}P_i$ in the ^{32}P assay system permits quantitation of the phosphatases that are present in the crude extracts.

Definition of Unit and Specific Activity. One unit of enzyme I is the amount that will catalyze the synthesis of 1 micromole of *N*-acetylmannosamine-6-P when incubated for 30 minutes at 37° under the conditions described above, in the presence of excess enzyme II and HPr. A similar definition applies to enzyme II. Specific activity is defined as the units of enzyme I, or II, per milligram of protein. The method of Lowry *et al.*[4] is used to determine protein concentration.

Purification Procedure

Growth of Organism. A number of *E. coli* strains, such as B, K12, contain the phosphotransferase system. Most of the kinetic studies were conducted with fractions obtained from *E. coli* K235.[5] The organism is grown in Todd-Hewitt broth supplemented with 1.5% glucose,[6] or in a mineral salts medium,[6] with either 0.4% glucose or 0.4% glycerol as the carbon source. With 600 ml quantities of medium, the cells are grown in 2-liter Erlenmeyer flasks on a New Brunswick Rotary Shaker at 37° (400 rpm). For larger quantities of cells, 14-liter New Brunswick fermentor jars are used, containing up to 10 liters of medium; here, the maximum yield of the phosphotransferase system is obtained when the culture is stirred at approximately 200 rpm, but without passage of air through the sparger. In both cases, it is essential to maintain the pH

[4] O. H. Lowry, N. J. Rosebrough, A. L. Farr, and R. J. Randall, *J. Biol. Chem.* **193**, 265 (1951).

[5] G. W. Barry and W. F. Goebel, *Nature* **179**, 206 (1957).

[6] The mineral salts medium contains per liter: 0.05 mole potassium phosphate, pH 7.3; 2 g $(NH_4)_2SO_4$; 0.2 g $MgSO_4 \cdot 7\ H_2O$; 0.5 mg $FeSO_4 \cdot 7\ H_2O$; 1 g Casamino acids, (Difco). After sterilization 1 mg of thiamine is added. Todd-Hewitt broth is available from Difco Laboratories, Detroit, Michigan.

between 7.0 and 7.5 by adding 6 *M* NaOH as required. The culture is permitted to grow until it reaches two-thirds of maximum growth and is then *rapidly chilled* to stop growth; the yield of HPr is greatly decreased if the cells reach the stationary phase of growth. After harvesting, the cells are washed twice with cold 1% KCl, and the wet paste is stored at −18°.

Purification of Enzymes I and II

Unless otherwise specified, all operations are conducted between 0° and 4°, and all buffers contain 5 micromoles EDTA, and 1 mg 2-mercaptoethanol per milliliter.

Step 1. Crude Extracts. The cells (1 g wet paste) are suspended in 20 ml of 0.025 *M* Tris-HCl buffer, pH 7.6, and ruptured either by sonic vibration, or by passage through a French Pressure Cell. The mixture is centrifuged at 10,000 *g* for 20 minutes, and the residue is discarded.

Step 2. Charcoal Treatment. The crude extract is treated with charcoal to remove HPr. Twenty milliliters of a 10% suspension of Darco G-60 charcoal is centrifuged for 5 minutes at 10,000 *g*, and the resulting precipitate is mixed with 20 ml of crude extract. The mixture is stirred for 5 minutes, then centrifuged at 10,000 *g* for 20 minutes; the residue is discarded.

Step 3. Ammonium Sulfate. A saturated solution of ammonium sulfate (Enzyme Grade, Mann Research Lab., Inc.) is adjusted with NaOH until a 1:4 dilution with water is at pH 7.6 when measured with a glass electrode. The ammonium sulfate solution (20 ml) is slowly added, with stirring, to 20 ml of the supernatant fluid obtained in step 2, the mixture is occasionally stirred for 20 minutes, and the precipitate is collected by centrifugation at 35,000 *g* for 20 minutes. Five ml of 0.01 *M* K phosphate buffer, pH 7.6, is immediately added to the residue, and the turbid suspension is passed through a 100 ml column of Sephadex G-25 previously equilibrated with the same buffer. The column is eluted with the phosphate buffer, and the protein fractions are combined.

Step 4. Cγ Alumina Gel. A suspension containing Cγ alumina gel (2% solids; Alumina Cγ aged (Nutritional Biochemicals Corp.) is prepared, and 30 ml is centrifuged for 5 minutes at 35,000 *g*. The residue is mixed with 15 ml of the slightly turbid fraction obtained in the previous step, and after 5 minutes with occasional stirring, the suspension is centrifuged for 10 minutes at 35,000 *g*. The supernatant fraction contains enzyme II, while enzyme I is adsorbed to the gel. The pellet is first washed with 30 ml of 0.05 *M* K phosphate buffer, pH 7.6, and enzyme I is then eluted by twice extracting the gel with 10 ml portions of 0.10 *M* K phosphate buffer, pH 7.6. The washing and extractions are performed by mixing the gel for 5 minutes with the buffers, followed by 10-minute centrifugations at 35,000 *g*. The extracts are combined and dialyzed against 0.01 *M* Tris-HCl buffer, pH 7.6, for 12 hours.

Step 5. DEAE-Cellulose Chromatography (Enzyme I). DEAE-cellulose is equilibrated with 0.01 *M* Tris-HCl, pH 7.6, and a column is prepared (2 × 15 cm) containing 60 ml of the cellulose packed by gravity filtration. The dialyzed gel eluate obtained in the previous step is transferred to the column, which is then eluted with 300-ml portions of 0.01 *M* Tris-HCl, pH 7.6, containing the increasing concentrations of KCl in the following sequence: 0.00, 0.05, 0.10, and 0.20 *M*. The 0.20 *M* eluate contains enzyme I, and is concentrated to 5 ml by pressure dialysis against 8 liters of 0.01 *M* Tris-HCl, pH 7.6.

Step 6. Purification of Enzyme II. The alumina gel supernatant fraction (step 4) contains enzyme II, which can be sedimented at 100,000 *g* (90 minutes). The pellet is washed with 20 ml 0.01 *M* Tris-HCl, pH 7.6, centrifuged, and resuspended in the same buffer. Enzyme II can also be purified by chromatography on DEAE-cellulose, phosphate form, previously equilibrated with 0.01 *M* K phosphate buffer, pH 7.6. The alumina gel supernatant fraction (15 ml) is placed on a 60-ml column of the cellulose (2 × 15 cm), and the column is eluted with 300 ml portions of phosphate buffer at the following concentrations: 0.01, 0.10, 0.20, and 0.30 *M*. The 0.30 *M* eluate contains enzyme II which is concentrated to 5 ml by pressure dialysis against 8 liters of 0.01 *M* Tris-HCl, pH 7.6. At this stage, enzyme II is sedimentable when subjected to centrifugation for 90 minutes at 100,000 *g*.

PURIFICATION OF ENZYMES I AND II

Fraction	Specific activity[a]			Yield (%)
	I + II	I	II	
Crude extract	0.30	—	—	100
Charcoal supernatant	0.82	—	—	62
50% Ammonium sulfate	2.2	—	—	51
Alumina gel				
Supernatant		0	16	39
Eluate		19	0	41
DEAE-cellulose				
Alumina gel supernatant		0	98	24
Alumina gel eluate		119	0	27

[a] Specific activity is defined as μmoles *N*-acetyl-mannosamine-6-P formed per milligram protein per 30 minutes. After the alumina gel step, either enzyme I or enzyme II is added in rate-limiting amounts and the other enzyme is added in excess. Zero values indicate no detectable activity (i.e., less than 0.05%).

The purification procedure for enzymes I and II is summarized in the table.

Purification of Heat-stable Protein (HPr)

These steps are conducted at room temperature, unless otherwise indicated.

Step 1. Crude Extract. The cells (2.3 kg wet paste) are suspended in 10 liters of 0.025 *M* K phosphate buffer, pH 7.6, without 2-mercaptoethanol or EDTA. The cells are ruptured by passage through a continuous flow sonic oscillator, centrifuged at 35,000 *g* for 30 minutes, and 500-ml aliquots of the supernatant fluid are heated for 15 minutes in a boiling water bath. The coagulated protein is removed by centrifugation and discarded.

Step 2. Acid Precipitation. The supernatant fluid of the preceding step is cooled to about 4°, and adjusted to pH 1.0 with conc. HCl solution. After 30 minutes, the resulting white precipitate is collected by centrifugation and washed once with cold 0.01 *M* HCl. The precipitate is then suspended in 2 liters of 0.5 *M* K phosphate buffer, pH 7.6, at room temperature, using a Waring blendor, and the suspension is stirred for 24 hours with a magnetic stirrer. After centrifugation at 35,000 *g* for 20 minutes, the precipitate is discarded, and the supernatant fluid is dialyzed against at least 4 changes of distilled water (10 liters each) for a total of 48 hours. The dialysis residue is concentrated to about 200 ml in a rotary evaporator, and dialyzed against 2-liter portions (4 changes) of 0.025 *M* K phospate buffer, pH 6.5.

Step 3. ECTEOLA-Cellulose. The cellulose derivative is converted to the phosphate form, adjusted to pH 6.5, and equilibrated with 0.025 *M* K phosphate, pH 6.5; a column of the *ECTEOLA*-cellulose is prepared by gravity filtration (6 × 40 cm) and contains 1 liter of packed cellulose. The dialyzate obtained in the preceding step (200 ml) is placed on the column and eluted with the 0.025 *M* phosphate buffer. The protein-containing fractions, detected by a modified biuret method,[4] are combined, concentrated in a rotary evaporator to approximately 200 ml, and dialyzed against 5 changes of 0.01 *M* Tris-HCl buffer, pH 7.6 (5 liters each).

Step 4. DEAE-Cellulose, First Column. The eluate obtained in the previous step is placed on a 1 liter column of DEAE-cellulose (5 × 30 cm; previously converted to the Cl^- form and equilibrated with 0.01 *M* Tris-HCl, pH 7.6). The column is then eluted with 3-liter portions of 0.01 *M* Tris-HCl, pH 7.6, containing the following concentrations of KCl 0.0, 0.050, and 0.10 *M*. The last fraction contains HPr, and is concentrated to 100 ml in the rotary evaporator, and dialyzed against 4 changes of 0.01 *M* Tris-HCl, pH 7.6 (5 liters each).

Step 5. DEAE-Cellulose, Second Column. A column containing 500 ml of DEAE-cellulose is pretreated as described in step 4; the HPr solution of the preceding step is transferred to the column. After the column has been washed with 0.01 *M* Tris-HCl buffer, pH 7.6, HPr is eluted using a linear gradient (total volume 6 liters) of KCl with the initial concentration of 0.0 *M* and the final concentration at 0.12 *M*, the gradient containing 0.01 *M* Tris-HCl at all points. The fractions are assayed for protein (absorbance at 280 mμ) and for HPr activity; inactive protein is eluted early after application of the gradient, while HPr is eluted in a single symmetrical peak at about 0.075 *M* KCl. The fractions containing HPr are combined, concentrated to 20 ml in a rotary evaporator, and dialyzed against 5 changes of distilled water (2 liters each).

Based on analysis by a modified biuret method,[4] HPr is purified between 8000- and 10,000-fold by the procedure described above. The final yield is approximately 10–15 mg.

Properties

Stability of Enzymes I, II, and HPr. Purified enzyme I is stable for approximately 1 week when stored at 0° and is not stable to freezing and thawing; enzyme II is also inactivated by freezing and thawing and is stable for only 24–48 hours at 0°.

HPr is stable at 100° at neutral pH for at least 20 minutes, at pH 1 at room temperature for several hours, but is unstable in 0.1 *M* NaOH for 60 minutes at room temperature. It is completely resistant to prolonged digestion and dialysis with the following nucleases: purified venom phosphodiesterase, venom 5′-nucleotidase, polynucleotide phosphorylase, pancreatic RNAse and DNAse. On the other hand, it is very labile to the action of the following proteinases: chymotrypsin, trypsin, papain, pepsin, and pronase. HPr is either irreversibly adsorbed or inactivated by charcoal, but not by mixed-bed ion exchange resins.

Properties of the Complete System. The purified enzyme system is active only with PEP as the phosphate donor (K_m, 6×10^{-4} *M*).[1] The optimum divalent metal ion is Mg^{++}, which can partially be replaced by Mn^{++}, Zn^{++}, and Co^{++}. The following ions are strongly inhibitory: Cu^{++}, Fe^{++}, and Ca^{++}. Other proteins cannot replace HPr. Preliminary experiments indicate that the components of the system ioslated from *E. coli* will substitute for the components of the system isolated from *B. subtilis* and other organisms.[2]

The sugar specificity of the system is confined to enzyme II, and there are at least three such enzymes present in the particulate fraction. The specificity of enzyme II is determined by the growth medium. For example, when a Todd-Hewitt glucose broth is used, active phosphate acceptors are of the D-gluco and D-manno configuration (hexoses, hexitols, hexosamines, *N*-acylhexosamines, methyl α-D-glucopyranoside), while a number of pentoses, pentitols, other hexitols, fructose, galactose, and disaccharides are inactive. When the cells are grown in mineral medium containing the inactive compounds as the carbon source, then enzyme II activity is demonstrable with these substrates. Enzyme II from cells grown on a glycerol mineral salts medium shows a broad substrate specificity, and the activity toward galactose can be increased at least 4-fold when the nonmetabolized compound D-fucose is added to the medium at 0.001 *M* concentration.[2] When the following sugars are used as substrates, the corresponding 6-phosphate esters are formed in the reaction (all *D*-sugars): glucose, mannose, glucosamine, mannosamine, *N*-acetyl- and *N*-glycolylglucosamine, and the corresponding *N*-acyl mannosamines, galactose, and methyl β-1-thiogalactoside.

The pH optimum for the complete system is approximately 7.4 for hexoses of the gluco and manno configuration, and 8.0 for those of the galacto configuration.

[73] Phosphoramidate-Hexose Transphosphorylase

By ROBERTS A. SMITH and MYNA C. THEISEN

$$KO_3PN^+H_3 + \text{glucose} \rightarrow \text{glucose 1-phosphate} + NH_3$$

Assay Method

Principle. The most convenient and sensitive assay method is based on the formation of glucose 1-phosphate by phosphoryl transfer from $^{32}P\text{-}NH_2$. The phosphorylated glucose is estimated after hydrolysis of the remaining $^{32}P\text{-}NH_2$ to $^{32}P_i$ and its subsequent removal from solution by precipitation of the phosphomolybdate complex according to the procedure of Sugino and Miyoshi.[1] Phosphoramidate is so extremely acid labile[2] that it can be converted to P_i virtually quantitatively while glucose 1-phosphate remains unaltered. This conversion is accomplished by the addition of trichloroacetic acid to terminate the reaction.

[1] Y. Sugino and Y. Miyoshi, *J. Biol. Chem.* **239**, 2360 (1964).

[2] T. Rathlev and T. Rosenberg, *Arch. Biochim. Biophys.* **65**, 319 (1956).

Reagents

Tris-chloride buffer, 1.0 *M* pH 8.7
^{32}P-phosphoramidate (^{32}P-NH_2), 0.1 *M*
Glucose, 0.4 *M*
Trichloroacetic acid, 25% (w/v)
$HClO_4$, 2.0 *N*
Ammonium molybdate, 0.08 *M*
Triethylamine chloride, pH 5.5, 0.8 *M*

Procedure. The assay mixture (0.8 ml) contains 0.1 ml of tris buffer, pH 8.7; 0.1 ml of ^{32}P-NH_2; 0.1 ml of glucose; and the enzyme preparation to be assayed. The reagents, excluding the enzyme, are preincubated at 37° for 5 minutes and the reaction is started by the addition of enzyme. After 15 minutes' incubation the reaction is stopped by the addition of 0.2 ml of 25% trichloroacetic acid and allowed to remain at 37° for 2 minutes. During this time the unreacted ^{32}P-NH_2 is converted to $^{32}P_i$ and is removed by the addition of 0.1 ml of $HClO_4$, followed by 0.25 ml of ammonium molybdate and finally by 0.1 ml of triethylamine hydrochloride. The reaction mixture is well mixed, and precipitation of the phosphomolybdate complex is permitted to proceed at room temperature for 10 minutes, after which it is removed by centrifugation. The radioactivity remaining in the supernatant fluid is a measure of the phosphoryl transfer. Because some preparations of ^{32}P-NH_2 contain small amounts of $^{32}PP_i$, it is always necessary to run a reagent blank with each determination.

One unit of phosphoramidate hexose transphosphorylase is defined as that amount of enzyme required to catalyze the transfer of 1 micromole of phosphoryl group per minute from P-NH_2 to glucose under the standard assay conditions.

Preparation of ^{32}P-NH_2. Phosphoramidate can be prepared essentially according to the method of Stokes,[3] and the ^{32}P-labeled material is prepared by the same method, starting from $H_3\,^{32}PO_4$. Ten millicuries of $H_3\,^{32}PO_4$ are introduced into a thick-walled test tube and water is removed by evaporation under a stream of dry air. When the sample is air dry (usually 2–3 hours), 0.5 ml of $POCl_3$ is introduced into the vessel and then frozen in dry ice. While the reaction mixture is frozen in the ice bath the thick-walled test tube is closed off with an oxygen flame and sealed carefully so that it will withstand considerable pressure. The closed vessel is then placed in a protective container and placed in an

[3] H. N. Stokes, *Am. Chem. J.* **15**, 198 (1893); *ibid* **16**, 123 (1894).

oven at 150° for at least 8 hours, during which exchange takes place.[4] After cooling, the sealed vessel is carefully broken open and the $^{32}POCl_3$ is carefully washed out with 0.50 ml of cold $POCl_3$ into a tared ground-glass round-bottom flask. For each mole of $^{32}POCl_3$, 2.1 moles of reagent grade phenol are added and the mixture refluxed for 2–2½ hours. Without further purification the reaction mixture is washed into a beaker containing a considerable excess of ammoniacal ethanol (concentrated ammonium hydroxide-ethanol, v/v) and allowed to react for 10 minutes at room temperature. The reaction mixture is then diluted 5-fold with cold distilled water and filtered on a Büchner flask. The crude ^{32}P-diphenylphosphoramidate can be recrystallized from ethanol at this point and then filtered and washed with 50% ethanol and dried over P_2O_5. For every gram of ^{32}P-diphenylphosphoramidate, 1.2 g of KOH and 2.5 ml of water are placed in a beaker with a boiling chip and gently boiled for 5 minutes. The reaction mixture is cooled in an ice bath and neutralized carefully with 50% acetic acid followed by the addition of approximately an equal volume of ethanol. After an hour in the ice bath the supernatant fluid is removed by filtration and the $^{32}P\text{-}NH_2$ (monopotassium salt) is collected and washed with ethanol followed by ether and dried over P_2O_5. Approximately a 75% yield of $^{32}P\text{-}NH_2$ is obtained by this procedure.

Purification Procedure

Step 1. Preparation of Extract. Escherichia coli, Crookes strain (ATCC No. 8739), grown according to the procedure of Korkes,[5] except that disodium succinate is substituted for cerelose in the medium, is harvested with a Sharples centrifuge and stored frozen until needed. Cell paste, 30 g suspended in 200 ml of 0.05 *M* Tris buffer pH 7.5, is treated for 15 minutes in three batches in a 12-kc sonic oscillator (Biosonik, Bronwill Scientific). The cell debris is removed by centrifugation for 20 minutes at 13,000 *g* in a Servall centrifuge at 0°. All subsequent fractionations are performed at or near 0°.

Step 2. Acid Treatment. The crude extract is adjusted to pH 1 (glass electrode) by the addition, with constant stirring, of the appropriate amount of 1 *N* HCl. The protein which precipitates at this pH is removed by centrifugation at 13,000 *g* for 20 minutes. The supernatant fluid from the acid treatment is then adjusted to pH 9 (glass electrode) with 1 *N* KOH, again with constant stirring, in an ice bath. The small

[4] R. D. O'Brien, "Toxic Phosphorus Esters," p. 347. Academic Press, New York, 1960.
[5] S. Korkes, Vol. I [77].

amount of protein precipitated with this treatment is also removed by centrifugation and discarded.

Step 3. Streptomycin Treatment. The supernatant fluid from the previous treatment is adjusted to pH 6 (glass electrode) with 1 N acetic acid and diluted with water so that the protein concentration does not exceed 15 mg/ml. About 0.1 volume of 2% streptomycin sulfate solution is added in order to remove the nucleic acids. The entire mixture is dialyzed overnight at 4° against two changes of at least 20 volumes of distilled water. The precipitate formed during the dialysis is removed by centrifugation at 13,000 g for 20 minutes.

Step 4. DEAE-Cellulose Chromatography. A column of DEAE-cellulose (5 × 30 cm) is converted to the bicarbonate phase with 2 liters of 1 N ammonium bicarbonate (adjusted to pH 9.5 with NH_4OH). After washing the column with water at 4° until the eluate gives off no gas bubbles when treated with acid, the column is then washed with 500 ml of 0.05 M $NaHCO_3$ (adjusted to pH 9.5 with 1 N NaOH). The dialyzed streptomycin-treated preparation of the enzyme is adjusted to pH 9.5 and applied to the column at a level not exceeding 1.25 mg of protein per milliliter of resin. After the column has been washed with 4 volumes of glass-distilled water at 4°, the enzyme elution is begun. Elution is accomplished using a linear gradient in which the mixing flask contains 3 liters of 0.05 M $NaHCO_3$, pH 9.5, and the reservoir contains 3 liters of 0.3 M potassium chloride in the same buffer. The enzyme usually emerges between 0.1 M KCl and 0.14 M KCl.

Properties

Stability. The *E. coli* phosphoramidate hexose transphosphorylase as reported by Fujimoto and Smith[6] is a remarkably stable enzyme when stored at either 4° or at −15°. Active fractions have been stored for as long as six years in the author's laboratory with no loss in activity. At 55°, and pH 7.5, the *E. coli* enzyme can be heated for more than 30 minutes with no loss in activity, but at 100° for 10 minutes approximately 60% of its activity is lost.

Most preparations prepared as outlined above still contain measurable phosphoramidase[6] and glucose 1-phosphate transphosphorylase.[7] In general the ratio of transfer to hydrolytic activity varies with different preparations and may be as low as 2 and as great as 10 or more.

Specificity. A variety of hexoses are phosporylated by the *E. coli* phosphoramidic hexose transphosphorylase. In decreasing order of acceptor ability these are: glucose, 2-deoxy-D-glucose, mannose, α-methyl-

[6] A. Fujimoto and R. A. Smith, *Biochim. Biophys. Acta* **56**, 501 (1962).

[7] A. Fujimoto, P. Ingram, and R. A. Smith, *Biochim. Biophys. Acta* **96**, 91 (1965).

D-glucoside, fructose, glucosamine, 3-*O*-methylglucose, 1,5-anhydro-D-glucitol. Except for sedoheptulose no other hydroxylic compound tested functions as a phosphoryl group acceptor at a significant rate. The initial product formed has been shown in the case of glucose, mannose, and fructose to be the hexose 1-phosphate. Presumably the phosphorylation of α-methyl-D-glucoside and 1,5-anhydro-D-glucitol results from the activity of glucose 1-phosphate transphosphorylase,[7] which contaminates most preparations, and a small but significant amount of reducing sugar in the substrate preparations. The K_m for glucose has been measured as 1.25×10^{-2} *M*.

A number of P-N compounds serve as phosphoryl donors, among which $P\text{-}NH_2$ and *N*-phosphorylglycine are the most effective. Several other *N*-phosphoryl amino acids as well as phosphohistidine have been shown to be effective phosphoryl donors. The K_m for $P\text{-}NH_2$ is 5.4×10^{-3} *M*.

Among a wide variety of inhibitors tested with this enzyme, fluoride ion has been shown to be the most effective. Its inhibition is markedly dependent on pH. At pH 8.7, 1.5×10^{-3} *M*, KF produces only 8% inhibition whereas at pH 6.5 the same concentration of KF results in 90% inhibition of the phosphoryl transfer reaction.

PURIFICATION OF PHOSPHORAMIDATE HEXOSE TRANSPHOSPHORYLASE FROM EXTRACTS OF SUCCINATE-GROWN *E. coli*

Fraction	Total protein (mg)	Total units	Specific activity[a]
1. Crude	10527	18.4	0.016
2. Fraction A (pH 1.0)	1769	30.6	0.183
3. Fraction B (pH 9.0)	1747	30.5	0.175
4. Fraction C (strept. SO_4)	1987	32.4	0.163
5. DEAE-cellulose (pooled peak)	0.274	6.2	22.7

[a] Units per milligram protein.

[74] Galactokinase

By M. R. HEINRICH and SALLY M. HOWARD

$$\alpha\text{-D-Galactose} + \text{ATP} \rightleftharpoons \alpha\text{-D-galactose-1-P} + \text{ADP}$$

Assay Method

Principle. The assays which are applicable to kinases in general have been used for this enzyme (see Vol. I [35]). The most satisfactory assay is the spectrophotometric procedure. The rate of ADP formation by galactokinase is measured, in the presence of pyruvate kinase and lactate dehydrogenase, by following the oxidation of DPNH at 340 mμ (see this volume [82]).

Reagents

KCl, 0.8 *M*
EDTA, 0.02 *M*, pH 7
$MgSO_4$, 0.1 *M*
ATP, 3 mg/ml
Phosphoenolpyruvate, 3.8 mg/ml
Lactate dehydrogenase (containing pyruvate kinase), 0.2 mg protein/ml
DPNH, 0.7 mg/ml
Potassium phosphate buffer, 0.16 *M*, pH 7.0
Galactose, 3 mg/ml

Procedure. One-tenth milliliter each of the first 7 reagents is pipetted into a cuvette of approximately 1 ml capacity and 1 cm light path, 0.005–0.1 ml of enzyme solution is added, and the volume is adjusted to 1.0 ml with buffer. Then 0.1 ml of galactose is added, the cuvette is mixed by inversion and placed in the spectrophotometer. In a volume of 1.1 ml, the change in number of micromoles of galactose is equal to the change in absorbance at 340 mμ divided by 5.65. The use of higher galactose concentrations will increase the velocity, up to about 1 mg of galactose in the mixture (5 m*M*). Substrate concentrations greater than this are inhibitory.

Definition of Unit and Specific Activity. One unit of galactokinase is that amount which phosphorylates 1 micromole of galactose per minute. Specific activity is expressed as units per milligram of protein.

Application of Assay Method to Crude Preparations. Enzymes which oxidize DPNH or produce ADP will interfere with the assay by indicating activity which is higher than the true value. Their presence can be detected and a correction made if the reaction rate is measured both before and after addition of galactose; the difference in these rates represents galactokinase activity. The presence of hexokinase does not interfere unless the galactose is contaminated with glucose. Hexokinase may be assayed by substituting glucose for galactose.

Source of Enzyme

Galactose-adapted *Saccharomyces fragilis,* strain C-106, has been used for the procedure described. Twenty strains of the yeast were tested and were found to have approximately the same enzyme content.

Growth of yeast. The medium[1] contains:

Galactose or lactose (autoclaved separately as 10% solution), 2%
$NH_4H_2PO_4$, 0.1%
$(NH_4)_2HPO_4$, 0.1%
KH_2PO_4, 0.2%
Difco yeast extract, 0.2%
Trace element solution, 1.0%: 2% $MgSO_4 \cdot 7\ H_2O$, 0.1% NaCl, 0.05% $FeSO_4 \cdot 7\ H_2O$, 0.05% $ZnSO_4 \cdot 7\ H_2O$, 0.05% $MnSO_4 \cdot 3\ H_2O$, 0.005% $CuSO_4 \cdot 5\ H_2O$, 1% of 0.1 N H_2SO_4.
A few drops of silicone antifoam are added if the culture is to be aerated with a stream of air.

The yeast is adapted to galactose by inoculating a small volume of the above medium from a stock slant and incubating the preparation overnight. This culture is used as inoculum for the main portion of medium, which is incubated 24–40 hours at 25–30°. Aeration may be provided by a stream of air, or by a rotary shaker for smaller volumes. The yeast is collected by centrifugation, and washed twice with distilled water. The yeast paste is spread on thin plastic film (Saran Wrap) in a 2-mm layer and allowed to dry at room temperature. Eight liters of growth medium yields about 20 g of dried yeast, which can be stored at —15° for several months without significant loss of galactokinase activity.

Purification Procedure[2,3]

Some purification of the *S. fragilis* enzyme was carried out by Leloir and Trucco (Vol. I [35]). Wilkinson[1] described the preparation of the enzyme from another type of yeast, and the enzyme from *E. coli* has been purified.[3a]

All steps except heat treatment are carried out at 0–5°. All solutions contain 0.001 M EDTA. Quantities are based on 100 g of dried yeast.

Extraction. The yeast is extracted 6–10 hours in 500 ml of 0.167 M diammonium phosphate with occasional stirring, and then centrifuged.

Heat Treatment. The extract is heated in a 50° water bath, with swirling, for 10 minutes, then rapidly chilled and centrifuged.

Ammonium Sulfate. The supernatant solution is diluted to 1 liter (20–30 mg protein/ml) and solid ammonium sulfate is added to 45% saturation (277 g). The pH is maintained above 6.5 with NH_4OH. The precipitate is discarded, and ammonium sulfate is added to 60% satura-

[1] J. F. Wilkinson, *Biochem. J.* **44**, 460 (1949).
[2] M. R. Heinrich, *J. Biol. Chem.* **239**, 50 (1964).
[3] S. M. Howard and M. R. Heinrich, *Arch. Biochem. Biophys.* **110**, 395 (1965).
[3a] J. R. Sherman and J. Adler, *J. Biol. Chem.* **238**, 873 (1963).

tion (113 g). The precipitate is collected and dissolved in 50 ml of 0.001 *M* EDTA.

pH 5 Treatment. Solid cysteine hydrochloride is added to a final concentration of 0.01 *M*, and the pH of the solution is adjusted to 5.0 with dilute acetic acid. The suspension is centrifuged briefly and the supernatant is immediately adjusted to pH 6.5 with NH_4OH.

DEAE-Cellulose Chromatography. This step has usually been carried out with small portions, approximately one-tenth of the total protein. After dialysis against 0.005 *M* ammonium phosphate buffer, pH 7.0, a portion of the enzyme solution is applied to a 22 × 2 cm column of DEAE-cellulose which has been equilibrated with the same buffer. The column is eluted with a linear gradient consisting of 375 ml of the above buffer in the mixing bottle, and 375 ml of 0.005 *M* ammonium phosphate–0.5 *M* KCl, pH 4.5, in the reservoir. Galactokinase appears just after the main protein peak, at an elution volume of about 400 ml; the pH of the effluent remains above 6.

Two additional procedures have been used in some preparations[2]: calcium phosphate gel, and removal of hexokinase with bentonite. These steps are more variable than the remainder of the procedure, and should be tested with small portions of the preparation. *Gel adsorption* may be carried out from a dilute protein solution in 0.01 *M* cysteine, at pH 6.5. Other proteins are usually adsorbed by a weight of calcium phosphate gel equal to the total weight of protein, and galactokinase is adsorbed by a second portion of gel of the same weight. The second portion of gel is eluted several times with neutral 0.1 *M* ammonium phosphate to recover the enzyme. *Bentonite* will adsorb hexokinase preferentially when added to an enzyme solution thoroughly dialyzed against 0.05 *M* ammonium phosphate, pH 7. A weight of bentonite (washed with buffer and water, and dried) equal to twice the weight of protein present is added to the solution, and the pH is adjusted to 5.2 with *M* acetic acid. The suspension is centrifuged, and the supernatant is immediately neutralized. The preparations have approximately equal hexokinase and galactokinase activities before treatment with bentonite.

The table summarizes a typical enzyme preparation, including all the steps described.

Properties

The bentonite supernatant contains no hexokinase or ATPase activity. Moving boundary electrophoresis at pH 6.6 showed two major components moving toward the anode, galactokinase activity being associated with the slower peak.

PURIFICATION PROCEDURE

Fraction	Volume (ml)	Protein (mg/ml)	Specific activity	Total units
Original extract	395	57.2	0.29	6600
Ammonium sulfate, 45–60% precipitate	119	87.8	0.42	4400
Calcium phosphate gel eluate	200	7.6	1.36	2300
DEAE-cellulose fractions, lyophilized	20	9.6	2.20	550
After bentonite	22	2.5	3.30	185

Metal ion requirement: Mg^{++} is required for activity; Mn^{++} or Zn^{++} can be substituted, but Zn^{++} is inhibitory above 0.01 *M*. Removal of the metal by dialysis reduces activity to zero. The equilibrium constant (ADP) (α-galactose-1-P)/(ATP) ($\alpha + \beta$-galactose) was found to be 26 at pH 7.00 and 25°, using a crude enzyme preparation.[4]

Stability. Galactokinase becomes increasingly unstable on purification. It is inactivated by organic solvents at all stages. In the crude extract galactokinase is relatively heat stable, but warming of purified fractions inactivates the enzyme. Cysteine gives some protection against inactivation by heat and storage. Addition of cysteine, however, frequently causes an initial decrease in activity which is recovered very slowly. Galactokinase is rapidly inactivated at pH 5 or below. Lyophilization is not uniformly successful; the enzyme is also unstable to repeated freezing and thawing and to prolonged frozen storage. Dilute solutions can be stored frozen or at 0–5° for short periods; for longer periods, the most successful method has been precipitation of galactokinase by addition of ammonium sulfate to 80% saturation; the precipitate is then suspended in a small volume of phosphate buffer and frozen. Before use, the enzyme can be desalted by dialysis or on a small column of Sephadex G-25.

Specificity. The K_m of galactokinase for solutions of D-galactose, determined by the spectrophotometric assay described here, is 1.1×10^{-3} *M*. The K_m for α-D-galactose, added to the assay mixture as a solid, is 0.53×10^{-3} *M*. β-D-Galactose is not phosphorylated but mutarotates to α-D-galactose, which is then phosphorylated.[3] No mutarotase activity is present in purified galactokinase, but the salts in the assay system catalyze rapid mutarotation. The sole product is α-D-galactose 1-phosphate, as shown by an ion exchange procedure capable of separating the α and β anomers.

2-Deoxygalactose is a poor substrate, reacting at about 2% the rate

[4] M. R. Atkinson, R. M. Burton, and R. K. Morton, *Biochem. J.* **78**, 813 (1961).

of galactose under comparable conditions. Phosphorylation of galactosamine has been reported for crude[5] and slightly purified[6] yeast galactokinase, but it does not appear to be a substrate for the preparation described here. Talose was also reported to react slightly.[6]

The following sugars are not phosphorylated in concentrations of 1.5×10^{-3} to 1.5×10^{-2} M and over a 10-fold range of enzyme concentration: galactitol, galacturonic acid, tagatose, L-fucose. The same compounds did not inhibit galactose phosphorylation when present at ten times the galactose concentration. Other sugars which were not tested as extensively, but gave no indication of phosphorylation or inhibition, were L-galactose, 6-*O*-methyl-D-galactose, galactaric acid, galactono-γ-lactone, lactose, glucose, glucosamine, and L-arabinose.

[5] C. E. Cardini and L. F. Leloir, *Arch. Biochem. Biophys.* **45**, 55 (1953).
[6] F. Alvarado, *Biochim. Biophys. Acta* **41**, 233 (1960).

[75] D-Allokinase

By F. J. SIMPSON and L. N. GIBBINS

D-Allose + ATP → D-allose 6-phosphate + ADP[1]

Assay Method

The method measures the rate at which ADP is formed by coupling with reactions catalyzed by pyruvate kinase and lactic acid dehydrogenase.[2,3] The reagents employed and procedure are the same as described for D-xylulokinase except that D-allose replaces D-xylulose as the substrate and 0.5 M KH_2PO_4–0.01 M EDTA buffer pH 6.5 replaces the Tris buffer.[4] Dilutions of the kinase for assay are made in 0.01 M KH_2PO_4–0.001 M EDTA buffer, pH 6.5. One unit is defined as the amount of enzyme that phosphorylates 0.01 micromole of allose per minute.

D-Allose may be prepared from D-ribose by cyanohydrin synthesis and reduction of the lactone,[5] or from sucrose by fermentation and reduction.[6]

[1] Issued as NRC No. 9056.
[2] M. J. Bessman, Vol. VI [20].
[3] L. N. Gibbins and F. J. Simpson, *Can. J. Microbiol.* **9**, 769 (1963).
[4] F. J. Simpson, this volume [81].
[5] F. L. Humoller, *in* "Methods in Carbohydrate Chemistry" (R. L. Whistler and M. L. Wolfrom, eds.), Vol. I, p. 102. Academic Press, New York, 1962.
[6] M. J. Bernaerts, J. Furnelle, and J. De Ley, *Biochim. Biophys. Acta* **69**, 322 (1963).

Purification Procedure

All operations are performed at 0–5° except where otherwise stated. The pH of fractions containing the enzyme is maintained between 6.5 and 7.0. Glass-distilled water or distilled water passed through a mixed-bed ion exchange resin (Amberlite MB-3) are used to prepare reagents. The method presented is a modification of a method previously described.[3] Although the final preparation is less pure, the recovery of kinase is greater and the steps can be completed in one day. Protein was determined by the method of Lowry *et al.*[7]

Step 1. Growth of Culture and Preparation of Cell Extract. Aerobacter aerogenes, PRL R3, is grown in a medium containing 0.7% K_2HPO_4, 0.3% KH_2PO_4, 0.05% sodium citrate dihydrate, 0.01% $MgSO_4 \cdot 7 H_2O$, 0.1% $(NH_4)_2SO_4$, 0.001% $FeSO_4 \cdot 7 H_2O$, 0.1% casein hydrolyzate (i.e., NZ-amine Type A, Sheffield Chemical Co., New York) and 0.5% D-allose. The D-allose is sterilized separately and aseptically added to the sterile basal medium. The culture is transferred from an agar slant (allose) to 11 ml of the above medium and after 8–12 hours of growth, 1 ml is used to inoculate ten 500-ml Erlenmeyer flasks containing 100 ml of medium. These flasks are incubated at 30° on a rotary shaker until exponential growth ceases (9–10 hours). The cells are harvested by centrifugation and washed once with buffer (0.01 *M* KH_2PO_4–0.001 *M* EDTA, pH 6.8). The cells are then resuspended in the above buffer (250 mg wet weight per milliliter) and broken by treatment for 2 minutes at 0–15° with a 110 watt, 20 kc/sec oscillator. The debris and unbroken cells are removed by centrifugation for 10 minutes at 25,000 *g*. The pellet is suspended in buffer as before and treated again. The debris removed by centrifugation after the second treatment is discarded, and the extracts are pooled (40 ml, pH 6.7).

Step 2. Protamine. A solution of 2% protamine adjusted to pH 5.0 with acetic acid is prepared. A portion of this (0.14 ml per milliliter of extract) is slowly added to the cell-free extract. The mixture is stirred throughout the addition and for a further 5 minutes before the precipitate is removed by centrifugation for 10 minutes at 25,000 *g*. The precipitate is discarded.

Step 3. Calcium Phosphate Gel.[8] The enzyme (42 ml, pH 6.6) is absorbed on calcium phosphate gel by adding with stirring 23.6 ml of gel (26 mg/ml) to the supernatant fraction. The mixing is continued for an additional 3 minutes and then the gel is recovered by centrifugation for

[7] O. H. Lowry, N. J. Rosebrough, A. L. Farr, and R. J. Randall, *J. Biol. Chem.* **193**, 265 (1951).

[8] S. P. Colowick, Vol. I, p. 97.

1 minute at 10,000 *g*. Twenty milliliters of buffer (0.015 *M* KH_2PO_4–0.001 *M* EDTA, pH 7.0) is mixed with the gel, then the gel is recovered by centrifugation and the liquid is discarded. The gel is mixed with 20 ml of a solution (1.5 *M* KH_2PO_4–0.002 *M* EDTA, pH 7.0) by rubbing the gel against the side of the centrifuge tube with a thick glass rod until a smooth homogenate is obtained. The mixture is then allowed to sit for 5 minutes prior to centrifugation. The gel is discarded and 0.1 g of NaCl is added to the supernatant liquid (22 ml). The chloride ion is used as an indicator (silver nitrate) in the next step for the presence of salt in the eluates from Sephadex.

Step 4. Gel Filtration. The day before use, 20 g of Sephadex, G50 coarse bead, are equilibrated with 0.002 *M* EDTA–0.002 *M* glycerol, pH 6.6, and placed in a column (4 × 16 cm). The phosphate in the eluate from the calcium phosphate gel is removed by absorbing the eluate on this column and washing the enzyme through with the above buffer. The flow rate is adjusted to 2.5 ml per minute, and fractions of 5 ml are collected. The first 55–65 ml through the column can usually be discarded; the enzyme appears in the next 60–70 ml and the chloride and phosphate soon after.

Step 5. Alumina Cγ.[8] Ten milliliters of Alumina Cγ (24 mg/ml) are added slowly with stirring to the chloride-free eluate (65 ml, pH 6.6) from the column of Sephadex. The gel is recovered by centrifugation at 10,000 *g* for 1 minute and washed with 20 ml of 0.0075 *M* KH_2PO_4–0.001 *M* EDTA, pH 7.0, then the enzyme is eluted with 10 ml of 0.08 *M* KH_2PO_4–0.001 *M* EDTA, pH 7.0. The elution takes about 15 minutes.

Step 6. Ammonium Sulfate. A saturated solution (room temperature) of ammonium sulfate is adjusted to pH 7.2 with concentrated ammonium hydroxide. Of this solution, 9.9 ml is added slowly, with stirring, to the eluate (11 ml, pH 6.6) from the alumina Cγ. This mixture, pH 6.6–6.7, is

PURIFICATION OF ALLOKINASE

Treatment	Kinase (units)	Specific activity (units/mg protein)	Kinase, % recovered	A-6-P isomerase, % recovered
Cell extract	4000	6	100	100
Protamine	3400	10	85	78
$CaPO_4$ gel	2100	18	52	26
Sephadex	1400	13	35	15
Alumina Cγ gel	1200	25	30	3
$(NH_4)_2SO_4$	900	55	22	Trace

allowed to sit at room temperature for 10 minutes. The precipitate is recovered by centrifugation at 25,000 *g* for 10 minutes and dissolved in 3 ml of buffer (0.01 *M* glycerol–0.01 *M* KH_2PO_4–0.001 *M* EDTA), pH 6.8.

The purification procedure is summarized in the table.

Properties

The allokinase is quite labile. About 30–50% of the activity is lost on storage overnight at 0°, slightly more at −20°; thus the purification should be completed in one day. The enzyme is most stable between pH 6.5 and 7.5, and most active at pH 6.5. The kinase requires magnesium ions for activity and is inactivated by *p*-chloromercuribenzoate at 1×10^{-5} *M*. K_s for allose is 0.98×10^{-3}. The preparation has some activity with glucose and ribose as substrates, but is usually free of allose 6-phosphate isomerase. The latter enzyme and the ATPase have a lower affinity for calcium phosphate gel than the allokinase. The isomerase is also less stable below pH 7 than above.

[76a] *N*-Acetyl-D-glucosamine Kinase

I. *Streptococcus pyogenes*

By S. S. Barkulis

ATP + *N*-acetyl-D-glucosamine or D-glucose → ADP + H^+
+ *N*-acetyl-D-glucosamine-6-P or D-glucose-6-P

A kinase from *Streptococcus pyogenes* which catalyzes ATP-dependent phosphorylation of D-glucose and *N*-acetyl-D-glucosamine at equal rates has been purified 1500-fold.[1] The ratio of the enzymatic activity on both substrates remained constant throughout the fractionation. Similarity in heat stability, *p*-hydroxymercuribenzoate inhibition, protection by either substrate, and lack of repression of enzymatic activity when bacteria were grown exclusively in medium supplemented with only one of the carbohydrates support the likelihood that the kinase activities are associated with one enzyme.

Assay Method

Principle. Kinase activity is measured by the rate of hydrogen ion liberated during the phosphorylation of substrate by ATP. This is con-

[1] L. D. Zeleznick, H. Hankin, J. J. Boltralik, H. Heymann and S. S. Barkulis, *J. Bacteriol.* **88**, 1288 (1964).

veniently determined in a constant-pH microtitrimetric procedure.[2] For each 1 mole of substrate and ATP that disappears, 1 mole of acid-stable phosphate and 1 mole of ADP are formed; the products of the phosphorylation are acetyl glucosamine 6-phosphate or glucose 6-phosphate.

Reagents

NaOH, 0.1 *N*
ATP, 0.1 *M*
$MgCl_2$, 1.0 *M*
EDTA, 0.1 *N*
NaF, 1.0 *N*
Glucose, 0.1 *M*
Acetyl glucosamine, 0.1 *M*

Procedure. Reactions are carried out in 4-ml volumes at 30° and pH 7.4 with 0.002 *M* ATP, 0.02 *M* $MgCl_2$, 0.002 *M* EDTA, 0.02 *M* NaF, enzyme, and 0.0015 *M* sugar. The pH is maintained at 7.4 by addition of 0.1 *N* NaOH from a microtitrator (Micro-Metric Instrument Co., Cleveland, Ohio). The assay may be conducted within a 3-minute period with recordings of NaOH utilization made at 15-second intervals. The reaction rate is a linear function of the amount of enzyme between 0.25 and 0.75 unit of enzymatic activity, and assays should be conducted on samples or dilutions of enzyme which are within these limits. Endogenous acid liberation in crude extracts due to phosphatase activity is measured in control vessels lacking substrate.

Glucokinase activity is also measured by spectrophotometric assay at 340 mμ of glucose 6-phosphate formed, in cuvettes with a 10-mm light path; the test solution (1 ml) contains 0.03 *M* Tris buffer, 0.006 *M* $MgCl_2$, 0.0006 *M* EDTA (pH 8.0), 0.0002 *M* NADP, and 1 μl of glucose 6-phosphate dehydrogenase (1.40 K units/ml).[3] This assay permits measurement of the rate of conversion of glucose in the presence of acetyl glucosamine.

Definition of Unit and Specific Activity. One unit of activity was defined as the amount of enzyme required to form 1 micromole of hydrogen ion per minute at 30° and specific activity (SA) as the number of units per milligram of protein.

Purification Procedure

Growth of Microorganisms. S. pyogenes strain S23, serotype 14, was grown according to Barkulis and Jones.[4] Inocula of 100 ml of log-phase

[2] M. Schwartz and T. C. Myers, *Anal. Chem.* **30,** 1150 (1958).
[3] See M. W. Slein, Vol. I, p. 299.
[4] S. S. Barkulis and M. F. Jones, *J. Bacteriol.* **74,** 207 (1957).

cells were used per 1 liter of medium contained in a 4-liter Erlenmeyer flask. The cells were harvested by centrifugation after 16 hours of stationary growth at 37° and were washed twice with distilled water. The yield of cells was 4–5 g (wet weight) per liter of medium. Similar kinase activity was found in cells grown on medium supplemented with *N*-acetyl-D-glucosamine instead of D-glucose.

Step 1. Crude Extract. Streptococci were suspended (20% wet weight/volume) in 0.05 *M* Tris buffer (pH 8.0) containing 0.01 *M* $MgCl_2$, 0.001 *M* EDTA, and 0.2 *M* glucose; 70 ml of the suspension were mixed with 15 ml of Ballotini No. 12 beads, and were added to the large head of a Raytheon sonic oscillator. The instrument was cooled with water and operated at 10 kc for 20 minutes. The resulting crude extract was centrifuged at 29,000 *g* at 4° for 30 minutes. The supernatant liquid was removed and centrifuged again in a Spinco Model L centrifuge at 105,000 *g* for 1 hour at 4°. The residue was discarded, and the clear yellow supernatant fluid was used as the starting material for fractionation. All subsequent operations, except as otherwise stated, were carried out at 0–4°.

Step 2. Streptomycin Treatment. Streptomycin sulfate (1 ml of a 20% solution) was added with stirring to each 100 ml of the supernatant fraction. The fine precipitate of nucleic acids was sedimented and discarded.

Step 3. First Ammonium Sulfate. Solid ammonium sulfate was added gradually to the supernatant liquid until the salt concentration was 65% of saturation. The resulting precipitate, containing most of the enzyme, was collected by centrifugation and was dissolved in the minimal volume of 0.005 *M* Tris buffer (pH 7.6) containing 0.001 *M* $MgCl_2$, 0.0001 *M* EDTA, and 0.2 *M* glucose or acetyl glucosamine. The same buffer mixture was used for redissolving the enzyme fractions obtained in the subsequent steps. The presence of either sugar substrate was essential for preventing the inactivation of the enzyme during purification.

Step 4. Heat Treatment. A considerable amount of inert protein was removed at this point by heating, performed in the presence of ammonium sulfate to avoid inactivation. Even with this precaution, approximately 25% of the enzymatic activity was lost. The dissolved enzyme from step 3 was brought to 13% $(NH_4)_2SO_4$ saturation by addition of the solid salt. The solution was placed into a bath (60°) with constant agitation and was kept there for 1 minute after it had attained the bath temperature. It was cooled rapidly in an ice bath and centrifuged at 29,000 *g* for 30 minutes to remove an inert precipitate.

Step 5. Second Ammonium Sulfate. The supernatant fluid was brought to 35% saturation by further addition of solid $(NH_4)_2SO_4$. The resulting precipitate, containing the major portion of the enzymatic activity, was collected by centrifugation and dissolved in buffer.

Step 6. Alcohol-Acetone Fractionation. Fractionation with alcohol and acetone was used to bring about substantial purification, but considerable inactivation of enzyme also took place. The maintenance of 0.2 *M* glucose or *N*-acetyl-D-glucosamine based on total volume was essential in this procedure. Sufficient sugar was added prior to each addition of alcohol and acetone to ensure a final concentration of 0.2 *M*. Absolute ethanol chilled to —40° was added drop by drop to the solution from step 5 until a concentration of 50% (v/v) was attained. The enzyme solution was maintained at 0° to —10° during this procedure. An inactive precipitate was removed by centrifugation. Acetone cooled to —40° was added to the ethanol supernatant fraction, which was maintained at —10° to —15°. When 10% acetone (v/v) was attained, the major portion of enzyme activity was precipitated. The protein recovered by centrifugation and dissolved in supplemented Tris buffer.

Step 7a, b, c. Additional Ammonium Sulfate Fractionations. Solid $(NH_4)_2SO_4$ was added to 18% saturation, and an inert precipitate was collected by centrifugation. The supernatant liquid was brought to 35% $(NH_4)_2SO_4$ saturation and allowed to stand for 3 hours at 0–4°. The resulting precipitate contained most of the enzymatic activity. It was

TABLE I
PURIFICATION OF KINASE FROM *S. pyogenes*

Step and fraction	Total protein (mg)	Total units	Specific activity (units/mg)	Recovery (%)
1. Crude extract	40,380	10,818	0.26	100
2. Supernatant after streptomycin treatment	19,872	9,000	0.45	83
3. First $(NH_4)_2SO_4$ fractionation (0–65%)	10,300	7,416	0.72	69
4. Supernatant after heat treatment	5,460	5,405	0.99	50
5. Second $(NH_4)_2SO_4$ fractionation (0–35%)	1,686	5,040	2.90	46
6. Alcohol-acetone fractionation	177.6	2,500	14.0	23
7a. Third $(NH_4)_2SO_4$ fractionation (18–35%)	62.0	2,180	35.0	20
7b. Fourth $(NH_4)_2SO_4$ fractionation (18–35%)	17.1	1,162	68.0	11
7c. Fifth $(NH_4)_2SO_4$ fractionation (18–35%)	4.6	1,160	250.0	11

centrifuged and redissolved in supplemented Tris buffer. This procedure was repeated twice (steps 7b and c), yielding a preparation purified 1000-fold in kinase activity. A critical feature for obtaining purification in steps 7a, b, and c was the dilution of the protein to concentrations below 5 mg/ml in supplemented Tris buffer before beginning the addition of $(NH_4)_2SO_4$.

The particular fractionation shown in Table I resulted in 1000-fold purification. Other fractionations, in which small variations of the procedure were employed, yielded preparations with a specific activity up to 400 units/mg or approximately 1500-fold purification. The ratio of glucokinase to *N*-acetyl-D-glucosamine kinase activity remained the same throughout.

A variety of chromatographic procedures were tried, and either failed to increase the specific activity of the enzyme or were inactivating. These included diethylaminoethyl cellulose, carboxymethyl cellulose, Amberlite CG 45, and starch columns. A three- to sixfold purification could be achieved at step 7a by use of Sephadex G-200. However, purification was not obtained from Sephadex gels of lower cross-linkage.

Properties

Enzyme Stability. The optimal substrate concentration for preserving enzymatic activity is 0.2 *M*. The enzymatic activity is stable for several months at —25°. At 4°, it is unstable in the absence of either substrate although ammonium sulfate in small amounts 2.5% is partially protective. Dialysis at 4° for 17 hours against 0.005 *M* Tris buffer pH 7.6 results in 92% loss of activity. In the presence of 0.1 *M* of either sugar there was a 33% loss of activity. Under the same conditions of dialysis but with 0.2 *M* substrate or 2.5% $(NH_4)_2SO_4$ added to the dialyzing medium all the enzymatic activity was recovered. Since this amount of ammonium sulfate does not cause any measurable buffering action in the titrimetric assay, small samples of enzyme were dialyzed in this manner to remove the protective sugar in order to assay for the two substrates on the same enzyme sample.

In the presence of substrate and 10–15% ammonium sulfate, the enzyme is stable at 60° for 1 hour; at 70°, it is completely inactivated in 3 minutes. The ratio of kinase activities does not change during denaturation at various temperatures and for varying periods of time.

Substrate Specificity. The enzyme catalyzes the phosphorylation of *N*-acetyl-D-glucosamine and D-glucose at equal rates. D-Glucosamine was phosphorylated by relatively crude preparations at a rate 0.4 times that observed for glucose or *N*-acetyl-D-glucosamine. Purified preparations are inactive on D-glucosamine and the substituted D-glucosamines in Table II. Maltose, D-galactose, *N*-acetyl-D-galactosamine, D-fructose, D-mannose, and *N*-acetyl-D-mannosamine are also inactive as substrates.

TABLE II
SUBSTITUTED D-GLUCOSAMINES NOT ACTIVE AS SUBSTRATES FOR THE KINASE

Sample No.	Position				
	1	2	3	4	6
1		$-NH_2$	$-OCH_3$		
2		$-NHC_4H_9$			
3	$-OCH_3$, α,β mixed	$-NH-Ac$			
4	$-OCH_3$, α	$-NH-Ac$	$-OCH_3$		
5		$-NH_2$	$-OCH_3$	$-OCH_3$	$-OCH_3$
6	$-OCH_3$, α,β mixed	$-NH-Ac$	(Mixture of mono- and di-*O*-methyl)		
7		$-NH_2$	$-OCH(CH_3)COOH$		
8		$-NH-Ac$	$-OCH(CH_3)COOH$		
9		$-NH_2$	$-OCH_2COOH$		

Activators and Inhibitors. *p*-Hydroxymercuribenzoate (PHMB) inhibited equally the enzymatic activity on both substrates; 1.25×10^{-6} *M* PHMB was the minimal concentration for complete inhibition. The substrates did not protect against this inhibition. Thioethanol protected the enzyme against PHMB inhibition only to the extent of neutralizing the unreacted PHMB, but was unable to reactivate the enzyme. Thioethanol protects the enzyme against an equimolar amount of PHMB added subsequently. However, complete inhibition of the enzyme with 1.25×10^{-6} *M* PHMB cannot be reversed by 2.5×10^{-2} *M* thioethanol.

In contrast, the PHMB inhibition of the kinase is reversed by cysteine. After prolonged (18 hours) periods of PHMB inhibition various quantities of DL-cysteine will completely reactivate the enzyme within 30 minutes.

pH. The pH optimum for the phosphorylation of both substrates lies between 7.0 and 8.0. Activity falls off rapidly above and below this range.

Kinetics. K_m values for the two substrates are 1.09×10^{-3} *M* for glucose and 0.77×10^{-3} *M* for acetyl glucosamine. When both substrates are present in the incubation mixture, the rate of phosphorylation is lower than the sum of the rates observed with each substrate alone, indicating an inhibition. Each substrate acts as competitive inhibitor of the other; the K_i for glucose is 1.8×10^{-3} *M* and the K_i for acetyl glucosamine is 0.9×10^{-3} *M*. All constants were determined from Lineweaver-Burk plots by the method of Dixon.

[76b] *N*-Acetyl-D-glucosamine Kinase

II. *Escherichia coli*

By CARLOS ASENSIO and MANUEL RUIZ-AMIL

ATP + *N*-acetyl-D-glucosamine or D-glucose → ADP + H^+ + *N*-acetyl-1-D-glucosamine-6-P or D-glucose-6-P

Assay Method

Principle. ASSAY A. The ADP produced in the reaction is estimated by coupling with a system of phosphoenol pyruvate, pyruvate kinase, lactic dehydrogenase, and NADH (see this volume [82]). The reaction is started with the addition of *N*-acetyl-D-glucosamine (NAGA), and is followed by the decrease in optical density at 340 mμ. A blank without NAGA must be included in order to estimate the ATPase activity usually present in the purified preparations here described. ATPase activity is particularly high in crude preparations, wherein this assay is inappropriate.

ASSAY B. Activity in crude preparations can be estimated by the procedure used by Leloir *et al.*[1] for the same enzyme from liver. In this procedure, *N*-acetylglucosamine kinase (NAGA-kinase) activity is measured by the disappearance of NAGA in the presence of ATP and Mg ions. A blank without ATP is used to correct for deacetylase activity. Residual NAGA is estimated by a modification[2] of the Morgan-Elson method for *N*-acetylhexosamines. Because of the high ATPase activity and the strong inhibition by ADP (see below) a great excess of ATP over NAGA in the assay system becomes necessary.

Reagents

A. Substrates and buffer at the following concentrations:
  NAGA, 1 m*M*
  ATP and $MgCl_2$, 3.3 m*M*
  Tris-HCl, pH 7.8, 66 m*M*
The auxiliary system (phosphoenol pyruvate, pyruvate kinase, lactic dehydrogenase, and NADH) as described in this volume [82].
B. NAGA, 0.01 *M*
  Tris buffer, 0.5 *M*, pH 7.8

[1] L. F. Leloir, C. E. Cardini, and J. M. Olavarría, *Arch. Biochem. Biophys.* **74**, 84 (1958).
[2] J. L. Reissig, J. L. Strominger, and L. F. Leloir, *J. Biol. Chem.* **217**, 959 (1955).

ATP, 0.1 *M*, neutralized
$MgCl_2$, 0.1 *M*
$ZnSO_4$, 5%
$Ba(OH)_2$, 0.15 *N*
Reagents for the determination of *N*-acetylhexosamines[2]

Procedure. Add 0.1 ml of the solutions NAGA, Tris buffer, ATP, and $MgCl_2$ to a small tube. Prepare another tube with the same reagents except ATP. Allow both tubes to reach 37° in a water bath and start the reaction by adding 0.1 ml of the enzyme preparation. Crude extracts (step 1) must be incubated for about 30 minutes to obtain adequate utilization of the sugar substrate. Stop the reaction by addition of 2.0 ml of $ZnSO_4$ and 2.0 ml of $Ba(OH)_2$ to each tube. Mix thoroughly and centrifuge at low velocity for a few minutes. Take out 0.5 ml of the supernatant for the estimation of residual NAGA.[2] Compare color yield against a standard with 0.1 micromole of NAGA. For color reading, any conventional spectrophotometer or photocolorimeter, at 545 or 585 mμ, can be used.

Definition of Unit and Specific Activity. One unit is defined as the amount of enzyme that catalyzes the phosphorylation of 1 micromole of NAGA per minute at 37°. Specific activity is expressed as units per milligram of protein. Protein is determined by ultraviolet absorption at 280 and 260 mμ.[3]

Purification Procedure

Culture Conditions. Escherichia coli, strain 86 NCTC Oxford,[4] was the source of this enzyme. The culture medium contained (per liter, in distilled water): Na_2HPO_4, 6.0 g; KH_2PO_4, 3.0 g; NH_4Cl, 1.0 g; $MgSO_4 \cdot 7H_2O$, 0.2 g; NaCl, 0.5 g; and glucose, 4.0 g. Glucose (4% solution) was autoclaved separately. The use of NAGA instead of glucose did not increase enzyme production. Twelve 2-liter Erlenmeyer flasks with 500 ml of the above mixture were inoculated (1% v/v) with an overnight culture of the bacteria and incubated for 20 hours at 37° in a gyratory shaker (G-25 model, New Brunswick Scientific Co.) at about 200 rotations per minute. Cells were harvested by centrifugation and washed twice with cold 0.1 *M* potassium phosphate, pH 7.0.

All the operations that follow were carried out at about 2° unless indicated otherwise.

Step 1. Preparation of the Crude Extract. Twelve grams of fresh

[3] O. Warburg and W. Christian, *Biochem. Z.* **310**, 384 (1941).
[4] Of other strains tested, NCTC 6121 was also a good source of NAGA-kinase, but not ATCC 9637 nor ATCC 11105, as assayed at the level of crude extracts.

bacteria were ground in a mortar with 36 g of alumina (Alcoa A-305) and extracted with 36 ml of a solution containing: 50 m*M* phosphate, 1 m*M* EDTA, and 0.1 m*M* NAGA, pH 7.6 (PEN solution). The homogenate was centrifuged at 20,000 *g* for 15 minutes and the sediment discarded. The supernatant was diluted with PEN solution to a concentration of 10 mg protein per milliliter (fraction I).

Step 2. Protamine Sulfate Treatment. To fraction I, 2% aqueous protamine sulfate solution was added up to 0.1 mg of protamine sulfate per milligram of protein. After it had stood for 10 minutes, the mixture was centrifuged at 27,000 *g* for 15 minutes and the sediment was discarded.

Step 3. Ammonium Sulfate Fractionation. A 90% saturated solution of ammonium sulfate was added to the above supernatant (fraction II) to obtain 60% saturation, and the solution was allowed to stand for 20 minutes and centrifuged at 27,000 *g* for 15 minutes. The supernatant was removed and the precipitate was dissolved with 6 ml of PEN solution. This fraction was filtered in two steps (3 ml each) by a Sephadex G-25 column (10 cm height, 2 cm width). Both filtrates were pooled and diluted with PEN solution up to the volume of the crude extract.

Step 4. Heat Treatment. The above fraction (fraction III) was heated for exactly 5 minutes at 65°, quickly cooled in an ice water bath, and kept frozen overnight. After it had been thawed, the solution was centrifuged for 15 minutes at 27,000 *g* and the sediment was discarded (fraction IV).

Step 5. DEAE-Cellulose Chromatography. DEAE-cellulose was treated as described[5] by Seubert and Remberger: (a) suspension in deionized water for several days, followed by discarding of ultrafine particles not sedimenting after 2 hours; (b) washing with 1 *M* NaOH and then water until the material is free of the alkali; (c) addition of 1 *M* NaCl until the Cl^- form is obtained; (d) washing with 1 *M* KOH and then adding enough water to remove the alkali. DEAE-cellulose treated in this way was packed in a column (6 cm height, 1 cm width) and equilibrated with PEN solution. Aliquots of 15 ml of fraction IV were separately added to the column and chromatographed as follows: (a) washing with 5 ml of 0.10 *M* phosphate, pH 7.6; (b) washing with 5 ml of 0.15 *M* phosphate, pH 7.6; (c) elution by passing two 5 ml portions of 0.25 *M* phosphate, pH 7.6 through the column. The second 5-ml eluate was the active fraction. All the phosphate solutions employed also contained 1 m*M* EDTA and 0.1 m*M* NAGA (as in the PEN solution). A flow rate of approximately 1 ml per minute was maintained during the

[5] W. Seubert and U. Remberger, *Biochem. Z.* **334**, 401 (1961).

whole operation. Finally, all the active eluates were pooled (fraction V).

Step 6. Ammonium Sulfate Precipitation. In the above fraction, solid $(NH_4)_2SO_4$ was dissolved to reach 60% saturation. After standing for half an hour, the light precipitate which formed was collected by centrifugation for 30 minutes at 27,000 *g* and dissolved in 1 ml of PEN solution for each 5 ml of fraction V. This constituted fraction VI.

With this procedure a purification of 125- to 150-fold, with a recovery of approximately 25% of the enzymatic activity, has been repeatedly achieved. Fraction VI is free of glucokinase (which is totally inactivated in the heat treatment[6]), and is essentially free (less than 0.5% of NAGA-kinase) of deacetylases. It does contain some ATPase activity (about 25% of the NAGA-kinase activity). A summary of the purification procedure is given in the table.

PURIFICATION PROCEDURE OF KINASE FROM *E. coli*

Fraction	Total protein (mg)	Total activity (units)	Specific activity	Yield (%)
I. Crude extract	700	18.6	0.026	(100)
II. Protamine sulfate	660	21.0	0.032	112
III. Ammonium sulfate and Sephadex	294	8.9	0.032	47
IV. Heat treatment	77	8.2	0.106	44
V. DEAE-cellulose	3.4	8.0	2.3	43
VI. Ammonium sulfate	1.2	4.6	3.8	25

Properties

Stability and Optimum pH. The purified preparations are stable for several months at −20°. The activity is not affected by treatment at pH 9.5 for 30 minutes at 30°.[7]

Activity at different pH values, shows a plateau from pH 7.5 to 9.5. This property is not shared by the NAGA-kinase from liver, which shows a sharp peak at about pH 7.2.[1]

Specificity and Kinetics. NAGA-kinase of *E. coli* has a high affinity and specificity for the sugar substrate. K_m (NAGA) $= 5 \times 10^{-5}\ M$. The enzyme shows no detectable activity on *N*-methyl-D-glucosamine, D-glucosamine (at pH 7.8), *N*-acetyl-D-mannosamine, and *N*-acetyl-D-galactosamine. K_i values are, $1 \times 10^{-3}\ M$ for *N*-methyl-D-glucosamine, ca. $0.01\ M$ for glucosamine and higher than $0.1\ M$ for the other two amino-sugars. The only marginal substrate found was D-glucose (V_{max} 15% of

[6] C. Asensio, *Rev. Españ. Fisiol.* **16,** Suppl. 2, 121 (1960).

[7] Buffer used: A mixture of 2-amino-2-methyl-1-propanol and acetic acid, 0.1 *M*.

that on NAGA; $K_m = 5 \times 10^{-3}\ M$). This activity is a genuine property of NAGA-kinase as shown by cross inhibition experiments, and therefore is unrelated to glucokinase [K_m (glucose) $= 3 \times 10^{-4}\ M$[8]], which is absent in the purified preparations. In contrast to this specificity, a NAGA-kinase recently obtained in a highly purified form from *Streptococcus pyogenes*[9] phosphorylates both substrates at the same maximal rate and has nearly the same affinity for both (K_m (NAGA) $= 7.7 \times 10^{-4}\ M$; K_m (glucose) $= 1.1 \times 10^{-3}\ M$). This enzyme is also inactive on glucosamine, *N*-acetyl-D-mannosamine, and *N*-acetyl-D-galactosamine.

The enzyme shows a less stringent requirement for the trinucleotide substrate. At the concentration of $2 \times 10^{-3}\ M$, relative velocities are: ATP = 1; ITP = 0.75; UTP, GTP, and CTP less than 0.05; K_m (ATP) $= 1.4 \times 10^{-4}\ M$; K_m (ITP) $= 2.5 \times 10^{-4}\ M$.

Inhibition by Products. ADP is a strong competitive inhibitor, with a K_i equal to the K_m for ATP. This may have physiological implications. On the contrary, NAGA-6-P does not inhibit significantly the reaction rate (less than 5% for a ratio NAGA-6-P: NAGA = 4).

[8] C. Asensio and A. Sols, *Intern. Abstracts Biol. Sci.*, Suppl., p. 125 (1958); *ibid.*, *Rev. españ. Fisiol.* **14**, 269 (1958).

[9] L. D. Zeleznick, H. Hankin, J. J. Boltralik, H. Heymann, and S. S. Barkulis, *J. Bacteriol.* **88**, 1288 (1964).

[77a] Phosphofructokinase

I. Skeletal Muscle

By K. H. LING, VERNER PAETKAU, FRANK MARCUS, and HENRY A. LARDY

Fructose 6-phosphate + ATP → fructose 1,6-diphosphate + ADP

Assay[1]

Principle. During the purification of the enzyme, phosphofructokinase (P-fructokinase) activity is determined by coupling with aldolase, triose phosphate isomerase, and α-glycerophosphate dehydrogenase. Each mole of fructose diphosphate formed by P-fructokinase leads to the oxidation of 2 moles of DPNH. The assay is carried out at pH 8.0 where the enzyme activity is maximal. Alternatively, the enzyme activity may be determined with a pH-stat,[2] or by measuring fructose diphosphate formed during a selected time of incubation.

[1] E. Racker, *J. Biol. Chem.* **167**, 843 (1947).

[2] J. E. Dyson and E. A. Noltmann, *Anal. Biochem.* **11**, 362 (1965).

Reagents[2a]

Tris-Cl, 0.20 *M*, pH 8.0 (33 m*M*)
ATP, 0.02 *M*, pH 7 (2 m*M*)
$MgSO_4$, 0.20 *M* (5 m*M*)
K_2 fructose 6-phosphate, 0.02 *M* (2 m*M*)
DPNH, 2.4 m*M* (0.16 m*M*)
KCl, 0.20 *M* (0.05 *M*)
Dithiothreitol, 0.10 *M* (Cleland's Reagent, CalBiochem) (1 m*M*)

Auxiliary enzyme solution is made up as follows: 0.25 ml of aldolase (10 mg/ml) and 0.05 ml of a mixture of triose phosphate isomerase and α-glycerophosphate dehydrogenase (10 mg/ml) are dissolved in 4.70 ml 0.01 *M* Tris-Cl, pH 8.0, containing 10 mg bovine serum albumin. The auxiliary enzymes are obtained from Boehringer and Soehne.

Dilution. The enzyme to be assayed is diluted to a final concentration of not lower than 1 μg P-fructokinase per milliliter in 0.1 *M* potassium phosphate, 2 m*M* in dithiothreitol, pH 8.0.

Procedure. A reaction mixture of 3.0 ml is made up as follows: 0.5 ml of Tris buffer, 0.3 ml of ATP, 0.075 ml of $MgSO_4$, 0.3 ml of fructose-6-P, 0.2 ml of DPNH, 0.75 ml of KCl, 0.03 ml of dithiothreitol, and 0.20 ml of auxiliary enzyme solution. The assay mixture described above is added to a 10-mm light path cuvette, and the volume is adjusted with water so that addition of the P-fructokinase will bring the total volume to 3.0 ml. The assay is run in a spectrophotometer thermostated at 28°. After a background rate of DPNH oxidation is obtained, P-fructokinase is added and the reaction rate, which is constant for several minutes following a brief lag, is recorded as change in optical density at 340 mμ. The units of P-fructokinase present are obtained by multiplying the net rate of DPNH oxidation (O.D. units per minute) by 0.24. The assay described is linear with P-fructokinase concentration up to an optical density change of 0.20 min^{-1}.

Units. By definition, one unit of P-fructokinase catalyzes the conversion of 1 micromole of fructose 6-phosphate to fructose diphosphate per minute at pH 8.0 and 28°. Protein is measured by the biuret method using bovine serum albumin as standard.[3] If Tris or ammonium sulfate is present, the biuret reaction is carried out on the material precipitated by 5% trichloroacetic acid.

[2a] Final concentrations are enclosed in parentheses.
[3] A. G. Gornall, C. J. Bardawill, and M. M. David, *J. Biol. Chem.* **177,** 751 (1949).

Purification[4]

The quantitative data given are typical for this procedure.

Step 1. Extraction. A large albino rabbit is decapitated after deep anesthesia (50 mg Nembutal i.v.). The muscle mass from the back and hind legs is removed and chilled in ice. Unless otherwise indicated, subsequent manipulations are carried out in a cold room at 4°. The muscle mass is passed through a meat grinder, weighed (650 g), and treated for 5 minutes with occasional stirring with 3 volumes of 0.03 *M* KF, 1 m*M* in EDTA. Then the mixture is homogenized for two 30-second periods in a Waring blendor and centrifuged for 60 minutes at 1300 *g* at 0° in an International SR-3 centrifuge. The supernatant fraction is centrifuged for 30 minutes at 10,000 *g*, yielding 1580 ml of extract, pH 6.6. Fraction one can be kept at 4° for at least 1 week without loss of P-fructokinase activity.

Step 2. Heating and Precipitation with Isopropanol. This step is carried out on two equal portions of fraction 1. 790 ml of fraction 1, in a 2-liter stainless steel beaker which has been mounted in a 45° water bath, is stirred continuously. When the temperature reaches 40°, 0.10 volume (79 ml) of isopropanol is added at a rate of 10 ml per minute. The solution, still immersed in the 45° bath, is stirred for an additional 10 minutes. The beaker is then transferred to a cold bath, cooled to 20° with stirring, and centrifuged at 20° for 20 minutes at 10,000 *g*.

The clear reddish supernatant solution (830 ml) is transferred to a second 2-liter stainless steel beaker, chilled to —4° with stirring in an alcohol bath, and 0.10 volume (83 ml) of ice-cold isopropanol is added at a rate of 10 ml per minute. The mixture is stirred in the —4° bath for an additional 20 minutes. The white precipitate collected by centrifugation for 20 minutes at 10,000 *g* at —4° is kept in the centrifuge bottles so that the final centrifugation of the second batch may be carried out in the same bottles. The precipitate is taken up in 45 ml (about 1/30 volume of fraction 1) of 0.1 *M* Tris-phosphate, pH 8.0, 0.2 m*M* in EDTA (fraction 2).

Step 3. Fractionation on DEAE-Cellulose. The Tris-phosphate buffers used in this step are expressed as molarity of Tris; they have been titrated to pH 8.0 with H_3PO_4, and are 0.2 m*M* in EDTA.

Fraction two is dialyzed against two changes of 0.1 *M* Tris-phosphate, 0.2 m*M* in fructose diphosphate. DEAE-cellulose (BioRad Cellex

[4] K. H. Ling, F. Marcus, and H. A. Lardy, *J. Biol. Chem.* **240,** 1893 (1965).

D, high capacity[5]) is prepared by the method of Peterson and Sober,[6] a column of 15 × 2.5 cm is made up and equilibrated with 0.1 *M* Tris-phosphate. After the sample has been applied and washed in, elution is carried out with 0.1 *M* Tris-phosphate until no further protein emerges (about 400 ml). The percentage transmission of the column effluent is continuously monitored at 280 mμ (Gilson Medical Electronics UV Absorption unit) and recorded. A large protein peak with no P-fructokinase activity is washed off the column in this step. Elution is then switched to 0.3 *M* Tris-phosphate to bring off a peak (80 ml) containing the P-fructokinase activity. The material in this peak is precipitated by the addition of solid ammonium sulfate to 66% saturation, allowed to equilibrate for 1 hour, and centrifuged for 20 minutes at 10,000 *g*. The enzyme may be dissolved in 14 ml 0.1 *M* potassium phosphate, pH 8.0, 1 m*M* in EDTA directly to yield a stable preparation of specific activity ca. 150.

Step 4. Crystallization. This is accomplished by the procedure of Parmeggiani and Krebs.[7] The material from step 3, after the precipitation with ammonium sulfate, is dissolved in a buffer of the following composition: 0.05 *M* Na glycerophosphate (25% α), 0.002 *M* EDTA, 0.004 *M* ATP, pH 7.0, keeping the protein concentration as high as possible (preferably 25–30 mg/ml). Any insoluble material is removed by filtration through a sintered glass filter. Saturated ammonium sulfate solution (pH adjusted to 7.0 with NH_4OH) is then added over a period of a week to a final concentration about 28% of saturation. The solution is set aside at 4°, and crystallization is observed after a few weeks. When no further material crystallizes, the suspension is centrifuged at very low speed, the mother liquor is removed, and the crystals are resuspended in 0.05 *M* potassium phosphate, saturated with ammonium sulfate, 0.1 m*M* in dithiothreitol and 0.1 m*M* in EDTA, pH 8.0. After a second low speed centrifugation the harvested crystals are resuspended in the same buffer.

Properties

Stability. At high protein concentrations (10–20 mg/ml) P-fructokinase prepared by this method is very stable in phosphate buffers at pH 8.0, losing only a small amount of activity over a period of months. The enzyme is not stable in phosphate buffer at pH 7.

The enzyme loses activity when diluted, the degree of inactivation increasing with dilution. At a concentration of 1 μg/ml, purified P-fruc-

[5] The results described were obtained with batch B-1897.

[6] See E. A. Peterson and H. A. Sober, Vol. V [1].

[7] A. Parmeggiani and E. G. Krebs, *Biochem. Biophys. Res. Commun.* **19**, 89 (1965).

tokinase retains 90% of its activity after 2 hours (pH 8.0, phosphate buffer, 1 mM dithiothreitol).

Cation Requirement. Muscle P-fructokinase has two cation requirements; Mg^{++}, which is typical of ATP-utilizing enzymes, and either potassium or ammonium ion. Sodium has only a limited capacity to replace potassium.

Specific Activity. The specific activity of the crystalline P-fructokinase obtained by this procedure is 180. This compares to a specific activity of 161 reported by Parmeggiani and Krebs.[7]

Sedimentation.[4] At pH 8 the protein sediments in a heterogeneous pattern which varies with protein concentration. However, in the presence of 2 M urea, the enzyme sediments as a single peak with $S_{20,w} = 13.7$, while retaining its activity.

Yield. The first extract contains 110 units of P-fructokinase per gram of muscle, of which 63% is recovered after DEAE-cellulose fractionation (step 3) and 44% is obtained in the first crop of crystalline enzyme with specific activity 180. This is a much higher yield than any other reported for this source.

Other Preparations. Parmeggiani and Krebs[7] have developed a procedure for the isolation of rabbit muscle P-fructokinase in which the yields are low but the final product is obtained in crystalline form. The procedure described here, involving the fractionation steps of Ling *et al.*,[4] and the crystallization step of Parmeggiani and Krebs,[7] gives good yields of crystalline enzyme of high activity.

Mansour has described a purification of guinea pig heart P-fructokinase[8] and Ramaiah *et al.*[9] have purified yeast P-fructokinase to a specific activity of 1.3.

Mansour *et al.*[10] have crystallized sheep heart P-fructokinase, obtaining a final specific activity of 120.

[8] T. E. Mansour, *J. Biol. Chem.* **238,** 2285 (1963).

[9] A. Ramaiah, J. A. Hathaway, and D. E. Atkinson, *J. Biol. Chem.* **239,** 3619 (1964).

[10] T. E. Mansour, N. W. Wakid, and H. M. Sprouse, *Biochem. Biophys. Res. Commun.* **19,** 721 (1965).

[77b] Phosphofructokinase

II. Heart Muscle

By TAG E. MANSOUR

$$\text{ATP} + \text{fructose-6-P} \rightarrow \text{ADP} + \text{fructose-1,6-di-P} + \text{H}^+$$

Assay Method

Principle. Fructose-1,6-di-P produced in the phosphofructokinase reaction can be determined enzymatically by coupling the enzyme with aldolase, triosephosphate isomerase, and α-glycerophosphate dehydrogenase systems.[1] The rate of disappearance of DPNH is followed spectrophotometrically at a wavelength of 340 mμ for 5 minutes. One micromole of fructose-1,6-di-P is equivalent to 2 micromoles of DPNH oxidized.

Reagents

1. Glycylglycine-KOH buffer, 0.5 *M*, pH 8.2
2. $MgCl_2$, 0.1 *M*
3. Bovine serum albumin, 0.1%
4. DPNH, 1 mg/ml
5. Cysteine HCl, 0.1 *M*, pH 8.2 (prepared fresh daily)
6. Fructose-6-P, 0.01 *M*, adjusted with KOH to pH 8.2
7. K-ATP, 0.01 *M*, pH 8.2
8. Glycylglycine-KOH buffer, 0.01 *M*, pH 7.5
9. Mixture of α-glycero-1-phosphate dehydrogenase (GDH) and triosephosphate isomerase, 10 mg/ml (specific activity: 67.8 GDH EU/mg·protein), purchased from Calbiochem in Los Angeles
10. Aldolase crystalline suspension, 10 mg/ml; purchased from Calbiochem in Los Angeles
11. Reduced glutathione, 0.1 *M*, adjusted with KOH to pH 8.0
12. Tris-HCl buffer, 1.0 *M*, pH 8.0

A stock reaction mixture (stock solution A) for 20 enzyme assays is freshly prepared by mixing the following: 2 ml of reagent 1; 0.2 ml of reagent 2; 2 ml of reagent 3; 1.6 ml of reagent 4; 2.8 ml of reagent 5; and 3.4 ml of distilled water. This solution may be used for 1 day if kept in ice. Substrate stock solution (stock solution B) is prepared by mixing the following: 0.2 ml of reagent 6; 1 ml of reagent 7 and 2 ml of H_2O.

[1] K.-H. Ling, W. L. Byrne, and H. Lardy, Vol. I [38].

Stock solution B can be stored frozen for a month. A mixture of aldolase, α-glycero-1-phosphate dehydrogenase, and triose phosphate isomerase (enzyme mixture) is prepared by mixing 1.88 ml of reagent 8; 0.02 ml of reagent 9; and 0.1 ml of reagent 10. This solution must be prepared fresh daily. Phosphofructokinase diluting fluid (PFK diluting fluid) is prepared by mixing the following: 10 ml of reagent 11, 10 ml of reagent 3, 1 ml of reagent 12; then the solution is made to 100 ml. The diluting fluid should be prepared at least once every week.

Procedure. Add the following to a 1-ml spectrophotometer cuvette: 0.6 ml of stock solution A; 0.16 ml of stock solution B; 0.05 ml of enzyme mixture; 0.29 ml of H_2O and appropriate concentration of phosphofructokinase diluted in 0.2 ml of PFK diluting fluid. The cuvette content is mixed and optical density is measured at 340 mμ every minute.

A unit of phosphofructokinase is the amount of enzyme that catalyzes the formation of 1 micromole of fructose-1,6-di-P per 1 minute. A change in optical density of 0.310 in a 5-minute period corresponds to enzyme activity of 0.005 unit.

Purification Procedure

Phosphofructokinase from the sheep heart as received in the laboratory from the slaughterhouse is largely inactive and is present in the insoluble fraction of the cell. The following procedure describes methods for activation and solubilization of the enzyme as well as its purification and crystallization. The methods are those of Mansour, Wakid, and Sprouse.[2,3] One preparation requires two working days exclusive of the time for crystallization. The steps are for a small batch of 300 g of heart muscle. This method has been successfully tried in the author's laboratory in 55 preparations. All operations are carried out at 0° to 4° unless otherwise indicated. Results of a typical preparation on a 300 g batch of heart are summarized in the table.

Tissue Preparation. Sheep hearts are obtained from a local slaughterhouse about 40 minutes after the animal is killed. The hearts are transported to the laboratory in crushed ice. Upon arrival (approximately 2 hours after the animal is killed) the hearts are freed of connective tissue and fat. The ventricles are cut into sections suitable for insertion into a meat grinder. The pieces are then rinsed with cold distilled water and blotted on absorbent paper towels. Every 2 or 3 pieces are wrapped in aluminum foil and stored at —18°. Tissues thus frozen for one month have no noticeable loss of activity.

[2] T. E. Mansour, N. W. Wakid, and H. M. Sprouse, *Biochem. Biophys. Res. Commun.* **19**, 721 (1965).

[3] T. E. Mansour, N. W. Wakid, and H. M. Sprouse, *J. Biol. Chem.* **241**, 1512 (1966).

PURIFICATION OF SHEEP HEART PHOSPHOFRUCTOKINASE[a]

Fraction No. and step	Volume (ml)	Total (units)	Protein (mg/ml)	Specific activity (units/mg)	Yield
I. Activated heart extract	1222	6780	2.55	2.18	100
II. Ethanol precipitate					
Without ATP	60	1650	—	—	—
With ATP	60	5220	11.5	7.54	77
III. Ethanol fraction extract	55	3300	2.67	22.4	49
IV. DEAE-cellulose eluate	70	2940	.566	74.1	43
V. Ammonium sulfate	1.9	2360	9.92	125.0	35

[a] The preparation summarized in this table was carried out in a 300-g batch as described in the text.

Step 1. Preparation of Heart Extracts. Chop the frozen heart muscle in small pieces with a bone scissors. Grind 300 g of frozen heart muscle without thawing in a previously chilled meat grinder. After weighing the ground muscle, transfer to a Waring blendor connected to a variable transformer. To the ground muscle add 1200 ml of a solution containing 0.01 M Tris-HCl, pH 8.0, and 0.002 M EDTA. Gradually increase the speed of the blendor to a final adjustment of 90 volts on the transformer and then homogenize the muscle for 1 minute. If the capacity of the Waring blendor is limited to 750 ml volume the previous step can be carried out in two lots. Centrifuge the homogenate in 250 ml polycarbonate bottles in a GSA Servall rotor at 9000 rpm (13,000 g) for 20 minutes. The supernatant fluid has no significant phosphofructokinase activity and is discarded. Transfer the sedimented material with a glass rod to a Waring blendor and suspend in 1200 ml of the same solution described above. Homogenize for 15 seconds in the Waring blendor. Centrifuge as described above for 20 minutes. The pinkish supernatant fluid which has no enzyme activity is discarded. Suspend the residue from the second centrifugation in a liter of solution with the following composition: 15 ml M Tris-HCl buffer, pH 8.0; 75 ml 0.1 M mercaptoethanol; 75 ml of M $MgSO_4$; 7.5 ml 0.1 M (K)ATP in 1 liter. After the residue is suspended, bring the final volume to 1500 ml with distilled water. The final composition of this suspension is therefore: 0.01 M Tris-HCl (pH 8.0); 0.005 M mercaptoethanol; 0.05 M $MgSO_4$ and 5×10^{-4} M ATP. The suspension is rehomogenized in the Waring blendor for 15 seconds. Pour the suspension in a 2-liter Erlenmeyer flask and place in a bath at 37°. Stir in the thermostatic bath for 25 minutes. Centrifuge in the 250-ml plastic tubes at the same speed indicated above for 30 minutes. The supernatant fluid (1222 ml) which contains high phosphofructokinase activity has a pH of about 7.4. The specific activity of the extract prepared by this method (fraction I) varies from 1.4 to 2.6 units/mg protein.

Step 2. Ethanol Fractionation. This step was adapted from the procedure of Ling *et al.*[1] for purification of skeletal muscle phosphofructokinase. Pour fraction I into a 2-liter chilled Erlenmeyer flask and place it in a bath maintained at —3°. Add gradually absolute ethanol cooled at —16° with continuous gentle stirring until a final alcohol concentration of 8% is reached (106 ml of absolute ethanol per 1222 ml of fraction I). Stir the mixture at —3° for 20 minutes. Centrifuge at 6000 rpm in the GSA Servall head (5800 *g*) for 10 minutes. Discard the residue. Collect supernatant fluid in a 2-liter Erlenmeyer flask and place in the cold bath. Add cold absolute ethanol again to the supernatant fluid to a final concentration of 13% (74.5 ml of cold ethanol to 1292 of supernatant fluid). Stir the mixture at —3° for 20 minutes and then centrifuge it at the same speed for 10 minutes. Discard the supernatant fluid. Suspend the residue in 60 ml (5% of the volume of the original heart extract) of a solution containing 0.01 *M* potassium phosphate buffer, pH 8.0, and 0.005 *M* mercaptoethanol. Homogenize the suspension in an all glass homogenizer. Phosphofructokinase in this fraction is not fully active. Addition of 10^{-4} *M* ATP to an aliquot of the concentrated enzyme after ethanol fractionation in the cold results in approximately 50–100% increase in enzyme activity. The specific activity of the enzyme after ethanol fractionation (fraction II) varies from 6.1 to 11.0 when assayed following ATP activation. ATP is not added to this fraction and is used here merely in an aliquot to assay for total phosphofructokinase activity. Fraction II can be stored overnight in the cold without significant loss of activity.

Step 3. Extraction of the Enzyme from the Ethanol Fraction. To the brownish opaque suspension of fraction II (60 ml) add 0.6 ml of 1 *M* KH_2PO_4 while stirring gently. This addition lowers the pH to approximately 6.7. Transfer the suspension to 30 ml polycarbonate tubes. Centrifuge in a Servall SS-34 rotor at 10,000 rpm (12,000 *g*) for 20 minutes. The supernatant fluid is discarded. Suspend the residue in a solution of 0.02 *M* $MgSO_4$, 0.01 *M* Tris-HCl (pH 8.6), 0.005 *M* 2-mercaptoethanol, 10^{-4} *M* ATP and 10^{-5} *M* fructose-1,6-di-P. Final volume is made up with this solution to 60 ml. The presence of ATP and fructose-1,6-di-P in this step as well as subsequent steps is essential to achieve maximum enzyme yield by stabilizing the enzyme.[4] Homogenize the

[4] N. Wakid and T. E. Mansour, *Mol. Pharmacol.* 1, 53 (1965).

suspension in an all glass homogenizer for 1 minute. Centrifuge enzyme suspension in the same type of tubes at 16,000 rpm (32,000 *g*) for 30 minutes. Collect the almost colorless supernatant fluid. The specific activity of this fraction varies from 21 to 39 units/mg.

Step 4. DEAE-Cellulose Chromatography. A batch of DEAE is prepared by suspending the dry powder in 0.1 *N* NaOH and then washing it exhaustively with distilled water. The fine particles are eliminated by decantation during the washing. The slurry is then made 1 *M* with respect to NaCl and rinsed exhaustively with distilled water. The final suspension is stored under refrigeration.

Prepare a column of DEAE-cellulose ($10.5\ cm^2 \times 10$ cm) by allowing the first 5 cm of the DEAE slurry to settle by gravity. Pack the remainder under pressure from a rubber bulb. Equilibrate the column with 100 ml of 1 *M* Tris-HCl, pH 8.6. Then wash the column with 250 ml of a solution with the following composition: 0.2 *M* Tris-HCl, pH 8.6; 0.005 *M* mercaptoethanol; 10^{-4} *M* ATP and 10^{-5} *M* fructose-1,6-di-P. Apply fraction III to the column and when it has completely descended add 20 ml of the 0.2 *M* Tris-HCl solution described above to the column. Wait until nearly all this solution has passed through the column. Prepare a linear gradient with 0.2 *M* and 0.9 *M* Tris-HCl buffer, pH 8.6, as limiting concentrations. This eluent has a total volume of 400 ml and contains 0.005 *M* mercaptoethanol, 10^{-4} *M* ATP and 10^{-5} *M* fructose-1,6-di-P at constant concentration throughout the gradient. Adjust rate of flow to 1.5–2.0 ml/min. Collect 5-minute fractions. Slow elution can result in poor recoveries. By the use of this procedure 80–85% of the activity applied to the column is eluted between 0.6 *M* and 0.7 *M* Tris-HCl buffer. Pool fractions with peak enzyme activity (70 ml). The pooled enzyme (fraction IV) has a specific activity which varies from 71 to 118.

Step 5. Ammonium Sulfate Fractionation. After fraction IV is measured it is transferred to a 200-ml chilled flask fitted with a magnetic stirrer and placed in ice. Add from a burette dropwise (2 ml/min), with continuous stirring, ammonium sulfate solution saturated at room temperature to bring the saturation to 0.42 (50.6 ml of ammonium sulfate solution per 70.0 ml of enzyme solution). Centrifuge the mixture in a Servall SS-34 rotor at 14,000 rpm (24,000 *g*) in a 30-ml polycarbonate tube for 10 minutes. Discard the residue. To the supernatant fluid add enough saturated ammonium sulfate solution dropwise to bring the saturation to 0.6 (52.6 ml of ammonium sulfate solution to 117 ml of supernatant fluid). Centrifuge at the same speed for 10 minutes. Discard the supernatant fluid. Dissolve the residue in a minimum volume of a solution of 0.05 *M* potassium phosphate buffer (pH 8.0); 0.001 *M* dithioerythritol[5]; 10^{-4} *M* ATP and 10^{-5} *M* fructose-1,6-di-P (final volume about 2 ml). This yields a water clear solution containing about 10 mg of protein per milliliter with a specific activity which varies from 120 to 157 units/mg protein (fraction V). This fraction is referred to as the "purified enzyme" or "purified phosphofructokinase." In order to maintain the stability of this fraction, dialyze against a solution of 0.05 *M* potassium phosphate buffer (pH 8.0); 0.001 *M* dithioerythritol; 0.01 *M* ATP; 10^{-5} *M* fructose-1,6-di-P and 0.001 *M* EDTA. After such treatment the enzyme loses only about 25% of its activity after 3 weeks. The major part of the lost activity can be recovered by dialyzing again against the same solution for 4 hours.

Crystallization. Dialyze fraction V containing about 10 mg of protein per milliliter against a solution composed of 0.3 saturated ammonium sulfate; 0.05 *M* K-phosphate buffer, pH 8.0; 10^{-2} *M* ATP; 10^{-5} *M* fructose-1,6-di-P; 0.001 *M* dithioerythritol and 10^{-3}*M* EDTA for 3–4 hours. Leave the enzyme in a conical centrifuge tube in ice in the cold room. By the next day a precipitate consisting of minute hexagonal crystals contaminated with amorphous material is formed. When the tube is left undisturbed for a few days the amorphous material is reduced and the crystals become more defined.

Enzyme crystals prepared by this procedure are insoluble in the regular phosphate buffer solutions used above. For solubilization the crystalline enzyme is isolated by centrifugation in the cold at 4000 *g*. The crystals are suspended in a solution of the following composition: 0.05 *M* K-phosphate buffer, pH 8.0; 10^{-2} *M* ATP; 10^{-5} *M* fructose-1,6-di-P; 0.001 *M* dithioerythritol and 10^{-3} *M* EDTA. The concentration of the protein should be approximately 2–10 mg/ml. The protein suspension is then placed in a water bath at 60° for 3 minutes and the suspension is stirred with a glass rod. The enzyme crystals dissolve after this time and the tube is then placed in ice.

The crystalline enzyme suspension is stable in the 0.3 saturated solution for at least a month. The enzyme, once solubilized, is not stable for more than a few days. The specific activity of the crystalline enzyme never exceeded that of the purified enzyme.

Properties

Electrophoresis in Starch Gel. Purified phosphofructokinase moves in the electrophoretic field toward the positive electrode as a diffuse patch. Maximum migration is achieved at pH 8.0 to pH 8.5. Because the en-

[5] The Cleland's reagent used was dithioerythritol (2,3-dihydroxy-1,4-dithiolbutane), which is available commercially from Cyclo Chemical Corporation.

zyme has the tendency to aggregate, some of the protein does not move in the starch gel.[3]

Ultracentrifugal Sedimentation Patterns. Both the purified and the crystalline enzyme have the tendency to aggregate when present in high concentrations. Ultracentrifugal sedimentation analysis of the crystalline enzyme shows a schlieren pattern with a single asymmetric peak with an $S_{20,w}$ value of 25.4. The schlieren pattern for the purified enzyme has a pattern characterized by a single minor peak and a major heavy peak. The $S_{20,w}$ value for the light component varies from 8 to 11 and that for the heavy component varies from 38 to 51. The sedimentation coefficient is concentration dependent. This can be shown on sucrose gradient. The lowest $S_{20,w}$ value is 15.2 when enzyme concentration is 0.03 mg/ml.

Kinetics.[6] The dissociation constant for fructose-6-P at pH 8.2 has been found to be 3.5×10^{-5} *M*. The K_m for ATP at the same pH when the ratio of ATP:Mg^{++} was fixed to 1:2 has been found to be 7.4×10^{-5} *M*. Concentrations of ATP between 2×10^{-4} *M* and 2×10^{-3} *M* are inhibitory in the presence of 0.05 *M* glycylglycine buffer, pH 6.9 or pH 7.2, and 5×10^{-5} *M* fructose-6-P. Cyclic 3′,5′-AMP, 5-AMP, 5-ADP, high fructose-6-P concentration or a more alkaline pH reverse ATP inhibition. When testing for ATP inhibition reagent 9 for the enzyme mixture should be prepared as follows: prepare the enzyme mixture as described above using twice the amount of enzyme; dialyze against 0.01 *M* glycylglycine buffer, pH 7.5, for 3 hours. This would ensure the removal of much of the ammonium sulfate present with the enzymes which reverses ATP inhibition.[6]

Acknowledgment

This work was supported by Research Grant AI04214 from the National Institute of Allergy and Infectious Diseases, and Research Career Development Award GM-K3-3848 from the Division of General Medical Sciences, U. S. Public Health Service.

[6] T. E. Mansour, unpublished observations.

[77c] Phosphofructokinase

III. Yeast

By ALBERTO SOLS and MARÍA L. SALAS

Assay Method

Principle. The phosphofructokinase reaction in the forward direction [Eq. (1)] can be assayed spectrophotometrically by coupling with aldolase [Eq. (2)], triosephosphate isomerase [Eq. (3)] and glycerolphosphate dehydrogenase [Eq. (4)] and following the oxidation of NADH:

$$\text{F-6-P} + \text{GTP} \rightarrow \text{FDP} + \text{GDP} \quad (1)$$
$$\text{FDP} \rightarrow \text{glyceraldehyde-3-P} + \text{dihydroxyacetone-P} \quad (2)$$
$$\text{Glyceraldehyde-3-P} \rightarrow \text{dihydroxyacetone-P} \quad (3)$$
$$\text{2 Dihydroxyacetone-P} + \text{2 NADH} \rightarrow \text{2 L-glycerol-3-P} + \text{2 NAD} \quad (4)$$

$$\text{F-6-P} + \text{GTP} + \text{2 NADH} \rightarrow \text{GDP} + \text{2 L-glycerol-3-P} + \text{2 NAD}$$

For the assay of yeast phosphofructokinase GTP is used as substrate preferentially to ATP to avoid allosteric inhibition by the latter.[1]

The enzyme can also be assayed in the reverse direction with FDP and GDP as substrates by coupling with glucose phosphate isomerase and glucose 6-phosphate dehydrogenase and following the reduction of NADP.[2]

Reagents

The composition of the assay mixture is shown in Table I (p. 438).

Procedure. To a cuvette with a 1-cm light path are added the components of the assay mixture and water to complete a final volume of 2 ml. The reaction is started with the addition of the preparation of phosphofructokinase (containing no more than about 10 milliunits enzyme), and is followed by the decrease in optical density at 340 mμ. A blank without GTP must be carried out, particularly with crude preparations of phosphofructokinase.

Definition of Unit and Specific Activity. The micromolar unit is used: the amount of enzyme that phosphorylates 1 micromole of F-6-P per minute in the above conditions at about 25°. For each mole of F-6-P phosphorylated, 2 moles of NADH are consumed in the above assay system. Protein is determined by ultraviolet absorption (at 280 and 260 mμ).[3] Specific activity is expressed as units per milligram of protein.

[1] E. Viñuela, M. L. Salas, and A. Sols, *Biochem. Biophys. Res. Commun.* **12**, 140 (1963). Alternatively, ATP can be used if the concentration is lowered to minimize inhibition while still approaching saturation of the substrate site. Thus, with 1 m*M* fructose-6-P, 0.1 m*M* ATP gives an initial rate some 70% of that given by 1 m*M* GTP; at the same concentration of fructose-6-P, 1 m*M* ATP, in the assay conditions here described, inhibits the native enzyme over 90%, whether GTP is present or not.

[2] With purified preparations free of myokinase, ADP may be used as substrate; in these conditions the addition of hexokinase (0.2 unit) and glucose (1 m*M*) doubles the yield of NADPH while preventing accumulation of inhibitory ATP.

[3] E. Layne, Vol. III [73].

TABLE I
ASSAY MIXTURE FOR PHOSPHOFRUCTOKINASE

F-6-P[a]	1 mM
GTP	1 mM
$MgCl_2$	5 mM
Potassium phosphate, pH 6.5	25 mM
Ethanethiol	5 mM
NADH	0.15 mM
Aldolase[b]	0.2 units (per 2 ml)
Glycerolphosphate dehydrogenase[b]	1 unit (per 2 ml)
Triosephosphate isomerase[b]	3 units (per 2 ml)

[a] For quantitative study of the inhibition by ATP in crude preparations containing glucose phosphate isomerase it is important to avoid marked changes in the concentration of F-6-P by isomerization during the assay. This can be prevented by the use of a 0.2 M neutralized solution of glucose-6-P with some 20 units of glucose phosphate isomerase per milliliter. Within 1 hour at room temperature this solution will be equilibrated to approximately 0.05 M F-6-P −0.15 M G-6-P.

[b] For work on allosteric effectors, diluted suspensions of enzymes in concentrated ammonium sulfate should be avoided, since NH_4^+ ions markedly counteract the inhibition of yeast phosphofructokinase by ATP.

Purification Procedure[4]

All operations are carried out at about 2° unless indicated otherwise.

Step 1. Extraction. Pressed bakers' yeast is ground in a mortar with twice its weight of alumina (Alcoa A-305) and the homogenate is diluted with 5 volumes of 10 mM $MgCl_2$.[5] This operation can be carried out efficiently in lots of 3 g yeast. Pool the homogenates from 30 g yeast, centrifuge at 20,000 g for 15 minutes, and discard the sediment.

Step 2. Acetone Fractionation. Dilute the extract with 10 mM $MgCl_2$ to about 15 mg protein per milliliter. Add acetone, precooled to −15°, slowly and with stirring, up to a concentration of 18%, while lowering gradually the temperature of the extract to −6°. Centrifuge at this temperature for 10 minutes at 20,000 g and discard the precipitate. Add to

[4] Developed in collaboration with Dr. E. Viñuela. Each fractionation tried was judged on the basis of recovery of catalytic activity *and* sensitivity to what appeared to be the main allosteric property: sensitivity to inhibition by ATP. Treatments that gave enzyme insensitive to inhibition by ATP were abandoned irrespective of the yield in catalytic activity. In the procedure here described there is some decrease in sensitivity to ATP, particularly after the chromatographic step, but qualitatively the purified preparation is sensitive to ATP and AMP, citrate, and NH_4^+ ions.

[5] For large scale preparation a suspension of yeast in 2 volumes of 10 mM $MgCl_2$ can be homogenized with the Ribi Cell Fractionator (I. Sorvall, Inc., Norwalk, Connecticut) with similar results.

the supernatant more acetone as before up to a concentration of 30%, lowering gradually the temperature to −8°. Centrifuge as before, at −8°, discard the supernatant, and dissolve the precipitate with 7 ml of 10 m*M* $MgCl_2$.

Step 3. Acid Precipitation. Dilute the acetone fraction to a protein concentration of 20 mg/ml, adjust the pH to 5.1 (measured at 2° without temperature correction) with 0.1 *M* acetic acid added slowly with stirring, centrifuge at 20,000 *g* for 10 minutes, and discard the supernatant. Suspend the precipitate in 7 ml of 10 m*M* $MgCl_2$–2 m*M* ethanethiol–10 m*M* potassium phosphate, pH 6.8, using a Potter-Elvehjem homogenizer to achieve a fine dispersion, and recentrifuge to discard the insoluble residue.

Step 4. DEAE-Cellulose Chromatography. Twenty-five grams of DEAE-cellulose is washed twice with 0.5 *N* NaOH, then twice with water, and finally with 0.2 *M* potassium phosphate, pH 6.8, until the washings have this pH. A diluted suspension in this buffer is carefully packed in a column of 2.5-cm diameter until a height of 6 cm packed cellulose is reached (volume 25 ml). The column is equilibrated with 10 m*M* $MgCl_2$ in 2 m*M* ethanethiol–0.5 m*M* ATP–0.75 *M* sorbitol–10 m*M* potassium phosphate, pH 6.8. The acid precipitate fraction is applied to the column, washing with 1.5 volumes of the equilibrium mixture. The column is eluted with increasing concentrations of $MgCl_2$ in the above mixture, using first 1.5 volumes of 25 m*M* $MgCl_2$ and discarding the eluate, and then 40 m*M* $MgCl_2$, at a flow rate of approximately 1 ml per minute. The latter eluate is collected in 5-ml fractions, and the samples containing more phosphofructokinase activity are combined.

Step 5. Ammonium Sulfate Precipitation. In the above eluate dissolve $(NH_4)_2SO_4$ for 80% saturation, allow to stand for half an hour, centrifuge at 20,000 *g* for 10 minutes, and discard the supernatant. Suspend the precipitate with a solution of $(NH_4)_2SO_4$ at 80% saturation up to a volume of 1 ml.

With this procedure a purification of 100- to 200-fold, with a yield of 30–50% is usually achieved. The preparation contains some hexokinase and glucose 6-phosphate dehydrogenase, but not (less than 0.5% of the activity of phosphofructokinase) glucose phosphate isomerase, aldolase, glyceraldehyde phosphate dehydrogenase, ATPase, or myokinase.[6] A summary of the purification procedure is given in Table II.

[6] The best preparations obtained had a specific activity about one-fifth that of crystallized muscle phosphofructokinase (see this volume [77a]). A precise comparison is difficult because of the many factors that affect the activity of these enzymes.

TABLE II
PURIFICATION PROCEDURE FOR YEAST PHOSPHOFRUCTOKINASE

Step	Volume (ml)	Total enzyme (units)	Protein (mg/ml)	Specific activity (units/mg)	Purification (-fold)	Yield (%)
Crude extract	122	290	35	0.068	—	100
Acetone	8.6	290	38	0.89	13	100
Acid	9.8	175	7	2.55	40	57
DEAE-cellulose	15	135	—	—	—	45
$(NH_4)_2SO_4$	1	125	10	12.5	183	43

Properties

The enzyme requires Mg^{++} ions for activity, preferably at concentration not less than that of the nucleotide substrate. Mn^{++} or Co^{++} ions are also effective.

The pH optimum for activity is about 6.5 with phosphate buffer and about 7.0 with imidazol buffer.

Kinetics. The effect of the concentration of one substrate on the rate of the phosphofructokinase reaction varies with the concentration of the other substrate, giving raise to parallel lines in the double reciprocal plots.[1,7]

Reversibility. The rate of the reverse reaction at pH 6.5 is about 10% that of the forward reaction.

Substrate Specificity. The enzyme can use a variety of nucleotide substrates. The following trinucleotides give approximately equal V_{max} and have apparent K_m values at 0.5 mM F-6-P as indicated: ATP, 0.02 mM; GTP, 0.1 mM; ITP, 0.2 mM; CTP, 0.4 mM; and UTP, 0.8 mM. In addition to F-6-P, the enzyme can phosphorylate F-1-P, with a V_{max} 0.005 that of the former. The K_m values for F-6-P and F-1-P, with 0.5 mM ITP, are 0.15 mM and 1 mM, respectively.

Metabolic Effectors. Yeast phosphofructokinase, obtained from resting yeast as indicated, is strongly inhibited by ATP, competitively respect to F-6-P. In the presence of inhibitory concentrations of ATP the enzyme follows second order kinetics respect to the concentration of F-6-P.[1,8]

[7] Muscle phosphofructokinase gives similar kinetics. The possibility of an enzyme-phosphate intermediate is very unlikely since the enzyme does not catalyze the reaction ATP + ADP* → ADP + ATP* [H. A. Lardy, *in* "The Enzymes" (P. D. Boyer, H. Lardy, and K. Myrbäck, eds.), 2nd ed. Vol. 6, Chapter 3 (1962)] nor the reaction GTP + ADP → GDP + ATP (unpublished results with a crystalline preparation of muscle phosphofructokinase kindly made available by Dr. H. U. Bergmeyer).

[8] In these conditions the activity with low concentrations of F-6-P can be increased by either FDP or 2,5-anhydroglucitol-6-P (a nonphosphorylatable analog

This inhibition is highly specific, in contrast with the relatively wide specificity for trinucleotide substrate and does not depend on excess of ATP over Mg^{++}. At 0.5 m*M* F-6-P (in phosphate buffer, pH 6.5 and with excess Mg^{++}), as little as 0.1 m*M* ATP is enough for 50% inhibition, while for a similar inhibition with GTP a concentration about 4 m*M* is required, and ITP, CTP, and UTP have even smaller apparent affinities as inhibitors. Apparently the enzyme has a regulatory site for ATP, not as a substrate but as end product of the glycolytic pathway of which phosphofructokinase is the first physiologically irreversible step. This allosteric inhibition by ATP can be very efficiently counteracted, competitively, by AMP.[9] Kinetic evidence suggests that AMP can occupy the regulatory site for ATP without being itself inhibitory for the reaction. The apparent affinities for ATP and AMP, as inhibitor and antiinhibitor, respectively, are markedly affected by the pH and the nature of the buffer. Thus, while within the physiological range (pH 6.5–7.0) the apparent affinity for AMP is considerably greater than that for ATP, above pH 8 the efficiency of the trinucleotides as inhibitors increases while that of AMP as antiinhibitor becomes small. ADP is a relatively weak inhibitor; inhibition of the reverse reaction by excess ADP-Mg, competitive with respect to FDP, suggests that it can act as an analog of ATP for inhibition at the allosteric site for ATP. Cyclic 3′,5′-AMP is inert in this system.

Citrate inhibits phosphofructokinase within the range of physiological concentrations (2–10 m*M* in yeast), particularly in the presence of inhibitory concentrations of ATP. Isocitrate is similarly inhibitory, but not α-ketoglutarate. Apparently citrate and ATP can cooperate at moderate concentrations of both to cause an inhibition, competitive with F-6-P, greater than either alone.[10]

NH_4^+ ions in the 5–50 m*M* range can activate phosphofructokinase counteracting competitively the inhibition by ATP. That is, NH_4^+ ions decrease the apparent affinity of the ATP inhibitory site for ATP without marked effect on either the V_{max} or the K_m for F-6-P of the uninhibited enzyme. The effect is independent of the pH in the 6.5–8.5 range. K^+ ions are ineffective.[11]

of F-6-P) at moderate concentrations; at high concentrations both compounds can inhibit the enzyme competitively with F-6-P.

[9] A. Ramaiah, J. A. Hathaway, and D. E. Atkinson, *J. Biol. Chem.* **239**, 3619 (1964).

[10] Inhibition of phosphofructokinase in yeast by citrate, whose level increases in aerobiosis and may be considered as the carbon end product of aerobic glycolysis, seems to be a key factor in the Pasteur effect: M. L. Salas, E. Viñuela, M. Salas, and A. Sols, *Biochem. Biophys. Res. Commun.* **19**, 371 (1965).

[11] The antiinhibitory effect of NH_4^+ ions, a key substrate for yeast growth, is likely to have physiological significance.

Reversible Loss of Sensitivity to Inhibition by ATP. Incubation of a crude yeast extract with fluoride (about 20 mM NaF) leads within minutes to the disappearance of the sensitivity of its phosphofructokinase to inhibition by ATP, without change in maximal rate. Two forms of yeast phosphofructokinase have been postulated from kinetic observations, an ATP-sensitive "*b*" form which seems to be the predominant form in resting yeast, and an ATP-insensitive "*a*" form.[12] The *a* form slowly reverts to the "b" form, particularly if diluted in the absence of fluoride, or after fractionation with acetone.

[12] E. Viñuela, M. L. Salas, and A. Sols, *Biochem. Biophys. Res. Commun.* **15**, 243 (1964).

[78] D-Arabinokinase[1]

By Wesley A. Volk

D-Arabinose + ATP → D-arabinose-5-P + ADP

Assay Method

Principle. Arabinose 5-phosphate and the adenine nucleotides were precipitated at the end of the reaction, and residual arabinose was measured with the orcinol reaction.

Reagents

- D-Arabinose, 0.01 M
- ATP, 0.1 M, pH 8.0
- $MgCl_2$, 0.2 M
- Glycylglycine, 0.1 M, pH 8.0
- NaF, 0.1 M
- $Ba(OH)_2$, 0.3 N
- $ZnSO_4$, 0.18 M
- Reagents for orcinol reaction (Vol. III [11])

Procedure. Into a 12-ml centrifuge tube were placed 0.2 ml of glycylglycine buffer, 0.1 ml of $MgCl_2$, 0.1 ml of ATP, 0.1 ml of NaF, 0.25 ml of D-arabinose, and enzyme to a total volume of 0.95 ml. The mixture was incubated for 15 minutes at 37°, and the reaction was stopped by the addition of 3.0 ml of $Ba(OH)_2$ and 3.0 ml of $ZnSO_4$. This step served to deproteinize the mixture as well as to precipitate residual nucleotides

[1] W. A. Volk, *J. Biol. Chem.* **237**, 19 (1962).

and any phosphorylated arabinose. After centrifugation, 1.0 ml of the supernatant solution was used to measure residual arabinose.

Definition of Unit and Specific Activity. One unit of D-arabinokinase is defined as the amount of enzyme which will phosphorylate 1 micromole of D-arabinose per minute at 37°. Specific activity is expressed as units per milligram protein. Protein was determined by the method of Lowry *et al.*[2]

Purification Procedure

Step 1. Growth of Culture and Preparation of Crude Extract. Propionibacterium pentosaceum, strain E14, was grown for 4–5 days in 2-liter Erlenmeyer flasks containing 1500 ml of a medium consisting of 0.5% yeast extract, 0.5% Bacto-peptone and 0.5% D-arabinose in a final concentration of 0.05 *M* phosphate buffer, pH 6.8. The cells were harvested by centrifugation, washed with distilled water, and disrupted in 0.1 *M* glycine buffer, pH 10.3 by treatment for 30 minutes in a Raytheon 10-kc sonic oscillator. After centrifugation the sedimented cells were disrupted one or two additional times in a similar manner. All subsequent operations were carried out at 0–2°.

Step 2. Protamine Precipitation. To 438 ml of crude cell-free extract (10 mg of protein per ml) was added 67 ml of 3.75 *M* $(NH_4)_2SO_4$, pH adjusted to 7.0, to yield a final concentration of 0.5 *M*. The pH was then adjusted to 6.0 by the addition of 2 *N* acetic acid. Although a small precipitate was present at this time, the solution was not centrifuged, and 87.5 ml of 1% protamine sulfate, pH 5.0, was added. The entire mixture was then dialyzed overnight against 18 liters of distilled water and centrifuged; precipitate was discarded.

Step 3. Ammonium Sulfate Fraction. To 585 ml of enzyme (5.5 mg protein per milliliter) was added 99.5 g of solid $(NH_4)_2SO_4$ to bring the final concentration of $(NH_4)_2SO_4$ to 1.2 *M*. After being stirred for 45 minutes, the mixture was centrifuged and the precipitate was discarded. Solid $(NH_4)_2SO_4$ (66 g) was added to the supernatant enzyme (580 ml) to bring the concentration to 1.9 *M*. The enzyme could be kept in this state at 0° for at least several weeks with essentially no loss in activity.

Step 4. Calcium Phosphate Gel Negative Adsorption. The 1.9 *M* suspension was centrifuged and the precipitate was dissolved in 75 ml of water (8.5 mg of protein per milliliter). To 70 ml of this enzyme solution was added 28 ml of $Ca_3(PO_4)_2$ gel (12 mg dry weight per milliliter). After stirring for 15 minutes, the suspension was centrifuged and the gel discarded.

[2] O. H. Lowry, N. J. Rosebrough, A. L. Farr, and R. J. Randall, *J. Biol. Chem.* **193**, 265 (1951).

Step 5. DEAE-Cellulose Column Fractionation. To 87 ml of the supernatant solution from the $Ca_3(PO_4)_2$ step (5.6 ml protein per milliliter) was added 313 ml of water. The solution was then poured on a DEAE-cellulose column (12 × 2.6 cm) which had been equilibrated with 0.001 *M* $(NH_4)_2SO_4$, pH 7.2. The enzyme was eluted with a linear gradient of $(NH_4)_2SO_4$ from 0.001 *M* to 0.6 *M*. Sample size was approximately 6.0 ml.

Step 6. First Pressure Dialysis. Fractions 93 through 115 were pooled (125 ml containing 1.1 mg of protein per milliliter) and were subjected to pressure dialysis[3] under 13 pounds per square inch against 1.2 *M* $(NH_4)_2SO_4$, pH unadjusted. The precipitate which resulted from 16 hours of dialysis (precipitate I) was collected by centrifugation and dissolved in 3.0 ml of 0.1 *M* glycylglycine buffer, pH 8.0.

Step 7. Second Pressure Dialysis. The supernatant solution from step 6 was again subjected to pressure dialysis against 1.6 *M* $(NH_4)_2SO_4$. After 16 hours the enzyme was centrifuged and the resulting precipitate (precipitate II) was dissolved in 2.0 ml of 0.1 *M* glycylglycine buffer, pH 8.0.

The table lists the results of these purification steps.

PURIFICATION OF D-ARABINOKINASE

Enzyme fraction	Volume (ml)	Protein (mg/ml)	Specific activity (× 10^2)	Total units	Yield (%)	Purification
Crude sonic extract	80	54.8	3.2	141	100	—
Protamine treated and dialyzed	585	5.5	3.3	106	75	1.05
$(NH_4)_2SO_4$, 1.2–1.9 *M*	75	8.5	16.2	103	73	5.1
$Ca_3(PO_4)_2$ gel supernatant	87	2.83	28.0	69	49	8.9
DEAE-cellulose (fractions 93–115)	125	1.1	40.0	50	36	12.6
Pressure dialysis, precipitate I	3.2	14.5	10.5	5	3.5	3.3
Pressure dialysis, precipitate II	2.2	2.03	367.0	16	11.4	116

Properties

During the purification the enzyme was quite unstable in solutions of low ionic strength. However, in the presence of 0.1 *M* $(NH_4)_2SO_4$ even such dilute solutions as those eluted from the DEAE-cellulose column maintained most of their activity for several days at 0°. Precipitate II

[3] G. Goldstein, I. S. Slizys, and M. W. Chase, *J. Exptl. Med.* **114,** 89 (1961).

still had 54% of its activity after 1 week at 0°, but had lost essentially all activity after 2 weeks at 0°.

The kinase requires Mg^{++} for activity. The K_m for Mg^{++} is 2.9×10^{-3} *M*. $MnSO_4$ (0.01 *M*) under the conditions of the assay resulted in full activity, but a similar concentration of $CaCl_2$ was inactive.

The effectiveness of several other trinucleotides serving as phosphate donors as compared to ATP is: GTP, 100%; ITP, 88%; UTP, 64%; and CTP, 51%. The K_m for ATP is 8.3×10^{-4} *M*, and the K_m for D-arabinose is 1.2×10^{-3} *M*.

pH Optimum. The enzyme exhibits maximum activity at pH 9.5 and approximately 75% of maximum activity at pH 8.0 and 10.5.

Specificity of the Enzyme. Even the most purified preparations contained varying amounts of phosphoarabinoisomerase and kinases which phosphorylate D- and L-ribulose and D-ribose. Disruption of the cells in glycine buffer, pH 10.3 inactivated much of the phosphoarabinoisomerase. Large scale preparations of D-arabinose 5-phosphate can be carried out in the presence of phosphoarabinoisomerase if the phosphorylation is carried out at pH 10. The isomerase is completely inactive at this pH[4] while the kinase has nearly maximum activity. Thus one can obtain D-arabinose 5-phosphate as the sole product using the 1.2–1.9 *M* $(NH_4)_2SO_4$ fraction of the enzyme if the pH of the phosphorylation is kept above 10.

Experiments in which D-arabinose was super-added to phosphorylation experiments with D- and L-ribulose showed no inhibition of ribulose phosphorylation, and it is concluded from these results that the enzyme or enzymes phosphorylating D- and L-ribulose are not identical to the D-arabinokinase. A similar experiment with D-arabinose and D-ribose did not show an additive effect. This result plus the fact that the ratio between D-arabinose and D-ribose activities remained quite constant throughout the purification led to the conclusion that D-arabinokinase phosphorylates D-ribose at about 80% of the rate at which it phosphorylates D-arabinose.

The enzyme was inactive under the conditions of the standard assay on the following substrates: D-glucose, D-galactose, D-mannose, D-fructose, D-xylose, and L-arabinose.

[4] W. A. Volk, *J. Biol. Chem.* **235**, 1550 (1960).

[79] D-Ribulokinase

By R. P. MORTLOCK and W. A. WOOD

D-Ribulose + ATP → D-ribulose 5-phosphate + ADP

Assay Method

Principle. Rate of ADP formation is determined by the coupled reaction sequence of Anderson and Wood[1] plotting NADH oxidation against time at 340 mμ.

D-Ribulose + ATP → D-ribulose 5-phosphate + ADP (1)
ADP + phosphoenolpyruvate → pyruvate + ATP (2)
Pyruvate + NADH + H^+ → lactate + NAD^+ (3)

Reagents

Tris-chloride buffer, 0.4 *M*, pH 7.5
$MgCl_2$, 0.1 *M*
Glutathione, sodium salt, 0.1 *M*
ATP, 0.1 *M*
Phosphoenolpyruvate, potassium salt, 0.025 *M*
NADH, 0.01 *M*
Lactic dehydrogenase, crystalline, rabbit muscle[2]
D-Ribulose, 0.15 *M*

Procedure. The reaction mixture contains 0.02 ml Tris-chloride buffer, 8 micromoles; 0.01 ml $MgCl_2$, 1 micromole; 0.015 ml glutathione, 1.5 micromoles; 0.005 ml ATP, 0.5 micromole; 0.01 ml phosphoenolpyruvate, 0.25 micromole; 0.01 ml NADH, 0.1 micromole; 0.02 unit of lactic dehydrogenase containing pyruvate kinase; 0.02 ml of D-ribulose, 2.0 micromoles; and the D-ribulokinase sample in a total volume of 0.15 ml.

Reaction rates with crude preparations of kinase must be corrected for adenosine triphosphatase, NADH oxidase, and D-ribulose reductase (ribitol dehydrogenase) activities by utilizing control cuvettes deficient in D-ribulose (NADH oxidase, ATPase control), deficient in ATP (D-ribulose reductase, NADH oxidase control) and deficient in D-ribulose and ATP (NADH oxidase control). For determination of kinase in the presence of high reductase activity, NADPH is substituted for NADH and the lactic dehydrogenase concentration is increased 10-fold. Glutathione must be omitted in this latter assay because of the presence of NADPH specific glutathione reductase in the lactic dehydrogenase preparation.

[1] R. L. Anderson and W. A. Wood, *J. Biol. Chem.* **237**, 1029 (1962).
[2] Worthington Biochemical Corp., Freehold, New Jersey. Containing pyruvate kinase.

Definition of Unit and Specific Activity. One unit of kinase is defined as that amount which will result in an absorbancy change of 1.0 per minute in the assay system described above. Specific activity is expressed in units per milligram protein. Protein is determined by the ratio of absorbances[3] at 280 and 260 mμ or by the method of Lowry *et al.*[4]

Purification Procedure

Cell Culture. For the preparation of extracts containing D-ribulokinase, *Aerobacter aerogenes* PRL-R3 may be cultured on either ribitol salts or D-arabinose-salts media.[5,6] Mutants constitutive for D-ribulokinase, coordinately induced with ribitol dehydrogenase,[7] may be readily isolated after growth on the selective substrate xylitol,[8] and extracts of such mutants normally contain 10-fold higher activity for D-ribulokinase than can be obtained by growth on either of the above carbohydrates.

Cells are harvested by centrifugation, washed once with cold, distilled water and broken by means of the French Pressure cell. The crude extract is diluted to a protein concentration of 10 mg/ml prior to the protamine sulfate treatment.

Step 1. Protamine Sulfate Treatment. All fractionation steps are carried out at 0–4° unless otherwise noted. Solid ammonium sulfate is added to 0.1 *M*, and one-tenth volume of protamine sulfate solution (40 mg/ml) is added dropwise with stirring; the mixture is centrifuged at 12,000 *g* for 10 minutes, and the pellet is discarded.

Step 2. First Ammonium Sulfate Fractionation and Heat Treatment. The supernatant liquid is adjusted to 1.9 *M* ammonium sulfate (by addition of solid ammonium sulfate) and centrifuged for 5 minutes; the pellet is dissolved in 0.001 *M* EDTA. This fraction is heated to 50° for 5 minutes and quickly cooled. After centrifugation, the supernatant solution is dialyzed overnight against 0.01 *M* Tris buffer (pH 7.5), containing 0.001 *M* EDTA.

Step 3. DEAE-Cellulose-Phosphate. After dialysis, the ammonium sulfate concentration is adjusted to 0.01 *M*, the fraction is passed through a DEAE-cellulose-phosphate column, and the proteins are eluted from the column with 0.03, 0.05, and finally 0.08 *M* potassium phosphate buffer (pH 7.5). A linear gradient from 0.005 to 0.2 *M* potassium phos-

[3] O. Warburg and W. Christian, *Biochem. Z.* **310,** 384 (1941).

[4] O. H. Lowry, N. J. Rosebrough, A. L. Farr, and R. J. Randall. *J. Biol. Chem.* **193,** 265 (1951).

[5] R. L. Anderson and W. A. Wood, *J. Biol. Chem.* **237,** 296 (1962).

[6] R. P. Mortlock, D. D. Fossitt, D. H. Petering, and W. A. Wood. *J. Bacteriol.* **89,** 129 (1965).

[7] R. P. Mortlock, D. D. Fossitt, and W. A. Wood, *Bacteriol. Proc.* p. 95, (1964).

[8] R. P. Mortlock and W. A. Wood, *Bacteriol. Proc.* p. 82, (1965).

phate (pH 7.5) may also be used for elution. The kinase is found in the eluate obtained with 0.08 *M* phosphate. Most of the ribitol dehydrogenase is found in the lower (0.05 *M*) eluate.

Step 4. Second Ammonium Sulfate Fractionation. If the protein concentration of the ribulokinase eluate is greater than 1 mg/ml, the next ammonium sulfate fractionation is carried out directly on the eluate. If the protein concentration is less than 1 mg/ml, the fraction is diluted to give a salt concentration of 0.02 *M* and passed through a small DEAE-cellulose-phosphate column; the kinase is eluted with several small portions of 0.1 *M* phosphate buffer. The eluate obtained by either method is fractionated with saturated ammonium sulfate (pH 7.0). The fraction precipitating between 0.45 and 0.54 saturation is collected by centrifugation and dissolved in distilled water.

Step 5. Alumina Gel Treatment. The fraction resulting from the last step is diluted to an ammonium sulfate concentration of 0.02 *M*, and alumina C_γ[9] is added (2 mg dry weight per milligram protein); the preparation is stirred for 5 minutes, then the gel is removed by centrifugation and eluted with 0.03 *M* phosphate buffer pH 7.5. The exact procedure for gel treatment may vary with different gel preparations and should be first tested using a small portion of the kinase fraction.

D-Ribulokinase has been purified by the above procedure from cells

PURIFICATION OF D-RIBULOKINASE[a]

Step	Protein (mg/ml)	Total units	Specific activity (units/mg of protein)	Kinase: dehydrogenase ratio
Crude extract[b]	10	30,000	0.945	0.096
Protamine supernatant fraction	8.5	33,000	1.10	0.103
Ammonium sulfate fraction	58.4	28,000	2.23	0.107
Heat treatment	21	25,200	5.98	0.179
Dialysis	16.8	24,080	6.52	0.218
DEAE-cellulose	1.5	13,200	31.3	5.55
Ammonium sulfate	3.5	8,510	47.9	5.25
Alumina C_γ gel[c]	0.58	8,450	205	13.3

[a] Data from R. P. Mortlock, D. D. Fossitt, D. H. Petering, and W. A. Wood, *J. Bacteriol.* **89**, 129 (1965).

[b] Extract was prepared from a constitutive mutant with limited D-ribulokinase activity. The growth medium was 2% peptone.

[c] Prepared as described in Vol. I [97].

[9] Prepared as described in Vol. I [97].

of *Aerobacter aerogenes* grown with ribitol and D-arabinose as substrates, from constitutive mutants, and from the high-activity constitutive mutants obtained after growth on the selective substrate xylitol.

A summary of the purification obtained is given in the table.

Properties

Specificity. Of the four 2-ketopentoses, the kinase is specific for D-ribulose as substrate.

Substrate Affinity. The K_m value for D-ribulose in $5.0 \times 10^{-4} M$.

pH Optima. The pH optima is 7.5 in Tris-chloride buffer.

[80] L-Ribulokinase

By N. L. LEE and ELLIS ENGLESBERG

L-Ribulose + ATP → L ribulose 5-phosphate + ADP

Assay Method

Principle. L-Ribulokinase can be assayed by several methods: (1) by measuring CO_2 evolution from bicarbonate manometrically[1,2]; (2) by coupling the reaction with pyruvate kinase and lactic dehydrogenase and following the disappearance of NADH spectrophotometrically[3]; and (3) by using radioactive L-ribulose and precipitating and assaying the radioactive L-ribulose 5-phosphate formed.[4,5] The second method cannot be used to assay crude extracts because of the presence of interfering substances, but it is very convenient for partially purified preparations and gives identical results as with the third assay method, for which a simplified procedure is given below.

Reagents

Tris-HCl buffer, 0.3 *M*, pH 7.6
GSH, 0.1 *M*, pH 7.6
EDTA, 0.1 *M*, pH 7.6
$MgCl_2$, 1.0 *M*
NaF, 0.5 *M*

[1] S. P. Colowick and H. M. Kalckar, *J. Biol. Chem.* **148**, 117 (1943).
[2] E. Englesberg, *J. Bacteriol.* **81**, 996 (1961).
[3] N. Lee and E. Englesberg, *Proc. Natl. Acad. Sci. U.S.* **48**, 335 (1962).
[4] Modified from B. L. Horecker, J. Thomas, and J. Monod, *J. Biol. Chem.* **235**, 1580 (1960).
[5] N. Lee and I. Bendet, unpublished data.

ATP, 0.2 *M* pH 7

L-Ribulose-1-^{14}C, 0.1 *M*, enzymatically prepared,[2] and purified by column chromatography. Specific activity = 1×10^4 cpm/micromole.

Procedure. Mix together 0.28 ml of Tris buffer, 0.08 ml glutathione (GSH), 0.04 ml EDTA, 0.04 ml $MgCl_2$, 0.08 ml NaF, 0.08 ml ATP, 0.2 ml ATP, 0.2 ml L-ribulose-1-^{14}C, and 0.2 ml H_2O. Use 0.1 ml of this mixture per assay. The enzyme is added to the reactive mixture, and after 4 minutes of incubation at 37° the reaction is stopped by the addition of 0.8 ml of absolute ethanol and 0.1 ml of 1 *M* barium acetate. The mixtures are placed in an ice bath for 15 minutes, and then the entire contents of each tube are transferred to a membrane filter (Millipore, HA) followed by six 1-ml washings with cold 80% ethanol. The wet filters are glued to planchets with paper cement, dried, and counted in a thin-window gas flow counter. The reaction is linear with time and with enzyme concentration up to 0.5 unit per reaction mixture.

Definition of Units and Specific Activity. One unit of L-ribulokinase is defined as that amount which phosphorylates 1 micromole of L-ribulose per minute. Specific activity is in units per milligram of protein.[6,7]

Purification Procedure

All operations are carried out at 0–4°.

Step 1. Growth of Culture and Preparation of Crude Extract. Escherichia coli B/r, ara A-2, an L-arabinose isomeraseless mutant,[8,9] is grown for 16 hours in 40 liters of casein hydrolyzate, L-arabinose (0.05%) medium,[2] in a 50-liter fermentor (New Brunswick, Fermacel), operating with a positive air pressure of 4–5 psi, sparging at a rate of 1.5–2.0 cu. ft. per minute and agitating at 300 rpm. A 10% inoculum is used consisting of an overnight aerated culture grown in the same medium as above but without L-arabinose. Foaming is controlled by Foamkil (Nutritional Biochemicals). The cells are harvested with a Sharples centrifuge. About 30 g dry weight of cells is obtained with each 40 liters of medium. The cells are resuspended with the aid of a Waring blendor in 300 ml of EDTA-GSH solution (EDTA 1 m*M*, GSH 1 m*M*, adjusted to pH 7.4 with NaOH), disrupted by treatment for 8–10 minutes in a 10-kc Raytheon sonic oscillator at 3° in 15-ml aliquots,

[6] O. H. Lowry, N. J. Rosebrough, A. L. Farr, and R. J. Randall, *J. Biol. Chem.* **193,** 265 (1951).

[7] See Vol. III [73].

[8] J. Gross and E. Englesberg, *Virology* **9,** 314 (1959).

[9] E. Englesberg, see this volume [3].

and centrifuged at 34,850 *g* for 1 hour; the supernatant solution (318 ml) is then collected.

Step 2. Removal of Nucleoproteins. To the supernatant solution, 15.9 ml of 1.0 *M* $MnCl_2$ is added with stirring. After 1 hour in an ice bath, the precipitate is removed by centrifugation at 34,850 *g* for 1 hour.

Step 3. Ammonium Sulfate Fractionation. The supernatant solution (280 ml) is adjusted to pH 7.6 with 1 *M* KOH. Solid ammonium sulfate (2 × crystallized, General Biochemicals) is added slowly over 1½ hours to bring the solution to 40% saturation at 0° (64.6 g). The precipitate formed is separated by centrifugation at 34,850 *g* for 1 hour and discarded. The supernatant (286 ml) is brought to 50% saturation with solid ammonium sulfate (an additional 16.8 g over 1½ hours, and the precipitate is collected by centrifugation at 34,850 *g* for 1 hour, dissolved in 35–40 ml of EDTA-GSH solution, and dialyzed overnight against 2 liters of EDTA-GSH. The dialysis fluid is then replaced by 2 liters of pH 7.6 column buffer (00.2 *M* potassium phosphate buffer, pH 7.6, 1 m*M* EDTA, and 1 m*M* GSH). After 3 hours, the enzyme solution is recovered and diluted to 50.0 ml with pH 7.6 column buffer.

Step 4. pH 7.6 DEAE-Cellulose Column Chromatography. DEAE-cellulose (Selectacel, Carl Schleicher & Schuell Co.) is pretreated with ten washings of equal volumes of 1 *N* NaOH, washed ten times with demineralized water, and neutralized with concentrated phosphoric acid. The DEAE-cellulose is then packed by gravity into a 4 × 30 cm column, washed with 0.1 *M* KH_2PO_4–K_2HPO_4 buffer, pH 7.6, to the same pH, and finally equilibrated in the cold with pH 7.6 column buffer by allowing five times the column volume of buffer to flow through the column. One-fourth of the material from step 3 is placed on a column, one-half column volume of buffer is added, and the enzyme is eluted with a convex NaCl gradient. The mixing bottle contains 1 liter of column buffer, and the reservoir contains the same buffer with 0.5 *M* NaCl. The first 450 ml of eluent is discarded, and the remainder is collected in 18–20 ml fractions. Protein concentrations are estimated spectrophotometrically.[7] The fractions containing the highest specific activities are pooled, then concentrated by addition of solid ammonium sulfate to 70% saturation. The precipitate is collected by centrifugation, dissolved in 10–12 ml of EDTA-GSH solution, and dialyzed against 2 liters of EDTA-GSH solution overnight. The dialysis fluid is then replaced with 2 liters of pH 6.5 column buffer (0.01 *M* KH_2PO_4–K_2HPO_4 buffer, pH 6.5; 1 m*M* EDTA; and 1 m*M* GSH) and dialyzed for 3 hours immediately before the next step.

Step 5. pH 6.5 DEAE-Cellulose Column Chromatography. The pH 6.5 DEAE-cellulose column is prepared in the same manner as above, except that it is washed and equilibrated to pH 6.5. One-half of the concentrated eluent from step 4 is placed on a 2.7 × 30 cm column, one-half column volume of buffer is added, and the enzyme is eluted with a convex NaCl gradient. The mixing bottle contains 500 ml of pH 6.5 column buffer, and the reservoir contains the same buffer with 0.5 *M* NaCl. The volume of each fraction is about 9 ml. Protein concentrations are determined as in step 4. The fractions containing the highest specific activities are pooled, then concentrated with 70% saturation with ammonium sulfate. After centrifugation, the pellets are dissolved in 3–4 ml of EDTA-GSH, and dialyzed overnight against 2 changes of EDTA-GSH, 600 ml each.

Step 6. Sephadex G-200 Column Chromatography. Sephadex G-200 beads, soaked overnight in pH 7.6 column buffer, are packed by gravity into a column seven times the volume of the concentrated eluent from step 5. Elution is accomplished using pH 7.6 column buffer. Forty to fifty fractions, 1.5 ml each, are collected. Proteins are determined as in step 4. All peak fractions contain approximately the same specific activity. The fractions containing the highest protein contents are pooled. Often there is no increase in specific activity after this step.

Step 7. Crystallization. Of the pooled Sephadex G-200 eluent, 3 ml is brought to 40% saturation with ammonium sulfate by the slow addition of 2.0 ml of 100% saturated ammonium sulfate solution (pH 7.0) with stirring. A slight cloudiness is removed by centrifugation with no loss of activity. The solution is then dialyzed against 45% saturated ammonium sulfate in the cold, and the concentration of ammonium sul-

Purification Procedure

Fraction	Volume (ml)	Protein (mg)	Total units	Specific activity	OD_{280}: OD_{260} ratio	Yield (%)
1. Crude extract	318	14,060	7260	0.515	—	100
2. $MnCl_2$ supernatant	280	—	6270	—	—	86.5
3. 40–50% ammonium sulfate fraction, dialyzed	50	1,625	3320	2.04	—	46.8
4. pH 7.6 DEAE-cellulose eluent, concentrated	60	163	1530	9.39	1.0	21.1
5. pH 6.5 DEAE-cellulose eluent, concentrated	9.5	73.1	981	13.4	1.4	13.5
6. Sephadex G-200 eluent (sum of two pools of different concentrations)	39.5	49.0	681	13.9	1.7	9.4
7. Second crystals	—	32.0	—	12.5	—	5.5

fate is slowly raised by evaporation. After a few days, the contents of the bag show some crystalline material amid amorphous precipitates. The precipitate is collected, redissolved in 45% saturated ammonium sulfate solution, and left to crystallize once more. The second crystals, after several weeks, contain two-thirds of the enzyme activity and two-thirds of the protein with no increase in specific activity.

A summary of the procedure is given in the accompanying table.

Properties

Molecular Weight Determination. The pooled eluent from the G-200 column appears homogeneous in disc electrophoresis,[10] immunoelectrophoresis[11] against rabbit antisera prepared against crude extract, and in the ultracentrifuge. Sedimentation studies performed on varying dilutions of the enzyme in pH 7.6 column buffer gives a value of $s^{\circ}_{20,w}$ of 6.205 with no evidence of dissociation of the molecule upon dilution. $D_{20,w}$ is found to be 5.28×10^{-7} cm^2/sec. The molecular weight obtained from these s and D values is 96,000, using a partial specific volume calculated from the amino acid composition. Osmotic pressure measurements give a number average molecular weight of 100,000 for L-ribulokinase.

Amino Acid Composition. After acid hydrolysis for 22 hours, the enzyme is found to be devoid of valine and methionine, and contains 154 amino acid residues (excluding tryptophan, which is lost during the acid hydrolysis) with a minimum molecular weight of 15,900.

Ultraviolet Light Absorption Spectra and Extinction Coefficient. The enzyme absorbs maximally at 280 mμ, with a shoulder at 290 mμ which probably indicates tryptophan. The specific absorbance of the enzyme at 280 mμ is 1.51 absorbancy units.

Effect of Heat and pH on Enzyme Stability. Purified L-ribulokinase is stable at pH 7.6 and above in 0.1 M Tris maleate buffer, even when stored at room temperature, but loses activity rapidly at acid pH's. In pH 7.6 column buffer, it retains complete activity at 60.5° even after 30 minutes. This heat stability is not observed with partially purified preparations.

Michaelis Constants for Substrates and Cofactor. K_m values are 1.11×10^{-4} M for L-ribulose, 0.76×10^{-4} M for ATP, and 1.30×10^{-3} M for Mg ion as $MgCl_2$. The enzyme also phosphorylates D-ribulose, the K_m being 2.70×10^{-4} M. The enzyme is inactive on L-xylulose, D-xylulose, D-ribose, D-fructose, and L-arabitol.[3]

pH Optimum. The enzyme has optimum activity at pH 7.6–7.8, with no activity below pH 7.[3]

[10] L. Ornstein and B. J. Davis, "Disc Electrophoresis," a preprint obtainable from the Distillation Product Industries (Division of Eastman Kodak Co.).

[11] A. J. Crowle, "Immunodiffusion." Academic Press, New York, 1961.

Molecular Activity. The molecular activity is 1700 moles of product formed per minute per mole of enzyme.

Occurrence. L-Ribulokinases have been purified from *Aerobacter aerogenes*[12] and *Lactobacilus pentosus.*[13] L-Arabinose-induced *E. coli* B/r and ara A-2 contain approximately 1–1½% and 3–4% L-ribulokinase, respectively.

[12] W. A. Wood and F. J. Simpson, see Vol. V [32a].
[13] D. P. Burma and B. L. Horecker, *J. Biol. Chem.* **231**, 1039 (1958).

[81] D-Xylulokinase

By F. J. Simpson

D-Xylulose + ATP → D-xylulose 5-phosphate + ADP[1]

D-Xylulokinase has been prepared from calf liver[2] and *Lactobacillus pentosus.*[3] The following procedure, a modification of one previously described,[4] provides a method for obtaining the enzyme on a rather large scale from *Aerobacter aerogenes.*

Assay Method

Principle. The convenient method of Kornberg and Pricer that measures the rate of formation of ADP is used.[5] The ADP is immediately rephosphorylated to yield ATP by pyruvate kinase acting on phosphopyruvate. This reaction is coupled to the reduction of pyruvate to lactate by lactic acid dehydrogenase, and the oxidation of NADH to NAD. The method is rapid and easily performed but suffers from the disadvantage that enzymes which oxidize NADH, produce ADP from ATP, or pyruvate from phosphopyruvate will interfere. Thus, a control without substrate is necessary. Other methods for measuring the activity of kinases also are applicable.[2-4]

Reagents

0.5 *M* Tris–0.5 *M* KCl–0.01 *M* EDTA buffer, pH 7.8
$MgCl_2$, 0.05 *M*
Phosphoenolpyruvate, 0.01 *M*, pH 7.0
ATP, 0.01 *M*, pH 7.0

[1] Issued as NRC No. 9055.
[2] G. Ashwell, Vol. V [22].
[3] P. K. Stumpf and B. L. Horecker, *J. Biol. Chem.* **218**, 753 (1956).
[4] F. J. Simpson and B. K. Bhuyan, *Can. J. Microbiol.* **8**, 663 (1962).
[5] A. Kornberg and W. E. Pricer, *J. Biol. Chem.* **193**, 481 (1951).

NADH, 0.003 *M*
D-Xylulose,[6] 0.01 *M*
Lactic acid dehydrogenase, 60 μg/ml,[7] diluted in 0.01 *M* KCl
D-Xylulokinase, diluted in 0.01 *M* Tris–0.001 *M* EDTA–0.001 *M* dithiothreitol buffer, pH 7.8

Procedure. Into a cuvette ($l = 1$ cm) are added 0.1 ml of buffer, 0.25 ml of a mixture (1 volume of each $MgCl_2$ and phosphoenolpyruvate and 0.5 volume of ATP) prepared daily, 0.03 ml of NADH, 0.1 ml of D-xylulose, 0.03 ml of lactic acid dehydrogenase, water, and D-xylulokinase. The final volume is 1.0 ml. The reaction is begun by adding the kinase and the rate of change in absorbance at 340 mμ is measured. The rate of oxidation of NADH is corrected for that of a control without substrate. One unit of activity is defined as the amount of enzyme required to phosphorylate 1 micromole of substrate per minute.

Purification Procedure

Step 1. Growth of Cells. Aerobacter aerogenes (PRL R3) is grown at 33° in a 5-gallon fermentor or in shaken flasks. The medium consists of 0.8% K_2HPO_4, 0.3% $NH_4H_2PO_4$, 0.05% Na_3 citrate·2 H_2O, 0.02% $MgSO_4 \cdot 7\ H_2O$, 0.005% $FeSO_4 \cdot 7\ H_2O$, 0.1% $(NH_4)_2SO_4$, 0.1% peptone or casein hydrolyzate (e.g., NZ-amine Type A, Sheffield Chemical Co. Inc., New York) and 1% D-xylose (technical). The sugar and the peptone are sterilized separately and aseptically added prior to inoculation. The fermentor is inoculated with 0.005 volume of an 8–10 hour culture grown on the same medium. During growth, the culture is aerated at the rate of 4 liters of air per minute and stirred at 400 rpm, or by incubating flasks on a rotary shaker. Growth is complete within 7–9 hours. Production of xylulokinase is concomitant with growth and ceases just before growth stops and the supply of xylose is exhausted. The cells may be harvested at this time or allowed to incubate for an additional 10 hours without much loss of enzyme. The culture, usually at pH 5.9–6.4, is adjusted to pH 7.5 with ammonium hydroxide and diluted with 0.5 volume of 0.005 *M* EDTA, pH 7.5; the cells are recovered by centrifugation. The cells are then washed with 0.001 *M* EDTA buffer, pH 7.5, and after recovery by centrifugation may be stored as a thick paste at −20°. The cells lose about 10% of their activity per month.

Step 2. Preparation of Cell Extract. The xylulokinase is conveniently extracted from the cells with glycine. One volume of the cell paste is

[6] F. J. Simpson, this volume [9].
[7] 2 × Crystallized lactic acid dehydrogenase prepared from rabbit muscle by the method of E. Racker [*J. Biol. Chem.* **196**, 347 (1952)] and available commercially contains sufficient pyruvate kinase to supply both enzymes for the assay.

suspended in 2 volumes of 3 *M* glycine–0.001 *M* EDTA (pH 7.5 with NH_4OH), and the suspension is mixed slowly with a stirrer or in Erlenmeyer flasks on a rotary shaker at 30° overnight (15–20 hours). In the first 2–3 hours the pH may fall and should be readjusted to pH 7.5–7.7. The next day the insoluble debris is removed by centrifugation at 11,000 *g* for 20 minutes. The precipitate is discarded. When stored at pH 7.5–8.0 at −20°, the glycine extract loses about 10% of its activity in a month.

Step 3. Heat Treatment. D-Xylose (1.5 g) is added to the cell extract (1000 ml), and the pH is adjusted to 8.0. Two aliquots, 500 ml each, in 1-liter stainless steel beakers, are heated with stirring in a boiling water bath to 65° and held at that temperature for 10 minutes in a second water bath. The preparations are then quickly cooled to 10°, and the precipitate is removed by centrifugation at 11,000 *g* for 20 minutes at 0–10°. Subsequent steps are conducted at 0–10°.

Step 4. Treatment with DEAE-Cellulose. DEAE-cellulose (0.95 meq/g) is washed with 1 *N* NaOH, then with water until neutral, and is equilibrated with 0.01 *M* Tris-HCl buffer, pH 7.8. The equilibrated cellulose is collected on a fritted glass Büchner funnel (e.g., No. 1 porosity, Gallenkamp, England) and rinsed with 0.001 *M* EDTA, pH 7.8. The cellulose can then be air dried and stored for future use. Fifty grams of this dry equilibrated cellulose are suspended with vigorous stirring in a liter of 0.001 *M* EDTA, pH 7.8, and mixed for not less than 10 minutes. This slurry is added to the enzyme preparation (850 ml) and mixed. The amount of DEAE-cellulose required to absorb the enzyme has been found to vary from 0.04 g to 0.08 g per milliliter depending on the lot. The efficiency of absorption can be readily tested during this step by filtering a few milliliters through a fritted glass funnel and determining the amount of enzyme in the filtrate. The preparation is mixed for 15–20 minutes then the DEAE-cellulose is recovered by filtration with vacuum on a fritted glass funnel. The cloudy filtrate is discarded and the DEAE-cellulose is washed by suspending it in 1.5 liters of buffer (prepared by adding 8.7 g of NaCl, 3.6 g of Tris, and 0.56 g of disodium EDTA to distilled water, adjusting the pH to 7.5 with HCl, and diluting to 1.5 liters). The suspension is vigorously stirred for 1 minute and then rapidly filtered. The D-xylulokinase is eluted by suspending the DEAE-cellulose in 1 liter of 0.1 *M* $MgCl_2$–0.02 *M* Tris–0.001 *M* dithiothreitol buffer, pH 6.0–6.2 and mixing slowly for 20 minutes. The DEAE-cellulose is removed by filtration and rinsed with 50 ml of eluting agent.

Step 5. Treatment with Alumina Cγ.[8] The eluate from the DEAE-

[8] S. P. Colowick, Vol. I, p. 97.

cellulose (1100 ml, pH 7.8) is mixed with 66 ml of alumina Cγ (24 mg/ml) for 3 minutes, and the gel is removed by centrifugation and discarded. About 10% of the enzyme is absorbed. The supernatant liquid is adjusted to pH 6.5 with dilute HCl, 110 ml of alumina Cγ is added, and the sample is mixed for 5 minutes to absorb the enzyme. The gel is recovered by centrifugation and washed with 220 ml of 0.002 *M* K_2HPO_4–0.001 *M* EDTA buffer, pH 9.0. The enzyme is eluted by thoroughly mixing the gel with 110 ml of 0.04 *M* potassium phosphate–0.001 *M* EDTA–0.001 *M* dithiothreitol buffer, pH 8.5. The gel is recovered by centrifugation and treated with the eluting agent again. If necessary, the two eluates are combined. This preparation, pH 8.0, is dialyzed overnight against cold flowing buffer (0.001 *M* Tris–0.001 *M* EDTA, pH 8.0). The extract amounts of alumina Cγ required for the negative and positive absorptions vary somewhat and have to be predetermined by trials on small aliquots of the eluate from the DEAE-cellulose.

Step 6. Treatment with Calcium Phosphate Gel.[8] Calcium phosphate gel (50 ml, 22 mg/ml) is added slowly to the dialyzed preparation (155 ml) while stirring and mixed for 5 minutes. The gel is recovered by centrifugation and the kinase is eluted with 30 ml of 0.027 *M* potassium phosphate–0.001 *M* EDTA–0.001 *M* dithiothreitol buffer, pH 8.5.

Step 7. Ammonium Sulfate Fractionation. Fifteen grams of ammonium sulfate is added to the gel eluate (32 ml) and after mixing for 5 minutes the precipitate is recovered by centrifugation at 20,000 *g* for 10 minutes. This precipitate is extracted with 5 ml of 48% saturated ammonium sulfate in 0.001 *M* EDTA, pH 7.5, for 5 minutes at 0°. The residue is recovered by centrifugation and the kinase extracted from it with 10 ml of 37% saturated ammonium sulfate in 0.001 *M* EDTA, pH 7.5. The undissolved material is removed by centrifugation. Four milliliters of saturated ammonium sulfate is added to the second extract.

Purification Procedure

Step	Total activity (units)	Specific activity (units/mg protein)	Yield (%)
Cell extract	29,000	1.0	100
Heat, 65°	22,600	1.4	78
DEAE-cellulose	20,600	9	71
Alumina C_γ gel	11,000	32	38
Calcium phosphate gel	9,000	40	31
Ammonium sulfate	6,700	95	23

The kinase is recovered by centrifugation and dissolved in 5 ml of 0.2 *M* glycylglycine–0.01 *M* EDTA buffer, pH 8.0, containing 10% glycerol.

Properties

The 95-fold purified kinase is free of D-ribulose 5-phosphate 3-epimerase, transketolase, transaldolase, glyceraldehyde 3-phosphate dehydrogenase, and D-xylose isomerase. The purified kinase is quite stable when stored at —20°. A preparation purified 60-fold retained 30% of its activity after four years. D-Xylulose, ATP, glycerol, and EDTA stabilize the enzyme.

Activators. The purified enzyme is inactive in the absence of either EDTA or a thiol such as reduced monosodium glutathione or dithiothreitol. One of these at 1 micromole per milliliter promotes full activity. $MgCl_2$ is required for activity: 50 micromoles per milliter are optimal whereas more than 80 micromoles inhibit. $MnCl_2$ is 90% and $NiSO_4$ 50% as effective as $MgCl_2$.

Specificity. The kinase is specific for D-xylulose. Other pentuloses, pentoses, pentitols, and all hexoses tested were not phosphorylated except for a slow rate on xylitol. ATP serves as the phosphate donor. ITP and GTP are only 25% and UTP 14% as effective. The K_s values for D-xylulose and ATP are 1.6×10^{-3} and 4.6×10^{-3} *M*, respectively. This kinase has been used to prepare D-xylulose 5-phosphate.

[82] L-Xylulokinase

By R. L. ANDERSON and W. A. WOOD

L-Xylulose + ATP → L-xylulose 5-phosphate + ADP

Assay Method

Principle. The continuous spectrophotometric assay is based on the following sequence of reactions[1]:

$$\text{L-Xylulose} + \text{ATP} \xrightarrow{\text{L-xylulokinase}} \text{L-xylulose 5-phosphate} + \text{ADP}$$

$$\text{Phosphoenolpyruvate} + \text{ADP} \xrightarrow{\text{pyruvate kinase}} \text{pyruvate} + \text{ATP}$$

$$\text{Pyruvate} + \text{DPNH} + \text{H}^+ \xrightarrow{\text{lactate dehydrogenase}} \text{lactate} + \text{DPN}^+$$

With pyruvate kinase and lactate dehydrogenase present in excess, the rate of L-xylulose phosphorylation is equivalent to the rate of DPNH oxidation, which is measured by the absorbance decrease at 340 mμ.

[1] R. L. Anderson and W. A. Wood, *J. Biol. Chem.* **237**, 1029 (1962).

Reagents

Tris-HCl or glycylglycine buffer, 0.2 *M*, pH 7.5
$MgCl_2$, 0.1 *M*
ATP, 0.05 *M*
Phosphoenolpyruvate, 0.05 *M*
Sodium glutathione (reduced), 0.15 *M*
DPNH, 0.01 *M*
L-Xylulose,[2] 0.15 *M*
Crystalline lactate dehydrogenase
Crystalline pyruvate kinase

Procedure. The following are added to a microcuvette with a 1.0-cm light path: 0.05 ml of buffer, 0.01 ml of $MgCl_2$, 0.01 ml of ATP, 0.01 ml of phosphoenolpyruvate, 0.01 ml of glutathione, 0.01 ml of L-xylulose, 0.005 ml of DPNH, excess lactate dehydrogenase, excess pyruvate kinase, L-xylulokinase, and water to a volume of 0.15 ml. The reaction is initiated by the addition of L-xylulokinase. A control cuvette minus L-xylulose measures adenosine triphosphatase and DPNH oxidase activities, which must be subtracted from the total rate. A control cuvette minus ATP should also be run to check for possible L-xylulose reductase activity. The reaction rates are most conveniently measured with a Gilford multiple sample absorbance recorder. The cuvette compartment should be thermostatted at 25°.

Evaluation of the Assay. When care is taken to see that the coupling enzymes are in excess, the velocity is constant with time and proportional to L-xylulokinase concentrations up to 0.3 unit per cuvette.

Definition of Unit and Specific Activity. One unit of enzyme is defined as the amount that effects an absorbance change of 1.0 per minute in the above assay. By calculation 1 unit thus would phosphorylate 1.4 micromoles of L-xylulose per hour. Specific activity (units per milligram of protein) is based on a spectrophotometric determination of protein.[3]

Purification Procedure[1]

Aerobacter aerogenes PRL-R3 was grown on an L-xylose–mineral medium as described,[4] except that the concentration of $FeSo_4 \cdot 7\ H_2O$ was 0.0005%.

Preparation of Cell Extracts. Cells harvested from 3.5 liters of me-

[2] O. Touster, *in* "Methods in Carbohydrate Chemistry" (R. L. Whistler and M. L. Wolfrom, eds.), Vol. I, p. 98. Academic Press, New York, 1962.
[3] O. Warburg and W. Christian, *Biochem. Z.* **310**, 384 (1941).
[4] R. L. Anderson and W. A. Wood, *J. Biol. Chem.* **237**, 296 (1962).

dium were suspended in 1 mM sodium ethylenediaminetetraacetate (pH 7.0) and broken in a Raytheon 10-kc sonic oscillator. After removal of the cellular debris by centrifugation at 31,000 g, the supernatant solution was used as the cell extract.

General Fractionation Precautions. All fractionation steps were performed at 0–4°. Because the unpurified enzyme was unstable to storage, the entire purification was carried out without interruption.

First Ammonium Sulfate Fractionation. The call extract was diluted to 200 ml with 1 mM sodium ethylenediaminetetraacetate (pH 7.0). The protein concentration was then 24.8 mg/ml, and the 280:260 mμ ratio was 0.68. Ammonium sulfate, 5.3 g, was added (0.2 M), followed by 40 ml of 2% protamine sulfate. The precipitate that formed was removed by centrifugation and discarded. To the supernatant (236 ml) was added 69.2 g of ammonium sulfate (55% of saturation). The precipitate that formed was collected by centrifugation and dissolved in water. This fraction (200 ml) contained 16 mg of protein per milliliter and had a 280:260 mμ ratio of 1.20.

Alumina Cγ Gel Adsorption and Elution. Alumina Cγ gel (Sigma, containing 10% solids), 16 ml, was added to the above fraction, stirred for 15 minutes, and centrifuged. The supernatant solution was treated with 9 ml more alumina Cγ gel and centrifuged; the resulting supernatant was discarded. The two gel centrifugates were washed separately with 50-ml portions of 0.01 M potassium phosphate buffer, pH 7.0. The first gel centrifugate was then eluted 6 times and the second gel centrifugate was eluted twice with 10- to 20-ml portions of 0.02 M potassium phosphate buffer, pH 7.0. The eight fractions were pooled to yield 112 ml of eluate with a protein concentration of 6 mg/ml and a 280:260 mμ ratio of 1.45.

Second Ammonium Sulfate Fractionation. Ammonium sulfate, 27.9 g, was added to the above fraction (45% of saturation). The resulting precipitate was collected by centrifugation and dissolved in water to yield 10 ml with a protein concentration of 13.2 mg/ml. The 280:260 mμ ratio was 0.99.

DEAE-Cellulose Chromatography. DEAE-cellulose (Brown Company, Berlin, New Hampshire) was treated with 0.5 M potassium phosphate buffer (pH 7.0), washed with water, and equilibrated with 0.05 M potassium phosphate buffer (pH 7.0). A column (1.2 $\times$ 10 cm) was prepared, and the above kinase fraction was added. The protein which adsorbed (about 50%) was eluted successively with 20-ml portions of 0.05, 0.08, 0.10, 0.12, and 0.14 M potassium phosphate buffer, pH 7.0. Five-milliliter fractions were collected. Fraction 12 was the most highly purified (520-fold), and contained 8.4% of the original activity. The 280:260 mμ ratio was 1.44 and the protein concentration was 0.16 mg/ml. A summary of the purification procedure is shown in the table.

PURIFICATION OF L-XYLULOKINASE

Fraction	Units	Units/mg	Fold purified	Yield (%)
Cell extract	200,000	40	1.0	100
Ammonium sulfate I	214,000	67	1.7	107
Alumina Cγ gel	180,000	340	8	90
Ammonium sulfate II	175,000	1,350	34	87
DEAE-cellulose 12	16,800	20,900	520	8.4 } 21
DEAE-cellulose 13	13,300	14,700	368	6.6 } 21
DEAE-cellulose 14	11,800	14,600	365	5.9 } 21

Properties

Specificity. Of 26 ketoses, aldoses, and polyols tested, only L-xylulose was phosphorylated ($K_m = 0.4$ mM). With 0.5 unit of 520-fold purified L-xylulokinase in the standard assay, phosphorylation of the following compounds could not be detected: D-xylulose, D- or L-glucose, D- or L-ribulose, D- or L-xylose, D- or L-arabinose, D-lyxose, D-ribose, D-fructose, L-sorbose, D-mannose, D-galactose, D-altrose, D- or L-fucose, L-rhamnose, L-erythrulose, D- or L-arabitol, xylitol, ribitol, and erythritol.

Activators. A divalent metal ion (Mg^{++}) is required for activity. Glutathione usually stimulates activity about 2-fold.

pH Optimum. Activity as a function of pH is maximal at about pH 7.5 and half-maximal at about pH 5.6.

[83] L-Fuculokinase[1]

By MOHAMMAD A. GHALAMBOR and EDWARD C. HEATH

L-Fuculose + ATP → L-fuculose 1-phosphate + ADP

Assay Method

Principle. L-Fuclokinase may be assayed most conveniently by measuring the amount of ADP formed in the reaction by the method of Kornberg and Pricer.[2] After the kinase has been inactivated by heat-

[1] This presentation is based on work previously published: E. C. Heath and M. A. Ghalambor, *J. Biol. Chem.* **237,** 2423 (1962).

[2] A. Kornberg and W. E. Pricer, *J. Biol. Chem.* **193,** 481 (1951).

ing, aliquots are removed and assayed for ADP by the coupled reactions of phosphopyruvate kinase and lactic acid dehydrogenase; the oxidation of DPNH in the latter reaction is determined spectrophotometrically at 340 mμ. Control incubation mixtures without L-fuculose must be included at all stages of purification in order to compensate for ATPase activity. The assay is of limited reliability in the crude extract because of the high ATPase activity present.

Reagents

L-Fuculose, 0.02 *M*. Prepare by the method of Barnett and Reichstein.[3] Purify by fractional crystallization of the *o*-nitrophenylhydrazone or by cellulose column chromatography.[1]
ATP, 0.1 *M*, pH 7
$MgCl_2$, 0.1 *M*
Glutathione, 0.1 *M*
Potassium fluoride, 1.0 *M*
Tris-chloride buffer, 0.5 *M*, pH 7.8
Phosphopyruvate, 0.05 *M*
DPNH, 0.02 *M*
Lactic acid dehydrogenase, 1.0 mg/ml. The two-times recrystallized preparation of Worthington Laboratories contains a sufficient amount of pyruvate kinase for the quantitative determination of ADP under the conditions described below.

Procedure. The incubation mixture contained 0.1 ml of L-fuculose, 0.05 ml of ATP, 0.05 ml of $MgCl_2$, 0.05 ml of glutathione, 0.005 ml of KF, 0.1 ml of Tris buffer, and appropriate amounts of L-fuculokinase in a total volume of 0.5 ml. After incubation at 37° for 10 minutes, the mixture was heated in a boiling water bath for 1 minute, cooled to room temperature, and centrifuged. ADP is determined in an aliquot of the supernatant fluid.

Definition of a Unit and Specific Activity. One unit of enzyme is defined as that amount required to form 1 micromole of ADP per minute. Specific activity is expressed as units per milligram of protein. Protein is determined by the method of Waddel.[4]

Purification Procedure

Step 1. Preparation of Cell-Free Extract. Escherichia coli 0111-B_4 (ATCC 12015) is grown in a medium that consists of the following (grams per liter): NH_4Cl, 5; Na_2HPO_4, 10; KH_2PO_4, 3; K_2SO_4, 1; NaCl,

[3] J. Barnett and T. Reichstein, *Helv. Chim. Acta* **20**, 1529 (1937).
[4] J. J. Waddel, *J. Lab. Clin. Med.* **48**, 311 (1956).

1; $MgSO_4 \cdot 7\ H_2O$, 0.2; $CaCl_2 \cdot 6\ H_2O$, 0.02; $FeSO_4 \cdot 7\ H_2O$, 0.001; L-fucose, 5. The sugar is autoclaved separately as a 50% solution and is added aseptically to sterile media. One liter of medium (contained in a 2-liter flask) is inoculated from agar slants (prepared from the same medium with 2% agar added) and incubated at 37° on a rotary shaker. The cells are harvested after 12–16 hours of growth and washed twice with cold 0.15 *M* KCl solution. Approximately 5 g of cell paste is obtained per liter of medium. Cell-free extracts are prepared by disrupting the cells in a Raytheon 10-kc sonic oscillator; 5–10 g batches of cells are suspended in 20 ml of cold 0.02 *M* phosphate buffer, pH 7, and sonicated for 15 minutes. The suspension is centrifuged at 25,000 *g* for 10 minutes, the residue is washed with 0.02 *M* phosphate buffer, pH 7, and the supernatant solutions are combined. All procedures are conducted between 0° and 4°.

Step 2. Protamine Sulfate. As excess protamine sulfate precipitates the enzyme, pilot tests with small aliquots of the crude extract are used to determine the optimal amount. In a typical experiment, 50 ml of crude extract is diluted with 0.02 *M* phosphate buffer, pH 7, to give a protein concentration of about 26 mg/ml. To the diluted extract is added, with gentle stirring, 30 ml of a 2% solution of protamine sulfate; the mixture is allowed to stand for 5 minutes. The suspension is centrifuged at 30,000 *g* for 10 minutes, and the supernatant solution (93 ml) is diluted with phosphate buffer to give a protein concentration of 2.2 mg/ml.

Step 3. Ammonium Sulfate. The diluted protamine sulfate supernatant solution (180 ml) is treated with 35 g of solid ammonium sulfate, the suspension is stirred for 5 minutes and centrifuged at 30,000 *g* for 5 minutes, and the small precipitate is discarded. The supernatant solution is treated with an additional 10 g of ammonium sulfate, and after stirring for 5 minutes, the suspension is centrifuged at 30,000 *g* for 10 minutes. The precipitate, containing most of the L-fuculokinase, is dissolved in phosphate buffer to yield a final volume of 9.25 ml.

The procedure is summarized in the table (p. 464).

Properties

Specificity. This preparation catalyzes the phosphorylation of a variety of ketoses with the following relative rates: L-fuculose, 100; D-ribulose, 38; D-xylulose, 36; D-fructose 45. L-Fucose, D-psicose, D-tagatose, L-sorbose, 6-deoxy-L-sorbose, L-rhamnulose, and L-ribulose are inactive. As the products of phosphorylation of those compounds other than L-fuculose have not been isolated, the position of the phosphate group in these products is not known.

PURIFICATION PROCEDURE

Fraction	Total volume (ml)	Protein (mg)	Specific activity (units/mg × 10^2)	Recovery (%)
1. Crude extract	50	1850	2.9	—[a]
2. Protamine sulfate	93	400	13.4	100
3. Ammonium sulfate[b]	9.25	42	96.6	74

[a] Interfering enzymes in the crude extract prevent accurate determination of total activity.

[b] This preparation is free of detectable amounts of L-fucose isomerase and contains only traces of ATPase.

Stability. The enzyme preparation is completely stable to storage at −16° for 2 weeks, although 20% of the activity is lost after storage for 4 months. Repeated freezing and thawing of the preparation results in considerable losses in activity.

Effect of pH. The pH optimum of the enzyme is approximately 8. The rates of the reaction at pH 7 and pH 8.5 are approximately 70% of maximum.

[84] L-Rhamnulokinase[1]

By T. H. CHIU and DAVID SIDNEY FEINGOLD

L-Rhamnulose + ATP → L-rhamnulose 1-phosphate + ADP

Assay

Principle. ADP released from ATP by the action of L-rhamnulokinase is determined by the following coupled reactions, catalyzed by the indicator enzymes pyruvate kinase and lactic dehydrogenase.[2]

ADP + PEP → ATP + pyruvate
Pyruvate + NADH + H^+ → lactate + NAD

The oxidation of NADH is followed photometrically.

Reagents

NADH, 0.6 m*M*, freshly prepared solution in 0.03 *M* Tris-HCl buffer, pH 8.5

[1] T. H. Chiu and D. S. Feingold, *Biochim. Biophys. Acta* **92**, 489 (1964).

[2] A. Kornberg and W. E. Pricer, Jr., *J. Biol. Chem.* **193**, 481 (1951).

L-Rhamnulose,[3] 0.13 M
Phosphoenolpyruvic acid, 0.05 M
ATP, 0.02 M, in 0.03 M Tris-HCl buffer, pH 8.5
Reduced glutathione, 0.01 M, freshly prepared solution
Ethylenediaminetetraacetic acid (EDTA), 0.01 M
$MgCl_2$, 0.1 M
Tris-HCl buffer, 0.03 M, pH 8.5
Lactic dehydrogenase (specific activity 40 units/mg protein) containing pyruvate kinase (Vol. I [67]) 40 mg/ml

Procedure. The following quantities of reagents (microliters) are added in the order given to a silica cuvette with 1.3-ml capacity and a 1-cm light path: 300 NADH, 2 lactic dehydrogenase, 50 reduced glutathione, 50 EDTA, 35 $MgCl_2$, 35 phosphoenolpyruvic acid, 450 Tris-HCl buffer, 35 ATP and 15 L-rhamnulose. The decrease of absorbancy at 340 mμ and 37° is followed for 5 minutes. A control is run without sugar.

Definition of Unit and Specific Activity. One unit of enzyme catalyzes formation of 1 micromole of ADP per minute at 37°. Specific activity is expressed as units per milligram protein.[4]

Preparation and Purification

Growth of Cells. Escherichia coli K40 is grown in the following medium (grams per liter): K_2HPO_4, 7.0; KH_2PO_4, 3.0; $(NH_4)_2SO_4$, 1.0; $MgSO_4 \cdot 7\ H_2O$, 0.1; L-rhamnose, 2.0. The sugar is sterilized by Seitz filtration and added aseptically to the sterile salts solution. Two liter flasks containing 1 liter of medium are inoculated with 100 ml of starter culture with a Klett reading of approximately 200 (Filter No. 42). After 12–16 hours of incubation on a rotary shaker at 37°, the cultures are in

[3] L-Rhamnose (30 g) is refluxed for 3 hours in 250 ml dry pyridine. The pyridine is removed by vacuum distillation; water is added, and vacuum distillation of the solution is repeated until no odor of pyridine is detectable. The sirupy mixture of L-rhamnose and L-rhamnulose is then loaded onto a column of powdered cellulose and separated by chromatography in butanone–acetic acid–H_2O; 75:25:10 (v/v). Carbohydrate-containing fractions are located by: (1) spotting on filter paper; (2) dipping into a solution freshly prepared by addition of 0.1 ml of saturated $AgNO_3$ solution to 20 ml acetone and just dissolving the precipitate by addition of water; (3) drying the paper in air; (4) spraying with 1 N ethanolic NaOH. L-Rhamnulose in the fractions is located by paper chromatography in the above solvent. Appropriate fractions are pooled, taken to dryness at 37°, dissolved in water, and any sediment is removed by filtration. L-Rhamnulose concentration is determined by the method of Dische and Borenfreund [*J. Biol. Chem.* **192,** 583 (1951), with 36.2 Klett units = 0.01 micromole L-rhamnulose].

[4] J. J. Waddel, *J. Lab. Clin. Med.* **48,** 311 (1956).

the stationary phase of growth. The cells are then harvested and washed twice with ice-cold 0.85% NaCl solution. Approximately 2 g of packed cells per liter of culture medium are obtained.

Preparation of Cell-Free Extracts. Ten-gram batches (wet weight) of cells are suspended in 25 ml of cold 0.02 *M* sodium and potassium phosphate buffer (pH 7.0) and sonicated for 15 minutes in a Raytheon 10-kc sonic oscillator cooled by circulation of ice-water. The suspension is centrifuged in the cold at 30,000 *g* for 20 minutes; the pellet is discarded. The supernatant fluid releases ADP from ATP when either L-rhamnose or L-rhamnulose is used as substrate, indicating the presence of L-rhamnose isomerase as well as L-rhamnulokinase. These enzymes are inducible, since they are not present in the crude enzyme solution obtained from cells grown in a medium in which the L-rhamnose is replaced by D-glucose.

$MnCl_2$ Treatment. (This and all subsequent operations are conducted at 0–4°.) The supernatant fluid from the previous step is diluted with 0.02 *M* phosphate buffer (pH 7.0), to give a protein concentration of 30 mg/ml, and 0.1 volume of 0.5 *M* $MnCl_2$ is added. The mixture is allowed to stand for 30 minutes and then centrifuged at 30,000 *g* for 10 minutes. The residue is washed with a small amount of 0.02 *M* phosphate buffer, and the washings are combined with the supernatant liquid.

$(NH_4)_2SO_4$ Fractionation I. To the supernatant solution (9.2 mg protein per milliliter) is added saturated $(NH_4)_2SO_4$ solution (pH 7.0) to 33% saturation. [In all $(NH_4)_2SO_4$ precipitations 5 minutes is allowed to elapse between addition of $(NH_4)_2SO_4$ and centrifugation.] The precipitate is discarded and the supernatant solution is brought to 55% saturation by addition of a saturated solution of $(NH_4)_2SO_4$ (pH 7.0). The precipitate, which contains most of the L-rhamnulokinase activity, is dissolved in a volume of 0.01 *M* phosphate buffer (pH 7.0), 0.01 *M* in respect to EDTA, equal to the volume of the crude enzyme solution. At this point the protein concentration is approximately 4 mg/ml.

Calcium Phosphate Gel Treatment. Calcium phosphate gel (Vol. I [11]) (12 mg dry weight per milligram protein) is added to the enzyme solution. The suspension is stirred occasionally for 15 minutes, and the gel is then removed by centrifugation.

$(NH_4)_2SO_4$ Fractionation II. The above supernatant solution is fractionated as previously between 35% and 50% saturation with $(NH_4)_2SO_4$. The precipitate is dissolved in sufficient 0.02 *M* phosphate buffer (pH 7), 0.01 *M* in respect to EDTA, to make the protein concentration approximately 1.5 mg/ml.

Alumina Gel Treatment. Alumina gel (Vol. I [11]) is added to the above solution (7 mg dry weight per milligram protein). After 5 minutes the gel is spun down and discarded. The supernatant is treated with a second portion of gel to adsorb the enzyme (9 mg dry weight per milligram protein). The mixture is stirred occasionally for 15 minutes, and the gel is then spun down and washed once with a volume of cold water equal to one-half the volume of the supernatant fluid in the adsorption step. The enzyme is eluted from the gel with 0.1 *M* sodium and potassium phosphate buffer (pH 7.6). Usually two elutions, each with one-half the volume of the supernatant fluid in the adsorption step, suffice to elute the bulk of the enzyme. The purification is summarized in the table.

PURIFICATION OF L-RHAMNULOKINASE

Fraction	Volume (ml)	Protein (mg/ml)	Specific activity[a] (units)	Total units	Purification (-fold)	% Recovery
Crude extract	15.0	30.00	0.20	90.0	1.0	100
$MnCl_2$	16.5	9.20	0.54	82.0	2.7	91
$(NH_4)_2SO_4$ fractionation I	15.0	3.90	0.92	54.0	4.6	60
$Ca_3(PO_4)_2$ gel	15.2	1.28	1.96	38.0	9.8	42
$(NH_4)_2SO_4$ fractionation II	5.0	1.44	4.05	29.0	20.3	32
Al_2O_3 gel	2.5	0.63	6.83	10.8	34.2	12

[a] Micromoles of ADP released at 37° per minute per milligram protein.

Properties

Substrate Specificity. Purified L-rhamnulokinase catalyzes the phosphorylation of a number of ketoses other than L-rhamnulose, with relative activities as follows: L-rhamnulose, 1.0; L-fuculose, 0.30; L-xylulose, 0.11; D-xylulose, 0.02; L-ribulose 0.01; D-fructose 0.01; D-psicose, 0; D-tagatose, 0; L-sorbose, 0. The structures of the phosphates formed have not been determined.

The kinase can use other nucleotides than ATP as phosphoryl donor. The ratio of activity with various nucleotides is: ATP, 1; UTP, 0.9; CTP, 0.7; GTP, 0.25; TTP, 0.10. K_m values at 37° are 8.2×10^{-5} *M* for L-rhamnulose and 1.1×10^{-4} *M* for ATP.

Activation by Metal Ions. The kinase requires certain divalent metal ions for maximum activity; K_m for Mg^{++} at 37° is 2.7×10^{-4} *M*. Purified preparations of the enzyme are stimulated 6-fold by 2.5×10^{-3} *M* Mg^{++}. The relative effect of the same concentration of other metal ions is as follows: Mg^{++}, 1.0; Mn^{++}, 1.0; Co^{++}, 0.90; Fe^{++}, 0.85; Ca^{++}, 0.77; Cu^{++}, 0.

Optimum pH. L-Rhamnulokinase has a rather sharp pH optimum at 8.5. Activity falls off rapidly on the basic side, more slowly the acid side.

Stoichiometry. One mole each of L-rhamnulose phosphate and ADP is formed for each mole of L-rhamnulose utilized in the reaction.

Preparation of L-Rhamnulose 1-Phosphate

A large-scale reaction is run with the following quantities of reactants (millimoles): L-rhamnulose, 2.2; ATP, 3.0; $MgCl_2$, 1; Tris-HCl (pH 8.5), 8.0; and 52.3 mg of enzyme protein in a total volume of 300 ml. After incubation at 37° for 1 hour, the reaction mixture is chilled in ice; analysis of an aliquot should indicate that all the L-rhamnulose has been phosphorylated.

The reaction mixture is fractionated on a column of Dowex 1 X-8 resin (formate form, 200–600 mesh) by gradient elution with 200 ml H_2O in a constant-volume mixing chamber and 1 liter of 0.4 N formic acid containing 0.1 M sodium formate in the reservoir. Fractions (15 ml each) containing ketose phosphate esters are pooled, concentrated at 37° to 60 ml, and adjusted to pH 6.4 with saturated $Ba(OH)_2$ solution. Then 4 volumes of ethanol are added and the solution is left at 0° for 1 hour. The precipitate is spun down and washed with cold 80% ethanol, washed with cold ether, and dried in vacuum. Contaminating ultraviolet-absorbing material is removed by treatment of an aqueous solution of the barium salts with charcoal (acid-washed Norit A) and subsequent reprecipitation of the barium salt as mentioned above. The barium salt is finally washed with cold 80% ethanol, cold ether, and dried in vacuum.

[85] Pyruvate Kinase: Clinical Aspects

By WILLIAM N. VALENTINE and KOUICHI R. TANAKA

Clinical Significance

A deficiency of pyruvate kinase in human erythrocytes resulting in a hereditary form of hemolytic anemia was first demonstrated in 1961.[1,2] Similar cases, while comparatively rare, have since been identified in the United States, England, Australia, various European countries, Japan, and Canada and in a Mexican child, in a Syrian infant, and in Umbrian families. The disease is genetically determined, homozygotes

[1] W. N. Valentine, K. R. Tanaka, and S. Miwa, *Trans. Assoc. Am. Physicians* **74,** 100 (1961).

[2] K. R. Tanaka, W. N. Valentine, and S. Miwa, *Blood* **19,** 267 (1962).

having a marked red cell pyruvate kinase deficiency and a variable but often severe anemia, while heterozygotes possess a partial deficiency in red cell pyruvate kinase activity not associated with anemia or symptomatic disease. The disorder is characterized by premature destruction of erythrocytes in subjects homozygous for the deficiency, anemia, reticulocytosis, enlargement of the spleen, often but not invariable requirement of repeated transfusions, and, as in other hemolytic anemias, by the frequent occurrence of gallstones developing at an early age. No abnormal hemoglobin has been demonstrated, spherocytic red cells are not a characteristic of the disorder, freshly harvested erythrocytes possess normal osmotic fragility, and splenectomy results in only partial if any benefit. When sterile defibrinated blood from affected subjects is incubated for 48 hours at 37°, autohemolysis is usually substantially greater than in the case of similarly incubated normal cells. In nearly all instances this is not correctable by addition of glucose or adenosine to the incubation medium. This pattern of autohemolysis (categorized in 1954 as type II by Selwyn and Dacie[3] to differentiate such cases from other hemolytic anemias with dissimilar behavior in the autohemolysis test) was noted prior to recognition of the underlying enzymatic deficiency, as was the fact that the red cells of affected individuals do not metabolize glucose normally. Still later, deGruchy and his colleagues[4] observed an abnormal accumulation of glycolytic intermediates in incubated erythrocytes, again pointing to a possible defect in glycolysis which has more recently been defined. Addition of neutralized ATP to the autohemolysis test system corrects the abnormality *in vitro*.

The disease is transmitted as an autosomal semirecessive trait. Parents, children, and approximately 50% of siblings of subjects with the disease have no clinical manifestations but have biochemically demonstrable partial red cell deficiency of the enzyme. Leukocytes have been found to have normal levels of pyruvate kinase activity.

The mature human erythrocyte lacks DNA, RNA, an intact Krebs cycle or cytochrome system, ribosomes, and the capacity for oxidative phosphorylation. Its small but definite energy requirements are satisfied largely by the conversion of glucose to lactate via the Embden-Meyerhof and hexose monophosphate shunt pathways. A severe deficiency in the important ATP generating reaction in glycolysis catalyzed by pyruvate kinase, and the consequent interference with the ability to oxidize DPNH at the following lactic dehydrogenase step presumably imposes a meta-

[3] J. G. Selwyn and J. V. Dacie, *Blood* **9**, 414 (1954).

[4] G. C. deGruchy, J. N. Santamaria, I. C. Parsons, and H. Crawford, *Blood* **16**, 1371 (1960).

bolic handicap leading to premature erythrocyte destruction in the body, hemolytic anemia, and the attendant clinical manifestations of the disease.

Methods of Preparation of Human Blood for Assays

The following directions are for the preparation of red and white cell-rich suspensions which may be utilized for various enzymatic assays. Fresh venous human blood is mixed in an approximate ratio of 4 parts of blood to 1 part of a solution containing 5% polyvinylpyrrolidone as sedimenting agent and 2.5% sodium citrate as anticoagulant. After erythrocyte sedimentation, the supernatant containing primarily leukocytes (and platelets) is removed and washed three times in cold 0.15 *M* NaCl. The sedimented red cells are washed likewise. Saline suspensions of erythrocytes in a concentration of about three million per microliter and leukocytes in a concentration between 30,000 and 40,000 per microliter are prepared. Quadruplicate counts for both the red and white cell-rich suspensions and for the contamination of the red cell-rich suspensions by white cells, and vice versa, are performed. Platelet contamination is disregarded. The leukocyte contamination in the erythrocyte suspension should be as low as possible, preferably less than 1500 per microliter. Assays performed with higher leukocyte contaminations are not always reliable because of the large correction necessary for the leukocyte contribution. Because the ratio of per cell activity for pyruvate kinase of leukocytes to erythrocytes is approximately 300 to 1, the activity of the white cell-rich suspension does not usually require correction for its erythrocyte contamination, unless the contamination is very high.

An alternative procedure, if only red cells are to be assayed, is to obtain blood in acid citrate dextrose solution (NIH Formula B) or with heparin as anticoagulant and to discard the buffy coat during each centrifugation of the washing process. Care should be taken not to remove too many of the top red cells (reticulocytes).

Assay Method

The method is adapted for human red cells and white cells from Bücher and Pfleiderer (Vol. I [66]).

Principle. The reader is referred to a discussion of the compound optical assay in which the pyruvate kinase reaction is coupled with that of lactic dehydrogenase (Vol. I [66]).

Reagents. The composition of the assay mixture is shown in Table I.

Triethanolamine buffer. The preparation of this buffer solution is described in connection with the enzyme α-glycerophosphate dehydrogenase (Vol. I [58]).

Lactic dehydrogenase (rabbit muscle, essentially free of pyruvate kinase). Dilute original solution just prior to use so that 0.1 ml contains 18 Enzyme Units. This dilute solution is stable for 1 day at 4°.

Phosphoenolpyruvic acid, crystalline, trisodium salt (tricyclohexylammonium salt may also be used).

ADP, DPNH, and phosphoenolpyruvic acid may be conveniently kept frozen at —20° in small ampuls.

Procedure. All reagents are pipetted in order, including DPNH, as in Table I into a cuvette (1-cm light path), which is placed in a spectrophotometer (at 37°, wavelength 340 mμ) for temperature equilibration.

TABLE I
ASSAY MIXTURE FOR PYRUVATE KINASE[a]

Reagents	Blank	Unknown RBC	Unknown WBC	Amount/cuvette
0.05 *M* Triethanolamine–HCl buffer, pH 7.5	0.5	0.5	0.5	25 micromoles
Glass-distilled or deionized water	1.5	1.2	1.4	—
2.25 *M* KCl	0.1	0.1	0.1	225 micromoles
0.24 *M* $MgSO_4$	0.1	0.1	0.1	24 micromoles
0.006 *M* ADP	0.2	0.2	0.2	1.2 micromoles
Lactic dehydrogenase	0.1	0.1	0.1	18 Enzyme units
DPNH (1.4 micromoles/ml)	0.4	0.4	0.4	0.56 micromoles
RBC or WBC homogenate	—	0.3	0.1	About 3.6×10^7 RBC or 1.6×10^5 WBC
0.045 *M* Phosphoenolpyruvic acid (trisodium salt)	0.1	0.1	0.1	4.5 micromoles

[a] Each silica cuvette in a final volume of 3 ml contains the reagents in the quantities listed in the table.

The following cell mixture is prepared and frozen and thawed three times (in dry ice–propylene glycol methyl ether mixture) just prior to addition to cuvette: 0.05 ml red cell or white cell suspension (see above), 0.4 ml of 0.05 *M* TEA-HCl buffer, and 0.8 ml 0.15 *M* NaCl. The reaction is started by pipetting the substrate phosphoenolpyruvic acid into the assay mixture. Readings are taken at timed intervals and recorded for 14 minutes on a manually operated spectrophotometer or on a multiple sample absorbance recorder.[5] The change in absorbance occurring between the 4th and 14th minute is usually used for calcula-

[5] Gilford Instrument Laboratories, Oberlin, Ohio.

tion, since experience indicated that the reaction is linear during this period.

The above conditions have been found to be suitable and provide reproducible results, but the conditions are not optimal. Increasing the concentration of ADP severalfold (5 micromoles per cuvette in above system) will result in greater activity. A hemolyzate of red cells in distilled water may be used, but complete hemolysis may not occur in pathological states.

Definition of Unit. Enzyme activity is expressed as micromoles of substrate utilized per minute per 10^{10} erythrocytes or leucocytes at 37° under the assay conditions stated.

Alternative Procedure. The enzymatic activity may also be assayed by measurement of the pyruvate produced from phosphoenolpyruvate. Pyruvate is measured by the color produced in reaction with 2,4-dinitrophenylhydrazine and alkalinization.[6] The compound optical assay method is preferable.

Application of assay to Qualitative or Screening Tests. The procedure detailed above has been used almost exclusively in clinical investigation. Brunetti and Nenci[7] have adapted the basic phosphoenolpyruvate to pyruvate reaction for screening purposes by using *o*-cresol red in an unbuffered system and noting a color change from yellow to purple-red as the pH increases from 6.5 to 8.5 during 30 minutes of incubation at 37°. Whole blood (0.01 ml) taken directly from the finger is used. Variations in hematocrit and white cell count may distort the results. Experience has been very limited with this method.

Interpretation of Results. Table II illustrates the findings in normal subjects, 9 homozygous subjects with the clinical disorder, and 47 heterozygous subjects in affected kindreds. Homozygous deficiencies are severe, but care must be taken (1) to exclude high white cell contamination from the red cell assay system to prevent distortion of results by leukocyte pyruvate kinase activity, (2) to recognize that blood containing transfused cells from normal donors will reflect the pyruvate kinase contribution of the donated blood, (3) to note that nonaffected subjects with hemolytic anemia, reticulocytosis, and erythrocytes of young mean cell age will exhibit *increased* pyruvate kinase activity in their red cells, and (4) to recognize that zero values do not necessarily exclude the presence of some enzyme activity since dilute hemolyzates are used in the assay system. Some overlap between pyruvate kinase activity of heterozygotes and either normal subjects on the one hand and ho-

[6] J. F. Kachmar and P. D. Boyer, *J. Biol. Chem.* **200,** 669 (1953).

[7] P. Brunetti and G. Nenci, *Enzymol. Biol. Clin.* **4,** 51 (1964).

mozygous individuals on the other occasionally occurs and may lead in certain instances to difficulties in definitive categorization. Since the assay system is, of course, sensitive to reactant concentrations and assay conditions, the values given in Table II apply only to the specified assay procedure. It is therefore important that each laboratory establish its own normals as a basis of interpretation.

TABLE II
PYRUVATE KINASE DEFICIENCY HEREDITARY HEMOLYTIC ANEMIA

		Pyruvate kinase[a]			
		RBC		WBC	
Subject	No.	Mean ± SD	Range	Mean ± SD	Range
Normals	40	2.65 ± 0.34	2.00–3.40	850 ± 195	547–1260
Heterozygotes	47	1.20 ± 0.32	0.63–1.73[b]	1018 ± 284	513–1883
Case 1. J.L.		0.00		729	
2. R.C.		0.18		697	
3. H.C.		0.14		627	
4. E.T.		0.30		1144	
5. S.H.		0.00		670	
6. K.H.[c]		0.81		621	
7. M.M.[c]		0.83		638	
8. W.N.		0.29		664	
9. S.R.[c]		0.71		810	

[a] Enzyme activity is expressed as micromoles of substrate utilized per 10^{10} erythrocytes or leucocytes at 37° under the assay conditions stated. (Molar absorbance of DPNH = 6.18×10^3.)
[b] Except for single value of 0.35.
[c] Not entirely free of transfused cells when studied.

[86] Acetol Kinase

By O. NEAL MILLER

The phosphorylation of acetol (monohydroxyacetone) can be demonstrated in rat- and rabbit kidney homogenates (reaction 1). This enzymatic activity has been demonstrated[1] by analyzing the phosphorylating reaction mixture (acetol kinase) for acetol phosphate (acetol-P) using the 1,2-propanediol phosphate (PDP) dehydrogenase (reaction 2)

[1] O. Z. Sellinger and O. N. Miller, *Biochim. Biophys. Acta* **36**, 266 (1959).

(see this volume [61]). The assay for acetol kinase is summarized by Eqs. (1) and (2).

$$\text{Acetol} + \text{ATP} \rightarrow \text{acetol-P} + \text{ADP} \quad (1)$$
$$\text{Acetol-P} + \text{NADH} + \text{H}^+ \rightleftharpoons \text{PDP} + \text{NAD}^+ \quad (2)$$

Materials

Preparation of Acetol. Acetol is prepared from commercially available acetol acetate as follows: Acetol acetate (25.0 g) is mixed with 15 ml of methanol containing 1 ml of concentrated HCl. The solution is refluxed for 5 minutes. Most of the methanol and methylacetate is distilled *in vacuo* with the use of a water aspirator. The residue is dissolved in 100 ml of ethyl ether and washed with 15 ml of saturated $NaHCO_3$ solution. The ether is removed by heating in a steam bath, and the residue is dried using anhydrous sodium sulfate. It distills at 15 mm *in vacuo* at 48–50°.

Preparation of Acetol-P. Acetol-P, prepared by the method previously described[2] is used to determine the enzymatic activity of PDP dehydrogenase used to assay for the acetol-P content of acetol kinase incubation mixtures.

The diethyl ketal of acetol is prepared by mixing 33.2 g of acetol acetate (b.p. 174–175°), 43 g of triethyl orthoformate, 3.6 ml of absolute ethanol, and two drops (about 0.1 ml) of concentrated sulfuric acid and allowing the mixture to stand at room temperature for 48 hours. The resulting dark purple solution is neutralized (pH 7.2) by the addition of 2 *N* NaOH and is then fractionated by vacuum distillation; 23.3 g of the diethyl ketal of acetol acetate (I) is collected as a colorless, water-insoluble liquid (b.p. 58–59°/4 mm). Calcium oxide (4.4 g) is suspended in 86 ml of water and 23.3 g of I is added with stirring to break the acetate ester linkage, the temperature of the mixture being maintained below 90°. After 15 minutes the flocculant precipitate is filtered off by suction and the somewhat viscous yellow filtrate is extracted with four successive 50-ml portions of diethyl ether. The water layer is discarded. The ether is removed using a water aspirator and the amber-colored residue is distilled *in vacuo* yielding 8.1 g of the colorless slightly sirupy diethyl ketal of acetol (II) (b.p. 62–64/8.5 mm).

Three grams of (II) is dissolved in 10 ml of dry pyridine (distilled over BaO, b.p. 114–115°). To this well chilled solution is added, dropwise, 7.1 g of diphenyl phosphorochloridate, and the mixture is kept cold with crushed ice for an additional 5 minutes. It is then stored at 4° for 12 hours. Water is then added to destroy the excess diphenyl phosphorochloridate and the solution is concentrated *in vacuo* using a water as-

[2] O. Z. Sellinger and O. N. Miller, *Biochim. Biophys. Acta* **29**, 74 (1958).

pirator (bath temperature, 45°). Additional water is added and the distillation is repeated until only a faint odor of pyridine persists. The resulting sirup is dissolved in 75 ml of benzene and the benzene layer is successively washed with 50 ml of cold water, 50 ml of cold 1 *N* HCl, 50 ml of cold 1 *M* $KHCO_3$ and lastly 60 ml of cold water, after which it is dried over anhydrous sodium sulfate overnight. The benzene is removed *in vacuo*, leaving 6.1 g of residue consisting of the crude diphenyl phosphoryl derivative of acetol diethyl ketal (III).

Three grams of (III) is dissolved in 125 ml of absolute ethanol, and to this is added 0.5 g of platinum oxide. Hydrogenation at slightly above atmospheric pressure is then carried out until 1450 ml of hydrogen has been absorbed (theory: 1450 ml). The catalyst is filtered off, 25 ml of water is added to the filtrate, and the solution is then brought to pH 9.5 with 1 *M* aqueous cyclohexylamine. The mixture is stored overnight at 20°. The solvent is removed *in vacuo* (bath temperature below 50°) yielding 1.4 g of the crude cyclohexylamine salt of acetol-*p* diethyl ketal (IV). One gram of (IV) is dissolved in 25 ml of warm absolute ethanol; the solution is filtered, and the ethanol is removed in a stream of air. The residue is suspended in acetone, quickly filtered by suction, and washed on the funnel with more acetone. The air-dry material weighs about 660 mg. This material is dissolved in 5 ml of water, and 8 ml of acetone is added; the solution is filtered by suction. Fifty milliliters of acetone and a few drops of a mixture of acetone–ether (1:1) are added to the filtrate. After about 5 minutes in the refrigerator, a crystalline precipitate of (IV) forms (needles). The product is collected by filtration and recrystallized by dissolving it in the least amount of water (about 1 ml) and adding acetone until, upon shaking, a turbidity persists. Four hundred milligrams of air-dried material, m.p. 147–149° (decomp.), is obtained.

A 51.3-mg sample of dicyclohexyl-ammonium acetol phosphate diethyl ketal is dissolved in 0.2 ml of water, 1 ml of a suspension of Dowex 50 (H form) resin (about 350 mg dry weight) is added; the solution is swirled for 30 seconds, filtered, made up to 5.0 ml with water, and incubated (pH 1.8) for 3 hours at 40°. It is found that this treatment completely hydrolyzes the ketal structure, without affecting the phosphate ester bond.

PDP dehydrogenase is provided by one of the preparations described in this volume [61].

Procedure

The acetol kinase incubation mixture contains acetol 1.2×10^{-2} *M*; glycine-NaOH buffer, pH 9.8, 1.2×10^{-2} *M*; potassium fluoride, 7.5×10^{-2} *M*, and ATP, 4×10^{-3} *M* and homogenate or other extract containing acetol kinase prepared in 0.15 M KCl. The total volume of reaction mixture is 3.0 ml. Incubations are carried out at 37° in capped, graduated (15 ml) centrifuge tubes. At the end of 30 minutes of incubation, 0.2 ml of 100% (w/v) trichloroacetic acid is added, and the volume is made up to 10 ml with distilled water. The control consists of the complete incubation mixture to which the trichloroacetic acid is added at "zero" time. Following centrifugation, 4 ml of the supernatant fluid is adjusted to pH 7.0 and the volume is made up to 5 ml. One-half milliliter of this solution is then added to a cuvette containing 0.45 micromole of NADH and 120 micromoles of tris(hydroxymethyl)aminomethane, pH 7.65. The oxidation of NADH is measured at 340 mμ in a suitable spectrophotometer, following the addition of a PDP dehydrogenase,[3] 3.5 mg protein of a Myogen A preparation. The control cuvette contains all components except NADH. The volume in each cuvette is 3.0 ml.

The kinetics of the reaction (2) are such that under the conditions described, the oxidation of NADH follows first-order kinetics for 5 minutes. Arbitrarily the amount of oxidation of NADH which takes place in 2 minutes is taken to indicate reduction due to added acetol phosphate from the reaction mixture. Units are usually expressed in micromoles of NADH oxidized per minute per gram of tissue, fresh weight; therefore,

$$\frac{\text{Decrease in absorbance} \times 0.48}{\text{Tissue equivalent (mg)} \times 2 \text{ (time)}} \times 1000 = \text{micromoles/g/min}$$

Properties of Acetol Kinase

The kinase enzyme activity is lost by dialysis for 5–48 hours and is not restored by adding Mg^{++} or thioglycolic acid. Mg^{++} at 3×10^{-3} M inhibits acetol kinase in the presence of 4×10^{-3} M ATP. The enzyme is stable to freezing and thawing. It thus appears that a high molar ratio of ATP to Mg^{++} favors the phosphorylation of acetol, and this point may be one of the distinguishing features which differentiates acetol kinase from glycerokinase and triokinase. Acetol kinase activity is easily demonstrated in kidney homogenates, but it is difficult to demonstrate in liver homogenates. Paradoxically, however, acetol kinase is demonstrable in liver extracts.

[3] Y. T. Baranowski, *Z. Physiol. Chem.* **260,** 43 (1939).

Section V

Aldolases

[87] Fructose Diphosphate Aldolase

By WILLIAM J. RUTTER, JAMES R. HUNSLEY, WILLIAM E. GROVES, JOHN CALDER, T. V. RAJKUMAR, and B. M. WOODFIN

Fructose 1,6-diphosphate (FDP) $\rightleftharpoons$ dihydroxyacetone phosphate (DHAP)
+ D-glyceraldehyde 3-phosphate (GAP)

Introduction

Fructose diphosphate aldolases may be segregated in two classes, based on their molecular and catalytic properties as well as their biological distribution.[1]

Class I aldolases resemble the mammalian muscle enzyme: the pH profile for catalysis of FDP cleavage and hydrogen exchange into DHAP is broad and congruent.[2] There is no evidence implicating metal ions in enzyme structure or activity. On the other hand, lysine residues are present in the active site, and a Schiff's base intermediate may be involved in the reaction mechanism.[3] Carboxypeptidase treatment results in partial loss of enzyme activity, and, in some instances at least, in an altered substrate specificity.[2]

Judging from the similarity of their sedimentation constants to those of the mammalian aldolases, the molecular weight may be approximately 150,000. Also, by analogy with the mammalian enzyme, the molecules are composed of at least three (perhaps four) subunits.

The properties of class II aldolases resemble those of the yeast enzyme.[1] The pH profile for catalysis of FDP cleavages is sharp and displaced from that of the hydrogen exchange activity.[2] Lysine groups are apparently not involved in catalysis; instead, there is evidence suggesting a functional role of divalent metal ions. Enzyme activity is inhibited by metal chelating agents.[2] In the instances examined, this is correlated with a requirement for a divalent metal ion (either firmly bound or dissociable). Monovalent ions (K^+, or NH_4^+) also must be present for optimal activity.[4] Carboxypeptidase treatment has little if any effect on class II aldolases.[2] From their sedimentation properties, a molecular weight similar to that of the yeast aldolase (approximately 70,000) may be inferred.[4] The number of subunits present in the mole-

[1] W. J. Rutter, *Federation Proc.* **23**, 1248 (1964).

[2] O. C. Richards and W. J. Rutter, *J. Biol. Chem.* **236**, 3185 (1961).

[3] E. Grazi, T. Cheng, and B. L. Horecker, *Biochem. Biophys. Res. Commun.* **7**, 250 (1962).

[4] W. E. Groves, Ph.D. Thesis, Univ. of Illinois, Urbana, 1962.

cules is unknown, but preliminary evidence suggests two chains may be present in the yeast enzyme.[5]

Class I aldolases are present in animals, plants, protozoans, and green algae; class II aldolases are present in bacteria, yeast, fungi, blue-green algae; both class I and II enzymes have been found in *Euglena* and *Chlamydomonas*.[1] Because of the gross difference in structure and properties, it has been proposed that class I and II aldolases are analogous, i.e., they have independent evolutionary origins.[4] A possible evolutionary relationship of FDP aldolases to other aldolases has been discussed.

The possibility of aldolase variants within a particular organism exists.[6] Three major aldolase variants have, indeed, been demonstrated in mammalian tissues. The classical muscle aldolase (prototype of class I aldolase) has been termed aldolase A; another variant found particularly in liver and kidney has been termed aldolase B; and still another aldolase found in brain, aldolase C, has recently been discovered.[7]

The isolation and properties of a number of aldolases have been described previously: rabbit muscle aldolase A (class I)[8]; pea aldolase (class I)[9]; *Aspergillus niger* aldolase (class II).[10] Here we describe the isolation and properties of additional class I and II aldolases: rabbit liver aldolase B (class I), and FDP aldolase obtained from yeast (class II) and *Clostridium perfringens* (class II).

I. Yeast

By WILLIAM J. RUTTER and JAMES R. HUNSLEY

The fructose diphosphate aldolase present in yeast is a class II aldolase. Warburg and Gawehn,[11] Richards and Rutter,[12] and Vanderheiden *et al.*[13] have reported methods of isolation of the enzyme from yeast. The present method is an adaptation of the procedure developed by Vanderheiden *et al.*[13] The inclusion of high concentrations of β-mercaptoethanol (BME) throughout the procedure greatly stabilizes the activity. Slight modification of the methodology has resulted in a simple,

[5] W. J. Rutter and J. Hunsley, unpublished observations.

[6] W. J. Rutter, B. M. Woodfin, and R. E. Blostein, *Acta Chem. Scand.* **17**, Suppl. 1, 226 (1963).

[7] T. V. Rajkumar, E. Penhoet, and W. J. Rutter, *Federation Proc.* **25**, 523 (1966).

[8] J. F. Taylor, Vol. I, p. 310.

[9] P. K. Stumpf, *J. Biol. Chem.* **176**, 233 (1948).

[10] V. Jagannathan, K. Singh, and M. Damodaran, *Biochem. J.* **63**, 94 (1956).

[11] O. Warburg and K. Gawehn, *Z. Naturforsch.* **9b**, 206 (1954).

[12] O. C. Richards and W. J. Rutter, *J. Biol. Chem.* **236**, 3177 (1961).

[13] B. S. Vanderheiden, J. O. Meinhart, R. G. Dodson, and E. G. Krebs, *J. Biol. Chem.* **237**, 2095 (1962).

reliable method for the isolation of relatively large quantities of the enzyme in crystalline form.

Assay Method

Principle. Several methods for the assay of FDP aldolase have been published. The most convenient methods trap the triosephosphates formed from FDP, either by cyanide,[14] hydrazine,[15] or coupling to a pyridine nucleotide-linked enzyme, e.g., α-glycerophosphate dehydrogenase.[16] The latter two methods are especially advantageous since direct kinetic measurements are obtained. Of these, the coupled enzymatic assay is usually preferable since it is more broadly applicable and the data are more readily expressed in specific quantitative terms (moles of substrate cleaved per minute). The assay employed here is the same as that used for aldolase B (see p. 492) except that potassium ions[12] and β-mercaptoethanol (BME)[5] are also included since they are required for optimal catalytic activity of yeast aldolase enzyme.

Reagents

Buffer-K-SH solution: Glycylglycine 0.1 *M* pH 7.5 (neutralized with NaOH), potassium acetate 0.2 *M*, BME 5×10^{-2} *M*. This solution should not be used more than 72 hours after the time of preparation. In practice BME is added to an aliquot of glycylglycine-potassium acetate sufficient for 1–2 days' assays.
FDP-Na (Sigma), 0.02 *M*, pH 7.5
NADH-Na (Sigma) 0.002 *M* in 0.001 *M* NaOH stored at 0°–4° for a period not to exceed 2 weeks[17]
α-Glycerolphosphate dehydrogenase-triosephosphate isomerase mixed crystals (Boehringer) 10 mg/ml, diluted 1:5 in distilled water at 0°

Procedure. One-half milliliter of the buffer-K-SH mixture, 0.1 ml FDP, 0.1 ml NADH, 10 μl glycerophosphate dehydrogenase–triosephosphate isomerase solution are diluted to 1.0 ml with distilled water in a 1.0-ml cuvette with a 1.0-cm light path. A suitable dilution of the enzyme (an aliquot containing 0.003–0.03 units of enzyme) is added in a negligible volume, i.e., 10 μl. After rapid mixing, the cuvette is placed in a spectrophotometer equipped with a water-circulating thermostat, and

[14] D. Herbert, H. Gordon, V. Subrahmanyan, and D. E. Green, *Biochem. J.* **34,** 1108 (1940).
[15] J. A. Sibley and A. L. Lehninger, *J. Biol. Chem.* **177,** 859 (1949).
[16] E. Racker, *J. Biol. Chem.* **167,** 843 (1947).
[17] O. H. Lowry, J. V. Passonneau, and M. K. Roch, *J. Biol. Chem.* **236,** 2756 (1961).

absorbance measurements (340 mμ) are made at 10–15 second intervals, or continuously (with the aid of a recorder) until a linear rate is obtained (3–5 minutes). It is assumed that the oxidation of 2 micromoles of NADH (12.44 absorbance units) reflects the cleavage of 1 micromole of fructose diphosphate under these conditions.

Definition of the Unit and Specific Activity. A unit of aldolase activity is defined as that amount of enzyme which catalyzes the aldol cleavage of 1 micromole of FDP per minute under the conditions described in the assay procedure. The specific activity is defined as units of enzyme activity per milligram of protein. The protein concentration may be determined spectrophotometrically by the method of Waddell[18] or by the method of Lowry *et al.*[19] These methods agree within 10% in both crude and semipurified solutions. The concentration of the pure enzyme may be determined from the following relationship: (absorbance 280 mμ)/1.06 = milligrams protein per milliliter (determined in a cuvette with 1.0-cm light path).

Purification Procedure

All procedures are carried out at 0–4° unless otherwise stated.

Step 1. Crude Autolyzate. Twenty pounds of Fleishman's yeast (Standard Brands, Inc.) or Budweiser yeast (Anheuser-Busch, Inc.) are crumbled in equal portions into two 14 liter stainless steel pails, and 2.4 liters of reagent grade toluene at 40–45° are added to each. The pails are then placed in a 45° water bath and their contents are stirred occasionally with a wooden paddle until the yeast liquefies (60–90 minutes). The containers are then removed from the water bath and allowed to stand at room temperature for 2 hours; they are then placed in an ice bath and their contents are rapidly cooled to 10° (20 minutes). The contents of both pails are then poured into a glass battery jar (approximately 20 liters) and mixed with 9.6 liters of cold glass-distilled water. The solutions are stirred for 1 hour at 2–4° and then allowed to stand at this temperature for 18 hours; during this time the phases separate. The bottom (aqueous) phase is carefully siphoned off and centrifuged at 14,600 *g* for 20 minutes. The aqueous extract (the dark, clear middle layer) is removed by means of a vacuum trap. This turbid extract is then mixed with 200 g of Hyflo Supercel (acid washed according to Noda and Kuby[20]), and the mixture is filtered through two sheets of Whatman No. 1 filter paper (32 cm diameter) in a large stainless steel Büchner

[18] W. J. Waddell, *J. Lab. Clin. Med.* **48**, 311 (1956).

[19] O. H. Lowry, N. J. Rosebrough, A. L. Farr, and R. J. Randall, *J. Biol. Chem.* **193**, 265 (1951).

[20] L. Noda and S. Kuby, *J. Biol. Chem.* **226**, 541 (1957).

funnel. To the filtrate is added enough BME and 0.1 M zinc sulfate to obtain a final concentration of 5×10^{-2} M and 1×10^{-4} M, respectively.

Step 2. Ammonium Sulfate Fractionation. To the above solution (11,735 ml) is added per liter, with rapid stirring, 314 g of enzyme grade ammonium sulfate (Mann) to bring the concentration to 50% saturation. The mixture is allowed to stand overnight at 4° and centrifuged at 14,600 g for 15 minutes. The supernatant is mixed with 200 g of acid-washed Hyflo Supercel and filtered through two sheets of Whatman No. 1 filter paper in a stainless steel Büchner funnel, as before. The filtrate (12,635 ml, approximately pH 5.5) is brought to pH 6.8 by the careful addition, with stirring, of cold concentrated ammonium hydroxide (approximately 1.0–1.5 ml per liter of solution). The ammonium sulfate concentration is then brought to 70% saturation by the addition with stirring of 140 g of enzyme grade ammonium sulfate per liter of solution. Immediately after the ammonium sulfate has dissolved, the suspension is centrifuged at 14,600 g for 20 minutes. To the supernatant is again added 200 g of acid washed Hyflo Supercel. The suspension is then filtered using a stainless steel Büchner funnel as before. It is essential that the filtration be completed within 1 hour.

To the very clear supernatant (12,220 ml) is added 75 g of ammonium sulfate per liter to bring the solution to 80% saturation. The suspension is allowed to stand overnight and centrifuged at 14,600 g for 30 minutes. The precipitate is dissolved in 100 ml 0.05 M BME to obtain a protein concentration of approximately 50 mg/ml.

Step 3. Crystallization from Ammonium Sulfate. To the above solution is added, with rapid stirring, 70 ml of saturated (room temperature) ammonium sulfate (total volume 298 ml). Crystals form immediately and are collected by centrifugation at 14,600 g for 30 minutes. The precipitate is dissolved in 60 ml of 0.05 M BME, and the solution is centrifuged at 14,600 g for 15 minutes to remove undissolved material. To the clear supernatant is added 23 ml (total volume 153 ml) of saturated ammonium sulfate (room temperature), and the crystallization is allowed to proceed overnight. The crystals are then centrifuged and resuspended in 75 ml 80% saturated ammonium sulfate containing 0.05 M BME (total volume 133 ml).

Properties

Catalytic Specificity and Kinetic Properties.[2] Yeast aldolase is specific for the substrate D-fructose 1,6-diphosphate. The K_m for EDP is approximately 3×10^{-4} M, and the V_{max} for aldol cleavage of this substrate is approximately 12,500 moles per minute per 70,000 g protein at 30°. The K_m for the triosephosphates DHAP and glyceraldehyde-3-

phosphate is approximately 2×10^{-3} M, and the maximum velocity for FDP synthesis from these substrates is approximately 38,000 moles per minute per 70,000 g protein at 30°.

The most favorable substrate analog, L-sorbose 1,6-diphosphate is cleaved at a rate about three orders of magnitude less than FDP. This molecule acts as a competitive inhibitor of FDP cleavage and has a K_I of 1.3×10^{-4} M. L-Sorbose 1-phosphate and D-fructose 1-phosphate are acted upon even more slowly and have K_I's of 2×10^{-4} M and 1×10^{-3} M, respectively. Glyceraldehyde is a much poorer substrate for aldol condensation than glyceraldehyde 3-phosphate. The K_m is 1×10^{-2} M, and the velocity of condensation is about four orders of magnitude lower than with the phosphorylated derivative. Fructose 6-phosphate is not acted upon significantly, but is a competitive inhibitor ($K_I = 3.8 \times 10^{-3}$ M). At the optimum pH the DHAP-^{3}H exchange reaction occurs at a rate approximately equivalent to the rate of FDP synthesis. In contrast, the enzyme-catalyzed ^{3}H exchange with acetol phosphate is at least three orders of magnitude lower.

Activators and Inhibitors. The activity of yeast aldolase is inhibited by ethylenediaminetetraacetic acid ($K_I = 5 \times 10^{-6}$ M) as well as other chelating agents such as dipyridyl,[21] *o*-phenanthroline,[17] or even cysteine at high concentrations.[17] The inhibitions observed are consistent with the finding of a divalent metal ion in the protein,[1] but there is no evidence for removal of zinc ions by the chelating agent. Thus, the partial reversal of inhibition of chelating agents by metal is probably due to an interaction of the metal ion with the chelating agent to form a complex that is noninhibitory. The inhibition by EDTA is at least partially competitive with the substrate.[16]

Potassium and ammonium ions specifically stimulate the catalytic activity of FDP aldolase. The optimal concentration of both ions is approximately 0.1 M. Sodium and rubidium ions have little effect on the system, and their presence does not affect the stimulation by potassium ions.[2]

The stability of the enzyme is greatly increased in the presence of high concentrations of BME (0.05 M). Though this level is necessary for optimal stability, it is inhibitory for catalytic activity. Therefore, routine assays are performed at 10^{-3} M BME.[5]

pH Optimum. The activity of yeast aldolase (FDP aldol cleavage or aldol condensation) is optimal at pH 7.0–7.5. The DHAP-^{3}H exchange activity, on the other hand, shows a rather sharp optimum at pH 6.0.[12] This behavior contrasts with the catalytic properties of the class I

[21] O. Warburg and W. Christian, *Biochem. Z.* **314,** 149 (1943).

aldolases A or B, in which case the profiles of aldol cleavage and hydrogen exchange activity are coincident. Potassium ions stimulate the exchange reaction at all pH values tested, and the optimum of the exchange system is not markedly altered by potassium; thus, the observed potassium stimulation cannot be ascribed to an adventitious alteration of the pH profile.[1]

Molecular Properties. The crystalline FDP aldolase behaves as an essentially homogeneous protein, both electrophoretically and ultracentrifugally. The electrophoretic mobility is -8.5×10^{-5} cm^2 volts^{-1} sec^{-1} and the sedimentation constant, $S_{20,w}$, is 5.4×10^{-13} sec. The molecular weight by sedimentation analysis has been calculated to be approximately 70,000.[12] Vanderheiden *et al.*[12] reported a molecular weight of 75,000. The crystalline protein contains approximately 1 mole of zinc per mole of enzyme.[12,13,22]

Carboxypeptidase treatment has no effect on the enzymatic activity (in contrast to the effects seen with class I aldolases) even though a number of amino acid residues are released.[12] The presence of two C-terminal leucines suggests the possibility of at least two subunits.[5]

Catalytic Activity. The markedly different general catalytic properties of the enzyme (divalent and monovalent metal requirement, the lack of inhibition of activity on treatment with sodium borohydride in the presence of substrate, markedly different pH profiles, the strong sulfhydryl requirement, etc.) suggest that the catalytically active sites may be basically different from that of aldolase A.[1] Nevertheless, the reaction catalyzed by the yeast enzyme, like that of the muscle enzyme, is an

TABLE I
ALDOLASE FROM YEAST

Fraction	Vol. (ml)	Total protein (mg)	Total units	Specific activity per mg protein	Recovery (%)
Autolyzate	11,735	272,000	1,290,000	4.7	100.0
50% ammonium sulfate supernatant	12,635	263,000	917,000	3.5	71.0
70% ammonium sulfate supernatant	12,220	133,000	418,000	3.1	32.5
80% ammonium sulfate precipitate	298	16,400	385,000	23.6	30.0
First crystallization	153	6,820	371,000	54.2	29.0
Second crystallization	133	1,270	137,000	108.0	11.0

[22] W. J. Rutter and B. L. Vallee, unpublished observations.

ordered sequence: viz., enzyme + FDP ⇌ glyceraldehyde 3-phosphate + enzyme-dihydroxyacetone phosphate (lacking a proton on the α-carbon) ⇌ enzyme + dihydroxyacetone phosphate.[23]

II. *Clostridium perfringens*

By WILLIAM E. GROVES, JOHN CALDER, and WILLIAM J. RUTTER

The fructose diphosphate (FDP) aldolase present in *Clostridium perfringens* is a class II aldolase.[1] Unlike the *Aspergillus niger*[24] and yeast aldolase[13,25,26] which contain tightly bound zinc, the *Clostridium perfringens* enzyme appears readily dissociable: Fe^{++} or Co^{++} must be added for optimal activity.[1,27] As with other known class II aldolases,[1] cysteine or other SH compounds stabilize enzyme activity.[1,4] The present purification procedure minimizes loss of enzyme activity by excluding oxygen where possible and carrying out most procedures in relatively high concentrations of β-mercaptoethanol.

Assay Method

Principle. The assay for aldolase is based on the estimation of dihydroxyacetone phosphate formed from FDP in the presence of limiting amounts of aldolase by conversion to α-glycerophosphate in the presence of NADH and excess triosephosphate isomerase and α-glycerophosphate dehydrogenase.[16,28] This assay may be employed for other class I and class II aldolases. For optimal catalytic activity of the *Clostridium perfringens* enzyme, cobaltous ions and potassium ions are included in the reaction mixture.[4]

Reagents

Buffer–Co^{++}–K^{+}–cysteine solution: Tris(hydroxymethyl)aminomethane 0.1 M, pH 7.5 acetic acid; potassium acetate, 0.2 M; $CoCl_2$, 1.4×10^{-3} M; L-cysteine-HCl, 2×10^{-4} M.

FDP-Na (Sigma) 0.02 M, pH 7.5.

NADH-Na (Sigma) 0.002 M in 0.001 M NaOH. The solution may be stored at 0° to 4° for periods up to 2 weeks.

[23] I. A. Rose, E. L. O'Connell, and A. H. Mehler, *J. Biol. Chem.* **240,** 1758 (1965).

[24] V. Jagannathan, K. Singh, and M. Damodaran, *Biochem. J.* **63,** 94 (1956).

[25] W. J. Rutter and K. H. Ling, *Biochim. Biophys. Acta,* **30,** 71 (1958).

[26] W. J. Rutter and B. L. Vallee, unpublished observations.

[27] R. C. Bard and I. C. Gunsalus, *J. Bacteriol.* **59,** 387 (1950).

[28] G. Beisenherz, H. J. Boltze, T. Bücher, R. Czok, K. H. Garbade, E. Meyer-Arendt, and G. Pfleiderer, *Z. Naturforsch.* **8b,** 555 (1953).

α-Glycerophosphate dehydrogenase-triosephosphate isomerase mixed crystals (Boehringer), diluted to 2 mg/ml in distilled water at 0°.

Procedure. One-half milliliter of the buffer–Co^{++}–K^{+}–cysteine mixture, 0.1 ml FDP, 0.1 ml NADH, and 10 μl glycerophosphate dehydrogenase-triosephosphate isomerase solution are diluted to 1.0 ml with distilled water in a 1.0-ml cuvette having a 1.0-cm light path. A suitably diluted sample of the enzyme (0.003–0.03 unit) is added in a negligible volume (10 μl). After rapid mixing, the cuvette is placed in a spectrophotometer equipped with a water-circulating thermostat at 28°, and changes in absorbance at 340 mμ are read until a linear rate is obtained (3–5 minutes). It is assumed that the oxidation of 2 micromoles of NADH (12.44 absorbance units) reflects the cleavage of 1 micromole fructose diphosphate under these conditions.

Definition of the Unit and Specific Activity. A unit of aldolase activity is defined as that amount of enzyme which catalyzes the aldol cleavage of 1 micromole of FDP per minute under the conditions described in the assay procedure. The specific activity is defined as units of enzyme activity per milligram of protein. The protein concentration is determined either by the spectrophotometric method of Warburg and Christian[29] or by the procedure of Lowry *et al.*[19] The relative values obtained by the two methods can vary considerably, especially if the 280:260 absorption ratio of the sample is <0.6. The method of Lowry is preferred, especially for crude fractions.

Purification Procedure

Growth and Harvest of Bacteria. Clostridium perfringens strain BP6K[30] is grown in a medium containing, per liter, tryptone, 10 g; yeast extract, 10 g; K_2HPO_4, 5 g; glucose, 10 g; cysteine, 2.5 mg; $FeSO_4 \cdot 7\ H_2O$, 10 mg; $CoCl_2 \cdot 5\ H_2O$, 2.1 mg; $MgSO_4 \cdot 7\ H_2O$, 100 mg; $MnSO_4 \cdot 1\ H_2O$, 1.1 mg; $ZnSO_4 \cdot 7\ H_2O$, 2.5 mg; $CuSO_4 \cdot 5\ H_2O$, 2.5 mg; NaCl, 5 mg. The medium is prepared in carboy (17.5 liters) quantities as follows: Appropriate quantities of tryptone, yeast extract, and K_2HPO_4 are added to 17.0 liters of distilled water and sterilized by autoclaving for 30 minutes at 15 psi pressure. The solution is kept anaerobic by immediately adding approximately 100 ml of a liquefied, presterilized mixture of one part petroleum jelly and one part paraffin (vaspar). The other components (glucose, cysteine, and salts) are sterile-filtered separately in 500, 10,

[29] O. Warburg and W. Christian, *Biochem. Z.* **310**, 384 (1941).

[30] Kindly provided by Dr. L. S. McClung, Department of Bacteriology, Univ. of Indiana, Bloomington, Indiana.

10 ml volumes, respectively, and steam heated for 20 minutes to reduce oxygen tension. The solutions are then mixed (total volume 17.5 liters) and inoculated with 1.0 liter of similar medium containing the organism in the log phase of growth (6–10 hours' growth under vaspar). The temperature is maintained at 37°, and the cells are harvested at the end of the log phase (approximately 6 hours) using a water-cooled (20°) air-driven Sharples centrifuge (40,000 rpm). The average yield of *Clostridium perfringens* cells was 6.7 g, wet weight, or 1.7 g, dry weight, per liter of medium. When stored frozen as a paste, the cells retain activity for at least 4 months.

Crude Extract. Fifty grams of frozen *Clostridium perfringens* cells are suspended in 100 ml of a solution of 0.05 *M* potassium phosphate, pH 6.5, —0.05 *M* β-mercaptoethanol and allowed to thaw overnight at 3°. The suspension is then placed in a bath at —10°, sonicated for 30 minutes at 9 amp (Branson 1250-watt sonifier). The solution is then centrifuged at 15,000 *g* at 0° for 30 minutes. The precipitate is discarded. The supernatant is diluted to a concentration of about 20 mg protein per milliliter (final volume, 195 ml).

Acetone Fractionation. The diluted crude extract (195 ml) is added to a 1-liter flask[31] in a —10° bath, and 130 ml of acetone–0.05 *M* β-mercaptoethanol (precooled to —50° by passage through a coil immersed in a Methyl Cellosolve–dry ice bath) is added just fast enough (5–10 ml/min) to prevent freezing. The solution (now 40% acetone) is stirred continuously for 45 minutes and then centrifuged at 15,000 *g* for 30 minutes at —10°. The precipitate is discarded and the supernatant (250 ml) is placed in a 2.0 liter flask.[31] One liter of acetone-0.05 *M* β-mercaptoethanol is added (10–20 ml/min) to bring the concentration to 80% acetone. After equilibration for 45 minutes, the solution is centrifuged at 15,000 *g* for 30 minutes at —10°, and the supernatant is discarded. The precipitate, kept at —10° by means of a sodium chloride-ice bath, is suspended in 110 ml of 0.01 *M* potassium phosphate, pH 6.5, —0.05 *M* β-mercaptoethanol (final temperature: —2° to —5°), and dialyzed overnight against 3–4 changes each of 4.0 liters of the same buffer at 0°.

pH 5.0 Precipitation. To the dialyzate (120 ml obtained from above, maintained at 3°) is added, with stirring, cold 1.0 *M* HCl to bring the pH to 5.0. The solution is stirred for 10 more minutes and centrifuged at

[31] The alcohol fractionation is carried out in a three-necked flask. The center neck contains a stirring shaft with an air tight fitting. One outer neck is fitted with a three-way stopcock for evacuating the flask and filling it with nitrogen. The third neck is fitted with a serum cap containing a 11-gauge needle through which acetone is added.

37,000 *g* for 20 minutes at 0°. The precipitate is discarded, and the supernatant is used for further fractionation.

Protamine Sulfate Precipitation. The supernatant (120 ml) is stirred at 3° while the pH is adjusted to 5.5 with cold 1.0 *M* ammonium hydroxide. A 4.0-ml sample of 1% solution of protamine sulfate in 0.01 *M* potassium phosphate, pH 5.5, —0.05 *M* β-mercaptoethanol is slowly added with stirring. The solution is then stirred for an additional 30 minutes at 3° and centrifuged at 37,000 *g* for 20 minutes at 0°. The precipitate is discarded.

Ammonium Sulfate Fractionation. The supernatant obtained from the protamine sulfate precipitation step (125 ml) is adjusted to pH 6.5 with cold 1.0 *M* ammonium hydroxide. Solid ammonium sulfate, 43.9 g, is added slowly with stirring to bring the solution to 55% saturation. After the ammonium sulfate has been added, the solution is stirred for at least 30 minutes at 3° and centrifuged at 37,000 *g* for 20 minutes at 0°. The residue is discarded.

The supernatant (135 ml) is brought to 80% saturation by adding 24.2 g ammonium sulfate. The suspension is centrifuged at 37,000 *g* for 20 minutes at 0°. The supernatant is discarded and the residue is dissolved in 10 ml ice cold 0.01 *M* potassium phosphate, pH 6.5, —0.05 *M* β-mercaptoethanol, through which nitrogen gas has been bubbled for 10 minutes prior to use. The solution (13 ml) is then dialyzed against two-liter changes (1 hour each) of 0.01 *M* potassium phosphate, pH 6.5, —0.05 *M* β-mercaptoethanol, bubbled with nitrogen. The dialyzate is centrifuged at 32,000 *g* for 20 minutes at 0° and the residue discarded.

Chromatography on DEAE-Sephadex. A DEAE-Sephadex[32] chromatographic column (7.1 $cm^2 \times 30$ cm.) is prepared by gravity packing at room temperature and washed with 0.01 *M* potassium phosphate, pH 6.5, —0.05 *M* β-mercaptoethanol at 3° until the effluent is approximately pH 6.5 and the conductivity is approximately 2 mmho (Radiometer conductivity meter). The enzyme solution is then carefully placed on the column and followed by 40 ml of the 0.01 *M* phosphate, pH 6.5–0.05 *M* β-mercaptoethanol buffer. A linear gradient from 0.01 *M* potassium phosphate to 0.5 *M* potassium phosphate is then applied with both limiting solutions (400 ml each) containing 0.05 *M* β-mercaptoethanol. Fractions of 5.0 ml are collected at a flow rate of 1.0–2.0 ml per minute.

[32] Fifty grams DEAE-sephadex A-25 are slurried in 1.0 liter 0.5 *M* acetic acid, filtered on a Büchner funnel, and washed (by suspending and filtering with a Büchner funnel) successively with 1.0 liter water, 1.0 liter 0.5 *M* NaOH, 1.0 liter water, 2 ×; 0.01 *M* potassium phosphate, pH 6.5 + 0.05 *M* β-mercaptoethanol (2 ×).

Aldolase activity appears in the range of 6.5 mmho to 10.5 mmho (0.12 M to 0.22 M potassium phosphate). The peak tubes 39 through 47 are pooled, and the enzyme is concentrated by bringing the solutions to 80% saturation with ammonium sulfate (0.56 g ammonium sulfate is added per milliliter). The suspension is centrifuged at 32,000 g for 20 minutes at 0°, and the pellet is resuspended in 80% ammonium sulfate —0.05 M β-mercaptoethanol. Such suspensions lose from 10% to 30% of enzyme activity per week when stored at 0°.

Properties

Catalytic Activity. The specific activity of this *Clostridium perfringens* aldolase preparation is as high as that of the crystalline aldolase A from rabbit muscle (a class I aldolase), but is only approximately one-quarter that of the crystalline yeast aldolase (another class II aldolase). The K_m for FDP (3×10^{-4} M) is independent of pH between 6.5 and 8.75.[4] The K_m of aldolase A from rabbit muscle markedly increases at pH values below 7.0.[33]

The enzyme catalyzes a dihydroxyacetone phosphate-^{3}H exchange reaction, but even at the pH optimum for exchange the rate of the exchange reaction is less than 20% of that of the FDP cleavage reaction.[4]

Specificity. Clostridium perfringens aldolase, like yeast aldolase, shows a high degree of specificity for fructose diphosphate. Fructose 1-phosphate and fructose 6-phosphate are acted upon at least three orders of magnitude less readily than FDP, but both act as competitive inhibitors of the enzyme ($K_I = 8.3 \times 10^{-3}$ M and 3.8×10^{-3} M, respectively).

pH Optimum. The pH profile for FDP cleavage exhibits a maximum at pH 7.5–7.8. The activity for the DHAP-^{3}H exchange reaction, on the other hand, is optimal at pH 6.0 to 6.3.[4] The pH profiles of activity, therefore, are similar to those of yeast aldolase[2]; however, the comparative rates (aldol cleavage/^{3}H-DHAP exchange) are about 1.7 for the yeast enzyme[2] and 5.0 for the *Clostridium perfringens* enzyme.[4]

Inhibitors and Activators. The enzyme is stimulated at least fivefold by the inclusion of approximately 10^{-2} M $CoSO_4$ in the assay medium.[4] A larger activation is elicited by 5×10^{-3} M Fe^{++} ions as measured by the colorimetric hydrazine assay (the presence of ferrous ions precludes assay by the coupled α-glycerophosphate dehydrogenase system).[4]

Clostridium perfringens aldolase activity, like other class II aldolases studied, is stimulated by monovalent cations.[4] Potassium and ammonium ions at optimal concentrations (10^{-1} M) stimulate the activity at least

[33] A. H. Mehler, *J. Biol. Chem.* **238,** 100 (1963).

fivefold. Sodium ions also stimulate the activity (up to threefold) at optimal concentrations (10^{-1} M). Lithium and rubidium ions are slightly inhibitory.

Clostridium perfringens aldolase is stabilized by sulfhydryl compounds (0.05 M β-mercaptoethanol is required for optimal stability[34]), moreover, an SH compound or a reducing agent such as ascorbic acid is required for optimum catalytic activity. In the presence of 7×10^{-4} M $CoSO_4$, the optimum concentration of cysteine is 2×10^{-5} M, of ascorbate, 10^{-5} M. Glutathione shows little, if any, effect at the concentrations tested.[4]

Inhibition by Chelating Agents. Enzyme activity is inhibited by chelating agents like ethylenediaminetetraacetic acid, *o*-phenanthroline and sodium pyrophosphate (the K_I's are 5×10^{-4} M, 8×10^{-4} M, and 5×10^{-3} M, respectively). The inhibition by EDTA is partially prevented by FDP.[4]

Molecular Properties. The sedimentation constant ($S_{20,w}$) estimated from sucrose density gradient centrifugation is 5.6×10^{-13} sec.[4] This value is similar to that for the yeast enzyme,[2] and it is, therefore, predicted that the molecular weight is also similar (70,000–75,000).

TABLE II
Clostridium perfringens Aldolase Purification Procedure

Fraction	Volume (ml)	Total protein (mg)	Total activity (units)	Specific activity (units/mg)	Recovery (%)
Sonicate	195	3980	4720	1.2	100
80% acetone	110	1870	3320	1.8	71
pH 5.0 supernatant	120	767	3260	4.3	70
Protamine sulfate supernatant	125	456	3410	7.5	72
80% $(NH_4)_2SO_4$ ppt.	13	239	2830	11.8	60
DEAE-sephadex combined eluate	40	47	1560	27.0	33

III. Aldolase B from (Adult) Rabbit Liver

By T. V. Rajkumar, B. M. Woodfin, and W. J. Rutter

Fructose 1,6-diphosphate (FDP) $\rightleftharpoons$ dihydroxyacetone phosphate (DHAP) + D-glyceraldehyde 3-phosphate (GAP)

Fructose 1-phosphate (F-1-P) $\rightleftharpoons$ dihydroxyacetone phosphate + D-glyceraldehyde

The aldolases present in vertebrate tissues, are class I aldolases.[1] Aldolase A and aldolase B have strong structural and catalytic similari-

[34] J. R. Hunsley, J. Calder, and W. J. Rutter, unpublished observations.

ties,[6] but are distinct entities since fingerprint patterns of chymotrypsin-trypsin digests are not identical. The enzymes can also be distinguished by their substrate specificity (the FDP:F-1-P activity ratio is about 50 and 1 for aldolases A and B, respectively), and immunological properties (antibodies prepared against aldolase B precipitate B and completely inhibit its catalytic activity, but do not cross react with aldolase A; conversely, antibodies prepared against aldolase A react with aldolase A, but not with B).

Liver aldolase was first crystallized from the bovine by Peanasky and Lardy.[35] The following purification procedure, developed for rabbit liver, can also be employed for isolation of the enzyme from livers of other species. A modification of the substrate elution technique of Pogell[36] is utilized in the procedure.

Assay Method

Principle. Limiting amounts of aldolase are incubated in an appropriate medium with excess glycerol phosphate dehydrogenase, triose phosphate isomerase, and NADH. Under these conditions, the triose phosphates formed from FDP or F-1-P are quantitatively converted to α-glycerol phosphate with the concomitant oxidation of a stoichiometric amount of NADH to NAD.[16,28] The latter conversion is followed spectrophotometrically by loss of absorption at 340 mμ. The quantity of NADH oxidized (as determined from the molar extinction coefficient of NADH) is equated with the triose phosphates formed, and in turn related to the moles of substrate cleaved during the incubation period (2 moles of NADH are oxidized for every mole of FDP cleaved, and 1 mole of NADH is oxidized for each F-1-P cleaved).

Reagents

Glycylglycine buffer adjusted to pH 7.5, 0.5 M
Fructose 1,6-diphosphate sodium salt, 0.02 M, pH 7.5 (Sigma)
Fructose 1-phosphate solution, 0.10 M, pH 7.5 (Sigma)
α-Glycerophosphate dehydrogenase–triosephosphate isomerase mixed crystals (Boehringer), 10 mg/ml
β-NADH, disodium salt (Sigma)

Procedure. The incubation mixture contains the following components at the concentrations indicated: glycylglycine 49 micromoles; NADH,

[35] R. J. Peanasky and H. A. Lardy, *J. Biol. Chem.* **233**, 365 (1958).
[36] B. M. Pogell, *Biochem. Biophys. Res. Commun.* **7**, 225 (1962).

0.1 micromole; glycerophosphate dehydrogenase–triosephosphate isomerase mixture, 50 μg; FDP (when present), 2 micromoles; F-1-P (when present), 10 micromoles; 0.003–0.03 units of enzyme; water to a volume of 1 ml. The assay is carried out at pH 7.5, 28°.

It is usually convenient to prepare, in sufficient quantity for one day's assays, a cocktail containing the components of the assay minus substrate and enzyme. (For 50 individual assays; 0.1 ml glycerophosphate dehydrogenase–triosephosphate isomerase mixture and 6 mg NADH are added to 4.9 ml glycylglycine buffer.) This mixture is kept at 0°, and aliquots are raised to 28° no more than 1 hour before use. The cocktail should not be used more than 24 hours after preparation.

For assay, 0.1 ml of the above cocktail, 0.1 ml FDP (or F-1-P) are diluted with water to 1.0 ml in a 1.0-ml cuvette having a 1-cm light path. The enzyme solution, suitably diluted, is then added in a negligible volume; e.g., 10 μl. After rapid mixing, the rate of decrease in absorbance at 340 mμ is measured spectrophotometrically (most conveniently by means of a recorder, but absorbance measurements at 30-second intervals for several minutes are also satisfactory). The extinction coefficient of NADH at 340 mμ is 6.22×10^6 cm^2 mole[37]; therefore, in this assay, absorbance changes of 12.44 and 6.22 correspond to the cleavage of 1 micromole of FDP and F-1-P, respectively. The rate of cleavage of substrate in micromoles per minute is calculated from the data.

Definition of the Unit and Specific Activity. A unit of aldolase B activity is defined as that amount of enzyme which catalyzes the aldol cleavage of 1 micromole of D-fructose 1,6-diphosphate, or 1 micromole of D-fructose 1-phosphate per minute under the conditions described in the assay procedure. The specific activity is defined as units of enzyme activity per milligram of protein. Protein in crude fractions is determined by the spectrophotometric procedure of Warburg and Christian,[29] or by the method of Lowry *et al.*[19] These methods usually agree within 10%, but the spectrophotometric method cannot be employed in extracts which are turbid or which contain a high level of hemoglobin. The concentration of pure aldolase B is accurately estimated by the following relationships (valid for cuvettes having 1-cm light path):

$$\frac{\text{Absorbance 280 m}\mu}{0.85} = \text{milligrams protein per milliliter}^{38}$$

[37] A. Kornberg and B. L. Horecker, *in* "Biochemical Preparations" (E. E. Snell, ed.), Vol. III, p. 27. Wiley, New York, 1953.

[38] A value of 0.85 was arrived at from the ratio of absorbance at 280 mμ (A_{280}) in a 1.0-cm light path to dry weight of protein; i.e., 0.85 absorbance units = 1.0 mg protein per milliliter.

Isolation Procedure

The following manipulations are carried out at 0° unless otherwise stated. The addition of ethanol, and pH adjustments are made with good mechanical stirring, but foaming is avoided.

Step 1. Crude Extract. About 16 or 17 adult rabbit livers (approximately 1500 g) are homogenized individually in three volumes (3 ml/1 g wet weight) of 25% glycerol containing 10^{-3} *M* EDTA, pH 8.0, for 10 seconds in a blendor at 3°. The pooled homogenate is centrifuged at 40,000 *g* for 45 minutes at 0° to remove particulate material. The supernatant (about pH 6.5) is adjusted to pH 7.1 by the careful addition of approximately 24 ml of 1 *N* NaOH, and then passed through cheesecloth to remove floating lipid material.

Step 2. Alcohol Fractionation. The supernatant from step 1 (4.1 liters) in a 12-liter stainless steel container equipped with mechanical stirring, is placed in a —10° bath. Ethanol, 4180 ml 95% (enough to raise the ethanol concentration to 48%[39]), precooled by passage through a glass coil immersed in a dry ice-Methyl Cellosolve bath at about —45° is added just fast enough (~40 ml/minute) to preclude freezing of the solution. After it has been stirred for approximately 30 minutes, the solution is centrifuged at 15,000 *g* for 15 minutes at —12°. The precipitate containing 10–15% of the total activity is discarded. The supernatant solution (approximately 7400 ml, —10°) is then brought to 60% ethanol concentration by adding 2530 ml 95% ethanol according to the procedure outlined above. After 30 minutes of stirring, the solution is centrifuged at 15,000 *g* for 15 minutes at —12°. The centrifuge cups are inverted over absorbent paper for 1–2 minutes to drain the alcohol solution from the precipitate. The latter is then dissolved in 450 ml 0.01 *M* Tris-Cl, 0.001 *M* EDTA, pH 7.4 by adding successive 50–100 ml aliquots serially to the centrifuge cups. The solutions are maintained near freezing by keeping the cups in a NaCl-ice bath at —10° during dissolution. The initial volume added to the precipitate (450 ml) and the final volume of the cloudy solution (530 ml) are carefully measured. The percentage of alcohol in the solution is estimated from the increased volume, assuming the volume of the precipitate is 60% ethanol and neglecting volume changes.[39] The solution is then centrifuged at 14,600 *g* for 10 minutes at 0°, and the precipitate is reextracted with 130 ml of the Tris-Cl EDTA solution as before; the precipitate is discarded and the

[39] The amount of 95% ethanol to be added was calculated using the following expression: $(v_1 \times a) + 95v_2 = (v_1 + v_2)b$, where v_1 = volume of extract, a = concentration of alcohol in v_1, v_2 = volume of 95% ethanol to be added, and b = final concentration of alcohol desired.

combined supernatants (637 ml, 2.2 units/ml, 8.9% ethanol) are usually carried through the procedure. The solution, however, may be stored frozen at this stage without appreciable loss of enzyme activity.

Step 3. Heat Treatment in the Presence of 10% Ethanol. The concentration of alcohol of the solution from step 2 is adjusted to 10% by adding approximately 10 ml of ethanol at —45°.[39] The solution is then placed in a stainless steel container and brought to 40°, incubated for 10 minutes, and then cooled immediately to 2° and centrifuged at 14,600 *g* for 10 minutes at 0°. The precipitate is discarded. The supernatant can be used immediately for further fractionation described below, or at this stage it can be stored either frozen for about a week, or at 3°–5° for 24–48 hours.

Step 4. Chromatography on Cellulose Phosphate.[40] Cellulose phosphate 0.86 meq/g capacity is prepared for chromatography as follows: 50 g of dry cellulose phosphate which passes through a 60-mesh screen is slurried in 2 liters of 2 *M* NaCl and filtered on a Büchner funnel. It is resuspended in 2 liters of 2 *M* NaCl and allowed to stand for 30 minutes and again filtered. The cellulose phosphate is then resuspended and washed four to five times in deionized water until the filtrate is free of chloride ions ($AgNO_3$ test). After each suspension, the slurry is allowed to stand for 15–20 minutes and the fines are decanted before filtering on the Büchner funnel. The residue is then washed twice with 2 liters of 95% ethanol (the suspension is allowed to stand at least ½ hour before filtration and each wash). The material is then suspended in glass-distilled water and allowed to stand overnight.

The cellulose phosphate is then filtered and washed successively with 2-liter batches of the following solutions: (a) 0.2 *M* Tris, (b) glass-distilled water, (c) 0.2 *M* Tris-Cl buffer, pH 5.5, (d) glass-distilled water, (e) 0.1 *M* Tris-Cl, pH 7.4, (f) glass-distilled water, (g) 0.01 *M* Tris-Cl, 0.001 *M* EDTA, pH 7.4.

The washed cellulose phosphate is slurried in 0.01 *M* Tris-Cl, 0.001 *M* EDTA, pH 7.4, and poured into a 4-cm glass column to a height of 40 cm (gravity packed at room temperature). The column is then placed in the cold room at 3°–5° and washed with the latter buffer until the conduc-

[40] If chromatography is carried out on a deaerated cellulose phosphate column in the presence of 10^{-2} *M* β-mercaptoethanol, aldolase B activity is eluted from the column sharply, but at a somewhat lower specific activity (approximately 0.5). The activity of the enzyme increases significantly on standing in the cold for several hours (to approximately 0.8). On crystallization from ammonium sulfate in the presence of β-mercaptoethanol by the dialysis procedure outlined above, the activity further increased and the specific activity of the final product (1.2–1.4) approached that obtained by the usual purification procedure. β-Mercaptoethanol does not affect the activity of the crystalline enzyme in solution.

tivity of the effluent is approximately 400 μmho (2500 ohms). The enzyme-containing solution obtained from step 4 (1455 units of activity, 8.9 g of protein in 10% ethanol) is then placed on the column and subsequently washed with 0.05 *M* Tris-Cl, 0.005 *M* EDTA, pH 7.4 at approximately 1 ml per minute. The hemoglobin band is eluted after approximately 1500 ml (about 10% of the aldolase activity is usually associated with it). A linear gradient elution with the substrate, FDP, by a modification of the procedure of Pogell[36] is then carried out. The mixing flask contains 1 liter of 0.06 *M* Tris-Cl, pH 7.4, and the reservoir 1 liter of 0.033 *M* Tris-Cl, pH 7.4, and 0.0025 *M* FDP sodium salt. The rate of elution is 0.5 ml/min, and fractions of 15 ml are collected. Enzyme activity appears between tubes 30 and 60 [the conductivity of these fractions ranged from 1.6 to 1.8 millimhos (555–625 ohms)]. The peak fraction contained 6.6 units/ml at a specific activity of 1.55. Those fractions which have a specific activity of one or greater than (representing about 40–50% of the activity applied on the column) are pooled.

Step 5. Crystallization from Ammonium Sulfate. The pooled fractions obtained from the cellulose phosphate column are brought to >70% saturation by adding 49 g ammonium sulfate for each 100 ml. After standing for about 12 hours at 0°, the turbid solution is centrifuged at 32,000 *g* for 30 minutes at 0°. The precipitate is dissolved in a minimal amount of cold 0.01 *M* EDTA, pH 7.5 (20–40 mg protein per milliliter) and centrifuged at 32,000 *g* for 10 minutes to remove undissolved (denatured) protein. The solution is then dialyzed in the cold for 8–12 hours against 500 ml of 10% saturated ammonium sulfate, 10^{-3} *M* EDTA, pH 7.5. At the end of this period, ammonium sulfate saturated at 0° containing 10^{-3} *M* EDTA, pH 7.5 was added dropwise and with stirring at about a rate of 0.1 ml per minute to the dialyzing solution. Fine crystals start to appear inside the dialysis bag after approximately 450 ml of saturated ammonium sulfate have been added (bringing the solution to approximately 55% ammonium sulfate saturation). Fifty milliliters of saturated ammonium sulfate (to 58% saturation) is then added as before, and the bag of crystals is left in the medium for 24 hours. The crystals are centrifuged at 32,000 *g* for 10 minutes, dissolved in a minimal amount of 0.01 *M* EDTA, pH 7.5, and recentrifuged to remove undissolved denatured protein. The solution is then dialyzed against 10% ammonium sulfate, and the crystallization repeated as before.

Properties

Catalytic Specificity. Aldolase B catalyzes the aldol cleavage of fructose diphosphate and fructose 1-phosphate at equivalent rates (460 moles/minute per 150,000 g protein). The activity of aldolase B toward

FDP is about an order of magnitude lower than that of aldolase A, and the activity toward F-1-P is about fourfold greater. The K_m's for FDP and F-1-P, dihydroxyacetone-phosphate, and glyceraldehyde-3-phosphate are 2×10^{-6} *M*, 8×10^{-4} *M*, 3.7×10^{-4} *M*, and 3×10^{-4} *M*, respectively.[6] Aldolase B catalyzes the reversible condensation of formaldehyde with dihydroxyacetone phosphate to erythrulose phosphate at a rate severalfold greater than that by aldolase A. The K_m for formaldehyde is approximately 10^{-2} *M*.[41] A number of aldehydes have been shown to inhibit the aldol cleavage of fructose 1-phosphate and fructose diphosphate by aldolase B. Some of these compounds are, no doubt, also substrates for the reaction.[42]

The kinetic parameters reported here for aldolase B differ from those reported[35] for the bovine liver enzyme. The major discrepancies are believed to be due to the different assays employed in the studies.

Inhibitors. Aldolase B, like aldolase A, is inhibited by a number of ions, but there is a considerable (perhaps metabolically significant) difference in the specificity of inhibition of the two enzymes. Aldolase B is inhibited by AMP > ADP and is not inhibited by ATP. On the other hand, aldolase A is competitively inhibited by ATP > ADP > AMP; magnesium reverses the ATP inhibition.[42] A number of aldehydes show inhibition of the aldol cleavage of fructose 1-phosphate and fructose diphosphate. The three-carbon aldehydes are more inhibitory than the two-carbon aldehydes for aldolase B; therefore, it has been suggested that this enzyme has a binding site for the C-6 methylene group.[42]

pH Optimum. In contrast to aldolase A, which has a broad pH optimum, aldolase B shows an optimum between 7.5 and 8.0 with FDP as a substrate. The optimum is even sharper near pH 7.5 with F-1-P as a substrate.[6] The pH profile of the ^{3}H-DHAP exchange reaction is similar to that of FDP cleavage.

Catalytically Active Sites. There are at least two DHAP binding sites per molecule protein as indicated by the formation of stably bound DHAP in the presence of borohydride at pH 5.0. The isolation of β-glyceryl lysine from the hydrolysis products of the bound DHAP-aldolase implicate lysine at or near the catalytically active site of the enzyme and suggest the formation of a Schiff base intermediate during the course of the reaction.[43] Kinetic studies suggest that the aldol cleavage of fructose 1-phosphate by aldolase B is an ordered reaction in which glyceraldehyde is released before the enzyme dihydroxyacetone-phosphate complex is dissociated.[42]

[41] O. C. Richards and W. J. Rutter, *J. Biol. Chem.* **236**, 3193 (1961).
[42] P. D. Spolter, R. C. Adelman, and S. Weinhouse, *J. Biol. Chem.* **240**, 1327 (1965).
[43] D. Morse, C. Y. Lai, B. L. Horecker, T. Rajkumar, and W. J. Rutter, *Biochem. Biophys. Res. Commun.* **18**, 679 (1965).

Molecular Properties

The crystalline protein is homogeneous by electrophoretic and ultracentrifugal criteria (electrophoretic mobility $= -0.91 \times 10^{-5}$ cm^2 volt^{-1} sec^{-1} in 0.015 M sodium phosphate, 0.01 M EDTA, pH 7.3; $S_{20,w} = 7.56 \times 10^{-13}$ sec; $D_{20,w} = 4.35 \times 10^{-7}$ cm^2 sec^{-1}). The partial specific volume is calculated to be 0.73 ml g^{-1}, and the molecular weight is 154,000.[6]

Acid treatment ($<$pH 3.0) of the protein changes the sedimentation constant to 2.1×10^{-13} sec. This value corresponds with the sedimentation constant of the subunits having an average molecular weight 50,000 formed from aldolase A under acidic conditions.

Aldolase B, like aldolase A, contains at least three carboxy terminal tyrosine residues. A kinetic analysis of the release of amino acids by carboxypeptidase suggests the presence of at least two different polypeptide chains in aldolase B, both of which are different from the chains present in aldolase A.[44]

Distribution. The substrate specificity and immunological reactivity of aldolase activity in crude extracts of mammalian tissues suggest they contain aldolase A and/or B instead of other aldolase variants.[45] Adult liver, intestines, and kidney are rich in aldolase B, but the earlier embryonic rudiments of these tissues contain primarily aldolase A.[46]

TABLE III

PREPARATION OF RABBIT LIVER ALDOLASE[a]

Fraction	Volume (ml)	Total protein (mg)	Total units	Specific activity, (units/mg protein)	Recovery (%)
Crude extract	4095	184,000	2700	0.015	100.0
48% ethanol supernatant	7600	56,500	1589	0.028	58.0
60% ethanol precipitate	365	9,140	1230	0.132	45.0
Filtrate after heat treatment	350	8,000	1135	0.142	42.0
Combined eluates from cellulose phosphate column	138	381	543	1.43	20.0
Crystallization	14.5	377	550	1.46	20.4

[a] Average of three experiments.

[44] T. Rajkumar and W. J. Rutter, unpublished observations.

[45] R. E. Blostein and W. J. Rutter, *J. Biol. Chem.* **238**, 3280 (1963).

[46] W. J. Rutter and C. S. Weber, *in* "Developmental and Metabolic Control Mechanisms and Neoplasia," Proc. 19th Ann. Symp. Fundamental Cancer Res. p. 195. Williams & Wilkens, Baltimore, Maryland, 1965.

[88] Transaldolase

By O. TCHOLA and B. L. HORECKER

Fructose 6-P + erythrose 4-P ⇌ sedoheptulose 7-P + glyceraldehyde 3-P

Introduction[1]

Candida utilis contains very high transaldolase activity, and a crystalline preparation of the enzyme has been obtained from this source.[2] However, some commercial preparations of *C. utilis* fail to yield the crystalline enzyme by the above procedure, despite the fact that they contain high transaldolase activity. This can now be attributed to the presence of three chromatographically distinct forms of the enzyme in extracts of either dried or frozen cells. These forms, which are readily separated on DEAE-Sephadex columns, will be referred to as types 1, 2, and 3, in their order of elution from the column. The crystalline enzyme previously described[2] belongs to type 3. The dried cells now available contain approximately 55% of type 1, 12% of type 2, and 30% of type 3, while frozen cells from the same source contain approximately 90% of the total activity as type 1 and only minor amounts of types 2 and 3. Attempts to convert type 1 to type 3 have been unsuccessful, and for this reason a new procedure has been developed for the isolation in pure form of type 1. In this chapter we will describe this procedure and also a new method for the isolation of crystalline type 3.

Assay Method

Principle. Glyceraldehyde 3-phosphate produced in the reaction is converted to dihydroxyacetone phosphate by triosephosphate isomerase. This product is measured with DPNH and α-glycerophosphate dehydrogenase. Under appropriate conditions the rate of oxidation of DPNH is proportional to the quantity of transaldolase present. The reaction is measured spectrophotometrically by the change in absorbance at 340 mμ.[3]

[1] Taken from a dissertation to be submitted by O. Tchola to the Sue Golding Graduate Division of Medical Sciences, Albert Einstein College of Medicine, in partial fulfillment of the requirements for the degree of Doctor of Philosophy.

[2] S. Pontremoli, B. D. Prandini, A. Bonsignore, and B. L. Horecker, *Proc. Natl. Acad. Sci. U.S.* **47**, 1942 (1961).

[3] P. T. Rowley, O. Tchola, and B. L. Horecker, *Arch. Biochem. Biophys.* **107**, 305 (1964).

Reagents

D-Fructose 6-phosphate (0.14 *M*). Dissolve 168 mg of D-fructose 6-phosphate, Ca salt 70% pure, in water. Add one drop of 2 *N* HCl and adjust the volume to 2 ml.

DPNH (0.01 *M*). Dissolve 10 mg of reduced diphosphopyridine nucleotide in 1.0 ml of 0.001 *N* NaOH.

Erythrose 4-P (0.01 *M*). Prepare D-erythrose 4-phosphate by the oxidation of glucose 6-phosphate with lead tetraacetate according to the procedure of Baxter *et al.*[4] Store the final product as a solution (0.01 *M* at pH 2–3) in the cold. Commercial preparations of erythrose 4-P may also be used.

α-Glycerophosphate dehydrogenase-triosephosphate isomerase. A mixture of the two enzymes (10 mg/ml) is commercially available.[5]

TEA-EDTA buffer. Triethanolamine-HCl buffer, 0.04 *M*, pH 7.6, is made 0.01 *M* with respect to EDTA.

Procedure. To 0.94 ml of TEA-EDTA buffer in a quartz cuvette add 0.02 ml of fructose 6-phosphate, 0.01 ml of DPNH, 0.02 ml of erythrose 4-phosphate, 0.001 ml of the suspension of α-glycerophosphate dehydrogenase-triosephosphate isomerase, and 0.01 ml of transaldolase, diluted to give a change in absorbance of 0.005–0.040 per minute. Measure the change in absorbance at 340 mμ at 25°. Run a control cuvette without transaldolase, and correct for the oxidation of DPNH in the absence of transaldolase.

Definition of Unit and Specific Activity. One unit is defined as the quantity of enzyme necessary to cleave 1 micromole of fructose 6-phosphate per minute under the conditions of the assay. Specific activity is expressed as units per milligram of protein. Determine protein by the turbidimetric method of Bücher.[6]

Analytical Column for Chromatography of Transaldolase. Suspend DEAE-Sephadex A-50 (Pharmacia Fine Chemicals Inc.) in water and allow to swell overnight. Remove the fine particles by repeated decantation. Wash the resin successively with 0.5 *N* HCl, water, 0.5 *N* NaOH, and water, using a large suction flask to filter. Transfer the washed resin to a large beaker, suspend in water, and add 0.5 *M* Na_2HPO_4 until the pH reaches 6.5. Then dilute the suspension until the Na_2HPO_4 concentration

[4] J. N. Baxter, A. S. Perlin, and F. J. Simpson, *Can. J. Biochem. Physiol.* 37, 199 (1959).

[5] Boehringer and Sons, Mannheim, Germany.

[6] T. Bücher, *Biochim. Biophys. Acta* 1, 292 (1947).

is 0.05 M. Allow the resin to settle, decant the supernatant fluid, and resuspend it twice in fresh 0.05 M phosphate buffer, pH 6.5. Store it in the cold. To prepare a column, decant the supernatant and dilute the stock DEAE-Sephadex suspension at room temperature with 4 volumes of 0.05 M phosphate buffer, pH 6.5. Allow the suspension to settle for 20 minutes, decant the fine particles and repeat the dilution and decanting procedure. Pack a column 0.8×26 cm with the freshly settled DEAE-Sephadex by a single addition of the slurry to the column which has previously been filled with buffer. Store the column in the cold overnight. To this column add 1 ml of transaldolase solution containing no more than 40 mg of protein which has previously been dialyzed against 0.05 M sodium phosphate buffer, pH 6.5. Elution is begun as soon as the enzyme is on the column, using a linear gradient prepared with 100 ml of 0.1 M KCl and 100 ml of 0.2 M KCl, both in 0.05 M phosphate buffer, pH 6.5. Collect fractions (2–4 ml) and analyze for transaldolase. Type 1 transaldolase can be expected to appear at 0.105 M KCl, type 2 at 0.13 M KCl, and type 3 at 0.15 M KCl. When the analytical column is applied to extracts, these are first concentrated with ammonium sulfate to contain 150–200 units per milliliter before they are placed on the column.

Purification Procedure for Transaldolase Type 1

Step 1. Extraction. All operations are carried out in the cold. Break a frozen cake of cells[7] into small pieces and, while still frozen, pass them through a household-type food and meat chopper. To each 300 g of the ground material add 60 ml of 0.5 M $NaHCO_3$ (at room temperature) and 120 ml of cold water. Blend the mixture in a Waring blendor at high speed for 6 minutes, not including the time required for the mixture to begin to revolve freely. To the suspension add 413 ml of cold water. Let the suspension stand 5 minutes and centrifuge in the large capacity head of the model PR-2 International centrifuge at 4500 rpm for 10 minutes at 0°. Repeat the extraction and centrifugation procedures 2 times and combine the supernatant solutions and store in the cold (Extract).

Step 2. Acetone Fractionation. Carry out this step in two batches. To 1 liter of extract at 4° add 5 N acetic acid (approximately 10 ml) until the pH is 4.8. Transfer the mixture to a freezing bath at —10° and set up for mechanical stirring. When the temperature reaches 1° add 300 ml of acetone, precooled in dry ice to —70°, slowly (approximately 1.5

[7] *Candida utilis,* frozen or dried at low temperature, can be purchased from the Lake States Yeast and Chemical Division of the St. Regis Paper Co., Rhinelander, Wisconsin.

minutes is required), stirring all the while. Centrifuge the suspension immediately for 10 minutes in a centrifuge cooled to —10°. To the supernatant solution add 250 ml of —70° acetone in the same way. Centrifuge again at once and discard the precipitate. To the supernatant solution add 500 ml of —70° acetone over a period of 3 minutes. After the last addition is complete, continue stirring the mixture for 12 minutes at —10° and then collect the precipitate by centrifugation. Dissolve the precipitate in 80 ml of cold water and adjust the pH to 7.0 with 0.35 ml. of 1 *N* NaOH. Repeat the acetone fractionation procedure with the remaining 1 liter of extract.

Combine the last solutions and adjust the volume to 167 ml. To this solution add 49.3 g of ammonium sulfate and collect the precipitate by centrifugation.[8] To the supernatant solution from the first ammonium sulfate precipitate add 30.2 g of ammonium sulfate, collect the precipitate by centrifugation, and dissolve it in a minimum volume of cold TEA-EDTA buffer (see Assay Procedure for the composition of this buffer). This solution can be stored for several weeks at —16° (acetone fraction).

Step 3. Calcium Phosphate Gel Adsorption. Combine two preparations at the stage of acetone fraction and dialyze at 2° against cold 5 m*M* sodium phosphate buffer, pH 6.5. Change the dialyzing solution (1 liter) once during the overnight dialysis. Add 52.4 ml of dialyzed acetone fraction to 240 ml of calcium phosphate gel,[9] 28 mg/ml, slowly with stirring. The precise amount of gel required is determined by small pilot experiments as the minimum amount necessary to adsorb all of the en-

[8] For the ammonium sulfate fractionations use Fisher Granulated ACS Grade ammonium sulfate. Allow each precipitate to stand in ice for 10 minutes before centrifugation. Collect the precipitates by centrifuging at 0° for 10 minutes in an International centrifuge, Model PR-2, at 10,000 rpm.

[9] Prepare the gel according to the procedure of Swingle and Tiselius[10] as modified by V. Massey (personal communication). Dissolve 450 g of pure sucrose in a total volume of 2 liters with distilled water. Add 75 g of finely powdered calcium oxide and stir for several hours until this is almost completely dissolved. Bring the temperature to 5° and keep the suspension at this temperature for the remaining operations. Add 50–55 ml of phosphoric acid with vigorous stirring until the pH is 9.5 (but not below). This addition should require about 1 hour. Add 2 liters of cold distilled water and stir for several hours. Allow the precipitate to settle, remove the supernatant solution, and wash the precipitate 8 times over a 4-day period. After the final washing, which is carried out with quartz-distilled water, 1 liter of quartz-distilled water is added and aliquots are taken for dry weight determination. The gel is then diluted with quartz-distilled water to 30 mg/ml and stored in the cold in a dark bottle. For the purification of transaldolase, the pH of the freshly prepared gel must be 6.2, and it is adjusted with 8.5% phosphoric acid. With 2- to 3-month-old gel (pH 6.9–7.0), pH adjustment does not appear to be necessary.

[10] S. M. Swingle, and A. Tiselius, *Biochem. J.* 48, 171 (1951).

zyme activity. Keep the suspension for 10 minutes in the cold, centrifuge briefly, and elute the enzyme from the gel with two 240-ml batches of cold 0.01 M sodium phosphate buffer, pH 7.0. Combine the 2 eluates (calcium phosphate gel eluate).

Step 4. Acid Ammonium Sulfate Fractionation. To the eluate (475 ml) add 178 g of ammonium sulfate. Adjust the pH to 5.0 with 1.3 ml of 2 N acetic acid. Keep the solution in the cold for 10 minutes and centrifuge. To the supernatant solution add 87.5 g of ammonium sulfate, collect the precipitate by centrifugation in 8 centrifuge tubes (40 ml capacity) and dissolve in a minimal amount of cold TEA-EDTA buffer. This solution is stable at $-16°$ for months (acid ammonium sulfate fraction).

Step 5. Chromatography on DEAE-Sephadex. The DEAE-Sephadex is prepared as described for the analytical column. Pack a column (1×19 cm) as before and store in the cold overnight. In the meantime, dialyze the ammonium sulfate fraction against 0.01 M KCl in 0.05 M sodium phosphate buffer, pH 6.5, with one change of dialyzate. Place the dialyzed solution (36 ml, containing 708 mg of protein) onto the column in the cold room and begin elution with 0.038 M KCl in 0.05 M phosphate buffer, pH 6.5. After 161 ml of effluent has been collected, change the eluting solution to 0.056 M KCl in 0.05 M phosphate buffer, pH 6.5. After 333 ml of effluent has been collected, change the eluting solution again, this time to 0.07 M KCl in the same phosphate buffer. Collect fractions (5–7 ml) at hourly intervals. Assay each fraction for protein (280 mμ absorption, assuming that one absorbance unit equals 1 milligram of protein) and for transaldolase. Combine the fractions with a specific activity of 64–80 (usually from 240 to 320 ml of effluent). Combine the fractions with a lower specific activity (32–64 units per milligram) separately. Concentrate the two enzyme fractions separately in a flash evaporator under reduced pressure with a bath temperature not exceeding 35°.[11] Centrifuge the concentrated solutions briefly at 2000 rpm and store the supernatant solutions at $-16°$ (DEAE-Sephadex fraction I and DEAE-Sephadex fraction II).

The purification procedure for type 1 is summarized in Table I.

Purification Procedure for Transaldolase Type 3

Step 1. Autolysis. To 200 g of low temperature dried *C. utilis* in a 2-liter conical flask, add 600 ml of 0.1 M $NaHCO_3$, cover the flask, and allow to autolyze at 25° for 7.5 hours. At the end of this time add 3 liters of cold water, centrifuge the mixture at 4500 rpm for 10 minutes (0°), and store the supernatant solution at 5° overnight (autolyzate).

[11] The effluents may also be concentrated by ammonium sulfate precipitation, but this is less convenient.

TABLE I
PURIFICATION PROCEDURE FOR TRANSALDOLASE TYPE 1

Fraction	Volume (ml)	Units/ ml	Total units	Protein (mg/ml)	Specific activity (units/mg)	Recovery (%)
Extract	2010	3.13	6300	5.60	0.56	—
Acetone fraction	21	248.00	5200	47.50	5.2	83
Calcium phosphate gel eluate	475	16.23	7700[a]	1.62	10.0	61
Acid ammonium sulfate fraction	15.6	436.0	6800	27.30	16.0	54
DEAE-Sephadex fraction I	12.5	216.0	2700[b]	2.77	78.0	27[b]
DEAE-Sephadex fraction II	9.5	284.0	2700	4.10	69.2	27

[a] At this point two acetone fractions were combined.
[b] 9900 units, specific activity 14, were placed on the column.

Step 2. Acetone Fractionation. The procedure is carried out in three batches, as described for type 1 (acetone fraction).

Step 3. Calcium Phosphate Gel Adsorption. The procedure is carried out as described for type 1 (calcium phosphate gel eluate).

Step 4. Acid Ammonium Sulfate Precipitation. To 325 ml of calcium phosphate gel eluate add 94.5 g of ammonium sulfate, dissolve, and centrifuge. To the supernatant solution add 34.8 g of ammonium sulfate, adjust the pH to 4.95 with 1.4 ml, of 1 *N* acetic acid, and centrifuge. To the supernatant solution add 52.4 g of ammonium sulfate and again centrifuge. Suspend the last precipitate in about 30 ml of cold 80% saturated ammonium sulfate solution. Centrifuge the suspension and dissolve the precipitate in a minimal amount of cold distilled water (at this point the ammonium sulfate saturation is approximately 0.45, as determined with a conductivity meter) (water extract).

Step 5. Crystallization. Clarify the suspension, if necessary, by centrifugation at 2000 rpm and store it in the cold for 3 days, during which time the enzyme will crystallize together with some amorphous material. Collect the precipitate by centrifugation for 10 minutes at 10,000 rpm at 0°. Remove the supernatant solution, which contains transaldolase type 1. Digest the precipitate in the cold for 90 minutes with 0.2 ml of distilled water. Centrifuge the mixture and quickly rinse the residue, which contains crystalline type 3 transaldolase, with 0.2 ml of cold water. Suspend the crystals in cold 60% saturated ammonium sulfate solution. The crystal suspension is stable for years (crystals).

Recrystallization. For this purpose crystals from a number of preparations having a specific activity of 24–64 are combined and centrifuged; the precipitate is dissolved in cold TEA-EDTA buffer. The solution at this point should contain 600–1000 units per milliliter and have an ammonium sulfate saturation of 0.60, determined with a conductivity meter. Adjust the clear solution to pH 5.25 with 1 N acetic acid and immediately centrifuge to remove the faint turbidity. Adjust the pH to 4.95 with 0.3 ml of 1 N acetic acid and allow to stand at room temperature from 4 to 5 hours. Collect the crystals by centrifugation and suspend them in 4 ml of cold 60% saturated ammonium sulfate solution (recrystallized product).

The purification procedure for type 3 is summarized in Table II.

TABLE II
PURIFICATION PROCEDURE FOR TRANSALDOLASE TYPE 3

Fraction	Volume (ml)	Units/ ml	Total units	Protein (mg/ml)	Specific activity (units/ mg)	Recovery (%)
Autolysate	3000	4.17	12,500	9.70	0.43	—
Acetone fraction	26.5	23.0	6,100	45.1	5.1	49
Calcium phosphate gel eluate	325	12.0	3,900	1.36	8.8	31
Water extract	4.8	563.0	2,700	36.18	15.3	22
First crystals	1.7	223.5	380	3.91	57	3
Recrystallized product	4.0	380.0	1,520[a]	4.75	80	49[a]

[a] 3100 units combined for recrystallization.

Properties

Transaldolase type 1 shows a K_m for D-fructose 6-phosphate of 5.3×10^{-4} M, and for D-erythrose 4-phosphate of 2×10^{-5} M. The corresponding K_m values for type 3 transaldolase are: 7.1×10^{-4} M and 1.9×10^{-5} M, respectively.

The purified proteins are homogeneous by sedimentation velocity analysis. The molecular weights, estimated from sedimentation equilibrium experiments, are 76,100 for type 1 and 65,900 for type 3.

Both enzyme types are stable in the neutral pH range between 0° and 25°.

[89] Transketolase: Clinical Aspects

By MYRON BRIN

Introduction

Transketolase is an enzyme in the metabolic pathway variously referred to as the glucose oxidative pathway, pentose phosphate pathway, pentose shunt, etc. The initial oxidation of glucose 6-phosphate in this system is mediated by glucose 6-phosphate dehydrogenase, and variations in the activity of this enzyme in human erythrocytes are significant in clinical medicine.[1]

Transketolase activity has been studied in human serum and erythrocytes by the measurement of either substrate utilization, or product formation, such as the formation of sedoheptulose phosphate (S-7-P) or hexose phosphate. While ribose 5-phosphate (R-5-P) was the substrate employed in the various assay methods, it should be noted that the transketolase enzyme has a dual requirement for both R-5-P and xylulose 5-phosphate (Xu-5-P) as reactants.

Blood Transketolase Activity

Serum

There is now a vast literature on the measurement of serum or plasma enzymes as aids in the diagnosis of human diseases.[2-4] The theoretical basis for this is their release from tissues which have been traumatized or otherwise inflamed by infection, chemical toxicity, or anaerobioses, or from certain malignancies.

Serum transketolase activity has been shown to vary in patients with liver disease and cancer. Mello[5] demonstrated increased R-5-P utilization in sera from patients with malignancies as compared with sera from healthy individuals or with diseases not involving the liver. Bruns *et al.*[6] observed increased S-7-P formation from R-5-P in serum in uremia and

[1] See P. A. Marks, this volume [25].

[2] "Enzymes in Blood," *Ann. N.Y. Acad. Sci.* **75,** 1–384 (1958).

[3] D. M. Greenberg and H. A. Harper, "Enzymes in Health and Disease." Thomas, Springfield, Illinois, 1960.

[4] J. H. Wilkinson, "An Introduction to Diagnostic Enzymology." Arnold, London, 1962.

[5] M. I. Mello, *Proc. Intern. Symp. Enzyme Chem. Tokyo Kyoto, 1957,* p. 106. Maruzen, Tokyo, 1958.

[6] F. H. Bruns, E. Dunwald, and E. Noltmann, *Biochem. Z.* **330,** 497 (1958).

viral hepatitis with somewhat decreased activity in diabetes. These diagnostic applications for human serum transketolase activity have not come into general use, however.

Erythrocytes

Erythrocyte transketolase activity (RBC-TK) has been used in the evaluation of thiamine (B_1) adequacy in man and experimental animals. Following the demonstration that the isolated transketolase enzyme was associated with thiamine pyrophosphate (TPP),[7,8] Brin and co-workers showed that methylene blue activated a recycling glucose oxidative pathway in rat erythrocytes and that the activity of this pathway was markedly reduced in red cells from B_1-deficient rats and man.[9–11] These studies were done by studying RBC-TK in intact cells, with the aid of ^{14}C-glucose. A hemolyzate assay was developed[12] and modified,[13] in which the sample was preincubated with and without TPP, and which used R-5-P as substrate. The "TPP-effect" was calculated from the pentose utilized and the hexose formed in the reaction. This index reflected the proportion of the transketolase enzyme which was not saturated with TPP. As the B_1-depletion progressed the RBC-TK decreased, and the TPP-effect increased in magnitude. These findings were both sensitive and specific for B_1-deficiency and permitted differentiation from diseases with similar clinical findings.[14,15] In similar studies, Bruns *et al.* in rats[6] and Dreyfus[16] in rats and man showed that B_1-deficient hemolyzates formed less S-7-P from R-5-P than did normal samples. While Brin[10,12] and Dreyfus[16] observed a TPP-effect in both man and in rats, Bruns *et al.*[6] did not in rats. (We have observed large variations between strains in obtaining a TPP-effect in rat erythrocytes.[15] The obtaining of a TPP-effect in deficient cells which have reduced RBC-TK, suggests that B_1-deficient red cells contain apotransketolase protein which is not fully saturated with TPP.[14,15] The effects of B_1-deficiency on RBC-TK parallel similar effects in other tissues.[13]

[7] B. L. Horecker and P. Z. Smyrniotis, *J. Am. Chem. Soc.* **75**, 1009 (1953).
[8] E. Racker, G. de la Haba, and I. G. Leder, *J. Am. Chem. Soc.* **75**, 1010 (1953).
[9] M. Brin, S. S. Shohet, and C. S. Davidson, *Federation Proc.* **15**, 224 (1956).
[10] M. Brin, S. S. Shohet, and C. S. Davidson, *J. Biol. Chem.* **230**, 319 (1958).
[11] S. J. Wolfe, M. Brin, and C. S. Davidson, *J. Clin. Invest.* **37**, 1476 (1958).
[12] M. Brin, M. Tai, A. S. Ostashever, and H. Kalinsky, *J. Nutr.* **71**, 273 (1960).
[13] M. Brin, *J. Nutr.* **78**, 179 (1962).
[14] M. Brin, *Ann. N.Y. Acad. Sci.* **98**, 528 (1962).
[15] M. Brin, *J. Am. Med. Assoc.* **187**, 762 (1964).
[16] P. M. Dreyfus, *New Engl. J. Med.* **267**, 596 (1962) [as discussed by M. Brin, *New Engl. J. Med.* **267**, 1265 (1962)].

Assay for Transketolase Activity

Preparation of Materials for Assay

Hemolyzate

Five milliliters of heparinized blood is adequate for multiple assays, although 1 ml of blood is sufficient for the semimicro method. The sample must be drawn before any oral or parenteral therapy. (The TPP effect is corrected very rapidly following parenteral therapy.) The RBC are packed by centrifugation in a graduated tube. The plasma and buffy coat are removed by suction, and a volume of distilled water equal to the volume of packed cells is added to the cells. The cells are suspended in the water, and the mixture is transferred to a vial and frozen with the vial tipped to prevent breakage. To prepare for immediate assay, freeze (in dry ice) and thaw three times to accelerate hemolysis. The hemolyzed sample is stable at room temperature for a number of hours, for 2–3 days in the refrigerator, and for at least 3 months when frozen.

Preparation of Reagents

a. "B" Buffer. Combine the following solutions in the proportions shown and adjust the final pH to 7.4.

0.9% NaCl, 40 ml (9 g/1000 ml)
1.15% KCl, 1030 ml (11.5 g/1000 ml)
1.75% K_2HPO_4, 200 ml (17.5 g + 18–20 ml of 1 *N* HCl per 1000 ml, to adjust to pH 7.4)
3.82% $MgSO_4 \cdot 7\ H_2O$, 10 ml

b. TPP (thiamine pyrophosphate)

TPP stock solution: 1 mg TPP per 1 ml in "B" buffer (25 ml). Keep frozen.
TPP working solution: 1 part of thawed stock TPP solution is diluted with 8 parts "B" buffer. Distribute 1–2 ml into small vials and freeze.

c. Substrate R-5-P. Place 3.24 g barium R-5-P in a 100-ml beaker and add 8.5 ml 1 *N* HCl. Stir until dissolved. Add 45 ml distilled water and distribute into four 40–50-ml centrifuge tubes. Rinse beaker and add washings to the tubes. To each of the 4 tubes add 2 ml of Na_2SO_4. Stir well, and allow to precipitate in the cold room for 15 minutes. When settled, add about 0.5 ml Na_2SO_4 dropwise to determine whether precipitation was complete. If no further turbidity occurs, centrifuge the tubes for 5 minutes at high speed. Decant supernatants of the 4 tubes into a 250-ml Erlenmeyer flask. Wash the $BaSO_4$ residues in each tube 3 times with 5 ml H_2O. Centrifuge after each washing and combine the supernatants in the Erlenmeyer flask. Adjust with 5 *N* KOH to pH 7.4. Filter if turbid. The volume will be approximately 125 ml; however, the actual volume must be determined. Assay for the concentration of ribose in this solution by the orcinol method as described below, and adjust the total volume so that the final concentration is 7.0 mg ribose per milliliter of solution. (R-5-P gives a different color equivalent per mole than ribose alone, with orcinol.) Distribute approximately 10 ml into a series of small vials and freeze. (When sodium ribose 5-phosphate is available, dissolve it in distilled water and standardize the solution to a final concentration of 7.0 mg ribose per milliliter as described.)

d. 7½% Trichloroacetic Acid. Dilute the contents of two ¼ lb bottles (113 g/bottle) to 3 liters with distilled water. Refrigerate.

e. Hexose Standard Solutions

Glucose stock solution (D-glucose): Make a stock solution of 1 mg D-glucose per milliliter in distilled water. Refrigerate.

Hexose working standard: Dilute 10 ml of the glucose stock solution to 100 ml with 7½% trichloroacetic acid to yield a solution of 100 μg/ml.

f. Anthrone Reagent. Dissolve 0.5 g anthrone[17] and 10.0 g thiourea in 66% H_2SO_4 by heating to 60°–70° (do not exceed 90°). Cool to room temperature and add additional 66% H_2SO_4 to make 1 liter. Refrigerate.

g. Pentose Standard Solutions

D-Ribose stock solution: Make a stock solution of 1 mg D-ribose per milliliter in distilled water. Refrigerate.

Pentose working standard. Dilute 1 ml of the ribose stock solution to 100 ml with distilled water to yield a solution of 10 μg/ml. Refrigerate.

h. Orcinol Reagent. Dilute 4.0 g orcinol[18] and 0.2 g $FeCl_3$ to 100 ml with distilled water. Add 30% HCl to a volume of 2 liters.

Procedure

1. *Macromethod*

a. Incubation Procedure. A unit of 4 test tubes is set up for each hemolyzate—the number of units depending upon the choice to run singlicate, duplicate or triplicate determinations. Label the 4 tubes A, B, and D and R, respectively, as indicated in Table I. (The "R" tube contains saline in the place of the hemolyzate.)

[17] Eastman Distillation Products Industries, Division of Eastman Kodak Company, Rochester, New York.

[18] Fisher Scientific Company, 633 Greenwich St., New York, N. Y. 10014.

TABLE I
INCUBATION PROCEDURE

Tube	Hemolyzate (ml)	"B" Buffer (ml)	TPP working solution[a] (ml)	Incubate at 38° (min)	R-5-P[a] (ml)	Incubate at 38° (min)	7½% TCA[a] (ml)
A	0.5	0.45	—	30	0.2	60	6.0
B	0.5	—	0.45	30	0.2	60	6.0
D	0.5	0.65	—	—	—	—	6.0
R	0.5 saline	0.45	—	—	0.2	—	6.0

[a] Addition is followed by mixing on Vortex.

The final volume for each test tube is 7.15 ml. Tube A contains hemolyzate without TPP. The tube B contains hemolyzate with TPP. Tube D serves as a blank in order to determine the amount of hexose and/or pentose endogenous to the tissue sample. Tube R contains substrate only and is used in the determination of the amount of ribose utilized in the reaction. Tubes A and B are initially incubated at 38° for 30 minutes (as shown in the table) to allow the enzyme to combine with the coenzyme, TPP, before the substrate is added. The substrate R-5-P is then added with shaking while noting the time and order of each addition. Exactly 60 minutes later, for each tube in turn, one adds the 6.0 ml of TCA with shaking in the same tube order and time interval as in the substrate addition. Centrifuge the tubes at a high setting for 10 minutes. The protein-free filtrates are used for the determination (1) of the hexose formed by the anthrone method, and (2) of the pentose utilized by the orcinol method. (The filtrates are stable for up to 5 days when refrigerated.)

TABLE II
PROCEDURE FOR DETERMINATION OF HEXOSE

Tube	Filtrate (ml)	Hexose (ml)[a]	7½% TCA (ml)	Anthrone,[b] cold (ml)	Boiling water bath (min)	Cold water bath (min)	Dark (min)
All incubation tubes	1.0	—	—	10.0	10	5	20
50-μg standard	—	0.5	0.5	10.0	10	5	20
100-μg standard	—	1.0	—	10.0	10	5	20
Blank	—	—	1.0	10.0	10	5	20

[a] 100 μg/ml.
[b] Addition is followed by mixing on Vortex.

b. Determination of the Hexose Formed in the Reaction. Hexose is determined in the filtrates of the tubes A, B, and D. The R tubes are *not* analyzed since they contain only the original substrate R-5-P. Standards are run in duplicate.

Place the tubes in the boiling water bath; start timing the reaction when the temperature has returned to 99°. Read the solutions against the blank set at 0 (zero) optical density (O.D.), in a spectrophotometer or colorimeter, at 620 mμ.

c. Determination of Pentose. Pentose is determined in the solutions of the A, B, D, and R tubes. Standards are run in duplicate.

TABLE III
PROCEDURE FOR DETERMINATION OF PENTOSE

Tube	Filtrate (ml)	Pentose[a] (ml)	H_2O (ml)	Orcinol[b] (ml)	Boiling water bath (min)	Cold water bath (min)
A, B, and D tubes	0.2	—	1.3	4.5	20	5
R tubes	0.1	—	1.4	4.5	20	5
5-μg standard	—	0.5	1.0	4.5	20	5
10-μg standard	—	1.0	0.5	4.5	20	5
Blank	—	—	1.5	4.5	20	5

[a] 10 μg/ml.
[b] Addition is followed by mixing on Vortex.

Place the tubes in the boiling water bath; start timing the reaction when the temperature has returned to 99°. Read the solutions against the blank set at 0 (zero) optical density (O.D.), in a spectrophotometer or colorimeter, at 670 mμ.

Calculations

Hexose. To calculate the amount of hexose formed per milliliter of hemolyzate per hour, a dilution factor for each sample must be determined:

$$\underset{\text{(ml hemolyzate used)}}{\frac{1}{0.5}} \times \underset{\text{(total ml per incubation tube)}}{\frac{7.15}{1}} \times \underset{\text{(ml filtrate used)}}{\frac{1}{1.0}} = 14.3$$

Calculate the average O.D. per microgram of hexose standard and call it SH. The dilution factor and O.D. per microgram of hexose are constant for each tube of the determination. Thus, 14.3 = a constant, KH, for all tubes of the determination (assuming that 1.0 ml filtrate was used for the determination of hexose in each case). The O.D. readings obtained from the A, B, and D tubes for each hemolyzate are referred to as A, B, and D, respectively.

(i) $(A - D) \times \text{KH}$ equals the micrograms hexose per milliliter of hemolyzate per hour formed during the incubation without TPP = TH_1

(ii) $(B - D) \times \text{KH}$ equals the micrograms hexose per milliliter hemolyzate per hour formed during the incubation with TPP = TH_2

(iii) $$\text{TPP-effect (\%)} = \frac{\text{TH}_2 - \text{TH}_1}{\text{TH}_1} \times 100$$

PENTOSE. To calculate the amount of pentose utilized per milliliter of hemolyzate per hour, a dilution factor for each sample read must be determined:

$$\underset{\text{(ml hemolyzate used)}}{\frac{1}{0.5}} \times \underset{\text{(total ml per incubation tube)}}{\frac{7.15}{1}} \times \underset{\text{(ml filtrate used)}}{\frac{1}{0.2}} = 71.5$$

Calculate the average O.D. per microgram of pentose standard and call it SP. The dilution factor and O.D. per microgram of pentose are constant for each tube of the determination. Thus 71.5/SP = a constant, KP, for all tubes of the determination. The O.D. readings which were obtained from the A, B, D, and R tubes for each hemolyzate are referred to as A, B, D, and R respectively. $2R + D$ = the amount of pentose originally present in each tube before incubation ($2R$ is used since only 0.1 ml R-5-P was added to the R tube whereas 0.2 ml R-5-P was added to the A, B, and D tubes).

(i) $(2R + D) - A \times \text{KP}$ equals micrograms of pentose utilized per milliliter of hemolyzate per hour for the incubation without TPP = TP_1.

(ii) $(2R + D) - B \times \text{KP}$ equals micrograms of pentose utilized per milliliter of hemolyzate per hour during the incubation with TPP = TP_2.

(iii) $$\text{“TPP-effect” (\%)} = \frac{\text{TP}_2 - \text{TP}_1}{\text{TP}_1} \times 100$$

2. *Semimicro Method*

a. This procedure is essentially the same as the macromethod except for the use of smaller volumes. Use the schema of Table IV.

TABLE IV
PROCEDURE FOR SEMIMICRO METHOD

Tube	Hemolyzate (ml)	"B" Buffer (ml)	TPP solution[a] (ml)	Incubation at 38°C (min)	R-5-P[b] (ml)	Incubation at 38°C (min)	TCA[c] (ml)
A	0.1	0.4	—	30	0.1	60	3.0
B	0.1	—	0.4	30	0.1	60	3.0
D	0.1	0.5	—	—	—	—	3.0
R	—	0.5	—	—	—	—	3.0

[a] TPP: Take 1 ml stock TPP + 16 ml buffer. Use 0.4 ml per tube.
[b] Substrate: Dilute 6.8 ml substrate to 10 ml with buffer. Use 0.1 ml per tube.
[c] TCA: Use 5% TCA.

b. Hexose and Pentose Determinations. These are done as described for the macromethod.

c. Calculations. Calculate with appropriate formulas. With the semimicro assay absolute enzyme activity values are somewhat higher, although TPP-effect values are comparable.

Notes

It is important to standardize the substrate as described in order to stabilize data on transketolase activity over a period of time in the laboratory. This is because one cannot saturate the enzyme with substrate, and therefore, the observed enzyme activity increases as the substrate concentration increases. Small variations in substrate concentration, however, will not markedly affect the TPP-effect, as this is a calculated value.

The use of hemolyzed erythrocytes for transketolase assay is recommended because (1) the hemolyzates can be stored for at least 3 months in the frozen state; (2) no special equipment or skills are necessary; (3) a therapeutic trial may be made *in vitro* by drawing an initial blood sample, administering 50 mg thiamine intramuscularly to the subject, and drawing another assay sample 2 hours later. The TPP-effect, if present, reverts to normal during the 2-hour interval.

Interpretation of the Assay

1. Transketolase is an enzyme within the erythrocyte, and as such is independent of nonspecific changes in the extracellular plasma. As B_1 deficiency becomes more severe (a) thiamine becomes limiting in the body cells, (b) the availability of the coenzyme for metabolic work becomes depleted of coenzyme, and therefore (c) the transketolase

activity diminishes. The "TPP-effect" measures the extent of depletion of the transketolase enzyme for coenzyme.[15]

2. Most normal individuals have an RBC–TK activity range of 850–1000 μg hexose per milliliter hemolyzate per hour. Duplicate determinations may vary by 5%, and repeat assays by 10%. No differences have been observed between males and females. The ranges of TPP-effect are as tabulated.[15]

Clinical thiamine condition	TPP-effect
Normal	0–15%
Marginally deficient	15–25%
Severely deficient (with clinical signs)	25 + %

3. Hexose activity of less than 800 μg hexose formed per milliliter hemolyzate per hour has been generally associated with a positive TPP-effect, i.e., 15% or higher.

4. The TPP-effect is generally eliminated within 2–4 hours after the parenteral administration of 50 mg of thiamine to the patient. Oral therapy is slower in action. Occasionally, however, samples with very low transketolase activity may be encountered (i.e., 200–400 μg hexose per milliliter hemolyzate per hour) in which the TPP-effect is effectively eliminated by the parenteral therapy, but the total enzyme activity may not return to normal values for 1–14 days after therapy. These samples are generally from very malnourished individuals, often with liver involvement, who may have an apotransketolase deficit in addition to a coenzyme depletion. Although this may appear to be a deterrent to the use of the assay, we feel that it is not, as the TPP-effect data are still applicable in that they reflect the proportion of the *available* transketolase apoenzyme which was depleted of thiamine pyrophosphate.

5. In our hands, the hexose activity and the TPP-effect hexose have been more reflective of clinical objective and subjective findings. Some laboratories have found the pentose data more useful.

Summary

Erythrocyte transketolase activity and the "TPP-effect" present a functional evaluation which is both sensitive to, and specific for, thiamine deficiency. It may be used to confirm clinical beriberi or other manifestations such as Wernicke's encephalopathy, to reveal a biochemical defect in marginal B_1 deficiency, or to differentiate B_1 deficiency from clinically similar diseases of other etiology.

[90] Phosphoketolase[1-3]

By MELVIN GOLDBERG, JUNE M. FESSENDEN, and EFRAIM RACKER

D-Xylulose-5-P + P → acetyl-P + D-glyceraldehyde-3-P + H_2O

Assay Method A

Principle. After deproteinization of the sample, the formation of glyceraldehyde-3-P is measured by the oxidation of DPNH in the presence of glyceraldehyde-3-P isomerase and α-glycero-P dehydrogenase. Only crude extracts are assayed by this method.

Reagents

Potassium phosphate, 0.1 *M* pH 6.0
Isomerase product (20 m*M* xylulose-5-P)[4]
Glutathione, 0.3 *M*, pH 6.0
$MgCl_2$, 0.1 *M*
Thiamine pyro-P, 0.03 *M*
Triethanolamine, 1 *M*, pH 7.5
0.001 *M* DPNH
Glyceraldehyde-3-P isomerase and α-glycero-P dehydrogenase mixture (Boehringer) diluted with H_2O to 0.2 mg/ml
Aldolase (Boehringer) diluted with H_2O to 1 mg/ml
Phosphoketolase, crude extract diluted to 200 μg/ml

Procedure. In a final volume of 0.3 ml the following solutions are pipetted into a test tube: 0.1 ml of potassium phosphate, 0.1 ml of isomerase product, 0.01 ml of glutathione, 0.01 ml of $MgCl_2$, 0.01 ml of thiamine pyro-P, 0.01 ml of mixture of glyceraldehyde-3-P isomerase and α-glycero-P dehydrogenase, 0.01 ml of aldolase, and 0.01–0.05 ml of diluted phosphoketolase. After incubation at 37° for 15 minutes, the mixture is placed in a boiling-water bath for 1 minute to stop the reaction. After centrifugation, 0.05–0.15 ml of the supernatant solution is assayed for the presence of fructose-1,6-di-P by addition to the following reagents in a Beckman quartz cuvette (10-mm light path) in a final volume of 1 ml: 0.05 ml of triethanolamine and 0.01 ml of DPNH. The

[1] E. Racker, Vol. V [29d]. Phosphoketolase from *Acetobacter xylinum*.
[2] B. L. Horecker, Vol. V [28d]. Phosphoketolase from *Lactobacillus plantarum*.
[3] J. Hurwitz, *Biochim. Biophys. Acta* **28**, 599 (1958); also M. L. Goldberg, J. M. Fessenden, and E. Racker, in preparation. Phosphoketolase in *Leuconostoc mesenteroides*.
[4] G. de la Haba and E. Racker, Vol. I [54].

optical density is measured at 340 mμ in a Gilford recording spectrophotometer. On addition of 0.01 ml of mixture of glyceraldehyde-3-P isomerase and α-glycero-P dehydrogenase plus 0.01 ml of aldolase the change in optical density is followed until no further changes are seen.

Assay Method B

Principle. This assay is based on the measurement of acetyl-P from xylulose-5-P in the presence of P_i and the enzyme. Acetyl-P is measured as hydroxamic acid.[5,6]

Reagents

Succinate, 0.2 *M*, pH 6.2
Potassium phosphate, 0.5 *M*, pH 6.1
Thioglycerol, 0.6 *M*
$MgCl_2$, 0.1 *M*
Thiamine pyro-P, 0.05 *M*
Isomerase product (20 m*M* xylulose-5-P)[4]
Phosphoketolase, diluted to 100 μg/ml before assay
$NH_2OH \cdot HCl$, 28%, brought to pH 6.4 with an equal volume of 15% NaOH
Sodium acetate buffer, 0.1 *M*, pH 5.4
HCl, 6 *N*
Trichloroacetic acid, 24%
$FeCl_3$, 20%, in 0.1 *N* HCl

Procedure. In a final volume of 0.5 ml the following solutions are pipetted into a test tube: 0.05 ml of succinate, 0.05 ml of potassium phosphate, 0.01 ml of thioglycerol, 0.01 ml of $MgCl_2$, 0.02 ml of thiamine pyro-P, 0.1 ml of isomerase product, and 0.01–0.1 ml of phosphoketolase. After 15 minutes at 37°, a 0.2 ml aliquot is removed, added to a test tube containing 0.3 ml of NH_2OH, 0.3 ml of acetate, and 0.1 ml of water, and kept at 23° for 10 minutes. Then 0.1 ml of HCl, 0.1 ml of trichloroacetic acid, and 0.1 ml of $FeCl_3$ are added; after 5 minutes the solutions are read at 540 mμ in the Beckman DU spectrophotometer with succinic anhydride as a standard.

Definition of Unit of Enzyme and Specific Activity. A unit of enzyme is defined as the amount of enzyme which catalyzes the formation of 1 micromole of acetyl-P per minute under the specified conditions. Specific activity is defined as units of enzyme per milligram of protein.

[5] F. Lipmann and L. C. Tuttle, *J. Biol. Chem.* **159,** 21 (1945).
[6] E. C. Wolff and S. Black, *Arch. Biochem. Biophys.* **80,** 236 (1959).

Cultivation and Harvest of Bacteria

Leuconostoc mesenteroides[7] is grown in a test tube in 10 ml of Bacto-A.P.T. broth (Difco)[8] supplemented with 0.35 μg/ml of thiamine-HCl. The medium is adjusted to pH 6.9 with HCl. After 8–12 hours of growth at 27°, the culture is transferred to an Erlenmeyer flask containing 50 ml of the same media and is grown for 12–15 hours. Fifteen liters of media in a 25-liter bottle are innoculated with the 50-ml culture. No attempt is made to maintain anaerobiosis. After 20–24 hours, the bacteria are harvested in a Sharples steam-driven continuous flow centrifuge at 4° at a flow rate of approximately 800 ml per minute, washed once with cold distilled water and stored at —55°. The yield is approximately 3.5 g wet weight of bacteria per liter of media.

Purification Procedure

All operations are carried out at 0°–4° unless otherwise stated; "Mann-enzyme grade" ammonium sulfate is used.

Step 1. Preparation of Bacterial Extract. Bacteria, 15 g, are suspended in 30 ml of cold distilled water and exposed to sonic oscillation for 60 minutes in a Raytheon sonic oscillator (250 watts, 10 kc) cooled by flowing ice water. After centrifugation at 78,000 *g* for 45 minutes, the clear, pale yellow supernatant solution is decanted; care is taken not to disturb the loosely packed precipitate. This crude extract can be stored at —55° for months with little loss of activity.

Step 2. pH Fractionation. The crude extract is diluted to 20 mg of protein per milliliter and, while being stirred, is brought to between pH 4.5 and 4.6 with the slow addition of 1 *N* acetic acid. If the pH of the extract falls below pH 4.45, phosphoketolase precipitates out and could not be recovered. The milky suspension is centrifuged for 45 minutes at 78,000 *g*, and the clear supernatant solution is carefully decanted. The supernatant solution, while being stirred in an ice bath, is quickly brought to between pH 5 and 5.1 with 1 *M* potassium phosphate pH 8.0.

Step 3. Streptomycin Treatment. Streptomycin (1.5 mg per milligram of protein) is dissolved in a volume of water equal to 0.2 volume of the supernatant solution, to which it is slowly added. The milky white mixture is kept for 2–4 minutes in ice, then centrifuged 15 minutes at 78,000 *g*. The clear supernatant solution is quickly carried to the next step.

Step 4. Concentration by Ammonium Sulfate. To each 100 ml of the

[7] We are indebted to Dr. W. A. Wood for the culture of *Leuconostoc mesenteroides* and for information on the properties of phosphoketolase from this organism.

[8] J. B. Evans and G. F. Niven, Jr., *J. Bacteriol.* **62**, 599 (1951). For the 15-liter cultures, the media were prepared in the laboratory.

supernatant solution being stirred gently, 0.25 ml of 12 M thioglycerol and 55.9 g of ammonium sulfate are added. After 15 minutes, the precipitate is collected by centrifugation for 20 minutes at 18,000 g and is dissolved in a minimal volume of 12 mM thioglycerol–10 mM sodium succinate (pH 5.0). After dialysis overnight at 4° against at least 120 volumes of the same buffer, the extract can be stored for weeks at —55° with little loss of activity. Steps 2–4 of the purification are always carried out on the same day.

Step 5. Fractionation with Protamine. The optimal amount of protamine to be added to the extract is determined on a small scale. The dialyzed extract is diluted to 9–10 mg of protein per milliliter. To small aliquots (0.4 ml) varying amounts (0.05–0.09 ml) of a 1% protamine sulfate solution (pH 5.0) are slowly added. These mixtures are kept in ice for 8 minutes with frequent stirring. The precipitate is removed by centrifugation for 5 minutes at 18,000 g, and the supernatant solution is immediately assayed.

To the remainder of the extract, gently stirred, an amount of protamine sufficient to precipitate 85–95% of the phosphoketolase is added. After 8 minutes at 0° the mixture is centrifuged at 18,000 g for 5 minutes, the supernatant solution is discarded, and the tubes are carefully wiped dry. The gummy precipitate is immediately extracted by homogenization with 0.5 volume (of the amount of dialyzed extract used) of a solution containing 10 mM sodium succinate, pH 5.0; 24 mM thioglycerol, and 48 mM ammonium sulfate. After centrifugation at 18,000 g for 5 minutes, the supernatant solution is carefully decanted and saved. The precipitate is reextracted with 0.2 volume (of the amount of dialyzed extract) of the same buffer. The precipitate is removed by centrifugation and discarded. The two supernatant solutions are combined and immediately carried to the next step. Throughout the protamine fractionation it is essential to work as fast as possible.

Step 6. First Ammonium Sulfate Fractionation. Solid ammonium sulfate (3.26 g) is added for each 10 ml of the combined extract. After 20 minutes, the precipitate is collected by centrifugation for 20 minutes at 18,000 g, dissolved in 12 mM thioglycerol–10 mM succinate buffer (pH 5.0) and labeled fraction A.

To each 10 ml of the supernatant solution from above, 1.1 g of ammonium sulfate is added. After 20 minutes at 0° the precipitate is collected and dissolved as described above (fraction B).

Most of the activity consistently appears in the second ammonium sulfate fraction. The enzyme is stable for many months when stored at —55°. Repeated freezing and thawing causes gradual loss of activity.

Step 7. Second Ammonium Sulfate Fractionation. Fraction B from above is diluted to 10 mg/ml with 10 mM succinate (pH 5.0). Two successive ammonium sulfate fractions are obtained by the addition of 3.14 g (fraction C) and 0.73 g (fraction D) of ammonium sulfate per 10 ml, respectively, and dissolved in 10 mM succinate (pH 5.0).

Step 8. Crystallization. Both fractions are diluted to a protein concentration of about 2 mg/ml with 10 mM succinate (pH 5.0), and cold unneutralized saturated ammonium sulfate solution is added, with constant stirring, until a very faint haze is visible. The solutions are stored in the refrigerator and examined periodically. At the end of 3 weeks crystals usually begin to appear.[9] The table shows a typical purification.

PURIFICATION OF PHOSPHOKETOLASE

Fraction	Volume (ml)	Protein (mg/ml)	Activity			Recovery (%)
			Units/ml	Total units	Specific activity	
Step 1. Crude extract	168	44.0	224	3760	0.51	100
Step 2. After pH fractionation	305	2.2	6	1815	2.7	48
Step 4. After ammonium sulfate concentration	11.2	24.7	123.5	1385	5.0	37
Step 6. After first ammonium sulfate fractionation						
Fraction A	1.0	35.4	216	216	6.1	5.7
Fraction B	1.2	51.9	617	740	11.9	20
Step 7. After second ammonium sulfate fractionation						
Fraction C	1.0	26.1	464	464	18.5	12
Fraction D	0.6	11.0	109	65	9.9	1.7

Properties

Specificity. Xylulose-5-P, fructose-6-P, glycolaldehyde, and hydroxypyruvate serve as substrate. Arsenate replaces phosphate, yielding acetate as the end product. Sulfhydryl compounds, such as thioglycerol, mercaptoethanol, cysteine, and glutathione, accept the acetyl group; the formation of thiolesters results. The rate of thiolester formation is only about 10–20% of the rate of acetyl-P formation.

Activators and Inhibitors. Phosphoketolase activity in the presence of sulfhydryl compounds is stimulated about 10–15% by 1 mM EDTA

[9] The crystallization is variable, and for most experiments the preparation was used after step 7.

and about 15–20% by 10 mM sodium borate. EDTA replaces the sulfhydryl compounds as a stimulant of the reaction.

The reaction is inhibited by glyceraldehyde-3-P and erythrose-4-P, but not by acetyl-P. An inhibition of close to 100% is observed with 2 mM *p*-hydroxymercuribenzoate at pH 8.0. About 83% of the activity is recovered on addition of 20 mM histidine. No inhibition is seen with *N*-ethyl maleimide, iodoacetic acid, or 5,5′-dithiobis(2-nitrobenzoic) acid.

Kinetic Properties. The optimal pH range is 5.8–7.6. The K_m for xylulose-5-P is 4.7×10^{-3} M, and for fructose-6-P, 2.9×10^{-2} M. With xylulose-5-P as substrate the K_m for P_i is 5.55×10^{-3} M, for Mg^{++}, 3.95×10^{-7} M, and for thiamine pyro-P, 4.04×10^{-7} M.

Stability. The enzyme is most stable between pH 4.5 and 5.5. For long-term storage 12 mM thioglycerol is added. The purified enzyme could be stored at —55° for several years with little loss of activity.

[91] 2-Keto-3-deoxy-6-phosphogluconic Aldolase (Crystalline)[1]

By H. Paul Meloche, Jordan M. Ingram, and W. A. Wood

2-Keto-3-deoxy-6-phosphogluconate ⇄ pyruvate + D-glyceraldehyde 3-phosphate

Assay

KDPG aldolase is assayed spectrophotometrically at 340 mμ. Pyruvate resulting from KDPG cleavage is reduced by NADH in the presence of added lactic dehydrogenase. The assay, carried out in microcuvettes (3 × 25 mm, 1-cm light path), in a total volume of 0.15 ml, is proportional to enzyme concentration over a range of velocities from near zero to more than 10 absorbance units per minute. One unit of KDPG aldolase is defined as an absorbance change of 1.0 per minute. This is equivalent to the cleavage of 0.0241 micromoles of KDPG per minute.

Reagents

1. Stock assay solution consisting of 4 mg of NaNADH; 0.25 ml of 1.0 M imidazole, pH 8; 0.05 ml of a commercial suspension of crystalline muscle lactic dehydrogenase; water to 5 ml
2. Na or K salt of KDPG, 0.05 M, pH 6

[1] Abbreviations used are: KDPG, 2-keto-3-deoxy-6-phosphogluconate (ic) and NADH, reduced nicotinamide-adenine-dinucleotide.

For assay, 0.05 ml of reagent 1, 0.015 ml of reagent 2, and 0.085 ml of water are added to a microcuvette. Upon addition of the aldolase sample, the linear absorbancy change per minute at 340 mμ is determined.

Enzyme Source

The mass cultivation of *Pseudomonas fluorescens,* especially in a fermentor, does not consistently produce high quality cells. The major problem seems to involve aeration control. This problem is still under study at this writing. However, acceptable cells can be obtained in a fermentor using the procedure and precautions described below.

Pseudomonas fluorescens A 3.12 is grown in a medium modified from that described elsewhere.[2] The organism must be grown in a fermentor, with a very high rate of air input, an efficient baffling system, and vigorous stirring. The medium composition for a 40-liter fermentor (about 35 liters of liquid) is: citric acid, 50 g; $MgSO_4 \cdot 7\ H_2O$, 30 g; $(NH_4)_2HPO$, 150 g; K_2HPO_4, 50 g; KH_2PO_4, 40 g; dextrose, 200 g; and $FeCl_3$ (anhydrous) 10–20 ppm (citrate and $MgSO_4 \cdot 7\ H_2O$ are combined first; the resulting chelation retards magnesium-ammonium phosphate formation). Growth of the organism is rapid enough that sterilization prior to inoculation is unnecessary. Of a 24-hour culture grown in liquid stock culture medium,[2] 100 ml is used to inoculate a 40-liter fermentor. (It is important that the inoculum culture be initiated from a fresh slant; transfer of slants is bimonthly.) The temperature of the fermentor must be maintained between 25 and 30° (the metabolic rate of the organism is high enough to produce significant quantities of heat). Foam is controlled with Antifoam A (Dow Corning) and growth is allowed to proceed 17–20 hours. Cells are collected by continuous flow centrifugation as rapidly as possible at 2°, stored either in the frozen or lyophilized state, and fractionated within 2 weeks. (One gram of fresh cells is equivalent to 0.25 g of lyophilized cells.) The cell paste must be *red,* and reasonably firm. *If the cells are white, discard the run.*

Purification of 2-Keto-3-deoxy-6-phosphogluconic Aldolase

Sodium or potassium phosphate buffer at pH 6 is used throughout this procedure. All steps are carried out at 0–4° unless otherwise specified, and centrifugations are performed at 18,000 *g* or higher. Protein is estimated by the ratios of absorbancies at 280 and 260 mμ.[3]

Preparation of Sonic Extract. The data shown in the table were ob-

[2] See this volume [116].
[3] E. Layne, Vol. III [73].

tained by fractionating cells from three 40-liter fermentor runs (530 g). The cell mass was suspended in 800 ml (1.5 volumes based on weight) of water and sonicated with stirring for 10 minutes at room temperature, a Branson Sonifier (Model S 110) being used. The maximum power output setting was used. Some warming of the preparation occurs, but cooling is not necessary. This preparation *is not* centrifuged at this point.

To the sonic extract is added 20 g of solid $MnCl_2 \cdot 4\ H_2O$ per liter of material (0.1 *M*), and the preparation is allowed to stir at room temperature for 15 minutes or until all the salt is dissolved. Then 1 *M* phosphate buffer (0.2 volume) is added and stirring is continued for another 15 minutes. Finally 0.2 volume of 2% protamine sulfate (pH 5.0–5.5) is added and after another 10 minutes of stirring, the preparation is centrifuged for 30 minutes in the cold. Essentially all the activity of the original sonic extract is found in the supernatant fraction. The ratio of absorbancies at 280 and 260 mμ should approach 0.9. These steps have been found necessary to make possible the successful further fractionation of the cells.

Acid Treatment at 40°. The temperature of the supernatant fraction is raised to 40°, and the pH is then adjusted to 2.5 with concentrated HCl. The preparation is allowed to stand at 40° for 10–20 minutes, and is then centrifuged for 30 minutes. The precipitate is discarded.

Ammonium Sulfate Fractionation. Without neutralization of acidity, the salt level is raised to 1.0 *M* by the addition of 140 g of solid ammonium sulfate per liter of supernatant. The preparation is centrifuged 30 minutes and the pellet is discarded.

The supernatant is brought to 2.0 *M* by the addition of 152 g of solid ammonium sulfate per liter. A very faint precipitate forms (if a heavy precipitate is found at this point, the enzyme cannot be crystallized), which is recovered by centrifugation for 30 minutes. The precipitate is dissolved in 10–20 ml of water per fermentor load of cells (in the data shown in the table, 40 ml of water was used). This is centrifuged to remove denatured protein, and is then adjusted to about 0.025 *M* with $MnCl_2$ by adding 0.025 ml of a 1.0 *M* solution per milliliter of enzyme preparation. The pH of the solution is about 4, and at this point the enzyme can be stored for a few days.

Calcium Phosphate Gel Treatment. The preparation from above is adsorbed onto calcium phosphate gel, using 5–6 mg of the gel per milligram of protein. Essentially all the activity is adsorbed. The gel is recovered by centrifugation. The activity is eluted from the gel using 0.2 *M* phosphate buffer, 1 ml of buffer being used per 15,000 units adsorbed. About 80% of the activity is recovered.

Concentration. The concentration of the eluted protein is about 1.2

mg/ml, making a quantitative activity recovery by ammonium sulfate precipitation difficult, as shown in the table. Better recovery of activity can be achieved by first pervaporating the preparation so that the protein level is about 2.5 mg/ml. The enzyme can then be precipitated by adding 0.373 g of ammonium sulfate per milliliter (2.5 *M*).

Crystallization. The pellet from the concentration step is dissolved in 0.050 *M* $MnCl_2$ so as to approximate 10 mg of protein per milliliter. To this is added 0.2–0.3 volume of saturated ammonium sulfate solution, until a reasonable turbidity occurs. Upon centrifugation, a brownish pellet should be found, which is discarded. Saturated ammonium sulfate solution is then added to the supernatant until faint turbidity once again occurs (if centrifuged the pellet would be white). The turbid solution is allowed to stand in the cold, and crystallization is usually complete overnight.

A summary of the purification steps and results obtained is shown in the table. The specific activity of crystalline enzyme is about 13,000 units per milligram of protein. The overall recovery of activity in crystalline form is about 30% of the initial activity. Highly purified enzyme must be dissolved in dilute (0.02 *M*) phosphate buffer for greatest stability and maximum activity, suggesting that phosphate affects the enzyme's active conformation.

PURIFICATION AND CRYSTALLIZATION OF 2-KETO-3-DEOXY-6-PHOSPHOGLUCONIC ALDOLASE

Step	Total activity (units)	Recovery (%)	Specific activity (units/mg protein)
Sonic extract	995,000	100	40
Acid treatment at 40°	852,000	86	60
Ammonium sulfate fractionation	722,000	73	3,500
Calcium phosphate gel eluate	560,000	56	8,100
Concentration	410,000	41	—
Crystallization	326,000	33	13,000

Properties of 2-Keto-3-deoxy-6-phosphogluconic Aldolase

Crystals of KDPG aldolase are well formed rhombohedrons measuring 10–20 μ across their long axis. The enzyme has an average molecular weight of 87,825 from sedimentation equilibrium experiments, and an S_{20} value of 4.5.[4] The aldolase is inactivated by borohydride when pyru-

[4] H. P. Meloche and W. A. Wood, *J. Biol. Chem.* **239,** 3515 (1964).

vate is present, a result indicating the formation of a Schiff's base.[5] One mole of pyruvate is bound per aldolase unit of molecular weight 43,500.[4] This suggests two active sites per molecule. The ϵ-amino group of a lysine, presumably in the active site, has been shown to be linked to the C-2 of pyruvate as a secondary amine, N_6-α-propionyllysine.[6] In addition, the enzyme is inactivated by dinitrofluorobenzene, 4 moles being stably fixed per mole of aldolase activity lost. This arylation occurs at four lysine residues which appear not to be involved in pyruvate binding, and the inactivation is inhibited by phosphate and KDPG[6] KDPG aldolase is also inactivated by bromopyruvate,[7] 2 moles being stably incorporated per mole of activity lost.[8] This alkylation has been found to occur at a cysteine residue, forming *S*-carboxyketomethyl cysteine.[8] In addition the aldolase is dissociated with guanidine-HCl into species having a minimum molecular weight of 49,500,[4] providing preliminary evidence that the enzyme may have two subunits. Although no metal requirements for the aldolase have been demonstrated, managanous ion partially protects the enzyme from inactivation on standing.[8]

The enzyme is induced in *Pseudomonas fluorescens* grown on glucose or gluconate. Virtually no aldolase is found in cells grown in salts media with glycerol or asparagine as carbon sources, nor in cells grown on peptone alone.[9] The enzyme has a pH optimum from 7.5 to 8.5, and the K_m of the enzyme for KDPG in the cleavage reaction is 0.11 m*M*.[10]

[5] E. Grazi, H. Meloche, G. Martinez, W. A. Wood, and B. L. Horecker, *Biochem. Biophys. Res. Commun.* **10,** 4 (1963).
[6] J. M. Ingram and W. A. Wood, *J. Biol. Chem.* **240,** 4146 (1965).
[7] H. P. Meloche, *Biophys. Res. Commun.* **18,** 277 (1965).
[8] H. P. Meloche, unpublished data, 1965.
[9] H. P. Meloche and B. Parekh, unpublished data, 1965.
[10] R. Kovachevich and W. A. Wood, *J. Biol. Chem.* **213,** 757 (1955).

[92] 2-Keto-3-deoxy-6-phosphogalactonic Acid Aldolase

By C. W. Shuster

2-Keto-3-deoxy-6-phosphogalactonate $\rightleftarrows$ D-glyceraldehyde 3-phosphate + pyruvate

Assay Method

Principle. One of the steps in the oxidative degradation of galactose is the cleavage of 2-keto-3-deoxy-6-phosphogalactonate (KDPGal) to pyruvate and glyceraldehyde 3-phosphate by an aldolase.[1] This enzyme

[1] M. Doudoroff and C. W. Shuster, *Bacteriol. Proc.*, p. 64 (1962).

is present in extracts of *Pseudomonas saccharophila* and other pseudomonads when grown with galactose as carbon source. The activity can be assayed by several methods: (1) by coupling the formation of pyruvate with lactic dehydrogenase, (2) chemical determination of pyruvate,[2] and (3) by coupling the formation of glyceraldehyde 3-phosphate with triose phosphate dehydrogenase. The first method is both rapid and convenient and may be applied to crude extracts of most microorganisms providing that particulate DPNH oxidase is removed by centrifugation.

Reagents

KDPGal, 0.02 *M*, pH 7.8, potassium salt (see below)
Tris-HCl buffer, 0.02 *M*, pH 8.0
Lactic dehydrogenase dissolved in Tris-HCl buffer in a concentration of 400 units/ml
DPNH, 5×10^{-3} *M*

Procedure. In a final volume of 3.0 ml the following solutions are placed in a spectrophotometer cell with a 1-cm light path: 2.8 ml Tris-HCl buffer, 0.05 ml lactic dehydrogenase solution, 0.05 ml KDPGal, 0.05 ml DPNH solution, and 0.05 ml enzyme solution to be tested.

After mixing, density readings at 340 mμ are taken at 30-second intervals for 5 minutes. Maximum rates are achieved immediately and are linear for several minutes.

Definition of Unit and Specific Activity. One unit of enzyme is defined as that amount which catalyzes the cleavage of 1 micromole of KDPGal per minute under the conditions of assay. The calculations are based on the assumption that 1 micromole of DPNH gives an optical density of 2.08. Cleavage of 1 micromole of KDPGal results in the oxidation of 1 micromole of DPNH. Specific activity is expressed as units of enzyme activity per milligram of protein.

Purification Procedure

Growth of Cells. Cultures of *P. saccharophila*[3] are grown with continuous aeration in liquid mineral media at 30° containing 2.5 g of galactose, 1 g NH_4Cl, 0.5 g of $MgSO_4 \cdot 7\ H_2O$, 50 mg $FeCl_3 \cdot 6\ H_2O$ per liter of 0.33 *M* KH_2PO_4–Na_2HPO_4 buffer at pH 6.8.[4] Cells in the exponential growth phase have a generation time of about 3 hours. Starting with a 2% inoculum from a mature culture, cells in the early stationary

[2] See Vol. III [66].
[3] For this preparation strain S-105 was used. Other cultures of *P. saccharophila*. presumably including ATCC 9114, would also contain the aldolase.
[4] J. De Ley and M. Doudoroff, *J. Biol. Chem.* **227**, 745 (1957).

phase are harvested by centrifugation after 18–20 hours of incubation at 30°. One liter of medium yields approximately 4 g of cell paste, which may be stored indefinitely at —16° without significant loss of enzyme activity.

Step 1. Preparation of Extracts. Extracts are prepared by sonic disruption in a 10kc Raytheon sonic oscillator or equivalent. Cells (160 g) are suspended in 1600 ml 0.05 *M* potassium phosphate buffer, pH 7.6 and disrupted in 65 ml portions by sonic treatment for 15 minutes. The suspension is centrifuged for 30 minutes at 30,000 rpm. The pooled supernatant solution contains between 6 and 8 mg protein per milliliter.

Step 2. Acid Precipitation. All operations in step 2 are carried out at 2–4°. Dilute the crude extract with 0.05 *M* phosphate buffer to a final protein concentration of 5 mg/ml and add 1 *M* acetic acid until the pH is exactly 5.0. After it has stood for 20 minutes, the acidified suspension is centrifuged for 30 minutes at 16,000 rpm. The precipitate containing approximately 50% of the total protein is resuspended in one-half the original volume of 0.05 *M* potassium phosphate buffer. Much of the precipitate is insoluble after the first acid precipitation, and the suspension should be clarified by centrifugation at 16,000 for 30 minutes. The protein solution is acidified again to pH 5.0, and the precipitate is dissolved in 600 ml of phosphate buffer. The purpose of acid precipitation is to separate KDPGal aldolase from a similar aldolase for the glucose analog, KDPG, which remains in the acid-soluble fraction. Activity of KDPG aldolase is reduced by 80% after the first acid precipitation. This step may be repeated 3 or 4 times with increasing resolution. Optimal precipitation of KDPGal aldolase is obtained with crude extracts having protein concentrations of no more than 5 mg/ml.

Step 3. Heat Treatment. The pH of the redissolved acid precipitate is adjusted to pH 7.6 with 1.0 *M* NaOH. The extract is heated with stirring to a temperature of 70° and maintained for 10 minutes at this temperature. Afterward the extract is cooled as rapidly as possible. After it has stood for 1 hour at 2–5°, the extract is centrifuged at 16,000 rpm and the precipitate is discarded.

Step 4. Ammonium Sulfate Fractionation. All operations in this step are carried out at 2–4°. Add to the heat-treated extract 300 ml (0.5 volume) of cold saturated ammonium sulfate previously adjusted to pH 7.0 with 2 *N* NH_4OH. After 20 minutes, centrifuge (16,000 rpm, 30 minutes), and discard the precipitate. Add 300 ml of neutral ammonium sulfate to the supernatant solution, and collect the precipitate by centrifugation. Dissolve the precipitate in 20 ml of 0.01 *M* phosphate buffer pH 7.6, and dialyze the fraction overnight against 2 liters of the same buffer.

Step 5. First DEAE-Cellulose Column. A DEAE-cellulose column, 3.5 × 22 cm, is equilibrated with 0.01 *M* phosphate buffer pH 7.6. The dialyzed extract is added, and the column is washed with 100 ml of phosphate buffer. Protein is eluted with a linear salt gradient consisting of 300 ml of 0.01 *M* phosphate buffer and 300 ml 0.07 *M* sodium chloride dissolved in phosphate buffer. The activity peak of KDPGal aldolase is eluted at a sodium chloride concentration of approximately 0.025 *M*. Fractions (10 ml) are collected, and those with activity are pooled and dialyzed against 0.01 *M* phosphate buffer.

Step 6. Second DEAE-Cellulose Column. A DEAE-cellulose column is prepared in 0.01 *M* phosphate buffer having the dimension of 2.5 × 35 cm. The dialyzed eluate from the first colümn is added to column 2 and eluted in a stepwise manner. Two hundred milliliters of 0.01 *M* phosphate buffer containing 0.01 *M* sodium chloride is passed through the column. Then 200 ml of 0.03 *M* sodium chloride in phosphate buffer is passed through the column. The KDPGal aldolase at its highest specific activity is eluted after 120 ml of the 0.03 *M* sodium chloride solution. KDPGal aldolase is not a heat-sensitive enzyme, and, although the acid precipitations and ammonium sulfate fractionations are carried out at low temperature, no advantage is to be gained by cooling either of the DEAE-cellulose columns or the fractions obtained from them. No activity is lost at room temperature for 6–8 hours. Results of these procedures are outlined in the table.

PURIFICATION OF KDPGAL ALDOLASE

Fraction	Total volume (ml)	Total protein (mg)	Total units	Specific activity (units/mg)	Recovery (%)
1. 10% sonic extract	1348	9420	983	0.104	—
2. Second acid precipitate	650	4030	621	0.154	63
3. Heat-treated supernatant	1400	1680	771	0.470	79
4. Ammonium sulfate (30–60%)	22	525	693	13.4	69
5. DEAE-cellulose column eluate	231	9	519	57.5	53
6. Second DEAE-cellulose column eluate	9.4	2.07	274	132.0	28

Properties

KDPGal aldolase exhibits typical Michaelis-Menton kinetics with no indication of either substrate or product inhibition. The K_m for KDPGal has been determined to be 5×10^{-4} *M*. The pH optimum is 7.8 in either Tris or phosphate buffers. Purified preparations of KDPGal aldolase exhibit a single homogeneous peak in the ultracentrifuge with a sedimentation constant of 3.84 S. Activity of frozen preparations declines about 20% in 2 weeks. This loss of activity is accompanied by the appearance of a 20.8 S peak in the ultracentrifuge accounting for 15–20% of the protein.

Substrate. The substrate used in the initial experiments on KDPGal aldolase[1] was synthesized chemically by Dr. L. Szabo, of the C.N.R.S. of the University of Paris. The product of Szabo's synthesis is only 50% pure by enzyme assay. It is possible to synthesize small quantities of KDPGal by phosphorylation of 2-keto-3-deoxygalactonate catalyzed by a kinase described by Wilkinson and Doudoroff.[5] The substrate for the kinase can be prepared in gram quantities by the method of De Ley and Doudoroff.[4] KDPGal synthesized by this method is isolated by the method of MacGee and Doudoroff.[6]

Specificity. KDPGal aldolase does not cleave KDPG, 3-deoxy-6-phosphogalactose, glucometasaccharinic acid, or keto-3-deoxy-L-arabonate, nor were these compound inhibitors of the reaction. Purified preparations of KDPGal aldolase react with KDPG to the extent of 1% of the total aldolase activity. Immunochemical experiments with antibody prepared to crystalline KDPG aldolase show that activity toward KDPG is the result of impurities in KDPGal aldolase. KDPGal aldolase does not cross-react with antibody to KDPG aldolase, and no residual KDPG aldolase activity is detected after reaction with antibody.

Equilibrium. The equilibrium of KDPGal aldolase measured at pH 6.8 at 25° is 3.7×10^{-3} *M* in direction of synthesis of KDPGal. The equilibrium constant does not vary with pH in this reaction.

Other Properties. KDPGal aldolase is an inducible enzyme of the galactose pathway and is present only in cells grown on galactose or galactose derivatives. The wild-type *P. saccharophila* grows well on galactose, but not on galactonic acid. Mutants utilizing galactonate may be isolated by incubating cells in media containing 0.1% galactonic acid for 3 or 4 days. These mutants grow equally well on galactose or galactonic acid, and have somewhat higher specific activity with galactonic acid as substrate. These mutants will utilize 2-keto-3-deoxygalactonic acid, but growth on this substrate is poor. Reciprocal transfer experiments between galactose derivatives and other substrates have not yielded mutants constitutive for KDPGal aldolase.

[5] J. F. Wilkinson and M. Doudoroff, *Science* **144,** 569 (1964).

[6] J. MacGee and M. Doudoroff, *J. Biol. Chem.* **210,** 617 (1954).

[93] 2-Keto-3-deoxy-D-glucarate Aldolase[1,2]

By DONALD C. FISH and HAROLD J. BLUMENTHAL

2-Keto-3-deoxy-D-glucaric acid
or → pyruvic acid + tartronic acid semialdehyde
5-Keto-4-deoxy-D-glucaric acid

Assay Method

Principle. The enzyme is usually assayed by disappearance of the substrate. The formyl pyruvate formed by cleavage of ketodeoxyglucarate with periodate forms a chromogen with thiobarbituric acid that is measured at 551 mμ.

Reagents

α-Keto-β-deoxy-D-glucarate, 0.00077 M. See this volume [13a] for the preparation of 5-keto-4-deoxy-D-glucarate or of a 85:15 mixture of this and 2-keto-3-deoxy-D-glucarate.
Potassium phosphate, 0.5 M, pH 7.8
$MgSO_4$, 0.07 M
Ketodeoxyglucarate aldolase
Trichloroacetic acid, 10%
Periodic acid, 0.025 M, in 0.125 N H_2SO_4
Sodium arsenite, 2%, in 0.5 N HCl
2-Thiobarbituric acid, 0.3%

Procedure. The incubation mixture (0.4 ml) in a 10 × 75 mm test tube contains 0.1 ml $MgSO_4$, 0.02 ml potassium phosphate buffer, 0.1 ml ketodeoxyglucarate, and about 0.0002 unit of aldolase preparation. After equilibration at 30°, the reaction is started by addition of the enzyme. The reaction is terminated after 10 minutes by addition of 0.1 ml trichloroacetic acid, and any protein precipitate is removed by centrifugation. A 0.2-ml sample is used to measure the disappearance of substrate by the periodate–thiobarbiturate test[3]; a zero time control must be included to measure the initial quantity of ketodeoxyglucarate. The assay is linear with respect both to time and to enzyme concentration until 40% of the substrate present at zero time has been utilized. Because this assay is based on substrate disappearance, it must first be established

[1] D. C. Fish and H. J. Blumenthal, *Bacteriol. Proc.* p. 110 (1963).
[2] H. J. Blumenthal and D. C. Fish, *Biochem. Biophys. Res. Commun.* **11,** 239 (1963).
[3] A. Weissbach and J. Hurwitz, *J. Biol. Chem.* **234,** 705 (1959).

that pyruvate, rather than α-ketoglutarate, and tartronate semialdehyde are indeed the products.[2,4] The pyruvate formation can be readily measured spectrophotometrically by coupling with lactic dehydrogenase and measuring the DPNH utilization.

Definition of Unit and Specific Activity. One unit is defined as the amount of enzyme that catalyzes the utilization of 1 micromole of ketodeoxyglucarate per minute under these conditions. An extinction coefficient of 60,000 at 551 mμ for ketodeoxyglucarate[4] in the periodate-thiobarbituric acid test is employed. The specific activity is expressed as units per milligram protein. Protein is determined either spectrophotometrically, by the method of Waddell[5] at 215 and 225 mμ, or nephelometrically with sulfosalicylic acid,[6] when the protein concentration is low or when the protein is in the presence of high concentrations of certain ions, such as Tris, acetate, or glutathione,[7] that interfere with the spectrophotometric method.

Purification Procedure

Culture Conditions. Escherichia coli strain CR63MA is grown aerobically for 13–15 hours on a rotary shaker at 37° with either 0.8% potassium acid glucarate or galactaric acid as the sole carbon source in a salts medium[8] adjusted with NaOH to pH 6.8. A 5% inoculum of a culture grown for 24 hours in the same medium is added to 950 ml of the medium in a 2000-ml Erlenmeyer flask. The cells are harvested by centrifugation and washed twice with 0.1 *M* sodium phosphate buffer, pH 6.8. The yield of cells grown on glucarate is about 3 g. The cells can be used immediately or stored at —14° for over a year.

Step 1. Preparation of Extracts. Five grams of cell paste is suspended in 50 ml of 0.15 *M* potassium phosphate buffer, pH 7.7, containing 0.003 *M* NaGSH, and disrupted for 15 minutes in a 10-kc Raytheon sonic oscillator. Cell debris is removed by centrifugation at 20,000 *g* for 5 minutes.

Step 2. Neutral Ammonium Sulfate Fractionation. An equal volume of neutral, saturated ammonium sulfate is added to the cell-free extract (50% saturation). The tubes are stirred slowly for 15 minutes and the precipitate is removed by centrifugation at 20,000 *g* for 5 minutes. The supernatant liquid is decanted, and to it is added an additional volume of neutral saturated ammonium sulfate to bring the solution to 65%

[4] See H. J. Blumenthal, D. C. Fish, and T. Jepson, this volume [13a].

[5] W. J. Waddell, *J. Lab. Clin. Med.* **48,** 311 (1956).

[6] See E. Layne, Vol. III, p. 447.

[7] J. B. Murphy and M. W. Kies, *Biochim. Biophys. Acta* **45,** 382 (1960).

[8] See H. J. Blumenthal, this volume [118].

saturation. The liquid is stirred for 15 minutes, then the precipitate is collected by centrifugation and the supernatant liquid is decanted. The precipitate is dissolved in 40 ml of 0.01 *M* potassium phosphate buffer, pH 7.7, containing 0.003 *M* NaGSH.

Step 3. DEAE-Cellulose Column Chromatography. DEAE-cellulose equilibrated with 0.1 *M* potassium phosphate buffer, pH 7.7, is used to pack a column 2.0 × 24 cm. The 50–65% ammonium sulfate enzyme fraction is placed on the column and then washed with 0.1 *M* potassium phosphate buffer, pH 7.7, until all the protein and nucleic acid which is not initially absorbed has been eluted, as judged by a return to a low and constant value of absorbance measurements at *ca.* 260 mμ (Canalco UV Flow Analyzer; Canal Industrial Corp.). A linear gradient of increasing phosphate buffer concentration at the same pH is then applied, and 7-ml fractions are collected. The reservoir contains 175 ml of 0.4 *M* potassium phosphate buffer, and the mixing chamber contains 175 ml of 0.1 *M* potassium phosphate buffer initially. The enzyme is eluted when the phosphate concentration approaches 0.32 *M*.

Step 4. Ammonium Sulfate Precipitation. The tubes containing the peak amounts of enzyme activity, usually about 10 tubes, are pooled, and the enzyme is precipitated by the addition of solid ammonium sulfate until the solution is saturated (approximately 65% of theory because of the high buffer concentration). It is necessary to add dilute NH_4OH during the addition of the solid ammonium sulfate to maintain the pH of the solution between 7.5 and 7.8. The saturated solution is allowed to stand for 1 hour with occasional stirring. The precipitate is collected by centrifugation at 20,000 *g* for 10 minutes and dissolved in 10 ml of 0.01 *M* potassium phosphate buffer, pH 7.7, containing 0.003 *M* NaGSH.

Step 5. Calcium Phosphate Gel Treatment. Aged calcium phosphate gel,[9] 16 mg/ml, is added to the enzyme in a ratio of 1.5 gel for every 2.25 mg protein. After 10 minutes, during which the suspension is stirred occasionally, the gel is sedimented by centrifugation for 2 minutes and the supernatant liquid is decanted. The aldolase is not adsorbed by the calcium phosphate gel, whereas some D-glucaric dehydrase and other proteins are adsorbed.

The purification procedure is summarized in the table (p. 532).

Properties

Effect of pH. The aldolase is most stable in potassium phosphate buffer, pH 6.0–7.5. Its stability decreases rapidly at either end of this range. The enzyme activity is maximal between pH 7.7 and 8.6 in potas-

[9] T. P. Singer and E. B. Kearney, *Arch. Biochem. Biophys.* **29**, 190 (1950).

Purification Procedure for Ketodeoxyglucarate Aldolase

Fraction	Units enzyme	Specific activity	Recovery (%)	Purification (-fold)	Aldolase to D-glucarate dehydrase ratio	Aldolase to tartronate semialdehyde reductase ratio
1. Crude extract	273	0.69	100	1	1.1	1.5
2. 50–65% $(NH_4)_2SO_4$ precipitate	168	2.01	62	2.9	1.6	1.2
3. DEAE-cellulose, pooled fractions	82	8.61	30	12.6	3.0	22.8
4. $(NH_4)_2SO_4$ precipitate	48	10.7	17.6	15.6	3.0	41.0
5. Negative $Ca_3(PO_4)_2$ gel	46	30.6	17	44.5	1010.0	52.0

sium phosphate buffer and between pH 7.4 and 8.0 in Tris-maleate buffer. High salt concentration inhibits activity. The enzyme does not cleave ketodeoxyglucarate in the following buffers: glycine-NaOH, collidine-acetate, barbital-acetate, imidazole-acetate, Tris-acetate, and lysine-HCl-NaOH.

Effect of Cations and Anions. There is an absolute cation requirement. Magnesium serves best, although cobalt, ferrous, manganous, and molybdenum cations, in decreasing order of effectiveness, may be used. When anions are tested, as their magnesium salts, only sulfate is non-inhibitory; acetate, chloride, nitrate, and bromide ions are inhibitory.

Effect of Inhibitors. Free sulfhydryl groups are probably not necessary for enzyme activity since *p*-chloromercuribenzoate, iodoacetate, and L-cysteine have almost no effect. In confirmation of the absolute cation requirement, fluoride, cyanide, EDTA, and pyrophosphate are all potent inhibitors of the aldolase. The cyanide inhibition can be reversed by Mg^{++} and Co^{++}, but not by Fe^{++} on Mn^{++}.

Five classes of sugar acids inhibit aldolase activity to various degrees. The order of effectiveness of the acids as inhibitors is hexaric $>$ hexuronic $=$ pentaric $>$ hexonic $=$ tetraric. The inhibition caused by D-glucarate, galactarate, and xylarate is due both to chelation and binding to the enzyme.

Specificity. The ketodeoxyglucarate aldolase exhibits stringent substrate specificity. There is essentially no activity with 2-keto-D-gluconate, 2-keto-3-deoxyoctonate, 2-keto-3-deoxy-6-phospho-D-gluconate, 2-keto-3-deoxyheptonate, 2-keto-3-deoxy-D-gluconate, 2-keto-3-deoxy-D-galactonate, 2-keto-3-deoxytartrate (oxalacetate), or FDP. The first three sugar acids listed inhibit aldolase activity on ketodeoxyglucarate competitively, while ketodeoxyheptonate exhibits a mixed type of inhibition.

The aldolase cleaves 5-keto-4-deoxy-D-glucarate, the product of the action of galactarate dehydrase on galactarate,[10] and a 85:15 mixture of 5-keto-4-deoxy- and 2-keto-3-deoxy-D-glucarate, the product of D-glucarate dehydrase on D-glucarate,[8] at the same rate. Hence the enzyme is best called an α-keto-β-deoxy-D-glucarate aldolase.

Substrate Affinity. The K_m for ketodeoxyglucarate is $2.5 \times 10^{-4}\, M$ under the standard assay conditions. Although the standard assay does not employ saturating levels of substrate, the assay is linear with respect to time and enzyme concentration until 40% of the substrate is cleaved.

Distribution. Like D-glucarate dehydrase[8] and galactarate dehydrase,[10] the ketodeoxyglucarate aldolase is found in extracts of a variety of bacteria, particularly enteric bacteria, with either D-glucarate or galactarate as the sole carbon source. The aldolase is not present when the cells are grown on a variety of other sugars or sugar acids. However, the aldolase is absent from hexarate-grown microbes, such as *Bacillus megaterium,* that form α-ketoglutarate and CO_2 from hexaric acids.[11]

Reversibility. The equilibrium constant cannot be measured under the usual aldolase assay conditions since the cleavage proceeds essentially to completion. However, the enzyme does catalyze a slow condensation of pyruvate with a variety of aldehydes.[1] As noted with 2-keto-3-deoxy-6-phosphogluconate aldolase,[12] the position of the equilibrium is influenced by the buffer used. The results of experiments with the ketodeoxyglucarate aldolase demonstrate that there is little agreement between buffer requirements for enzymatic activity in the forward and reverse direction. For example, the condensation of pyruvate with glycolaldehyde to yield 2-keto-4,5-dihydroxy-L-valeric acid in collidine-acetate buffer, pH 7.8, proceeds at about ten times the rate of the same condensation in Tris-maleate, pH 7.4. However, there is no measurable cleavage of ketodeoxyglucarate in the collidine-acetate buffer whereas the cleavage by aldolase is rapid in Tris-maleate buffer.

On the basis of the characterization of the condensation products from pyruvate with glycolaldehyde, glyoxylate, D- and L-glyceraldehyde,

[10] See H. J. Blumenthal and T. Jepson, this volume [119].

[11] H. J. Blumenthal and T. Jepson, *Bacteriol. Proc.* p. 82 (1965).

[12] R. Kovachevich and W. A. Wood, *J. Biol. Chem.* **213,** 757 (1955).

the ketodeoxyglucarate aldolase catalyzes a stereospecific polarization of the aldehyde in such a way that the hydroxyl group which is formed upon condensation is located on the left side of the 2-keto-3-deoxy onic acid when it is drawn in the Fischer projection.[13]

[13] D. C. Fish, Ph.D. Thesis, Univ. of Michigan, 1964.

[94] 3-Deoxyoctulosonic Acid Aldolase[1]

By MOHAMMED A. GHALAMBOR and EDWARD C. HEATH

3-Deoxyoctulosonic acid ⇄ pyruvic acid + D-arabinose

Assay Method

Principle. The most convenient method of assay of 3-deoxyoctulosonate aldolase is to measure the disappearance of the substrate colorimetrically with the thiobarbituric acid reagents.[2] An alternate procedure for determination of this activity with the purified preparation of 3-deoxyoctulosonate aldolase (after DPNH oxidase has been removed) is to couple the reaction to lactic acid dehydrogenase in the presence of DPNH and to measure the formation of pyruvate continuously.

Reagents

Potassium 3-deoxyoctulosonate, 0.05 *M*. Prepare either enzymatically from phosphopyruvate and D-arabinose 5-phosphate,[3,4] chemically,[4] or isolate and purify from cell wall lipopolysaccharide preparations.[4]
Phosphate buffer, pH 7, 0.2 *M*
Trichloroacetic acid, 25%
DPNH, 0.01 *M*
$MgCl_2$, 0.1 *M*
Lactic acid dehydrogenase, 5 mg/ml, (2×) crystalline, obtained from Worthington Laboratories

Procedure. Incubation mixtures contain 0.1 ml of phosphate buffer, 0.01 ml of potassium 3-deoxyoctulosonate, water, and aldolase to a final

[1] This presentation is based on work that was previously published in preliminary form: M. A. Ghalambor and E. C. Heath, *Biochem. Biophys. Res. Commun.* **11,** 288 (1963).
[2] D. Aminoff, *Biochem. J.* **81,** 384 (1961).
[3] D. H. Levin and E. Racker, *J. Biol. Chem.* **234,** 2532 (1959).
[4] See this volume [14].

volume of 0.5 ml. All components of the reaction, except enzyme, are mixed; the reaction is initiated by the addition of enzyme (0.1–0.2 ml crude extract) and immediately, a zero-time aliquot (0.05 ml) is withdrawn from the mixture and transferred to a tube containing 0.025 ml of trichloroacetic acid. The remainder of the incubation mixture is placed in a 37° water bath; 0.05 ml aliquots are removed at 5, 10, and 20 minute intervals and transferred, as described above, to tubes containing trichloroacetic acid. The trichloroacetic acid-inactivated aliquots are centrifuged, and 0.05-ml aliquots of the supernatant fluids are analyzed for 3-deoxyoctulosonate content with the thiobarbituric acid reagents. The disappearance of thiobarbituric acid-reactive material is a measure of 3-deoxyoctulosonate aldolase activity and is linear with regard to enzyme concentration throughout the assay period under the conditions described above. For assay of the purified preparation, incubation mixtures are prepared in cuvettes containing 0.05 ml of potassium 3-deoxyoctulosonate, 0.1 ml of phosphate buffer, 0.01 ml of DPNH, 0.01 ml of $MgCl_2$, and 0.005 ml of lactic acid dehydrogenase in a final volume of 0.5 ml. Zero-time absorbancies at 340 $m\mu$ are determined, 0.005–0.01 ml of purified aldolase preparation (appropriately diluted) is added, and the decrease in absorbancy at 340 $m\mu$ is followed continuously for a period of 10 minutes.

Definition of Unit and Specific Activity. One unit of 3-deoxyoctulosonate aldolase activity is defined as the amount that catalyzes the cleavage (3-deoxyoctulosonate disappearance or pyruvate formation) of 1 micromole of 3-deoxyoctulosonate per minute. Specific activity is expressed as units per milligram of protein. Protein is determined by the method of Waddell.[5]

Purification Procedure

Step 1. Preparation of Cell-Free Extract. Aerobacter cloacae[6] is grown in a medium that consists of the following (g/l): NH_4Cl, 5; Na_2HPO_4, 10; KH_2PO_4, 3; K_2SO_4, 1; NaCl, 1; $MgSO_4 \cdot 7\ H_2O$, 0.2; $CaCl_2 \cdot 6\ H_2O$, 0.02; $FeSO_4 \cdot 7\ H_2O$, 0.001; glucose, 0.25; yeast extract, 1.0;

[5] W. J. Waddell, *J. Lab. Clin. Med.* **48**, 311 (1956).

[6] This culture was originally obtained from Dr. Saul Roseman, Department of Biology, The Johns Hopkins University, Baltimore, who had isolated it by an enrichment technique. Using the procedure described in this paper for cultivation of a variety of other organisms (*Escherichia coli,* strains 0111, B, and K-12, as well as *Salmonella typhimurium, Salmonella adelaide,* and *Salmonella typhi*), followed by extraction of these cells provided crude preparations that exhibited 3-deoxyoctulosonate aldolase activities comparable to that obtained from *Aerobacter cloacae.*

and synthetic potassium 3-deoxyoctulosonate,[7] 5.0. Glucose is autoclaved separately as a 50% solution. A 10% solution of potassium 3-deoxyoctulosonate is sterilized by filtration. Each of the sugars is added aseptically to the remainder of the sterile growth medium. One liter of medium (contained in a 2-liter flask) is inoculated with 25 ml of an overnight culture of the organism and incubated at 37° on a rotary shaker for 18–24 hours. The cells (approximately 6 g) are harvested, washed with cold 0.15 *M* KCl, resuspended in 20 ml of water, and sonicated for 10 minutes. The suspension is centrifuged at 25,000 *g* for 10 minutes, and the supernatant fluid is retained. All procedures are conducted between 0° and 4°.

Step 2. Protamine Sulfate Precipitation. To 24 ml of crude extract are added 33 ml of 0.025 *M* phosphate buffer, pH 7, and 3 ml of 2% protamine sulfate solution. The suspension is stirred for 5 minutes, and the precipitate is removed by centrifugation at 25,000 *g* for 10 minutes and discarded.

Step 3. Ammonium Sulfate Fraction I. The protamine sulfate supernatant solution (58 ml) is treated with 13 g of solid ammonium sulfate, stirred for 5 minutes, and centrifuged at 25,000 *g* for 10 minutes. The precipitate is discarded and to the supernatant solution (66 ml) is added 12 g of solid ammonium sulfate; the suspension is stirred for 5 minutes and centrifuged at 25,000 *g* for 10 minutes; the precipitate is dissolved in phosphate buffer to a total volume of 6 ml.

Step 4. pH 5 Dialysis. A flask containing 250 ml of 0.01 *M* acetate buffer, pH 5.1, is placed in a salt–ice mixture and stirred until the temperature of the solution reaches 0°. The ammonium sulfate fraction from the preceding step is placed in dialysis tubing, chilled to 0°, and placed in the acetate buffer solution which is stirred with a magnetic bar for 4 hours. It is essential that this step is done exactly in this manner in order to maintain the temperature between 0° and 2°; if the temperature rises significantly about 2° during this procedure, rapid inactivation of the enzyme occurs. After dialysis, the suspension is centrifuged at 25,000 *g* for 10 minutes and the precipitate is discarded. The volume of the supernatant solution is 7.2 ml.

Step 5. Calcium Phosphate Gel. The supernatant solution (7.2 ml) from the preceding step is diluted with 14.4 ml of water and then treated with 14.4 ml of calcium phosphate gel (14 mg/ml). The suspension is stirred for 10 minutes and centrifuged at 10,000 *g* for 5 minutes; the calcium phosphate gel is successively washed with 10-ml portions of

[7] The crude synthetic product (this volume [14]) that contains a mixture of isomers of 3-deoxyoctulosonate (approximately 70% of which corresponds to natural 3-deoxyoctulosonate) was used in the growth medium.

water, 0.005 *M* phosphate buffer, pH 7, 0.01 *M* phosphate buffer, pH 7, and 0.015 *M* phosphate buffer, pH 7. Most of the 3-deoxyoctulosonate aldolase activity is obtained in the 0.01 *M* phosphate buffer eluate; this fraction is retained and used for further purification.

Step 6. Ammonium Sulfate Fraction II. The gel eluate obtained in the preceding step (10 ml) is treated with 4.4 g of solid ammonium sulfate, stirred for 5 minutes, and centrifuged at 35,000 *g* for 15 minutes. The precipitate is dissolved in 0.025 *M* phosphate buffer, pH 7, to a final volume of 1.4 ml.

The purification procedure is summarized in the table.

PURIFICATION PROCEDURE FOR 3-DEOXYOCTULOSONIC ACID ALDOLASE

Fraction	Total volume (ml)	Protein (mg)	Specific activity (units/mg × 10^2)	Recovery (%)
1. Crude extract	24	500	9.3	100
2. Protamine sulfate	60	250	21	100
3. Ammonium sulfate I	6	110	43	85
4. pH 5 Dialysis	7.2	60	85	94
5. Calcium Phosphate gel	10	7.6	250	40
6. Ammonium sulfate II	1.4	3.2	565	34

Properties

Specificity. The preparation of 3-deoxyoctulosonate aldolase is essentially specific for 3-deoxyoctulosonate and exhibits identical kinetics regardless of the source of substrate (prepared enzymatically, chemically, or isolated from cell wall lipopolysaccharide). A variety of other 2-keto-3-deoxy-onic acids have been tested as substrate for the aldolase preparation with the following results (millimicromoles substrate disappearance): 3-deoxyoctulosonate, 140; and 3-deoxy-D-*arabino*-heptulosonate, 3-deoxy-D-*erythro*-hexulosonate, and *N*-acetyl neuraminate, <6.

The specificity of the aldolase in the direction of condensation is less precise than in the direction of cleavage. Thus, incubation mixtures containing aldolase, phosphate buffer, pyruvate (0.08 *M*), and various aldoses (0.04 *M*) were assayed for the formation of thiobarbituric acid-reactive material with the following results (millimicromoles of thiobarbituric acid-reactive material): D-arabinose, 125; D-ribose, 36; and D-arabinose 5-phosphate, L-arabinose, D-xylose, D-lyxose, and *N*-acetyl-D-mannosamine, <3.

Stability. Various preparations of 3-deoxyoctulosonate aldolase are relatively stable. The crude extract is stable indefinitely when stored at —20°; the first ammonium sulfate fraction (step 3) and the second ammonium sulfate fraction (step 6) exhibited only slight (10–20%) loss when stored at —20°, even when the preparations are repeatedly thawed and refrozen several times over a period of one year.

Other Properties. Regardless of the direction in which the reaction is conducted, analysis of incubation mixtures of 3-deoxyoctulosonate aldolase with substrate concentrations of 1.5×10^{-2} *M*, indicates that the equilibrium mixture consists of 12% of 3-deoxyoctulosonate and 88% of pyruvate and D-arabinose.[8] The K_m for 3-deoxyoctulosonate has been estimated to be 6×10^{-3} *M*. The pH optimum of the reaction is 7.0 in phosphate buffer.

[8] Higher substrate concentrations (0.2–0.4 *M*) result in an alteration in the ratio of the various components of the aldolase reaction at equilibrium; this factor is used to advantage in the preparation of 3-deoxyoctulosonate-^{14}C from pyruvate-^{14}C and D-arabinose with 3-deoxyoctulosonate aldolase (see Vol. VIII [14]).

[95] L-Fuculose 1-Phosphate Aldolase[1]

By MOHAMMAD A. GHALAMBOR and EDWARD C. HEATH

L-Fuculose 1-phosphate ⇄ dihydroxyacetone phosphate + L-lactaldehyde

Assay Method

Principle. L-Fuculose 1-phosphate aldolase is assayed by determination of dihydroxyacetone phosphate with glycerol 1-phosphate dehydrogenase and DPNH[2]; the amount of DPNH utilized is measured spectrophotometrically at 340 mμ and is proportional to dihydroxyacetone phosphate in the mixture. Because of the presence of DPNH oxidase in the cruder preparations, these fractions are assayed in two steps: the aldolase preparation is incubated with L-fuculose 1-phosphate, and after inactivation of the enzyme, dihydroxyacetone phosphate is determined in an aliquot. In more purified preparations (step 4. Ammonium Sulfate Fraction), the amounts of contaminating enzymes have been sufficiently reduced to permit coupling of the aldolase reaction directly with the glycerol 1-phosphate dehydrogenase system; the rate of formation of dihydroxyacetone phosphate is determined spectrophotometrically in a continuous manner.

[1] This presentation is based on work previously published: M. A. Ghalambor and E. C. Heath, *J. Biol. Chem.* **237**, 2427 (1962).

[2] Racker, E., *J. Biol. Chem.* **167**, 843 (1947).

Reagents

L-Fuculose 1-phosphate, 0.02 *M*. Prepare with L-fuculo kinase (see this volume [83]) and isolate as the barium salt as previously described.[3] Appropriate amounts of the solid barium salt are dissolved in water (a trace of acetic acid is sometimes necessary to dissolve the salt) and treated with 10% molar excess of potassium sulfate. The barium sulfate precipitate is removed by centrifugation and washed with a small amount of water; the combined supernatant solutions are adjusted to the desired volume with water.

Tris-chloride buffer, pH 7.2, 0.1 *M*

Potassium fluoride, 1.0 *M*

DPNH, 0.01 *M*

Glycerol 1-phosphate dehydrogenase, 5 mg/ml (the crystalline enzyme preparation from Boehringer Mannheim Corporation was used; the specific activity of this preparation is 36 units/mg).

Procedure. For the two-step assay method, incubation mixtures contain 0.025 ml of L-fuculose 1-phosphate, 0.1 ml of Tris buffer, 0.005 ml of potassium fluoride, and 0.025–0.05 ml of aldolase solution in a final total volume of 0.2 ml. Control incubation mixtures contain all the components except L-fuculose 1-phosphate. After 5–20 minutes incubation at 37°, the tubes are placed in a boiling water bath for 1 minute, cooled, and centrifuged. Aliquots of the supernatant fluids are placed in cuvettes that contain 0.01 ml of DPNH, 0.1 ml of Tris buffer, and water to a total volume of 0.4 ml. Initial absorbancies at 340 mμ are determined, 0.005 ml of glycerol 1-phosphate dehydrogenase solution is added, and the absorbancy is followed until no further change occurs (about 5 minutes). The difference between the initial and final absorbancies is used to calculate the amount of dihydroxyacetone phosphate present in the sample.

In the single step assay procedure, incubation mixtures prepared in cuvettes contain 0.025 ml of L-fuculose 1-phosphate, 0.1 ml of Tris buffer, 0.005 ml of potassium fluoride, 0.01 ml of DPNH, and 0.005 ml of glycerol 1-phosphate dehydrogenase and water in a final volume of 0.4 ml. Control cuvettes contain all the components except L-fuculose 1-phosphate. After obtaining initial absorbancies at 340 mμ, L-fuculose 1-phosphate aldolase is added to all the cuvettes, and absorbancies are determined continuously with a spectrophotometer.

Definition of a Unit and Specific Activity. One unit of activity is defined as that amount required to form 1 micromole of dihydroxy-

[3] E. C. Heath and M. A. Ghalambor, *J. Biol. Chem.* **237**, 2423 (1962).

acetone phosphate per minute. Specific activity is defined as units per milligram of protein. Protein was determined by the method of Waddell.[4]

Purification Procedure

Step 1. Preparation of Cell-Free Extracts. Escherichia coli 0111-B_4 (ATCC 12015) is grown and cell-free extracts are prepared as previously described [83]. All procedures are conducted between 0° and 5°.

Step 2. Protamine Sulfate Treatment. A 10-ml aliquot of the crude extract (36 mg protein per milliliter) is diluted with 0.02 *M* phosphate buffer, pH 7, to a final protein concentration of 20 mg/ml. This solution is treated with 2 ml of a 2% solution of protamine sulfate and allowed to stand for 5 minutes; the precipitate is removed by centrifugation at 25,000 *g* for 10 minutes.

Step 3. Calcium Phosphate Gel I. The supernatant fluid, 19 ml, is diluted with 38 ml of phosphate buffer and slowly added, with stirring, to 36 ml of calcium phosphate gel suspension (10 mg solids per milliliter). The mixture is stirred occasionally over a period of 5 minutes and centrifuged at 2000 *g* for 5 minutes; the residue is discarded.

Step 4. Ammonium Sulfate Fractionation. To the supernatant fluid is added 40 ml of 1 *M* acetate buffer, pH 5.5, followed by 12 ml of water and 32 g of solid ammonium sulfate. The mixture is stirred until the ammonium sulfate dissolves and then allowed to stand for 5 minutes. After centrifugation at 25,000 *g* for 5 minutes, the inactive precipitate is discarded. Ammonium sulfate, 23 g, is slowly added with continuous stirring to the supernatant fluid, the mixture is stirred for 5 minutes and centrifuged at 25,000 *g* for 5 minutes. The resulting precipitate is dissolved in 1.5 ml of phosphate buffer (6.0 mg protein per milliliter).

Step 5. Dialysis. The ammonium sulfate fraction is dialyzed against

TABLE I

SUMMARY OF PURIFICATION PROCEDURE FOR L-FUCULOSE 1-PHOSPHATE ALDOLASE

Fraction	Specific activity ($\times 10^2$)	Recovery (%)
1. Crude extract	—[a]	—
2. Protamine sulfate	3.1	100
3. Calcium phosphate gel I	5.7	90
4. Ammonium sulfate	34	75
5. Dialysis	169	78
6. Calcium phosphate gel II	400	70

[a] Interfering enzymes in the crude extract prevent accurate determination.

[4] W. J. Waddell, *J. Lab. Clin. Med.* **48**, 311 (1956).

300 volumes of 0.01 *M* acetate buffer, pH 5.5, for 3 hours. The resulting precipitate was removed by centrifugation at 25,000 *g* for 5 minutes and discarded. The supernatant fluid contains 1.3 mg protein per milliliter.

Step 6. Calcium Phosphate Gel II. The dialyzed fraction from the preceding step is diluted with 3 ml of phosphate buffer to give a final concentration of 0.42 mg of protein per milliliter and then mixed with 1.5 ml of calcium phosphate gel suspension. The mixture is stirred for 5 minutes and centrifuged at 3000 *g* for 5 minutes; the supernatant fluid is retained.

The purification procedure is summarized in Table I.

Properties

Specificity. L-Fuculose 1-phosphate aldolase is essentially specific for L-fuculose 1-phosphate. Thus, the relative activity (formation of dihydroxyacetone phosphate) of the enzyme with a variety of ketose 1-phosphates is as follows: L-fuculose 1-phosphate, 100; D-fructose 1,6-diphosphate, 4; 6-deoxy-L-sorbose 1-phosphate, 5; L-sorbose 1-phosphate, 0; D-fructose 1-phosphate, 4; and D-ribulose 1,5-diphosphate, 6. While the enzyme is specific for dihydroxyacetone phosphate in the direction of condensation, each of a variety of aldehydes will serve as substrate as follows (relative rates): L-lactaldehyde, 100; D-lactaldehyde, 73; formaldehyde, <0.5; acetaldehyde, <0.5; glycolaldehyde, 10; glyceraldehyde, 33; L-glyceraldehyde 3-phosphate, <0.5.

Thus, the specificity of the aldolase appears to require a ketose 1-phosphate exhibiting *cis* configuration at carbons 3 and 4 in the direction of cleavage. As shown in Table II, the same specificity is maintained

TABLE II
CONDENSATION PRODUCTS WITH VARIOUS ALDEHYDES

Substrates	Product isolated
Dihydroxyacetone phosphate	
+ L-Lactaldehyde	L-Fuculose 1-phosphate
+ L-Glyceraldehyde	L-Tagatose 1-phosphate
+ D-Glyceraldehyde	D-Psicose 1-phosphate
+ Glycolaldehyde	D-Ribulose 1-phosphate

in the condensation reaction between dihydroxyacetone phosphate and the various aldehydes that serve as substrate in this reaction.

Thus, each of the products of condensation of dihydroxyacetone phosphate with the various aldehydes is a ketose 1-phosphate with *cis* configuration of hydroxyls at carbons 3 and 4 (to the right of the carbon chain when viewed in the Fischer projection formula).

Stability. The enzyme is relatively stable when stored at $-16°$; only 10–20% of the activity is lost over a period of several months of storage with repeated freezing and thawing of the preparation.

Effect of pH. Optimal activity of the enzyme is obtained in the range from pH 6.9 to 7.2; approximately 50% of the maximal activity is observed at pH 6.5 and 8.

Other Properties. The equilibrium constant for the reaction L-fuculose 1-phosphate → dihydroxyacetone phosphate + L-lactaldehyde is 4.6×10^{-4} *M*, and the K_m for L-fuculose 1-phosphate is 7×10^{-4} *M*.

[96] L-Rhamnulose 1-Phosphate Aldolase

By YASUYUKI TAKAGI

L-Rhamnulose 1-phosphate ⇌ dihydroxyacetone phosphate + L-lactaldehyde

Assay Method

Principle. The disappearance of L-rhamnulose 1-phosphate added as substrate is measured colorimetrically by the cysteine–sulfuric acid test for the ketohexoses[1]; or the reaction is followed by determining either dihydroxyacetone phosphate spectrophotometrically with α-glycerolphosphate dehydrogenase and DPNH,[2] or lactaldehyde colorimetrically.[3]

Reagents

L-Rhamnulose 1-phosphate, 0.01 *M*
KCN, 0.5 *M*
Glycine buffer, 0.4 *M*, pH 9.3
Trichloroacetic acid, 10%
Cysteine hydrochloride, 25%, freshly prepared
H_2SO_4, 75% by volume

Preparation of L-Rhamnulose 1-Phosphate. L-Rhamnulose was prepared chemically by refluxing L-rhamnose with pyridine according to the method of Barnett and Reichstein.[4] Then 200 micromoles of L-rhamnulose was incubated with partially purified L-rhamnulokinase (see this volume [84]) in the presence of 500 micromoles of ATP, 1 millimole of $MgCl_2$, 5 millimoles of KCl, and 2 millimoles of Tris buffer (pH 7.3) in a total

[1] Z. Dische and A. Devi, *Biochim. Biophys. Acta* **39**, 140 (1960).
[2] E. Racker, *J. Biol. Chem.* **167**, 843 (1947).
[3] S. B. Barker and W. H. Summerson, *J. Biol. Chem.* **138**, 535 (1941).
[4] J. Barnett and T. Reichstein, *Helv. Chim. Acta* **20**, 1529 (1937).

volume of 100 ml. After 60 minutes, the reaction was terminated by the addition of 10 ml of cold acetic acid, and nucleotides were removed with charcoal (Norit A). The reaction mixture was then adjusted to pH 7.2 and treated with 20 ml of 1 *M* barium acetate; four volumes of cold ethanol were added with gentle stirring. The suspension was kept at 0° overnight, and the resulting precipitate was collected by centrifugation at 2°, washed three times with 80% ethanol, and dried *in vacuo* over KOH. The precipitate thus obtained was dissolved in a small volume of 0.05 *M* acetic acid and converted to the sodium salt with a slight excess of Na_2SO_4. $BaSO_4$ formed was removed by centrifugation, and the supernatant solution was adjusted to pH 6.6.

Procedure. The incubation mixture (1.0 ml) contained 0.5 micromole of L-rhamnulose 1-phosphate, 50 micromoles of KCN, 40 micromoles of glycine buffer, enzyme, and water. After 10 minutes of incubation at 37°, the reaction was stopped by the addition of 1.0 ml of trichloroacetic acid solution and centrifuged. To a 0.5-ml aliquot of the deproteinized supernatant fluids was added 0.1 ml of cysteine hydrochloride solution and 6 ml of H_2SO_4.[1] The mixture was shaken and left at room temperature for 2 hours. Intensity of color was then read at 402 and 380 mμ, and the amount of rhamnulose phosphate was calculated by using the difference of optical densities at these two wavelengths.

Definition of Unit and Specific Activity. One unit of enzyme is defined as the quantity which causes the cleavage of 1 micromole of rhamnulose phosphate in 10 minutes. Specific activity is expressed as units per milligram of protein. Protein was determined by the turbidimetric method of Bücher.[5]

Purification Procedures

This is from the work of Sawada and Takagi.[6] *Escherichia coli,* strain B, was grown in a medium containing 0.8 g of KH_2PO_4, 0.3 g of $(NH_4)_2HPO_4$, 0.4 g of $(NH_4)_2SO_4$, 0.5 g of yeast extract, and 0.5 g of L-rhamnose in a final volume of 100 ml. The final pH of the medium was adjusted to 7.0. After 10 hours of incubation with vigorous shaking the cells were harvested by centrifugation and washed twice with cold distilled water; the packed cells were stored at —10°.

Preparation of Extract. Frozen cells, 2.5 g, were ground for 5 minutes with 2.5 g of alumina, and the resulting paste was extracted with 12 ml of cold distilled water. The insoluble residue was then removed by centrifugation at 6000 *g* for 15 minutes (extract, 9 ml).

[5] T. Bücher, *Biochim. Biophys Acta* **1,** 292 (1947).
[6] H. Sawada and Y. Takagi, *Biochim. Biophys. Acta* **92,** 26 (1964).

Protamine Fractionation. To the supernatant fluid was added 3 ml of a 1% protamine sulfate solution with stirring, and after 20 minutes the precipitate was removed by centrifugation at 10,500 *g* for 10 minutes (protamine fraction, 11 ml).

Alumina Cγ. Pilot tests with small volumes were always carried out to determine the maximal quantity of gel required to adsorb most of the inactive protein and not the enzyme. In a typical experiment, 11 ml of the protamine fraction was mixed with 1.8 ml of alumina gel Cγ (24 mg/ml), and allowed to stand at 3° for 20 minutes. The gel was removed by centrifugation at 6000 *g* for 10 minutes. The activity was confined to the supernatant fraction (alumina gel Cγ fraction, 11 ml).

Ammonium Sulfate Fractionation. To the alumina gel fraction was added an equal volume of cold saturated ammonium sulfate solution, and the mixture was kept at 2° for 30 minutes with occasional stirring. The resulting precipitate was collected by centrifugation at 6000 *g* for 10 minutes and dissolved in 7 ml of 0.005 *M* glycylglycine buffer (pH 7.0) (ammonium sulfate fraction, 7 ml).

This final fraction represents an approximately fifteenfold purification over the crude extract, with a yield of about 50%. The method has proved to be very reproducible in the hands of the author, and the enzyme loses no activity after storage for several months at —10°C. However, this fraction is still slightly contaminated with L-rhamnose isomerase, L-rhamnulokinase, and hexose diphosphate aldolase.

PURIFICATION PROCEDURE FOR L-RHAMNULOSE 1-PHOSPHATE ALDOLASE

Fraction	Units	Specific activity (units/mg protein)
Extract	32	0.2
Protamine fraction	30	0.8
Alumina gel C_γ fraction	29	2.5
Ammonium sulfate fraction	15	3.0

Properties

pH Optimum. The reaction proceeds at maximal velocity at approximately pH 9.3 in glycine buffer, with a considerable decrease in rate at higher or lower pH values. The K_m value for rhamnulose phosphate is estimated to be 1.0×1.0^{-2} *M*.

Cation Requirement. The addition of divalent cations to any of the purified enzyme preparations results in no stimulation of the reaction. Inclusion of Mg^{++} or Mn^{++} into the assay mixtures has no effect on the reaction velocity, but the presence of Co^{++} or Ni^{++} inhibits the activity at

$1 \times 10^{-3} M$ and $5 \times 10^{-3} M$ by 50 and 95%, respectively. The enzyme is unaffected by the addition of *p*-chloromercuribenzoate.

Reversibility. Formation of rhamnulose phosphate is indicated from the fact that incubation of L-lactaldehyde and dihydroxyacetone phosphate with the aldolase followed by treatment of the reaction product with acid phosphatase, yielded a sugar reacting with L-rhamnose isomerase and chromatographing on paper identically with rhamnulose. Incubation of these two C_3 substrates with hexose diphosphate aldolase results in no detectable formation of rhamnulose phosphate. On the other hand, the reaction product from D-lactaldehyde instead of L-isomer added as substrate is presumed to be 6-deoxy-D-sorbose from the comparison of the mobility in paper chromatography with the value reported in the literature for the possible diastereoisomers. Thus rhamnulose phosphate aldolase appears to require or produce the *trans*-configuration of hydroxyl groups at C-3 and C-4 of the ketose 1-phosphate, but is distinct from hexose diphosphate aldolase. Therefore rhamnulose phosphate aldolase, hexose diphosphate aldolase, and L-fuculose 1-phosphate aldolase[7] may be analogous in that each yields one of four possible diastereoisomers at C-3,4 in the condensation reactions.

[7] M. A. Ghalambor and E. C. Heath, *J. Biol. Chem.* **237**, 2427 (1962).

[97a] Deoxyribose 5-Phosphate Aldolase

I. *Lactobacillus plantarum*

By P. Hoffee, O. M. Rosen, and B. L. Horecker

Deoxyribose 5-phosphate ⇌ D-glyceraldehyde 3-phosphate + acetaldehyde

Assay Method[1]

Principle. The amount of deoxyribose 5-phosphate formed in the condensation of acetaldehyde and D-glyceraldehyde 3-phosphate is measured by the diphenylamine method of Dische.[2]

Reagents

Potassium maleate buffer, 1.0 *M*, pH 6.3
Redistilled acetaldehyde, 0.18 *M* (stock solution stored as 4 *M*)
Sodium fructose diphosphate, 0.05 *M*, pH adjusted to 6.0

[1] W. E. Pricer, Jr. and B. L. Horecker, *J. Biol. Chem.* **235**, 1292 (1960).
[2] Z. Dische, *Mikrochemie* **8**, 4 (1930).

Rabbit muscle fructose diphosphate aldolase. A solution containing 1.13 mg/ml is prepared from the crystal suspension.

Enzyme. Dilute enzyme to give optical density reading in the activity test of between 0.100 and 1.00 at 595 mμ.

Diphenylamine reagent. Dissolve 1.0 g diphenylamine in 100 ml glacial acetic acid and add 2.75 ml concentrated sulfuric acid.

Procedure. Place 0.1 ml of buffer solution, 0.1 ml of acetaldehyde, 0.1 ml of fructose diphosphate solution, 0.1 ml of muscle aldolase solution, and sufficient water in a tube to make a total volume of 1.00 ml. Add enzyme to begin the reaction. Incubate the reaction mixture at 37° for 10 minutes. Stop the reaction by the addition of 2 ml of diphenylamine reagent. Cover the tubes with marbles and heat for 10 minutes in a boiling water bath. Cool, and read absorbance at 595 mμ. Run control tubes without enzyme or without acetaldehyde.

Definition of Unit and Specific Activity. One unit of enzyme is defined as the quantity required to give a change in absorbance of 1.000, equivalent to 0.775 micromole of deoxyribose 5-phosphate. Specific activity is expressed as units per milligram of protein. Before the ammonium sulfate step (see Purification) protein is determined by the method of Lowry *et al.*[3]; after ammonium sulfate extraction, protein is determined by the method of Bücher.[4]

Preparation of Enzyme[5]

Growth Medium. Prepare a growth medium containing (per liter) 4 g of yeast extract, 20 g of nutrient broth, 20 g of sodium acetate·4 H_2O, 0.02 g of $MnCl_2$, and 8 g of fructose.

Growth of Cells and Induction. Inoculate 2-liter batches of growth medium in 2-liter flasks with 20 ml of an 8-hour subculture of *Lactobacillus plantarum* strain 124-2 (ATCC 8041) and incubate without aeration for 16 hours. Dilute the cultures in the morning with an equal volume of fresh medium, allow to grow for 1 hour, and then add sterile deoxyribose to a final concentration of 0.07%. When the deoxyribose is about 80% utilized, as measured with the diphenylamine reaction (3–4 hours), chill the cells and harvest in a refrigerated Sharples centrifuge. Store the cells as a frozen paste at —10°.

Purification Procedure

Step 1. Preparation of Cell Extract. Extract frozen induced cells by one of the following three methods: (a) Transfer 5 g of cells (wet

[3] O. H. Lowry, N. J. Rosebrough, A. L. Farr, and R. J. Randall, *J. Biol. Chem.* **193**, 265 (1951).

[4] T. Bücher, *Biochim. Biophys. Acta* **1**, 292 (1947).

[5] The procedure is based on previous publications.[1,7]

weight) together with 30 ml of 0.05 *M* NH_4OH-HCl buffer, pH 9.5, and 30 g of washed glass beads to a 10-kc Raytheon sonic oscillator and sonicate for 45 minutes at 40°. Centrifuge the mixture and wash the residue with 10 ml of buffer. (b) Grind 30 g of cells (wet weight) in a cold mortar with 60 g of Alumina A-301. Take up the paste in 120 ml of 0.05 *M* NH_4OH-HCl buffer, pH 9.5, and centrifuge. Wash the residues 3 times with 10 ml of buffer and combine the supernatant fluid and washings. (c) Suspend 50 g of cells (wet weight) in 250 ml of NH_4OH-HCl buffer, add 250 g of washed glass beads and blend in a Waring blendor for four 5-minute periods in a cold room at —10°. Centrifuge the mixture and wash the residues twice with 10 ml of buffer. Again treat the final residue in the Waring blendor in the same way and combine the supernatant fluid and washings. The specific activity of the extracts varies from 14 to 45 units per milligram protein, depending on the level of induction but not on the method of extraction. The Waring blendor treatment provides a method for extracting large amounts of cells in a relatively short time.

Step 2. Protamine Step. Adjust the extract to pH 7.0 with 5 *N* HCl and treat with a 2% solution of protamine sulfate (Lilly). The amount of protamine added should precipitate less than 2% of the activity at pH 7.0; determine this by an initial titration on small amounts of extract. After 10 minutes at 0°, remove the precipitate by centrifugation at 10,000 rpm. Adjust the supernatant fluid to pH 9.5 and keep at 0° for 2 hours. Collect the fine precipitate which forms by centrifugation at 17,500 rpm and dissolve it in 0.5 *M* potassium maleate buffer, pH 6.3. Dilute this solution 5-fold with water and centrifuge. Wash the precipitate with water and combine the supernatant fluid and washings.

Step 3. pH Precipitation. Dilute the protamine fraction with potassium maleate buffer, 0.02 *M*, pH 6.3, to 1 mg protein per milliliter, and adjust the pH to 9.5. After 10 minutes at 0°, collect the precipitate by centrifugation at 17,500 rpm and dissolve in 0.5 ml of 0.5 *M* potassium maleate buffer, pH 6.3. Dilute the solution with 10 ml of water and centrifuge at 10,000 rpm. Adjust the supernatant fluid to pH 8.1, keep for 10 minutes at 0°, and then adjust to pH 9.5. Keep this solution at 0° for 2 hours. Collect the precipitate by centrifugation at 17,500 rpm, dissolve it in 0.5 ml of 0.5 *M* potassium maleate, pH 6.3, and dilute with 10 ml of water. Repeat the precipitations at pH 8.1 and 9.5 three times in the same way.

Step 4. Ammonium Sulfate Extracts. Treat the final solution of the pH 9.5 precipitate (5 ml) with 3.15 g of solid ammonium sulfate to 90% saturation. After 30 minutes at 0°, centrifuge the suspension at 17,500 rpm. Extract the precipitate with three 0.2-ml portions each of 65%, 60%, 55%, and 50% saturated solutions of ammonium sulfate at pH 7.8, respectively. Assay each extract for enzyme activity and protein. Combine those having a specific activity of 4000 or greater.

Step 5. Zinc Crystals. To the combined solution add 1 *M* potassium maleate buffer, pH 6.3, until it is clear and then adjust to pH 7.8. Add $ZnCl_2$ to a final concentration of 10^{-3} *M*. In the cold, a silky precipitate forms. Set the solution at 4° for 2 days. Dissolve the crystals which have formed in 1 *M* potassium maleate buffer, pH 6.3, and adjust the solution to pH 7.8. Add $ZnCl_2$ to 10^{-3} *M* and then solid ammonium sulfate, slowly, until a slight turbidity appears, keeping the solution at pH 7.8 with small additions of 1 *M* NH_4OH. Cool the slightly turbid solution slowly to 4° and allow it to stand for 2 days to permit crystal formation. The final ammonium sulfate concentration should be about 50% of saturation, as tested with a conductivity meter.

The purification procedure is summarized in the table.

Purification Procedure for Deoxyribose Phosphate Aldolase of *Lactobacillus plantarum*

Fraction	Total volume (ml)	Units/ml	Total units	Protein (mg/ml)	Specific activity (units/mg)	Recovery (%)
1. Extract	450	95	43,000	2.7	35	—
2. Protamine fraction	43	630	27,000	1.6	400	63
3. pH precipitate	5	3130	15,650	1.08	2900	36.5
4. Ammonium sulfate extracts	2.5	2080	5,200	0.4	5200	12
5. Zinc crystals	1.0	4700	4,700	0.58	8000	11

Properties

Specificity. The enzyme shows greatest activity with acetaldehyde but will also react with propionaldehyde; with the latter substrate the product is 2-methyl-2-deoxypentose phosphate.[6] The other reactant, D-glyceraldehyde 3-phosphate, can be replaced by L-glyceraldehyde 3-phosphate, D-erythrose 4-phosphate, glyceraldehyde phosphate, and D-ribose 5-phosphate as well as by free glyceraldehyde and D-erythrose.[1]

Activators and Inhibitors. A carboxylic acid is required for maximum product formation.[7] Full activity is observed in the presence of maleate, succinate, glutarate, propionate, malonate, isobutyrate, or acetate. The enzyme is resistant to such sulfhydryl reagents as iodoacetic acid and *N*-ethylmaleimide.

[6] O. M. Rosen, P. Hoffee, and B. L. Horecker, *J. Biol. Chem.* **240,** 1517 (1965).
[7] P. Hoffee, O. M. Rosen, and B. L. Horecker, *J. Biol. Chem.* **240,** 1512 (1965).

Effect of pH. The purified enzyme shows a relatively sharp pH optimum with maximum activity at pH 6.3.

Equilibrium Constants. The cleavage of deoxyribose 5-phosphate to yield acetaldehyde and glyceraldehyde 3-phosphate is freely reversible with a K_{eq} of 2×10^{-4} *M*.

[97b] Deoxyribose 5-Phosphate Aldolase

II. Liver

By D. P. GROTH

Deoxyribose 5-phosphate ⇌ D-glyceraldehyde 3-phosphate + acetaldehyde

Deoxyribose aldolase has been purified from *E. coli*[1] and *L. plantarum*.[2] The procedures employed are not suitable for purification of the aldolase from rat liver, since the liver enzyme is very unstable.[3] With very mild conditions, however, the liver aldolase can be conveniently obtained in good yield.

Assay Methods

Principle. The metabolism of deoxyribose 5-phosphate to acetaldehyde and glyceraldehyde 3-phosphate is measured either colorimetrically with diphenylamine[4] when the crude preparations are assayed or spectrophotometrically with a modified procedure of Racker[5] when purified preparations are assayed. A very sensitive and specific assay can be employed with the highly purified preparations. In this assay the rate of incorporation of carbon-14 labeled acetaldehyde into deoxyribose 5-phosphate is measured.

Reagents

Tris buffer, 0.05 *M*, pH 7.5, containing 0.01 *M* EDTA
Sodium citrate, 0.1 *M*, pH 7.5
Deoxyribose 5-phosphate, 0.016 *M* pH 7.5
DPNH, 0.005 *M*
Acetaldehyde-^{14}C, 0.02 *M*, specific activity 5×10^5 cpm per micromole

[1] E. Racker, *J. Biol. Chem.* **196**, 347 (1952).
[2] W. E. Pricer, Jr. and B. L. Horecker, *J. Biol. Chem.* **235**, 1292 (1960).
[3] D. P. Groth, *Federation Proc.* **24**, 666 (1965).
[4] G. Ashwell, Vol. III [12].
[5] E. Racker, Vol. I [56].

DL-Glyceraldehyde 3-phosphate, 0.02 M, pH 6.5

Crystalline yeast alcohol dehydrogenase, 5 mg/ml, (Sigma lyophilized), containing bovine serum albumin, 1 mg/ml

Enzyme diluted with 0.005 M Tris buffer, pH 7.5, containing 0.001 M EDTA and 0.005 M 2-mercaptoethanol to obtain 100–1000 Racker[11] units per milliliter

Procedure. (a) WITH DIPHENYLAMINE. Add the following reagents to 8 × 70 mm test tubes: Tris buffer, 30 μl; sodium citrate, 10 μl; deoxyribose 5-phosphate, 10 μl; distilled water, 200 μl; diluted enzyme, 50 μl. Incubate at 22° for 30 minutes. Stop the reaction by the addition of 100 μl of 3 M perchloric acid. After centrifugation for 5 minutes at 1500 rpm the residual deoxyribose 5-phosphate is estimated in the supernatant according to Dische.[6] With crude enzyme preparations, an unincubated control is assayed at each enzyme concentration.

(b) WITH ALCOHOL DEHYDROGENASE. Add the following reagents to a 1-cm silica cuvette of 0.5-ml capacity: Tris buffer, 30 μl; sodium citrate, 10 μl; deoxyribose 5-phosphate, 10 μl; alcohol dehydrogenase, 10 μl; DPNH, 10 μl; distilled water, 180 μl; and diluted enzyme, 50 μl. A cuvette in which the deoxyribose 5-phosphate is omitted serves as control for each enzyme preparation. Absorbancy readings at 340 mμ are taken at 30-second intervals for 5 minutes at 22°.

(c) WITH ^{14}C-LABELED ACETALDEHYDE. Add the following reagents to 8 × 70 mm test tubes: Tris buffer, 15 μl; sodium citrate, 5 μl; glyceraldehyde 3-phosphate, 10 μl; acetaldehyde-^{14}C, 10 μl; distilled water, 60 μl; and diluted enzyme 50 μl. Incubate at 22° for 10 minutes. Stop the reaction by the addition of 50 μl of 0.2 M formic acid. Dilute the mixture with 3 ml of distilled water and apply to a 0.6 × 2 cm column of Dowex 1-formate. After addition of the sample, wash the column with 6 ml of distilled water. Enzymatically synthesized deoxyribose 5-phosphate-^{14}C is then eluted from the column with 2 ml of 1 M sodium formate. The ^{14}C is measured with a suitable liquid scintillation technique.

Definition of Unit and Specific Activity. Protein concentration is determined by the biuret method up to the hydroxylapatite step in the purification (below) and then spectrophotometrically.[7] Units are expressed according to Racker.[5]

Application of Assay Methods to Crude Tissue Preparations. The diphenylamine method appears to be valid for crude mammalian tissue homogenates from which the large particulates have been removed by a 10-minute centrifugation at 10,000 g.

[6] Z. Dische, *Mikrochemie* 8, 4 (1930).

[7] E. Layne, Vol. III [73].

Purification Procedure[3]

All steps in the procedure were carried out at 2–4°, and all solutions contained 0.005 *M* 2-mercaptoethanol and 0.001 *M* EDTA *unless stated otherwise.*

Step 1. Crude Extract. Seventy grams of rat liver (100–150 g animals are fasted overnight) is cut into small pieces and homogenized in 200 ml of 0.05 *M* Tris, pH 7.2, in a glass homogenizer with a tight-fitting Teflon pestle.

Step 2. Separation of Active Cell Fraction. Essentially all the activity of the homogenate is found to be localized in the soluble fraction not sedimented by centrifugation at 30,000 rpm for 90 minutes in a number 30 rotor in the Spinco Model L preparative ultracentrafuge.

Step 3. Ammonium Sulfate Fractionation. The soluble fraction is adjusted to 25% saturation by the slow addition of saturated ammonium sulfate, pH 7.4. The precipitate is collected by centrifugation and discarded. The ammonium sulfate concentration in the supernatant fraction is raised to 35% of saturation, and the resulting precipitate, after collection by centrifugation, is dissolved in 30 ml of 0.05 *M* Tris, pH 7.2. The solution is dialyzed twice for 45 minutes against 2-liter portions of 0.001 *M* phosphate buffer, pH 7.5, containing 0.005 *M* ammonium sulfate. The preparation at this stage can be stored frozen at —20° for at least 1 month with little loss in activity.

Step 4. Chromatography on DEAE-Cellulose.[8] The ammonium sulfate fraction above is thawed, centrifuged to removed denatured protein, if necessary, and diluted with 500 ml of 0.001 *M* phosphate buffer, pH 7.5, containing 0.005 *M* ammonium sulfate. The diluted solution is applied to a 3 $\times$ 25 cm DEAE-cellulose column with a flow rate of 200 ml per hour. The column is eluted with 200 ml of 0.001 *M* phosphate buffer, pH 7.5, containing 0.05 *M* ammonium sulfate and then with 125 ml of 0.001 *M* phosphate buffer, pH 7.5, containing 0.1 *M* ammonium sulfate. Discard these fractions. The column is then eluted with an additional 150 ml of the buffered 0.1 *M* ammonium sulfate solution. The majority of the aldolase activity appears in this fraction. All elutions are carried out at a flow rate of 150–200 ml per hour. The aldolase activity is precipitated by the slow addition of 150 ml of saturated ammonium sulfate, pH 7.2,

[8] Whatman DE 50, after repeated cycling through 1 *N* NaOH and 1 *N* HCl, is sedimented several times by gravity in a large volume of water to remove fines. The very coarse cellulose particles are removed by homogenization in a Waring blendor for 1 minute. The column is packed under 2 psi pressure with the DEAE suspended in distilled water and is equilibrated with 9 liters of 0.001 *M* phosphate buffer, pH 7.5, containing 0.005 *M* ammonium sulfate.

over a period of 30 minutes. The precipitate is collected by centrifugation and dissolved in 30 ml of 0.001 *M* phosphate buffer, pH 7.5, containing 0.005 *M* ammonium sulfate. This solution may be frozen and stored indefinitely at —20°, although freezing results in the loss of 40–50% of the activity.

Step 5. Hydroxylapatite Chromatography.[9] Hydroxylapatite is prepared according to Tiselius *et al.*[10] sedimented repeatedly by gravity from a large volume of 10^{-3} *M* sodium phosphate buffer, pH 6.8, to remove fines, and stored in the cold. The column is prepared as follows: A suspension of hydroxylapatite is poured into a 2 × 80 cm glass column to give a final bed height of 70 cm. The column is equilibrated with 0.01 *M* potassium phosphate buffer, pH 8.0, at a flow rate of 20–30 ml per hour. Fifteen milliliters of the DEAE fraction above are concentrated to 1–2 ml by ultrafiltration against 500 ml of 0.001 *M* phosphate buffer, pH 7.5, containing 0.005 *M* ammonium sulfate and 0.001 *M* sodium ascorbate. The concentrated enzyme solution is applied to the hydroxylapatite column, and elution is carried out first with 70 ml of 0.01 *M* phosphate buffer, pH 8.0, and second with 300 ml of 0.03 *M* phosphate buffer, pH 8.0. Column flow rate is maintained at 15 ml per hour. The aldolase activity appears in the final 30–35 ml of the 0.03 *M* phosphate buffer eluate. The active fractions are concentrated by ultrafiltration as above. The concentrated enzyme solution is then rechromatographed on hydroxylapatite, which increases the specific activity approximately 30–40% with little loss in enzyme activity. Freezing and storage at —20° for several months results in the loss of 50–70% of the activity, whereas the activity in the unfrozen solution has a half-life of approximately 1 week at 2°.

Step 6. Rechromatography on DEAE-Cellulose. The aldolase from step 5 still contains small amounts of triosephosphate isomerase activity. This activity is removed as follows: a 0.8 × 10 cm column of DEAE-cellulose is equilibrated with 0.001 *M* phosphate buffer, pH 7.5, containing 0.005 *M* ammonium sulfate and 0.001 *M* sodium citrate. Two-tenths milliliters of hydroxylapatite enzyme is diluted with 9 ml of the same buffer and 0.25 ml of 0.016 *M* deoxyribose 5-phosphate is added. After incubation at 22° for 30 minutes, the mixture is applied to the DEAE-cellulose column. Elution is carried out at room temperature with successive 10-ml portions of the following solutions: (a) 0.001 *M* phosphate buffer, pH 7.5, 0.001 *M* citrate, 0.05 *M* ammonium sulfate; (b) 0.001 *M* phosphate buffer, pH 7.5, 0.001 *M* citrate, 0.1 *M* ammonium sulfate; (c)

[9] EDTA is omitted from the solutions used for chromatography of the enzyme on hydroxylapatite.

[10] A. Tiselius, S. Hjerten, and Ö. Levin, *Arch. Biochem. Biophys.* **65**, 132 (1956).

0.001 M phosphate buffer, pH 7.5, 0.001 M citrate, 0.2 M ammonium sulfate. One-milliliter fractions of each solution are collected every 2–3 minutes. Isomerase-free deoxyribose aldolase is eluted in milliliters 4–7 of the 0.2 M ammonium sulfate fraction. These active fractions are used as such. The activity has a half-life of 24–48 hours at 2°.

The purification procedure is summarized in the table. Step 5 is repeated twice and step 6 five times to completely purify the aldolase from 70 g of liver.

PURIFICATION PROCEDURE FOR DEOXYRIBOSE ALDOLASE OF LIVER

Fraction	Volume (ml)	Aldolase activity (units)[a]	Protein concentration (mg/ml)	Specific activity (units/mg)	Recovery (%)
1. Crude extract	280	83,000	59	5	100
2. High speed supernatant	200	62,000	30	10	75
3. Ammonium sulfate fractionation	30	61,000	67.7	64	74
4. DEAE-cellulose chromatography	30	46,000	7.9	820	55
5. Hydroxylapatite chromatography	1.0	33,000	3.3	9,800	40
6. DEAE-cellulose	20	8,300	0.038	11,000	10

[a] A unit of enzyme activity is defined as a change in absorbancy at 340 mμ of 0.001 per minute at 22°. When the diphenylamine procedure is employed, this absorbancy change is calculated from the measured rate of metabolism of deoxyribose 5-phosphate.

Properties

Contamination with Other Enzymes. The purified enzyme is free from triosephosphate isomerase, glyceraldehyde 3-phosphate dehydrogenase, alcohol dehydrogenase, and glycerophosphate dehydrogenase.

Specificity. The purified enzyme has no measurable activity with deoxyribose 1-phosphate, deoxyribose, purine deoxyribonucleotides, and purine deoxyribonucleosides. Optimal activity with deoxyribose 5-phosphate in either Tris or phosphate buffer is at pH 7.4–7.5.

Activators. The purified enzyme is activated markedly by di- and tricarboxylic acids.[11]

Equilibrium.[2] The equilibrium constant for reaction (1) is approximately 2×10^{-4} at pH 7.5 and 22°.

Molecular Weight. The apparent minimal molecular weight of the

[11] N. Jiang and D. P. Groth, *J. Biol. Chem.* **237**, 3339 (1962).

aldolase is 253,000 as determined by the sucrose density gradient technique of Martin and Ames.[12]

Kinetic Properties.[3] The K_m values for the substrates and activators are 1.7×10^{-4} *M*, 2.67×10^{-4} *M*, 2.0×10^{-4} *M*, 1×10^{-4} *M*, 2×10^{-4} *M*, 5×10^{-4} *M* for deoxyribose 5-phosphate, acetaldehyde, glyceraldehyde 3-phosphate, citrate, succinate, and glutarate, respectively.

[12] R. G. Martin and B. N. Ames, *J. Biol. Chem.* **236,** 1372 (1961).

Section VI

Isomerases and Epimerases

[98a] Phosphoglucose Isomerase

I. Rabbit Muscle (Crystalline)[1]

By ERNST A. NOLTMANN

Glucose 6-phosphate ⇌ fructose 6-phosphate

Assay of Enzymatic Activity

Enzyme activity is measured either in the forward direction

glucose 6-phosphate → fructose 6-phosphate

by a coupled recording pH-stat assay, or in the reverse direction

fructose 6-phosphate → glucose 6-phosphate

employing a coupled spectrophotometric assay. The frequently used resorcinol method of Roe,[2] which analyzes for fructose 6-phosphate colorimetrically after chemically stopping the enzyme reaction, has not been found by the author to be very satisfactory, particularly when only one or a small number of determinations have to be made at one time, as is the case in enzyme purification.

Spectrophotometric Assay.[3] The production of glucose 6-phosphate is followed by the change in absorbance at 340 mμ produced by NADPH in a coupled enzyme system with fructose 6-phosphate as substrate and glucose 6-phosphate dehydrogenase as the indicator enzyme. The reaction mixture (2.0 ml) has the following composition (final concentrations): 0.1 *M* Tris, pH 8.0; 1.7 m*M* fructose 6-phosphate; and 0.5 m*M* NADP. Sufficient glucose 6-phosphate dehydrogenase is added to produce a low steady state level of glucose 6-phosphate in the coupled enzyme system, i.e., conditions where the isomerase velocity is equal to the glucose 6-phosphate dehydrogenase velocity (cf. Kahana *et al.*[4]). After all the glucose 6-phosphate (which may be present as an impurity in the fructose 6-phosphate solution) is oxidized, the reaction is recorded for several more minutes in order to obtain a rate for possible phosphoglucose

[1] D-Glucose 6-phosphate ketol-isomerase, EC 5.3.1.9.

The experimental work described in this article was supported in part by Grants AM 07203 from the U.S. Public Health Service and GB 2236 from the National Science Foundation and by Cancer Research Funds of the University of California.

The author wishes to acknowledge the permission of the Journal of Biological Chemistry to make quotations from his original publication.[3]

[2] J. H. Roe, *J. Biol. Chem.* **107,** 15 (1934).

[3] E. A. Noltmann, *J. Biol. Chem.* **239,** 1545 (1964).

[4] S. E. Kahana, O. H. Lowry, D. W. Schulz, J. V. Passonneau, and E. J. Crawford, *J. Biol. Chem.* **235,** 2178 (1960).

isomerase contamination in the glucose 6-phosphate dehydrogenase preparation employed. The isomerase reaction is then initiated by addition of 5 μl of suitably diluted enzyme sample and the time required to change the absorbance at 340 mμ by 0.1 is measured. When isomerase-free crystalline glucose 6-phosphate dehydrogenase is used, $1/\Delta t$ is directly proportional to the phosphoglucose isomerase concentration, whereas with the commercial dehydrogenase, the measured isomerase activity has to be corrected for isomerase present in the glucose 6-phosphate dehydrogenase aliquot, by subtraction of the rate obtained in the prereaction period. An arbitrary enzyme unit has been defined as the amount of enzyme per 1 ml of reaction mixture which, under the conditions described, requires the time of 1 minute to cause an increase in absorbance of 0.1. Multiplied by 0.0161 (derived from the molar absorbance index for NADPH of $6.22 \times 10^3 M^{-1}$ cm^{-1} at 340 mμ) the value obtained is equal to the micromoles of glucose 6-phosphate formed per minute per milliliter of assay reaction mixture.

Crystalline glucose 6-phosphate dehydrogenase free of phosphoglucose isomerase is prepared as described previously[5] and in this volume [23]. Commercial analytical grade glucose 6-phosphate dehydrogenase with low phosphoglucose isomerase content (e.g., Boehringer) is also satisfactory but usually requires correction of the overall velocity measured in the reaction mixture for the phosphoglucose isomerase contamination in the auxiliary enzyme. NADP of sufficient purity is available commercially. Difficulties may be encountered in obtaining high purity fructose 6-phosphate from commercial manufacturers. For absolute rate measurements, contributions by sugar phosphates other than fructose 6-phosphate should not exceed 1%. Highly purified fructose 6-phosphate preparations have been described by Hines and Wolfe,[6] Bonsignore *et al.*,[7] and Borrebaek *et al.*[8] The substrate is to be used in its sodium or potassium form. Concentrations of substrate and coenzyme solutions are determined by enzymatic assay.

pH-Stat Assay.[9] Formation of fructose 6-phosphate produced in the forward reaction from glucose 6-phosphate can be measured conveniently through coupling with phosphofructokinase and titration with dilute base of the hydrogen ions produced in the sequence:

$$\text{Fructose 6-phosphate} + ATP^{4-} \rightarrow \text{fructose 1,6-diphosphate} + ADP^{3-} + H^{+}$$

[5] E. A. Noltmann, C. J. Gubler, and S. A. Kuby, *J. Biol. Chem.* **236**, 1225 (1961).
[6] M. C. Hines and R. G. Wolfe, *Biochemistry* **2**, 770 (1963).
[7] A. Bonsignore, S. Pontremoli, G. Mangiarotti, A. De Flora, and M. Mangiarotti, *J. Biol. Chem.* **237**, 3597 (1962).
[8] B. Borrebaek, S. Abraham, and I. L. Chaikoff, *Anal. Biochem.* **8**, 367 (1964).
[9] J. E. Dyson and E. A. Noltmann, *Anal. Biochem.* **11**, 362 (1965).

Base equivalents required to maintain constant pH are stoichiometrically related to moles of substrate converted at pH 8.5 or above. The pH-stat assay has been described in detail[9] for the Radiometer Titrator–Titrigraph combination. The assay is performed in a thermostatted microtitration vessel with 2 m*M* NaOH as the titrant introduced into the reaction mixture by means of a precision-bore syringe of 0.5-ml total capacity. The standard reaction mixture contains final concentrations of 5 m*M* ATP, 5 m*M* magnesium sulfate, 0.08 *M* KCl, 3 m*M* glucose 6-phosphate, and 4–9 units of phosphofructokinase (diluted with 0.05 *M* cysteine of pH 8.5 and preincubated at room temperature for 5 minutes), adjusted to pH 8.5 in a total volume of 3.0 ml. After pH equilibration the reaction is initiated by addition of a 10- to 50-μl aliquot of phosphoglucose isomerase diluted in glass-distilled water to give approximately 0.1–0.2 unit per aliquot. Rate measurements corresponding to a base consumption of 0.02–0.35 μeq per minute take approximately 3 minutes when 30 rpm motors are used for both the pen and the chart drive of the Titrigraph recorder, with 2% gearing for the pen drive (60% of recorder scale per minute) and 2.5 mm per revolution (7.5 cm per minute) for the chart drive. Various degrees of sensitivity and accuracy can be obtained by proper choice of motor speeds for pen and chart drives. Daily blank titrations are required to obtain small but significant corrections for nonenzymatic hydrolysis of ATP and seepage of KCl from the calomel electrode.

Phosphofructokinase free of ATPase and other phosphatases, with a specific activity of 145–160 units per milligram of protein at 30° has been prepared in the author's laboratory by modification of an unpublished method of J. F. Morrison. Several equally good or better preparations are now described in the literature[10, 11] (cf. this volume [77]).

Phosphofructokinase used as auxiliary enzyme has to be dialyzed prior to use to remove ammonium sulfate or other inorganic salts whose addition may cause base consumption which would be registered by the instrument as an apparent reaction rate. It is also desirable to keep phosphofructokinase aliquots below 25 μl, or, if it is required to use larger aliquots, to reduce the concentration of cysteine. The buffering capacity of too much cysteine will otherwise cause irregularities in the recorded reaction rate curves. With care, the pH-stat assay permits rate measurements of phosphoglucose isomerase with an accuracy of $\pm$ 3% or better.

Comparison of the Spectrophotometric and the pH-Stat Assay. Both assays record initial velocity measurements under the specified conditions

[10] A. Parmeggiani and E. G. Krebs, *Federation Proc.* **24**, 284 (1965).
[11] K.-H. Ling, F. Marcus, and H. A. Lardy, *J. Biol. Chem.* **240**, 1893 (1965).

and have been found to give approximately equal values for the same enzyme sample, if proper temperature control is maintained. It should be kept in mind, however, that both assays measure initial velocity at substrate saturation under the conditions of assay and not true V_{max}. The spectrophotometric assay is somewhat dependent on the degree of purity of the fructose 6-phosphate employed, with the rate slightly affected by 6-phosphogluconate produced from glucose 6-phosphate impurities in the preequilibration period, and possibly also by other inhibitors[12] in commercial fructose 6-phosphate preparations. 6-Phosphogluconate produced during the assay does not interfere because the initial portion of the recorded reaction rate curve permits determination of the *initial* velocity. The suitability of the pH-stat assay, on the other hand, for crude enzyme preparations has to be investigated in each case, with possible interference coming from ATPase or other phosphatases. In the opinion of the author, however, both assays described are superior to a one-point chemical stop procedure depending on the colorimetric analysis of fructose 6-phosphate.

Enzyme units given in this chapter, except where otherwise indicated, are equivalent to micromoles of substrate converted per minute at 30° under standard assay conditions. Specific activity is expressed as units per milligram of protein. Protein is determined with the biuret procedure of Gornall *et al.*[13] or, at stages after elution from the DEAE-cellulose, by spectrophotometric measurement of the absorbance at 280 mμ. The biuret factor for 10.0 ml total reaction volume is 32 mg of protein per absorbance unit for a 10-mm light path at 540 mμ.[14] The absorbance of 1 mg of enzyme per milliliter at 280 mμ (10-mm light path) in 0.01 *M* phosphate of pH 7.0 is 1.32, as calibrated against the biuret assay.[14]

Isolation Procedure

A method for partial purification of rabbit muscle phosphoglucose isomerase was described in Vol. I [37] by M. W. Slein, who also suggested distinguishing this enzyme from phosphomannose isomerase which catalyzes the conversion of fructose 6-phosphate to the other epimeric

[12] Significant inhibition has been observed in the author's laboratory with commercial fructose 6-phosphate preparations of particularly poor quality. This inhibition could not have been caused by 6-phosphogluconate since the glucose 6-phosphate content of these preparations, as determined by enzymatic assay, was not unusually high. For a comparison of the fructose 6-phosphate content of several commercial preparations designated as fructose 6-phosphate, refer to the paper by M. K. Schwartz and O. Bodansky, *Anal. Biochem.* **11,** 48 (1965).

[13] A. G. Gornall, C. J. Bardawill, and M. M. David, *J. Biol. Chem.* **177,** 751 (1949).

[14] Unpublished experiments, 1964

aldohexose 6-phosphate. The following procedure, which, except for the extraction step, is essentially that previously reported,[3] describes the isolation of the enzyme in crystalline form.

Unless otherwise stated, all manipulations are carried out in a cold room at 2–4° or in an ice bath. Required amounts of ammonium sulfate to obtain a certain degree of saturation, ammonium sulfate concentrations in dissolved ammonium sulfate-precipitated pellets, and volumes of solvent additions are calculated as described previously[5] and in this volume [23].

Fraction I. Two rabbits are killed by a blow on the neck, bled and skinned. Back and leg muscles are excised and passed through a chilled meat grinder. The minced muscle is then homogenized in 100-g portions in a Waring blendor, each batch for 1 minute with 300 ml of 0.01 *M* KCl. The homogenates are combined and are stirred for 30 minutes with a heavy duty, stainless steel propeller-blade, mechanical stirrer. Thereafter, the cell debris is removed by centrifugation (15 minutes at 15,000 *g*). The volume of the opaque solution obtained (fraction I) is approximately equal to that of the KCl used.

Fraction II. A solution of 1 *M* zinc acetate is added to give a final concentration of 0.03 *M*, and the pH is adjusted with ice-cold 1 *N* NaOH to 8.5. After 1–2 hours the thick, white suspension is centrifuged for 15 minutes at 15,000 *g* (e.g., Servall GSA rotor), and the precipitate is suspended in 0.25 *M* ammonium citrate of pH 8.5, (0.2 volume of fraction I) and is kept overnight at 0–4°. A considerable precipitate appears during that period and is removed the next day by centrifugation (30 minutes at 15,000 *g*). The precipitate is washed with 0.04 volume of fraction I of 0.25 *M* ammonium citrate, pH 8.5. Both extracts are combined and the slightly turbid, reddish solution is fraction II.

Fraction III. Solid ammonium sulfate is added to fraction II to give 0.33 saturation and the precipitate is removed by centrifugation (20 minutes at 15,000 *g*). The supernatant liquid is brought to 0.63 saturation and the precipitate is collected (30 minutes at 15,000 *g*). The pellet is dissolved in 0.125 volume of fraction II of 0.05 *M* EDTA, pH 8.0, and the solution is dialyzed (Visking thin-wall seamless cellulose tubing, 36/32-inch inflated diameter) overnight against three to four 6-liter portions of 0.005 *M* magnesium acetate or more conveniently against an equivalent volume by flow dialysis.[5] Any insoluble material is removed by centrifugation, and the clear, red solution is called fraction III.

Fraction IV. The protein concentration of fraction III is determined and diluted to 15 mg/ml with 0.005 *M* magnesium acetate. The pH (approximately 6.5–7.0) is brought to 8.6 with ice-cold 0.5 *N* NH_4OH (approximately 0.007–0.01 v/v of solution is required). In a −10° bath, cold 95% ethanol is slowly added with efficient stirring to bring the ethanol concentration to 34% (with respect to 95% ethanol). Care must be exercised that the temperature in the solution does not exceed 0°. The solution is centrifuged (15 minutes at 15,000 *g*) at a running temperature setting of —15°, 20 minutes after the last addition of the ethanol. The precipitate is discarded and the ethanol concentration in the supernatant fluid is slowly increased to 64% (of 95% ethanol). After centrifugation (15 minutes at 15,000 *g*, —15°), the pellet is suspended in 0.1 *M* magnesium acetate (0.25 volume of fraction III after dilution to 15 mg/ml). The volume is measured, the suspension is centrifuged (20 minutes at 15,000 *g*), and the clear dark yellow extract is fraction IV.

Fraction V. The protein concentration of fraction IV should be approximately 15 mg/ml. The ethanol concentration in fraction IV is estimated by assuming the concentration in the volume increase due to the suspended pellet to be equal to the ethanol concentration in the previous fraction. Fraction IV is then brought to 40% (v/v) with acetone (A. R. grade) by counting the ethanol present in fraction IV with two-thirds of its percentage toward the total of 40% for the acetone. The acetone is chilled in a —10° bath and is added slowly in the —10° bath to fraction IV with care not to exceed 0° in the mixture. The solution is then equilibrated at 0° in an ice bath and centrifuged for 15 minutes at 15,000 *g* with the running rotor temperature set at —5°. The heavy precipitate is discarded, and the acetone concentration in the supernatant liquid is increased at 50% (v/v). After 10 minutes the precipitate is collected by centrifugation (15 minutes at 15,000 *g*, —5°) and dissolved in 0.5 volume of fraction IV of 0.05 *M* magnesium acetate. The slightly opalescent, pale yellow solution is designated as fraction V.

Fraction VI. The protein concentration of fraction V is determined and diluted to 7 mg/ml with 0.05 *M* magnesium acetate. The pH is adjusted to 7.4, if necessary, with 0.2 *M* acetic acid and bentonite (Fisher Scientific Co.) is added with vigorous stirring, at a ratio of 4 g per gram of protein. Stirring is continued for exactly 10 minutes, and the adsorbent is removed immediately by centrifuging for 30 minutes at 15,000 *g*. The clear supernatant liquid (fraction VI) is significantly lighter in color than fraction V.

Fraction VII. To concentrate the dilute protein in the bentonite supernatant liquid, an equal amount of chilled acetone is slowly added to fraction VI in a —10° bath with care to avoid raising the temperature in the sample above 0°. Approximately 10 minutes after the last addition of acetone the milky solution is centrifuged (15 minutes at 15,000 *g*); the small pellet, which may have a pinkish tint, is dissolved in 0.005–0.01 volume of fraction VI of 0.2 *M* Tris, pH 8.0, and is dialyzed overnight against at least a 1000-fold volume of 0.01 *M* sodium phosphate, pH 7.0. The clear, almost colorless dialyzate is fraction VII.

Crystallization. The protein concentration of the dialysate should not be lower than 30 mg/ml. Neutralized, saturated ammonium sulfate[15] solution is added to 0.45 saturation, which will cause a sudden, considerable increase in viscosity. The sample is immediately centrifuged for 5 minutes at 25,000 *g* to remove traces of insoluble material. Thereafter, the ammonium sulfate concentration is cautiously brought to 0.48–0.50 saturation until the sample turns slightly opalescent. Simultaneously, on gentle swirling, a silky luster can be observed, whose intensity increases within the next hours. In the course of the following days, the ammonium sulfate saturation is gradually increased to about 0.08 above the saturation at which crystallization initiated, at a rate of not more than 0.01–0.02 per day. Within 2 days after reaching the final saturation, approximately 80% of the active protein will have crystallized. For recrystallization the crystalline suspension is centrifuged for 1 hour at 25,000 *g*, the mother liquor is removed with a transfer pipette, and the crystals are dissolved in 0.01 *M* phosphate, pH 7.0 (one-fourth to one-third the volume of the crystalline suspension). The ammonium sulfate saturation in the dissolved pellet is estimated, and addition of neutralized, saturated ammonium sulfate solution is made as before. Further crystallizations are carried out in the same manner.

The first two mother liquors may show a pink tint whereas additional ones are colorless. Also, packed crystalline pellets after centrifugation except for the first one are pure white in appearance. Usually, three to five crystallizations are required until constant specific activity is obtained, as may be seen from the table, which summarizes data for a typical preparation. The procedure, without consideration of the time required for crystallization and recrystallizations, can be carried through in 3 days, the first night being used for precipitation of inert protein from the ammonium citrate extract of the zinc acetate-precipitated pellet, the second night for dialysis prior to ethanol fractionation, and the night after the third day for dialysis in preparation for the crystallization. However, if desired, the procedure can be interrupted at almost any stage without loss in activity as long as care is exercised not to leave the enzyme in high concentrations of organic solvent without lowering the temperature accordingly. The procedure as described here has been reproduced in the author's laboratory by several individuals, with results essentially as shown in the table, which is in good agreement with that

[15] For crystallization, Mann Assayed Ammonium Sulfate, Special Enzyme Grade, is preferred over the regular A. R. Grade salt.

FRACTIONATION OF PHOSPHOGLUCOSE ISOMERASE (PREPARATION GCRM-16[a,b])

Fraction	Volume (ml)	Total protein (mg)	Total activity (units[c] $\times 10^{-3}$)	Specific activity (units/mg protein)	Purification		Recovery (%)	
					Overall	Over preceding step	Overall	From preceding step
I. KCl extract	3,440	42,000	445	10.6	—	—	(100)	(100)
II. Zinc acetate ppt	1,020	17,650	352	20.0	1.9	1.9	79	79
III. $(NH_4)_2SO_4$ ppt, 0.63 saturation, after dialysis	244	8,390	384	45.8	4.3	2.3	82	103
IV. Ethanol ppt, 64%	144	2,260	267	118	11	2.6	60	73
V. Acetone ppt, 50%	76	1,400	222	159	15	1.3	50	83
VI. Bentonite supernatant	187	563	138	245	23	1.6	31	62
VII. Second acetone ppt, after dialysis	6.0	436	135	310	29	1.3	30	98
Crystallizations								
1. Crystals	6.0	203	109	533	50	1.7	25	81
2. Crystals	17.3	154	108	702	66	1.3	24	98
3. Crystals	14.6	143	109	758	71	1.1	25	104
4. Crystals	14.3	125	92	736	69	1.0	21	84
5. Crystals	9.2	115	87	760	72	1.0	20	95
5. Mother liquor	15.5	4	3	646	—	—	—	—

[a] Initially, 1180 g of ground rabbit skeletal muscle.
[b] Carried out by G. C. Chatterjee.
[c] Micromoles of glucose 6-phosphate formed per minute from fructose 6-phosphate at 30° measured by the spectrophotometric assay.

given in the original description[3] of the isolation procedure. Differences in the absolute values reflect differences in the assay conditions used.

Properties

Work on the physical, chemical, and kinetic properties of the crystalline enzyme has only recently been started and final data are not yet available. Sedimentation velocity studies[16] yielded values of 7.2 S and 5.1 D for $S_{20,w}$ and $D_{20,w}$, respectively, indicating a molecular weight of 130,000–140,000.

[16] G. C. Chatterjee and E. A. Noltmann, unpublished experiments, 1965.

[98b] Phosphoglucose Isomerase

II. Mammary Gland

By F. J. Reithel

Glucose 6-phosphate $\rightleftharpoons$ fructose 6-phosphate

Assay Method

Principle. The determination of the fructose 6-phosphate formed from glucose 6-phosphate by the Roe method[1] is preferable for most purposes. In this method a derivative formed from fructose in acid reacts with resorcinol to form a pink color. An alternative method[2] is a linked assay system employing phosphofructose kinase. It is also possible to determine the speed of the reaction in the opposite direction by using glucose 6-phosphate dehydrogenase. The disadvantage of the latter method is the requirement for highly purified assay enzyme and substrate.

Reagents

Tris-acetate buffer, 0.1 *M*, pH 8.0 (0.1 *M* with respect to acetate), which is 10^{-3} *M* with respect to glucose 6-phosphate

Resorcinol, 0.1% reagent grade, in 95% ethanol

Hydrochloric acid, concentrated

Enzyme. Dilute the enzyme solution with distilled water or 0.1 *M* Tris-acetate buffer to obtain a concentration of about 5 units/ml (see definition of unit below).

[1] J. H. Roe, *J. Biol. Chem.* **107**, 15 (1934).

[2] S. E. Kahana, O. H. Lowry, D. W. Schulz, J. V. Passonneau, and E. J. Crawford, *J. Biol. Chem.* **235**, 2178 (1960).

Procedure. To 20 × 150 mm test tubes is added 4.0 ml of the substrate solution; the tubes are then allowed to come to temperature equilibrium in a 30° bath. The reaction is initiated by introducing 0.01 ml of enzyme solution with a micropipette or a Lucite holder. The reaction is stopped after 1, 2, and 5 minutes by blowing in 1.0 ml concentrated HCl. To each tube is then added 4.0 ml of the resorcinol reagent plus 11.0 ml concentrated HCl (by burette). The color is developed by heating to 80° for 10 minutes (total elapsed time). After rapid cooling to room temperature, the optical densities are measured at 530 mμ. Highly purified fructose 6-phosphate produces 95% of the optical density of an equimolar quantity of fructose. This correction must be applied when fructose is used as a standard.

Definition of Unit and Specific Activity. A unit of enzyme is that amount which catalyzes the formation of 1.0 micromole of fructose 6-phosphate per minute. Specific activity is expressed as units per milligram of protein.

Purification Procedure

Either lactating or nonlactating bovine mammary gland may be used, but the former is preferable because of higher activity. Either fresh or deep-frozen glands may be used.[3]

Step 1. Preparation of Extract. Frozen gland trimmed of fat is cut into very thin slices (about 2 mm) with a commercial meat slicer and suspended in a volume of water corresponding to half the weight of the gland used. Mixing is best done by hand. After about 30 minutes' extraction the liquid extract is separated by filtration through cheesecloth. The pulp is washed again by resuspending it in one-half the former volume of cold water; the mass is filtered again through cheesecloth. The combined filtrates are centrifuged clear (800 *g*, 30 minutes).

Step 2. Ammonium Sulfate Fractionation. The enzyme is precipitated at a concentration of 2.8 *M* ammonium sulfate and a pH of 5.2 (430 g salt added per liter of extract). The precipitate is filtered off and suspended in 1.84 *M* ammonium sulfate at pH 7.3. The volume is equivalent to half the weight of the gland extracted. The enzyme is extracted by stirring about 30 minutes in the cold room and filtering. The filtrate is made 3.4 *M* in ammonium sulfate (205 g per liter of 1.84 *M* extract), and the precipitate is allowed to settle. The precipitate is collected by centrifugation, then it is dissolved in a minimal volume of 10^{-3} *M* Tris-acetate buffer, pH 7.3, and dialyzed free of ammonium sulfate. All steps are to be carried out at 4°.

[3] A. Baich, R. G. Wolfe, and F. J. Reithel, *J. Biol. Chem.* **235**, 3130 (1960).

Step 3. DEAE-Cellulose Column. The enzyme solution is percolated through a 6 × 12 cm column of DEAE-cellulose previously equilibrated with 0.001 *M* Tris-acetate buffer, pH 7.3. The enzyme is not adsorbed by DEAE-cellulose, but contaminating proteins are.

Step 4. Hydroxylapatite Column. The eluate from the last step is next added to a 4 × 26 cm column of hydroxylapatite equilibrated with 10^{-3} *M* Tris-acetate, pH 7.3. The column is developed with liter volumes of 0.02, 0.03, 0.04, and 0.05 *M* sodium phosphate buffer, pH 7.3. Most of the activity will be found in the 0.04 *M* eluate.

Step 5. Carboxymethyl Cellulose Column. The eluate from the preceding step is freed of phosphate ions by dialyzing it against 10^{-3} *M* Tris-acetate, pH 7.3. After concentration by ultrafiltration, the solution is introduced on a 4 × 26 cm column of carboxymethyl cellulose. The activity is eluted with 0.06 *M* Tris-acetate buffer, pH 7.3.

Step 6. Zone Electrophoresis. The enzyme solution is dialyzed against 0.05 *M* Tris-acetate, pH 7.75, and concentrated. The enzyme is applied in a minimal volume to a 3.5 × 24 cm starch column in a Flodin-Porath apparatus.[4] Electrophoresis under these conditions provides good separation in 24–30 hours at 15 ma. The eluate contains an enzyme which exhibits a single boundary in free boundary electrophoresis or sedimentation by the ultracentrifuge.

A summary of this procedure and its results are provided by the table.

PURIFICATION PROCEDURE FOR MAMMARY GLAND PHOSPHOGLUCOSE ISOMERASE

Fraction	Protein (mg)	Total units	Specific activity
Extraction (10 liters)	80,000	100,000	1.2
Ammonium sulfate fractionation	10,500	50,000	4.8
DEAE-cellulose	2,300	35,000	15
Hydroxylapatite	220	30,000	140
Carboxymethyl cellulose	42	30,000	710
Zone electrophoresis	24	28,000	1160

Properties

Specificity. In addition to the substrates noted, glucose 6-sulfate also appears to be acted upon. The enzyme is not known to have contaminants, but does exhibit mutarotase activity. It is this activity that lessens the utility of rotation measurements for assay purposes. There is no

[4] P. Flodin and J. Porath, *Biochim. Biophys. Acta* **13**, 175 (1954).

detectable conversion of glucose to fructose catalyzed by this enzyme even at very high substrate concentrations.

pH. The maximum initial velocity is attained close to pH 8 but the exact value depends on the substrate used.[5] The activity drops by half if the pH is lowered to 6.5 or raised to 10.

Physical Constants. The molecular extinction coefficient at 280 $m\mu$ is 4.0×10^4. At 30° the molecular activity was found to be 4.8×10^4 moles G-6-P/min/mole enzyme assuming a molecular weight of 125,000. The equilibrium constant for F-6-P:G-6-P = 0.32. The pH independent maximal initial velocity is 5×10^4 for G-6-P and 9.7×10^4 for F-6-P. The Michaelis constant is 1.2×10^{-4} for G-6-P and 7×10^{-5} for F-6-P at pH 8. The $S_{20,w}$ extrapolated to zero protein concentration is 7.09, and the molecular weight determined by sedimentation equilibrium methods is 125,000. The pycnometric value for the partial specific volume is 0.75 at 25°.

Stability. The activity of purified phosphoglucose isomerase does not diminish over a period of several weeks when it is kept in ammonium sulfate solutions as a slurry at cold-room temperatures. In solution at pH 7 the activity decreases slowly.

[5] M. C. Hines and R. G. Wolfe, *Biochemistry* **2**, 770 (1963).

[99] Phosphohexose Isomerase: Clinical Aspects

By Oscar Bodansky and Morton K. Schwartz

Introduction

Phosphohexose isomerase (PHI) catalyzes the reaction: glucose-6-phosphate (G-6-P) $\rightleftharpoons$ fructose-6-phosphate (F-6-P). Elevations of this enzyme activity have been found in the serum of patients with cancer,[1-9]

[1] O. Bodansky, *Cancer* **7**, 1191 (1954).
[2] O. Bodansky, *Cancer* **7**, 1200 (1954).
[3] O. Bodansky, *Cancer* **8**, 1087 (1955).
[4] O. Bodansky, *Cancer* **10**, 865 (1957).
[5] M. A. Schwartz, M. West, W. S. Walsh, and H. J. Zimmerman, *Cancer* **15**, 346 (1962).
[6] M. A. Schwartz, W. S. Walsh, M. West, and H. J. Zimmerman, *Cancer* **15**, 927 (1962).
[7] M. West, M. A. Schwartz, W. S. Walsh, and H. J. Zimmerman, *Cancer* **15**, 931 (1962).
[8] A. Rose, M. West, and H. J. Zimmerman, *Cancer* **14**, 726 (1961).
[9] C. O. Tan, J. Cohen, M. West, and H. J. Zimmerman, *Cancer* **16**, 1373 (1963).

hepatitis,[10,11] myocardial infarction,[12] various myopathies,[13] leukemia,[14,15] and other diseases. In metastatic carcinoma, elevations are associated with the progression of the disease and decrease with its remission, as judged by clinical observation and by biochemical criteria, such as urinary calcium excretion, serum calcium levels, or liver function tests.[2,3,16-18] Elevation in the serum PHI may be frequently an early herald of a worsening in the patient's metastatic disease. In contrast to several other enzymes that have recently been studied, electrophoretic and immunochemical studies have indicated the probable existence of only one molecular form.[19,20]

The PHI activity of other body fluids, such as cerebrospinal fluid and pleural and peritoneal fluid, has also been investigated in some detail.[21-23] The PHI activity in cerebrospinal fluid has been reported to be a more sensitive indicator than LDH or glutamic-oxalacetic transaminase (GOT) of the presence of secondary and primary tumors of the central nervous system and of meningitis and meningoencephalitis caused by pyogenic, viral, or yeast organisms.[21] Brauer *et al.* noted that the ratio of PHI activity in serum to that in the effusion was elevated in 72% of patients with neoplastic disease and in about 45% of patients with benign effusions due to heart failure or cirrhosis.[23] Study of PHI activity in gastric juice has yielded no distinctive diagnostic information.[24]

Assay Procedure

Principle. Fructose reacts in solution with resorcinol and HCl at a temperature of about 80° to form a red-colored compound. This is the

[10] O. Bodansky, S. Krugman, R. Ward, M. K. Schwartz, J. P. Giles, and A. M. Jacobs, *A.M.A. J. Diseases Children* **98,** 166 (1959).
[11] H. F. Bruns and W. Jacob, *Klin. Wochschr.* **32,** 104 (1954).
[12] R. J. Bing, A. Castellanos, and A. Siegel, *J. Am. Med. Assoc.* **164,** 647 (1957).
[13] G. Schapira, J. C. Dreyfus, F. Schapira, and J. Kruh, *Am. J. Phys. Med.* **32,** 313 (1955).
[14] L. G. Israels and G. E. Delory, *Brit. J. Cancer* **10,** 318 (1956).
[15] M. C. Blanchaer, P. T. Green, J. P. Maclean, and M. J. Hollenberg, *Blood* **13,** 245 (1958).
[16] G. F. Joplin and K. Jegatheesan, *Brit. Med. J.* **I,** 827 (1962).
[17] M. M. Griffith and J. C. Beck, *Cancer* **16,** 1032 (1963).
[18] M. K. Schwartz, E. Greenberg, and O. Bodansky, *Cancer* **16,** 583 (1963).
[19] M. N. Lipsett, R. B. Reisberg, and O. Bodansky, *Arch. Biochem. Biophys.* **84,** 171 (1959).
[20] M. K. Schwartz and O. Bodansky, *Am. J. Med.* **40,** 231 (1966).
[21] H. G. Thompson, Jr., E. Hirschberg, M. Osnos, and A. Gellhorn, *Neurology* **9,** 545 (1959).
[22] K. Hulanicka, R. Arend, and M. Orlowski, *Arch. Neurol.* **8,** 194 (1963).
[23] M. J. Brauer, M. West, and H. J. Zimmerman, *Cancer* **16,** 533 (1963).
[24] D. D. Piper, M. L. Macon, F. L. Broderick, B. H. Fenton, and J. E. Builder, *Gastroenterology* **45,** 614 (1963).

well known Seliwanoff reaction for ketose sugars and consists in the formation, first, of hydroxymethylfurfural and, secondly, in the condensation of this compound with resorcinol to form a cherry red colored product or products.[25] Heating under the usual conditions of this reaction, 15 minutes at 80° in 6 *N* HCl, results in the hydrolysis of F-6-P to fructose and inorganic phosphate to the extent of 40–50%.[26] Fructose, as well as the unhydrolyzed F-6-P, react to the same extent with HCl and resorcinol on the basis of the fructose content.[27] The method for PHI to be described here is based on the conversion of G-6-P to F-6-P at pH 7.4 and 37°.

Reagents. All solutions of organic compounds should be covered with a few drops of toluene and kept in the refrigerator.

Tris buffer, 0.1 *M*, pH 7.4. Dissolve 12.1 g tris(hydroxymethyl)-amino methane in about 800 ml water. Adjust pH to 7.4 with 10 *N* HCl. Dilute to 1000 ml.

Disodium D-glucose 6-phosphate, 0.03 *M* (stock solution). Dissolve 5.0 g of a good grade of this compound[28] which contains 1.5 moles of H_2O per mole of compound (molecular weight, 331) in about 450 ml H_2O. If necessary, adjust pH to 7.40 and bring final volume to 500 ml with water. The exact molarity of the stock solution should be determined by analyzing for inorganic and total organic phosphorus by the method of Fiske and SubbaRow.[29]

G-6-P, 0.00625 *M*, buffered. Dilute 104 ml of 0.03 *M* stock G-6-P solution to 500 ml with 0.1 *M* Tris buffer. If the molarity of the stock solution is not precisely 0.03 *M*, then correspondingly larger or smaller amounts of this stock solution are used.

HCl, 10 *N*. Mix 1500 ml concentrated HCl with 300 ml distilled water.

0.1% resorcinol–95% ethanol. Dissolve 1 g resorcinol in 1000 ml of 95% ethanol.

D-Fructose standard, 0.0012 *M*. Dissolve 216 mg of D-fructose in 1000 ml distilled water.

Trichloroacetic acid, 5%

NaCl, 0.85%

[25] W. A. Ekenstein and J. J. Blankona, *Ber. Deut. Chem. Ges.* **43**, 2355 (1910).

[26] M. K. Schwartz and O. Bodansky, unpublished observations, 1964.

[27] M. K. Schwartz and O. Bodansky, *Anal. Biochem.* **11**, 48 (1965).

[28] We have found that a good grade of this compound is available from the Sigma Chemical Co., St. Louis, Missouri.

[29] C. H. Fiske and Y. SubbaRow, *J. Biol. Chem.* **66**, 375 (1925).

Enzyme. Serum or other biological fluid. After collection, PHI in the serum is stable for at least 10 hours at room temperature, for up to 21 days in the refrigerator, and for at least one year in the deepfreeze.[30]

Procedure. Into a small test tube pipette 4 ml of buffered 0.00625 *M* substrate and 0.8 ml of 0.85% NaCl. Mix well and place in water bath at 37° for 3 minutes. To start the reaction add 0.2 ml of serum which has been brought to 37°. Mix rapidly. Within the first 15–30 seconds of the reaction withdraw 2.00 ml mixture and add to 2.00 ml of 5% TCA. Filter. At 30 minutes after the beginning of the reaction, withdraw a second aliquot of 2.00 ml of the reaction mixture, add to 2.00 ml 5% TCA, and filter. Place 2.00 ml of each filtrate into tubes of approximately 20–25 ml capacity. Add 2.00 ml of the resorcinol reagent and 6.00 ml of the HCl reagent. At the same time set up one blank containing 2 ml of H_2O and a set of standard solutions containing 0.1 ml, 0.3 ml, 0.5 ml, and 0.7 ml of the standard fructose solution in 2 ml of water. To the blank and each of these standards, add 2.0 ml of the resorcinol reagent and 6.0 ml of the HCl reagent. Mix thoroughly the contents of each tube containing the standard or unknown solution. Place the tubes in a water bath at 80° ± 1° for 15 minutes. The tubes are then cooled for 3 minutes in water at room temperature and are read at 490 mμ in a suitable photoelectric colorimeter. The cherry red color is stable for about 2 hours.

The absorbance at the start of the reaction (zero time) is subtracted from the absorbance at 30 minutes. The difference is read against the calibration curve which shows absorbance values for the reaction of 0–840 millimicromoles of D-fructose in a total volume of 10 ml.

The final concentration of G-6-P in the reaction mixture is 0.005 *M*. Since the ratio of G-6-P to F-6-P at equilibrium is 76:24,[27] a maximum of 0.0012 *M* F-6-P or 1200 millimicromoles of F-6-P per milliliter of reaction mixture may be formed and may be present in the 2 ml of TCA filtrate. The reaction is of zero order for about 50% of the time course[31] or for about 600 millimicromoles of F-6-P formed per milliliter reaction mixture. Should the amount formed be more than 600 millimicromoles, the assay should be repeated with a more dilute sample of serum or of other biological fluid.

Definition of a Unit. The activity is defined as milliunits, that is, the millimicromoles of F-6-P formed per minute per milliliter of reaction mixture by 1 ml of serum or other biological fluid under the conditions described above. The calibration curve would ordinarily yield a value for

[30] M. K. Schwartz and O. Bodansky, *Methods Med. Res.* **9,** 5 (1961).
[31] O. Bodansky, *J. Biol. Chem.* **202,** 829 (1953).

the amount of F-6-P or fructose formed per milliliter of reaction mixture by 0.04 ml of serum in 30 minutes. The calibration curve may also be prepared so as to translate the absorbance readings directly into milliunits of activity by multiplying the F-6-P values on the abscissa by $1.00/0.04 \times 1/30$ or 0.833. Figure 1 shows these alternate forms of ex-

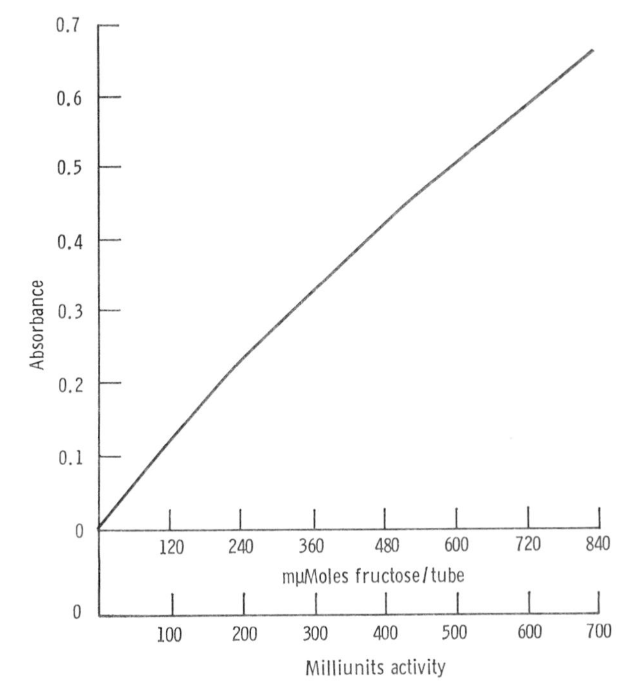

FIG. 1. Illustrative calibration curve for PHI activity. Ordinates represent absorbance at 490 mμ in a Coleman Junior spectrophotometer with cuvettes, 17 mm in internal diameter. Abscissas represent millimicromoles of F-6-P in 10 ml of colored solution (upper row of figures) or, alternatively, milliunits of PHI activity (lower row of figures).

pression when a Coleman Junior spectrophotometer with cuvettes, 17 mm in internal diameter, was employed.

Alternate Methods

Three other methods of assaying PHI activity deserve consideration. The first is based on coupling the reactions F-6-P $\rightleftharpoons$ G-6-P, and G-6-P + TPN $\rightleftharpoons$ G-6-phosphogluconate + TPNH in the presence of excess G-6-P dehydrogenase. For example, Wu and Racker's assay system contained Tris buffer, pH 7.4, 50 mM; TPN, 0.15 mM; $MgCl_2$, 20 mM; F-6-P, 2 mM and G-6-P dehydrogenase, 0.2 unit in a volume of 3 ml.[32] If this coupled reaction is employed, G-6-P dehydrogenase free of contaminating enzymes, such as PHI or phosphatase, and 95–100% enzymatically pure F-6-P should be used. However, the purity of commercially available preparations of F-6-P varies widely and a good preparation is not readily available. We have also found difficulties in obtaining G-6-P dehydrogenase preparations that are free of isomerase and other enzymes.[27]

The method originally proposed by Bodansky[1] was essentially the same as that described here. However, Veronal-acetate was used as buffer and, because of the scarcity of G-6-P, a concentration of 0.002 M of this substrate was employed. In conjunction with the use of this low concentration of substrate, a new method was devised for expressing enzyme activity in terms of the substrate changed at any stage of the reaction.[31,33] The activity was defined as the reciprocal of that concentration of serum, expressed as milliliters per milliliter of reaction mixture, that would cause the formation of 25 μg of fructose as F-6-P in 30 minutes at pH 7.4 and 37°. This point was within the zero-order portion of the reaction. The values were calculated from a calibration curve based on the use of a commercially available preparation of F-6-P that, according to the literature[34] and our own experience,[31] gave an average of approximately 61–62% of the color value of a molecularly equivalent amount of fructose in the resorcinol-HCl reaction. It has been shown that such preparations were consistently impure and that pure preparations of F-6-P give the same color values as molecularly equivalent concentrations of fructose.[35,36]

The units obtained by this earlier procedure can be converted to yield values under the conditions of the method proposed in this paper. Multiplication by 0.61 corrects for the impurity of the former F-6-P standards, and multiplication of the unitage obtained at 0.002 M G-6-P by 1.34 yields the reaction velocity at 0.005 M G-6-P.[31] Division by 180 and again by 30 converts micrograms formed in 30 minutes to micromoles formed per minute, and multiplication by 25 and by 1000 yields millimicromoles formed by 1 ml of serum per milliliter reaction mixture. These operations may be combined into one factor, 3.79, which, when

[32] R. Wu and E. Racker, *J. Biol. Chem.* **234,** 1029 (1959).

[33] O. Bodansky, *J. Biol. Chem.* **205,** 731 (1953).

[34] W. W. Umbreit, R. H. Burris, and J. E. Stauffer, *in* "Manometric Techniques in Tissue Metabolism." Burgess, Minneapolis, Minnesota, 1949.

[35] S. E. Kahana, O. H. Lowry, D. W. Schulz, J. V. Passonneau, and E. J. Crawford, *J. Biol. Chem.* **235,** 2178 (1960).

[36] E. A. Noltmann, *J. Biol. Chem.* **239,** 1545 (1964).

multiplied by the old units, yields milliunits of PHI activity by the present procedure. For example, a value of 21 units obtained by the previous method[1] corresponds to 79 milliunits by the present procedure.

An automated method in which a continuous flow instrumental system of analysis is employed has also been described by the present authors.[37] The essentials of such a system have been described in this volume [55]. The flow rates are arranged so that the concentrations in the reaction mixture correspond to those in the manual method proposed here. The complete reaction mixture passes through an incubation bath at 37° for a precisely measured interval of about 10 minutes. At the end of this period, the reaction is stopped by combination with 50% H_2SO_4 and 0.1% resorcinol in 50% ethylene glycol. This mixture is heated for about 2 minutes at 95° and is then passed into a cuvette. The absorbance at 505 mμ appears as a deflection on a suitable recorder. Calibration curves are developed under the same conditions with known amounts of fructose.

Interpretation of the Assay. Using the method described in this paper, we have obtained a mean value of 77 milliunits, SD = 42 milliunits, for 13 presumably normal adults. This is in excellent agreement with the value, 79 milliunits, SD = 27 milliunits, obtained by conversion of the results obtained in our previous study of 87 normal adults.[1] The values obtained in normal persons by several groups of investigators[12,14-17] are in generally good agreement with the values obtained by Bodansky's original method[1] and, if converted, would also agree with the present normal values.

The frequency and extent of elevations of serum PHI in various diseases are shown by several representative studies, listed in the table. For example, in 23 patients studied at various stages after myocardial infarction, all showed elevations above normal on day 1, but this incidence decreased rapidly so that 60% of the determined values were within normal range on day 3.[12] In infectious hepatitis in children, the maximal elevations attained were above normal in all cases, and 48% of the elevations were more than 4 times the upper limit of normal.[10]

Of the various ubiquitous enzymes that have been studied in neoplastic disease, the PHI is more frequently elevated than any other enzyme in serum.[5-9] Thus, of 119 patients with carcinoma of the gastrointestinal tract, the PHI was elevated in 74% of the cases, in contrast to frequencies of 70% elevations for aldolase, of 53% for LDH, and of 35% for GOT.[5] In 88 patients with carcinoma of the head, neck, or esophagus, the PHI was elevated in 51% of the cases, the aldolase in 45%, the LDH in 21%, and the GOT in only 16% of the cases.[6] The PHI was most fre-

[37] M. K. Schwartz, G. Kessler, and O. Bodansky, *Ann. N.Y. Acad. Sci.* **87**, 616 (1960).

Elevation of Serum PHI in Various Diseases

Disease	Normal range	1.1–2.0 × normal range	2.1–4.0 × normal range	4.1–6.0 × normal range	6.1–10.0 × normal range	>10.0 × normal range
	Percentage of cases having values in					
Myocardial infarction[a]						
Day 1	0	7	43	21	7	21
Day 2	33	33	22	11		
Day 3	60	20	20			
Extensive hepatic metastasis[b]	17	36	29	7	8	3
Infectious hepatitis in children[c]	0	29	24	24	24	—
Lymphocytic leukemia[d]	77	15	0	8	—	—
Granulocytic leukemia[d]	7	36	22	29	7	—

[a] R. J. Bing, A. Castellanos, and A. Siegel, *J. Am. Med. Assoc.* **164**, 647 (1957).
[b] C. O. Tan, J. Cohen, M. West, and H. J. Zimmerman, *Cancer* **16**, 1373 (1963).
[c] O. Bodansky, S. Krugman, R. Ward, M. K. Schwartz, J. P. Giles, and A. M. Jacobs, *A.M.A. J. Diseases Children* **98**, 166 (1959).
[d] M. C. Blanchaer, P. T. Green, J. P. Maclean, and M. J. Hollenberg, *Blood* **13**, 245 (1958).

quently elevated in metastatic carcinoma of the liver, as has already been indicated. In 284 such cases, the PHI was elevated in 84%, as contrasted with 75% frequency of elevations for aldolase, 69% for LDH, and 51% for GOT.[9]

The sequential use of PHI determinations has proved to be of particular value in following the course of neoplastic disease. Decreases in elevated values have been shown by several groups of investigators[16–18] to be associated with clinical improvement, and, as has already been noted, a slight rise from a normal level may be the first herald of recrudescence of neoplastic growth.

[100] Phosphoglucosamine Isomerase from *Proteus vulgaris*

By Henry I. Nakada

$$\text{Glucosamine-6-P} + H_2O \rightarrow \text{fructose-6-P} + NH_3 \quad (1)$$

Phosphoglucosamine isomerase has also been isolated from *Escherichia coli*[1,2] and animals.[3] The enzyme has been called glucosamine-6-P de-

[1] J. B. Wolfe, B. B. Britton, and H. I. Nakada, *Arch. Biochem. Biophys.* **66**, 333 (1957).
[2] D. A. Comb, and S. Roseman, *J. Biol. Chem.* **232**, 807 (1958).
[3] L. F. Leloir and C. E. Cardini, *Biochem. Biophys. Acta* **20**, 33 (1956).

aminase (see Vol. V [56, 57]). However, because the mechanism of action follows the Lobry de Bruyn–Van Eckenstein rearrangement, as does phosphoglucoisomerase, this enzyme has been named as an isomerase to indicate its chemical behavior.

Assay Method

Principle. Fructose-6-P formation can be followed spectrophotometrically by measuring the formation of NADPH at 340 mμ in the presence of phosphoglucoisomerase and glucose-6-P dehydrogenase. Crude and partially purified extracts and most commercial preparations of glucose-6-P dehydrogenase contain sufficient phosphoglucoisomerase to convert fructose-6-P to glucose-6-P rapidly. Highly purified preparations of phosphoglucosamine isomerase can be measured by the addition of both phosphoglucoisomerase and glucose-6-P dehydrogenase.

The enzyme can also be measured by following the disappearance of glucosamine-6-P (method of Levvy and McAllan[4]), the formation of fructose-6-P (Dische and Borenfreund[5]), or the formation of ammonia (Braganca *et al.*[6]).

Reagents and Procedure. Solutions in the indicated quantities were added to a quartz cuvette having a 1-cm light path:

K Phosphate buffer, 0.05 *M*, pH 7.2	0.25 ml
$MgSO_4$, 0.1 *M*	0.05 ml
Cysteine, 0.2 *M*	0.05 ml
NADP, 10 m*M*	0.05 ml
Glucose-6-P dehydrogenase, 6 units/ml	0.005 ml
Glucosamine-6-P, 10 m*M*	0.10 ml
Enzyme	0.10 ml

Each cuvette was brought up to a volume of 1.4 ml with water. In every experiment, a control cuvette containing all components except the substrate was employed. The reaction was started by adding substrate and, after mixing, absorbance readings are taken at 30-second intervals or recorded using a double-beam spectrophotometer.

Definition of a Unit and Specific Activity. One unit of enzyme is defined as the amount which catalyzes the formation of 0.1 micromole of fructose-6-P per 10 minutes. Specific activity is expressed as units per milligram protein. The calculations are based on the assumption that the absorbance of NADPH is 6.22 per micromole.

[4] G. A. Levvy and A. McAllan, *Biochem. J.* **73,** 127 (1959).
[5] Z. Dische and E. Borenfreund, *J. Biol. Chem.* **192,** 583 (1951).
[6] B. M. Braganca, J. H. Quastel, and R. Schucker, *Arch. Biochem. Biophys.* **52,** 18 (1954).

Purification Procedure

Growth of Bacteria. All cultures were grown in a liquid medium of the following final composition: 1% Bacto-proteose peptone; 0.5% sodium chloride; 0.5% glucose; 0.5% glucosamine. The pH was adjusted to 7.4 with potassium hydroxide. A 50% solution of glucose and a 30% solution of glucosamine HCl were sterilized by Seitz filtration. All other components, glassware, and equipment were sterilized by autoclaving. Thirty liters of culture medium, contained in three 10-liter round-bottom flasks, were routinely used.

Starter cultures were prepared by inoculating flasks containing 100 ml medium with *P. vulgaris* 13315 GL grown on nutrient agar slants. The flasks were shaken at ambient temperature for 12 hours. To each of the large round-bottom flasks was added 600 ml of bacteria-containing medium. Dow Corning Antifoam was used to prevent foaming during the growth period of 48 hours. The flasks were aerated by the passage of acid and alkali washed air through the cultures during the growth period. The cells were harvested by use of the Servall RC-2 continuous flow centrifuge, washed twice with 0.05 *M* potassium phosphate buffer, pH 7.2, and stored in a freezer at −15°.

Step 1. Crude Extract. Approximately 10 g (wet weight) of frozen *P. vulgaris* 13315 GL were ground manually in a frozen mortar with an equivalent weight of 200-mesh ground glass. Soluble proteins were extracted by stirring the ground cell paste with about 50 ml potassium phosphate buffer, 0.05 *M*, pH 7.2 for 20 minutes followed by centrifugation at 15,000 rpm for 25 minutes at 0–4° in a Servall RC-2 refrigerated centrifuge. The precipitate was discarded. Unless otherwise indicated, all subsequent steps were carried out at 0–5° and centrifugation was accomplished in the manner just described.

Step 2. First Ammonium Sulfate Fraction. Finely ground ammonium sulfate was added with stirring to the crude, cell-free extract to 70% saturation. The pH was adjusted to 7.6 with 2 *N* sodium hydroxide. The suspension was stirred for 30 minutes, then centrifuged. The supernatant fluid was discarded, and the precipitate was dissolved in about 50 ml water.

Step 3. Removal of Nucleic Acids. Excess nucleic acids were removed by the addition of 0.05 volume of 1 *M* manganous chloride. The solution was dialyzed overnight against potassium phosphate buffer, 0.005 *M*, pH 8.0. After centrifugation, the supernatant fluid was checked for enzymatic activity and for nucleic acid and protein content.[7] If the

[7] O. Warburg and W. Christian, *Biochem. Z.* **310**, 384 (1941).

nucleic acids were above 10%, additional manganous chloride was added until an acceptable level of nucleic acids was attained (below 10%).

Step 4. Fractionation on DEAE-Cellulose. The method used was based on that of Peterson and Sober.[8] DEAE-cellulose (alkali and acid washed, then equilibrated with potassium phosphate buffer, 0.005 *M*, pH 8.0) was packed under pressure (10–15 psi) in a column (1 cm I.D.) to a height of about 12 inches. The enzyme (step 3) in 0.005 *M* potassium phosphate buffer, pH 8.0, was absorbed on the column. Proteins were eluted by passage of 100 ml each of the following potassium phosphate buffers: (I) 0.005 *M*, (II); 0.05 *M*, pH 7.6; (III) 0.1 *M*, pH 7.2; (IV) 0.15 *M*, pH 7.0; (V) 0.2 *M*, pH 6.8; and (VI) 0.2 *M*, pH 6.5. Sodium chloride was added to buffer VI to a concentration of 0.2 *M*.

The flow rate was adjusted to approximately 0.5 ml per minute, and approximately 10 ml fractions were collected. The protein content of each fraction was measured by readings at 280 mμ. The peak eluted by pH 7.0 buffer contained phosphoglucosamine isomerase, and the protein eluted by pH 6.5 buffer contained phosphoglucoisomerase. The purification procedure is summarized in the table.

PURIFICATION OF PHOSPHOGLUCOSAMINE ISOMERASE FROM *Proteus vulgaris*

Fraction	Volume (ml)	Protein (mg/ml)	Units/ml	Specific activity (units/mg protein)
Crude	62	11.8	4.8	0.41
Step 3	54	0.78	5.2	6.67
DEAE-cellulose eluate	40	0.054	4.41	81.8

The overall yield was relatively good. The purification was about 190-fold over the crude extract. Because of the low protein concentration, the tubes of the peaks were pooled and concentrated. The activity dropped considerably during the concentration process but was stable to freezing thereafter.

Properties

Specificity. The enzyme was not active with glucosamine, *N*-acetylglucosamine, or *N*-acetylglucosamine-6-P.

Activators and Inhibitors. No cofactor or ion requirement was noted. *N*-Acetylglucosamine-6-P was not an activator for this system, even when low concentrations of glucosamine-6-P were used. 2-Deoxyglucose-

[8] E. A. Peterson and H. A. Sober, *J. Am. Chem. Soc.* 78, 751 (1956).

6-P, an inhibitor of phosphoglucoisomerase, had no influence on the reaction rate.

Effect of pH. The pH optimum was found to be 7.2 and was fairly broad. This differs from the *E. coli* enzyme which proceeded at maximum velocity at pH 7.8. No differences in reaction rate were noted when phosphate, Tris chloride, or sodium acetate buffers were used.

Effects of Substrate Concentration. The Michaelis-Menten constant for this enzyme was found to be 1.25×10^{-2} *M*.

Reversibility. Although studied under a variety of conditions such as widely differing substrate concentrations and pH, no indication of reversibility was noted.

[101] L-Rhamnose Isomerase

By GOETZ F. DOMAGK and RONALD ZECH

L-Rhamnose ⇆ L-rhamnulose

L-Rhamnose isomerase is an enzyme inducible in a great variety of microorganisms by growing them in the presence of L-rhamnose.[1] The method described below uses *Lactobacillus plantarum* as the source of the enzyme.[2]

Assay Method

Principle. The assay procedure is based on the appearance of the 2-keto sugar L-rhamnulose, which is measured by the cysteine-carbazole reaction of Dische and Borenfreund.[3]

Reagents

L-Rhamnose, 0.05 *M*. Dissolve 18.2 mg of L-rhamnose in 2.0 ml of water. Commercial rhamnose gives a rather high blank in the cysteine–carbazole test and should be recrystallized from water before use.

Tris-HCl buffer, 0.1 *M*, pH 8.5

Potassium thioglycolate, 0.4 *M*. 0.27 ml of freshly distilled thioglycolic acid (b_{12} 101°) are dissolved in water, the pH is adjusted to 6.5 by the addition of KOH, and the volume is made up to 10 ml.

[1] R. G. Eagon, *J. Bacteriol.* **82**, 548 (1961).

[2] G. F. Domagk and R. Zech, *Biochem. Z.* **339**, 145 (1963).

[3] Z. Dische and E. Borenfreund, *J. Biol. Chem.* **192**, 583 (1951).

$MnCl_2$, 0.01 *M*
Cysteine·HCl, 0.2% solution (prepared fresh daily)
Sulfuric acid, 70% solution
Carbazole in ethanol, 0.10% solution. Commercial carbazole is recrystallized from xylene or sublimed in vacuum.

Procedure. In a regular test tube (16 mm diameter) are pipetted: 0.5 ml of buffer, 0.05 ml of thioglycolate, 0.01 ml of $MnCl_2$, and 0.05 ml of rhamnose. Water is added to bring the final volume (including the enzyme) to 0.80 ml. The incubation is started by the addition of enzyme. A blank is obtained by running an incubation without the enzyme. After incubation for 15 minutes at 37° the reaction is stopped by the addition of 0.20 ml of cysteine, followed by 5.0 ml of sulfuric acid. Before the addition of 0.20 ml of carbazole the contents of the test tube have to be mixed thoroughly, so that no turbidity occurs upon the addition of the carbazole solution. Mix again and leave at room temperature for 60 minutes. The extinction at 546 mμ is then measured in 1-cm cuvettes.

In the authors' laboratory it proved to be most convenient to add cysteine, sulfuric acid, and carbazole from burettes. All volumes of the incubation sample and of the final assay can be reduced if the enzyme is to be used for the micro assay of L-rhamnose.[4]

Definition of Unit and Specific Activity. One unit is defined as the amount of enzyme which produces an optical density of 1.00 under the assay conditions given above. Specific activity is the number of units of activity per milligram of protein, as determined by the biuret method[5] or, in dilute solutions, by the optical method.[6] Under our assay conditions 0.10 micromole of chromatographically pure L-rhamnulose, which has not yet been crystallized, gives an extinction of 0.535 at 546 mμ. Our working units can be converted to International Enzyme Units (micromoles per minute) by applying a factor of 0.0125.

Purification Procedure

Growth of Bacteria. Lactobacillus plantarum, ATCC 8041, was grown at 37° in a synthetic medium[7] containing 1% Difco Nutrient Broth, 0.4% Difco Yeast Extract, 1% sodium acetate, 0.02% $MgSO_4 \cdot 7\ H_2O$, 0.001% NaCl, 0.001% $FeSO_4 \cdot 7\ H_2O$, 0.001% $MnSO_4 \cdot 4\ H_2O$, 1% fructose, and 0.3% rhamnose. The sugars were autoclaved separately and

[4] L. Krueger, O. Lüderitz, J. L. Strominger, and O. Westphal, *Biochem. Z.* **335,** 548 (1962).
[5] G. Beisenherz, H. J. Boltze, T. Bücher, R. Czok, K. H. Garbade, E. Meyer-Arendt, and G. Pfleiderer, *Z. Naturforsch.* **8b,** 555 (1953).
[6] O. Warburg and W. Christian, *Biochem. Z.* **310,** 384 (1941).
[7] P. K. Stumpf and B. L. Horecker, *J. Biol. Chem.* **218,** 753 (1956).

added to the medium before inoculation. The bacteria should be subcultured in this medium several times before a 5% inoculum is used to start the main culture. After 22 hours at 37° the cells were collected by centrifugation and were washed once with cold 0.9% NaCl. These resting cells retain their isomerase activity for many months if stored frozen at —15°.

Extraction. One gram of washed cells and 3 g of alumina A303 were ground in a mortar for 10 minutes at 0°. The paste was then extracted, using 20 ml of 0.05 *M* potassium phosphate buffer pH 6.8, and centrifuged: the residue was reextracted twice, using the same buffer. The combined supernatants gave 40.5 ml of crude extract.

First DEAE-Sephadex Chromatography. The crude extract was put on a column of DEAE-Sephadex A50 coarse (37 × 120 mm) equilibrated with 0.05 *M* potassium phosphate, pH 6.8; 850 ml of 0.05 *M* potassium phosphate, pH 6.8, was used for the removal of contaminating proteins, and subsequently the isomerase was eluted using 650 ml of the same buffer containing 0.03 *M* NaCl. The eluate was concentrated approximately 10-fold using a rotatory flash evaporator at 35°; the concentrate was then dialyzed for 20 hours against 1 liter of cold distilled water (renewed 3 times).

TEAE-Cellulose Chromatography. The dialyzed material was put on a column of TEAE-cellulose (10 × 50 mm) equilibrated with 0.01 *M* Tris buffer, pH 7.4. Then the column was washed with 0.01 *M* Tris containing 0.05 *M* NaCl; no isomerase was removed by this treatment. Ninety per cent of the isomerase was eluted with the first 4 fractions (10 ml each) when the column was washed with 0.01 *M* Tris, pH 7.4, containing 0.2 *M* NaCl. These fractions were combined and dialyzed at +2° for 3 hours (first against 1 liter of distilled water, then against 1 liter of 0.05 *M* potassium phosphate pH 6.8).

Second DEAE-Sephadex Chromatography. The dialyzed enzyme was now put on a column of DEAE-Sephadex A50 coarse (10 × 100 mm) equilibrated with 0.05 *M* potassium phosphate pH 6.8, and chromatography was carried out using a NaCl gradient. The mixing chamber contained 500 ml of 0.05 *M* potassium phosphate pH 6.8, and the same buffer containing 0.03 *M* NaCl was in the reservoir. Ten-milliliter fractions were collected; those containing isomerase (numbers 7–22) were pooled and concentrated by means of a rotatory flash evaporator.

The table summarizes the total purification.

Properties

Cofactors and Inhibitors. The enzyme can be activated by various divalent cations, of which Mn^{++} in a concentration of 1.2×10^{-4} *M* was found to be the most effective. The isomerase activity can be decreased

PURIFICATION PROCEDURE FOR L-RHAMNOSE ISOMERASE

Step	Volume (ml)	Units of Isomerase	Specific activity	International enzyme units	Recovery (%)
Crude extract	40	1520	10	19	100
1st DEAE-Sephadex	650	1190		15	80
Concentration, dialysis	105	1045	160	13	69
TEAE-cellulose	50	845	920	10.6	56
2nd DEAE-Sephadex	45	575	2550	7.2	38

by the addition of SH-blocking reagents, whereas sulfhydryl reagents activate the enzyme.

pH Optimum. The pH optimum was found to be at 8.5, with a steep decrease of activity on the alkaline side, and about 50% of the optimum activity left at pH 6.0.

Substrate Affinity and Equilibrium. The K_m value for L-rhamnose was determined to be 7×10^{-4} *M*. The equilibrium of the reaction was reached at 57% rhamnose and 43% rhamnulose.

Substrate Specificity. No reaction was found with the following substrates: D-rhamnose, L-fucose, D-mannose, D-arabinose, L-arabinose, D-xylose, and L-xylose. However, L-mannose was isomerized; its K_m value, as determined in our experiments, was found to be 5×10^{-3} *M*. Ascarylose (3-deoxy L-rhamnose) gave a very faint red color after the reaction; this might be due to the poor response of 3-deoxy ketoses in the cysteine–carbazole reaction.[8]

Other Sources of the Enzyme

L-Rhamnose isomerase had been first demonstrated in extracts from rhamnose-grown *E. coli*[9,10] and *Pasteurella pestis*.[11] The *E. coli* enzyme has been purified by Tecce and DiGirolamo,[9] who found cobalt ions to be the best activator, and by Takagi and Sawada.[12]

[8] G. F. Domagk and H. W. Schiwara, unpublished observations, 1965.
[9] G. Tecce and M. Di Girolamo, *Giorn. Microbiol.* **1**, 286 (1956).
[10] D. M. Wilson and S. Ajl, *J. Bacteriol.* **73**, 410, 415 (1958).
[11] E. Englesberg, *J. Bacteriol.* **74**, 8 (1958).
[12] Y. Takagi and H. Sawada, *Biochim. Biophys. Acta* **92**, 10 (1964).

[102] D-Arabinose Isomerase

By R. P. Mortlock

D-Arabinose ⇌ D-ribulose

Assay Method

Principle. Two assay methods may be employed. The first utilizes the cysteine–carbazole test for ketopentoses[1] to determine the quantity of D-ribulose formed after incubation of D-arabinose and enzyme.[2] The second spectrophotometrically measures the formation of D-ribulose with excess ribitol dehydrogenase as the rate of NADH oxidation at 340 mμ.

Colorimetric Assay

Reagents

D-Arabinose, 1.0 *M*
$MnCl_2$, 0.1 *M*
Cacodylate buffer, 0.5 *M* pH 7.0

Procedure. The assay mixture contains cacodylate buffer, pH 7.0 (500 micromoles), manganous chloride (10 micromoles), D-arabinose (100 micromoles), and isomerase in a total volume of 2.0 ml. The mixture is incubated at 37° for 20 minutes, and 0.1-ml samples are removed for determination of ketopentose by the cysteine–carbazole test. Reagent mixtures minus enzyme must be sampled to correct for color development due to D-arabinose alone.

Definition of Unit and Specific Activity. One unit of enzyme is equal to that amount which catalyzes the formation of 1 micromole of D-ribulose per hour under the above conditions. Specific activity is expressed as units per milligram protein. Protein is measured by the Lowry test[3] or by determination of ratio of absorbancy at 280 and 260 mμ.[4]

Spectrophotometric Assay

Reagents

D-Arabinose, 1.0 *M*
$MnCl_2$, 0.1 *M*

[1] Vol. III [12].
[2] R. L. Anderson and W. A. Wood, *J. Biol. Chem.* **237,** 296 (1962).
[3] O. H. Lowry, N. J. Rosebrough, A. L. Farr, and R. J. Randall, *J. Biol. Chem.* **193,** 265 (1951).
[4] O. Warburg and W. Christian, *Biochem. Z.* **310,** 384 (1941).

Tris buffer, 0.2 *M*
NADH, 0.01 *M*
Ribitol dehydrogenase

Procedure. The assay mixture contains 0.01 ml D-arabinose (10 micromoles), 0.01 ml $MnCl_2$ (1 micromole), 0.005 ml NADH (0.05 micromoles), 0.05 ml Tris buffer, pH 7.0 (10 micromoles), 10 units ribitol dehydrogenase, and isomerase in a total volume of 0.2 ml. D-Ribulose formed by the isomerization of D-arabinose is reduced to ribitol, and rate of decrease in absorbancy at 340 mμ due to oxidation of NADH is measured. This assay cannot normally be used with crude isomerase preparations due to high NADH oxidase activity.

Definition of Unit and Specific Activity. One unit of enzyme in the spectrophotometric assay is defined as that amount which catalyzes a decrease in absorbancy of 1.0 per minute under the conditions described above. Specific activity is expressed in units per milligram. One unit determined by the spectrophotometric assay is equivalent to two units determined using the cysteine–carbazole test.

Purification Procedure

Growth of Cells. D-Arabinose isomerase may be purified from *Aerobacter aerogenes,* PRL-R3, grown on the following inorganic salts medium: 1.35% $Na_2HPO_4 \cdot 7\ H_2O$, 0.15% K_2HPO_4, 0.3% $(NH_4)_2SO_4$, 0.02% $MgSO_4 \cdot 7\ H_2O$, 0.0005% $FeSO_4 \cdot 7\ H_2O$, and 0.5% D-arabinose (autoclaved separately). Higher enzyme yields can be obtained using L-xylose as the substrate,[5] and mutants constitutive for the enzyme may be isolated using L-xylose as a selective substrate.[6] Cells are harvested by centrifugation in the cold, washed once with 0.2 volume of cold distilled water, and broken by means of the French pressure cell. All subsequent steps are carried out at 0–4°. The broken-cell suspension is centrifuged for 10 minutes at 15,000 *g,* and the clear supernatant is diluted with water to a final concentration of 10 mg protein per milliliter.

Step 1. Protamine Sulfate Treatment. Solid ammonium sulfate is added to the extract to a final concentration of 0.1 *M.* A solution of protamine sulfate (20 mg/ml) is added dropwise with slow stirring, using 1 ml of protamine sulfate solution for each 10 ml of extract. The mixture is centrifuged as before, and the precipitate is discarded.

Step 2. Ammonium Sulfate Fractionation. Solid ammonium sulfate is added to the supernatant from the protamine sulfate step, and protein fractions are collected by centrifugation. The fraction precipitating between 1.9 and 3.0 *M* ammonium sulfate is dissolved in 0.001 *M* EDTA

[5] R. P. Mortlock and W. A. Wood, *J. Bacteriol.* **88,** 835 (1964).
[6] R. P. Mortlock, D. Fossitt, and W. A. Wood, *Bacteriol. Proc.* p. 95 (1964).

(pH 7.0), 0.001 *M* mercaptoethanol and 0.1 *M* Tris-chloride buffer (pH 7.0). The pH is adjusted to 7.0 if necessary, and the fraction are dialyzed against 0.001 *M* Tris-chloride buffer (pH 7.0) with EDTA and mercaptoethanol in the concentrations indicated above.

Step 3. DEAE-Cellulose. After dialysis, the salt concentration (ammonium sulfate) is adjusted to 0.02 *M* using a precalibrated Barnstead conductivity meter. The fraction is passed through a DEAE-cellulose column (5 cm × 13 cm) equilibrated against 0.02 *M* phosphate buffer, pH 7.0. Most of the enzyme activity is not absorbed. After passage through the column the spectrophotometric assay may be employed.

Step 4. Alumina-Cγ. Alumina Cγ is added (1 mg gel per milligram protein) mixed for 1 minute, and removed by centrifugation.

Step 5. Second Ammonium Sulfate Fractionation. The supernatant from the alumina-Cγ treatment is fractionated by the addition of saturated ammonium sulfate (pH 7.0). The protein fraction precipitating between 60 and 65% of saturation represents a fifteenfold purification over the crude extract.

Enzyme Properties

The enzyme displays a pH optimum at 7.0. The Michaelis constant for D-arabinose at pH 7.0 is 2.2×10^{-1} *M*. Manganese ion is required for optimal activity. In addition to catalyzing the isomerization of D-arabinose to D-ribulose, the purified fraction is capable of catalyzing the isomerization of L-xylose to L-xylulose and L-fucose to L-fuculose.

[103] D-Arabinose 5-Phosphate Isomerase[1]

By Wesley A. Volk

D-Arabinose 5-phosphate ⇄ D-ribulose 5-phosphate

Assay Method

Principle. The amount of ketopentose formed from D-arabinose 5-phosphate was determined at the end of the reaction period by the cysteine–carbazole method.

Reagents

Glycylglycine, 0.1 *M*, pH 8.0
D-Arabinose 5-phosphate, 0.01 *M*
Reagents for cysteine-carbazole determination (Vol. I [52] and Vol. III [12])

Procedure. Into a standard colorimeter tube is placed 0.3 ml of glycyl-

[1] W. A. Volk, *J. Biol. Chem.* **235**, 1550 (1960).

glycine buffer, 0.2 ml of D-arabinose 5-phosphate, and enzyme to a total volume of 0.6 ml. After incubation at 37° for 15 minutes, the enzymatic reaction is stopped by the addition of the cysteine–carbazole reagents. After an additional 30 minutes' incubation at 37°, the color is read at 540 mμ. A tube from which substrate is withheld until after the addition of the H_2SO_4 reagent serves as the control.

Definition of Unit and Specific Activity. One unit of phosphoarabinoisomerase is defined as the amount of enzyme which will isomerize 1 micromole of D-arabinose 5-phosphate to D-ribulose 5-phosphate in 1 minute. Specific activity is expressed as units per milligram protein. Protein was determined by the method of Lowry *et al.*[2]

Purification Procedure

Step 1. Growth of Culture and Preparation of Crude Extract. Propionibacterium pentosaceum strain E214 was grown for 4–5 days in 2-liter Erlenmeyer flasks containing 1500 ml of a medium consisting of 0.5% yeast extract, 0.5% Bacto-peptone, and 0.5% D-arabinose in a final concentration of 0.05 *M* phosphate buffer, pH 6.8. The cells were harvested by centrifugation, washed with 0.001 *M* EDTA, pH 7.4, suspended in 0.001 *M* EDTA, pH 7.4, and disrupted for 30 minutes in a 10 kc Raytheon sonic oscillator. After centrifugation, the sedimented cells were disrupted two additional times. All subsequent operations were carried out at 0–2°.

Step 2. Protamine Precipitation. To 284 ml of the crude cell-free extract was added 250 ml of 0.001 *M* EDTA, pH 7.4. Protamine sulfate (213 ml of 0.5% solution, pH 5.0) was added slowly with stirring. After 15 minutes, the suspension was centrifuged and the precipitate was discarded.

Step 3. 2.0 M Ammonium Sulfate Fractionation. The supernatant solution (685 ml) from the protamine treatment was brought to 2.0 *M* by the addition of 761 ml of 3.75 *M* $(NH_4)_2SO_4$, pH 7.0. After the preparation had stood overnight, the precipitate was collected by centrifugation, and the supernatant solution was discarded.

Step 4. 1.2–2.0 M $(NH_4)_2SO_4$ Fractionation. The above precipitate was dissolved in 0.001 *M* EDTA, pH 7.4. Residual $(NH_4)_2SO_4$ was determined with Nessler reagent,[3] and 23 ml of 3.75 *M* $(NH_4)_2SO_4$, pH 7.0, were added to bring the $(NH_4)_2SO_4$ concentration to 1.2 *M*. The suspension was centrifuged and the resulting precipitate was discarded.

[2] O. H. Lowry, N. J. Rosebrough, A. L. Farr, and R. J. Randall, *J. Biol. Chem.* **193**, 265 (1951).

[3] O. Folin and H. Wu, *J. Biol. Chem.* **38**, 81 (1919).

The supernatant solution (119 ml) was brought to 2.0 *M* by the addition of 53.5 ml of 3.75 *M* $(NH_4)_2SO_4$, pH 7.0. This precipitate was dissolved in 50 ml of 0.001 *M* EDTA, pH 7.4, and dialyzed for 3 hours against 8 liters of 0.001 *M* EDTA, pH 7.4.

Step 5. Calcium Phosphate Negative Adsorption. The dialyzed extract (51 ml) was treated with 51 ml of $Ca_3(PO_4)_2$ gel (10 mg dry weight per milliliter). After 15 minutes, the gel suspension was centrifuged and the gel was discarded.

Step 6. 1.6–2.2 M $(NH_4)_2SO_4$ Fractionation. The enzyme solution from the $Ca_3(PO_4)_2$ gel step was fractionated by the addition of 3.75 *M* $(NH_4)_2SO_4$, pH 7.0, and the protein precipitating between 1.6 *M* and 2.2 *M* was saved. This precipitate was dissolved in 20 ml of glass-distilled water and dialyzed for 4 hours against 8 liters of glass distilled water. The final $(NH_4)_2SO_4$ concentration of the enzyme after dialysis was 0.04 *M*.

Step 7. DEAE-Cellulose Column Fractionation. The dialyzed enzyme (17.5 ml) was absorbed on a DEAE-cellulose column (1.8 × 8.0 cm) which had been equilibrated with 0.1 *M* phosphate buffer and washed with 500 ml of distilled water before use. After absorption of the enzyme, a linear gradient of phosphate buffer, pH 7.0, from 0 to 0.2 *M* was passed through the column. The column eluate was collected in approximately 10-ml fractions. The active enzyme was eluted in fractions 30 through 55 with the peak activity in fraction 42.

A summary of this procedure is presented in the table; it includes protein concentration and specific activity of the enzyme during purification.

PURIFICATION OF PHOSPHOARABINOISOMERASE

Enzyme fraction	Volume (ml)	Protein (mg/ml)	Specific activity (units/mg protein)	Total units	Yield (%)	Purification (-fold)
Crude extract	284	18.8	0.13	704	100	
Protamine treated	685	3.8	0.21	548	78	1.6
2.0 *M* $(NH_4)_2SO_4$ ppt	100	6.3	1.00	650	92	7.8
1.2–2.0 *M* $(NH_4)_2SO_4$ ppt	50	8.3	0.88	367	52	6.7
Supernatant fluid from $Ca_3(PO_4)_2$ gel	87	1.7	1.51	223	32	11.4
1.6–2.2 *M* $(NH_4)_2SO_4$ ppt	20	5.6	1.79	200	28	13.6
Tube 42 from DEAE-cellulose	10	0.044	8.18	—	—	62
Total recovery from DEAE-cellulose (tubes 30–55)	240	0.048	5.40	62	8.8	40.7

Properties

Stability and Characteristics of the Enzyme. The phosphoarabinoisomerase appears to be sensitive to heavy-metal inactivation; therefore, all operations were carried out in 0.001 *M* EDTA, pH 7.4, in glass-distilled water. Active fractions from a DEAE-cellulose column were pooled (106 ml containing 32 units per milliliter of enzyme); after dialysis against 0.001 *M* EDTA, pH 7.4, for 19 hours under 13 pounds of pressure, the final volume was 23 ml and contained 85 units/ml. This dialyzed enzyme (containing 0.08 mg of protein per milliliter) was stored at 0° and assayed at intervals. After 14 days, 62 units/ml remained. The K_m for D-arabinose 5-phosphate is 1.98×10^{-3} *M*.

pH Optimum. The optimum pH for the activity of the phosphoarabinoisomerase is 8.0; however, the enzyme retains approximately 90% of its activity at pH 7.5 and 8.5.

Specificity and Inhibitors. Phosphoarabinoisomerase is not active on the following substrates: L-arabinose, D-arabinose, D-xylose, D-ribose, D-ribose 5-phosphate, glucose 6-phosphate. None of the above compounds exerted any inhibitory effect on the isomerization of D-arabinose 5-phosphate when present in the assay mixture at 10 times the concentration of D-arabinose 5-phosphate.

The enzyme was not inhibited by 0.0025 *M* iodoacetate, but 0.0012 *M* *p*-chloromercuribenzoate caused a 31% inhibition.

Equilibrium of Isomerization. At 37° equilibrium is established when 22.8% of the D-arabinose 5-phosphate is converted to D-ribulose 5-phosphate, yielding an equilibrium constant of 0.295. The equilibrium is shifted toward the formation of D-ribulose 5-phosphate as the temperature of the enzymatic reaction is increased.

[104] D-Xylose Isomerase

By KEI YAMANAKA

D-Xylose ⇄ D-xylulose

This enzyme has been described by Cohen[1] and by Slein[2] briefly. The method of purification described here is more reproducible.

[1] S. S. Cohen, see Vol. I [52].
[2] M. W. Slein, see Vol. V [43].

Assay Method

The method commonly used is based on the determination of the xylulose formed in the reaction by cysteine–carbazole test.[3]

Reagents

Maleate buffer, 0.05 *M*, pH 6.6
$MnCl_2$ (or $MnSO_4$), 0.01 *M*
D-Xylose, 0.1 *M*
H_2SO_4 solution prepared by mixing 190 ml of H_2O and 450 ml of concentrated H_2SO_4
Cysteine-HCl, 1.5%. Keep in a refrigerator.
Carbazole, 0.12%, in 95% ethanol.

Procedure. The reaction mixture (1.0 ml) contains 0.5 ml of maleate buffer, 0.05 ml of $MnCl_2$ (or $MnSO_4$), and 0.01–0.1 ml of enzyme preparation with adequate dilution. After equilibration for 5 minutes at 35°, 0.05 ml of D-xylose is added; the mixture then incubated at 35° for 10 minutes. The reaction is stopped by adding 0.05 ml of 50% trichloroacetic acid. A blank with heat-inactivated enzyme is included in every determination. The cysteine–carbazole reaction[3] is carried out with color development at 35° for 20 minutes. Since xylose also gives color, it is necessary to have a zero time determination or to subtract the color produced from the same amount of xylose.

Definition of Unit and Specific Activity. One unit is defined as the amount of enzyme required to produce one micromole of D-xylulose in 10 minutes of incubation. Specific activity is expressed as unit per milligram of protein.

Spectrophotometric Assay. D-Xylose isomerase activity can be determined as NADH oxidation in a coupled assay involving a D-xylulokinase, pyruvate kinase, and lactate dehydrogenase system (see L-arabinose isomerase[4]). It can also be determined in a coupled assay with xylitol dehydrogenase.[2,5]

Sources

D-Xylose isomerase activity has been demonstrated in the extracts of the following organisms grown on D-xylose: *Pseudomonas hydrophila*,[6]

[3] Z. Dische and E. Borenfreund, *J. Biol. Chem.* **192,** 583 (1951); see also Vol. III [12].
[4] K. Yamanaka and W. A. Wood, see this volume [106].
[5] O. Touster and G. Montesi, see Vol. V [34].
[6] R. M. Hochster and W. A. Watson, *J. Am. Chem. Soc.* **75,** 3284 (1953); *Arch. Biochem. Biophys.* **48,** 120 (1954).

Lactobacillus plantarum,[7] lactic acid bacteria,[8] *Pasteurella pestis*,[2,9] *Salmonella typhosa*,[10] *Aerobacter aerogenes*,[11] *Pediococcus pentosaceus*,[12] *Candida utilis*,[13] and *Brevibacterium pentoso-amino acidicum*.[14] Many heterofermentative lactic acid bacteria, such as *Lactobacillus brevis*, *L. mannitopous*, *L. gayonii*, *L. pentoaceticus*, *L. fermenti*, *L. lycopersici*, and *Leuconostoc mesenteroides* produce D-xylose isomerase from D-xylose medium.[8] Small amounts of this enzyme are also produced during growth on L-arabinose. *Lactobacillus xylosus*, strain Katagiri and Kitahara, also produces D-xylose isomerase from glucose, fructose, and D-xylose media. Yield of D-xylose isomerase by *L. brevis* is higher than from the above-mentioned lactic acid bacteria.

Purification Procedure Used with Lactobacillus brevis[15]

Growth Medium.[16] The culture medium is composed of 1% peptone, 0.2% yeast extract, 1% sodium acetate, 0.05% $MnSO_4 \cdot 4\ H_2O$, 0.02% $MgSO_4 \cdot 7\ H_2O$, and 0.01% $CoCl_2 \cdot 6\ H_2O$. The mixed salt solution is autoclaved separately. D-Xylose isomerase from this Mn^{++}-fortified medium is about 10 times higher in specific activity and 30 times higher in total units than is obtained with the basal medium. As carbon sources, a mixture of 1% D-xylose and 0.1% glucose are sterilized separately and added just before inoculation. For stab cultures, the same medium contains 2% agar and solid $CaCO_3$ is added.

Culture of Bacteria. *Lactobacillus brevis* is inoculated into 8 ml of the above medium and incubated overnight at 32°. The whole inoculum is transferred to 400 ml of the same medium. After 24 hours of growth, the entire culture is then transferred to 10 liters of the medium and incubated for 16 hours. Cells are harvested and washed with water or dilute Tris buffer.

Step 1. Preparation of Cell-Free Extract. Wet cells, 110 g obtained from 46 liters of medium, were disrupted in small portions by grinding with 300 g of levigated alumina. The enzyme is extracted with 1.7 liters

[7] S. Mitsuhashi and J. O. Lampen, *J. Biol. Chem.* **204,** 1011 (1953).

[8] K. Yamanaka, *Bull. Agr. Chem. Soc. Japan* **22,** 299 (1958); **24,** 305, 310 (1960) (in English).

[9] M. W. Slein, *J. Am. Chem. Soc.* **77,** 1663 (1955).

[10] E. S. Kline and L. S. Baron, *Arch. Biochem. Biophys.* **66,** 128 (1957).

[11] R. P. Mortlock and W. A. Wood, *J. Bacteriol.* **88,** 838 (1964).

[12] W. J. Dobrogosz and R. D. DeMoss, *J. Bacteriol.* **85,** 1356 (1963).

[13] M. Tomoeda and H. Horitsu, *Agr. Biol. Chem. Japan* **28,** 139 (1964) (in English).

[14] Y. Hirose and K. Yamada, *Agr. Biol. Chem. Japan* **25,** 757 (1961) (in English).

[15] K. Yamanaka, *Agr. Biol. Chem. Japan* **27,** 271 (1963) (in English).

[16] K. Yamanaka and T. Higashihara, *Agr. Biol. Chem. Japan* **26,** 162 (1962); K. Yamanaka, *ibid.* **27,** 265 (1963) (in English).

of 0.02 *M* Tris buffer, pH 7.0. The precipitate of alumina is removed by centrifugation or decantation.

Step 2. Manganese Treatment. To the extract (1.5 liters) is added 75 ml of 1 *M* $MnCl_2$. Ammonia was then added to adjust the pH to 6.8–7.2. After standing, the precipitate is removed in the Spinco super-centrifuge using a batch rotor at 15,000 rpm for 30 minutes (1.52 liters). For a large-scale treatment, the remaining alumina and the Mn precipitate may be removed by filtration.

Step 3. Ammonium Sulfate Fractionation. The manganese fraction is treated with 1.08 kg of ammonium sulfate and the precipitate is collected by centrifugation or by filtration. The precipitate (3.7 g wet weight) is dissolved in 100 ml of 0.01 *M* Tris buffer, pH 7.0 (ammonium sulfate I), and refractionated with ammonium sulfate. The precipitate obtained between 0.45 and 0.95 of saturation is collected, dissolved in Tris buffer, and dialyzed overnight against 0.01 *M* Tris buffer, pH 7.0, containing $MnSO_4$ at 5×10^{-3} *M* (ammonium sulfate II).

Step 4. Heat Treatment. The enzyme solution in the presence of 5×10^{-3} *M* of $MnSO_4$ is immersed to a bath at 80°. When the temperature reaches 53°, the flask is transferred to a bath at 55° and maintained for 10 minutes. The flask is then cooled in an ice-bath and the coagulated proteins are removed by centrifugation (volume = 90 ml). The activity is further fractionated by adding solid ammonium sulfate to 0.5 and to 0.85 saturation (ammonium sulfate III).

Step 5. Column Chromatography on DEAE-Sephadex G-50. The enzyme can be chromatographed on DEAE-cellulose at pH 7.0 by the usual procedure. The enzyme is eluted between 0.15 and 0.2 *M* KCl by gradient elution. DEAE-Sephadex G-50 is also useful. DEAE-Sephadex in a column (2×34 cm) is equilibrated with 0.02 *M* Tris-acetate buffer, pH 7.0, and the protein is chromatographed in the same buffer with increasing KCl molarity up to 0.4 *M*. Active fractions obtained by elution at 0.15 *M* KCl are pooled and the enzyme is precipitated by addition of ammonium sulfate. The purification procedure is summarized in the table.

Properties[17]

Effect of pH. The pH optimum lies between pH 5.9 and 7.0. The maximum activity is obtained between pH 5.9 and 6.6 in maleate buffer, whereas Tris buffer is inhibitory.

Stability. Enzyme is completely stable between pH 6.6 and 8.2 at 20° for 3 hours. Manganese ions effectively protect against the thermal

[17] K. Yamanaka, *Mem. Fac. Agr., Kagawa Univ.,* No. 16 (1963) (in English).

PURIFICATION OF D-XYLOSE ISOMERASE

Fraction	Protein (mg)	Units	Specific activity (micromoles/mg/10 min)
Crude extract	7500	85,000	11.3
Mn-treated fraction	4900	99,500	20.4
Ammonium sulfate I (0 to 1.0 saturation)	3700	76,000	20.7
Ammonium sulfate II (0.45–0.95 saturation)	1920	68,500	35.6
Heated fraction	990	39,600	40.0
Ammonium sulfate III (0.5–0.85 saturation)	580	34,400	59.4
DEAE-Sephadex	109	10,400	94.8

inactivation. The enzyme preparation can be stored with $5 \times 10^{-3}\,M$ of $MnSO_4$ at 5°.

Substrate Specificity and Affinity. D-Glucose and D-ribose are also converted to give D-fructose and D-ribulose, respectively, but the affinities for both aldoses are very low. Ketose is not formed from D-mannose, D-galactose, 2-deoxyglucose, L- and D-arabinose, L-xylose, gluconate, glucuronate, D-sorbitol (NAD or NADP), D-mannitol (with NAD), or xylitol (with NAD or NADP). The Michaelis values in the presence of excess Mn^{++} are $5 \times 10^{-3}\,M$ for D-xylose (pH 7.0, 40°) and $0.92\,M$ for D-glucose (pH 7.0, 50°).

Metal Requirement. This enzyme requires Mn^{++} for its activity specifically. The dissociation constant for manganese ions is $1.7 \times 10^{-5}\,M$ with D-xylose or $1.4 \times 10^{-5}\,M$ with D-glucose.[17]

Inhibitors. Chelating agents, such as ethylenediaminetetraacetic acid or *o*-phenanthroline inhibit the activity, but it is fully recovered by adding Mn^{++}. Ni^{++}, Zn^{++}, Cu^{++}, Cd^{++}, Fe^{++}, and Fe^{+++} strongly inhibit isomerization of both D-xylose and D-glucose. Cysteine neither inhibits nor stimulates these activities. Xylitol inhibits both isomerase activities competitively. K_i values are $1.5 \times 10^{-3}\,M$ with D-xylose or $4.5 \times 10^{-3}\,M$ with D-glucose. D-Sorbitol also inhibits both enzymatic reactions competitively.

Activation Energy. It was calculated as 15,100 cal per mole for D-xylose and 24,600 cal per mole for D-glucose.

Remarks on the Identity of D-Xylose Isomerase and D-Glucose Isomerizing Activity.[17] The pH-activity curves for isomerization of D-xylose or D-glucose are identical. The D-xylose isomerase activity was inhibited by glucose and the glucose isomerizing activity was also inhibited by D-xylose. Eight strains of heterofermentative lactic acid bacteria produce both the D-xylose isomerase and the glucose isomerizing activity from D-xylose medium in a constant ratio, and attempts to separate them using fractionations with ammonium sulfate or acetone at several pH values, by column chromatography, or by vertical electrophoresis did not succeed. The ratio of D-xylose isomerase and glucose isomerizing activity remained constant in various thermal inactivation experiments.[17] These results strongly suggest that the conversion of glucose to fructose may be catalyzed by the D-xylose isomerase.

In addition to the above isomerase found in lactic acid bacteria, several so-called glucose isomerase-producing bacteria have been found, mainly in Japan, and some of them require arsenate for isomerization of glucose. However, arsenate inhibits strongly the production of D-xylose isomerase and also inhibits the activities of D-xylose isomerase and of the glucose-isomerizing reaction *of L. brevis*. There is no positive evidence for the presence of specific glucose isomerase.

[105] D-Lyxose Isomerase

By R. L. ANDERSON

D-Lyxose ⇄ D-xylulose

Assay Method

Principle. With D-lyxose as the substrate, the assay is based on the appearance of ketopentose, as determined with the cysteine–carbazole reagents described by Dische and Borenfreund.[1]

Reagents

Sodium cacodylate buffer, 0.2 *M*, pH 7.0
D-Lyxose, 0.1 *M*
$MnCl_2$, 0.1 *M*
Cysteine·HCl, 1.5%
H_2SO_4, 25 *N*
Carbazole, 0.12%, in 95% ethanol

Procedure. To a test tube are added 0.3 ml of buffer, 0.20 ml of D-lyxose, 0.1 ml of $MnCl_2$, D-lyxose isomerase, and water to a volume of 1.0 ml. The reaction is initiated by the addition of D-lyxose isomerase. After incubation at 25° for 10 minutes, 0.2 ml of cysteine·HCl and 6.0 ml of sulfuric acid are added sequentially with mixing, followed im-

[1] Z. Dische and E. Borenfreund, *J. Biol. Chem.* **192**, 583 (1951).

mediately by 0.2 ml of carbazole. After incubation at 35° in a water bath for exactly 20 minutes, the absorbance at 540 mμ is measured with a colorimeter. The increase in absorbance above a control without enzyme is a measure of isomerase activity.

Discussion. The velocity is constant with time and proportional to enzyme concentration up to 1 unit. D-Ribulose (as the *o*-nitrophenylhydrazone) is available commercially and may be used as a standard in the cysteine–carbazole procedure in place of xylulose.[2] Color development with ribulose is complete in about 15 minutes, whereas with xylulose it is about 73% complete in 20 minutes and 100% complete in about 2 hours.

Definition of Unit and Specific Activity. A unit of D-lyxose isomerase is defined as the amount that catalyzes the isomerization of 1 micromole of D-lyxose per hour under the conditions described. Specific activity (units per milligram of protein) is based on a spectrophotometric determination of protein.[3]

Purification Procedure[4]

A strain of *Aerobacter aerogenes*, PRL-R3, selected for the ability to grow on D-lyxose without a lag[4] was grown aerobically at 30° for 18–24 hours and harvested by centrifugation. The medium consisted of 1.35% $Na_2HPO_4 \cdot 7\ H_2O$, 0.15% KH_2PO_4, 0.3% $(NH_4)_2SO_4$, 0.02% $MgSO_4 \cdot 7\ H_2O$, 0.0005% $FeSO_4 \cdot 7\ H_2O$, and 0.5% D-lyxose (autoclaved separately). Extracts were prepared by suspending washed cells in 1 m*M* sodium ethylenediaminetetraacetate (EDTA) (pH 7.0) and treating them for 2–4 minutes in a Raytheon 10-kc sonic oscillator circulated with ice water. The supernatant fluid resulting from 15-minute centrifugation at 31,000 *g* was used as the cell extract. All steps in the fractionation procedure were carried out at 0–4°.

Ammonium Sulfate Fractionation. Ammonium sulfate, 3.7 g, was dissolved in 140 ml of cell extract containing 16 mg of protein per milliliter with a 280:260 mμ ratio of 0.60. Twenty-eight milliliters of 2% protamine sulfate was then added with stirring, and the precipitate that formed was removed by centrifugation and discarded. To the supernatant solution (165 ml) was added 19.9 g of ammonium sulfate (60% of saturation), and the precipitate that formed was collected and dissolved in 10 ml of 1 m*M* EDTA (pH 7.0). This fraction contained 30 mg of protein per milliliter and had a 280:260 mμ ratio of 1.17.

Sephadex G-100 Chromatography. A portion (7 ml) of the above fraction was placed on a column (3.5 × 22 cm) of Sephadex G-100 and

[2] O. Touster, *in* "Methods in Carbohydrate Chemistry" (R. L. Whistler and M. L. Wolfrom, eds.), Vol. I, p. 98. Academic Press, New York, 1962.

[3] O. Warburg and W. Christian, *Biochem. Z.* **310,** 384 (1941).

[4] R. L. Anderson and D. P. Allison, *J. Biol. Chem.* **240,** 2367 (1965).

eluted with distilled water. Five-milliliter fractions were collected, and those which contained most of the activity were pooled. This solution (38 ml) contained 1.26 mg of protein per milliliter and had a 280:260 mμ ratio of 1.65.

DEAE-Cellulose Chromatography. DEAE-cellulose (Bio-Rad Cellex D, exchange capacity = 0.95 milliequivalents per gram) was pretreated as recommended by Peterson and Sober[5] and equilibrated with 4 m*M* sodium cacodylate—1 m*M* EDTA (pH 7.0) in a column 3.5 cm in diameter and 4.5 cm high. A portion (20 ml) of the above fraction was added to the column and eluted with 250 ml (5-ml fractions) of the same buffer containing NaCl in a linear gradient from 0 to 0.8 *M*. The fractions containing the highest specific activity were pooled to yield 28 ml of 70-fold purified D-lyxose isomerase. The protein concentration was 0.09 per milliliter and the 280:260 mμ ratio was 0.79.

Alumina Cγ Gel Adsorption and Elution. Alumina Cγ gel (Sigma Chemical Co., 4.2% solids) was mixed with an equal volume of 1 m*M* EDTA (pH 7.0), and 0.52 ml of the mixture was suspended in a 20-ml portion of the above D-lyxose isomerase fraction. After centrifugation, the supernatant was discarded and the centrifugate was resuspended in 20 ml of 10 m*M* potassium phosphate–1 m*M* EDTA (pH 7.0). The suspension was centrifuged and the supernatant discarded. The centrifugate was resuspended in 3.0 ml of 100 m*M* potassium phosphate–1 m*M* EDTA (pH 7.0) and centrifuged. A portion (2 ml) of the supernatant solution was placed on a Sephadex G-25 column (1.5 × 16 cm) and eluted with distilled water. This yielded 2.0 ml of 130-fold purified D-lyxose isomerase with a protein concentration of 0.09 mg/ml and a 280:260 mμ ratio of 1.63. Correcting for the portions of each fraction not further purified, the recovery was 6% of the activity of the cell extract. A summary of the purification procedure is given in the table.

PURIFICATION OF D-LYXOSE ISOMERASE

Fraction	Units[a]	Units/mg	Fold purified	Recovery (%)
Cell extract	7950	3.5	1.0	100
Ammonium sulfate	3530	12	3.4	44
Sephadex G-100	3000	44	12	38
DEAE-cellulose	1730	252	72	22
Alumina C_{γ} gel	463	462	130	6

[a] Micromoles of D-lyxose isomerized per hour; corrected for the portion of each fraction not further purified.

[5] See Vol. V [1].

Properties[4]

Substrate Specificity. D-Lyxose isomerase has been found only in cells grown on D-lyxose; D-mannose and D-fructose do not serve as inducers. However, the purified enzyme isomerizes D-mannose to D-fructose. The K_m values for D-lyxose and D-mannose are 3.6 mM and 10 mM, respectively. The V_{max} is 2.4 times greater with D-lyxose than with D-mannose. D-Ribose, D- or L-xylose, D- or L-arabinose, D-glucose, D-glucose 6-phosphate, and D-mannose 6-phosphate are not isomerized at a detection level of 0.5% the rate on D-lyxose.

Activators and Inhibitors. Activity is nil in the absence of a metal ion. At 1 mM concentrations, the relative activity with various metal salts is as follows: $MnCl_2$, 100; $FeCl_3$, 78; $FeSO_4$, 18; $CaCl_2$, 14; $CoCl_2$, 8; $MgCl_2$, 0; $ZnSO_4$, 0; $CuSO_4$, 0. The K_m for Mn^{++} is 4 μM. Activity is nil in the presence of 10 μM *p*-chloromercuribenzoate; the activity cannot be restored with the further addition of reduced glutathione, and reduced glutathione was never found to be protective or stimulatory during fractionation or assay. Iodoacetate, oxidized glutathione, 2,6-dichlorophenolindophenol, or $K_3Fe(CN)_6$ at 1 mM and pH 7.0 showed no inhibition in the standard assay.

Effect of pH. The pH optimum is about 7.0. Activity as a function of pH varies with the buffer, being half maximal at pH 6.3 in maleate and at pH 4.9 in cacodylate.

Molecular Weight. The sedimentation coefficient as determined by density gradient centrifugation is 3.7 to 3.8 S, suggesting a molecular weight of about 40,000.

[106] L-Arabinose Isomerase

By K. YAMANAKA and W. A. WOOD

$$\text{L-Arabinose} \rightleftarrows \text{L-ribulose}$$

This enzyme has been described by Cohen[1] briefly and by Smyrniotis[2] and by Horecker[3,4] in more detail.

[1] S. S. Cohen, see Vol. I [52].

[2] P. Z. Symrniotis, see Vol. V [42].

[3] B. L. Horecker, *in* "Methods of Enzymatic Analysis" (H. U. Bergmeyer, ed.), p. 178. Academic Press, New York, 1963.

[4] E. C. Heath, B. L. Horecker, P. Z. Smyrniotis, and Y. Takagi, *J. Biol. Chem.* **231**, 1031 (1958).

Assay Methods[2]

Principle. L-Arabinose isomerase activity is assayed spectrophotometrically at 30° using a standard spectrophotometer equipped with an absorbancy converter, automatic cuvette positioner, and recorder.[5] The assay system is based on the following sequences of reactions:

$$\text{L-Arabinose} \xrightarrow{\text{L-arabinose isomerase}} \text{L-ribulose}$$

$$\text{L-Ribulose} + \text{ATP} \xrightarrow{\text{L-ribulokinase}} \text{L-ribulose 5-phosphate} + \text{ADP}$$

$$\text{Phosphoenolpyruvate} + \text{ADP} \xrightarrow{\text{pyruvate kinase}} \text{pyruvate} + \text{ATP}$$

$$\text{Pyruvate} + \text{NADH} \xrightarrow{\text{lactate dehydrogenase}} \text{lactate} + \text{NAD}$$

Reagents

Maleate buffer, 0.4 *M*, pH 6.9
$MgCl_2$, 0.1 *M*
ATP, 0.1 *M*, adjusted to pH 6.9
Phosphoenolpyruvate, 0.025 *M*, adjusted to pH 6.9
Sodium glutathione
NADH, 0.01 *M*
L-Arabinose, 0.1 *M*
Purified L-ribulokinase, free from L-arabinose isomerase[6]
Lactate dehydrogenase-pyruvate kinase, crystalline (Worthington)

All solutions are prepared separately and kept in a deepfreeze. A reagent stock solution is then prepared by mixing 0.8 ml of maleate buffer, 0.4 ml of $MgCl_2$, 0.2 ml of ATP, 0.4 ml of phosphoenolpyruvate, 18 mg of glutathione, 0.4 ml of NADH, 0.05 ml of lactate dehydrogenase, and water to a total volume of 4.0 ml. This mixture can be kept frozen and prepared fresh every 3 days.

Procedures. To a microcuvette of 0.5-ml capacity and 1-cm light path are added 0.1 ml of reagent mixture, 2 units of purified L-ribulokinase, 2 micromoles of L-arabinose, suitable amounts of L-arabinose isomerase preparation, and water to a volume of 0.15 ml. Blanks are run for NADH oxidase with buffer and NADH, and for ATPase activity by omitting L-ribulokinase and L-arabinose. L-Ribulokinase activity may be determined by the same method by adding L-ribulose instead of L-arabinose. NADH oxidase and ATPase activities are subtracted from the total activity. A linear relationship between the rate of NADH

[5] W. A. Wood and S. R. Gilford, *Anal. Biochem.* **2**, 589, 601 (1961).
[6] See this volume [80]. See also F. J. Simpson and W. A. Wood, *J. Biol. Chem.* **230**, 473 (1958).

oxidation and amount of enzyme is obtained between 0 and 0.3 optical density units per minute. This corresponds to 0–0.067 unit of isomerase.

Alternate Assay. For crude preparations, the isomerase activity cannot be determined by a spectrophotometric method owing to the strong interference of NADH oxidase and ATPase. In that case, the following procedure is followed:

The reaction mixture (1.0 ml) contains 0.5 ml of 0.05 *M* maleate buffer, pH 6.9, for *Aerobacter* enzyme or of 0.05 *M* Tris-HCl buffer, pH 8.0, for *Lactobacillus* enzyme, 0.05 ml of 0.1 *M* $MnCl_2$ (or $MnSO_4$), and 0.01–0.1 ml of enzyme preparation with adequate dilution. After incubation at 35° for 5 minutes, the reaction is started by adding 0.05 ml of 0.1 *M* L-arabinose, and incubation is continued at 35° for 10 minutes. L-Ribulose is determined by the cysteine–carbazole procedure[2] using L-ribulose-*o*-nitrophenylhydrazone as a standard.

Definition of Unit and Specific Activity. A unit of enzyme is defined as the amount required to effect an absorbancy change of 1.0 per minute at 340 mμ under these conditions. Specific activity is expressed as units per milligram of protein. For the colorimetric procedure, see Vol. V [42] and the procedure for D-xylose isomerase.[7]

Sources

L-Arabinose isomerase activity has been found in the cells grown on L-arabinose for the following organisms: *Lactobacillus plantarum*,[4,8] some strains of homolactic acid bacteria related to *L. plantarum*,[9] and of *Aerobacter aerogenes*.[10,11] This enzyme is also produced in *L. plantarum* grown on D-xylose.[2] However, many of the heterofermentative lactic acid bacteria, such as *L. gayonii*, *L. mannitopous*, *L. buchneri*, *L. brevis*, *L. pentoaceticus*, and *L. lycopersici*, and some of the *Pediococcus* group also produce L-arabinose isomerase from the media containing glucose, fructose, galactose, D-xylose, L-arabinose, or malt extract.[9] The glucose-grown cells of *L. gayonii* contain L-arabinose isomerase but no L-ribulokinase and NADH oxidase. Induction of L-arabinose isomerase by *Pediococcus pentosaceus*[12] and by *Lactobacillus plantarum*[13] was reported.

[7] K. Yamanaka, see this volume [104].
[8] J. O. Lampen, *Proc. Am. Chem. Soc.*, 44c (1954).
[9] K. Yamanaka, *Bull. Agr. Chem. Soc. Japan* **24**, 310, (1960) (in English); *Mem. Fac. Agr., Kagawa Univ.*, No. 16 (1963) (in English).
[10] F. J. Simpson, M. J. Wolin, and W. A. Wood, *J. Biol. Chem.* **230**, 457 (1958).
[11] R. P. Mortlock and W. A. Wood, *J. Bacteriol.* **88**, 838 (1964).
[12] W. J. Dobrogosz and R. D. DeMoss, *J. Bacteriol.* **85**, 1350, 1356 (1963).
[13] M. Chakravorty, *Biochim. Biophys. Acta* **85**, 152 (1964).

Purification Procedures from *Aerobacter aerogenes*, PRL-R3

Growth Medium. The medium is composed of 0.2% $(NH_4)_2SO_4$, 0.7% K_2HPO_4, 0.3% KH_2PO_4, 0.01% $MgSO_4$, 0.2% yeast extract, 0.2% peptone, 0.05% $MnCl_2 \cdot 4\ H_2O$, and 0.2% L-arabinose. $MgSO_4$, $MnCl_2$, and L-arabinose solutions are sterilized separately in concentrated form and are added just before inoculation. Manganese ions form a precipitate with the phosphate buffer, but L-arabinose isomerase is produced in greater amounts than in the same medium without manganese.

Culture of Bacteria. The bacterium is inoculated into 5 ml of medium in a test tube and cultured for one night at 30° with shaking. One-tenth milliliter of culture is then transferred to 500 ml of the same medium in a Fernbach flask and grown for 20–24 hours on a rotary shaker. Two-hundred and fifty milliliters of this culture is added to 20 liters of the medium in a glass carboy equipped with an aeration device, and the culture is incubated for 16 hours at 20°. The cells are harvested in a Sharples centrifuge and washed with water.

Preparation of Cell-Free Extract. The cells are suspended in $10^{-3}\ M$ ethylenediaminetetraacetic acid (EDTA), pH 7.0, and disrupted in a 10-kc sonic oscillator for 10 minutes. The cell-free extract was obtained by centrifugation.

Protamine Treatment. The extract is diluted to a protein concentration of 10 mg/ml by addition of $10^{-3}\ M$ EDTA, pH 7.0. Solid ammonium sulfate is added to 0.2 M in final concentration, and 0.2 volume of 2% protamine sulfate is added to precipitate the nucleic acids. The precipitate is removed by centrifugation (protamine supernatant).

First Ammonium Sulfate Fractionation. Solid ammonium sulfate is added to give 45% saturation. The precipitate is removed by centrifugation, and additional ammonium sulfate is added to give 80% saturation. The precipitate was dissolved in 0.01 M Tris-acetate buffer, pH 7.0, and dialyzed overnight against the same buffer containing $10^{-3}\ M$ EDTA (ammonium sulfate I).

Heat Treatment. $MnCl_2$ solution is added to the above fraction to give $5 \times 10^{-3}\ M$ concentration. The fraction in a flask is immersed in a water bath at 80° and stirred until the temperature of the solution reaches 49°. It is then transferred to a bath at 50° for 10 minutes. After cooling, the precipitate is removed by centrifugation. Solid ammonium sulfate is added to the clear supernatant to obtain the precipitate between 45 and 75% saturation. The precipitate was collected, dissolved, and dialyzed against 0.01 M Tris-acetate buffer, pH 7.0, containing $5 \times 10^{-3}\ M$ $MnCl_2$. NADH oxidase activity was almost completely removed in this step (heat fraction I).

Acetone Fractionation. Acetone at —20° is slowly added while the temperature of the fraction is slowly lowered to —5°. Enzymatic activity is recovered mainly in the precipitate obtained between 20–30% and 30–40% acetone (acetone fraction). The precipitates were dissolved in 0.01 *M* Tris-acetate buffer, pH 7.0, and incubated at 55° for 5 minutes in the presence of 5×10^{-3} *M* of $MnCl_2$. The precipitate was removed by centrifugation (heat fraction II).

DEAE-Cellulose Chromatography. Enzyme was absorbed onto a column of DEAE-cellulose (3.8 × 25 cm) which was equilibrated with 0.02 *M* Tris-acetate buffer, pH 7.0, and eluted by a linear gradient of 0 to 0.4 *M* of NaCl in the same buffer. Enzyme was recovered at around 0.2 *M* of NaCl. The active fractions (nos. 72–85) were pooled, and solid ammonium sulfate was added to obtain the precipitates between 45 and 70% saturation. This preparation is almost free of L-ribulokinase activity (DEAE-cellulose fraction I).

Sephadex G-100 Chromatography. The enzyme fraction is absorbed on a column of Sephadex G-100 (2.3 × 25 cm). The active fractions were collected by eluting 0.05 *M* NaCl (Sephadex fraction).

Second DEAE-Cellulose Chromatography. Rechromatography on a column of DEAE-cellulose (1.4 × 22 cm) was carried out by the procedure described in the first DEAE-cellulose chromatography. The active fractions were combined and concentrated by the addition of solid ammonium sulfate (DEAE-cellulose fraction II).

The purification procedure is summarized in Table I. The final preparation represents an 83-fold enrichment of the specific activity and is essentially free from NADH oxidase, ATPase, L-ribulokinase, D-ribulokinase, D-xylulokinase, ribitol dehydrogenase, L-ribulose 5-phosphate 4-epimerase, D-ribulose 5-phosphate 3-epimerase, and phosphoketolase activities.

TABLE I
PURIFICATION OF L-ARABINOSE ISOMERASE FROM *A. aerogenes* (SPECTROPHOTOMETRIC ASSAY)

Fraction	Protein	Units × 10^3	Specific activity (units/mg of protein)
Crude extract	56,200	628	11
Ammonium sulfate I	20,350	468	23
Heat fraction I	10,300	620	60
Ammonium sulfate II	8,440	577	68
Acetone fraction	3,900	299	76
Heat fraction II	1,890	281	148
DEAE-cellulose fraction I	720	203	282
Sephadex G-100 fraction	460	201	439
DEAE-cellulose fraction II	153	141	923

Purification Procedure for *Lactobacillus gayonii*

The procedure given is that described by Smyrniotis,[2] but with some modifications.

Growth Medium. A manganese-fortified medium yielded about 20 times more L-arabinose isomerase than the basal medium without manganese.[14] The composition of the medium is as follows: 1% peptone, 0.2% yeast extract, 1% sodium acetate, 0.05% $MnSO_4 \cdot 4\ H_2O$, 0.02% $MgSO_4 \cdot 7\ H_2O$, 0.001% NaCl, 0.1% L-arabinose, and 0.5% glucose.

Culture of Bacteria. See D-xylose isomerase.[7] For maximum production of the isomerase by *L. gayonii*, incubation is carried out for about 40 hours with glucose or 16 hours with L-arabinose at 32° for 20 liters of medium.

Preparation Procedures. See Vol. V [42]. Further purification can be obtained by using heat treatment and by column chromatography on DEAE-cellulose procedures for *A. aerogenes* (above). The purification procedure is summarized in Table II.

TABLE II
PURIFICATION OF L-ARABINOSE ISOMERASE FROM *L. gayonii*
(BASED UPON THE COLORIMETRIC ASSAY)

Fraction	Protein (mg)	Units	Specific activity (micromoles/mg/10 min)
Crude extract	255	1670	5.9
Mn-treated fraction	162	2115	13.1
Ammonium sulate fraction (0.6–0.9 saturfation)	24	1110	46.6
Acetone fraction	15	1024	85.5
DEAE-cellulose fraction	6	770	140

Properties

These properties are shown by the purified preparation from *A. aerogenes* unless otherwise noted.

Effect of pH. The maximum activity was attained at pH 6.4–6.9 (*A. aerogenes*), 7.8–8.2 (*L. gayonii*) at 35° for 10 minutes of incubation.

Stability. Enzyme is stable between pH 6.6 and 9.5 at 30° for 2 hours. Manganese ions protect against the thermal inactivation.

Substrate Specificity and Affinity. Ketoses were produced from L-arabinose, D-galactose, or D-fucose, but not from other pentoses or from hexoses. The apparent Michaelis values from the Lineweaver-Burk plots

[14] K. Yamanaka and T. Higashihara, *Agr. Biol. Chem. Japan* **26,** 162 (1962) (in English).

are $3.3 \times 10^{-2} M$ for L-arabinose, 0.27 M for D-fucose, and 0.37 M for D-galactose (pH 6.9, 38°, 10 minutes). The ratio of relative activities toward these sugars (L-arabinose, D-fucose, and D-galactose) was 1:0.22: 0.23 (500 micromoles of substrate, 50°, 30 units of isomerase). These three sugars have the same configuration from C-1 to C-4. However, L-lyxose and L-ribose were not isomerized by this enzyme.[15] Eight strains of heterofermentative lactic acid bacteria grown on L-arabinose had the same ratio of isomerase activities on L-arabinose and D-galactose.[16] These activities were not separable by ammonium sulfate fractionation or by chromatography on DEAE-cellulose. The optimum pH for isomerization of L-arabinose, D-galactose, and D-fucose is identical at pH 6.4–6.9. From these observations, it may be concluded that this enzyme has one site with a high affinity for L-arabinose, but low affinity for D-galactose and D-fucose. The isomerase thus has the affinity for sugars with an L-*cis* hydroxyl configuration at C-3 and C-4.

Metal Requirement. The isomerase from both sources required Mn^{++} specifically for activity.

Inhibitors. L-Arabitol and ribitol inhibit competitively the isomerization of all three aldoses. With L-arabinose as substrate, the K_i values for L-arabitol or ribitol are about $2.3 \times 10^{-3} M$ and $3.5 \times 10^{-4} M$, respectively.

[15] K. Yamanaka and W. A. Wood, unpublished observation, 1964.

[16] K. Yamanaka, unpublished observation, 1962.

[107] 4-Deoxy-L-*threo*-5-hexosulose Uronic Acid Isomerase[1]

By JACK PREISS

$$CHO(CH_2OH)_2CH_2COCOOH \rightleftharpoons CH_2OHCOCHOHCH_2COCOOH$$

4-Deoxy-L-*threo*-5-hexosulose uronic acid	3-Deoxy-D-*glycero*-2,5-hexodiulosonic acid
(I)	(II)

Assay Method

Principle. The most convenient assay for the isomerization of compound (I) to compound (II) proved to be the spectrophotometric determination of the reduction of the latter by DPNH at 340 mμ in the presence of an excess of 3-deoxy-D-*glycero*-2,5-hexodiulosonic acid reductase. The preparation of this enzyme, free from contamination with

[1] J. Preiss and G. Ashwell, *J. Biol. Chem.* **238**, 1577 (1963).

isomerase and suitable for use in this assay, is described in this volume [41].

Reagents

Potassium phosphate buffer, 1 *M*, pH 7.0
Compound (I), 0.01 *M*. This compound is prepared by enzymatic degradation of polygalacturonate by polygalacturonate lyase.[2]
DPNH, 0.01 *M*
3-Deoxy-D-*glycero*-2,5-hexodiulosonic acid reductase, 3.9 units/ml

Procedure. A reaction mixture containing 0.02 ml of phosphate buffer, 0.05 ml of compound (I), and 0.0024–0.015 unit of enzyme in a total volume of 0.20 ml was incubated at 37° for 10 minutes. The enzyme was added last to initiate the reaction. The reaction was terminated by heating the incubation mixture for 30 seconds at 100° followed by quick cooling. An aliquot, 0.10 ml, was transferred to a cuvette of 1-cm light path containing 0.05 ml of potassium phosphate, pH 7.0, 0.01 ml of DPNH, and water in a final volume of 0.95 ml. The reaction was initiated by the addition of 50 μl of dehydrogenase, and the decrease in absorption at 340 mμ was recorded for 20–25 minutes, after which the reaction was complete. Control tubes in which either the isomerase or compound (I) were omitted from the original incubation mixture were also run at the same time.

Definition of Unit and Specific Activity. One unit is defined as that amount of enzyme required to catalyze the isomerization of 1 micromole of compound (I) in 1 minute. Specific activity is defined as units of enzyme per milligram of protein. Protein concentration was determined by the method of Lowry *et al.*[3]

Purification Procedure

The pseudomonad, ATCC 14968, was originally isolated by Dr. J. D. Smiley on the basis of its ability to utilize alginic acid as the sole carbon source. It was grown aerobically on a minimal medium consisting of ammonium nitrate, 1%; dibasic potassium phosphate, 1.5%; monobasic sodium phosphate, 0.5%; magnesium sulfate heptahydrate, 0.1%; and sodium polygalacturonate, 0.11%. After 5 days of growth at room temperature (25°), the bacteria were harvested in a Sharples centrifuge. The bacterial paste, bright orange in color, was stored frozen at —15°. Approximately 1 g of cells (wet weight) was obtained per liter culture fluid.

[2] J. Preiss and G. Ashwell, *J. Biol. Chem.* **238**, 1571 (1963).
[3] O. H. Lowry, N. J. Rosebrough, A. L. Farr, and R. J. Randall, *J. Biol. Chem.* **193**, 265 (1951).

Step 1. Crude Extract. The frozen bacterial paste was suspended in 2 volumes of 0.1 *M* potassium phosphate buffer, pH 7.5, containing 0.1% cysteine-HCl (or 0.01 *M* GSH), and was disrupted by sonic vibration for 20 minutes in a 10 kc Raytheon oscillator. The broken cell mixture was centrifuged at 10,000 rpm for 15 minutes, and the supernatant solution was used as the starting material for purification of the enzyme. All ensuing operations were carried out at 0–3°.

Step 2. Streptomycin Sulfate Fractionation. To 20 ml of the crude extract, 7.5 ml of a 5% solution of streptomycin sulfate was added slowly and with continuous stirring. After 10 minutes, the suspension was centrifuged at 10,000 rpm, and the precipitate was discarded.

Step 3. Ammonium Sulfate Fractionation. To 24 ml of the supernatant fluid was added 20 ml of a saturated solution of ammonium sulfate. After 10 minutes the precipitate was removed by centrifugation, and 13.5 g of solid ammonium sulfate was added. The suspension was stirred continuously for 10 minutes and centrifuged, and the precipitate was dissolved in 10 ml of 0.05 *M* Tris-HCl buffer, pH 7.5, containing 0.02 *M* mercaptoethanol and 0.001 *M* EDTA. The resulting solution was dialyzed overnight against 500 ml of the same buffer solution.

The purification is summarized in the table.

PURIFICATION OF THE REDUCTASE

Step	Volume (ml)	Activity (units)	Protein (mg/ml)	Specific activity (units/mg protein)
1. Crude extract	20	33.3	53.0	0.63
2. Streptomycin sulfate fractionation	25	26.0	35.0	0.74
3. Ammonium sulfate fractionation	13	48.0	11.0	4.36

Properties

Stability. The ammonium sulfate fraction was quite stable and has been kept at 0–3° for at least 1 month without loss of activity.

pH Optimum. The isomerase reaction proceeds with maximal velocity in phosphate buffer in the narrow range of pH 7.0–7.5.

Equilibrium. In phosphate buffer, pH 7.0, the equilibrium ratio of compound (II) to compound (I) was found to be about 1.1:1.0. In borate buffer, pH 8.0, the equilibrium ratio of compound (II) to compound (I) was about 2.7.

Occurrence of Enzyme. The enzyme does not appear to be present in extracts of glucose-grown cells.[1]

[108] D-Ribulose 5-Phosphate 3-Epimerase

By W. T. WILLIAMSON and W. A. WOOD

D-Ribulose 5-phosphate ⇄ D-xylulose 5-phosphate

Assay Method

Principle. The assay method determines the rate of formation of D-xylulose 5-phosphate in a coupled system involving excess phosphoketolase, triose phosphate isomerase and α-glycerol phosphate dehydrogenase. Thus, the rate of epimerization is measured spectrophotometrically at 340 mμ as the rate of NADH oxidation.

Reagents

Imidazole buffer, 0.5 *M* pH 7.0
$MgCl_2$, 0.05 *M*
NaGSH, 0.1 *M*
NADH (neutralized), 0.02 *M*
Thiamine pyrophosphate (neutralized), 0.01 *M*
D-Ribulose 5-phosphate, 0.1 *M* (see this volume [9, 10])
Na_2ASO_4, 0.001 *M*, pH 7.0
α-Glycerol phosphate dehydrogenase–triose phosphate isomerase (crystalline commercial), phosphoketolase[1] free of 3-epimerase.

Procedure. The complete reaction mixture in a volume of 0.2 ml contained imidazole buffer (10 micromoles), $MgCl_2$ (0.5 micromoles), NADH (2.0 micromoles), thiamine pyrophosphate (0.1 micromole), NaGSH (1.0 micromole), D-ribulose 5-phosphate (1.0 micromole), Na_2AsO_4 (0.1 micromole), and 1 μl of a slurry of crystalline triose phosphate isomerase, and 2.5 units of phosphoketolase. The assay was initiated by the addition of a suitable dilution of 3-epimerase. The velocity was linear over a range of 0.1–1.2 units of epimerase.

Definition of a Unit of Epimerase. One unit of enzyme is defined as the amount which produced an absorbancy change of 1.0 per minute in the above assay. Specific activity was in the usual units and was based upon the protein method of Warburg and Christian.[2]

[1] R. Votaw, W. T. Williamson, L. O. Krampitz, and W. A. Wood, *Biochem. Z.* **338,** 756 (1963). Also see this volume [90].
[2] O. Warburg and W. Christian, *Biochem. Z.* **310,** 384 (1941).

Purification Procedure

Crude Extract. Five-hundred grams of dried yeast (Standard Brands, 20–40) was slowly added to 1600 ml of 0.1 *M* $NaHCO_3$ with stirring. The mixture was slowly stirred for 4 hours at room temperature and centrifuged to remove cellular debris. The precipitate was washed twice with 1 liter of 0.1 *M* $NaHCO_3$, and these washings were combined with the crude extract. The extract was cooled to 4°–5° during centrifugation and all subsequent steps were performed at this temperature.

Protamine Fractionation. Solid ammonium sulfate was added to 0.1 *M* concentration and 0.1 volume of 2.0% protamine sulfate, pH 5, was added slowly with stirring. The mixture was centrifuged, and the precipitate was discarded.

Ammonium Sulfate I. Solid ammonium sulfate was added to 70% of saturation. The precipitate obtained upon centrifugation was dissolved in sufficient cold water to give an ammonium sulfate carry-over of 0.1 *M*.

Acetone Fractionation. One volume of 80% acetone (redistilled) at —20° was added, while the temperature of the enzyme fraction was maintained near 0°. The precipitate recovered by centrifugation was discarded. An additional 2 volumes of 80% acetone were added to the supernatant solution. The precipitate recovered by centrifugation was immediately dissolved in water, and solid ammonium sulfate was added to 50% of saturation. The precipitate was discarded, and the ammonium sulfate concentration was raised to 70% of saturation. The precipitate was dissolved in a minimum volume of cold water, and more water was added until the ammonium sulfate concentration was 0.5 *M*.

Heat Step. The solution was heated with stirring in an 80° water bath. When the temperature reached 60°, the solution was removed from the bath, held for 1 minute, and then cooled rapidly in a salt–ice bath. The copious precipitate was removed by centrifugation, and the supernatant solution was treated with ammonium sulfate to give a final concentration of 2.5 *M*.

Extraction with Ammonium Sulfate Solutions. The precipitate from the above step was extracted with 100 ml portions of 2.5 *M* and 2 *M* ammonium sulfate. Usually, most of the activity was in the 2.5 *M* extract. Solid ammonium sulfate was added to 3 *M*; the precipitate was collected by centrifugation and dissolved in 100 ml of 0.01 *M* phosphate buffer, pH 7.0.

Chromatography on DEAE-Cellulose. The above solution at pH 7.0 was added to a column of DEAE-cellulose (2 cm × 15 cm) which had been previously equilibrated at pH 7.4 with 0.01 *M* phosphate buffer. The epimerase was eluted with a linear gradient of phosphate buffer, pH

7.4, between 0 and 0.5 M. The epimerase eluted at a concentration of 0.2 M phosphate.

Fractionation with Alumina Cγ. The combined fractions from the DEAE-cellulose column were dialyzed against 7 liters of cold distilled water for 16 hours. Alumina Cγ was then added at the rate of 1 mg dry weight per milligram of protein to absorb the activity. The epimerase was eluted by stirring with 13 ml of phosphate buffer, pH 7.0, for 2 minutes.

As shown in the table, a 675-fold purification was obtained with a 4% yield of activity.

PURIFICATION OF D-RIBULOSE 5-PHOSPHATE 3-EPIMERASE

Step	Volume (ml)	Total activity (units $\times 10^{-3}$)	Recovery (%)	Protein (mg/ml)	Specific activity	Purification (-fold)
Crude extract	2580	1032	100	33	12	1
Protamine sulfate	2820	1130	100	33	2.5	—
Ammonium sulfate (0–70% saturation)	900	1040	100	48	24	2
Acetone fractionation	540	1000	97	—	—	—
Ammonium sulfate (50–70% saturation)	320	625	60	16	120	10
Heat step	294	710	68	4.6	535	44
Ammonium sulfate (0–2.5 M extract)	115	420	41	2.8	1340	111
Ammonium sulfate (3 M precipitate)	50	440	42	4.3	2025	169
Ammonium sulfate (50–60% saturation)	10.5	172	17	6.0	2740	228
DEAE-cellulose	108	122	12	0.19	6610	550
Alumina C_γ eluate	13	39	4	0.37	8100	675

Properties[3]

Based upon tritium incorporation data, the epimerization of D-ribulose 5-phosphate to D-xylulose 5-phosphate proceeds with the equilibrium incorporation of 1 μatom of hydrogen per micromole of ketopentose 5-phosphate. The tritiated D-xylulose 5-phosphate was converted to tritiated glycerol phosphate by the action of phosphoketolase, α-glycerol phosphate, and NADH. Also, the tritiated D-xylulose 5-phosphate, in the presence of phosphoketolase, glyceraldehyde 3-phosphate dehydrogenase,

[3] W. T. Williamson, D. D. Fossitt, and W. A. Wood, unpublished studies, 1963.

lactic dehydrogenase, pyruvate, and NAD yielded lactate of the same specific activity. Thus, the incorporation of tritium was at carbon 3 of D-xylulose 5-phosphate. No cofactors have been found for the epimerase, and ethylenediamine tetraacetate does not inhibit the reaction.

The purified epimerase appeared to be homogeneous in the ultracentrifuge, and had a Z average molecular weight of 45,800 and a weight average molecular weight of 45,900.

The pH optimum was at 8.0 when assayed in both 0.05 M potassium phosphate buffer and in 0.05 M tris buffer. The purified enzyme was stable when stored for periods of at least 2 months in 0.5 M ammonium sulfate at —20°.

109 Aldose-1-Epimerase from *Escherichia coli*

By Kurt Wallenfels and Klaus Herrmann

$$\alpha\text{-D-Aldose} \rightleftarrows \beta\text{-D-aldose}$$

Assay Method

Principle. A polarimetric method has been used to assay purified fractions of aldose-1-epimerase.[1] For assay based on the more rapid oxidation of β- than α-D-glucose by bromine solution,[2] and glucose oxidase,[3] see R. Bentley.[4] Very rapid determinations of even small amounts of aldose-1-epimerase can be done by means of a combined, kinetic, optical assay[5] with the β-specific D-galactose dehydrogenase from *Pseudomonas saccharophila*[6] and DPN.

$$\alpha\text{-D-Galactose} \rightarrow \beta\text{-D-galactose}$$
$$\beta\text{-D-Galactose} + \text{DPN}^+ \rightarrow \rightarrow \text{D-galactonic acid} + \text{DPNH} + \text{H}^+$$

This assay is applicable to crude cell-free extracts of *Escherichia coli* as well as to highly purified preparations.

Reagents

α-D-Galactose, 0.1%. Immediately before use the solid α-anomer is dissolved in water at 4°.

[1] D. Keilin and E. F. Hartree, *Biochem. J.* **50,** 331 (1952).

[2] D. S. Bhate and R. Bentley, *Federation Proc.* **16,** 154 (1957).

[3] A. S. Keston, *Federation Proc.* **17,** 253 (1958).

[4] R. Bentley, see Vol. V [24].

[5] K. Wallenfels, K. Herrmann, and G. Kurz, *Abstr. 6th Intern. Congr. Biochem., New York 1964,* p. 342. I.U.B. Vol. 32, New York, 1964.

[6] K. Wallenfels and G. Kurz, *Biochem. Z.* **335,** 559 (1962). See also K. Wallenfels and G. Kurz, this volume [22].

Potassium sodium phosphate buffer, 0.033 *M*, pH 6.8
DPN^+, 3.3×10^{-4} *M*, in buffer
β-D-Galactose dehydrogenase, 1500–2000 units/ml, concentration about 1 mg/ml[6]
Aldose-1-epimerase, 2000–3000 units/ml[7]

Procedure. One-tenth milliliter of DPN^+ solution, 2.6 ml of phosphate buffer, 0.1 ml of aldose-1-epimerase and 0.1 ml of β-D-galactose dehydrogenase solutions are incubated in a photometer cell at 25 ± 0.1° for 10–15 minutes. The reaction is started by addition of 0.1 ml of α-D-galactose at time zero. The optical density (366 mμ) E_{sample} is read 100 seconds later. A blank determination (E_{blank}) has to be done in the same way, but instead of 0.1 ml aldose-1-epimerase, 0.1 ml of phosphate buffer is added. The difference of the optical densities ($\Delta E = E_{sample} - E_{blank}$) is used for the calculation of the activity.

Definition of Unit. One unit of aldose-1-epimerase is the amount of protein that catalyzes the transformation of 1 micromole of α-D-galactose to β-D-galactose per minute. Further conditions: temperature 25°; 0.033 *M* phosphate buffer, pH 6.8; substrate saturation of enzyme.

As the determination of activity of aldose-1-epimerase at substrate saturation is impossible by means of the above procedure, V_{max} (the maximum velocity of the reaction at substrate saturation) is read from a Lineweaver-Burk plot determined by polarimetric assays, using twice-distilled water as a solvent (phosphate buffer has an inhibitory effect on aldose-1-epimerase).[8] For the amount of protein that shows a $\Delta E = 0.1$ in 100 seconds in the above procedure, the V_{max} from the Lineweaver-Burk plot is calculated to be 2080 micromoles per minute.

Calculation of Activity. The product of $\Delta E \times 20{,}800$ gives directly the activity in units per milliliter of the original enzyme solution.

Purification Procedure

E. coli K12, is cultivated on synthetic medium, using glycerol as a carbon source; the cells are harvested just before leaving the log-phase of growth by centrifugation in a Sharples supercentrifuge. The moist cells are lyophilized.

Methods for preparation of the cell-free extract and removal of the nucleic acids are the same as described for the β-galactosidase from *E. coli*.[9] After those two steps of purification both enzymes, β-galactosidase and aldose-1-epimerase, are separated from one another by fractional precipitation with ethanol.

[7] K. Wallenfels and K. Herrmann, *Biochem. Z.* **343**, 294 (1965).
[8] K. Wallenfels, K. Herrmann, and F. Hucho, *Biochem. Z.* **343**, 307 (1965).
[9] K. Wallenfels, see Vol. V [23].

To the supernatant (2770 ml) of the ethanol precipitation, 91850 ml of purified cold ethanol (—15°) is added slowly with constant stirring while the mixture is held at —35°. The suspension is allowed to warm up to —15° and to stand for several hours at —15°; it is then centrifuged for 60 minutes at 2500 *g* and —15°. The supernatant is rejected, and the precipitate is freed from ethanol as completely as possible. It is then suspended in 50 ml of 0.033 *M* phosphate buffer pH 6.8, stirred for 2 hours, and centrifuged for 30 minutes at 9000 *g* and 0°. The precipitate is discarded. The clear yellow supernatant is dialyzed for 2 days against 0.001 *M* phosphate buffer pH 6.8 (cellophane tubes of Visking Corporation are used); 69 ml of solution C is obtained.

First Precipitation with Ammonium Sulfate. To solution C, 8 ml of 1 *M* phosphate buffer pH 5.37 and 3 ml of water are added to a final concentration of 10 mg of protein per milliliter and 0.1 *M* buffer. Then 53 ml of saturated ammonium sulfate solution (pH 6.8, 4°) is added slowly with stirring. The suspension is allowed to stand for 24 hours at 0–4°. After centrifugation for 20 minutes at 20,000 *g* and 0°, the precipitate is discarded. To the supernatant, 67 ml of saturated ammonium sulfate solution is added slowly with stirring. The suspension is allowed to stand for 2 hours at 0°. Then it is centrifuged for 20 minutes at 20,000 *g* and 0°; the supernatant is discarded. The precipitate is washed twice with 10 ml of 60% saturated ammonium sulfate solution, once with 10 ml of 55% saturated ammonium sulfate solution, and finally once with 10 ml of 50% saturated ammonium sulfate solution. The supernatants are removed as completely as possible. After each centrifugation the sides of the centrifuge tubes are freed of liquid with filter paper. The precipitate is dissolved in 5 ml of 0.033 *M* phosphate buffer pH 6.8, the clear solution is dialyzed for 2 days against 0.001 *M* phosphate buffer pH 6.8, 13.9 ml of solution D is obtained.

Adsorption on Calcium Phosphate Gel.[10] To solution D 7 ml of calcium phosphate gel is added slowly with stirring at 0°. After 20 minutes the suspension is centrifuged for 10 minutes at 5000 *g* and 0°. The supernatant is discarded. The precipitate is suspended in 6.5 ml of 0.02 *M* phosphate buffer pH 6.8, stirred for 20 minutes at 0°, then centrifuged for 10 minutes at 5000 *g* and 0°. The supernatant is solution E_1; the precipitate is resuspended in 6.5 ml of 0.02 *M* phosphate buffer pH 6.8. It is stirred for 20 minutes at 0°, then centrifuged for 10 minutes at 5000 *g* and 0°. The supernatant is solution E_2; the precipitate is discarded. Solution E_1 and solution E_2 form solution E (13 ml).

Second Precipitation with Ammonium Sulfate. To solution E, 1.5 ml

[10] D. Keilin and E. F. Hartree, *Proc. Roy. Soc.* **B124**, 397 (1938).

of 1 M phosphate buffer pH 5.37 is added. Then 17 ml of saturated ammonium sulfate (pH 6.8; 4°) is added slowly with stirring. The suspension is allowed to stand for 1 hour; then it is centrifuged for 20 minutes at 20,000 g and 0°. The supernatant is removed as completely as possible and discarded. The precipitate is dissolved in 1 ml of 0.033 M phosphate buffer pH 6.8. It is dialyzed for 2 days against 0.001 M phosphate buffer pH 6.8. The yield is 3.6 mg of protein in 1.2 ml of solution F.

A summary of the purification procedure is given in the table.

PURIFICATION OF ALDOSE-1-EPIMERASE FROM *E. coli* K12

Step	Volume, (ml)	Total activity (units $\times 10^{-4}$)	Total protein[a] (mg)	Specific activity (units/mg)	Yield (%)	Purification (-fold)
Cell-free extract	990	413	13,900	297	100	1
$MnAc_2$ precipitation	1150	314	6220	505	76	1.7
Protamine sulfate precipitation	1220	283	5070	558	69	1.9
Ethanol precipitation	69	169	452	3740	41	12
First $(NH_4)_2SO_4$ precipitation	13.9	120	39.6	30,300	29	102
Adsorption on Ca-phosphate gel	13.0	109	11.1	98,600	26.5	332
Second $(NH_4)_2SO_4$ precipitation	1.2	89	3.6	248,000	21.7	835

[a] In the first 4 steps by biuret method using the factor 16.5 and the 546 mμ Hg-line. In the last three steps by measuring the optical density at 280 mμ and a factor of 1.08 (ml · mg^{-1}).

Properties

Aldose-1-epimerase from *E. coli* has a molecular weight of about 30,000[11]; pH-optimum is 6.5; K_m value for galactose is 6.45×10^{-3} M, and for glucose is 1.66×10^{-3} M. The turnover number of the purified preparation is 7.5×10^6 moles galactose per minute per mole enzyme, as calculated from the V_{max} value for galactose. For more detailed discussion of properties see Wallenfels *et al.*[8]

[11] K. Wallenfels, K. Weber, and K. Herrmann, unpublished observations (1965).

[110] UDP-*N*-Acetyl-D-glucosamine 2′-Epimerase

By CHAVA TELEM SPIVAK and SAUL ROSEMAN

UDP-*N*-acetylglucosamine + $H_2O \rightarrow$ UDP + *N*-acetyl-D-mannosamine

Assay Method

Principle. The product, *N*-acetyl-D-mannosamine is determined by a modification of the Morgan-Elson procedure.[1,2] With the purified enzyme, UDP can be determined by coupling the reaction with nucleotide diphosphokinase, PEP, lactic dehydrogenase, and DPNH.

Reagents

UDP-*N*-acetylglucosamine (UDPAg), potassium or sodium salt, 0.02 *M*
Magnesium sulfate, 0.50 *M*
Tris-HCl buffer, 2.00 *M*, pH 7.5
Dried Dowex 1-carbonate form resin, 20–40 mesh, and dried Dowex 50-hydrogen form resin, 20–40 mesh, are prepared immediately prior to use by washing with H_2O, then with acetone, and air-dried.
Potassium borate, 0.80 *M*, pH 9.0

Diluted DMAB reagent, prepared immediately before use. DMAB reagent: 10 g of *p*-dimethylaminobenzaldehyde is dissolved in 30 ml of glacial acetic acid and 12.5 ml of 10 *N* HCl; glacial acetic acid is added to a final volume of 100 ml. This concentrated reagent is stable for 30 days at 2° or for several months at —10°. Just before use, the concentrated reagent is diluted tenfold with glacial acetic acid.

Procedure. The incubation mixtures contain the following in final volumes of 0.050 ml: 0.005 ml of UDPAg, 0.005 ml of magnesium sulfate, 0.005 ml of Tris buffer, 0.01–0.04 units of enzyme. After incubation at 37° for 20 minutes the reaction is stopped by heating at 100° for 2 minutes. Control incubation mixtures contain heat-inactivated enzyme in place of active enzyme, or contain no substrate, or are incubated 0 minute. A few grains each (approximately 20–30 mg) of Dowex 1-carbonate and Dowex 50-hydrogen are added, followed by 0.10 ml of water in order to increase the volume. The addition of resin is necessary because the presence of salts such as ammonium sulfate interferes with the color

[1] J. L. Reissig, J. L. Strominger, and L. F. Leloir, *J. Biol. Chem.* **217**, 959 (1955).
[2] C. T. Spivak and S. Roseman, *J. Am. Chem. Soc.* **81**, 2403 (1959).

reaction. The tubes are shaken and the mixture is allowed to react with the resin for a few minutes. The resin and protein are removed by centrifugation, and 0.10 ml of the supernatant fluid is used for the estimation of N-acetyl-D-mannosamine. To the aliquot containing 0.005–0.030 micromole of N-acetyl-D-mannosamine is added 0.025 ml of potassium borate. The tubes are heated at 100° for 12 minutes and immediately chilled in ice. Diluted DMAB reagent (0.60 ml) is added, and the tubes are shaken thoroughly, heated for 10 minutes at 37°, and cooled in water at about 10°. Absorbancies are determined at 590 mμ.

Definition of Unit and Specific Activity. A unit of enzyme is the quantity that produces 1.0 micromole of N-acetyl-D-mannosamine in 20 minutes under the conditions described above. Specific activity is defined as the units of enzyme per milligram of protein.

Purification Procedure

All operations are conducted between 0 and 4°. Enzyme solutions contain the following at all stages: $2 \times 10^{-3} M$ EDTA, pH 7.5, and $5 \times 10^{-4} M$ uridine. These compounds are also present in all reagents used during the purification.

Step 1. Preparation of Crude Extract. Livers are obtained from exsanguinated male or female adult rats. Fresh rat liver (25 g) and 50 ml of 0.005 M K phosphate buffer pH 7.5 are homogenized for 1 minute in a Waring blendor, and centrifuged at 35,000 g for 15 minutes. The precipitate is discarded, and the supernatant fluid represents the crude extract.

Step 2. Removal of Nucleic Acids. A 2% solution of polymyxin sulfate (11 ml) adjusted to pH 7.5 with ammonia, is added with stirring to 55 ml of the crude extract. After standing for 5 minutes, the mixture is centrifuged at 35,000 g for 15 minutes, and the residue is discarded.

Step 3. First Ammonium Sulfate. To 60 ml of the supernatant liquid from step 2, 34.9 ml of saturated ammonium sulfate solution, pH 7.0, are added slowly and with stirring. After 20 minutes, with occasional stirring, the mixture is centrifuged at 10,000 g for 10 minutes, the supernatant fluid is discarded, the centrifuge tubes are carefully wiped to remove remaining liquid, and the paste is resuspended in 30 ml of 0.002 M phosphate buffer, pH 7.5. This fraction is designated PAS.

Step 4. Adsorption and Elution from Calcium Phosphate Gel. To 30 ml of PAS is added, with stirring, 15 ml of aged calcium phosphate gel suspension (16 mg of calcium phosphate per milliliter). After 5 minutes, the mixture is centrifuged at 5000 g for 3 minutes, and the supernatant liquid is discarded. The gel is washed with 30-ml portions of 0.005 M K phosphate buffer pH 7.5, containing ammonium sulfate; the first and second portions are 5% saturated with respect to the ammonium sulfate, and the third and fourth are 15% saturated.

After each suspension in the wash solution, the mixture is occasionally stirred for 5 minutes, and centrifuged at 5000 *g* for 3 minutes. The first two gel washes are discarded and the latter two combined and adjusted to 45% of saturation with respect to ammonium sulfate by the addition of 33 ml of saturated ammonium sulfate solution. After standing for 20 minutes, the mixture is centrifuged for 5 minutes at 9000 *g* and the supernatant liquid discarded. The paste is dissolved in 5 ml of 0.005 *M* phosphate buffer, pH 7.5.

Step 5. Adsorption and Elution from DEAE-Cellulose. In this step, the enzyme retains stability only if the medium contains 2-mercaptoethanol. Therefore, all solutions added to the enzyme contain 1% 2-mercaptoethanol in addition to the reagents noted above. The DEAE-cellulose (type 20; Brown and Co., Berlin, New Hampshire) is washed with a solution containing 0.10 *M* potassium phosphate buffer, pH 7.0, and 0.02 *M* potassium chloride, until the supernatant fluid is pH 7.0. Before use, the DEAE-cellulose is washed with several volumes of 0.005 *M* potassium phosphate buffer pH 7.5. The concentrated calcium phosphate gel eluate is dialyzed for 2 hours against 600 ml of 0.005 *M* phosphate buffer, pH 7.5, and the dialyzed enzyme (5 ml) is placed on a small column containing 6 ml of DEAE-cellulose packed by gravity filtration. The column is eluted, successively, with 20 ml of 0.005 *M* phosphate buffer, pH 7.5, and with 20 ml portions of the buffer containing increasing concentrations of potassium chloride as follows: 0.03 *M*, 0.05 *M*, and 0.10 *M*. The enzyme is eluted in the last fraction.

Step 6. Second Ammonium Sulfate. Saturated ammonium sulfate solution (20 ml) is added to 20 ml of the DEAE-cellulose eluate. The mixture is allowed to stand for 15–20 minutes and then centrifuged. The supernatant fluid is discarded and the precipitate is redissolved in 2 ml of 0.005 *M* phosphate buffer, pH 7.5.

PURIFICATION PROCEDURE

Step	Total units	Yield (%)	Specific activity (units/mg protein)
Crude extract	43.2	100	0.085
Polymyxin supernatant	42.0	97	0.16
Ammonium sulfate I	37.5	87	1.1
Calcium phosphate gel	22.5	52	4.2
DEAE-cellulose	12.0	27	8.0
Ammonium sulfate II	9.8	23	12.0

Properties

Specificity. The enzyme is specific for UDP-N-acetylglucosamine. UDP-glucose, N-acetyl-D-glucosamine, N-acetyl-D-glucosamine 1-phosphate, and UDP-N-acetylgalactosamine are inactive.

Stability. The purified enzyme is extremely unstable, often losing all activity within a few hours, and it is therefore advisable to carry out the entire purification procedure at one time. However, in the presence of uridine at $5 \times 10^{-4} M$, UDP at $5 \times 10^{-5} M$ and of various analogs of these compounds (but not UMP), the fraction obtained at the calcium phosphate gel step maintains well over 50% of its activity when stored for several days at 2°.

Optimum pH. The enzyme shows maxima of activity at pH 7.7 with Tris-HCl buffer and pH 6.7 using imidazole buffer. A double peak encompassing these two optima is observed when Tris-maleate buffer is used over the entire pH range.

Activators and Inhibitors. A threefold increase in activity is observed in the presence of magnesium ions, whereas barium and calcium ions have no effect and the presence of manganese, zinc, cupric, and ferric ions cause inhibition of N-acetyl-D-mannosamine formation. N-Ethyl maleimide and p-chloromercuribenzoate also cause inhibition of activity.

Reversibility. The reaction is not detectably reversible under conditions where 0.25% reversibility would be measurable.

Kinetic Properties. The K_m value for UDPAg is $2 \times 10^{-3} M$. In several enzyme preparations, continuous kinetic studies (using the coupled assay system) indicated a short, but detectable, lag period prior to the formation of UDP. This evidence, in addition to the double pH peak, suggests the possibility that the overall reaction is the result of two discrete steps, epimerization, followed by hydrolysis.

Section VII

Phosphatases

[111] Glucose-6-phosphatase

By ROBERT C. NORDLIE and WILLIAM J. ARION

$$\text{Glucose-6-P} + H_2O \rightarrow \text{glucose} + P_i$$
$$PP_i + H_2O \rightarrow 2\ P_i$$
$$PP_i + \text{glucose} \rightarrow \text{glucose-6-P} + P_i$$
$$\text{Nucleoside-5'-tri- or diphosphate} + \text{glucose} \rightarrow \text{glucose-6-P} + \text{nucleoside-5'-di- or monophosphate}$$
$$\text{Sugar or polyol phosphate} + \text{glucose} \rightleftarrows \text{glucose-6-P} + \text{sugar or polyol}$$

A method for assay of glucose-6-P hydrolysis has been described in an earlier volume.[1] However, it recently[2-6] has been demonstrated that this enzyme also catalyzes PP_i hydrolysis and a considerable number of quite active transphosphorylation reactions. In addition to those materials described here, a variety of other compounds also serve as substrates for the hydrolytic[7] and phosphotransferase[4,7] reactions catalyzed by this enzyme. Assays for these newly characterized activities, as well as a modified method for measuring glucose-6-P hydrolysis, are described.

Assay Methods

Reagents

Sodium cacodylate buffers, 0.10 *M*, pH 5.2, 5.5, 6.0, and 6.5
Sodium or potassium glucose-6-P, 0.15 *M*, pH 6.5
Sodium pyrophosphate, 0.15 *M*, pH 5.5 and 6.0.
Glucose, 0.90 *M*, or other[4,7] sugar or polyol
Sodium or potassium CTP, CDP, deoxyCTP, ATP, ADP, GTP, GDP, and ITP, 0.15 *M*, pH 5.2.
Potassium mannose-6-P,[8] 0.15 *M*, pH 6.0
Sucrose, 0.25 *M*; 6 *M* KOH, 0.01 *M* TPN; 0.1 *M* Tris buffer, pH 8.0
Sodium deoxycholate, 2.0% (w/v), pH 7
Perchloric acid solution, 12% (w/v)
Trichloroacetic acid solution, 10% (w/v)

[1] See Vol. II [83].
[2] R. C. Nordlie and W. J. Arion, *J. Biol. Chem.* **239**, 1680 (1964).
[3] W. J. Arion and R. C. Nordlie, *J. Biol. Chem.* **239**, 2752 (1964).
[4] R. C. Nordlie and W. J. Arion, *J. Biol. Chem.* **240**, 2155 (1965).
[4a] R. C. Nordlie and J. F. Soodsma, *J. Biol. Chem.* **241**, 1719 (1966).
[5] M. R. Stetten, *J. Biol. Chem.* **239**, 3576 (1964).
[6] M. R. Stetten and H. L. Taft, *J. Biol. Chem.* **239**, 4041 (1964).
[7] See J. Ashmore and G. Weber, *Vitamins Hormones* **17**, 91 (1959), and references cited therein.
[8] See Vol. III [20].

"Reagent A," prepared by dissolving 19 ml 60% (w/v) perchloric acid and 17.04 g Na_2SO_4 in distilled water and diluting to 200 ml

"Reagent B_1," prepared by dissolving 40 g ammonium molybdate in 210 ml distilled water

"Reagent B_2," prepared by mixing 40 ml of methanol, 1.2 ml *n*-hexanol, and 8.8 ml of distilled water

Procedures. Sodium cacodylate buffers[9] are employed in all assay systems because this compound buffers effectively over the range pH 5–7.4 and exerts no specific effect on the reactions. Sodium acetate buffer also may be employed in the phosphotransferase and inorganic pyrophosphatase assay systems, but it is not suitable for the glucose-6-P hydrolysis assay. Citrate recently has been found to inhibit competitively with respect to phosphate substrates, below pH 6.[10]

1. GLUCOSE-6-P HYDROLYSIS. The assay mixture contains 0.6 ml (60 micromoles) cacodylate buffer, pH 6.5, 0.2 ml (30 micromoles) glucose-6-P solution, and sufficient distilled water to bring the volume to 1.5 ml after addition of enzyme. Tubes containing the assay mixture are preincubated for 5 minutes, the reaction is initiated by addition of enzyme preparation, incubations[11] are carried out, and 1.0 ml of trichloroacetic acid solution is added to terminate the reaction. In this and all other assay procedures, enzyme preparations also are added to a series of "zero-time" control reaction mixtures *after* addition of acid. Denatured protein is sedimented by centrifuging the mixture for 8 minutes at 1800 rpm in an International Model CL clinical centrifuge. Suitable aliquots (usually 0.5 or 1.0 ml) of supernatant solution are transferred to test tubes and assayed for P_i.[12] Glucose-6-P hydrolysis is calculated as the difference in P_i present in incubated and "zero-time" control tubes. Alternatively, glucose production may be determined enzymatically. In this case, reactions are terminated by the addition of 0.15 ml of 6 *M* KOH solution. The samples are neutralized by the addition of predetermined aliquots of perchloric acid solution, denatured protein is sedi-

[9] H. L. Segal and M. E. Washko, *J. Biol. Chem.* **234,** 1937 (1959).

[10] R. C. Nordlie and D. G. Lygre, *Federation Proc.* **25,** 219 (1966).

[11] Incubations for assay of all activities are carried out at 30° for 10 minutes in a shaking, thermostatically regulated water bath.

[12] The Fiske-SubbaRow P_i determination is described in Vol. III [115]. This procedure has been followed, except that 0.4 ml of reducing agent, prepared as described by R. M. Flynn, M. E. Jones, and F. Lipmann [*J. Biol. Chem.* **211,** 791 (1954)] is added to each tube. Under these conditions, no interference by glucose-6-P (or citrate buffer, if employed) [see J. Imsande and B. Ephrussi, *Science* **144,** 854 (1964)] was noted.

mented as above, and 1.0-ml aliquots of clear, supernatant solution are transferred to test tubes and assayed for glucose by the glucose oxidase micromethod.[13]

2. PP$_i$ HYDROLYSIS. The assay mixture contains 0.6 ml (60 micromoles) cacodylate buffer, pH 5.5, 0.2 ml (30 micromoles) PP$_i$ solution, enzyme preparation, and distilled water in 1.5 ml. Incubations and P$_i$ assays are carried out as for glucose-6-P hydrolysis studies. Δ micromoles PP$_i$ = (micromoles P$_i$ in incubated assay mixture — micromoles P$_i$ in "zero-time" control mixtures)/2.

3. PHOSPHOTRANSFERASE ACTIVITIES. (a) Phosphoryl group transfer from PP$_i$, nucleoside-5'-tri- or diphosphate, or mannose-6-P to glucose. Assay mixtures contain 0.6 ml (60 micromoles) cacodylate buffer of desired pH,[14] 0.3 ml (270 micromoles) glucose solution, 0.2 ml (30 micromoles) PP$_i$, nucleotide,[15] or mannose-6-P solution, and sufficient distilled water to bring the volume to 1.5 ml including enzyme. After a 5-minute preincubation of assay mixtures, reactions are initiated by addition of enzyme preparation. Ten minutes later, 0.5 ml of ice-cold perchloric acid solution is added to inactivate the enzyme. The tubes are centrifuged for 10 minutes at 1800 rpm in the International Clinical centrifuge, and predetermined aliquots of KOH solution[16] are added to neutralize the acid. The contents of the tubes (less sedimented protein) are transferred quickly to a second series of tubes, which are cooled in ice and then centrifuged at 4° for 20 minutes at 2000 *g* to remove the precipitated potassium perchlorate. (The preliminary centrifugation and transfer may be omitted, and KOH solution added immediately to perchloric acid-containing tubes, if partially purified enzyme preparations are employed; with liver homogenates, however, turbidity is encountered occasionally if the procedure described above is not followed). Aliquots (usually 0.5 or 1.0 ml) of the clear, supernatant solution obtained following sedimenta-

[13] Commercial preparations of glucose oxidase ("Glucostat Special") are available from Worthington Biochemical Corporation, Freehold, New Jersey. Details of the glucose assay method are given in literature supplied with the enzyme. See also Vol. I [45].

[14] Both pH 5.5 and 6.0 have been used routinely for PP$_i$-glucose phosphotransferase assays, pH 5.2 for nucleotide-glucose assays, and pH 6.0 for mannose-6-P-glucose phosphotransferase assays. All phosphate-containing substrates are adjusted to buffer pH.

[15] Five rather than 30 micromoles of some nucleotide compounds have been employed where economic factors dictate.

[16] The required volume of KOH solution is determined as follows: A series of identical reaction mixtures is prepared, supplemented with 0.5-ml aliquots of 12% (w/v) perchloric acid solution, and titrated with 6 *M* KOH to pH 8. Approximately 0.15–0.20 ml of base is required. This titration must be repeated periodically.

tion of the potassium perchlorate are transferred to 1-cm quartz cuvettes, and glucose-6-P is measured enzymatically with glucose-6-P dehydrogenase by a modification of the method of Noltmann *et al.*[17] A reference molar absorbance index of $6.22 \times 10^3 M^{-1}$ cm^{-1} is utilized in calculating glucose-6-P concentrations.[18]

$$\mu\text{moles glucose-6-P produced per 1.5 ml assay mixture} = \frac{(\text{A340(I)} - \text{A340(C)}) \times (2.000 + v') \times 0.482}{v}$$

where A340(I) is observed absorbancy at 340 mμ due to glucose-6-P-dependent reduction of TPN in assays of aliquots of phosphotransferase reaction mixtures incubated with microsomal enzyme, A340(C) is the comparable value from assays for glucose-6-P in "zero-time" control samples, v is milliliters of neutralized phosphotransferase reaction mixture assayed with dehydrogenase system, and v' is the ml of KOH solution required for neutralization of the perchloric acid-containing phosphotransferase mixtures.

(b) Phosphoryl group transfer from PP_i to polyols or sugars other than glucose.[19] The assay mixture contains 0.6 ml (60 micromoles) of cacodylate buffer, pH 5.5 or 6.0, 0.1 ml (15 micromoles[20]) PP_i solution containing $^{32}PP_i$ equivalent to 5 to 10×10^5 cpm,[21] 0.3 ml (270 micromoles) sugar or polyol solution, and sufficient distilled water to bring to 1.5 ml including enzyme preparation. Carry out incubations, terminate reactions with perchloric acid, neutralize with KOH solution, and centrifuge as described for the PP_i-glucose reaction. Transfer 1.2-ml aliquots of supernatant solution to small (16×125 mm) test tubes, add 0.4 ml of 5 M HCl solution, and heat in a boiling water bath for 7 minutes to hy-

[17] E. Noltmann, C. J. Gubbler, and S. A. Kuby, *J. Biol. Chem.* **236**, 1225 (1961). Assay mixtures contain aliquots of neutralized phosphotransferase mixture, 2.0 ml (200 micromoles) Tris buffer, pH 8.0, 0.1 ml (1 micromole) TPN solution, an excess of glucose-6-P dehydrogenase, and distilled water in 3.0 ml. (Mg^{++} is omitted to preclude turbidity due to precipitation of PP_i or nucleotide complexes at the alkaline pH.) Glucose-6-P dehydrogenase, available as a suspension in saturated ammonium sulfate solution from C. F. Boehringer and Son, Mannheim and New York, catalyzes the reduction of 1 micromole of TPN per micromole glucose-6-P present. See also Vol. III [19].

[18] Pabst Laboratories Circular OR-17, Pabst Brewing Co., Milwaukee, 1961, pp. 2 and 38.

[19] The P_i extraction procedure is a modification of the altered method of J. B. Martin and D. M. Doty *Anal. Chem.* **21**, 965 (1947) described by A. B. Falcone and P. Witonsky *J. Biol. Chem.* **239**, 1954 (1964). See also Vol. VI [32] and [39].

[20] Fifteen rather than 30 micromoles PP_i are employed to avoid overloading the P_i extraction system used in the assay.

[21] Sodium $^{32}PP_i$ (specific activity 5 Mc/millimole) is available from Volk Radiochemical Co., Skokie, Illinois.

drolyze selectively the remaining PP_i.[22] Cool; add 1.0 ml-aliquots to large (20 × 175 mm) test tubes and supplement with 2.0 ml of "Reagent A," 1.5 ml of "Reagent B_2," and 1.5 ml of "Reagent B_1," in that order.[23] Extract with two 6-ml aliquots of *n*-hexanol, each time mixing vigorously with a Vortex mixer and then aspirating and discarding the upper, $^{32}P_i$-containing layer. Filter the lower, aqueous layer through Whatman No. 42 paper. Transfer suitable aliquots (usually 1.0 or 2.0 ml) to planchets, dry, and count with a Geiger-Mueller counter. Aliquots of "zero-time" control samples are carried through the hydrolysis and extraction procedures. Other aliquots of "zero-time" control samples also are taken through all steps, with the exception that they are not heated after addition of HCl. Radioactivity (cpm/aliquot) observed with the hydrolyzed "zero-time" control preparations is subtracted from all other cpm/aliquot readings. Amount (micromoles) of phosphate ester formed then is calculated from the following relationship:

$$\frac{\mu\text{moles PP}_i\text{ added/1.5 ml}}{\text{c.p.m./aliquot of unhydrolyzed "zero-time" sample (i.e., cpm }^{32}\text{PP}_i\text{ per aliquot)}} = \frac{\mu\text{moles sugar or polyol phosphate formed/1.5 ml}}{\text{cpm per aliquot of incubated, hydrolyzed, extracted sample}}$$

These reactions also may be studied indirectly by measuring inhibition of the PP_i-(or nucleotide-)glucose phosphotransferase reaction by various sugars and polyols.[2-4a]

Unit of Enzymatic Activity. One unit of activity is that amount of enzyme catalyzing the hydrolysis of 1 micromole of substrate, or formation of 1 micromole of phosphate ester, per minute.

Enzyme Preparations

Liver[24] *Homogenates.* Decapitate rats; quickly remove livers, blot, weigh, place in ice-cold 0.25 *M* sucrose solution, and grind for 1.5 minutes at 0° in a Potter-Elvehjem homogenizer operating at 600 rpm. Dilute to 15 ml per gram wet liver with the sucrose[25] solution. Supplement 9 parts of homogenate with 1 part 2% deoxycholate solution. (For assay of nucleotide-glucose phosphotransferase reactions, activation by the

[22] See Vol. III [115].

[23] Reagents "B_1" and "B_2" must not be mixed until added to sample to be extracted, or the high concentration of molybdate (4 times that used in the usual method) will precipitate in a matter of hours when combined with methanol reagent "B_1" in the absence of additional water.

[24] The procedure outlined for liver also is applicable to rat kidney preparations.[4a] However, the activities also present in guinea pig or rabbit intestinal mucosa[10] are not activated by deoxycholate.

[25] We have found the activities stable in sucrose solution, the use of which simplifies the control of reaction mixture pH at the rather acidic levels required.

detergent is essential.[4] Distilled water may be substituted for the deoxycholate solution if it is desired to assay the other activities in the absence of the bile salt.) Preparations may be frozen,[26] or placed in ice for immediate assay. Such preparations contain approximately 10 mg of protein per milliliter. One-tenth-milliliter aliquots conveniently are employed for enzymic assays.

Isolated Microsomes.[27] Isolate microsomes from homogenates[24] (not supplemented with deoxycholate) by differential centrifugation. Remove nuclei and mitochondria and sediment the microsomal fraction by centrifugation for 3,000,000 *g*-minutes (conveniently at 30,000 rpm for 38 minutes in the Spinco No. 30 rotor). Wash once with sucrose solution. Suspend microsomes in sucrose solution (9 ml per gram wet liver), treat with detergent as for homogenates, and assay 0.1-ml aliquots for enzymatic activity.

Partially Purified Enzyme Preparations. As described in earlier reviews,[1,7,28] glucose-6-phosphatase is a microsomal enzyme which has resisted attempts to obtain a satisfactory purification. An number of significant efforts, not previously reviewed, to solubilize and purify this enzyme recently have been carried out. Moderate enrichments have been obtained by differential centrifugation of microsomal preparations after treatment with detergents[9,29,30] or proteolytic enzymes,[29] and by fractional precipitation of detergent-dispersed microsomal preparations with acetone in the presence of added Mg^{++}.[31]

We[2] have achieved moderate success in our laboratory by the following procedure involving fractional ammonium sulfate precipitation[32] of deoxycholate-dispersed rat liver microsomal preparations: Isolate the microsomal fraction from rat liver homogenates and wash, as described above. Take the microsomal pellet up in 1.0 ml per gram wet liver of 0.25 *M* sucrose solution, containing 0.2% (w/v) sodium deoxycholate. By slowly adding the powdered salt with stirring, bring the ammonium sulfate concentration to 30% of saturation. After 5 minutes, remove the precipitate by centrifuging for 10 minutes at 79,000 *g*. In a series of similar steps, bring the ammonium sulfate concentration of the supernatant fraction to 40 and then 50% of saturation and collect the sedi-

[26] G. T. Cori and C. F. Cori, *J. Biol. Chem.* **199,** 661 (1952).

[27] See Vol. I [3].

[28] W. L. Byrne, *in* "The Enzymes" (P. D. Boyer, H. Lardy, and K. Myrbäck, eds.), 2nd Ed., Vol. 5, p. 73. Academic Press, New York, 1961.

[29] M. Görlich and E. Heise, *Nature* **197,** 698 (1963).

[30] L. Ernster, P. Siekevitz, and G. E. Palade, *J. Cell Biol.* **15,** 541 (1962).

[31] M. C. Ganoza, Doctoral dissertation, Duke Univ., 1964 (Univ. Microfilms, Inc., Ann Arbor, Michigan).

[32] See Vol. I [10].

ments after each addition. Resuspend the fraction precipitating between 40 and 50% saturation of ammonium sulfate in a volume of sucrose solution equal to one-half the original volume of deoxycholate-treated microsomes. By this means an approximate doubling of specific activity, relative to detergent-treated microsomes, is obtained. More importantly this treatment effects a substantial separation of glucose-6-phosphatase and associated activities from acid phosphatase and magnesium-stimulated inorganic pyrophosphatase and ATPase, which precipitate mainly above 50% ammonium sulfate saturation. The isolated fraction contains between 40 and 50% of recovered glucose-6-phosphatase activity, and is stable for at least 6 months when frozen at —15°. Before assaying, melt frozen preparations at 0° in an ice–water mixture and dilute 1 part enzyme with 4 parts of ice-cold 0.25 *M* sucrose solution. Employ 0.1-ml aliquots for enzymatic assays.

With the various preparations and reaction mixtures described, activity is a linear function of incubation time for at least 15 minutes and also is linearly proportional to protein concentration.

[112a] Fructose-1,6-diphosphatase

I. Rabbit Liver (Crystalline)

By SANDRO PONTREMOLI

Fructose 1,6-phosphate + H_2O → D-fructose 6-phosphate + P_i

Assay Method

TPN Reduction Method

This spectrophotometric method is based on the conversion of FDP to G-6-P which reduces TPN in the presence of D-glucose 6-phosphate dehydrogenase. This test depends on the presence of a large excess of D-glucose 6-phosphate dehydrogenase and D-glucose 6-phosphate isomerase so that the rate of reduction of TPN, as measured in a spectrophotometer at 340 mμ is proportional to the concentration of FDPase.

The activity of fructose-1,6-diphosphatase can be tested at pH 7.5 or 9.1 (the latter is the actual pH in the specified test system).

Reagents

FDP, 0.01 *M*. Prepare a solution of the sodium salt at pH 7.4

Glycine buffer, 0.2 *M*, pH 9.4, or 0.2 *M* triethanolamine buffer, pH 7.5

$MnCl_2$, 0.01 *M*
TPN, 0.01 *M*
D-Glucose 6-phosphate isomerase, 1 mg/ml
D-Glucose 6-phosphate dehydrogenase, 1 mg/ml. Both enzymes have been purchased from Boehringer and Söhne, Germany.
Enzyme. Dilute the solution to be tested to a concentration of 1–5 units per milliliter with water (see definition below).

Procedure. Place 0.2 ml of buffer, 0.68 ml of distilled water, 0.1 ml of $MnCl_2$, 0.01 ml of FDP, 0.005 ml of D-glucose 6-phosphate dehydrogenase, 0.001 ml of D-glucose 6-phosphate isomerase in a cell having a 1-cm light path. Take readings at 340 mμ at 1-minute intervals after the addition of 0.005 ml of enzyme.

Definition of Unit and Specific Activity. A unit of enzyme is defined as that amount which causes a change in optical density of 1.000 per minute (room temperature 21°–23°) under the above conditions. Specific activity is expressed as units per milligram of protein. Protein is determined by the ratio of absorbancies at 260 and 280 mμ, the method being standardized against a known dry weight of dialyzed crystalline FDPase. One milligram dry weight of crystalline diphosphatase per milliliter has absorbancies at 280 mμ and 260 mμ of 0.890 and 0.530, respectively.

Phosphate Liberation Method

The hydrolysis of sedoheptulose 1,7-diphosphate (SDP) is measured following the liberation of inorganic phosphate by the procedure of Fiske and SubbaRow.[1]

Reagents

SDP, 0.015 *M*, sodium salt (see Vol. III [30])
Glycine buffer, 0.2 *M*, pH 9.4, or 0.2 *M* triethanolamine buffer, pH 7.5
$MnCl_2$, 0.035 *M*

Procedure. Place 0.05 ml of SDP, 0.1 ml of buffer, 0.05 ml of $MnCl_2$, and 0.3 ml of distilled water in a centrifuge tube. Add the enzyme in an amount sufficient to release 0.05–0.1 micromole of P_i in 10 minutes at room temperature (21°–23°). The reaction is stopped by the addition of 0.1 ml of 5 *N* sulfuric acid. The protein precipitate is removed by centrifugation. An aliquot of the supernatant solution is analyzed for the inorganic phosphate by the procedure of Fiske and SubbaRow. For a di-

[1] C. H. Fiske and Y. SubbaRow, *J. Biol. Chem.* **66,** 375 (1925).

rect comparison of the phosphatase activity with the two substrates, the same procedure is used for FDP except that the concentration of $MnCl_2$ is 1×10^{-3} *M*, and FDP, 10×10^{-3} *M*, is added instead of SDP.

Definition of Unit and Specific Activity. One unit of the enzyme is defined as the amount which will liberate 1 micromole of P_i per minute under above conditions. Specific activity is expressed as units per milligram of protein.

Purification Procedure[1a]

Step 1. Extract of Acetone Powder. Freshly collected rabbit livers, if not used immediately, were frozen and stored at —30°. Seven frozen livers (800 g) were broken into pieces and homogenized for 2 minutes in a Waring blendor with 4 volumes (w/v) of cold acetone. The resulting suspension was filtered on a large Büchner funnel, and the residue was blended again for 1 minute with acetone as before.

The defatted tissue was spread and dried at room temperature and finally stored in a vacuum desiccator. To obtain the crude extract, the dried powder (150 g) was extracted for 10 minutes with 6 volumes of 10^{-2} *M* sodium phosphate buffer, pH 8.0, with continuous stirring. The suspension was centrifuged at 20,000 *g*, and the precipitate was discarded (fraction I, volume = 740 ml).

Step 2. Acid Fractionation. The supernatant solution was adjusted to pH 3.7 with 5 *M* lactic acid. After centrifugation for 10 minutes at 20,000 *g*, the supernatant solution was brought to pH 6.5 with 5 *N* NaOH. The heavy precipitate was then removed by centrifugation for 10 minutes at 20,000 *g* (fraction II, volume = 625 ml).

Step 3. Ammonium Sulfate Fractionation. The resulting supernatant solution was brought to pH 4.2 with 5 *N* acetic acid, and 2 *M* acetate buffer, pH 4.2, was added to a final concentration of 10^{-2} *M* (volume = 650 ml). This fraction was treated with 180 g of ammonium sulfate; after 10 minutes, the suspension was centrifuged and the precipitate discarded. To the supernatant solution were added 18.9 g of ammonium sulfate, the precipitate was discarded, and the supernatant solution was treated with 39.7 g of ammonium sulfate. The precipitate was collected and dissolved in water (fraction III, volume = 160 ml).

Step 4. Heat Fractionation. In order to reach a suitable protein concentration (8–10 mg/ml) fraction III was diluted with water to 175 ml. The diluted solution was heated in a water bath at 50° for 5 minutes, then chilled to 0°; the precipitate was removed by centrifugation at 30,000 *g* for 5 minutes (fraction IV, volume = 175 ml).

[1a] All operations were carried out at 0° unless otherwise stated.

Step 5. CM-Cellulose Column I. Fraction IV was dialyzed for 4–5 hours against $5 \times 10^{-3} M$ malonate buffer, pH 6.0, and then applied to a column (2×13 cm) of CM-cellulose equilibrated with $5 \times 10^{-3} M$ malonate buffer, pH 6.0. The column was then washed with the equilibrating buffer until the effluent was free of protein (measured by absorbance at 280 mμ). A solution of fructose 1,6-diphosphate ($5 \times 10^{-4} M$) in $5 \times 10^{-3} M$ malonate buffer, pH 6.0, was then applied for the elution of FDPase.[2] Fractions of 5 ml volume were collected at a flow rate of 1.5 to 1.7 ml per minute. Fractions containing FDPase were combined and concentrated to a small volume at low pressure in a rotatory evaporator (fraction V, volume = 6 ml).

Step 5. (Alternate) CM-Cellulose Column. Fraction IV was dialyzed for 4–5 hours against $5 \times 10^{-3} M$ malonate buffer, pH 6.3, and then applied to a column (2×13 cm) of CM-cellulose equilibrated with $5 \times 10^{-3} M$ malonate buffer, pH 6.3. The column was then washed with the equilibrating buffer and then with $5 \times 10^{-3} M$ malonate buffer, pH 6.8, until the effluent was free of protein (measured by absorbance at 280 mμ). A solution of fructose 1,6-diphosphate ($5 \times 10^{-4} M$) in $5 \times 10^{-3} M$ malonate buffer, final pH 6.8, was then applied for the elution of FDPase.[2] Fractions of 5 ml volume were collected at a flow rate of 1.5–1.7 ml per minute. Fractions containing FDPase were combined and concentrated to a small volume at low pressure in a rotatory evaporator (fraction V, volume = 6 ml).

With this modification it is possible to avoid the second CM-cellulose column since the specific activity of the enzyme, after step 5 is 90 and the total units are around 2000.

Step 6. CM-Cellulose Column II. Fraction V was dialyzed against $5 \times 10^{-3} M$ malonate buffer, pH 6.0, and chromatographed in a column (1×13 cm) of CM-cellulose using the same conditions described for the first column, except that the concentration of fructose 1,6-disphosphate for elution was 1×10^{-3}. The active fractions were concentrated as before (fraction VI, volume = 3.5 ml).

Step 7. Chromatography on Sephadex G-75. Fraction VI was chromatographed on a Sephadex G-75 column (2×40 cm), previously equilibrated with $5 \times 10^{-3} M$ malonate buffer, pH 6.0, at a flow rate of 2 ml per hour. Fractions containing the major part of FDPase activity were combined and concentrated to a small volume as described for the effluent from CM-cellulose (fraction VII, volume = 1 ml).

[2] B. M. Pogell, *in* "Fructose 1,6-Diphosphatase and Its Role in Gluconeogenesis," (R. W. McGilvery and B. M. Pogell, eds.), p. 89. American Institute of Biological Sciences, Washington, D. C., 1961. See also this volume [2].

Step 8. Crystallization of FDPase. Fraction VII (1 ml containing 1.200 units) was placed in a small dialysis bag and dialyzed against a solution containing 65% saturated ammonium sulfate which was 0.01 *M* with respect to $MgSO_4$ and EDTA. The pH was 7.5. After several days a suspension of fine needles was formed. The crystals were collected by centrifugation, the supernatant liquid was decanted, and the crystals were redissolved in 5×10^{-2} malonate buffer, pH 6.0. The specific activity of the solution was equal to that of the starting material. The overall purification was approximately 1000-fold with a yield of about 8%.

PURIFICATION OF RABBIT LIVER FDPASE[a]

Fraction and step	Total units[b]	Recovery (%)	Specific activity[c]
I. Acetone powder extraction	11,000	—	0.15
II. Acid fractionation	8,750	78	0.70
III. Ammonium sulfate fractionation	4,550	41	3.14
IV. Heat fractionation	4,400	40	9.6
V. CM-cellulose columns I	2,100	19	62
VI. CM-cellulose column II	1,200	11	105
VII. Sephadex G-75 effluent	1,200	11	140
VIII. Crystals	876	8	140

[a] Enzymatic assay at pH 9.1.
[b] Absorbancy changes per minute.
[c] Units per milligram of protein.

Properties

Specificity. As measured from the rate of P_i liberation (second assay method) glycine buffer, pH 9.4) SDP is dephosphorylated at 70% the rate of FDP. The relative activities with FDP or SDP as substrates do not change significantly throughout the purification and even in the crystalline form of the enzyme. A number of monophosphate and diphosphate esters, such as D-glucose 6-phosphate, D-glucose 1-phosphate, D-fructose 1-phosphate, D-fructose 6-phosphate, D-ribose 5-phosphate, D-ribulose 5-phosphate, D-xylulose 5-phosphate, D-ribulose 1,5-diphosphate, sedoheptulose 1-phosphate, and sedoheptulose 7-phosphate, have been tested as substrates for the crystalline enzyme. Under the conditions of the standard assay (method 2, glycine buffer pH 9.4) with substrate concentrations equal to 2×10^{-3} *M* none of these compounds was cleaved at a significant rate.

Activators and Inhibitors. Mg^{++} or Mn^{++} is required for activity. The enzyme is insensitive to diisopropylfluorophosphate.[3] Dinitrophenylation of one cysteine residue per mole of enzyme increases the catalytic activity at pH 7.5 with both FDP and SDP. The effect is most apparent when Mn^{++} is the activating cation.[4]

The enzyme contains 20 titratable sulfydryl groups with different reactivity for *p*-hydroxymercuribenzoate. The binding of 4–6 moles of *p*-hydroxymercuribenzoate per mole of enzyme increases the catalytic activity with both FDP and SDP. The reaction with all 22 groups leads to inactivation of the enzyme.

Proteolytic activation. The autolysis of liver extracts produces a 3- to 4-fold increase in the total amount of FDPase activity.[5] The same effect can be obtained by incubation of purified enzyme preparations with papain at pH 5.5.[6] At 26° the papain increases the rate of hydrolysis with FDP and SDP. With longer digestion the activity toward FDP returns to nearly the original level, while that toward SDP is reduced to very low levels. Similar, but reversible, changes are found to occur by treatment with urea.

Site of Cleavage of Fructose 1,6-Diphosphate. On hydrolysis of FDP it is the O—P bond which is split, leading to the formation of ^{18}O-labeled inorganic phosphate when the experiment is carried out in $H_2{}^{18}O$. The reaction mechanism is, in this respect, analogous to that observed with other specific or nonspecific phosphatases.[3]

Attempts to label the enzyme either with ^{32}P or with $FD^{32}P$ have been unsuccessful.

pH Optimum. The pH optimum of the FDPase is largely dependent on the concentration of Mg^{++} or Mn^{++} and on the nature of the buffer.[7] Histidine has been shown to shift the pH optimum of FDPase of rat liver from 9.3 to 6.5.[8] In our conditions, with Mn^{++} as activator, the hydrolysis of either FDP or SDP shows an optimum around pH 9.3 with a smaller peak for SDP around pH 7.3. With Mg^{++} the maximum rate of hydrolysis of FDP is achieved at pH 9.3, while for SDP the maximum

[3] S. Pontremoli, S. Traniello, B. Luppis, and W. A. Wood, *J. Biol. Chem.* **240,** 3459 (1965).

[4] S. Pontremoli, B. Luppis, W. A. Wood, S. Traniello, and B. L. Horecker, *J. Biol. Chem.* **240,** 3464 (1965).

[5] B. M. Pogell and R. W. McGilvery, *J. Biol. Chem.* **197,** 293 (1952).

[6] L. C. Mokrasch and R. W. McGilvery, *J. Biol. Chem.* **221,** 909 (1956).

[7] R. W. McGilvery, *in* "Fructose 1,6-Diphosphatase and Its Role in Gluconeogenesis" (R. W. McGilvery and B. M. Pogell, eds.), p. 3. American Institute of Biological Sciences, Washington, D. C., 1961.

[8] H. G. Hers and E. Eggermont, *in* "Fructose 1,6-Diphosphatase and Its Role in Gluconeogenesis" (R. W. McGilvery and B. M. Pogell, eds.), p. 14. American Institute of Biological Sciences, Washington, D. C., 1961.

rate is obtained around pH 7.3; the activity at pH 9.3 being about 40% of that at pH 7.3.[4]

Substrates and Activators Affinity. In the presence of Mg^{++} the K_m for FDP is $4.3 \times 10^{-6} M$ at pH 9.1 and less than $10^{-6} M$ at pH 9.1 and less than $10^{-6} M$ at pH 7.5. In the presence of Mn^{++}, K_m is $2.6 \times 10^{-6} M$ at pH 9.1 and less than $10^{-6} M$ at pH 7.5.

The K_m for Mg^{++} is $4 \times 10^{-4} M$ both at pH 9.1 and 7.5; the K_m for Mn^{++} is $2 \times 10^{-5} M$ at pH 7.5 and $12 \times 10^{-5} M$ at pH 9.1.[4]

Stability. The purified enzyme can be stored for months in the frozen state without appreciable loss of activity.

Homogeneity and Molecular Weight. After the Sephadex step, the enzyme preparation is homogeneous in the polacrilamide disc gel electrophoresis at pH 9.1. Crystalline FDPase has been shown to move as a single component in the ultracentrifuge; the molecular weight is approximately 130,000.

Acid Dissociation into Subunits. FDPase in phosphate-HCl buffer 0.01 *M*, pH 1.8, dissociates into two subunits as judged from the sedimentation pattern in a sucrose density gradient. About 20% restoration of the enzymatic activity is obtained by dilution of the dissociated protein with malonate buffer 0.01 *M* pH 6.5.

Comments. AMP inhibition of FDPase has been reported for FDPase from many sources: rat liver[9,10], rat kidney,[11] frog muscle.[12] Taketa and Pogell[13] have found that this inhibition is reversible and noncompetitive. Among the nucleotides tested, only deoxyadenosine 5'-monophosphate produces a significant inhibition of the enzyme being almost as potent as AMP. The K_i (for 50% inhibition) has been determined at 30° and pH 7.5 for AMP and found equal to 0.11 m*M*.

[9] E. A. Newsholme, *Biochem. J.* **89**, 38P (1963).
[10] K. Taketa and B. M. Pogell, *Biochem. Biophys. Res. Commun.* **12**, 229 (1963).
[11] J. Mendicino and F. Vasarhely, *J. Biol. Chem.* **238**, 3528 (1963).
[12] M. Salas, E. Viñuela, J. Salas, and A. Sols, *Biochem. Biophys. Res. Commun.* **17**, 150 (1964).
[13] K. Taketa and B. M. Pogell, *J. Biol. Chem.* **240**, 651 (1965).

[112b] Fructose-1,6-diphosphatase

II. *Candida utilis*

By O. M. Rosen, S. M. Rosen, and B. L. Horecker

Fructose 1,6-phosphate + H_2O → D-fructose 6-phosphate + P_i

Assay Method

Principle. Fructose 6-P produced in the reaction is converted to glucose 6-P by phosphohexoisomerase. In the presence of glucose 6-phosphate dehydrogenase this product reduces TPN. The production of TPNH is measured spectrophotometrically at 340 mμ.

Reagents

Sodium fructose 1,6-diphosphate, 0.01 *M*
Glycine buffer, 0.2 *M*, adjusted to pH 9.5 with 1.0 *N* KOH
$MgCl_2$, 0.1 *M*
TPN, 0.05 *M*
Glucose 6-phosphate deyhdrogenase, crystalline suspension in $(NH_4)_2SO_4$, 5 mg/ml
Phosphohexoisomerase, crystalline suspension in $(NH_4)_2SO_4$, 2 mg/ml

Procedure. Place into a quartz cell (1-cm light path) in a final volume of 1 ml: 0.2 ml glycine buffer, 0.01 ml $MgCl_2$, 0.01 ml TPN, 0.001 ml each of glucose 6-phosphate dehydrogenase and phosphohexoisomerase and the aliquot of enzyme to be measured. Record the absorbance at 340 mμ at 1-minute intervals until no further increase occurs. Add fructose diphosphate (0.02 ml) and record the absorbance at 340 mμ at 1-minute intervals. Use the interval between 6 and 10 minutes to calculate the rate.

Definition of Unit and Specific Activity. A unit of enzyme is defined as the amount which converts 1 micromole of fructose diphosphate to fructose 6-phosphate in 1 minute at room temperature. Specific activity is expressed as units per milligram of protein. Protein is determined by the method of Lowry *et al.*[1]

Application of the Assay Method to Crude Enzyme Preparations. The spectrophotometric assay is applicable to crude cell extracts provided the rate is corrected for TPN reduction which is not dependent upon added fructose diphosphate. In cases where the reaction continues in

[1] O. H. Lowry, N. J. Rosebrough, A. L. Farr, and R. J. Randall, *J. Biol. Chem.* **193,** 265 (1951). See also Vol. III, p. 448.

the absence of added substrate, it is necessary to measure the rate of change of absorbance in a control cell lacking fructose diphosphate. With pure preparations of enzyme the release of inorganic phosphate[2] from fructose diphosphate can be measured.

Purification Procedure[3]

Dried *Candida utilis* is obtained commercially from the Lake States Yeast Corporation, Rhinelander, Wisconsin. Unless otherwise stated, all procedures are carried out in the cold and all centrifugations are performed at 20,000 *g* in a refrigerated centrifuge at 0°.

Step 1. Autolyzate. Suspend 135 g of dried *C. utilis* in 400 ml of 0.2 *M* potassium phosphate buffer, pH 7.5, containing 1.0 m*M* β-mercaptoethanol. Allow to autolyze at 4° for 48 hours. Centrifuge the suspension and collect the pink supernatant fluid containing the enzymatic activity. This can be stored at —20° for several weeks.

Step 2. Acid-Heat Precipitation. Dilute the autolyzate with an equal volume of distilled water to achieve a protein concentration of approximately 25 mg/ml (diluted autolyzate). Adjust the pH to 5.0 by adding 5.0 *M* lactic acid, dropwise, with mechanical stirring. Centrifuge the suspension and discard the flocculant precipitate. Heat the supernatant fluid to 55° in a water bath which is initially at 60°. Maintain the solution at 55° for 5 minutes, cool rapidly to 4° and centrifuge for 10 minutes to remove precipitated protein (acid-heat fraction).

Step 3. Ammonium Sulfate Fractionation I. Again adjust the pH of the supernatant fluid to 5.0 with a few drops of 5.0 *M* lactic acid and add reagent grade ammonium sulfate (17.6 g per 100 ml of solution) with stirring. Allow the solution to stand in ice for 10 minutes, remove the precipitate by centrifugation, and treat the supernatant solution with additional ammonium sulfate (6.2 g per 100 ml of solution). Discard this second precipitate and treat the supernatant fluid with 9.7 g of ammonium sulfate per 100 ml of solution. Collect the last precipitate which contains the enzyme activity and suspend it in 15 ml of 55% saturated (2°) neutralized ammonium sulfate (pH 6.8–7.4) (ammonium sulfate fraction I).

Step 4. Ammonium Sulfate Fractionation II. Centrifuge the suspension for 5 minutes, discard the supernatant fluid, and repeat the extraction twice in the same way. Then, extract the pellet successively with 10-ml aliquots of 50% neutralized ammonium sulfate. Determine the

[2] C. H. Fiske and Y. SubbaRow, *J. Biol. Chem.* **81,** 629 (1929). See also Vol. III, p. 843.

[3] O. M. Rosen, S. M. Rosen, and B. L. Horecker, *Arch Biochem. Biophys.* **112,** 411 (1965).

fructose diphosphatase activity of each fraction, and pool those fractions containing specific activities equal to or greater than 1.7 (ammonium sulfate fraction II). Add solid ammonium sulfate to bring the saturation to approximately 80% (51 g/100 ml) and collect the precipitate by centrifugation. Dissolve this in 15 ml of 0.005 *M* sodium malonate buffer, pH 5.7, and pass the solution through a Sephadex G-25 column (3 cm $\times$ 15 cm) previously equilibrated with the same 0.005 *M* malonate buffer. Wash the enzyme through with malonate buffer and test fractions of the eluate both for enzymatic activity and with a conductivity meter to assure separation of enzyme from ammonium sulfate. The active fractions (approximately 40 ml) are pooled.

Step 5. Phosphocellulose Column Fractionation. Prepare a phosphocellulose (Schleicher and Schuell) column by suspending the resin in 0.005 *M* sodium malonate buffer, pH 5.7, and readjusting to pH 5.7 with 5.0 *N* NaOH. Pour the slurry into a 2.5 $\times$ 12 cm column and wash extensively with buffer. Dilute the enzyme solution with the malonate buffer (approximately 150 ml) to achieve a protein concentration of 1.0 mg/ml and apply to the column at a rate of 3 ml per minute. When all the enzyme solution has been applied, wash the column with 0.005 *M* malonate buffer until the absorbance of the effluent at 280 mμ is less than 0.01. Then wash the column with 0.01 *M* Na malonate buffer, pH 5.7, until the absorbance at 280 mμ of the effluent is again less than 0.01.

Decrease the column flow rate to 1 ml per minute and elute the enzyme with a solution of 2.0 m*M* sodium fructose diphosphate in 0.01 *M* sodium malonate buffer, pH 5.7.[4] Collect fractions (5 ml) and pool those with specific activities of 33 and above. The enzyme will appear in a broad peak between fractions 10 and 30 (phosphocellulose fraction).

Immediately concentrate the enzyme by negative pressure dialysis against air, or by lyophilization, to a protein concentration of 0.5–1.0 mg/ml. This solution can be stored frozen for several months without loss of activity.

Step 6. Crystallization. Precipitate the concentrated enzyme solution by addition of solid ammonium sulfate (5 g/10 ml) and resuspend the pellet in a volume of 30% saturated neutralized ammonium sulfate such that the final protein concentration is 2–3 mg/ml. Add $MgCl_2$ to a final concentration of 1.0 m*M* and follow this with saturated (2°) neutralized ammonium sulfate solution added dropwise at room temperature until the solution becomes faintly turbid. A silky luster should develop during the ensuing few hours at room temperature (crystals). Microscopic observation will reveal fine, long crystals. Harvest these by centrifugation at

[4] B. M. Pogell, *Biochem. Biophys. Res. Commun.* **7**, 225 (1962).

room temperature, dissolve in water or 0.005 M malonate buffer, pH 5.7, and store the solution frozen.

PURIFICATION PROCEDURE FOR ENZYME FROM *C. utilis*

Step	Total volume (ml)	Units/ml	Total units	Protein (mg/ml)	Specific activity units/mg	Recovery (%)
1. Diluted autolyzate	400	2.50	1000	25	0.1	—
2. Acid-heat fraction	350	2.86	1000	7.5	0.33	100
3. Ammonium sulfate fraction I	15	59.0	880	59	1.0	88
4. Ammonium sulfate fraction II	50	9.0	450	3.6	2.5	45
5. Phosphocellulose fraction	150	1.67	250	0.03	50	25
6. Crystals	1.2	167	200	2	83	20

Properties[3]

Specificity. Candida fructose diphosphatase is specific for fructose diphosphate. It shows no activity with a variety of triose, pentose, and hexose phosphates including sedoheptulose diphosphate, fructose 1-phosphate, and ribulose diphosphate.

Metal Requirement. Enzymatic activity is dependent upon the addition of Mg^{++} (1.0 mM). Mn^{++} (0.5 mM) may be substituted for Mg^{++} but is less effective.

pH Optimum. The pH optimum in glycine buffer lies between pH 9.0 and 9.5. Addition of 0.5 mM EDTA (ethylenediaminetetraacetic acid) enhances the activity at pH 9 by 50–100%; in the presence of this substance activity appears in the neutral pH range approximately equal to that at pH 9, with an optimum between pH 7.5 and pH 8.0. The activity in this range is measured in 0.04 M triethanolamine buffer. These effects of EDTA can be demonstrated in the autolyzate as well as with the purified enzyme.

K_m. The K_m for fructose diphosphate at pH 9.5 is 1×10^{-5} M. At pH 7.5 in the presence of EDTA it is approximately 2×10^{-6} M.

Molecular Weight. The molecular weight is approximately 98,000, as determined by equilibrium sedimentation.[5] Values of 117,000 ± 15% have been obtained using the sucrose gradient technique.[6]

Inhibitors. No inhibition of enzymatic activity is observed with fructose (0.1 M), fructose 6-P (1.0 mM), fructose 1-P (1.0 mM), or

[5] D. A. Yphantis, *Biochemistry* **3**, 297 (1964).
[6] R. G. Martin and B. N. Ames, *J. Biol. Chem.* **236**, 1372 (1961).

inorganic phosphate (0.01 M). Fe^{++}, Cu^{++}, Zn^{++}, and Hg^{++} do not inhibit at concentrations of 0.1 mM in the presence of 1.0 mM Mg^{++}, whereas Pb^{++} (0.1 mM) and Ca^{++} (0.1 mM) inhibit 80–100%, Ag^{+} (0.1 mM) inhibits 30%, and fluoride (0.01 M) inhibits 50–60%.

AMP inhibits enzymatic activity at concentrations of 0.05–0.5 mM. This inhibition is most marked at pH 7.5 in the presence of EDTA. Under these conditions, 0.1 mM AMP produces 80% inhibition of enzymatic activity.

[112c] Fructose-1,6-diphosphatase

III. *Euglena gracilis*

By A. A. APP

Fructose 1,6-phosphate + H_2O → D-fructose 6-phosphate + P_i

Assay Method

Principle. The colorimetric assay measuring the liberation of inorganic phosphate may be conveniently used.[1]

Reagents

Tris buffer, 0.5 M, pH 8.4
$MgCl_2$, 0.1 M
Na_2EDTA, 0.016 M
Fructose-1,6-diphosphate, sodium salt 4×10^{-3} M

Procedure. The following reagents are present in a total volume of 1 ml: 100 micromoles Tris (pH 8.4), 10 micromoles $MgCl_2$, 2 micromoles FDP, 1.6 micromoles EDTA, and the enzyme solution to be tested.[2] After incubation at 30° (1–3 minutes, depending on the level of activity) the reaction is stopped with 5% trichloroacetic acid and a suitable aliquot of the supernatant assayed for P_i.[3] Blanks consist of all components except FDP. One unit of fructose diphosphatase activity is defined as the liberation of 1 micromole of P_i per minute.

Protein was estimated by absorbency or by the Lowry procedure.[4] As long as consideration is given to phosphatase activity from other

[1] E. Racker and E. A. R. Schroeder, *Arch. Biochem. Biophys.* **74**, 326 (1958).
[2] A. A. App and A. T. Jagendorf, *Biochim. Biophys. Acta* **85**, 427 (1964).
[3] H. H. Taussky and E. Shorr, *J. Biol. Chem.* **202**, 675 (1953).
[4] O. H. Lowry, N. J. Rosebrough, A. L. Farr, and R. J. Randall, *J. Biol. Chem.* **193**, 265 (1961). See also Vol. III, p. 448.

than alkaline fructose-1,6-diphosphatase, the assay may be employed with crude *Euglena* extracts.

Purification Procedure

Euglena gracilis, strain Z, was grown aseptically in 9-liter carboys for 6–9 days in white light at 25°. The cultures were aerated through capillary tubing and grown in a previously described medium consisting of 0.5% proteose peptone, 0.2% yeast extract, and 0.1% sodium acetate.[5] The cells were harvested in a Sorvall continuous flow centrifuge, washed and resuspended in water, then broken by sonication for 3 minutes with a Raytheon 10-kc oscillator, Model DF 101.

The sonicate from 230 g (wet weight) of *Euglena* was centrifuged for 15 minutes at 20,000 *g*. The pellet was resuspended in water, the mixture was recentrifuged, and the combined supernatants were dialyzed overnight against water at 4°. The dialyzed extract was centrifuged at 20,200 *g* for 15 minutes and the pH of the supernatant was adjusted to 5.1 with 0.1 *M* citric acid. This acidified supernatant was held at 60° for about 10 minutes, or until heavy precipitation occurred. The precipitate was removed by centrifuging, the supernatant solution was then cooled to 3° and acidified to pH 4.35 with more 0.1 *M* citric acid. The second precipitate was also removed by centrifugation, and the pH of the supernatant solution was raised to 4.95 by addition of sodium citrate. One milliliter of a protamine sulfate solution (5.0 mg/ml, pH 5.5) was then added for every 20 ml of enzyme solution. Precipitation was allowed to occur for 25 minutes at 4°, and the precipitate was removed by centrifuging at 20,200 *g* for 15 minutes. The supernatant solution was retained and its pH readjusted to 4.95 if necessary.

The next step was chromatography on a DEAE-cellulose column. The cellulose derivative was cleaned as described by Peterson and Sober, formed into a 3.25 × 13.0 cm column, and equilibrated at 4° with 0.01 *M* citrate buffer (pH 5.5).[6] Approximately 380 ml of the enzyme solution was then passed through the column. Gradient elution of the enzyme was accomplished using a mixing vessel with 1 liter of 0.01 *M* citrate buffer at pH 5.5 and a reservoir containing 0.01 *M* citric acid. In this procedure a great deal of protein was removed before the fructose diphosphatase began to appear in the effluent at approximately pH 4.8. The most active fractions from this first run were pooled and rechromatographed with an identical procedure using a 2 × 12.75 cm column and a 700 ml mixing vessel. Elution profiles from this second column showed a close association between fructose diphosphatase activity and the remaining material.

[5] G. Brawerman and N. Konigsberg, *Biochim. Biophys. Acta* **43**, 374 (1960).
[6] E. A. Peterson and H. A. Sober, see Vol. V [1].

PURIFICATION OF FRUCTOSE DIPHOSPHATASE FROM *Euglena gracilis*[a]

Fraction	Protein, (mg/ml)	Fructose diphosphatase (units/ml)	Specific activity (units/mg protein)	Total recovery (%)
Dialyzed, centrifuged	20.0	3.25	0.16	100
Heated, centrifuged	4.0	2.80	0.70	85
pH precipitation, centrifuged	2.75	2.08	0.76	64
Protamine sulfate, centrifuged	2.37	2.25	0.95	69
First DEAE-column	0.130	3.65	28.0	46
Second DEAE-column	0.100[b]	7.2[b]	72[b]	31

[a] Reproduced from A. A. App and A. T. Jagendorf, *Biochim. Biophys. Acta* **85**, 427 (1964).
[b] Best specific activity.

The table shows the yield of enzyme activity and specific activities of the fractions obtained by this purification procedure. Although the highest specific activity found in this experiment was 72, in some experiments specific activities as high as 105–128 were obtained, indicating a maximum purification of 800-fold. Unfortunately, the low total yield of protein at this stage precluded attempts at further purification or even determining the degree of purity by ordinary physical chemical procedures. The highly purified fructose diphosphatase is stable at pH 4.95 in 0.01 M citrate and may be stored at $-20°$ for months without loss of activity.

Properties

pH Optimum and Substrates. The purified enzyme has a pH optimum for activity of 8.25. It is highly specific in that no hydrolysis of the following phosphorylated sugars at a concentration of 2 micromoles/ml can be detected: fructose 1-phosphate, glucose 1-phosphate, fructose 6-phosphate, glucose 6-phosphate, ribulose 5-phosphate, or DL-glycerophosphate.

Inhibitors and K_m. An effective inhibitor of *Euglena* fructose diphosphatase is PCMB, giving almost complete inhibition at $5 \times 10^{-5}\,M$. Iodoacetate has no effect, and sodium fluoride is only weakly inhibitory, requiring $2 \times 10^{-2}\,M$ concentration to even approach 50%.

A Lineweaver-Burke analysis of activity as a function of substrate concentration showed a K_m of $3 \times 10^{-4}\,M$ for fructose diphosphate. In these studies the magnesium concentration was varied with the FDP in order to maintain a constant Mg^{++}:substrate ratio.

Activators. It was observed that optimum enzyme activity requires a distinct ratio of magnesium to substrate rather than a given magnesium concentration.[2] Thus differing optimum curves are obtained for activity versus magnesium concentration at three different FDP levels (2, 5, and 10 micromoles/ml, respectively). However, when these data are recalculated on the basis of the Mg^{++}:FDP ratio, the optimum ratios coincide. The optimum Mg^{++}:FDP ratio appears to be between 20 and 25 moles of Mg^{++} per mole of FDP. This optimum ratio was independent of enzyme concentration over a fivefold range.

Under certain conditions, EDTA will enhance the activity. The optimum concentration of EDTA for enhancement apparently depends on the Mg^{++}:FDP ratio.

[113] Alkaline Phosphatase (Crystalline)[1]

By M. MALAMY and B. L. HORECKER

$$\text{R-O-P} + H_2O \rightarrow \text{P} + \text{ROH}$$

Assay Method

Principle. The assay is based on the formation of *p*-nitrophenol in the hydrolysis of *p*-nitrophenylphosphate. This is measured spectrophotometrically at 420 $m\mu$.

Reagents

p-Nitrophenylphosphate: 1.0 m*M* in 1 *M* Tris buffer, pH 8

Procedure. To 1 ml of the buffered *p*-nitrophenylphosphate solution add the enzyme solution and read the absorption at 420 $m\mu$ in a Beckman DU spectrophotometer. Keep the cell compartment at 27°. Calculate the units of enzyme on the basis of the amount required to liberate micromole of *p*-nitrophenol in 1 hour using a molar absorptivity coefficient for *p*-nitrophenol of 1.32×10^4. The specific activity is the number of units per milligram of protein under the conditions of the test. Determine protein by the turbidometric method of Bücher.[2] One milligram per milliliter of the purified protein gives an absorbance reading of 0.72 at 278 $m\mu$.[1,3]

[1] M. H. Malamy and B. L. Horecker, *Biochem. J.* **3**, 1893 (1964).
[2] T. Bücher, *Biochim. Biophys. Acta* **1**, 292 (1947).
[3] D. J. Plocke, C. Levinthal, and B. L. Vallee, *Biochemistry* **1**, 373 (1962).

Purification Procedure

Selection of Strain. Use *Escherichia coli* C4F1 or a suitable strain constitutive for alkaline phosphatase. Test the strain by streaking the cells on nutrient agar and examine the colonies for alkaline phosphatase with a drop of nitrophenylphosphate solution.[4] Select colonies which turn yellow within a few minutes and use them as the inoculum for cells for the preparation of enzyme.

Growth of Cells. Inoculate five 2-liter flasks containing 500 ml of peptone–glucose–salts medium with 2 ml of an exponential growing culture of *E. coli* K12-C4F1. Incubate the cultures at 37° on a high-speed rotary shaker for 16–20 hours until a final turbidity of about 1.3 at 590 mμ in the Beckman DU spectrophotometer is reached. The suspension of the intact bacteria should test for about 14 units of phosphatase per milliliter. Harvest the cells by centrifugation at 4° and wash the pellets with several 100-ml portions of 0.01 *M* Tris buffer, pH 8.

Preparation of Spheroplasts.[5] Resuspend the washed cells to a final volume of 1 liter with a solution of 20% sucrose, which is also 0.033 *M* with respect to Tris buffer at pH 8. The cell suspension should have a turbidity of about 3.4 at 590 mμ as measured in a Beckman spectrophotometer. Keep the suspension in a 2-liter flask at room temperature and stir slowly with a magnetic stirring bar. Add 10 ml of 0.01 *M* EDTA, pH 8, and 10 mg of crystalline egg white lysozyme (Nutritional Biochemicals Corp.). Test samples (0.1 ml) at intervals for completeness of spheroplast formation by diluting each to 1 ml with distilled water and measuring the final turbidity at 490 or 590 mμ in the Beckman DU spectrophotometer. Compare this reading with that obtained before the addition of lysozyme. When more than 90% of the cells have become osmotically sensitive at about 9 minutes, centrifuge the suspension at 8500 rpm for 15 minutes at 4°. Remove the supernatant fluid and wash the residue remaining with a small portion (ca. 100 ml) of the sucrose-Tris buffer solution. The supernatant solution should contain a total of 234,000 units of enzyme and 260 mg of protein. Discard the wash solution if it contains less than 10% of the enzyme activity.

Chromatography. Prepare a suspension of DEAE-cellulose (Bio-Rad Co., Calbiochem) by stirring 20 g of the dry powder into 1 liter of 0.05 *M* NaCl. Stir the slurry for several minutes, allow it to settle, and remove the fine particles by decantation. Repeat the washing with 0.05 *M* NaCl several times and acidify the final suspension by adding a quantity

[4] H. Echols, A. Garen, S. Garen, and A. Torriani, *J. Mol. Biol.* 3, 325 (1961).

[5] An alternate procedure for release of enzyme, which avoids the use of lysozyme, has been described.[6]

[6] H. C. Neu and L. A. Heppel, *Biochem. Biophys. Res. Commun.* 17, 215 (1964).

of HCl equivalent to the exchange capacity of the gel. For 20 g of DEAE-cellulose 17.4 ml of 1 N HCl is required. Use this suspension, which is slightly acid (about pH 5), to pack a column 2.6×5.5 cm. Maintain pressure in the system by means of a constant pressure cautery bulb. Wash the packed column with 0.05 M NaCl under pressure.

Allow the entire sucrose-Tris buffer supernatant fraction to drip slowly onto the column from a reservoir supported above the top of the column. Keep the flow rate slow and allow the process to continue overnight in the cold room (4°). When the entire fraction has been adsorbed, wash the column with 20 ml of 0.05 M NaCl to remove the remaining sucrose-Tris buffer solution. Collect fractions (4 ml) on an automatic fraction collector at 10-minute intervals. Elute the phosphatase activity from the column with 0.125 M NaCl; it should appear in 4 tubes between 20 ml and 36 ml of eluent (DEAE-cellulose eluate). These 4 tubes should contain a total of about 65 mg of protein, as calculated from the absorbance readings at 278 mμ.

Crystallization. To the column fractions add $MgCl_2$ to 0.01 M and bring the solution to 50% saturation by the addition of solid ammonium sulfate, 0.29 g/ml. Centrifuge the solution in the clinical centrifuge at room temperature to remove gas bubbles and adjust to about pH 8 with 2 N NaOH. Add saturated ammonium sulfate at pH 8, dropwise, until the solution becomes faintly turbid (about 61% saturation). Allow the turbid suspension to stand at room temperature for 1 hour; it will increase in turbidity. Place the suspension in an ice bath for several minutes, whereupon the turbidity should disappear almost entirely. Then allow the solution to warm slowly to room temperature by placing the tube in a 5-liter insulated water bath at 0°. As the solution warms, the turbidity reappears and the suspension develops a silky sheen when viewed against a dark background.

The purification is summarized in the table.

PURIFICATION PROCEDURE FOR ALKALINE PHOSPHATASE

Fraction	Total enzyme units (μmoles/hr/ml)	Volume (ml)	Specific activity (μmoles/hr/mg)
1. Intact cells	36,200[a]	2500	—
2. Sucrose-Tris buffer supernatant	234,000	1000	910
3. DEAE-cellulose eluate	142,000	18	2190
4. Crystals[b]	4,700[b]	1[b]	2900[b]

[a] The alkaline-phosphatase activity of intact cells was generally only a fraction of that obtained with cell extracts.

[b] These values were obtained with an aliquot (1 ml) of fraction 3.

Properties

Specificity. The alkaline phosphatase preparations obtained by this procedure show little or no nuclease activity but will hydrolyze a large variety of phosphomonoesters.[7]

Stability. The suspensions of crystals are stable at room temperature for many months. They can be stored at 0° but are not stable to freezing.

Inhibitors. Cysteine or thioglycolic acid are inhibitory at about 10^{-3} to 10^{-4} *M*. This inhibition is reversed by the addition of Zn^{++} ions.[8]

[7] L. A. Heppel, D. R. Harkness, and R. J. Hilmoe, *J. Biol. Chem.* **237**, 841 (1962).
[8] D. J. Plocke and B. L. Vallee, *Biochemistry* **1**, 1039 (1962).

[114] 2-Phosphoglycerol Phosphatase

By N. Appaji Rao and C. S. Vaidyanathan

I. 2-Phosphoglycerol Phosphatase (Acid) from Wheat Germ

Assay Method

Reagents

Sodium acetate-acetic acid buffer, 0.1 *M*, pH 5.7
Magnesium chloride solution, 0.2 *M*
2-Phosphoglycerol, 0.1 *M* solution

Procedure. For the routine assay of phosphatase activity the following components were mixed in a final volume of 1.5 ml; 75 micromoles of sodium acetate buffer, pH 5.7, 20 micromoles of $MgCl_2$, 6 micromoles of the substrate (2-phosphoglycerol), and approximately 0.5 unit of enzyme. After 15 minutes incubation at 37°, the reaction was stopped by the addition of 1 ml of *M* $HClO_4$, and any insoluble material was removed by centrifugation. Aliquots of the supernatant fluids were tested for P_i by the method of Fiske and SubbaRow.[1]

Enzyme Unit and Specific Activity. A unit of enzyme activity is the amount of enzyme which liberates 1 micromole of P_i under the standard conditions of assay. Specific activity is defined as the enzyme units per milligram of protein.

[1] C. H. Fiske and Y. SubbaRow, *J. Biol. Chem.* **66**, 375 (1925).

Purification

The method of purification is that of Joyce and Grisolia.[2] All operations during fractionation were carried out at 0° and all centrifugations were for 10 minutes, unless indicated otherwise. A speed of 5500 *g* in a Lourdes centrifuge with a rotor No. 3RA was used through Fraction II. In the succeeding steps, centrifugations were at 4000 *g* in an International Centrifuge. The $(NH_4)_2SO_4$ solutions were saturated at pH 5.5.

Wheat germ was extracted with 4 volumes of water for 30 minutes, and the preparation was centrifuged. To each liter of the supernatant fluid (crude extract) were added 20 ml of *M* $MnCl_2$ and the mixture was centrifuged. To each liter of the supernatant fluid (Mn^{++} fraction) 538 ml of a saturated $(NH_4)_2SO_4$ solution (35% saturation) was added. The insoluble material was removed by centrifugation. To the supernatant fluid 788 ml of saturated $(NH_4)_2SO_4$ was added (57% saturation). While swirling in 65–70° water bath, the mixture was heated to 60° and held at this temperature for 2 minutes. The preparation was rapidly chilled to 8° and centrifuged. The precipitate was suspended in water (one-third that of the Mn^{++} fraction), and was then centrifuged for 30 minutes. The supernatant fluid (fraction II), may be stored frozen. For each new bentonite and enzyme preparation, a preliminary titration for adsorption is required. One-milliliter samples of fraction II containing about 8 mg of protein per milliliter were adsorbed for 7 minutes with 0.2, 0.25, and 0.3 ml of washed bentonite suspension (60–70 mg/ml) and water to a final volume of 1.5 ml. The samples were centrifuged at 2000 *g* and the supernatant fluids were assayed. The best conditions obtained in the preliminary bentonite treatment were followed in purifying the remainder of the preparation. The supernatant fluid is fraction III. To each liter of this fraction, 272 g of solid $(NH_4)_2SO_4$ was added (47% saturation). The insoluble material was discarded after centrifugation and 97 g of $(NH_4)_2SO_4$ was added to the supernatant fluid (61% saturation). The mixture was centrifuged and the precipitate was dissolved in water (0.05 volume of fraction III) to give fraction IV. This fraction may be stored frozen. Fraction IV was adjusted to about 4.5 mg of protein per milliliter. To each milliliter of fraction IV, 0.11 ml of 0.2 *M* EDTA, 0.05 ml of saturated ammonium sulfate, and 1.75 volume of methanol (measured and added at −20°) were added. After centrifugation the precipitate was suspended in the minimal volume of water, and the insoluble material was discarded after centrifugation. The supernatant fluid was dialyzed for approximately 11 hours against 100 volumes of 0.01 *M* EDTA (five

[2] B. K. Joyce and S. Grisolia, *J. Biol. Chem.* **235**, 2278 (1960).

changes). The dialyzed material (fraction V) may be frozen overnight. Fraction V was percolated through a DEAE-cellulose column (approximately 1 × 7 cm), about one-fourth the volume of the enzyme solution. The elution was carried out with successive one-column volumes of Tris buffer, pH 7.4, at increasing molarities (0.01, 0.02, 0.04, and 0.08 *M*). The highest specific activity usually was eluted with approximately 0.02 *M* solutions. The best fractions were dialyzed with three successive changes for 3 hours against 100 volumes of 0.01 *M* EDTA. Samples were lyophilized to dryness and can be stored as such in the cold or dissolved in the minimal volume of water and kept frozen. The best fraction was fraction VI. A summary of the purification procedure is shown in Table I.

TABLE I
PURIFICATION OF 2-PHOSPHOGLYCEROL PHOSPHATASE (ACID) FROM WHEAT GERM

Fraction	Total volume (ml)	Total activity (units)	Total protein (mg)	Specific activity (units/mg protein)	Yield (%)
Crude extract	2630	106,000	126,000	0.84	100
Fraction II	1020	75,000	7,850	9.6	71
Fraction III	1450	53,500	2,180	24.5	42
Fraction IV	95.5	34,400	898	34.4	26.5
Fraction V	28.4	15,650	22.4	700	14.8
Fraction VI	1.0	4,150	1.4	3000	3.9[a]

[a] The specific activity and recovery rates for Fraction VI were probably higher than indicated. When this fraction was diluted in 0.5% serum albumin and tested (final concentration of serum albumin in the assay, 25 μg), it showed 50% higher activity. This was most likely due to enzyme inactivation during dilution to low protein concentration. The serum albumin had no phosphatase activity.

Properties

Specificity. The enzyme is nonspecific. However, it appears to have very little diesterase activity. 3-Phosphoglycerate 2,3-diphosphoglycerate, 2-phosphoglycerate, ATP, and 2-phosphoglycerol were all attacked at about the same rate.

Effect of Ions. The enzyme was inhibited completely by 0.3 micromole of Hg^{++} in contrast to the stimulation of muscle 2,3-diphosphoglycerate phosphatase.[3,4] Mg^{++} stimulated the activity (32% at 1.3×10^{-2} *M*) whereas Mn^{++} had only a slight stimulating effect. Pb^{++} at this concen-

[3] S. Rapoport and J. Luebering, *J. Biol. Chem.* **189**, 683 (1951).
[4] B. K. Joyce and S. Grisolia, *J. Biol. Chem.* **233**, 350 (1958).

tration inhibited the enzyme 75%, Ag^+ ($8 \times 10^{-3} M$), Zn^{++} and Cu^{++} ($1.2 \times 10^{-2} M$) completely inhibited the activity.

II. 2-Phosphoglycerol Phosphatase (Ferric Ion-Activated) from Mung Beans

Assay Method

Reagents

Sodium Veronal-acetate buffer, 0.02 *M*, pH 8.5
Sodium salt of 2-phosphoglycerol, 0.1 *M* solution in water
Ferric sulfate, 0.02 *M* solution
Trichloroacetic acid, 20% solution

Procedure. The reaction mixtures, unless otherwise stated, contained 10 micromoles per milliliter of Veronal-acetate buffer, pH 8.5, 20 micromoles per milliliter of sodium 2-phosphoglycerol, 2.9 micromoles per milliliter of ferric sulfate, and 0.5 ml of enzyme in a total volume of 2.5 ml. After temperature equilibration the reaction was started by the addition of the enzyme. The reaction mixture was incubated at 49° for 15 minutes, and the reaction was stopped by the addition of 1.0 ml of 20% trichloroacetic acid. The mixture was centrifuged at 2500 *g* for 10 minutes, 1.0-ml aliquots were used for phosphate estimation by the method of Fiske and SubbaRow.[1]

Units and Specific Activity. One unit of enzyme activity is defined as the amount of enzyme which liberates 1 μg of P_i in 15 minutes under standard assay conditions. Specific activity is the number of enzyme units per milligram protein.

Purification of Enzyme

The method adopted is that of Appaji Rao *et al.*[5] A total of 20 g of freshly powdered mung bean seeds (passing 40 mesh) was extracted with 200 ml of water for 6 hours in the cold (0–4°) and centrifuged at 4000 *g* for 15 minutes. The extract (step 1, Table II) was dialyzed against water in the cold for 18 hours with repeated changes of water, and the precipitate obtained on centrifugation at 4000 *g* for 15 minutes was discarded. The supernatant was fractionated by the addition of solid ammonium sulfate; the precipitate appearing between 0.40 and 0.55 saturation was dissolved in 25 ml of water and dialyzed free of ammonium sulfate in the cold. The dialyzed extract was centrifuged at 4000

[5] N. Appaji Rao, H. R. Cama, S. A. Kumar, and C. S. Vaidyanathan, *J. Biol. Chem.* **235**, 3353 (1960).

g for 20 minutes and made up to 50.0 ml (step 2). To the supernatant was added 5.0 ml of 0.02 *M* acetic acid in the cold, and the precipitate obtained on centrifugation at 4000 *g* for 10 minutes was discarded. The supernatant was dialyzed against water in the cold for 18 hours, and the clear dialyzed extract was used as the enzyme (step 3). An 80-fold purification with 81% recovery was achieved.

TABLE II
PURIFICATION OF Fe^{3+}-REQUIRING 2-PHOSPHOGLYCEROL PHOSPHATASE (ALKALINE) FROM MUNG BEAN

Step	Volume (ml)	Protein (mg)	Total activity (units)	Specific activity (units/mg protein)	Yield (%)
1. Extract	180	3840	15,750	4	100
2. $(NH_4)_2SO_4$ fraction	50	330	13,500	41	86
3. Supernatant from acetic acid treatment	55	40	12,810	320	81

Properties

The enzyme functioned optimally at a pH of 8.5, at 49°, and at a ferric ion concentration of 2.9 micromoles per milliliter. In the presence of ferric ions under the standard conditions of assay, the enzyme was specific for 2-phosphoglycerol. The K_m of the enzyme for 2-phosphoglycerol was 6.2 micromoles per milliliter. Heavy metal ions like Mn^{++}, Cu^{++}, Zn^{++}, Fe^{++}, and Hg^{++} were highly inhibitory; but PCMB and iodoacetate were without effect. In the absence of ferric ions, the enzyme was activated by aspartic acid, histidine, cysteine, and glutathione and in the presence of ferric ions, by cysteine and glutathione.

[115] Phosphoglycolate Phosphatase

By DONALD E. ANDERSON and N. E. TOLBERT

$$\begin{array}{l} COOH \\ | \\ CH_2OPO_3H_2 \end{array} \rightarrow \begin{array}{l} COOH \\ | \\ CH_2OH \end{array} + H_3PO_4$$

Assay Method

Principle. Inorganic phosphate (P_i) is determined colorimetrically after enzymatic hydrolysis of phosphoglycolate. Cacodylate buffer is only slightly inhibitory and it does not interfere with the P_i determination. Maleic acid is a good alternate buffer.

Reagents

Sodium phosphoglycolate, 0.01 *M*, pH 6.3
$MgSO_4$, 0.01 *M*
Cacodylate buffer, 0.20 *M*, pH 6.3
Trichloroacetic acid, 10%
Reagents for determination of P_i[1]

Phosphoglycolate has been synthesized from phosphoglycerate by controlled permanganate oxidation[2,3] or by phosphorylation of glycolic acid.[4] The tricyclohexylammonium salt of phosphoglycolate (General Biochemicals) is treated with Dowex 50 (H^+) and then the free phosphoglycolic acid solution is adjusted to pH 6.3 with base.

Procedure. A solution of 0.5 ml of cacodylate buffer, 0.6 ml of $MgSO_4$, 1.0 ml of phosphoglycolate, and 0.8 ml of water are equilibrated in a 30° water bath for 5 minutes. The reaction is started by addition of 0.10–0.01 ml of enzyme which has been kept at 0°. After 10 minutes, the reaction is stopped by addition of 1 ml of 10% trichloroacetic acid and P_i is determined on an aliquot after removal of denatured protein by centrifugation. Two controls, one without enzyme and one without substrate, are run to correct for P_i that is not liberated in the reaction.

Protein is determined in crude plant homogenates by the Kjeldahl method. In partially purified preparations protein concentration is estimated by optical density at 280 mμ in a 1-cm light path. A factor of 1 is used to convert spectrophotometric readings to milligrams protein per milliliter.

Definition of Unit and Specific Activity. One unit is defined as the amount of enzyme that catalyzes the formation of 1 μg of phosphorus in 10 minutes. Specific activity is units per milligram of protein.

Purification Procedure

The procedure below was developed with field-grown 'Maryland Mammoth' tobacco leaves. The enzyme from leaves of spinach, wheat, and alfalfa is also stable to acetone precipitation.

Step 1. Preparation of Crude Extract. About 30 to 40 leaves are harvested and washed; the base of the cut stems is placed in shallow

[1] C. H. Fiske and Y. SubbaRow, *J. Biol. Chem.* **66**, 375 (1925).
[2] W. Kiessling, *Ber.* **68**, 243 (1935).
[3] K. E. Richardson and N. E. Tolbert, *J. Biol. Chem.* **236**, 1285 (1961).
[4] I. Zelitch, *J. Biol. Chem.* **240**, 1869 (1965).

water. The leaves are chilled in a 2° room and all subsequent operations are performed at about 2°. The midribs are removed and the blades weighed. The leaves are homogenized in a large Waring blendor for 2 minutes with twice the weight of distilled water. The resulting slurry is squeezed through a double layer of cheesecloth. The extract is clarified by centrifugation at about 15,000 *g* for 20 minutes. The pH of this extract is about 5.7.

Step 2. First Acetone Fractionation. A volume of reagent grade acetone at 0° equal to 40% of the crude extract volume is added very slowly. The acetone is siphoned through 0.7 mm I.D. Teflon tubing over a period of 1½ hours into the extract while it is being stirred by a magnetic stirrer and kept at 0° by an ice bath. The mixture is centrifuged at about 15,000 *g* for 15 minutes, and the precipitate is discarded. Cold acetone equal to 20% of the volume of the original extract is then added as before to precipitate a protein fraction containing the enzyme. This precipitate is removed by centrifugation at 15,000 *g* for 8 minutes and the acetone supernatant is drained and rinsed as completely as possible from the centrifuge cups. The enzyme is dissolved in a volume of cold 0.02 *M* cacodylate buffer, pH 6.3, equal to one-eighth of the volume of the original extract. The enzyme solution is centrifuged again for 8 minutes to remove insoluble protein.

Step 3. Second Acetone Fractionation. To the aqueous extract from the first acetone fractionation is added 0° acetone (25 ml per 100 ml of extract) as before. The mixture is centrifuged at 5900 *g* for 8 minutes, and the precipitate is discarded. Then acetone (20 ml per 100 ml of starting extract) is added as before, and the mixture is centrifuged at 5900 *g* for 8 minutes to precipitate the enzyme. After careful removal of as much of the acetone solution as possible, the precipitate is dissolved in a volume of cold 0.02 *M* cacodylate buffer, pH 6.3, equal to 40% of the volume of the preparation from step 2. The enzyme solution is centrifuged again for 8 minutes to remove insoluble protein.

Step 4. Third Acetone Fractionation. The enzyme preparation is adjusted to pH 5.7. The volume is recorded as the basis for the volumes of acetone to be added. A volume of cold acetone equal to 30% of the enzyme preparation is added slowly by pipette while the mixture is stirred by hand. The precipitate is discarded by centrifugation at 5900 *g* for 8 minutes. To the supernatant fluid 10% more acetone is added to precipitate the enzyme. After centrifugation at 5900 *g* for 8 minutes, the acetone supernatant is drained and the sides of the bottles are rinsed with water. The enzyme is dissolved in 0.02 *M* cacodylate buffer, pH 6.3, of one-quarter the volume of the preparation from step 3.

A summary of the purification procedure is given in the table.

Fraction	Total units	Yield (%)	Specific activity	Purification
1. Crude Extract	236,600	100	5.6	1
Centrifuged extract	203,500	86.0	11	2
2. First Acetone	152,800	64.6	93	17
3. Second Acetone	153,100	64.9	383	68
4. Third Acetone	95,700	40.5	916	164

The enzyme from tobacco leaves has been alternately purified by a sequence of ammonium sulfate precipitation, calcium phosphate gel adsorption and elution and chromatography on DEAE-cellulose.[3] However, ammonium sulfate treatment alters the pH optimum for the enzyme from tobacco leaves and inactivates the enzyme from wheat or alfalfa leaves.

Properties

Heat Stability in Crude Extracts. When tobacco leaves are ground by cold mortar and pestle as much as 95% of the enzyme is inactivated by incubation at 30° for 1 hour. Several methods produce an enzyme preparation stable at 30°. Grinding the leaves in a Waring blendor for 2 minutes stabilizes the phosphatase. If the leaves are ground by mortar and pestle and the extract stored at 0° for 6–24 hours, the enzyme becomes stable. The curve of stability versus time is sigmoidal. The enzyme is stabilized also by the addition of *cis*-aconitate ($4 \times 10^{-3}\,M$) or of substrate. The stable enzyme as isolated from wheat sap, without any additions, contains *cis*-aconitic acid associated with it.

Storage Characteristics. Crude extracts prepared by the Waring blendor are stable at 2° for at least 24 hours. Acetone precipitated enzyme redissolved in cacodylate buffer is stable at 2° for at least 2 weeks. The enzyme at various stages of purification by acetone is unstable to freeze-thawing. The initial acetone precipitate may be dried at room temperature under about 15 mm of Hg pressure without sacrificing activity, and the dried precipitate is stable at room temperature. Rapid drying of the acetone precipitate at less then 1 mm Hg destroys much of the activity.

General Properties. Studies on highly purified enzyme obtained after chromatography on DEAE-cellulose indicate that the phosphatase is specific for phosphoglycolate.[3] The pH optimum is 6.3, and Mg^{++} is generally used as a cofactor although Zn^{++} functions equally well. This phosphatase is not present in roots or etiolated leaves,[5] but we have

[5] Y. L. Yu, N. E. Tolbert, and G. M. Orth, *Plant Physiol.* **39**, 643 (1964).

always observed it in the leaves of plants (about 8 tested) and in extracts of algae (*Chlorella* and *Chlamydomonas*). The enzyme appears to be lightly bound to the chloroplasts,[5] and for this reason 9–14% of it is lost with the chloroplasts during centrifugation to clarify the crude extract (see table).

Section VIII

Dehydrases

[116] 6-Phosphogluconic Dehydrase[1]

By H. PAUL MELOCHE and W. A. WOOD

$$\text{6-Phosphogluconate} \xrightarrow{M^{++}} \text{2-keto-3-deoxy-6-phosphogluconate} + H_2O$$

Assay

6-Phosphogluconic dehydrase is assayed spectrophotometrically at 340 mμ. The KDPG formed by the dehydrase is cleaved by added KDPG aldolase, and the resulting pyruvate is reduced by NADH in the presence of added lactic dehydrogenase. The assay, carried out in microcuvettes (3×25 mm, 1-cm light path), in a total volume of 0.2 ml, is proportional over a range of velocities up to 1.0 absorbance unit per minute. One unit of dehydrase is defined as an absorbance change of 1.0 per minute. This is equivalent to the dehydration of 0.032 micromoles of 6-PG per minute.

Reagents

Imidazole, pH 8.0, 200 m*M*
NaNADH, 2 m*M*
NaGSH, 30 m*M* (freshly prepared)
$MnCl_2$, 10.0 m*M* (freshly prepared)
Na 6-phosphogluconate, pH 7, 40 m*M*
Commercial crystalline muscle lactic dehydrogenase diluted 1:10 in water
KDPG aldolase, 2000 units/ml (see this volume [91])

For assay, 0.01 ml of each of the above reagents is added to a microcuvette and brought to a total volume of 0.2 ml with distilled water (0.13 ml). Upon the addition of the dehydrase sample, the linear rate of absorbance change per minute at 340 mμ is determined (an initial lag is observed).

Enzyme Source

Pseudomonas fluorescens A 3.12 is maintained on slants of the following composition: $(NH_4)_2HPO_4$, 4 g; K_2HPO_4, 4 g; $MgSO_4 \cdot 7\ H_2O$, 1 g; citric acid, 1 g; potassium gluconate, 5 g; $FeCl_3$ (anhydrous), 10 mg; agar, 20 g; and distilled water, 1000 ml. Growth occurs overnight at room temperature, and cultures are stored under refrigeration. Potassium

[1] Abbreviations used are: KDPG, 2-keto-3-deoxy-6-phosphogluconate(ic); 6-PG, 6-phosphogluconate(ic); NADH, reduced nicotinamide-adenine-dinucleotide; GSH, reduced glutathione.

gluconate solution at pH 6 is prepared by neutralizing the δ-glucono-lactone with about 1.25 equivalents of solid KOH.

For mass culture, inocula are prepared in the above media from which agar has been omitted. The medium is dispensed at the rate of 100 ml per 250-ml flask and autoclaved. The inoculum is incubated 24 hours at room temperature with shaking.

In mass culture, cells are grown in a medium composed of: $(NH_4)_2HPO_4$, 3 g; K_2HPO_4, 1.5 g; $MgSO_4 \cdot 7\ H_2O$, 0.5 g; glucose, 6 g; $FeCl_3$ (anhydrous), 10–20 ppm; and water, 500 ml. The ammonium and potassium phosphates are dissolved in 250 ml of water in 2.7 liter Fernbach flasks (solution A). The glucose and magnesium sulfate are dissolved in 250 ml of water in a 1000-ml Erlenmeyer flask (solution B). Solutions A and B are autoclaved 30 minutes at 121°, and then solution B is added to solution A. Prior to inoculation, $FeCl_3$ (unsterile) is added. A 0.33% inoculum is used, and the culture is grown 18 hours on a rotary shaker. For mass culture in a fermentor, the medium and conditions described for the preparation of KDPG aldolase are employed.[2]

Purification of 6-Phosphogluconic Dehydrase

The results of purification and separation of 6-phosphogluconic dehydrase from KDPG aldolase are shown in the table. The procedure is

PURIFICATION OF 6-PHOSPHOGLUCONIC DEHYDRASE

Step	Total activity (units)	Recovery (%)	Specific activity (units/mg protein)	Dehydrase to aldolase activity ratio
Crude extract	4150	100	0.58	0.093
Protamine sulfate	4820	—	0.83	0.072
Ammonium sulfate I	3540	85	5.1	0.45
Acetone I	3500	84	6.2	0.88
Ammonium sulfate II	1800	43	6.6	2.2
Acetone II	1650	40	6.7	6.1
Ammonium sulfate III	1180	28	8.2	8.4
Acetone III	1040	25	9.9	13
Storage	925	22	11.4	40

essentially a series of alternate ammonium sulfate and acetone fractionations.

Acetone-dried cells (10 g) are suspended in 100 ml of 0.05 M phosphate, pH 8.0, containing 1 mM $MnCl_2$. All steps are carried out in this

[2] H. P. Meloche, J. M. Ingram, and W. A. Wood, see this volume [91].

buffer. Protein is estimated from the absorbance ratio at 280 and 260 mμ.[3] The cells are disrupted by sonic oscillation, and the cell debris is removed by centrifugation. This and all steps are carried out at 2° unless otherwise specified.

Protamine Sulfate Fractionation. The crude preparation is diluted with buffer to give 14 mg of protein per milliliter. Ammonium sulfate is added to a final concentration of 0.1 M (13 g per liter), followed by 0.2 volume of 2% protamine sulfate (salmine) at pH 5.5. The resulting precipitate is removed by centrifugation and discarded.

Ammonium Sulfate I. To the supernatant is added 20.3 g of ammonium sulfate (1.5 M) per 100 ml. The resulting precipitate is discarded. To the supernatant is added 9.23 g of ammonium sulfate (1.5–2.1 M) per 100 ml. The precipitate is collected and, after carefully washing the centrifuge tubes with distilled water, the pellet is dissolved in 10 ml of buffer for each 100 ml of solution used at the start of this step.

Acetone I. All fractionations with acetone are carried out at −14° with 80% acetone-water (v/v). To the preparation from above is slowly added with stirring 3.75 volumes of acetone. The precipitate is removed by centrifugation at −20° and, after carefully draining the tubes, the enzyme is dissolved in buffer using the same volume present going into this step. Insoluble protein is removed by recentrifugation, and the supernatant is diluted with buffer to 10 mg of protein per milliliter.

Ammonium Sulfate II. To the above is added 31.9 g of ammonium sulfate (0–2.1 M) per 100 ml. The precipitate is collected by centrifugation and, after the centrifuge tubes have been carefully washed with water, the pellet is dissolved in buffer using 0.2 times the volume present going into this step. This preparation is used below.

Acetone II. Again 3.75 volumes of acetone are added as above. The precipitate is collected by centrifugation and dissolved in 10 ml of buffer. Insoluble protein is removed by centrifugation, and the supernatant is diluted with buffer to contain 10 mg of protein per milliliter and used in the next step.

Ammonium Sulfate III. Then 3.19 g of ammonium sulfate is added per 10 ml of solution and the precipitate is collected by centrifugation. After the centrifuge tubes have been carefully washed, the pellet is dissolved in buffer using 0.67 times the volume present at the beginning of this step. This preparation is used below.

Acetone III. Finally, 3.75 volumes of acetone are added with stirring at −14°. The precipitate is collected, dissolved in 5 ml of buffer and recentrifuged.

[3] E. Layne, Vol. III [73].

Storage. The enzyme preparation is frozen in a dry ice–acetone bath and stored at —20°. Subsequently, after thawing, insoluble protein is removed by centrifugation.

The residual aldolase activity slowly decays and the dehydrase is stable for at least 4 months if kept frozen. It is suggested that after use, the dehydrase preparation be rapidly refrozen in a dry ice bath. The overall purification is 20-fold, and the relative increase in the dehydrase to aldolase activity ratio is 430-fold. The final ratio of absorbancy at 280 mμ and 260 mμ is 0.9.

Properties of 6-Phosphogluconic Dehydrase

General. 6-Phosphogluconic dehydrase catalyzes the dehydration of 6-phosphogluconate by removing the elements of water between carbon atoms 2 and 3 to form KDPG.[4] The reaction has been shown to involve the formation of enol-KDPG which spontaneously rearranges to the keto form with the stable incorporation of a proton from water.[5] This is in contrast to cobamide coenzyme-requiring dehydrases which convert diols to the corresponding carbonyl-deoxy compounds by a mechanism involving the intramolecular shift of a hydride ion.[6-8]

6-Phosphogluconate is completely converted to KDPG by the enzyme. No evidence of reversibility (that is, conversion of KDPG to 6-phosphogluconate) has been obtained. This is probably due to the fact that the immediate substrate of a back reaction would be the enol form of KDPG, not the keto form.[5] Indeed, there is evidence that KDPG itself does not exist as the keto form, but as the hemiketal.[5,9] The K_m of the enzyme for 6-phosphogluconate is 0.6 m*M*.[4]

6-Phosphogluconic dehydrase activity exhibits a requirement for divalent cations, the enzyme being activated by ferrous, manganous, or magnesium ions. The manganous ion was most efficient of the three, a concentration of 0.5 m*M* affording maximum activation.[4] Further activation was achieved with 10 m*M* reduced glutathione.[4] The enzyme is partially inactivated by iodoacetate and cyanide.[4] EDTA and *p*-chloromercuribenzoate completely and irreversibly inactivate the dehydrase.[4]

[4] R. Kovachevich and W. A. Wood, *J. Biol. Chem.* **213**, 745 (1955).
[5] H. P. Meloche and W. A. Wood, *J. Biol. Chem.* **239**, 3505 (1964).
[6] A. M. Brownstein and R. H. Abeles, *J. Biol. Chem.* **236**, 1199 (1961).
[7] R. H. Abeles and H. A. Lee, Jr., *J. Biol. Chem.* **236**, 2347 (1961).
[8] K. L. Smiley and M. Sobolov, *Arch. Biochem. Biophys.* **97**, 538 (1962).
[9] J. MacGee and M. Doudoroff, *J. Biol. Chem.* **210**, 617 (1954).

[117] D-Glucosaminic Acid Dehydrase

By J. M. MERRICK and SAUL ROSEMAN

CO_2^- — NH_2 — CH_2OH ⟶ CO_2^- — C=O — H—C—H — CH_2OH + NH_3

D-Glucosaminic acid → 2-Keto-3-deoxy-D-gluconic acid

Assay Method

Principle. The dehydrase catalyzes the conversion of D-glucosaminic acid to 2-keto-3-deoxy-D-gluconic acid (KDG) and NH_3. The α-keto acid is measured as its semicarbazone according to the procedure of MacGee and Doudoroff.[1]

Reagents

- Glucosaminic acid, 0.2 M, adjusted to pH 8.0 (prepared as described below)
- Potassium phosphate buffer, 0.6 M, pH 8.0
- Pyridoxal phosphate, 0.0032 M
- 2-Mercaptoethanol, 0.2 M
- Semicarbazide reagent: 1% semicarbazide hydrochloride in 1.5% sodium acetate trihydrate solution.

Procedure. The reaction mixture contains 0.05 ml of glucosaminic acid (10 micromoles), 0.025 ml of buffer (15 micromoles), 0.005 ml of pyridoxal phosphate (0.016 micromole), 0.01 ml of 2-mercaptoethanol (2 micromoles), and enzyme to a final volume of 0.125 ml. The incubation is conducted at 37° for 5 minutes and stopped by heating for 3 minutes at 100°. Control mixtures include one with heat-inactivated enzyme and one without enzyme. Aliquots of the reaction mixtures are removed, diluted to 1.0 ml, mixed with 1 ml of the semicarbazide reagent, and incubated for 15 minutes at 37°. After the addition of 3 ml of water, semicarbazone formation is measured at 250 mμ.

Definition of Unit and Specific Activity. One unit of enzyme is defined as the quantity that yields 1 micromole of α-keto acid in 5 minutes under

[1] J. MacGee and M. Doudoroff, *J. Biol. Chem.* **210,** 617 (1954).

the above conditions. The specific activity is defined as the number of units of enzyme per milligram of protein. Protein is determined by a nephelometric micromethod.[2]

Purification Procedure[3]

An unidentified organism, isolated by an enrichment culture technique, has been deposited at the Northern Regional Research Laboratories, Peoria, Illinois. The organism, designated as NRRL-P-826, is grown in a medium containing 2.5 g of glucosaminic acid, 2.0 g of NH_4Cl, 6.0 g of Na_2HPO_4, 3.0 g of KH_2PO_4, 5.0 g of NaCl and 0.1 g of $MgSO_4$ per liter. A 2.5% inoculum is used, and cells are grown at 37° on a New Brunswick Rotary Shaker for 24 hours. The bacteria are harvested at 4° by centrifugation and washed with 0.15 *M* KCl solution. Approximately 2.5 g (wet weight) of cells are obtained per liter of growth medium. The remaining steps are performed at 0–5°. Either freshly harvested cells (2.5 g), or cells frozen and stored at —18° for periods of less than 1 month, are suspended in 8 ml of 0.15 *M* KCl; extracts are prepared by vigorous shaking with 8 g of fine glass beads.[4] After centrifugation at 16,000 *g* for 10 minutes, the debris is washed twice with 6 ml portions of 0.15 *M* KCl and the supernatant fluid and washings are combined. The resulting crude extract contains about 8100 units of enzyme with a specific activity averaging 127. The extract (19 ml) is treated with 5 ml of a 2% solution of protamine sulfate; the mixture is stirred for 10 minutes at 0°, then centrifuged at 16,000 *g* for 10 minutes. The supernatant fluid is dialyzed against 0.01 *M* KCl for 3–6 hours, and the dialysis residue is centrifuged to remove insoluble matter. The yield of enzyme in the supernatant fluid is about 98%, with a specific activity averaging 563.

Diethylaminoethyl cellulose (previously equilibrated with a solution containing 0.02 *M* KCl and 0.01 *M* potassium phosphate buffer, pH 7.0, followed by exhaustive washing with water) is added to 22 ml of the supernatant fluid until at least 75% of the enzyme is retained by the cellulose. The supernatant fluid is discarded, and the cellulose is washed several times with cold water. Finally, the enzyme is eluted by treating the cellulose twice with 20-ml portions of a solution containing 0.01 *M* potassium phosphate buffer, pH 7.1, and 0.08 *M* KCl. The specific ac-

[2] P. B. Hawk, B. L. Oser, and W. H. Summerson, "Practical Physiological Chemistry," 12th ed. McGraw-Hill (Blakiston), New York, 1947.

[3] J. M. Merrick and S. Roseman, *J. Biol. Chem.* **235**, 1274 (1960).

[4] Superbrite glass beads, No. 115 regular, Minnesota Mining and Manufacturing Company. The Nossal technique [P. M. Nossal, *Australian J. Exptl. Biol. Med. Sci.* **31**, 583 (1953)] was used to prepare the extracts.

tivity ranges between 700 and 1200 at this point, and the yield varies between 65 and 80%. Purified preparations show variable stability when stored at —18°; some preparations lose activity in a few days whereas others remain active for a few weeks.

Enzymatic Preparation of KDG[3]

Glucosaminic dehydrase as prepared above may be used in this preparation, or a less purified fraction may be prepared as follows:

Five grams of cells are extracted in 2.5 g batches as described above. The crude extracts are combined and the debris is not further extracted. After centrifugation at 16,000 *g* for 10 minutes, the supernatant fluid (10 ml) is treated with 1 ml of aged calcium phosphate gel containing 16 mg of solids per milliliter. After 15 minutes, with occasional stirring, the gel is removed by centrifugation. The supernatant fluid (10 ml) is treated with 2.26 g of solid ammonium sulfate, stirred for 15 minutes and centrifuged for 15 minutes at 16,000 *g*. The precipitate is discarded and the supernatant fluid is adjusted to 70% of saturation with respect to ammonium sulfate. The resulting precipitate is dissolved in 9.0 ml of phosphate buffer, 0.1 *M* pH, 6.6. The following mixture (860 ml final volume) is incubated for 2 hours at 37°: 8.6 ml of the enzyme solution; 2.7 g of glucosaminic acid adjusted to pH 7.4; 260 ml of 0.1 *M* potassium phosphate buffer, pH 7.4. The reaction is followed by the semicarbazide assay, and at the end of the incubation period approximately 91% of the glucosaminic acid is converted to keto acid. The pH is adjusted to 2–3 with phosphoric acid, and the mixture is passed through 100 ml of Dowex 50, hydrogen form resin (200–400 mesh). The eluate and washings are neutralized with $Ca(OH)_2$. Calcium phosphate is removed by centrifugation, and the supernatant fluid is concentrated under reduced pressure. The pH of the concentrated solution is maintained at 7.2 by adding $Ca(OH)_2$ until no change is detected over a period of several hours (the pH presumably decreases owing to hydrolysis of the lactone ring of the acid). Finally, the solution is concentrated to a syrup and triturated with methanol, giving 2.47 g of crude calcium salt as a white solid; about 94% of the keto acid present in the incubation mixture is isolated. The salt is crystallized by dissolving it in the minimum quantity of water, concentrating to a syrup, and adding several drops of acetone. Crystallization is difficult in the presence of excess H_2O.

Properties of the Enzyme

Cofactors. When crude extracts were stored at —18° for 6 months, glucosaminic acid dehydrase activity was substantially decreased. Protamine, followed by ammonium sulfate fractionation, gave preparations that were essentially inactive unless pyridoxal phosphate and 2-mercaptoethanol were added. Glutathione, thioglycolate, and cysteine were much less effective than 2-mercaptoethanol. AMP, an apparent cofactor with some amino acid dehydrases, showed no effect in this system. The concentration of pyridoxal phosphate required for maximum enzyme activity is approximately 1.5×10^{-5} *M*.

Specificity. The reaction catalyzed by glucosaminic acid dehydrase is analogous to the reactions catalyzed by serine and threonine dehydrases. The purified preparation described above showed little activity with D- and L-isomers of serine and threonine, the maximum being 2.2% of the activity exhibited with glucosaminic acid. The K_m for glucosaminic acid is 8.8×10^{-3} *M*.

Effect of pH. In phosphate buffer, maximum activity is observed between pH 7.7 and 8.1; in Tris-HCl, the maximum is between 8.3 and 8.7.

Stoichiometry and Reversibility Studies. The enzyme converts D-glucosaminic acid to stoichiometric quantities of 2-keto-3-deoxy-D-gluconic acid and NH_3. Attempts to show the reverse reaction, or to incorporate ^{15}N from $^{15}NH_3$ into glucosaminic acid in the presence of the enzyme were not successful.

[118] D-Glucarate Dehydrase[1]

By HAROLD J. BLUMENTHAL

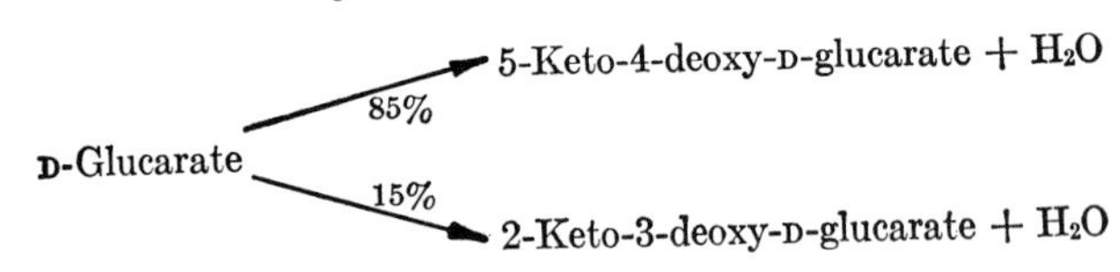

Assay Methods

Principle. The ketodeoxyglucarates resulting from the dehydration of D-glucaric (saccharic) acid are cleaved by periodate to yield formylpyruvic and glyoxylic acids. The chromogen formed by the reaction of thiobarbituric and formylpyruvic acids is measured at 551 mμ.

Reagents

Disodium D-glucarate (saccharate), 0.02 *M*. Commercial sources of glucarate are frequently impure. See this volume [13] for the

[1] H. J. Blumenthal, *Bacteriol. Proc.* p. 159 (1960).

method of purification of this substrate as the crystalline dicyclohexylammonium salt.

Tris-HCl buffer, 0.8 *M*, pH 8.0

$MgSO_4$, 0.08 *M*

D-Glucaric dehydrase. Make dilutions of the enzyme in 0.1% Triton X-100 (Rohm and Haas) to help prevent surface denaturation.

Trichloroacetic acid, 10%

Periodic acid, 0.025 *M*, in 0.125 *N* H_2SO_4

Sodium arsenite, 2.0%, in 0.5 *N* HCl

2-Thiobarbituric acid, 0.3%

Procedure. The incubation mixture (0.4 ml) in a 10×75 mm test tube contains 0.1 ml disodium D-glucarate, 0.05 ml Tris-HCl buffer, 0.05 ml $MgSO_4$, and about 0.01 unit of D-glucaric dehydrase preparation. The reactions, which are started by addition of enzyme to the temperature-equilibrated mixture, are terminated after an incubation period of 10 minutes at 30° by addition of 0.1 ml of trichloroacetic acid. If a precipitate forms, it is removed by centrifugation, and a 0.2-ml sample is used for the periodate–thiobarbituric acid test.[2] If increased amounts of substrate are used in the incubations, samples of 0.1 ml or smaller are analyzed so that periodate is not limiting. If the absorbancy of the chromogen at 551 mμ is too high, the analysis must be rerun using a smaller sample. Dilutions of the chromogen do not yield the expected absorbancy.

Ordinarily, only a zero time control is needed, trichloroacetic acid being added before the enzyme. However, if the fractions contain galactarate dehydrase[3] with galactarate as a stabilizer, then a control lacking substrate must be incubated for the 10 minutes and substrate added after the trichloroacetic acid. The difference between the glucarate dehydrase assay and the control lacking substrate determines the true glucaric dehydrase values. There is good proportionality between enzyme activity and enzyme concentration when up to 0.02 unit of enzyme is assayed.

Definition of Unit and Specific Activity. One unit of enzyme is defined as the amount of enzyme that catalyzes the formation of 1 micromole of ketodeoxyglucarate per minute under these conditions. In the periodate-thiobarbituric acid test[2] for ketodeoxyglucarate, an extinction coefficient of 60,000[3] at 551 mμ is employed. Specific activity is expressed as units per milligram of protein. Protein is determined spectrophotometrically by the method of Waddell[4] at 215 and 225 mμ.

[2] A. Weissbach and J. Hurwitz, *J. Biol. Chem.* **234,** 705 (1959).
[3] H. J. Blumenthal, D. C. Fish, and T. Jepson, see this volume [13a].
[4] W. J. Waddell, *J. Lab. Clin. Med.* **48,** 311 (1956).

Purification Procedure

Culture Conditions. Escherichia coli strain CR63MA is grown aerobically for 16 hours on a rotary shaker at 37° with either 0.8% D-glucarate or galactarate as the sole carbon source in a salts medium adjusted to pH 6.8 with NaOH. The medium contains the following in 1 liter of deionized water: 8.0 g of galactaric acid or potassium acid D-glucarate, 16.5 g of anhydrous dibasic sodium phosphate, 1.5 g of monobasic potassium phosphate, 2.0 g of ammonium sulfate, 0.2 g of magnesium sulfate heptahydrate, 10 mg of calcium chloride dihydrate, and 50 μg of ferrous sulfate heptahydrate. A 5% inoculum of organisms grown for 24 hours in the same medium is used to initiate growth. After harvesting by centrifugation, the cells are resuspended in, and washed twice with, 0.05 M KCl. The average yield of packed cells is 3.6 g per liter of medium. Cells prepared under these conditions can be stored at —20° for several years with retention of most of their D-glucarate dehydrase activity.

Step 1. Preparation of Crude Extract. To each gram of cells, 10 ml of 0.05 M Tris-HCl, pH 7.5, containing 0.003 M NaGSH is added and the suspension is disrupted by a 10-minute treatment in a 10-kc Raytheon sonic oscillator. The supernatant fluid obtained after centrifugation for 10 minutes at 21,000 g is used as the crude cell-free extract. The enzyme activity in this fraction is fairly stable at —20°, with considerable amounts of activity remaining after storage for over a year.

Step 2. Fractionation with Ammonium Sulfate. To each 18 ml of extract is added 12 ml of saturated ammonium sulfate which had been neutralized with ammonia. After 15 minutes, with occasional stirring, the 0–40% ammonium sulfate precipitate is removed by centrifugation. To the 30 ml of supernatant liquid is added 10 ml of neutral saturated ammonium sulfate, and the 40–55% precipitate is removed by centrifugation after 15 minutes. The precipitate is dissolved in 3 ml of 0.01 M Tris-HCl, pH 7.5, containing 0.003 M NaGSH. Ketodeoxyglucarate aldolase[5] and tartronate semialdehyde reductase can be recovered from the supernatant solution by adjusting it to 65% saturation with more neutral saturated ammonium sulfate.

Step 3. Fractionation with Acetone. To each 5.5 ml of the 40–55% ammonium sulfate fraction in a 50-ml stainless steel centrifuge tube is added 0.05 ml of 5 M potassium acetate. With the preparation immersed in ice, 1.5 ml of cold acetone is added during constant stirring. The centrifuge tube is transferred to a bath at —9°, and 3.0 ml of acetone

[5] D. C. Fish and H. J. Blumenthal, see this volume [93].

is slowly added. After 10 minutes, the 0–45% acetone fraction is recovered by centrifugation for 10 minutes at 21,000 g at —10°. The supernatant solution is transferred to a second stainless steel centrifuge tube in the —9° bath, and 1.0 ml of cold acetone is added. After 10 minutes, the 45–50% acetone fraction is collected by centrifugation as before, and this precipitate is dissolved in 2.0 ml 0.01 M Tris-HCl, pH 7.5, containing 0.003 M NaGSH.

Step 4. Column Chromatography on XE-64 Resin. Amberlite XE-64 (IRC-50) resin, equilibrated with 0.2 M potassium phosphate buffer, pH 5.4, is packed into a column 0.9 × 19 cm. The 45–50% acetone fraction (5.5 ml; *ca.* 18 mg protein) is applied to the column and washed in with 10 ml of the buffer used for equilibration. The column is then connected to a reservoir containing 0.2 M potassium phosphate buffer, pH 5.7, and 5-ml fractions are collected every 6 minutes. There is a large peak of protein and nucleic acids that is not adsorbed and comes off immediately. Tubes 28–40 contain glucarate dehydrase with a specific activity of 18.5 or greater. These fractions do not contain any α-keto-β-deoxy-D-glucarate aldolase[5] or galactarate dehydrase[6] although they still contain some tartronate semialdehyde reductase. For storage, the fractions are neutralized with 4 N KOH.

There is an alternate method of column chromatography which yields comparable results but takes longer to run. A sample of the same size is placed on a 2 × 22 cm column of DEAE-cellulose equilibrated with 0.15 M KCl containing 0.01 M Tris-HCl, pH 7.5, and 0.001 M $MgCl_2$. The column is washed at a rate of 1 ml/minute with the equilibrating solvent; 12-ml fractions are collected until the major protein and nucleic acid peak is completely eluted (200 ml) as judged by a return of the absorbancy of the eluate at *ca.* 260 mμ to the initial levels. Then a linear gradient of increasing KCl concentration is applied to the column. The mixing chamber contains 200 ml of 0.15 M KCl and the reservoir the same volume of 0.35 M KCl, both solutions containing 0.01 M Tris-HCl, pH 7.5, and 0.001 M $MgCl_2$. The dehydrase appears in about five tubes when the KCl concentration reaches 0.3 M.

The purification procedure is summarized in the table (p. 664).

Properties

Effect of pH. The enzyme exhibits maximal activity between pH 7.5 and 8.5 in Tris-HCl buffer, at pH 7.5 in barbital-acetic acid buffer, and at pH 7.2 in Tris-maleate buffer. The dehydrase is moderately stable at

[6] H. J. Blumenthal and T. Jepson, see this volume [119].

PURIFICATION PROCEDURE FOR D-GLUCARATE DEHYDRASE

Fraction	Units/ml	Specific activity	Recovery (%)	Purification (-fold)
1. Crude extract	9.4	0.625	100	1
2. 40–55% $(NH_4)_2SO_4$ precipitate	20.6	1.28	75.8	2
3. 45–50% Acetone precipitate	18.1	5.40	51.8	8.6
4. XE-64 Column, fractions 28–40	0.81	20.1	27.7	32.1

pH 5–6 so that fractionation at those pH values is permissible. However, it is most stable near pH 7.5.

Substrate Specificity. The D-glucarate dehydrase is quite specific, acting only upon D-glucaric and L-idaric acids, the latter at only about one-half the rate of D-glucarate. The list of compounds that will not serve as substrates includes galactaric acid, D-mannaric acid,[7] D-idaric acid, xylaric-acid, D-, L-, and *meso*-tartaric acid, D-glucosaminic acid, DL-threonine, L-serine, D-gulonic acid, D-gluconic acid, D-galactonic acid, L-galactonic acid, α-D-glucoheptonic acid, and 6-phospho-D-gluconic acid. This specificity complements that of the labile galactarate dehydrase, also present in cells grown on either D-glucarate or galactarate, which utilizes galactarate but not D-glucarate.[6]

Effect of Metals and Inhibitors. The D-glucarate dehydrase requires a divalent cation, $8 \times 10^{-2} M$ Mg^{++} serving best. Mn^{++} $(2 \times 10^{-4} M)$ is less than one-third as effective as Mg^{++}, while the same concentration of Co^{++} is about one-tenth as effective. Higher concentrations of Mn^{++} and Co^{++}, but not Mg^{++}, are inhibitory. The D-glucarate dehydrase is completely inhibited when $2 \times 10^{-2} M$ EDTA is added to the standard D-glucarate dehydrase assay, a concentration that does not affect galactarate dehydrase.[6]

The enzyme requires SH groups for activity, judging by its sensitivity to *p*-chloromercuryphenylsulfonic acid. Even at $10^{-5} M$, this compound inhibits over 90% of the enzyme activity.

When the effect of a number of compounds that are not substrates for D-glucarate dehydrase, but are structurally related to the substrate, are tested along with D-glucarate at 0.0025 M concentration, galactaric, D-idaric, and D-tartaric acids all inhibit the enzyme activity between 30 and 40%. The L-and *meso*-tartaric acids are only one-half as effective an inhibitor as D-tartaric acid.

Reversibility. The enzyme reaction is irreversible, and D-glucarate

[7] Care must be taken to include a control employing boiled enzyme since there is a nonenzymatic formation of ketodeoxyhexaric acid under alkaline conditions from mannarodilactone [D. Heslop and F. Smith, *J. Chem. Soc.* p. 577 (1944)].

can yield ketodeoxyglucarate quantitatively. Indeed, this is the best method to determine the ϵ_{551} of 60,000 for ketodeoxyglucarate[3] in the periodate-thiobarbituric acid procedure.

Substrate Affinity. The K_m for D-glucarate is 8×10^{-4} *M* under the standard assay conditions.

Product. The product of the dehydrase reaction with D-glucarate as the substrate is an approximately 85:15 mixture of 5-keto-4-deoxy- and 2-keto-3-deoxy-D-glucarate.[3,8] Apparently, the enzyme can dehydrate glucarate in either the 2,3- or 4,5-positions, the latter reaction being favored. With L-idarate as the substrate for this dehydrase, only 5-keto-4-deoxy-D-glucarate can be formed no matter which pair of hydroxyl groups is involved in the dehydration.

Distribution. Both D-glucarate dehydrase and galactarate dehydrase are induced by growth of *E. coli* on either D-glucarate or galactarate. This has also been demonstrated in *Erwinia carotovora, Aerobacter aerogenes, E. freundii, Paracolobactrum* sp., *Klebsiella pneumoniae, Bacillus subtilis,* and *B. megaterium.* D-glucarate dehydrase appears to be the first step in the metabolism of D-glucarate in all bacteria, including those that use a ketodeoxyglucarate aldolase[5] forming pyruvate and tartronate semialdehyde, and microbes such as *B. megaterium,* which yield CO_2 and α-ketoglutarate from D-glucarate.[9]

[8] D. C. Fish and H. J. Blumenthal, *Bacteriol. Proc.* p. 192 (1961).
[9] H. J. Blumenthal and T. Jepson, *Bacteriol. Proc.* p. 82 (1965).

[119] Galactarate Dehydrase[1]

By Harold J. Blumenthal and Theressa Jepson

Galactaric acid → 5-keto-4-deoxy-D-glucaric acid + H_2O

Assay Method

Principle. The ketodeoxyglucarate resulting from the dehydration of galactaric (mucic) acid by galactarate dehydrase is cleaved by periodate to yield formylpyruvic and glyoxylic acids. The chromogen formed by the reaction of thiobarbituric acid and formylpyruvic acid is measured at 551 mμ.

Reagents

Disodium galactarate (mucate), 0.01 *M*
Tris-acetate, 0.8 *M*, pH 8.0

[1] H. J. Blumenthal and T. Jepson, *Biochem. Biophys. Res. Commun.* **17**, 282 (1964).

Na_2EDTA, 0.02 M
Enzyme: about 0.05 unit enzyme/ml
Trichloroacetic acid (TCA), 10%
Periodic acid, 0.025 M, in 0.125 N H_2SO_4
Sodium arsenite, 2.0%, in 0.5 N HCl
2-Thiobarbituric acid, 0.3%

Procedure. The incubation mixture (0.4 ml) in a 10 × 75 mm test tube contains 0.1 ml of disodium galactarate, 0.05 ml of Tris-acetate, 0.05 ml of EDTA, and about 0.01 unit of galactarate dehydrase preparation. The assay is started, after temperature equilibration at 30°, by addition of enzyme, and is terminated, usually after 10 minutes, with 0.1 ml of 10% trichloroacetic acid. If a precipitate forms it is removed by centrifugation, and 0.2-ml sample is used for the periodate–thiobarbituric acid test.[2] A zero time control, wherein the TCA is added before the enzyme, and a no substrate control, wherein galactarate is added only after the 10 minute incubation has been terminated with trichloroacetic acid, are routinely included. The no substrate controls are necessary since galactarate is used to stabilize the enzyme. The enzyme bound galactarate is usually utilized in about 20 minutes.

Definition of Unit and Specific Activity. One unit is defined as the amount of enzyme that catalyzes the formation of 1 micromole of ketodeoxyglucarate per minute under these conditions. For ketodeoxyglucarate[3] in the periodate–thiobarbituric acid test, an extinction coefficient of 60,000 at 551 mμ is employed. The specific activity is expressed as units per milligram of protein. Protein is determined nephelometrically with sulfosalicyclic acid.[4]

Purification Procedure

Culture Conditions. Escherichia coli strain CR63MA is grown aerobically for 16 hours on a rotary shaker at 37° with 0.8% galactaric acid as the sole carbon source in a salts medium adjusted to pH 6.8 with NaOH. A 5% inoculum of a culture grown for 24 hours in the same medium is used. The medium contains 8.0 g of galactaric acid, 16.5 g of anhydrous dibasic sodium phosphate, 1.5 g of monobasic potassium phosphate, 2.0 g of ammonium sulfate, 0.2 g magnesium sulfate heptahydrate, 10 mg of calcium chloride dihydrate, and 50 μg of ferrous sulfate heptahydrate in each liter of deionized water. Five minutes before removal of the cells from the shaker and harvesting, 300 micromoles of

[2] A. Weissbach and J. Hurwitz, *J. Biol. Chem.* **234**, 705 (1959).
[3] H. J. Blumenthal, D. C. Fish, and T. Jepson, see this volume [13a].
[4] E. Layne, see Vol. III, p. 447.

ferrous sulfate is added to each liter of culture medium and shaking is resumed.

Step 1. Preparation of Extracts. After they have been harvested by centrifugation in the cold, the cells are resuspended in, and washed twice with, 0.02 *M* Na_2 galactarate. The average yield of packed cells is 3.6 g per liter of medium. Cells prepared under these conditions can be stored at —20° for several months with retention of much of their galactarate dehydrase activity. To each gram of cells, 10 ml of 0.02 *M* Na_2 galactarate is added, and the suspension is disrupted for 10 minutes in a 10-kc Raytheon sonic oscillator. If no purification beyond the crude extract is desired, the supernatant obtained after a 10-minute centrifugation at 21,000 *g* is used directly. Otherwise, 1.0 ml of 4.0 *M* sodium acetate is added directly to the uncentrifuged extract, followed immediately by 60 mg of finely powdered protamine sulfate (Nutritional Biochemical Corp.), and stirred until the powder is fully dissolved. The suspension is then centrifuged 10 minutes at 21,000 *g*, and the supernatant is immediately used for further purification.

Step 2. Potassium Acetate Fraction. To each 10 ml of the supernatant solution, 15 ml of 9.0 *M* potassium acetate, pH 7.3, is added rapidly with stirring, followed immediately by addition of 10 ml of 5.0 *M* potassium acetate, pH 5.7 (final pH 6.3). After centrifugation for 10 minutes at 21,000 *g*, the supernatant fluid is discarded and the pellet is dissolved in 9.8 ml of 0.02 *M* Na_2 galactarate. A partial desalting is accomplished by addition of 0.2 ml of 0.1 *M* $CaCl_2$. After 15 minutes, the precipitate of insoluble calcium salt is removed by centrifugation for 10 minutes at 8000 *g*.

Step 3. Calcium Phosphate Gel Treatment. Each milliliter of the supernatant fluid is treated with 0.1 ml (3.2 mg solids) of calcium phosphate gel.[5] The suspension is stirred for 5 minutes, then centrifuged; the gel is discarded.

Step 4. XE-64 Treatment. To each 4.0 ml of the supernatant fluid is added 4.0 ml of a suspension of Amberlite XE-64 anion exchange resin (*ca.* 85% solids, v/v) equilibrated with, and stored in, 0.05 *M* potassium acetate, pH 5.7. The preparation is stirred for 5 minutes; the resin is removed by centrifugation for 5 minutes at 1100 *g* and then successively treated, in the same manner, with 4.0 ml each of 0.02 *M* Tris-acetate, pH 6.0, and 0.02 *M*, 0.05 *M*, 0.1 *M* and 0.2 *M* Tris-acetate, pH 7.5. The galactarate dehydrase in these fractions is very labile and has to be used immediately.

The purification procedure is summarized in the table (p. 668).

[5] T. P. Singer and E. B. Kearney, *Arch. Biochem. Biophys.* **29**, 190 (1950).

PURIFICATION PROCEDURE FOR GALACTARATE DEHYDRASE

Fraction	Galactarate dehydrase Units/ml	Specific activity	Purification	Recovery (%)	D-Glucarate dehydrase (Units/ml)
1. Crude extract	3.19	0.38	1	100	9.32
2. K acetate precipitate	1.58	3.0	8	49.0	0.53
3. $Ca_3(PO_4)_2$ supernatant	1.28	6.0	16	40.1	0.19
4. XE-64 fractions					
(a) Supernatant	0.64	17.0	44	20.0	0.0
(b) 0.02 *M* Tris-acetate, pH 6.0	0.26	22.6	59	8.1	0.0
(c) 0.02 *M* Tris-acetate, pH 7.5	0.11	9.6	25	3.4	Trace
(d) 0.05 *M* Tris-acetate, pH 7.5	0.05	4.0	10	1.5	0.004
(e) 0.10 *M* Tris-acetate, pH 7.5	0.02	1.6	4	0.6	0.028
(f) 0.20 *M* Tris-acetate, pH 7.5	0.01	0.5	1	0.3	0.09

Properties

Stability. The cell-free enzyme is very labile although galactaric acid provides some degree of stabilization. The crude enzyme, isolated from galactarate-grown cells, is active for only a few minutes in its absence. No method for enzyme reactivation has been found.

Although galactarate dehydrase is also induced when cells are grown in a medium with D-glucarate replacing galactarate, the enzyme cannot readily be demonstrated unless a metal (preferably ferrous ion) is added to the growing cells about 5 minutes prior to harvest and the cells are sonicated in the presence of galactarate. Yet, no stabilization or requirement by the enzyme for metals can be shown.

Effect of Metals and Inhibitors. There is no metal requirement for galactarate dehydrase. Its enzyme activity is the same in the presence or absence of EDTA or in the presence of seventeen cations, including divalent metals such as magnesium, calcium, iron, or cobalt. This is a convenient and important difference from D-glucarate dehydrase,[6] which requires magnesium. Once the bound galactarate that is used to stabilize the enzyme is dehydrated by the enzyme, potassium cyanide completely inhibits further enzyme activity. Under the same conditions, pyrophos-

[6] H. J. Blumenthal, see this volume [118].

phate results in an 35% inhibition of further activity whereas sodium azide, sodium fluoride, or sulfhydryl reagents have no effect.

Effect of pH. The reaction is most rapid at pH 8.0, with relatively little activity below pH 7.0 or above pH 9.0. The enzyme reacts equally well in Tris-acetate, Tris-HCl, diethanolamine-acetate, glycine, lysine, or glutamate buffers. Phosphate, citrate, and cacodylate buffers are inhibitory.

Specificity. The enzyme, which can be completely separated from D-glucarate dehydrase, is specific for galactarate. It cannot utilize D-glucarate, mannarate, and either D- or L-idarate. Detailed studies on enzyme specificity are hampered by the lability of the enzyme in the absence of galactarate.

Nature of Reaction. When galactarate is examined by the criteria of Hirschmann,[7] it does have reflective, but not rotational symmetry. Evidence that this is indeed the case is obtained by isolating ^{14}C-labeled glycerate and unlabeled pyruvate following the further metabolism of galactarate-1-^{14}C by crude extracts of galactarate-grown *E. coli.*[1] The results provide evidence that galactarate is an asymmetric molecule that is dehydrated to form only 5-keto-4-deoxy-D-glucarate. This contrasts with the D-glucarate dehydrase[6] which dehydrates D-glucarate at either of two sites, although 5-keto-4-deoxy-D-glucarate is also the major product.[3]

Substrate affinity. At pH 8.0 and 30° the K_m for galactarate is about 4×10^{-4}.

Distribution. Both galactarate dehydrase and D-glucarate dehydrase are induced by growth of *E. coli* on either galactarate or D-glucarate. This has also been demonstrated in *Erwinia carotovora, Aerobacter aerogenes, Escherichia freundii, Paracolobactrum* sp., *Klebsiella pneumoniae, Bacillus subtilis,* and *B. megaterium.* Galactarate dehydrase appears to be the first step in the metabolism of galactarate in all bacteria, including the enteric bacteria, that use a ketodeoxyglucarate aldolase[8] forming pyruvate and tartronate semialdehyde and in microbes, such as *B. megaterium,* which yield CO_2 and α-ketoglutarate from galactarate.[9]

[7] H. Hirschmann, *J. Biol. Chem.* **235,** 2762 (1960).
[8] D. C. Fish and H. J. Blumenthal, see this volume [93].
[9] H. J. Blumenthal and T. Jepson, *Bacteriol. Proc.* p. 82 (1965).

[120] Enolase from Yeast and Rabbit Muscle

By E. W. WESTHEAD

$$\text{D-CH}_2\text{OH}\cdot\text{CHOPO}_3\text{H}_2\cdot\text{CO}_2\text{H} \rightleftarrows \text{H}_2\text{O} + \text{CH}_2\text{:COPO}_3\text{H}_2\cdot\text{CO}_2\text{H}$$

The preparation of enolase from brewers' yeast was described in the first volume of this series by T. Bücher ten years ago.[1] Since then, the preparation of that enzyme has been improved greatly, and convenient methods for the preparation of enolase from rabbit muscle have been published. Most of the published work on enolase, in fact, has appeared since Bücher's chapter was written.

Assay Method

Principle. In the presence of enolase, D-glyceric acid 2-phosphate is dehydrated to the unsaturated pyruvic acid phosphate. The equilibrium constant is influenced by pH and other factors, but at pH 7.8, in the presence of 0.4 N KCl and 8×10^{-3} M Mg^{++}, the ratio of pyruvic acid phosphate to glyceric acid 2-phosphate is 4.9; at pH 6.7 it is 4.1.[2] Pyruvic acid phosphate has a broad absorption band in the ultraviolet region, with a maximum near 215 mμ, while glyceric acid phosphate absorbs very little above 220 mμ. This difference in absorbancy was made the basis of an assay for enolase by Warburg and Christian,[3] who measured the increase in absorbancy at 240 mμ during the forward reaction. These authors fitted the data to a first-order plot to obtain a rate constant, but more recent workers have measured the initial velocity of the reaction by drawing a tangent at $t = 0$ to the curve of absorbancy *versus* time. Initial rates are easier to measure at 230 mμ than at 240 mμ; the sensitivity of the assay is doubled, and there is a helpful increase in the length of linear portion of the rate curve. The initial velocity measurement obviates assumptions about the correct form of the rate equation and, by minimizing the effects of secondary reactions, makes it possible to assay enolase in crude systems.

Reagents. Since the mammalian and yeast enzymes, at least, differ markedly in their pH optima, a single assay system is not ideal for enzymes from different sources. However, all enolases reported so far may be assayed conveniently, if not optimally, by either of the two assay systems given.

[1] T. Bücher, see Vol. I, p. 427.
[2] F. Wold and C. E. Ballou, *J. Biol. Chem.* **227,** 301 (1957).
[3] O. Warburg and W. Christian, *Biochem. Z.* **210,** 384 (1942).

The yeast enzyme is assayed at 30° in a solution of the following composition[4]:

D-Glyceric acid-2-phosphate	2×10^{-3} *M*
Magnesium acetate	1×10^{-3} *M*
Tris-acetic acid buffer, pH 7.8	0.05 *N* in acetate
EDTA	10^{-5} *M*

The standard assay for the rabbit muscle enzyme[5] is run at 25° in a solution containing:

D-Glyceric acid 2-phosphate	1×10^{-3} *M*
Magnesium ion (acetate or sulfate)	1×10^{-3} *M*
KCl	0.40 *N*
Imidazole-HCl buffer, pH 6.7	0.05 *M* in imidazole
EDTA	10^{-5} *M*

Substrate Preparation. The DL form of glyceric acid 2-phosphate may be synthesized by the method of Kiessling[6] or the pure D-isomer may be made by the method of Ballou and Fischer.[7] More conveniently, the barium salt of the D-isomer may be purchased.[8] Since the L isomer is without effect on the reaction (F. Wold, personal communication) it may be ignored when the racemic mixture is used. Substrate preparations from the different sources have given the same rates in comparative assays when they have been prepared carefully. It is important to remove all barium ion and to keep the substrate cold, especially when it is in acidic solution.

Barium is best removed by swirling a suspension of the salt in water with twice the equivalent amount of a cation-exchange resin (such as Dowex 50) in the hydrogen form. The white grains of the barium salt should dissolve in a few minutes. The solution should then be decanted slowly through a small column containing another equivalent of the same resin. The resins are washed two or three times with water, and the combined eluates are brought to pH 7 with KOH.

To assay the substrate, a small aliquot is diluted in the buffer used for the yeast enzyme assay. Upon addition of enough enolase to bring the reaction to completion within a few minutes, a 1×10^{-3} *M* solution of the substrate will show an increase in absorbancy at 240 mμ of 1.0 unit.

[4] E. W. Westhead and G. McLain, *J. Biol. Chem.* **239**, 2464 (1964).
[5] A. Holt and F. Wold, *J. Biol. Chem.* **236**, 3227 (1961).
[6] W. Kiessling, *Ber.* **68**, 243 (1935).
[7] C. E. Ballou and H. O. L. Fischer, *J. Am. Chem. Soc.* **76**, 3188 (1954).
[8] CalBiochem, Los Angeles, California, or Boehringer and Söhne, Mannheim, Germany.

Unsatisfactory substrate solutions are often responsible for low enzyme activities. In the presence of the complete assay mixture used for the muscle enzyme, the substrate is not stable. Double strength solutions of substrate and the mixture of other components are stored separately and combined only a few hours before use. The complete assay solution should then be kept on ice until shortly before use. The high KCl content of this assay mixture minimizes the sensitivity of the reaction rate to variations in the concentrations of all other components of the solution.

In the yeast enzyme assay solution, the substrate appears stable indefinitely when frozen, and for at least several hours at room temperature.

Procedure. A measured volume of substrate solution is put into a silica cuvette in a thermostatted cell holder or compartment of a spectrophotometer. The wavelength scale is set to 230 mμ, and the instrument is zeroed with a suitable blank. When temperature equilibrium has been reached, the reaction is started by adding an accurately measured aliquot of enolase solution with an efficient stirring device. If the reaction is to be followed manually, the final enzyme concentration should be near 0.1 μg/ml; if a logarithmic recorder is available, the enzyme concentration can be increased tenfold. A tangent is drawn to the curve of A_{230} *vs.* time, and the rate is expressed as absorbancy units per minute. Despite the apparently subjective element involved in drawing a tangent, the assay is very reproducible with a little practice.

Specific Activity. The specific activity may be expressed conveniently as the initial rate of change in absorbancy at 230 mμ (in units per minute) divided by the concentration of added protein (as absorbancy at 280 mμ due to added protein). Thus if 10 μl of a protein solution with an absorbancy at 280 mμ of 0.30 is added to a cuvette containing 1.0 ml of assay solution, the calculated protein concentration is 0.0030; if an initial rate of 0.60 per minute were found, the specific activity would be 200. In these units, the specific activity of pure yeast enolase is 620 in the pH 7.8 buffer at 30°. For the pure muscle enzyme, the specific activity is 250 in the pH 6.7 buffer at 25°. Conversion of these values to Enzyme Units (E.U.)[9] may be made with the extinction coefficients for the substrate[2] and for the two enzymes.[3,5] The corresponding values are 200 E.U./mg and 96 E.U./mg for the yeast and muscle enzymes, respectively.

Remarks. The reaction conditions given are those generally used in the literature. They can be varied significantly without lowering the

[9] "Report of the Commission on Enzymes of the International Union of Biochemistry." Macmillan (Pergamon), New York, 1961.

enzymatic activity, but the relationships among the optimal concentrations of salt, substrate, metal, and buffer ions are complex, and any alterations should be made carefully. In the yeast system at least, Tris and imidazole buffers are interchangeable. The yeast enzyme may be assayed at other temperatures by using the relationship, $\log(V_{30}/V) = 2.75 \times 10^{-3} (1/T - 1/T_{30})$ to correct to the rate at 30°.[10] The validity of the assay in crude extracts of *Escherichia coli*[11] and rabbit brain[12] has been established by adding pure enolase to such extracts and measuring the increase in activity. Since the initial increase in absorbance due to pyruvic acid phosphate production is followed by a decrease in such systems, it is essential to measure *initial* rates.

Preparation of the Enzymes

General Remarks. The muscle enzyme is perhaps somewhat easier to prepare than the yeast enzyme, and seems to be stable indefinitely as the crystalline suspension in ammonium sulfate. Yeast enolase, however, is far more stable in solution and more stable to freeze-drying. It is recommended as the enzyme of choice when a supply of enolase is to be kept available as a reagent. The solution should be stored at pH 8 to pH 8.5 in 10^{-4} *M* Mg^{++} solution in the presence of a drop of toluene. Solutions have been kept this way for nearly a year without loss of activity.

Both enzymes are sensitive to inhibition by heavy metals. Any dialysis tubing used should be washed until odorless and colorless by heating in several changes of dilute sodium carbonate solution containing 10^{-3} *M* EDTA. Glassware used for dilute solutions of pure enzyme should first be rinsed with fuming nitric acid and distilled water.

Preparation of Yeast Enolase

The enzymes from brewers' and bakers' yeast (*Saccharomyces cerevisiae*) appear to be identical,[4] but bakers' yeast, obtained in damp 1-pound cakes, is the more convenient starting material. The procedure given is that of Westhead and McLain,[4] modified in a few details. All operations, including pH measurements, are done near 0°.

Step 1. Toluene Autolysis and Extraction. A 1-pound cake of yeast is crumbled and added to 500 ml of toluene at 70–80°. The mixture is stirred well and adjusted to 38–40° by placing it in a water bath. Within 20 minutes the yeast should liquefy; at this point, the excess toluene, which contains much lipid, can be poured off to reduce the volume for subsequent centrifugation. The temperature of the mixture is held at

[10] E. W. Westhead and B. G. Malmström, *J. Biol. Chem.* **228,** 655 (1957).
[11] B. von Hofsten, *Physiol. Plantarum* **14,** 177 (1961).
[12] E. W. Westhead and H. Boerstling, unpublished data.

38–40° for another 2 hours, during which time the mixture should be stirred frequently. The container is then placed in ice, or in a cold-room, and 500 ml of cold distilled water is added. The mixture is stirred for at least ½ hour and centrifuged cold at not less than 800 g for 20 minutes. The aqueous layer between the lipid and the sediment contains the enzyme; it is removed by siphoning, or if the sediment is firmly compacted, by pouring it out from under the jellied lipid layer. The carryover of some sediment or lipid can be ignored. At least 500 ml of extract should be obtained, even from the rather dry bakers' yeast.

Step 2. Acetone Fractionation. A volume of reagent grade acetone equal to 0.5 volume of extract is measured at room temperature and cooled to below 0°; at the same time; an equal volume of acetone is cooled for the second precipitation. The extract is stirred vigorously while the first volume of acetone is poured in. The resulting precipitate settles well at 1000 g in 15 minutes. The supernatant is saved and the second volume of prechilled acetone is added to it as before. The suspension is centrifuged as above, and this time the precipitate is saved. It is dissolved immediately in 100–200 ml of ice-cold water. If necessary, the preparation may be held overnight at this stage without loss of activity but with slight formation of multiple components.

Step 3. Ethanol Fractionation. The pH of the above solution is adjusted to pH 4.8 with 2 N acetic acid, and the volume is measured. Two quantities of 95% ethanol are now measured at room temperature, then chilled to well below 0°. The volume of ethanol for the first precipitation is equal to 0.54 times the volume of the pH 4.8 solution; for the second precipitation it is equal to 0.73 times the volume of the pH 4.8 solution. The smaller volume of ethanol is added to the stirred aqueous solution slowly enough to keep the temperature below 10°. The precipitate is removed by centrifugation (1000 g, 10 minutes) and discarded. The second volume of ethanol is then added in the same way as the first. This precipitate, which settles very rapidly upon centrifugation, contains the enzyme. After the supernatant is drained off, the precipitate is dissolved in 100 ml of cold Tris-acetate buffer, 10^{-4} M in acetate, pH 8.4, which also contains 10^{-4} M magnesium acetate. The pH is adjusted to 8.4 with 2 N NH_3 solution, and the solution is centrifuged for 10 minutes at 8000 g to clarify it. At this point, the solution should have an absorbancy at 280 mμ of 10–20, with a specific activity of 160–180. This crude enzyme may be stored overnight in the presence of toluene without harm, or it may be freeze-dried and stored indefinitely in a refrigerator.

Step 4. Removal of Inactive Protein with DEAE-Cellulose. The next two steps are routinely carried out near 0° in ice buckets. Cellulose ion-exchangers are first cleaned in the usual way, by washing alternately with

0.5 N HCl and 0.5 N NaOH until the wash solutions are colorless. Following the final wash in NaOH, they are washed with 0.1 N sodium acetate, and then suspended in that solution. The DEAE-cellulose suspension is adjusted to pH 8.3 $\pm$ 0.1 with acetic acid and drained by vacuum filtration. After being rinsed with distilled water, it may be dried or stored cold in a tightly closed jar containing toluene. Phosphorylated cellulose is treated and stored similarly except that the sodium acetate suspension is adjusted to pH 6.0 before the water wash.

The clear enzyme solution from the previous step is diluted to an absorbancy at 280 mμ of 10–12. At the same time, it is made 1×10^{-4} M in magnesium acetate and 1.0 to 1.2×10^{-4} M in EDTA; this is important. DEAE-cellulose (dry or damp) is added in small portions (0.2 g dry) to the stirred solution. Throughout the addition, pH is monitored and held at 8.3 with 2 N ammonia or acetic acid. After each addition a small cone of filter paper is put part way under the surface of the liquid, and in a few seconds enough filtrate is obtained to allow an aliquot (20 μl) to be taken for concentration determination. The aliquot is diluted in a suitable volume of water in a silica cuvette, and the absorbancy is read at 280 mμ. After the protein concentration has been reduced by 30% it is best to assay each aliquot. For this purpose dilution should be made in assay buffer (not containing substrate) to eliminate any lag period in the assay due to contamination with inhibitory metals. The addition of DEAE-cellulose should be continued until the specific activity reaches 320. It may exceed this value at times, but there is increasing risk of adsorbing enolase. Should this happen, the cellulose should be removed and added to a subsequent batch of crude enolase whereupon the enolase will be displaced by the more strongly bound proteins. If an excessive amount of DEAE-cellulose is required for this step (the slurry should not become thick), the ionic strength is probably too high and a 2:1 dilution with 10^{-4} M EDTA may help. The recovery of activity in this step should be nearly 100%.

Step 5. Adsorption of Enolase on Phosphorylated Cellulose. After removal of the DEAE-cellulose by gentle suction filtration, the solution is adjusted to pH 6.0 with 2 N acetic acid. Portions of phosphorylated cellulose are added until a 5-μl aliquot of the undiluted supernatant causes an initial rate of only 0.2–0.4 per minute in 1 ml of the standard assay solution. After each addition of phosphorylated cellulose, the pH must be readjusted to 6.0. When the enolase has been adsorbed, the slurry is filtered with strong suction and the cellulose is pressed to remove as much solution as possible. The filter cake is then resuspended in 10^{-4} M EDTA solution, *and the pH readjusted to 6.0* before filtering again. Enolase will be lost above or below pH 6. The cellulose may be washed with water to remove EDTA, but it is important to check the pH of the suspension or to assay the filtrate before discarding it.

Step 6. Removal of Enolase from Phosphorylated Cellulose. The pressed-dry filter cake is suspended in a fairly small volume of water, and $2\,M$ ammonia solution is added to raise the pH of the slurry to pH 8.6. After the mixture has been stirred for 10 minutes, the pH is checked and the mixture is filtered with gentle suction to avoid foaming. The filter cake is washed well with $10^{-4}\,M$ ammonia, and the combined eluates are assayed. With this procedure, the activity should be between 600 and 620. If it is not, incubation for an hour in $10^{-3}\,M$ magnesium acetate, at pH 8.6 and 30° may raise the activity to this range. Unless metal-free enzyme is needed, the solution should be made $10^{-4}\,M$ in magnesium ion before storage. For maximum stability the pH of the enzyme solutions must be between pH 8.0 and 8.6. The enzyme may be freeze-dried very successfully if it is frozen rapidly and if it is subsequently dissolved in ice-cold water or buffer.

This method has been used successfully to prepare enzyme from 50 g and from 5 kg of yeast. If centrifuge capacity is limited, part of the material can be kept up to twice as long in either the toluene autolysis or the water extraction stages without deleterious effects. Multiple enzymatic components that arise during storage of the yeast or during prolonged preparations of the enzyme can be tested for and isolated by chromatography on TEAE-cellulose. The enzyme is pure by the criteria of ultracentrifugation, gel-electrophoresis, and specific activity. It may be recrystallized readily as the inactive mercury salt by the procedure of Warburg and Christian,[3] but no increase in purity is achieved, and the specific activity is usually lowered.

Preparation of Rabbit Muscle Enolase

This is the procedure of Winstead and Wold[13] which differs in details from the original preparation.[5] The crystallization step is a modification, suggested to us by Dr. Wold, which we have found valuable.

Step 1. Preparation of the Aqueous Extract. Fresh rabbit muscle, or partly thawed frozen muscle,[14] is put through a meat grinder, then homogenized in a high speed blender with twice its weight of $10^{-3}\,M$ EDTA, pH 7. The homogenate is stirred for an hour at low temperature, then centrifuged at 8000 g for 15 minutes. The supernatant is decanted through gauze or glass wool to remove fat. At this point, the volume of extract should equal the volume of EDTA solution added.

[13] J. A. Winstead and F. Wold, *in* "Biochemical Preparations" (A. Mahley, ed.), Vol. 12. Wiley, New York, 1965.

[14] Available in U.S.A. from Pel-Freeze Biologicals, Rogers, Arkansas.

Step 2. Acetone Fractionation. For each liter of extract, 440 ml of reagent grade acetone at $-10°$ is measured and stirred rapidly into the extract. The mixture is centrifuged at $-5°$ for 10 minutes at 8000 g, and the precipitate is discarded. An additional 460 ml of acetone per liter of original extract is added to the supernatant. The suspensions is centrifuged as above, for 20 minutes, and the supernatant is thrown away. After the precipitate has been drained to remove solvent it is dissolved in a small volume of cold 0.05 M imidazole HCl buffer containing 10^{-3} M $MgSO_4$ at pH 7.8. The acetone fractionation must be carried out quickly to prevent loss of activity, and with large volumes it has been found advantageous to work with half the total extract at a time.

Step 3. Heat Treatment. The concentration of protein and $MgSO_4$ is adjusted with the above imidazole buffer and a concentrated $MgSO_4$ solution to give an absorbancy at 280 mμ of 20 and a concentration of $MgSO_4 \cdot 7\ H_2O$ of 5.0 g/100 ml. This solution must be incubated at 37° for 5–10 hours to remove all traces of acetone before the heat step. Unless all acetone is removed, complete loss of activity may occur during the heat step or later during the ammonium sulfate fractionation. When no acetone odor remains, the solution is brought rapidly to 55° by constant stirring in a 70° water bath. The solution is kept at exactly 55° for 5 minutes, then cooled in an ice bath to below 10°. It is then centrifuged at 8000 g for 20 minutes, and the active supernatant is poured through glass wool to remove floating material.

Step 4. Ammonium Sulfate Fractionation. The volume of the supernatant from the last step is measured and a quantity of reagent grade ammonium sulfate equal to 242 g per liter of supernatant is weighed out. This is then added to the solution at 0° with efficient stirring, slowly enough to prevent accumulation of solid ammonium sulfate on the bottom of the container. After removal of the inactive precipitate by centrifugation at 8000 g, 0° for 20 minutes, the volume of supernatant is measured. A further addition of 280 g of solid ammonium sulfate per liter of this supernatant is made as before. The precipitate is collected by centrifugation and saved for the next step.

Step 5. Crystallization. A solution is prepared by dissolving 475 g of analytical grade ammonium sulfate, plus a few crystals of $MgSO_4$, in a liter of 10^{-4} M EDTA. By the addition of concentrated ammonia solution, the pH is brought to 7.8 at room temperature. A series of about 7 dilutions of this solution is made, each one containing 2.5% more water than the previous one, and they are chilled to 0° before use.

The precipitate from the second ammonium sulfate addition (step 4) is suspended in the most concentrated of the above solutions (use about 100 ml of solution per kilogram of muscle). The suspension is stirred very thoroughly for about 10 minutes and then centrifuged at 10,000 *g* for 20 minutes. The supernatant is decanted into a clean container that can be stoppered tightly, and the precipitate is resuspended in a similar volume of the next most concentrated ammonium sulfate solution. It is stirred and centrifuged as before. This process is continued for each concentration of ammonium sulfate and then to each container is added 0.01 volume of saturated ammonium sulfate solution. The containers are closed tightly and allowed to stand at 0°. Within minutes or hours crystals begin to form in two or three of the middle fractions, and overnight excellent yields of crystalline enolase are obtained. The solutions are centrifuged to separate out the crystals which are then dissolved in small volumes of cold 10^{-3} *M* EDTA solution, pH 7.8. The activities are measured and all solutions with specific activities over 200 at 25° are combined for recrystallization.

Step 6. Recrystallization. Saturated ammonium sulfate at pH 7.8 containing 1 g of magnesium sulfate per liter is added at 0° to the above enzyme solution until the solution is very slightly turbid. The solution is centrifuged immediately at 8000 *g* for 10 minutes, and the supernatant is poured off. If distinct crystal formation does not begin within 30 minutes after centrifugation, saturated ammonium sulfate should be added dropwise until turbidity again occurs. Crystallization should be essentially complete overnight, and the yield of once-recrystallized enzyme should be over half a gram per kilo of rabbit muscle. The crystals can be stored near 0° as a suspension in 70% saturated ammonium sulfate containing 10^{-3} *M* EDTA for several months without noticeable loss of activity.

Preparations of this enzyme based solely on ammonium sulfate fractionations have been described in detail by Malmström[15] and in outline by Czok and Bücher.[16] The enzyme is homogeneous by the criteria of ultracentrifugation and electrophoresis.[5,15]

Molecular Forms of Enolase

At present, there appear to have been at least three types of enolase discovered. The yeast enzyme alone is devoid of sulfhydryl and disulfide groups.[17] It is a single polypeptide chain of 67,000 molecular weight[17] and shows maximum catalytic activity at pH 7.8.[5] The Michaelis constant is 2×10^{-4} with all activating metal ions.

The rabbit muscle enzyme has two (dissociable) peptide chains with a combined molecular weight of about 82,000.[5,15] It has 12 cysteine residues, no cystine,[15] and is maximally active at pH 6.7.[5] Purified

[15] B. G. Malmström, *Arch. Biochem. Biophys.*, Suppl. 1, 247 (1962).

[16] R. Czok and T. Bücher, *Adv. Protein Chem.* **15**, 315 (1960).

[17] B. G. Malmström, J. R. Kimmel, and E. L. Smith, *J. Biol. Chem.* **234**, 1108 (1959).

enolase from ox brain[18] resembles the rabbit muscle enzyme in its molecular properties and its activation by metal ions. It has the same activation energy as the yeast enzyme, but this property has not been measured for the muscle enzyme. Partially purified enolase from rabbit brain[12] has a pH dependence of activity identical to that from muscle and otherwise behaves similarly. However, a significant part (5–7%) of the rabbit brain enzyme is firmly bound to the mitochondrial fraction and is released only upon disruption of the mitochondria by mechanical or osmotic force. Partially purified enolases from fish muscle also appear to resemble the rabbit enzyme, but they show species specific patterns of electrophoretic isozymes.[19] Both fish muscle enolase and chicken muscle enolase have been crystallized in Wold's laboratory and have been found very similar to the rabbit enzyme. Details of these preparations have not yet been published.

The third type of enolase, from potato,[20] has been purified but not studied extensively. It contains sulfhydryl groups and is reported to be very unstable in solution. It differs from the second class of enolases in being inhibited by very low concentrations of *p*-mercuribenzoate. A partially purified enolase from the snail, *Busycon,* is also inhibited by concentrations of *p*-mercuribenzoate or 5,5′-dithio-bis(2-nitrobenzoate) in the range of 10^{-7} to 10^{-6} *M*.[21]

All enolases studied, including that from *E. coli,*[11] require divalent metals for activity, Mg^{++} giving the highest rates. No other naturally occurring substrates have been found for the enzyme and the structural requirements for a substrate or inhibitor are stringent.[22]

The most recent review of work on this enzyme is Malmström's.[23] Significant publications since this review, that are otherwise not cited in this chapter are on the inactivation of yeast enolase by bromoacetate[24]; on the reversible denaturation of the yeast enzyme[25-27]; and on the metal ion interactions with the yeast enzyme.[28-30]

[18] T. Wood, *Biochem. J.* **91,** 453 (1964).
[19] H. Tsuyuki and F. Wold, *Science* **146,** 535 (1964).
[20] H. Boser, *Z. Physiol. Chem.* **315,** 163 (1959).
[21] D. P. Hanlon and E. W. Westhead, unpublished observations.
[22] F. Wold and C. E. Ballou, *J. Biol. Chem.* **227,** 313 (1957).
[23] B. G. Malmström, *in* "The Enzymes" (P. D. Boyer, H. Lardy, and K. Myrbäck, eds.), 2nd ed., Vol. 5, p. 471. Academic Press, New York, 1961.
[24] J. M. Brake and F. Wold, *Biochemistry* **1,** 386 (1962).
[25] W. C. Deal, W. J. Rutter, V. Massey, and K. E. Van Holde, *Biochem. Biophys. Res. Commun.* **10,** 49 (1963).
[26] A. Rosenberg and R. Lumry, *Biochemistry* **3,** 1055 (1964).
[27] E. W. Westhead, *Biochemistry* **3,** 1062 (1964).
[28] M. Cohn, *Biochemistry* **2,** 623 (1963).
[29] D. P. Hanlon and E. W. Westhead, *Biochim. Biophys. Acta* **96,** 537 (1965).
[30] J. M. Brewer and G. Weber, *Federation Proc.* **24,** 285 (1965).

[121] L-(+)-Tartaric Acid Dehydrase

By RONALD E. HURLBERT and WILLIAM B. JAKOBY

$$\begin{array}{c} COOH \\ | \\ H-C-OH \\ | \\ HO-C-H \\ | \\ COOH \end{array} \rightarrow \begin{array}{c} COOH \\ | \\ C=O \\ | \\ CH_2 \\ | \\ COOH \end{array} + H_2O$$

Active L-(+)-tartaric acid dehydrase is composed of four monomeric protein units which are reversibly associated.[1,2] Although the enzyme has not been significantly purified, empirical findings concerning the association phenomenon are included.

Assay

Principle. The reaction is followed by treatment of the ketoacid formed with 2,4-dinitrophenylhydrazine and measurement at 540 mμ of the chromogen resulting from the addition of base. Although the product is a β-ketoacid and thus subject to rapid decarboxylation, both oxalacetate and pyruvate yield a chromogen with approximately the same extinction coefficient by the method described.

Reagents

Tris-chloride, *M*, at pH 8.5
Potassium L-(+)-tartrate, 0.2 *M*
2,4-Dinitrophenylhydrazine, 1% in 2 *N* HCl
NaOH, 2.5 *N*

Procedure. The following reagents are added in the indicated order with the total volume brought to 0.5 ml with water: Tris-chloride, 100 μl; potassium L-(+)-tartrate, 100 μl; appropriate quantities of enzyme. Prior to the addition of enzyme, each tube (1 × 10 cm) is flushed with helium for 4 minutes. The stoppered tubes are incubated for 5 minutes at 25° and the reaction is terminated with 0.5 ml of 2,4-dinitrophenylhydrazine. Hydrazone formation is maximal after 20 minutes, at which time 4 ml of 2.5 *N* sodium hydroxide are added and the color determined in a Klett-Summerson colorimeter with a No. 54 filter. The rate of enzymatic activity is directly proportional to time and to protein concentration under conditions in which less than 0.4 μmole of product is formed.

[1] R. E. Hurlbert and W. B. Jakoby, *Biochim. Biophys. Acta* **92**, 202 (1964).
[2] R. E. Hurlbert and W. B. Jakoby, *J. Biol. Chem.* **240**, 2772 (1965).

Definition of Unit. A unit of activity is defined as that quantity of enzyme resulting in the formation of 1 micromole of keto acid in 5 minutes. Specific activity is expressed as the number of units per milligram of protein. Protein was determined by a biuret method[3] with crystalline bovine serum albumin as a standard.

Purification Procedure

Growth. The organism[4] is a strain of *Pseudomonas putida* isolated by the enrichment culture technique with L-(+)-tartrate as sole carbon source. Growth is allowed by the following medium[5]: Potassium L-(+)-tartrate, 3 g; NH_4Cl, 1 g; $MgSO \cdot 7\ H_2O$, 0.05 g; $CaCl_2 \cdot 2\ H_2O$, 0.05 g; $FeCl_3$, 0.05 g; Difco yeast extract, 0.025 g; in 1 liter of 0.01 *M* sodium phosphate buffer at pH 7.2. Large-scale cultures are produced in the same medium with aeration at room temperature in 5-gallon carboys. After 12–16 hours of growth, cells are harvested at 3° with a Sharples centrifuge and washed with water. The frozen cell paste, stored at −85° for one year, does not lose tartaric acid dehydrase activity.

Preparation of Active Enzyme. Step 1. Cell paste is suspended in 3 times its volume of 0.01 *M* Tris-chloride, pH 7.0, containing 5 m*M* glutathione and 5 m*M* cysteine. The extract, obtained by passing this suspension through a cooled French pressure cell (Aminco), is centrifuged at 2° for 30 minutes at 16,000 *g*, and the residue is discarded (specific activity: 8.8).

Step 2. All further procedures are conducted at 25° unless otherwise noted. The extract is diluted with an equal volume of 0.01 *M* Tris-chloride, pH 7.0, containing 50 m*M* glutathione and 50 m*M* cysteine. Ammonium sulfate and RNase are added with stirring to a concentration of 7 g/100 ml and 2 mg/100 ml, respectively; the resulting solution is incubated under anaerobic conditions for 60 minutes; the presence of ammonium sulfate appears to stimulate RNase action as determined by the absorbance of the acid-soluble fraction at 260 mμ. After this incubation period, additional ammonium sulfate (24.5 g/100 ml) is added while a stream of helium is blown over the surface of the mixing vessel in order to minimize contact with air. After 15 minutes the precipitate is removed by centrifugation at 0° and discarded. The material precipitating upon the further addition of 14 g of the salt per 100 ml, the 45–65% ammonium sulfate fraction, is collected by centrifugation and dissolved in 0.01 *M* Tris-chloride, pH 7.0, containing 50 m*M* glutathione and 50 m*M* cysteine (specific activity: 24).

On several occasions, a further three-fold increase in specific activity

[3] T. E. Weichselbaum, *Am. J. Clin. Pathol.* **10**, 40 (1946).
[4] Available as No. 17642 from the American Type Culture Collection.
[5] M. Shilo, *J. Gen. Microbiol.* **16**, 472 (1957).

was obtained upon elution from DEAE-cellulose columns, but with yields of less than 10%. Concentration and recombination of various fractions did not account for lost activity.

Preparation of "Inactive Enzyme." The enzyme is inactivated by incubation of the ammonium sulfate fraction for 15 minutes in 0.02 M EDTA at 0° with stirring in an open beaker. The preparation is then passed through a Sephadex G-50 column (2 × 20 cm) previously equilibrated with 0.01 M Tris-chloride at pH 7.0, and eluted with the same solution.

Properties

The enzyme is specific for the dehydration of L-(+)-tartaric acid. Two distinct K_m values for L-(+)-tartrate are obtained, 8×10^{-4} M and 2×10^{-3} M, when special precautions are taken to eliminate air. One K_m, 2×10^{-3} M is found in the presence of air. Neither D-(−)-tartrate nor mesotartrate serve as substrate although the latter is a competitive inhibitor. Under anaerobic conditions two K_i values, 2×10^{-3} M and 8×10^{-3} M, are found for mesotartrate. In the presence of air the K_i is 8×10^{-3} M.

Optimum enzyme activity is attained at pH 8.5, and half-maximal activity at pH 7.6 and 9.4. The ammonium sulfate fraction, when stored under reduced pressure at −20° and pH 7.0, is stable for several months and can be repeatedly thawed and frozen without significant loss in activity, provided that care is taken to minimize contact with air.

Activation.[2] Inactive enzyme, prepared as described, may be reactivated by the addition of either GSH or cysteine plus ferrous ions; reactivation is dependent on time, temperature, and protein concentration. The requirement for reduced glutathione or cysteine cannot be met by a variety of other mercaptans, reducing agents, and cofactors. The requirement for iron is also specific.

The conversion of inactive to active protein is a linear function of the 4th power of the protein concentration, which has been interpreted as due to the extreme instability of all aggregates of monomers (S_{20} = 3.0 S; mol. wt. = 39,000) except that which includes 4 monomers, i.e., the active enzyme (S_{20} = 6.9 S; mol. wt. = 145,000). Optimal conditions for activation include incubation at pH 7.0 for 90 minutes at 25° in a helium atmosphere and in a medium containing the following: ferrous sulfate, 1 mM; glutathione, 50 mM; cysteine, 50 mM. Activity is tested by addition of aliquots of the products of such an incubation mixture to the assay system outlined above for L-(+)-tartaric dehydrase activity.

[122] Lactyl-CoA Dehydrase

By R. L. Baldwin and W. A. Wood

Acrylyl-CoA + $H_2O \rightleftharpoons$ lactyl-CoA

Assay Method[1]

Principle. The enzyme can be assayed spectrophotometrically with acrylyl-CoA as substrate. The rate of lactyl-CoA formation is determined continuously by conversion of this product to free lactate in the presence of acetate and CoA transphorase. The lactate is then oxidized with a muscle lactic dehydrogenase, NAD, diaphorase dye system. The rate of reduction of 2-(*p*-iodophenyl)-3-(*p*-nitrophenyl) tetrazolium chloride (INT) is determined at 500 mμ.

Reagents

- Tris-acetate buffer, 0.2 *M*, pH 8.0
- INT, 0.2% solution
- NAD, 0.15 *M*
- Acrylyl-CoA,[2] 0.005 *M*
- Muscle lactic dehydrogenase
- Coenzyme A transphorase[1]
- Diaphorase

Procedure. The assay mixture contains Tris-acetate buffer, pH 8.0, 20 micromoles; INT, 20 μg; NAD, 1.5 micromoles; acrylyl-CoA, 0.1 micromole; muscle lactic dehydrogenase, 0.5 unit; diaphorase, 1.0 unit; CoA transphorase, 1.0 unit, and lactyl-CoA dehydrase in a final volume of 0.30 ml. Either lactyl-CoA dehydrase or acrylyl-CoA can be added last. Often in crude enzyme preparations, it is desirable to preincubate the assay mixture without acrylyl-CoA for several minutes until the endogenous rate of dye reduction decreases. Coenzyme A transphorase and diaphorase from *Peptostreptococcus elsdenii* are fairly heat stable whereas lactyl-CoA dehydrase is not; hence, heated (55° for 10 minutes) crude extracts of *P. elsdenii* can replace purified CoA transphorase and diaphorase in the assay system. The reaction proceeds linearly with time for 4–6 minutes and is linear with enzyme concentration when the reaction rate is below 0.025 absorbancy unit per minute. Crude extracts contain very high levels of crotonase and β-hydroxybutyryl-CoA dehy-

[1] R. L. Baldwin, W. A. Wood, and R. S. Emery, *Biochim. Biophys. Acta* **97**, 202 (1965).

[2] See Vol. III, p. 931.

drogenase, which can reduce INT in the presence of acrylyl-CoA. Hence, in crude extracts it is essential that only the activity requiring lactic dehydrogenase for expression is considered.

Definition of Unit and Specific Activity. One unit of enzyme will catalyze the formation of 1.0 μmole of lactyl-CoA per minute at 25°, and specific activity is expressed as units per milligram of protein. The millimolar extinction coefficient for INT was determined to be 20.4×10^3. This value is somewhat higher than reported by Hirsch *et al.*[3] but similar to that reported by Pennington[4] for INT in a nonaqueous system.

Purification Procedure

Preparation of Crude Extract. P. elsdenii, strain B 159,[5] was grown on 1 liter of medium containing 1% corn steep liquor, 1.4% sodium DL-lactate, and 0.02% mercaptoethanol (see Walker[6] for details of the procedure). After growth at 38°, the cells were collected by centrifugation, washed twice with cold 0.02 *M* potassium phosphate buffer (pH 7.5), and disrupted by sonic oscillation. The suspension of disrupted cells was centrifuged at 20,000 *g* for 30 minutes to remove cellular debris. The crude extracts thus obtained were stored at —15°C.

Nucleic Acid Precipitation. The protein concentration of the crude extract was adjusted to approximately 15 mg/ml with 0.05 *M* potassium phosphate buffer (pH 7.5), and solid ammonium sulfate was added to a final concentration of 0.20 *M*. The nucleic acids were then precipitated by the dropwide addition of 0.20 volume of 1.0% protamine. The enzyme is easily precipitated with protamine and, hence, this step must be run carefully.

First Ammonium Sulfate Fractionation. Solid ammonium sulfate was added to the supernatant from the protamine step to 50% saturation, centrifuged, and the precipitate discarded. The supernatant was adjusted to 65% saturation with solid ammonium sulfate, centrifuged, and the precipitate dissolved and adjusted to twice the original volume with water.

Fractionation with Calcium Phosphate Gel. Calcium phosphate gel was added in a stepwise fashion until about 80% of the enzyme was adsorbed. The gel was then washed with 0.07 *M* phosphate buffer, pH 7.5, and then the enzyme was eluted from the gel by washing twice with

[3] C. A. Hirsch, M. Rasminsky, B. D. Davis, and E. C. C. Lin, *J. Biol. Chem.* **238,** 3770 (1963).
[4] R. J. Pennington, *Biochem. J.* **80,** 649 (1961).
[5] Obtained from Dr. Marvin Bryant, Department of Dairy Science, University of Illinois, Urbana, Illinois.
[6] D. J. Walker, *Biochem. J.* **69,** 524 (1958).

volumes of 0.12 *M* phosphate buffer, pH 7.5, equal to about twice the volume of the gel.

Second Ammonium Sulfate Fractionation. The eluates from the calcium phosphate gel were combined and solid ammonium sulfate added to a final concentration of 55% saturation. The precipitate was discarded and the ammonium sulfate concentration adjusted to 70% saturation. The precipitate was collected and resuspended in 0.5 *M* phosphate buffer, pH 7.5.

A summary of the procedure is shown in the accompanying table.

PURIFICATION PROCEDURE

Fraction	Volume (ml)	Total activity (units)	Yield (%)	Protein (mg)	Specific activity (units/mg protein × 10^2)
Crude extract	15	2.76		480	0.58
Protamine	35	2.58	93	420	0.61
First ammonium sulfate	60	2.20	80	126	1.75
Calcium phosphate gel	17	1.50	54	37	4.05
Second ammonium sulfate	10	1.10	40	16	6.90

Properties

The enzyme has not been purified sufficiently to study in detail the kinetics of the reaction and the effects of inhibitors and activators. The pH optimum for the reaction appears to be between 7.5 and 8.0. The activity observed with crotonyl-CoA as substrate is 75% that observed with acrylyl-CoA. EDTA does not inhibit, and sometimes stimulates, activity. The enzyme is relatively stable in crude extracts stored at —15°. However, it becomes very unstable during purification. For example, approximately 10% of the enzyme activity is lost per hour when the enzyme is allowed to stand at 0° after the first ammonium sulfate step.

[123] Dioldehydrase[1]

By ROBERT H. ABELES

$$CH_3—CHOH—CH_2OH \xrightarrow{DBCC^2} CH_3—CH_2CHO$$
$$CH_2OH—CH_2OH \xrightarrow{DBCC} CH_3—CHO$$

Assay of Dioldehydrase

Principle. The assay is based upon the colorimetric determination of propionaldehyde.[3] The assay is applicable to crude systems provided that no other enzymes are present which convert propanediol to a carbonyl compound. Interference by such reactions can frequently be eliminated by treating crude extracts with charcoal prior to assay.

Reagents

K_2HPO_4–KH_2PO_4, 0.2 *M*, pH 8.0
DBCC, 100 μg/ml
HCl, 2 *N*
2,4-Dinitrophenylhydrazine reagent: 25 mg 2,4-DNPH, 0.2 ml conc. HCl in 25 ml absolute methanol (aldehyde-free)
80% pyridine, 20% H_2O
Alcoholic KOH: 80 ml 40% KOH, 320 ml methanol

Enzyme. The enzyme is diluted in 0.01 *M* K_2HPO_4 containing 2% of 1,2-propanediol. The optimal concentration range is 0.05–0.50 units/ml. One unit is the amount of enzyme which produces 1 micromole of propionaldehyde per minute under the assay conditions.

Procedure. The assay mixture consists of 0.2 ml buffer, 0.2 ml DBCC, 0.2 ml enzyme, and 0.4 ml H_2O. The reaction is started by the addition of DBCC and is carried out at 37° for 10 minutes. The reaction is stopped by the addition of 0.1 ml 2 *N* HCl. The blank is of the same composition as the assay mixture, except that 2 *N* HCl is added prior to the addition of DBCC.

After the addition of 2 *N* HCl, 1.0 ml of the 2,4-DNPH reagent is added. The reaction mixture is thoroughly mixed and kept at room temperature for 20 minutes; 5.0 ml of 80% pyridine is then added fol-

[1] R. H. Abeles and H. A. Lee, Jr., *J. Biol. Chem.* **236,** 2347 (1962); A. M. Brownstein and R. H. Abeles, *ibid.* **236,** 1199 (1961); and also H. A. Lee, Jr. and R. H. Abeles, *ibid.* **238,** 2367 (1963).
[2] DBCC = dimethylbenzimidazoyl-cobamide coenzyme.
[3] H. Bohme and O. Z. Winkler, *Anal. Chem.* **142,** 1 (1954).

lowed by 1.0 ml of alcoholic KOH. After mixing, the assay mixture is allowed to stand for 5 minutes and clarified by centrifugation in an International Clinical Centrifuge. It is convenient to carry out the assay in heavy-wall Pyrex tubes which can be centrifuged. After centrifugation the color intensity is read in Klett colorimeter with a No. 54 filter (540 mμ). The blank is subtracted from the reading. Under these conditions, 1 micromole of propionaldehyde gives a Klett reading of 710. The assay was linear with time and proportional to the amount of dehydrase.

Assay of DBCC

Principle. Under the conditions of the assay, the amount of propionaldehyde produced is proportional to the amount of DBCC added. Since the assay is carried out with an excess of enzyme, hydroxo- and cyanocobalamin at concentrations up to twice that of the coenzyme do not interfere with the coenzyme assay. The useful range of the assay is from 0.001 μg to 0.012 μg DBCC per assay mixture. The system is specific for derivatives of vitamin B_{12} which function as coenzymes of dioldehydrase.

Reagents

Enzyme: 10 units/ml in 0.01 *M* K_2HPO_4 containing 2% 1,2-propanediol
DBCC standard 0.004–0.008 μg/ml
All other reagents, as in enzyme assay.

Procedure. The assay mixture consists of 0.2 ml K_2HPO_4-KH_2PO_4 buffer, 0.2 ml of enzyme, DBCC solution to be assayed and H_2O to give a final volume of 1.0 ml. The assay is carried out at 37° and is allowed to proceed for 30 minutes. The reaction is stopped by the addition of 0.1 ml of 2 *N* HCl. The assay is then completed as in the enzyme assay.

Purification Procedure

Reagents

Tris (free base), 0.01 *M*
Solution A: 0.01 *M* K_2HPO_4–0.05 *M* $(NH_4)_2SO_4$–2% propanediol adjusted to pH 8.0 with KOH
Solution B: Same as solution A except for the addition of 0.50 *M* $(NH_4)_2SO_4$
Solution C: 0.01 *M* K_2HPO_4–2% 1,2-propanediol

Preparation of Cell-Free Extract. Aerobacter aerogenes ATCC 8724 are grown in 20 L carboys without aeration at 35° for 15–18 hours. The medium consists of: K_2HPO_4 80 g; KH_2PO_4, 20 g; yeast extract (Amber Laboratories, Milwaukee, Wisconsin), 20 g; $(NH_4)_2SO_4$, 24 g; $MgSO_4 \cdot 7 H_2O$ 4 g; glycerol, 300 g; 1,2-propanediol, 50 ml; distilled water to 20 liters.[4] After harvesting, cells from 100 liters of medium are suspended in 500–600 ml 0.01 *M* Tris (free base) and centrifuged. The resulting cell paste is suspended in distilled water and lyophilized. Lyophilized cells, when stored at —10°, maintain full enzyme activity for at least 2 months.

To prepare a cell-free extract, 30 g of lyophilized cells is added to 300 ml of water and blended for 20 seconds in a Waring blendor. The suspension is then cooled to 5°, and the pH is adjusted to 8.7 with 40% KOH and stirred for 10–15 minutes. The pH is readjusted during this time, if necessary. The suspension is then sonicated in a 500-watt ultrasonic disintegrator [Measuring and Scientific Equipment Ltd. (MSE)]. Sonication is frequently interrupted to prevent the temperature from rising above 15°. After sonication, 30 g Darco G-60 charcoal suspended in 115 ml H_2O and 8 ml 1,2-propanediol are added. The mixture is stirred for 15 minutes and then centrifuged at 20,000 *g* for 90 minutes. The supernatant is lyophilized. The lyophilized material is stable, when stored at —15° for at least one month.

Protein Fractionation. All steps are carried out at 5–10°. Eight grams of lyophilized cell extract is suspended in 240 ml of solution A. The supernatant is lyophilized. The lyophilized material is stable, when solution (fraction E-1), 106 ml of a 2% protamine sulfate (Sigma Chemical Company or California Corporation for Biochemical Research) dissolved in solution A is added. If necessary, the protamine solution is adjusted to pH 8.0 with KOH. After stirring for 15 minutes, the suspension is centrifuged at 20,000 *g* for 75 minutes, and the residue is discarded. The supernatant is adjusted to pH 8.7 with KOH, and $(NH_4)_2SO_4$ is added slowly with continuous stirring. Three grams of $(NH_4)_2SO_4$ are added per 10 ml of fraction E-2. Throughout the $(NH_4)_2SO_4$ addition, the pH is not allowed to drop below 7.9; if necessary, the pH is raised by addition of KOH. After standing for 30 minutes, the mixture is centrifuged for 60 minutes at 20,000 *g*. The sediment is suspended in 12 ml of solution A. The suspension (fraction E-3) is stirred for 20 minutes and then centrifuged for 2 hours at 30,000 g. The supernatant fluid (fraction E-4) contains appreciable activity but is not

[4] Not all batches of yeast extract are suitable. Occasionally, although the crude extract contains dioldehydrase activity, the enzyme cannot be purified by the procedure described here. This difficulty is due to the yeast extract. We have, however, been unable to establish what component of the yeast extract is responsible for this difficulty. The suitability of a given batch of yeast extract is determined by preliminary trials.

further purified. This fraction is extremely stable and suitable for coenzyme assays after dialysis. The residue is suspended in 8 ml of solution B. After a uniform suspension is obtained, it is centrifuged for 2 hours at 30,000 *g*. The supernatant fluid (fraction E-5) is discarded, and the residue resuspended in 4 ml of solution C and dialyzed overnight against 2 liters of solution C. The dialyzed suspension is centrifuged for 2 hours at 30,000 *g*, and the residue is discarded; 0.2 ml of a solution containing 400 mg K_2HPO_4/ml is added slowly with stirring to the supernatant fluid (fraction E-6). After 1 hour, the suspension is centrifuged for 15 minutes at 10,000 *g*. The supernatant fluid (fraction E-7) is discarded, and the residue is dissolved in 1 ml of 0.003 *M* K_2HPO_4–0.10 *M* propanediol and allowed to stand for 24–36 hours before it is assayed. The solution (E-8) maintains full activity for several weeks when stored at 5–7°. Freezing leads to loss of activity.

PURIFICATION OF DIOLDEHYDRASE

Fraction	Total protein (mg)	Specific activity (units)	Yield (%)
Crude extract (E-1)	3620	0.24	100
Protamine supernatant (E-2)	2460	0.27	77
Suspended ammonium sulfate precipitate (E-3)	1100	0.57	72
Supernatant (E-4)	990	0.18	20
Wash (E-5)	51	0.41	
Dialyzate (E-6)	19	13	28
K_2HPO_4 supernatant (E-7)	14	3.9	
K_2HPO_4 precipitate (E-8)	4.5	47	24

Properties

Adenyl and benzimidazoyl cobamide coenzymes, at saturation, are as active as DBCC in this system. In addition to a cobamide coenzyme, a monovalent metal ion such as K^+, NH_4^+, or Ti^+ is required for full activity. Dioldehydrase is specific for 1,2-propanediol and ethylene glycol. Butylene glycol, glycerol, ribitol, ethanolamine, and mercaptoethanol are not acted upon. Incubation of dioldehydrase with DBCC in the absence of substrate leads to irreversible inactivation. Hydroxo- and cyanocobalamin also inhibit irreversibly. This inhibition can be partially prevented by DBCC. Glycolaldehyde and glyoxal at levels as low as 0.02 m*M* inhibit irreversibly in the presence of DBCC. The enzyme shows no definite pH optimum. No change in rate is observed between pH 9.5 and 6.5. Beyond this range the enzyme is unstable.

Section IX

Miscellaneous Enzymes

[124] Glyoxylate Carboligase from *Escherichia coli*

By Naba K. Gupta and Birgit Vennesland

$$2\ CHO \cdot COO^- + H^+ \rightarrow CHO \cdot CHOH \cdot COO^- + CO_2 \quad (1)$$

$$2\ \text{Glyoxylate} + H^+ \rightarrow \text{tartronic semialdehyde} + CO_2 \quad (2)$$

Glyoxylate carboligase is known primarily as an adaptive enzyme formed by microorganisms grown on two carbon substrates which are metabolized by way of glyoxylate. This group of enzymes all require thiamine pyrophosphate and Mg^{++}. In addition, the enzyme from *Escherichia coli*[1,2] has been shown to be a flavoprotein.[3]

Assay Method

Principle. The enzyme may be assayed by carrying the reaction out in unbuffered medium at pH 6.5, and titrating the acid consumption by adding standard acid to maintain a constant pH.[2] This should be done under anaerobic conditions to avoid autoxidation of the reaction product. The manometric assay is preferred because of the greater ease of controlling the reaction conditions. This latter method involves the measurement of the rate of CO_2 formation from added glyoxylate under standard conditions.

Reagents

- Sodium glyoxylate, 0.3 *M*
- H_2SO_4, 4 *N*
- Potassium phosphate buffer of pH 7.3, 1 *M*
- Cysteine hydrochloride, 0.05 *M*, freshly prepared without neutralization
- Thiamine pyrophosphate, approximately 6×10^{-3} *M*. Dissolve 3 mg of thiamine pyrophosphate chloride per ml H_2O.
- Magnesium chloride, approximately 0.03 *M*. Dissolve 6 mg of $MgCl_2 \cdot 6\ H_2O$ per ml H_2O.
- Enzyme, freshly diluted with 0.2 *M* potassium phosphate buffer, pH 7.3 to obtain 4 to 6 units/ml

Procedure. The enzyme assay is carried out in double-arm Warburg flasks at 30° under an atmosphere of argon gas. To the main compartment, add 0.1 ml phosphate buffer, 0.1 ml of thiamine pyrophosphate

[1] G. Krakow and S. S. Barkulis, *Biochim. Biophys. Acta* **21,** 593 (1956).
[2] G. Krakow, S. S. Barkulis, and J. A. Hayashi, *J. Bacteriol.* **81,** 509 (1961).
[3] N. K. Gupta and B. Vennesland, *J. Biol. Chem.* **239,** 3787 (1964).

solution, 0.1 ml of $MgCl_2$ solution, 0.1 ml of cysteine hydrochloride solution, 0.1 ml of suitably diluted enzyme preparation, and 1.4 ml of water. The vessels are gassed with argon for 10 minutes with shaking. After another 10 minutes of equilibration, the reaction is started by tipping 0.1 ml sodium glyoxylate solution from one side arm. The reaction is allowed to proceed with shaking for 10 minutes, and stopped by addition of 0.3 ml of the H_2SO_4 solution from the other side arm, and the pressure is read after 5 minutes. The CO_2 released is calculated from the net pressure increase. A correction for bicarbonate in the reagents is applied by subtracting the pressure change observed in a duplicate vessel containing all components except the enzyme.

Definition of Unit and Specific Activity. A unit of enzyme is defined as that amount which produces 1 micromole of CO_2 in 1 minute under the assay conditions. Specific activity is defined as units per milligram of protein, determined by the method of Lowry *et al.*[4]

Purification Procedure

Growth of Bacteria. E. coli, Crookes strain,[2] was grown in a medium containing $(NH_4)_2SO_4$, 12 g; Na_2HPO_4, 72 g; KH_2PO_4, 36 g; $MgSO_4 \cdot 7 H_2O$, 2.4 g; NaCl, 6.0 g; glycolic acid, 34 g; NaOH, 13 g; tap water, 1 liter; and distilled water, 12 liters.

The bacterial growth from a freshly cultured nutrient slant is transferred to 50 ml of the liquid medium and incubated at room temperature with shaking for 24 hours. This culture is transferred to 300 ml of liquid medium and incubated at room temperature, with shaking, for another 24 hours, then transferred to 3 liters of liquid medium and incubated at 35° with vigorous aeration. After overnight growth, the heavy suspension is transferred to 36 liters of liquid medium and incubated at 35° with vigorous aeration for another 24 hours. The cells are harvested by centrifugation, washed once with 250 ml of 1% KCl solution and centrifuged again. The packed cells may be stored in the frozen state. The yield is about 1–1.25 g wet weight of bacteria per liter of culture.

Fractionation of the Enzyme. The table summarizes the results of a typical fractionation procedure. This procedure consistently gives a yellow protein preparation with a specific activity of 25–30 units per milligram of protein. Unless otherwise mentioned, all operations are performed at 0–4°.

Step 1. Crude Extract. About 43 g of wet frozen cells are suspended in

[4] O. H. Lowry, N. J. Rosebrough, A. L. Farr, and R. J. Randall, *J. Biol. Chem.* **193,** 265 (1951).

200 ml of 1% KCl solution and ruptured in a sonic oscillator (Sonifier, Branson Instruments Company, Model LS 75). Sonication is done in 60-ml batches (at 4.5 amp) for a total of 7 minutes per batch, with interruptions at frequent intervals to keep the temperature of the suspension below 15°. The extract is clarified by centrifugation for 1 hour at 27,000 *g* in a Servall angle-head centrifuge, and the clear supernatant is dialyzed overnight against 4 liters of 0.01 *M* phosphate buffer of pH 6.7.

Step 2. Ammonium Sulfate Fractionation. Solid ammonium sulfate is added slowly to the extract to bring it to 35% saturation. The suspension is stirred for 30 minutes and then centrifuged at 27,000 *g* for 10 minutes. The precipitate is discarded. Additional solid ammonium sulfate is added to bring the supernatant to 55% saturation. After 30 minutes of stirring, the precipitate, which contains most of the enzyme, is collected by centrifugation as described above, dissolved in 0.01 *M* phosphate buffer of pH 6.7, and dialyzed against 4 liters of the same buffer.

Step 3. Heat Treatment. The dialyzed solution (about 100 ml) is mixed with 0.25 volume of 1 *M* phosphate buffer of pH 6.1. For each milliliter of this solution, add 3 mg of thiamine pyrophosphate, 3 mg of $MgCl_2$, and 1 mg of cysteine hydrochloride. The solution is then distributed in 15-ml portions into 25-ml thin-walled test tubes, which are held in a water bath, maintained at 70°, for 5 minutes. The tubes are then transferred to an ice bath, and cooled for 5 minutes. The slurry is clarified by centrifugation for 10 minutes at 27,000 *g* and the inactive precipitate is discarded. Solid ammonium sulfate is added to the yellow supernatant solution to 55% saturation, and the protein precipitated after 30 minutes of stirring is collected by centrifugation, dissolved in 0.01 *M* phosphate buffer and dialyzed as before. This step consistently gives a larger number of enzyme units than are present after step 2.

Step 4. Protamine Sulfate Treatment. A freshly prepared 1% solution of protamine sulfate is adjusted to pH 6.0 with dilute KOH solution and added to the enzyme solution in the proportion of 1.25 mg of protamine sulfate per 20 mg of protein. After 10 minutes of stirring, the precipitate is removed by centrifugation. The enzyme is precipitated from the supernatant solution by adding an additional 1.25 mg of protamine sulfate per 20 mg of protein. After 10 minutes of stirring, the precipitate is collected by centrifugation and dissolved in 0.2 *M* phosphate buffer, pH 6.1. The enzyme is recovered by addition of ammonium sulfate to 45% saturation. After 30 minutes of stirring, the yellow precipitate is recovered by centrifugation, dissolved in 0.01 *M* phosphate buffer at pH 6.7 and dialyzed against the same buffer.

Purification of Glyoxylate Carboligase from *E. coli*

Fraction	Volume (ml)	Protein (mg)	Specific activity (units/mg protein)	Yield (%)
Sonicated extract (from 43 g of cells)	212	3965	0.3	100
After first ammonium sulfate precipitation	115	1920	0.55	89
After heat treatment and second precipitation	33	152	12.4	159
After protamine fractionation and ammonium sulfate precipitation	11	41	25.4	88

Properties of the Enzyme

Stability of the Enzyme. Solutions of the enzyme prepared as described above retain their activity for a day or two at 4°, after which a slow progressive loss of activity sets in, accompanied by increasing development of turbidity. This denaturation can be retarded by storing in 0.2 *M* phosphate buffer of pH 6.1, with added cysteine (1 mg of cysteine hydrochloride per milligram concentrated enzyme solution). The addition of cysteine (or an equivalent SH compound) is required to prevent inactivation on subsequent precipitation with $(NH_4)_2SO_4$. The cysteine-protected enzyme can be stored in the frozen state, or as a lyophilized powder, with little loss in activity.

Effect of pH, Salts, and Sulfhydryl Reagents. The enzyme has maximum activity at pH 7.5, somewhat more alkaline than the pH used for the assay. A high salt concentration is stimulatory; thus 10^{-2} *M* NaCl, KCl, or NH_4Cl increase the activity of the enzyme in the assay system by 33%.

Particularly after purification, the enzyme shows a marked dependence on SH stabilizers such as cysteine; 10^{-4} *M* *p*-hydroxymercuribenzoate gives about 50% inhibition when added to the assay system in the absence of cysteine; 5×10^{-3} *M* sodium iodoacetate has no effect, 10^{-4} *M* Cu^{++} and Hg^{++} inhibit almost completely.

Flavin Content. The absorption spectrum of the purified carboligase preparation shows three maxima located at 275, 380, and 445 mμ with a distinct secondary maximum at 467 mμ. The absorption bands in visible light are due to the prosthetic group, FAD. The purified protein contains 7.3–8 millimicromoles of FAD per milligram of protein. The average molar absorptivity (*E*) at 445 mμ is 14×10^3 M^{-1} cm^{-1}, calculated on a flavin basis. The ratio of the absorbance at 275 mμ to that at 445 mμ ranges from 7.5 to 8.5. The active enzyme shows little flavin fluorescence.

Resolution of the Enzyme

The carboligase preparation can be resolved into FAD and a colorless apoenzyme by the acid ammonium sulfate procedure of Warburg and Christian.[5] The following resolution procedure is a modification described by Strittmatter.[6]

One milliliter of enzyme solution, containing 3.75 mg protein of specific activity 25 is mixed with 1 ml of 3 *M* KBr, and cooled in an ice bath. A solution of acid ammonium sulfate is prepared by adding 0.5 ml of 1 *N* H_2SO_4 to 7.5 ml of saturated ammonium sulfate, and likewise cooled. One milliliter of this solution is added drop by drop, with stirring to the solution containing enzyme in concentrated KBr. The white precipitate which forms is separated by centrifugation in the cold for 5 minutes at 8000 *g*. The yellow supernatant solution is poured off, and the liquid adhering to the inside of the centrifuge tube is removed with adsorbent paper. For reconstitution experiments, the white precipitate is dissolved in 5 ml of 0.2 *M* phosphate buffer, pH 7.3, containing 0.5 mg per milliliter of thiamine pyrophosphate, and 0.5 mg per milliliter of cysteine hydrochloride. A small amount of insoluble material is removed by brief centrifugation in the cold, and 0.1-ml aliquots of the clear supernatant solution are tested in the standard assay system with varying amounts of added FAD. If the resolution procedure has been carried out carefully, and without unnecessary delay, the apoenzyme shows only about 5% of its original activity in the absence of added FAD; and almost full recovery of the initial activity is obtained when 1.2 millimicromoles of FAD is added to the assay mixture. The reconstitution of the apoenzyme with FAD follows Michaelis-Menten kinetics. K_m, for FAD $= 2.0 \times 10^{-7}$ *M*. The apoenzyme is very unstable and loses most of its activity on overnight storage at 4°.

Mechanism

The mechanism of glyoxylate carboligase of *E. coli* has been shown to involve the formation of 2-hydroxymethylthiamine pyrophosphate as an intermediate,[7] as previously demonstrated for the glyoxylate carboligase of *Pseudomonas* subspecies.[8]

No evidence has been obtained to show that the FAD of carboligase functions by being reduced and oxidized during the enzyme-catalyzed reaction. The substrate, glyoxylate, does not bleach the flavin. If sodium

[5] O. Warburg and W. Christian, *Biochem. Z.* **298,** 150 (1938).

[6] P. Strittmatter, *J. Biol. Chem.* **236,** 2329 (1961).

[7] G. Kohlhaw, B. Deus, and H. Holzer, *J. Biol. Chem.* **240,** 2135 (1965).

[8] L. Jaenicke and J. Koch, *Biochem. Z.* **336,** 432 (1962).

hydrosulfite is added under anaerobic conditions, the color of the enzyme is bleached rather slowly. Addition of O_2 results in recovery of the absorption bands. Reduction by hydrosulfite is accompanied by substantial loss of enzyme activity. Reoxidation by O_2 restores the activity.[9] This reversible loss of activity on addition of hydrosulfite can be demonstrated by addition under argon of 0.2 mg sodium hydrosulfite to the assay system.

[9] N. K. Gupta and B. Vennesland, *Arch. Biochem. Biophys.* **113**, 255 (1966).

[125] Inositol 1-Phosphate Synthetase and Inositol 1-Phosphatase from Yeast

By Frixos Charalampous and I-Wen Chen

Glucose 6-phosphate → inositol 1-phosphate
Inositol 1-phosphate → inositol + P_i

Note on Nomenclature

The purified enzyme preparation from yeast described in the present article contains the two enzyme systems responsible for the biosynthesis of inositol from glucose-6-P.[1] The first enzyme system catalyzes the conversion of glucose-6-P to inositol 1-phosphate, and is called inositol 1-phosphate synthetase; the second enzyme, which catalyzes the hydrolysis of inositol 1-phosphate to inositol and inorganic phosphate, is referred to as inositol 1-phosphatase. This nomenclature is not intended to imply that the synthetase represents the action of a single enzyme, nor that the phosphatase is strictly specific for inositol-1-P.

The term inositol and the various inositol phosphates refer to *myo*-inositol and its phosphates.

Assay Method for Inositol-1-P Synthetase

Principle. The most convenient method of assaying this enzyme is based on the determination of the radioactive inositol 1-P formed enzymatically from ^{14}C-labeled glucose-6-P. The radioactive inositol-1-P present in the incubation mixture is quantitatively converted to free inositol by the action of purified alkaline phosphatase, and the radioactive inositol is isolated by paper chromatography and counted.

[1] I. W. Chen and F. C. Charalampous, *Biochem. Biophys. Res. Commun.* **19**, 144 (1965).

Reagents

0.02 *M* uniformly labeled ^{14}C-glucose-6-P (Nuclear Chicago Corp.) neutralized with Tris (specific activity, 5×10^5 cpm per micromole).

DPN, 8.0 m*M*

NH_4Cl, 0.14 *M*, in 0.5 *M* Tris-acetate, pH 8.0

Enzyme. The solution contained 3–4 mg of protein per milliliter (800–1100 units/ml). Bacterial alkaline phosphatase. This was a chromatographically purified preparation, purchased from Worthington Biochemical Corporation. It contained 2.8 mg of protein per milliliter and had a specific activity of 30,000 units.

Procedure. The incubations are carried out in test tubes containing 0.05 ml of glucose-6-P, 25 μl each of DPN and NH_4Cl, 0.05 ml of enzyme solution, and water to a final volume of 0.25 ml. The tubes are incubated for 1 hour at 29° and the reaction is stopped by immersing the tubes in boiling water for 1 minute. After removal of precipitated proteins, 25 μl of the supernatant fluid is incubated with 10 μl of alkaline phosphatase and 0.175 ml of 0.05 *M* Tris-acetate (pH 8.0) for 1 hour at 29°. This incubation time is adequate to allow complete dephosphorylation of all radioactive phosphorylated compounds. At the end of the incubation carrier inositol (10 μg) is added, and the mixture is demineralized by passage through a column of 0.3×1.5 cm each of Dowex 1-X8 (acetate form) and amberlite IR-120 (H^+ form). The Dowex 1 is placed on top of the amberlite. The demineralized solution is concentrated to 0.1 ml and treated with 0.1 ml of 0.15 *M* $Ba(OH)_2$ at 100° for 15 minutes. After dilution with 1.3 ml of water the mixture is demineralized as described above using 4 times more of each resin. The column eluate is concentrated to a small volume and chromatographed on Whatman No. 1 paper using acetone:water (85:15, v/v) as solvent. The inositol area is cut out and counted in a liquid scintillation counter using liquifluor in toluene, or it can be eluted with water and counted in a windowless gas-flow counter.

Assay Method for Inositol 1-Phosphatase

Principle. The method is based on the colorimetric determination of P_i released enzymatically from L-*myo*-inositol-1-P.

Reagents

0.01 *M* L-*myo*-inositol-1-P (purchased from CalBiochem) neutralized with Tris to pH 7.0

$MgCl_2$, 0.012 *M*, in 0.5 *M* Tris-acetate, pH 7.7
Enzyme: the same solution used in the synthetase assay

Procedure. The incubation is carried out in test tubes containing 0.1 ml of inositol-1-P, 25 μl of $MgCl_2$, 30 μl of enzyme solution, and water to a final volume of 0.25 ml. At the end of 30 minutes' incubation at 29°, the reaction is stopped by immersing the tubes in a boiling water bath for 1 minute; the preparation is cooled and centrifuged. Aliquots from the supernatant fluid are used in the colorimetric determination of inorganic phosphate.[2] Similar incubation mixtures containing heat-inactivated enzyme were used as controls.

Purification Procedure

The following procedure yields a purified preparation which contains both the synthetase and the phosphatase activities. All procedures are carried out at 2–4°.

Step 1. Preparation of Crude Extract. Candida utilis (*Torulopsis utilis*) Y-900, obtainable from the American Type Culture Collection, is grown as previously described.[3] One liter of culture medium contained: (a) glucose, 20 g; (b) salts—NH_4NO_3 2.0 g, KH_2PO_4 0.35 g, $MgSO_4 \cdot 7 H_2O$ 0.25 g, $ZnSO_4 \cdot 7 H_2O$ 1.33 mg, $MnSO_4 \cdot H_2O$ 2.33 mg, $CuSO_4 \cdot 5 H_2O$ 2.93 mg, Na_2SO_4 1.79 mg, $(NH_4)_6MO_7O_{24} \cdot 4 H_2O$ 0.037 mg; (c) iron solution—$FeSO_4 \cdot 7 H_2O$ 0.93 mg; (d) vitamins—calcium pantothenate 2.5 mg, pyridoxine·HCl 0.25 mg, thiamine·HCl 0.25 mg, biotin 0.025 mg. The individual solutions of glucose, salt, and iron were sterilized by usual autoclaving for 20 minutes. The vitamins were sterilized by filtration through standard bacteriological filters.

The yeast was maintained on slants of Sabouraud dextrose agar containing 1% Bacto yeast extract (Difco). Large-scale cultures were grown in Erlenmeyer flasks having a volume 10 times larger than the culture volume. The inoculum was obtained from a smaller, 24-hour culture, and enough was added to give a turbidity of about 30 Klett units. The flasks were shaken on a gyratory shaker for 18 hours at 28°.

The harvested cells are washed three times with 10 volumes each of ice-cold demineralized water. The washed cells (20 g wet weight) are suspended in 40 ml of 0.02 *M* phosphate buffer, pH 7.2, containing 5×10^{-4} *M* GSH, and the suspension is passed through a French pressure cell at a pressure gauge reading of 5000 pounds per square inch. The resulting cell-free extract is centrifuged at 900 *g* for 10 minutes, and the precipitate is discarded since it contains no activity.

Step 2. 100,000 g Supernatant. The supernatant fluid obtained from

[2] C. H. Fiske and Y. SubbaRow, *J. Biol. Chem.* **66**, 375 (1925).
[3] I. W. Chen and F. C. Charalampous, *J. Biol. Chem.* **239**, 1905 (1964).

step 1 is centrifuged for 3 hours at 100,000 *g* in the Spinco ultracentrifuge. The resulting supernatant fluid is dialyzed overnight against 0.005 *M* Tris-chloride, pH 7.2.

Step 3. Protamine Sulfate Fractionation. To the dialyzed enzyme solution from step 2 is added, dropwise with stirring, an amount of a 2% protamine sulfate solution calculated to give a ratio of nucleic acid to protamine sulfate equal to 1 (w/w). The mixture is allowed to stand for 10 minutes in an ice bath, and is then centrifuged for 10 minutes at 20,000 *g*. The clear supernatant fluid contains all the enzymatic activity.

Step 4. Ammonium Sulfate Fractionation. The protein concentration of the solution from step 3 is adjusted to 15 mg/ml by the addition of the calculated amount of 0.005 *M* Tris-chloride, pH 7.2. In order to protect the SH groups of the enzyme system, GSH is added to the enzyme solution to give a final concentration of 0.1 m*M*. For every 100 ml of enzyme solution 31.3 g of ammonium sulfate is added slowly, with stirring, at 2° (50% saturation). The suspension is allowed to stand for 15 minutes in an ice bath, and the precipitate is removed by centrifugation for 10 minutes at 20,000 *g*. It is free of enzymatic activity and is discarded. To 100 ml of the supernatant fluid 6.6 g of ammonium sulfate is added (60% saturation), and the precipitate is collected as described above. Finally, for every 100 ml of supernatant fluid thus obtained 21.9 g of ammonium sulfate is added to give 90% saturation. After standing for 30 minutes in an ice bath the precipitate is collected by centrifugation for 30 minutes at 100,000 *g*. It is dissolved in the minimal amount of 0.005 *M* Tris-chloride (pH 7.0), and is dialyzed for 3 hours against 1000 volumes of the same buffer; the dialysis solution is changed once at 1.5 hours. Throughout the ammonium sulfate fractionation the pH of the enzyme solution is maintained at 7.2 by the dropwise addition of 4 *N* NH_4OH.

Step 5. Adsorption on Calcium Phosphate Gel. The dialyzed enzyme solution is diluted with 0.005 *M* Tris-chloride (pH 7.0) to give a protein concentration of about 10 mg/ml and is placed in an ice bath. The suspension of the calcium phosphate gel (58 mg dry weight per milliliter) is added slowly, with stirring, in an amount calculated to give a ratio of gel (milligrams dry weight) to protein (mg) equal to 1.4. After stirring the mixture for 10 minutes, it is centrifuged at 900 *g*. The clear supernatant fluid containing most of the enzymatic activity (gel-unadsorbed fraction of Table I) is treated once more with calcium phosphate gel so that the ratio of gel to protein is now 2. The mixture is stirred and centrifuged as before, and the precipitated calcium phosphate gel, which has adsorbed more than 70% of the enzymatic activity, is used in the following step.

Step 6. Elution from Calcium Phosphate Gel. Selective elution of the enzyme system from calcium phosphate gel is achieved with phosphate buffer by carefully controlling the volume, the molarity, and the pH of the buffer. After many preliminary experiments it was found that the following procedure gives the best and most reproducible results: for every 100 mg (dry weight) of calcium phosphate gel used to adsorb the enzyme system, 4 ml of 0.05 *M* phosphate buffer (pH 7.0) is added, and the mixture is stirred gently for 15 minutes in an ice bath. The supernatant fluid obtained after centrifugation at 900 *g* contains most of the enzymic activity. It is water clear and has a pale straw color at a protein concentration of 5 mg/ml.

Step 7. Dialysis against EDTA. The enzyme solution is dialyzed against EDTA according to the method of Racker *et al.*[4] since it was desirable to remove any bound thiamine pyrophosphate. Finally, EDTA and KCl are removed by dialysis for 6 hours against 1000 volumes of 0.005 *M* Tris-chloride (pH 7.0) containing 0.1 m*M* GSH; the dialysis solution is changed once at 3 hours. The enzyme solution thus obtained is water clear and colorless at a protein concentration of 6 mg/ml.

Table I summarizes the results of the purification. Starting with about 20 g wet weight of yeast, the entire procedure takes about 4 days, and can be conveniently interrupted at the completion of the ammonium sulfate fractionation. The ammonium sulfate fraction can be dialyzed

TABLE I
PURIFICATION OF THE INOSITOL SYNTHESIZING SYSTEM[a]

Step	Total protein (mg)	Specific activity[a] (units/mg protein)	Total activity (units[b])	Yield (%)
1. Crude extract	1560	2.3	3,588	—
2. 100,000 *g* supernatant	966	29.0	28,000	100
3. Protamine sulfate	955	29.0	27,600	95
4. Ammonium sulfate fraction	242	81.0	19,600	70
5. Calcium phosphate gel-unadsorbed fraction	138	100.0	13,800	49
6. Calcium phosphate gel eluate	54	135.0	7,300	26
7. Dialysis against EDTA	54	252.0	13,600	48

[a] Throughout the purification the inositol-1-P synthetase and the inositol 1-phosphatase were assayed together rather than separately, using the assay described under "alternative assay method." The specific activity values refer to this combined activity of the two enzymes.

[b] One unit is the amount of enzyme which catalyzes the formation of 1 millimicromole of inositol per hour.

[4] E. Racker, G. de la Haba, and I. G. Leder, *J. Am. Chem. Soc.* **75,** 1010 (1953).

overnight instead of 3 hours, or it can be kept in an ice bath until next day when the purification is resumed.

Properties

Cofactors. The inositol-1-P synthetase requires DPN, which gives half-maximal activity at 0.1 mM. TPN, FAD, FMN, thiamine pyrophosphate, cyanocobalamine, and pyridoxal phosphate cannot substitute for DPN.

The inositol 1-phosphatase requires Mg^{++} ions, which give half-maximal activity at 0.4 mM. Mn^{++} ions at an optimal concentration of 1.3 mM are half as effective as Mg^{++} ions. Zn^{++} and Co^{++} ions cannot substitute for Mg^{++} ions.

Activators. The synthetase is activated 5-fold by NH_4^+ ions at a concentration of 14 mM, while K^+ ions can only partially substitute for NH_4^+ ions.

pH Optima. The optimal pH for the synthetase is 8.0, and that of the phosphatase is 7.7. In both cases the buffer was Tris-acetate.

Kinetic Properties. The reaction velocity of the synthetase reaches a maximum at a glucose-6-P concentration of 5 mM, and the K_m is 1.5 mM. In the case of phosphatase maximal velocity is reached at an inositol-1-P concentration of 4 mM, and the K_m is 1.67.

Inhibitors. The two enzyme systems are inhibited by sulfhydryl reagents, various phosphorylated compounds and heavy metals. The effects of these inhibitors are summarized in Table II.

Specificity. Besides inositol-1-P the phosphatase will hydrolyze other phosphorylated compounds. The rates of hydrolysis, expressed as percentage of the rate obtained with L-*myo*-inositol-1-P, are as follows: *myo*-inositol-2-P 40; (—)-inositol-3-P 77; glucose-6-P 76; glucose-1-P 83; fructose-6-P 109; fructose-1-P 208; 6-phosphogluconate 74; β-glycerophosphate 112; *p*-nitrophenyl phosphate 123.

Since the purified enzyme preparation contains small amounts of phosphoglucomutase and glucose phosphate isomerase, it was not possible to test the specificity of the synthetase toward various hexose phosphates.

Stability. The purified enzyme preparation retains all its synthetase and phosphatase activities for at least 40 days if kept frozen at —20° in the presence of 0.1 mM GSH. Repeated thawing and freezing of the enzyme solution causes progressive inactivation of the synthetase, but not of the phosphatase. The rate of heat inactivation of the synthetase is much faster than that of the phosphatase. Heating at 60° for 8 and 15 minutes destroys 52 and 90% of the synthetase, respectively; under the same conditions the phosphatase is inactivated by 11 and 25%, respectively.

TABLE II
ENZYME INHIBITION BY VARIOUS COMPOUNDS

Inhibitor	Concentration (m*M*)	Inhibition (%)	
		Synthetase	Phosphatase
2-Deoxyglucose-6-P	2.00	75.0	16.0
2-Deoxyglucose	4.00	19.0	—
6-Phosphogluconate	4.00	69.0	26.0
Ribose-5-P	5.00	77.0	34.0
Erythrose-4-P	1.00	14.0	—
CMB[a]	0.10	76.0	—
PMN[a]	0.30	57.0	33.0
$AgNO_3$	0.02	77.0	37.0
$HgCl_2$	0.05	73.0	28.0
$CuSO_4$	0.50	10.0	31.0
$ZnCl_2$	1.00	33.0	89.0
$CoCl_2$	2.70	20.0	20.0
KF	3.00	0.0	50.0
KH_2PO_4 (pH 7.7)	10.00	4.0	38.0

[a] CMB, *p*-chloromercuribenzoate; PMN, phenylmercuric nitrate.

Alternative Assay Method. If one wishes to assay the overall conversion of glucose-6-P to inositol, one adds to the incubation mixture of the synthetase assay $MgCl_2$ at 1.2 m*M* final concentration. The inositol is then isolated as described previously *without* prior incubation with alkaline phosphatase.

The various assay methods have been used very satisfactorily with crude soluble extracts from yeast and rat testis.

[126] Glucosamine 6-Phosphate *N*-Acetylase

By E. A. DAVIDSON

[Glucosamine 6-phosphate: $CH_2OPO_3^{2-}$, HO, HO, $^{+}NH_3$, H, OH] $+ CH_3\overset{O}{\overset{\|}{C}}-SCoA \longrightarrow$ [*N*-acetylglucosamine 6-phosphate: $CH_2OPO_3^{2-}$, HO, HO, $N(H)-C(=O)-CH_3$, H, OH] + CoASH

Assay Method

The reaction may be followed either by the disappearance of absorbancy at 230 mμ representing cleavage of the thiolester bond of acetyl

CoA or alternatively by measurement of the formation of *N*-acetylglucosamine 6-phosphate. For routine purposes, the latter assay is employed.[1]

Reagents

Glucosamine 6-phosphate, 0.08 *M*. Crystalline material is the dipolar ion; dissolve in water and adjust to pH 6.6 with 2 *N* NH_2OH. Prepared chemically by phosphorylation of glucosamine with polyphosphoric acid[2] or enzymatically by phosphorylation with ATP in the presence of hexokinase.[3]
Acetyl coenzyme A, 0.02 *M*. Most conveniently prepared by acylation of CoA with acetic anhydride.[4]
Potassium phosphate buffer, 0.1 *M*, pH 6.6
Tris-HCl buffer, 0.1 *M*, pH 7.4
Protamine sulfate, 2%

Procedure. The following components are present in an incubation mixture in a final volume of 0.35 ml; glucosamine 6-phosphate, 8 micromoles; acetyl CoA, 2 micromoles; pH 6.6 phosphate buffer, 10 micromoles.

An appropriate aliquot is then withdrawn for acetyl hexosamine determination according to the procedure of Reissig *et al.*[5] Under these conditions rates of formation of acetyl glucosamine 6-phosphate will be linear for at least 40 minutes. The assay is linear with enzyme concentration as well as time. A five-fold variation in enzyme concentration will give satisfactory results under the described assay conditions.

Preparation of Enzyme

The hexosamine 6-phosphate *N*-acetylase is widely distributed in both mammalian and lower animal sources as well as plant systems. Active preparations have been reported from several species of molds and fungi, group A hemolytic streptococci, yeast, liver, kidney, and muscle.[1,6,7] The procedure described below is for *Neurospora crassa* enzyme.[1]

Neurospora crassa strain 5297a may be obtained from the American Type Culture Collection, 12301 Parklawn, Rockville, Maryland, 20852. The organism is grown in shake culture for 18 hours on a sucrose-salts

[1] E. A. Davidson, H. J. Blumenthal, and S. Roseman, *J. Biol. Chem.* **226**, 125 (1957).
[2] J. Distler, J. M. Merrick, and S. Roseman, *J. Biol. Chem.* **230**, 497 (1958).
[3] D. H. Brown, *Biochim. Biophys. Acta* **7**, 487 (1951).
[4] S. Ochoa, see Vol. I, p. 685.
[5] J. Reissig, J. L. Strominger, and L. F. Leloir, *J. Biol. Chem.* **217**, 959 (1955).
[6] L. F. Leloir and C. E. Cardini, *Biochim. Biophys. Acta* **12**, 15 (1953).
[7] D. H. Brown, *Biochim. Biophys. Acta* **16**, 429 (1955).

medium.[8] After this period of time the mycelia are harvested by filtration, washed several times with distilled water and lyophilized. The lyophilized mycelia may be stored at —18 degrees *in vacuo* for as long as one year without loss in activity of extract prepared from them.

Purification

All steps are carried out at 0–4 degrees. Of lyophilized mycelia, 1.4 g is suspended in 20 ml of 0.1 *M*, pH 7.4, Tris buffer in a Waring blendor; 1.5 g of 60 μ diameter glass beads[9] are added and the resulting suspension homogenized for three min with efficient cooling. A jacketed Waring blendor permitting good temperature control is employed for this step. It is possible to prepare as many as 16 crude extracts and to combine them for further purification. The glass beads and cellular debri are removed by centrifugation at 2000 *g* for 10 minutes.

The cloudy tan supernatant is treated with 3.8 ml of a 2% solution of protamine sulfate for each 10 ml of extract. The solution is stirred for 10 minutes in an ice bath, and the precipitated material is removed by centrifugation as above.

The resulting clear tan supernatant is adjusted to pH 4.8 by dropwise addition of 6 *N* acetic acid and then placed in a water bath maintained at 80°. The solution is continually stirred until the internal temperature reaches 60°; the solution is maintained at that temperature for a period of 10 minutes and then cooled in an ice bath. After readjustment of the pH to 6 with potassium hydroxide solution, precipitated material is removed by centrifugation for 15 minutes at 3000 *g*. The supernatant resulting from this step is treated with 1.1 volumes of mixed-bed resin (either Amberlite MB 3 or Dowex 501 is suitable for this purpose) stirred for 20 minutes in an ice bath, the resin is removed by filtration, and the resulting solution is lyophilized.

The lyophilized protein is redissolved at a concentration of 10 mg/ml in 0.1 *M* phosphate buffer, pH 6.6. Any insoluble material is removed by centrifugation, and the fraction precipitating between 50 and 70% ammonium sulfate saturation is harvested by centrifugation at 12,000 *g* for 20 min. This precipitate is dissolved in the same phosphate buffer, dialyzed for 24 hours against the buffer and may be stored in the frozen state. The enzyme at this stage represents approximately a 300-fold purification over the crude extract with a 50 per cent yield. This preparation is stable for several months when stored in the deep freeze. Details of purification are summarized in the table.

[8] G. W. Beadle and E. L. Tatum, *Am. J. Botany* **32**, 678 (1945).

[9] These should be designated pavement marking beads, 60 μ diameter and may be obtained from Minnesota Mining and Manufacturing Corporation.

PURIFICATION OF GLUCOSAMINE 6-PHOSPHATE *N*-ACETYLASE

Step	Specific activity[a]	Yield (%)
Crude Extract	2.8	100
Protamine sulfate	9.4	78
Heat, pH (4.8, 60°)	140	72
Resin	180	65
Ammonium sulfate	700	54

[a] Assay conditions as described in the text. Expressed as micromoles substrate converted per milligram of protein per hour.

Properties and Specificity

The pH optimum of the purified enzyme is fairly broad between 6 and 7.1. There are no cofactor requirements and the enzyme appears relatively stable toward iodoacetate although parachloromercuribenzoate is strongly inhibitory. The Michaelis constant for glucosamine 6-phosphate was calculated as $7 \times 10^{-4}\ M$ and that for acetyl CoA at approximately $5 \times 10^{-4}\ M$. The purified enzyme exhibits no acetylating activity toward glucosamine, galactosamine, paranitroaniline, several amino acids, or mannosamine 6-phosphate.[1] Galactosamine 6-phosphate is fully reactive as a substrate for the enzyme, and the ratio of activities toward this substrate coincides very closely with the ratio toward glucosamine 6-phosphate throughout the purification procedure described above.[2] Accordingly, it seems likely that these activities are catalyzed by the same enzyme, but this has not been completely resolved at this time. It is not known if the enzyme is specific for the α or β anomer of the hexosamine phosphate. Purification of this enzyme from other sources has also been noted with most preparations exhibiting heat stabilities similar to those indicated here. The mammalian enzyme is relatively less stable at acid pH but retains stability at pH 6 at 60 degrees for 10 minutes.

The acetylation of glucosamine *in vitro* was initially reported by Chou and Soodak in preparations from pigeon liver.[10] Subsequent to this, Leloir and Cardini, and Brown reported activities against either glucosamine or glucosamine 6-phosphate in preparations from *N. crassa* and yeast.[6,7] From the above data it seems likely that the primary acetylating enzyme is that for the hexosamine 6-phosphate; activity observed with the free amino sugar probably reflects the relatively nonspecific aromatic amine acetylase.

[10] C. Chou and M. Soodak, *J. Biol. Chem.* **196,** 105 (1952).

[127] Galactose 1-Phosphate Uridyl Transferase

By JARY S. MAYES and R. G. HANSEN

UDP-Glc + Gal-1-P ⇄ UDP-Gal + Glc-1-P

Gal-1-P uridyl transferase catalyzes the reversible reaction shown above which is one of the important steps in galactose metabolism. It is also this enzyme which is lacking in galactosemia, a hereditary disorder of galactose metabolism.

The partial purification and properties of Gal-1-P uridyl transferase from calf liver and yeast have been presented[1,2] and are discussed in an earlier volume of this series.[3] In the present section a modified and extended procedure for fractionation of calf liver, leading to a more purified enzyme, and an additional procedure for the partial purification of the enzyme from human red blood cells are described.[4]

Assay Method

Principle. The assay is similar to the method previously described[3] and is based on the rate of Glc-1-P formed in the transferase reaction. Glc-1-P formation is followed by the rate of increase in optical density at 340 mμ which results from TPNH formation in a reaction coupled with phosphoglucomutase and Glc-6-P dehydrogenase.

Reagents

- UDP-Glc, sodium salt, 0.02 *M*
- Gal-1-P, sodium salt, 0.02 *M*
- TPN, 0.02 *M*
- $MgCl_2$, 0.02 *M*
- A solution containing: 0.5 mg/ml crystalline phosphoglucomutase (Calbiochem) and 2.5 units/ml Glc-6-P dehydrogenase (Type V, Sigma Chemical Co.)
- Glycine buffer, 0.1 *M* pH 8.7, containing 0.01 *M* mercaptoethanol
- Gal-1-P uridyl transferase-diluted to give a change in optical density between 0.008 and 0.02 per minute

Procedure. An appropriate amount of buffer to make a final volume of 0.50 ml was first added to each cuvette. Then 0.01 ml of UDP-Glc,

[1] K. Kurahashi and E. P. Anderson, *Biochim. Biophys. Acta* **29**, 498 (1958).
[2] K. Kurahashi and A. Sugimura, *J. Biol. Chem.* **235**, 940 (1960).
[3] E. S. Maxwell, K. Kurahashi, and H. M. Kalckar, see Vol. V [20].
[4] J. S. Mayes, Ph.D. Thesis, Michigan State University, 1965.

Gal-1-P and TPN; 0.02 ml of $MgCl_2$; 5 μg of phosphoglucomutase; and 0.025 units of Glc-6-P dehydrogenase are added. The reaction is started by mixing in the appropriate amount of transferase and is followed at 25° with a spectrophotometer equipped with an automatic cuvette changer and recording attachment.[5,6] After a brief initial lag period, the reaction is linear with time.

In crude preparations 6-phosphogluconic acid dehydrogenase is present; and if not measured and corrected for, the activity of transferase is overestimated. 6-Phosphogluconic acid dehydrogenase was therefore measured independently according to the principle of Horecker and Smyrniotis[7] with 0.2 micromole of TPN and 6-phosphogluconic acid, 0.4 micromole of $MgCl_2$ an appropriate amount of extract, and 0.1 *M* glycine buffer, pH 8.7, to a total volume of 0.50 ml. This procedure was employed for most of the assays of transferase and compensated for in calculating the activity. Also in crude preparations a pyrophosphatase is present which hydrolyzes UDP-Glc to Glc-1-P. Therefore, a control without Gal-1-P was necessary to correct for this activity.

Definition of Unit and Specific Activity. A unit is defined as the amount of enzyme that causes the formation of 1 micromole of products per minute under the conditions described. The calculation of product is based on a molar extinction coefficient of 6.22×10^3 for TPNH at 340 mμ. Specific activity is units per milligram of protein which was determined by the method of Lowry *et al.*[8]

Purification Procedure from Calf Liver

Calf liver was obtained from a local slaughterhouse as soon as possible after death of the animal, sliced in small pieces, packed in dry ice, and kept frozen until ready for use. Purification of the enzyme was accomplished in the following manner with all steps being carried out at 0–4° and with distilled-deionized water. Centrifugation was for 20 minutes at 24,000 *g*.

Step 1. Preparation of Crude Extract. Frozen calf liver, 1 kg, was diced and extracted with 2 liters of 0.03 *M* KOH–0.005 *M* EDTA Na_4 in a large Waring blendor for 2 minutes. The extract, which was originally pH 8.6, was adjusted to pH 6.8 with 6.0 *N* acetic acid; and 400 ml of 2%

[5] W. A. Wood and S. R. Gilford, *Anal. Biochem.* **2**, 589 (1961).

[6] W. A. Wood and S. R. Gilford, *Anal. Biochem.* **2**, 601 (1961).

[7] B. L. Horecker and P. Z. Smyrniotis, *J. Biol. Chem.* **193**, 371 (1951); also see Vol. I [42].

[8] O. H. Lowry, N. J. Rosebrough, A. L. Farr, and R. J. Randall, *J. Biol. Chem.* **193**, 265 (1951).

protamine sulfate was added. After it was stirred for 15 minutes, the mixture was centrifuged.

Step 2. Ammonium Sulfate Fractionation. To the supernatant from step 1, solid ammonium sulfate was added slowly until 50% saturation (0.298 g/ml). After it was stirred for 30 minutes, the solution was centrifuged and the supernatant discarded. The precipitate was dissolved in a minimum amount of cold water and dialyzed for 3 hours against two changes of 10 liters of water.

Step 3. Calcium Phosphate Gel Adsorption. To the dialyzate of step 2, calcium phosphate gel[9] (25–35 mg/ml) was added until the gel to protein ratio was 0.8. After it was stirred for 10 minutes, the solution was centrifuged. To the supernatant, four times the original amount of calcium phosphate gel was added; the mixture was stirred for 2 hours and centrifuged. The precipitated gel was suspended in 1500 ml of 0.02 *M* phosphate buffer, pH 8.0; and after elution overnight, the mixture was centrifuged.

Step 4. DEAE-Cellulose Chromatography. The supernatant from step 3 was placed directly on a DEAE-cellulose[10] column (30×2 cm). The activity was eluted with a gradient of phosphate buffer (pH 7.0). A gradient system similar to that described by Hurlbert *et al.*[11] was employed. Elution was started with water in the mixing flask (500 ml) and 0.05 *M* buffer in the reservoir and continued until the enzyme started to come through. The elution of the enzyme was completed with 0.08 *M* buffer in the reservoir.

Step 5. Concentration by DEAE-Cellulose. Tubes from step 4 containing the enzyme were combined and diluted with 1.5 volumes of cold water. The enzyme was adsorbed on a small column (7×1 cm) of DEAE-cellulose then eluted with 0.16 *M* phosphate buffer, pH 7.3.

Step 6. Ammonium Sulfate Fractionation. Tubes from step 5 containing the enzyme were combined and solid ammonium sulfate added until 30% saturation (0.168 g/ml). After it was stirred for 15 minutes, the solution was centrifuged. Ammonium sulfate was again added to the supernatant until 50% saturation (0.119 g/ml). After 15 minutes of

[9] For preparation, see Vol. I [11].

[10] The DEAE-cellulose was washed with 1 *N* NaOH until the filtrate was clear and then with water until neutral. For the material for columns, the DEAE-cellulose was then converted to the phosphate salt by washing with 0.5 *M* Na_3PO_4 and then again with water until the excess phosphate was removed. For material to remove hemoglobin, the DEAE-cellulose was then washed with 0.5 *M* phosphate buffer (pH 7.0) and with water until the excess phosphate was removed. For washing, also see Vol. V [1].

[11] R. B. Hurlbert, H. Schmitz, A. F. Brumm, and V. R. Potter, *J. Biol. Chem.* **209**, 23 (1954).

stirring, the solution was again centrifuged. The supernatant was discarded and the precipitate dissolved in a small amount of 0.05 *M* phosphate buffer (pH 7.0) which contained 0.01 *M* mercaptoethanol.

The purification of the preparation is shown in Table I. An overall purification of about 500-fold was achieved with a yield of about 23%. The procedure has been repeated several times with similar results. In the calcium phosphate gel step, it is important for best results to have the gel and protein concentration about the same as given. Also, the activity from the DEAE-cellulose column should be concentrated as soon as possible because the enzyme is unstable in dilute solution.

TABLE I
PURIFICATION OF CALF LIVER ENZYME

Step	Volume (ml)	Total units	Protein (mg)	Specific activity (units/mg protein)
Crude extract (after protamine sulfate)	2340	2000	80,000	0.025
Ammonium sulfate I	785	1870	41,900	0.045
Calcium phosphate gel	1600	1120	910	1.23
DEAE-cellulose (column)	635	780	89	8.79
DEAE-cellulose (concentration)	35	530	49	10.8
Ammonium sulfate II	2.2	480	39	12.4

Properties

Stability. The enzyme is fairly stable at pH 7.0 and 8.0, but not at pH 6.0 where almost half of the activity was lost in 18 hours. Sulfhydryl compounds stabilize the enzyme somewhat.

Specificity. In addition to the normal nucleotide sugars, UDP-Glc and UDP-Gal, the enzyme has some activity with the other pyrimidine diphosphate sugars (8.7% with TDP-Glc and 0.3% with CDP-Glc), but showed no detectable activity with purine diphosphate sugars (ADP-Glc, GDP-Glc, and IDP-Glc). The purified enzyme was active with Glc-1-P and Gal-1-P, but there was no activity detectable with the other sugar phosphates tested, namely, Man-1-P, Xyl-1-P, Gal-6-P, and Fru-1-P.

Activators and Inhibitors. All sulfhydryl compounds tested activated and stimulated the enzyme to about the same extent (30–50%). Mercaptoethanol usually appeared to be slightly more active than either glutathione or cysteine. The enzyme has no demonstrable requirement for divalent metals. In fact, all divalent metals that were tested were inhibitory including Cu^{++}, Mg^{++}, Mn^{++}, and Ca^{++}.

Purification Procedure from Human Red Blood Cells

The red blood cells from 4 pints of blood, obtained from the regional blood center, were washed twice with physiological saline (0.9%) and held frozen in the refrigerator overnight to rupture the cells. The cells were thawed and diluted with an equal volume of water to form the hemolyzate. Other conditions for purification were the same as for the calf liver.

Step 1. The purpose of this step is to remove the hemoglobin. The procedure is a modification of that described by Hennessey *et al.*[12] An equal volume of a thick slurry of DEAE-cellulose was added to the hemolyzate; the mixture was stirred for 1 hour, then poured into a large fritted glass funnel. The DEAE-cellulose on the funnel was washed with 0.0025 *M* phosphate buffer (pH 7.0) until the red color was removed. The DEAE-cellulose was then removed from the funnel, activity eluted with 0.5 *M* KCl for 1 hour, and filtered.

Step 2. To the filtrate of step 1, solid ammonium sulfate was added until 50% saturation (0.298 g/ml). The solution was stirred for 20 minutes, then centrifuged. The supernatant was discarded, and the precipitate was dissolved in water. The solution was dialyzed for 3 hours against two changes of water of 10 liters each.

Step 3. To the dialyzate of step 2, calcium phosphate gel was added until the gel to protein ratio reached 1.2. The solution was stirred for 15 minutes, then centrifuged. Eight times the original amount of gel was added to the supernatant. The solution was stirred for 2 hours, then again centrifuged. The precipitated gel was eluted with 0.05 *M* phosphate buffer (pH 8.0) for 2 hours and centrifuged.

TABLE II
PURIFICATION OF RED BLOOD CELL ENZYME

Step	Volume (ml)	Total units	Protein (mg)	Specific activity (unit/mg protein)
Hemolyzate	2020	35.9	323,000	0.00011
DEAE-cellulose	2150	22.4	11,400	0.0020
Ammonium sulfate I	270	19.0	4,780	0.0040
Calcium phosphate gel	340	11.9	220	0.053
Ammonium sulfate II	10	8.6	160	0.058

[12] M. A. Hennessey, A. M. Waltersdorph, F. M. Huennekens, and B. W. Gabrio, *J. Clin. Invest.* **41,** 1257 (1962).

Step 4. To concentrate the supernatant of step 3, solid ammonium sulfate was added until 60% saturation (0.390 g/ml); the solution was stirred for 15 minutes, then centrifuged. The supernatant was discarded, and the precipitate was dissolved in a small amount of 0.05 *M* phosphate buffer (pH 7.0) which contained 0.01 *M* mercaptoethanol.

A summary of the purification procedure is given in Table II. An overall purification of about 500-fold with a yield of about 24% was achieved. The procedure has been repeated several times with similar results. Several additional steps, such as DEAE-cellulose chromatography and Alumina Cγ gel fractionation, were attempted without success. Instability of the enzyme in dilute solutions is a problem. The above procedure gave a concentrated enzyme that was free of 6-phosphogluconic acid dehydrogenase and pyrophosphatase which cause trouble in the assay.

[128] Galactose 1-Phosphate Uridyl Transferase (Clinical Aspects)

By R. G. Hansen and Jary S. Mayes

$$\text{Gal-1-P} + \text{UDP-Glc} \xrightleftharpoons{\text{transferase}} \text{UDP-Gal} + \text{Glc-1-P} \quad (1)$$

$$\text{UDP-Glc} + 2\,\text{NAD} \xrightleftharpoons{\text{dehydrogenase}} \text{UDP-GA} + 2\,\text{NADH} + 2\,\text{H}^+ \quad (2)$$

$$\text{UDP-Gal} \xrightleftharpoons{\text{epimerase}} \text{UDP-Glc} \quad (3)$$

Introduction

Galactosemia, a hereditary disorder in humans, is characterized by a very low level or a complete absence of the enzyme galactose 1-phosphate uridyl transferase (transferase) in blood cells,[1] liver, and other tissues. In general, the symptoms of the fully manifested disease are jaundice, cataracts, failure to grow, and excess galactose, amino acids, and protein in the urine. If the disease is not detected early and treated in the infant, mental retardation and death may occur.

Measurement of transferase in blood cells is the most direct way of establishing or excluding a diagnosis of galactosemia. Further, by measuring the enzyme in blood cells, it has been possible to demonstrate that heterozygotes on the average have one-half the enzyme level of normal

[1] H. M. Kalckar, E. P. Anderson, and K. J. Isselbacher, *Biochim. Biophys. Acta* **20**, 262 (1956).

individuals; thus, an autosomal recessive mode of inheritance has been established.[2–7] In addition to both parents, some of the relatives of the galactosemic possess transferase at levels which are intermediate between the diseased and normal individual. Since there is an overlap of about 5% in transferase levels between these groups, it is not possible to classify individuals as heterozygotes with absolute certainty using chemical tests by presently available methods. For clinical purposes, however, a study of the transferase levels in the patient's blood or in various members of the family will indicate the possibility of galactosemia with almost complete certainty. Population and family studies have been used to assess the frequency of the galactosemia heterozygote and from this, to calculate an expected rate of occurrence of the disease.[8] Thus, chemical methods for both recognition of the disease and for identifying the heterozygote are of the utmost importance. As the treatment of galactosemia is to give the patient a galactose-free diet at the earliest possible age, methods for screening large populations of infants are also desirable.

A number of procedures which are suitable for establishing the disease with certainty (see the table) have been developed to detect the enzyme in blood cells. Some of these are inherently complex and not suited to routine or large-scale screening procedures. The detection of the heterozygote is more difficult, and various modifications of the spectrophotometric determination of UDP-Glc consumption by red cells[9] or the manometric procedures[3,6,10] have been used. The spectrophotometric procedure of Bretthauer *et al.*[2] for the quantitative assay of the transferase for diagnosing the disease and detecting the heterozygote will be described in detail here.

[2] R. K. Bretthauer, R. G. Hansen, G. Donnell, and W. R. Bergren, *Proc. Natl. Acad. Sci. U.S.* **45**, 328 (1959).
[3] H. N. Kirkman and E. Bynum, *Ann. Human Genet.* **23**, 117 (1959).
[4] G. N. Donnell, W. R. Bergren, R. K. Bretthauer, and R. G. Hansen, *Pediatrics* **25**, 572 (1960).
[5] D. Y. Hsia, M. Tannenbaum, J. A. Schneider, I. Huang, and K. Simpson, *J. Lab. Clin. Med.* **56**, 368 (1960).
[6] V. Schwarz, A. R. Wells, A. Holzel, G. M. Komrower, and I. M. Simpson, *Ann. Human Genet.* **25**, 179 (1961).
[7] N. J. Brandt, A. Froland, M. Mikkelsen, A. Nielsen, and N. Tolstrup, *Lancet* **ii**, 700 (1963).
[8] R. G. Hansen, R. K. Bretthauer, J. S. Mayes, and J. H. Nordin, *Proc. Soc. Exptl. Biol. Med.* **115**, 560 (1964).
[9] E. P. Anderson, H. M. Kalckar, K. Kurahashi, and K. J. Isselbacher, *J. Lab. Clin. Med.* **50**, 469 (1957).
[10] V. Schwarz, *J. Lab. Clin. Med.* **56**, 483 (1960).

METHODS FOR MEASURING GAL-1-P URIDYL TRANSFERASE IN RED CELLS

Principle	Applications			Reference
	Detection of disease	Detection of heterozygote	General screening	
Spectrophotometric				
UDP-Glc consumption	+	+		*a*
UDP-Glc consumption	+			*b*
UDP-Glc consumption	+			*c*
UDP-Glc consumption	+	+		*d*
UDP-Gal formation	+			*e*
Manometric				
Oxygen consumption	+	+		*f*
Oxygen consumption	+	+		*g*
Oxygen consumption	+	+		*h*
Radioactive substrates				
Galactose	+			*i*
Galactose	+			*j*
Galactose	+			*k*
Galactose	+			*l*
Galactose-1-P	+			*m*
Visual				
Dye-linked	+		+	*n*

[a] R. K. Bretthauer, R. G. Hansen, G. Donnell, and W. R. Bergren, *Proc. Natl. Acad. Sci. U.S.* **45,** 328 (1959).
[b] E. P. Anderson, H. M. Kalckar, K. Kurahashi, and K. J. Isselbacher, *J. Lab. Clin. Med.* **50,** 469 (1957).
[c] J. H. Nordin, R. K. Bretthauer, and R. G. Hansen, *Clin. Chim. Acta* **6,** 578 (1961).
[d] N. Tolstrup, *Nord. Med.* **67,** 742 (1962).
[e] E. S. Maxwell, H. M. Kalckar, and E. Bynum, *J. Lab. Clin. Med.* **50,** 478 (1957).
[f] H. N. Kirkman and E. Bynum, *Ann. Human Genet.* **23,** 117 (1959).
[g] V. Schwarz, A. R. Wells, A. Holzel, G. M. Komrower, and I. M. Simpson, *Ann. Human Genet.* **25,** 179 (1961).
[h] V. Schwarz, *J. Lab. Clin. Med.* **56,** 483 (1960).
[i] A. N. Weinberg, *Metabolism* **10,** 728 (1961).
[j] E. Eggermont and H. G. Hers, *Clin. Chim. Acta* **7,** 437 (1962).
[k] A. Robinson, *J. Exptl. Med.* **118,** 359 (1963).
[l] M. London, J. H. Marymont, and J. Fuld, *Pediatrics* **33,** 421 (1964).
[m] W. G. Ng, W. R. Bergren, and G. N. Donnell, *Clin. Chim. Acta* **10,** 337 (1964).
[n] E. Beutler, M. Baluda, and G. N. Donnell, *J. Lab. Clin. Med.* **64,** 694 (1964).

Assay Method

Principle. The method is based on the UDP-Glc consumption test of Anderson *et al.*[9] with suitable modifications to measure the transferase quantitatively. Red cells are incubated with Gal-1-P and UDP-Glc in suitable excess, the reaction is stopped, and the residual UDP-Glc is determined spectrophotometrically with UDP-Glc dehydrogenase by measuring the NADH which is formed. For an added check on the procedure, it is possible to measure the UDP-Gal which is formed in the reaction by addition of a purified UDP-Gal-4-epimerase,[11] thus re-forming UDP-Glc which may again be quantitated with UDP-Glc dehydrogenase.

Reagents

(1) Buffer, glycine, 1.0 M and 0.1 M (pH 8.7)
1.0 M, 7.5 g glycine and 7 ml 1 N NaOH to 100 ml with water
0.1 M, 1 volume of 1.0 M and 9 volumes of water.

(2) UDP-Glc, 3 micromoles/ml
18 mg of the disodium salt dissolved in 10 ml of water

(3) Gal-1-P, 20 micromoles/ml
79.0 mg of the barium salt dissolved in 5.0 ml of water
25.2 mg of Na_2SO_4 are added; and after centrifuging, the supernatant is adjusted to 10 ml with water; or 74.0 mg of dipotassium salt (2 H_2O) dissolved in 10 ml of water.

(4) NAD, 10 micromoles/ml
6.6 mg to 1 ml with water

(5) UDP-Glc dehydrogenase purified through step 5 by the procedure of Strominger *et al.*[12]

(6) UDP-Gal-4-epimerase

(7) Working solutions, made fresh daily
(A) 2.0 ml each of solutions (1) and (2) and 1 ml of solution (3)
(B) 2.0 ml each of solutions (1) and (2) and 1 ml of water

Procedure. PREPARATION OF SAMPLES. Blood is drawn by venipuncture into tubes with oxalate as an anticoagulent, and placed in ice until centrifuged at 0° in a clinical centrifuge. As the enzyme is labile, the assay should be completed as soon as possible; and care should be taken to keep the samples at ice temperatures, and if necessary, stored at —10°. After the plasma has been decanted, the erythrocytes are washed three times with cold saline (0.9%) and frozen. A hemolyzate is prepared by thawing the frozen erythrocytes and adding an equal volume of distilled water.

INCUBATION OF HEMOLYZATE WITH SUBSTRATES. Two tubes are prepared

[11] E. S. Maxwell, H. M. Kalckar, and E. Bynum, *J. Lab. Clin. Med.* **50,** 478 (1957).
[12] J. L. Strominger, E. S. Maxwell, J. Axelrod, and H. M. Kalckar, *J. Biol. Chem.* **224,** 79 (1957).

for each hemolyzate. To one tube in an ice bath, 0.2 ml of solution A is added; to a second tube, 0.2 ml of solution B is added. Two tenths milliliter of hemolyzate (equal volume of water and red cells) is added to each tube. They are then mixed and placed in a water bath at 37°. After incubation for 10 minutes, the tubes are placed with shaking in a boiling water bath for 1.0 minute, then chilled in ice and centrifuged.

SPECTROPHOTOMETRIC DETERMINATION OF RESIDUAL UDP-GLC. Aliquots of the supernatant (usually 0.01 ml) are taken from each tube and placed in reaction cuvettes containing 0.20 ml NAD and glycine buffer (0.1 *M*, pH 8.7) to make a final volume of 0.5 ml. Sufficient UDP-Glc dehydrogenase is added with mixing to complete the reaction in less than 15 minutes. After the optical density change ceases, UDP-Gal-4-epimerase is added and the further reduction of NAD is followed. This new reduction of NAD should be equivalent to the initial UDP-Glc disappearance (Fig. 1).

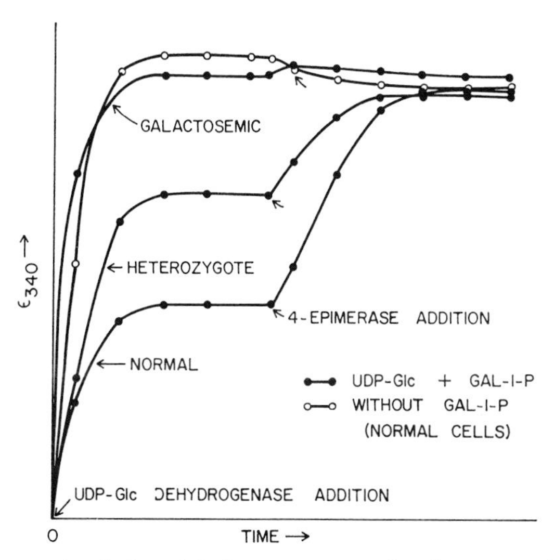

FIG. 1. The spectrophotometric measurement of galactose 1-phosphate uridyl transferase in hemolyzates of red cells. The hemolyzates were incubated with an excess of UDP-Glc and Gal-1-P. Residual UDP-Glc was then quantitated with the specific dehydrogenase from calf liver. UDP-Gal formed in the reaction was converted to UDP-Glc by addition of a C_4 epimerase and quantitated with the UDP-Glc dehydrogenase.

Calculation. The extinction coefficient for NADH at 340 mμ, 6.22 micromoles/ml, is the basis for the quantitative calculation. For a light path of 1 cm, reaction volume of 0.5 ml, blood dilution at the first incubation of 1–4, an aliquot for analysis of 0.01 ml, an incubation time of 10 minutes and with 2 micromoles of NADH formed per mole of UDP-Glc the following applies:

$$(\Delta\epsilon_B - \Delta\epsilon_A) \times \frac{4 \times 100 \times 6}{6.22 \times 2 \times 2} = \text{micromoles UDP-Glc/ml cells/hr}$$

or

$$(\Delta\epsilon_B - \Delta\epsilon_A) \times 97 = \text{micromoles UDP-Glc/ml cells/hr}$$

[129] Formate-Pyruvate Exchange System: *Streptococcus faecalis*

By N. P. Wood

$$CH_3COCOO^- + H^{14}COO^- \rightleftharpoons CH_3CO^{14}COO^- + HCOO^-$$

Assay Method

Principle. The formate-pyruvate exchange system catalyzes the exchange of formate-^{14}C with the carboxyl group of pyruvate. The velocity of the reaction is followed by measuring the increase in radioactivity of pyruvate which is recovered as the pyruvate-2,4-dinitrophenylhydrazone. This reaction is closely associated with the phosphoroclastic reactions of pyruvate.[1]

Reagents

Potassium pyruvate, 0.5 M
Sodium formate-^{14}C, 0.5 M, 0.67 μC/ml
Potassium phosphate, 0.6 M, pH 7.5
2,4-Dinitrophenylhydrazine, 1.5 M, in 18 N H_2SO_4
Reducing solution containing 0.01 M $FeSO_4$ and 0.03 M 2,3-dimercaptopropanol

Procedure. The reaction system contains, in a total volume of 1.0 ml: 50 micromoles of potassium pyruvate, 50 micromoles of sodium formate-^{14}C (0.067 μC), 60 micromoles of potassium phosphate (pH 7.5), 0.1 ml of reducing solution, and 1 unit of enzyme. The system is incubated for 15 minutes at 37°. Prior to the addition of extract and reducing solution, the reaction mixture is cleared of oxygen by evacuation and flushing with helium.[2]

The reaction is terminated by addition of 0.5 ml of 5 N H_2SO_4. Two milliliters of H_2O is added and the protein is removed by centrifugation. To the supernatant solution 0.2 ml of 2,4-dinitrophenylhydrazine solu-

[1] See Vol. I [73].
[2] M. O. Oster and N. P. Wood, *J. Bacteriol.* **87**, 104 (1964).

tion is added, and the pyruvate-2,4-dinitrophenylhydrazone is then recovered on a filter paper disk by vacuum filtration (Tracerlab filter tower). The phenylhydrazone on the disk is washed twice with water and dried under a heat lamp. The phenylhydrazone and disk are weighed and counted in a gas flow counter (D-47 Nuclear Chicago Corp.). Corrections are made for self absorption and coincidence.

Definition of Unit and Specific Activity. One unit is defined as that amount of enzyme which catalyzes the incorporation of ^{14}C (0.1 μC/ml) to produce 1 cpm per micromole of pyruvate per minute of reaction under specified reaction and counting conditions.[3] Specific activity is expressed as units per milligram of protein. Protein is measured by the method of Lowry *et al.*[4]

Application of Assay Method to Intact Cells. Cell suspensions will exchange formate-^{14}C provided that the reaction mixture is supplemented with 15 mg of yeast extract.[5] The mixture containing 0.4 mg cell suspension is incubated for 90 minutes at 37° in N_2 or helium atmosphere. By adding pyruvate to the growth medium and supplementing the reaction mixture with 0.01 *M* sodium acetate and 0.01 *M* histidine, the requirement for yeast extract can be replaced partially.

Purification Procedure[3]

The formate exchange enzyme is unusually sensitive to oxygen. Therefore stringent precautions are required to preserve activity (see this volume [1] for general procedures and details). All steps in the purification are carried out at 0°–4°. Reagents are prepared in deionized water.

Culture Methods. Streptococcus faecalis, strain 10C1 ATCC No. 11700, is grown in a medium consisting of 1% yeast extract (Basamin-bact, Anheuser Busch, Inc., St. Louis, Missouri), 0.4% K_2HPO_4, and 0.2% sodium citrate. The medium is inoculated with 2% of a 12-hour culture grown in a portion of the same medium and incubated without agitation at 37° for 12 hours. The cells are washed with distilled water and used immediately or stored at −20° under helium.[2]

Step 1. Crude Extract. To prepare cell-free extracts of *S. faecalis* the cells are suspended in 3 volumes of 0.033 *M* potassium phosphate, pH 6.0, and reducing solution is added, as a complex, to a final concentration of 0.003 *M* 2,3-dimercaptopropanol and 0.001 *M* $FeSO_4$. The suspension is then passed through a French pressure cell; the air within the pressure cell is replaced with helium prior to treatment. Cell debris and

[3] N. P. Wood, P. Paolella, and D. Lindmark, *J. Bacteriol.* in press.

[4] O. H. Lowry, N. J. Rosebrough, A. L. Farr, and R. J. Randall, *J. Biol. Chem.* **193**, 265 (1951).

[5] N. P. Wood and D. J. O'Kane, *J. Bacteriol.* **87**, 97 (1964).

fine particles are removed by centrifugation of the extract at 20,000 *g* for 60 minutes in capped tubes containing an atmosphere of helium.

Step 2. Removal of Nucleic Acids. Saturated ammonium sulfate, pH 7.0, is added to 20% of saturation; the solution is centrifuged at 20,000 *g*. Two per cent protamine sulfate solution, pH 5.0, is prepared in water that has been previously boiled and cooled under an atmosphere of helium; reducing solution is added as a complex to a concentration of 0.003 *M* 2,3-dimercaptopropanol and 0.001 *M* ferrous sulfate. The protamine solution is added by syringe to an extent of 20% of the volume (10–12 mg of extract per milliliter).

Step 3. Dialysis. The solution containing extract, ammonium sulfate, and protamine is dialyzed for 4 hours with gentle stirring in 100 volumes of 0.033 *M* potassium phosphate, pH 6.0, which contains 0.001 *M* $FeSO_4$ and 0.003 *M* 2,3-dimercaptopropanol. The precipitate is removed by centrifugation and discarded. The 280:260-mμ ratio is 0.8–0.91.

Step 4. Ammonium Sulfate. Fractionation. A saturated solution of ammonium sulfate is boiled and then cooled under helium. The solution is adjusted to pH 7.5 and reducing solution is added to provide a concentration of 0.003 *M* 2,3-dimercaptopropanol and 0.001 *M* $FeSO_4$. The dialyzed protamine solution is adjusted to pH 7.5 with 0.5 *N* KOH, and saturated ammonium sulfate is added to 50% of saturation. The solution is stirred for 20 minutes and the precipitate is removed by centrifugation and discarded. The supernatant solution is brought to 70% of saturation with saturated ammonium sulfate and the solution is stirred for 20 minutes and then centrifuged at 20,000 *g* for 30 minutes. The precipitate was redissolved in 0.005 *M* potassium phosphate, pH 7.5, containing 0.001 *M* $FeSO_4$ and 0.033 *M* 2,3-dimercaptopropanol to give one-tenth the volume of the original supernatant solution.

Step 5. Dialysis. The solution containing the 50–70% ammonium sulfate fraction is dialyzed for 6 hours, with gentle stirring, against two changes of 0.005 *M* potassium phosphate, pH 7.5 (100 volumes) containing 0.001 *M* $FeSO_4$ and 0.003 *M* 2,3-dimercaptopropanol.

Step 6. DEAE-Sephadex Column. Twenty-five grams of DEAE Sephadex-A-25 are suspended in 2.5 liters of distilled-deionized water. The water is removed by filtration and the material is washed with 1 liter of 0.5 *N* HCl and rinsed twice with equal volumes of deionized water. The Sephadex is then washed with 1 liter of 0.5 *N* NaOH and twice with 1 volume (1 liter) of deionized water. It is then treated with 100 ml of 0.6 *M* KH_2PO_4 and washed with 100 ml each of 0.6 *M*, 0.3 *M*, 0.1 *M*, and 0.01 *M* potassium phosphate, pH 7.5. After a final wash with 1 liter of 0.005 *M* potassium phosphate, pH 7.5, the Sephadex is then suspended in 1 liter of the same buffer and then boiled to remove oxygen. The material is cooled under helium, and 2,3-dimercaptopropanol is added to a final concentration of 0.003 *M*. The slurry is packed carefully in a 95 × 1.5 cm column (Pharmacia) and two bed volumes of 0.005 *M* potassium phosphate, pH 7.5, containing 0.003 *M* 2,3-dimercaptopropanol are passed through the column. To allow sufficient time for the 2,3-dimercaptopropanol to react with oxygen, the buffer-dimercaptopropanol solution is kept in the column for 24 hours at 1° and then replaced with buffer-dimercaptopropanol solution (4 void volumes) until the O-R potential has obtained maximum negative reading. After 24 hours, the buffer-dimercaptopropanol solution is replaced by 4 void volumes of fresh buffer-dimercaptopropanol solution. When the effluent remains steady, at —250 mv (calomel electrode), the sample (5 milliliters) is then added and the column is developed by gradient elution with 0.005 *M* and 0.6 *M* potassium phosphate, pH 7.5, containing 0.003 *M* 2,3-dimercaptopropanol. The flow rate of the column is 1 ml per minute. The effluent is collected in test tubes containing 0.5 ml of solution, which will provide a final concentration of 0.001 *M* $FeSO_4$ and 0.003 *M* 2,3-dimercaptopropanol and 5×10^{-4} *M* $MnSO_4$.

The purification procedure is summarized in the table.

PURIFICATION PROCEDURE

Step	Volume[a] (ml)	Total protein (mg)	Total units	Yield (%)	Specific activity (units/mg protein)	E_h^0 (volts)
Crude extract	60	390	416	—	1.07	−0.21
Protamine supernatant, dialyzed	89	268	428	100	1.54	−0.24
$AMSO_4$ (0.50–0.70 saturation), dialyzed	10	30	190	46	6.35	−0.60
DEAE-Sephadex A-25 column	10	1.96	165	40	84	−0.21

[a] Where aliquots of the total fraction were used, the figures have been corrected back to the original volume.

Properties[3]

Distribution. The formate exchange reaction has been studied with extracts of *Escherichia coli*,[6-9] *Micrococcus lactilyticus*,[10] *Clostridium butylicum*, *C. pasteurianum*, *C. kluyveri*,[8] and *Staphylococcus aureus*.[7]

[6] H. Chantrenne and F. Lipmann, *J. Biol. Chem.* **187**, 757 (1950).

[7] H. J. Strecker, *J. Biol. Chem.* **189**, 815 (1951).

[8] R. F. Asnis and M. C. Glick, *Bacteriol. Proc.* p. 113 (1956).

[9] M. F. Utter, F. Lipmann, and C. Werkman, *J. Biol. Chem.* **158**, 521 (1945).

[10] N. G. McCormick, E. J. Ordal, and H. R. Whiteley, *J. Bacteriol.* **83**, 899 (1962).

Specificity. The purified enzyme will exchange formate-^{14}C with α-ketobutyrate and α-ketoglutarate at a velocity of 11% and 1%, respectively, of pyruvate exchange. α-Ketovalerate, α-ketoisovalerate, and α-ketocaproate exchange formate-^{14}C at negligible rates or not at all. The K_m (Lineweaver and Burk) for pyruvate, α-ketobutyrate, and formate is 0.025 *M*, 0.027 *M*, and 0.073 *M*, respectively.

Effect of Buffers and pH. The optimal pH of exchange is approximately 7.5, and when the O-R potential is sufficiently negative, the enzyme is active between pH of 6.0 and 8.5. Under the same O-R conditions potassium phosphate buffer does not inhibit or stimulate the reaction. Tris, maleate, imidazole, 2,4,6-collidine, and glycylglycine inhibit the exchange reaction by crude extracts.[2]

Activators and Inhibitors. The exchange reaction is not inhibited by 0.1 *M* EDTA. Phosphate, lipoate, and CoA are not required. Avidin has no effect, but arsenite, arsenate, and PCMB are inhibitory. A Mn^{++} and thiamine diphosphate requirement for the formate exchange reaction in *E. coli* has not been studied in the *S. faecalis* system.[7]

Stability. The enzyme is rapidly and irreversibly inactivated by oxygen. When the O-R potential of the environment is kept sufficiently negative, wet cells (cell paste) can be stored 2–3 months. Crude extracts stored 4 days at —20° show a 10–20% loss of activity. Ammonium sulfate fractions can be stored as a precipitate without loss of activity. Enzyme activity in DEAE-Sephadex effluent is lost rapidly when stored in an ice bath at zero to 1° or frozen at —20°, but 80% of the activity remained when the effluent fraction was stored for 3 days in 30% (w/v) potassium tartrate at these temperatures.

The enzyme is inactivated when the extract is heated at 45° for 25 minutes or for 10 minutes at 55°.[2]

Other Properties. Pyruvate is not degraded by purified extracts.[3,7] The enzyme is colorless and has a molecular weight of over 200,000.

Author Index

Numbers in parentheses are reference numbers and indicate that an author's work is referred to although his name is not cited in the text.

A

B

C

D

H

I

J

K

N

O

P

Q

R

S

T

Y

Z

Subject Index

A

B

C

D

G

H

J

K

M

N

O

P

Q

R

S

T

U

V

W

X

Y

Z